Nanotechnology

Volume 1

Nanotechnology

Volume 1

WM Breck

CBS Publishers & Distributors Pvt Ltd

New Delhi • Bengaluru • Chennai • Kochi • Kolkata • Mumbai
Hyderabad • Jharkhand • Nagpur • Patna • Pune • Uttarakhand

Nanotechnology
Volume 1

ISBN: 978-81-239-2842-5

Copyright © Publisher

First Edition: 2016
Reprint: 2017

Published by Satish Kumar Jain and produced by Varun Jain for

CBS Publishers & Distributors Pvt Ltd
4819/XI Prahlad Street, 24 Ansari Road, Daryaganj, New Delhi 110 002, India.
Ph: 23289259, 23266861, 23266867   Website: www.cbspd.com
Fax: 011-23243014   e-mail: delhi@cbspd.com; cbspubs@airtelmail.in.
Corporate Office: 204 FIE, Industrial Area, Patparganj, Delhi 110 092
Ph: 4934 4934   Fax: 4934 4935   e-mail: publishing@cbspd.com; publicity@cbspd.com

Branches

- **Bengaluru:** Seema House 2975, 17th Cross, K.R. Road,
  Banasankari 2nd Stage, Bengaluru 560 070, Karnataka
  Ph: +91-80-26771678/79   Fax: +91-80-26771680   e-mail: bangalore@cbspd.com
- **Chennai:** 7, Subbaraya Street, Shenoy Nagar, Chennai 600 030, Tamil Nadu
  Ph: +91-44-26680620, 26681266   Fax: +91-44-42032115   e-mail: chennai@cbspd.com
- **Kochi:** Ashana House, No. 39/1904, AM Thomas Road, Valanjambalam,
  Ernakulam 682 016, Kochi, Kerala
  Ph: +91-484-4059061-65   Fax: +91-484-4059065   e-mail: kochi@cbspd.com
- **Kolkata:** 6/B, Ground Floor, Rameswar Shaw Road, Kolkata-700 014, West Bengal
  Ph: +91-33-22891126, 22891127, 22891128   e-mail: kolkata@cbspd.com
- **Mumbai:** 83-C, Dr E Moses Road, Worli, Mumbai-400018, Maharashtra
  Ph: +91-22-24902340/41   Fax: +91-22-24902342   e-mail: mumbai@cbspd.com

Representatives

- **Hyderabad**   0-9885175004   • **Jharkhand**   0-9811541605   • **Nagpur**   0-9021734563
- **Patna**   0-9334159340   • **Pune**   0-9623451994   • **Uttarakhand**   0-9716462459

Printed at India Binding House, Noida, UP, India

Preface

Nanoscience occurs at the intersection of traditional science and engineering, quantum mechanics, and the most basic processes of life itself. Nanotechnology encompasses how we harness our knowledge of nanoscience to create materials, machines and devices that will fundamentally change the way we live and work.

The world of material science is witnessing a revolution in the exploration of the matter at the small scale. The overall societal impact of nanotechnology is expected to be greater than the integrated chip since it is applicable to many more areas than electronics. The next two decades are bound to see many more changes due to nanotechnology than the changes that have occurred in the entire past century.

Nanotechnology is emerging fast, with revolutionary new products already on the horizon. The list of products that are already being envisioned is mind-boggling with applications ranging from medical breakthroughs, and revolutionary products to the mundane. The industry and the market are gearing for an unprecedented surge in the growth of this field.

Material science/technology is one of the areas with the high probability of producing early returns. Scientists manipulate individual atoms and molecules to build things with almost unimaginable precision.

This is just the beginning. As the technology improves, we will be able to move hundreds and thousands of atoms, to precisely construct objects of virtually any size, shape and material.

The first volume of nanotechnology is divided into two sections—general considerations and application of nanocomposites.

Chapter 1 is devoted to basic concepts of nanotechnology and discusses tools, techniques and applications of nanotechnology. Chapter 2 deals with quantum chemistry which describes the electronic behaviour of atoms and molecules as pertaining to their reactivity. Chapter 3 focus on solid state physics and discusses crystal structure and electronic properties of materials. Chapter 4 is devoted to fullerene which is any molecule composed entirely of carbon in the form of a hollow sphere, ellipsoid or tube. Chapter 5 concentrates on carbon nanotubes which are allotropes of carbon with a cylindrical nanostructure. Chapter 6 focuses on nanowire which is a nanostructure, with the diameter of the order of a nanometer (10^{-9} meters). Chapter 7 discusses nanomechanics which is a branch of nanoscience studying fundamental mechanical (elastic, thermal and kinetic) properties of physical systems at nanometer scale. Chapter 8 deals with micro- and nano-electromechanical system (MEMS) and describes MEMS basic process and manufacturing technologies. Chapter 9 focuses on supramolecular chemistry which refers to the area of chemistry beyond the molecules and focuses on the chemical systems made up of a discrete number of assembled molecular subunits and components. Chapter 10 concentrates on microfluidics which deals with the behaviour, precise control and manipulation of fluids that are geometrically constrained to a small, typically sub-millimeter scale. Chapters 11 and 12 are devoted to self-assembled monolayer and nanostructured advanced materials. Chapter 13 deals with devoted to gas phase cluster and discusses the design and development of functional molecules, nanomaterials and nanodevices. Chapter 14 concentrates on nanophotonics or nano-optics which is the study of the behaviour of light on the nanometer scale. Thus it deals with optics, or the interaction of light with particles or substances at deeply subwavelength scales.

Chapter 15 is devoted to nanomaterial synthesis and application. Chapter 16 focuses on oxide nanoprecursors: a technological perspective. Chapter 17 concentrates on core/shell nanoparticles. Chapter 18 focuses on kinetics and energetics in nanolubrication and discusses their various aspects. Chapter 19 focuses on MEMS packaging and discusses its various types, functions and multichip packaging. Chapter 20 deals with nanosensors which are biological, chemical or surgery, sensory points which convey information about nanoparticles to the macroscopic world.

Section II focuses on nanocomposites and their applications. Chapter 21 is devoted to nanocomposites which are made from two or more different materials having a wide difference in physical/mechanical properties. Chapter 22 concentrates on clay based nanocomposites and discusses their flammability properties, and ceramic-metal nanocomposites. Chapter 23 focuses on processing and characterisation of nanoceramic composites. Chapter 24 deals with polyolefin, polypropylene and polysterene nanocomposites. Chapters 25 and 26 focuses on polyester clay nanocomposites and nylon-6 nanocomposites. Chapter 27 concentrates on synthesis of inorganic nanomaterials and composites. This chapter discusses synthesis, physical and chemical methods along with industrial applications of nanocomposites. Chapter 28 is devoted to biologically derived synthetic nanocomposites. Chapter 29 focuses on metal and oxide nanocomposites. Chapters 30 and 31 concentrate on sustainable flame retardant nanocomposites and nondestructive testing of nanocomposites by optical techniques. Chapter 32 deals with nanoclay for polymer composites, paints, cosmetics and waste-water treatment. Finally chapter 33 focuses on hard and tough nanocomposite coatings.

This reference textbook is essential reading for all students, teachers, professionals, researchers and industrialists involved with all branches of engineering including material sciences. It is also a valuable source of information for industrialists and those preparing for or already associated with material science, etc. The reference textbook also caters to the requirement of the syllabus prescribed by various Indian universities for undergraduate and postgraduate students pursuing these courses. Constructive suggestions are always welcome from users of this book.

Diagrams, figures, tables and index supplement the text. All the topics have been covered in a cogent and lucid style to help the reader grasp the information quickly and easily.

WM Breck

Contents at a Glance

Preface   v

SECTION I

General Considerations   *1–342*

  1. Basic Concepts of Nanotechnology   3–22
  2. Quantum Chemistry   23–33
  3. Solid-state Physics   34–49
  4. Fullerene   50–58
  5. Carbon Nanotubes   59–89
  6. Nanowire   90–101
  7. Nanomechanics   102–112
  8. Micro- and Nano-electromechanical System   113–121
  9. Supramolecular Chemistry   122–157
10. Microfluidics   158–167
11. Self-assembled Monolayer   168–185
12. Nanostructured Advanced Materials   186–199
13. Gas Phase Cluster   200–217
14. Nanophotonics   218–231
15. Nanomaterial Synthesis and Application   232–260
16. Oxide Nanoprecursors: A Technological Perspective   261–275
17. Core/Shell Nanoparticles   276–285
18. Kinetics and Energetics in Nanolubrication   286–303
19. MEMS Packaging   304–329
20. Nanosensor   330–342

SECTION II

Nanocomposites and their Applications   *343–519*

21. Nanocomposites   345–361
22. Clay Based Nanocomposites   362–373
23. Processing and Characterisation of Nanoceramic Composites   374–387
24. Polyolefin, Polypropylene and Polystyrene Nanocomposites   388–406
25. Polyester-clay Nanocomposites   407–416

26. Nylon-6 Nanocomposites 417–422

27. Chemical Synthesis of Inorganic Nanomaterials and Composites 423–430

28. Biologically Derived Synthetic Nanocomposites 431–469

29. Metal and Oxide Nanocomposites 470–475

30. Sustainable Flame Retardant Nanocomposites 476–480

31. Nondestructive Testing of Nanocomposites by Optical Techniques 481–487

32. Nanoclays for Polymer Nanocomposites, Paints, Cosmetics and Waste-water Treatment 488–502

33. Hard and Tough Nanocomposite Coatings 503–519

References 521

Index 523–529

Contents

Preface   v

Contents at a Glance   vii

SECTION I

General Considerations   *1–342*

1. Basic Concepts of Nanotechnology   **3–22**

Introduction   3
Origin   3
Fundamental Concepts   4
    Simple to Complex: A Molecular Perspective   5
    Molecular Nanotechnology   5
Current Research   6
    Nanomaterials   6
    Bottom-up Approaches   6
    Top-Down Approaches   7
    Functional Approaches   8
    Biomaterials   8
    Speculative   10
Tools and Techniques   11
Applications   12
Implications   13
    Health and Environmental Concerns   13
    Regulation   13
Energy Applications of Nanotechnology   14
    Consumer Products   15
    Economic Benefits   15
Applications of Nanotechnology   16
    Medicine   16
    Chemistry and Environment   18
    Energy   18
    Information and Communication   19
    Heavy Industry   21
    Consumer Goods   21

2. Quantum Chemistry   **23–33**

Introduction   23
Electronic Structure   23
    Wave Model   24

Valence Bond	24
Molecular Orbital	24
Density Functional Theory	24
Chemical Dynamics	24
Adiabatic Chemical Dynamics	24
Non-adiabatic Chemical Dynamics	25
Quantum Chemistry and Quantum Field Theory	25
Computational Chemistry	25
Concepts	26
Methods	27
Interpreting Molecular Wave Functions	30
Software Packages	30
Quantum Chemistry Computer Programs	30
Quantum Electrochemistry	33

3. Solid-state Physics **34–49**

Introduction	34
Crystal Structure and Properties	34
Electronic Properties	35
Quasicrystal	36
Mathematical Description	36
Physics of Quasicrystals	37
Material Science of Quasicrystals	37
Spin Glass	38
Magnetic Behaviour	38
Model of Sherrington and Kirkpatrick	38
Non-ergodic Behaviour, and Applications	39
Superconductivity	39
Elementary Properties of Superconductors	40
Theories of Superconductivity	43
Classification	44
Semiconductor Device Fabrication	45
Wafers	45
Processing	45
Wafer Test	47
Device Test	47
Die Preparation	48
Packaging	48
List of Steps	48
Hazardous Materials	49

4. Fullerene **50–58**

Introduction	50
Prediction and Discovery	50
Variations	51

Buckyball 51
    Boron Buckyball 52
    Variations of Buckyballs 52
Carbon Nanotubes 53
Properties of Fullerenes 54
    Aromaticity 54
    Chemistry 54
    Solubility 55
    Quantum Mechanics 56
    Safety and Toxicity 56
    Superconductivity 57
Fullerite (Solid State) 58

5.   **Carbon Nanotubes** **59–89**

Introduction 59
Types of Carbon Nanotubes and Related Structures 59
    Single-walled 59
    Multiwalled 60
    Torus 62
    Nanobud 62
Properties 63
    Strength 63
    Kinetic 63
    Electrical 64
    Optical 64
    Thermal 64
    Defects 64
    One-dimensional Transport 64
    Toxicity 65
Synthesis 65
    Arc Discharge 66
    Laser Ablation 66
    Chemical Vapour Deposition (CVD) 66
    Natural, Incidental, and Controlled Flame Environments 67
Potential and Current Applications 67
    Structural 68
    In Electrical Circuits 68
    As Paper Batteries 69
    As a Vessel for Drug Delivery 69
    Current Applications 69
Boron Nitride Nanotubes 71
    Composites Containing BN 71
    Carbon Nanotubes in Photovoltaics 72
    Carbon Nanotube Composites in the Photoactive Layer 72

Carbon Nanotubes as a Transparent Electrode 73
CNTs in Dye-Sensitised Solar Cells 74
Optical Properties of Carbon Nanotubes 75
Electronic Structure of Carbon Nanotube 75
van Hove Singularities 76
Kataura Plot 77
Optical Absorption 77
Luminescence 78
Raman Scattering 80
Selective Chemistry of Single-walled Nanotubes 82
Structure and Reactivity 82
Sidewall Functionalisation 82
Reaction Mechanism 85
Reversibility of Diazonium Chemistry 86
Chemical Separation of Metallic and Semiconducting SWNTs 86
Carbon Nanotube Chemistry 86
Multiwalled Carbon Nanotubes 86
Single-walled Carbon Nanotubes 88
Colossal Carbon Tube 88
Silicon Nanotubes 88
Synthesis 88
Applications 88
Lithium Ion Batteries 89

6.   Nanowire **90–101**

Introduction 90
Synthesis of Nanowires 91
Suspension 91
VLS Growth 91
Solution-phase Synthesis 92
Physics of Nanowires 92
Conductivity of Nanowires 92
Welding Nanowires 93
Uses of Nanowires 93
Molecular Wires 93
Materials 94
Structure 94
Conduction of Electrons 94
Use of Nanowires in Molecular Electronics 95
Fabrication of Nanowires at Surfaces 95
Atom Chains, the Ultimate Nanowires 98
Solar Nanowires Promise Efficient, Low-cost Solar Power 100
Nanowire Battery 101

7.   Nanomechanics   **102–112**

Introduction   102
Molecular Machine   103
    Historical Insight and Studies   103
    Modern Insights and Studies   104
DNA Machine   106
DNA Nanotechnology   107
    Tile-based Arrays   108
    DNA Origami   108
    DNA Polyhedra   109
    DNA Nanomechanical Devices   109
    Stem Loop Controllers   109
    Applications   109
Geometric Phase   110
    Examples of Geometric Phases   111

8.   Micro- and Nano-electromechanical System   **113–121**

Introduction   113
Materials for MEMS Manufacturing   113
    Silicon   113
    Polymers   114
    Metals   114
MEMS Basic Processes   114
    Deposition Processes   114
    Physical Deposition   114
    Chemical Deposition   114
    Patterning   115
    Lithography   115
    Diamond Patterning   115
MEMS Manufacturing Technologies   117
    Bulk Micromachining   117
    Surface Micromachining   117
    High Aspect Ratio (HAR) Silicon Micromachining   117
    Applications   118
    Research and Development   118
    Industry Structure   119
MEMS Thermal Actuator   119
Micro-opto-Electromechanical Systems   119
Nanoelectromechanical Systems   119
    Importance for AFM   120
    Approaches to Miniaturisation   120
    Materials   120
    Future of NEMS   121

9. Supramolecular Chemistry   **122–157**

Introduction   122
Control of Supramolecular Chemistry   125
    Thermodynamics   125
Concepts in Supramolecular Chemistry   125
    Molecular Self-assembly   125
    Molecular Recognition and Complexation   126
    Template-directed Synthesis   126
    Mechanically-interlocked Molecular Architectures   126
    Dynamic Covalent Chemistry   126
    Biomimetics   126
    Imprinting   126
    Molecular Machinery   127
Building Blocks of Supramolecular Chemistry   127
    Synthetic Recognition Motifs   127
    Macrocycles   127
    Structural Units   127
    Photo-/Electro-chemically Active Units   128
    Biologically-derived Units   128
Applications   128
    Material Technology   128
    Catalysis   128
    Medicine   128
    Data Storage and Processing   128
    Green Chemistry   129
    Other Devices and Functions   129
Categories and Sub-categories of Supramolecular Chemistry   129
    Calixarene   129
    Carcerand   130
    Catenane   131
    Cavitand   131
    Crown Ether   131
    Cryptand   132
    Cryptophane   132
    Crystal Engineering   133
    Cucurbituril   133
    Cyclodextrin   135
    Dendrimer   136
    Dynamic Covalent Chemistry   136
    Endohedral Fullerene   137
    Foldamer   138
    Folding (Chemistry)   138
    Host-Guest Chemistry   138

Hydrogen Bond   139
Hydrophobic Effect   139
Inclusion Compound   140
Intercalation (Chemistry)   140
Macrocycle   141
Mechanical Bond   142
Mechanically-interlocked Molecular Architectures   142
Metallacrown   142
Micelle   143
Molecular Borromean Rings   144
Molecular Encapsulation   145
Molecular Tweezers   145
Molecular Knot   146
Molecular Imprinting   146
Molecular Machine   147
Molecular Self-Assembly   149
Molecular Sensor   150
Molecular Shuttle   151
Noncovalent Bonding   151
Porphyrin   152
Resorcinarene   153
Rotaxane   153
Stacking (Chemistry)   155
Supermolecule   156
Supramolecular Assembly   156
Supramolecular Polymers   157
Topoisomer   157

10. Microfluidics   **158–167**

Introduction   158
Microscale Behaviour of Fluids   158
Key Application Areas   159
    Continuous-flow Microfluidics   159
    Digital (Droplet-based) Microfluidics   160
    DNA Chips (Microarrays)   160
    Molecular Biology   160
    Acoustic Droplet Ejection (ADE)   161
    Fuel Cells   161
Digital Microfluidics   161
    Working Principle   161
    Implementation   161
Lab-On-A-Chip   162
    Chip Materials and Fabrication Technologies   163
    LOCs and Global Health   164

Nanofluid: Engineering the Fluid   165
     Preparation   166
     Experimental Investigation on Thermal Conductivity of Nanofluids   166
     Flow, Convection and Boiling   167

11. Self-assembled Monolayer   **168–185**

Introduction   168
Types of SAMs   169
Preparation of SAMs   169
Characterisation of SAMs   170
     Kinetics   170
     Defects   170
     Nanoparticle Properties   170
Self-Assembled Monolayers in Organic Chemistry   170
     Preparation   171
     Characterisation   171
     Reactivity   172
Mixed Monolayer Coverage on Gold Nanoparticles for Interfacial Stabilisation of
     Immiscible Fluids   174
Applications of SAMs   178
     Passivation of Surfaces: Protection from Corrosion   178
     Fabrication Using Patterned SAMs as Resists   178
     Fabrication on Curved Surfaces   179
     Directed Assembly of Materials on the Surface of a Patterned SAM   180
     Optical Systems   180
     Fabrication of Colloids   180
     Organic Chemistry   180
     Applications in Other Areas   181
Current Issues   184

12. Nanostructured Advanced Materials   **186–199**

Introduction   186
Prologue   186
Quantum Structures: Synthesis, Characterisation, Manipulation, and Assembly   187
Nanoworld is Different   189
     Size Effects   189
     Threshold Size Effects from a Single Particle to Collective Phenomena   191
     Metal–Nonmetal Transition in Finite Systems   191
     Coulomb Blockade   192
     Confinement   193
     Size Effects for Transport in Nanostructures   195
     Nanoscale Regime   196
     Nanoelectronics, Nano-optoelectronics, and Information Nanoprocessing   196
     Perspectives in Nanoscience and Nanotechnology   198

13. Gas Phase Cluster   **200–217**

Introduction   200
Molecular Clusters and Interaction Forces   202
Design and Development of Functional Molecules, Nanomaterials, and Nanodevices   202
    Ionophores/Receptors   203
    Organic Nanotubes and Nanowires   203
    Catalytic Residues in Enzymatic Reaction   204
    Left-handed Helix of Polypeptides   204
    Nanomechanical Device   205
Gas Cluster Ion Beam   205
    Process   205
    Industrial Applications   206
Cluster Chemistry   206
    Applications of Clusters in Catalysis   206
    Electronic Structure   207
Excursions in Cluster Science   208
    From Fission to Coulomb Explosion   209
    Ultraintense Laser — Cluster Interactions   212
    Nuclear Fusion Driven by Cluster Coulomb Explosion   213
    Perspectives   216

14. Nanophotonics   **218–231**

Introduction   218
Components of a Nanophotonic System   219
    Waveguide   219
    Coupler   220
    Optical Switch   221
    Photodetector   221
    Electro-Optic Modulator   222
    Wavelength-Division Multiplexing   223
    Amplifier   224
    Laser   225
    Optical Circulator   231

15. Nanomaterial Synthesis and Application   **232–260**

Introduction   232
Uniformity   233
Properties   234
    Synthesis   236
    Colloids   237
    Morphology   238
    Characterisation   238
Nanoscale Iron Particles   239
    Chemistry   239

Magnetic Nanoparticles 239
  Properties 239
Nanoshell Particles: Synthesis, Properties and Applications 241
  Synthesis of Nanoshell Particles 242
  Surface Plasmon Resonance of Metal Nanoparticles and Metallic Nanoshells 249
  Properties of Nanoshell Particles 251
  Applications of Nanoshell Particles 254
Nanotoxicity: Threat Posed by Nanoparticles 257
  Safety 259
Conclusion 260

16. Oxide Nanoprecursors: A Technological Perspective **261–275**

Introduction 261
Synthesis of Oxide Nanoparticles 261
  Mechanical Method 262
  Physical Methods 262
  Chemical Methods 263
Size Determination of Nanoparticles 265
Results 265
  Efficiency of Sol-gel Derived Nanostructured γ-alumina Porous Spheres as an
    Adsorbent in Liquid Chromatography 266
  Efficiency of Pd Impregnated Sol-gel Derived γ-alumina Porous Spheres as Catalyst 266
  Synthesis of High T_c Superconductor $YBa_2Cu_4O_8$ (1–2–4) Under the Condition of
    Normal Oxygen Pressure 268
  Densification of Lead Zirconium Titanate (PZT) 269
  Aqueous-sol Derived Thin Films 269
Synthesis of Low-agglomerated Nanoprecursors in the ZrO_2-HfO_2-Y_2O_3 System 270
  Theory 270
  Experimental 272
Results and Discussion 272

17. Core/Shell Nanoparticles **276–285**

Introduction 276
Experimental Procedure 276
Results and Discussion 277
Future Work 278
X-ray Diffraction on Core-Shell Nanoparticles for a Precise Structure Determination 278
Tailoring Magnetic Properties of Core/Shell Nanoparticles 280

18. Kinetics and Energetics in Nanolubrication **286–303**

Introduction 286
From Bulk to Molecular Lubrication 288
  Hydrodynamic Lubrication and Relaxation 288
  Boundary Lubrication 289
  Stick Slip and Collective Phenomena 289

Thermal Activation Model of Lubricated Friction   291
Functional Behaviour of Lubricated Friction   294
Thermodynamical Models Based on Small and Nonconforming Contacts   295
Limitation of the Gaussian Statistics — The Fractal Space   298
Fractal Mobility in Reactive Lubrication   299
Metastable Lubricant Systems in Large Conforming Contacts   300
Conclusion   303

19. MEMS Packaging   **304–329**

Introduction   304
Functions of MEMS Packages   304
   Mechanical Support   305
   Protection from Environment   305
   Electrical Connection to Other System Components   306
   Thermal Considerations   306
Types of MEMS Packages   307
   Metal Packages   307
   Ceramic Packages   308
   Thin-film Multilayer Packages   308
   Plastic Packages   309
   Package-to-MEMS Attachment   309
   Chip Scale Packaging   310
   Multichip Packaging   315
   Plastic Packaging (PEMs)   319
Generic Surface Micromachining Module for MEMS Hermetic Packaging at
   Temperatures Below 200°C   322
   Requirements   322
   Experimental Procedure   324
   Results and Discussion   325

20. Nanosensor   **330–342**

Introduction   330
Predicted Applications   330
Existing Nanosensors   331
Production Methods   331
Economic Impact   332
Infrared Temperature Sensors   332
   Gold-Flecked Nanosensor Detects Poisonous Mercury   332
   Smarter, Faster Nanosensor   333
Chemical Nanosensor Development and Characterisation   334
Nanosensors for Aqueous Environments   334
   Research Plan   334
Chemical Sensors and Nanotechnology   335
Nanosensors and Devices for Space and Terrestrial Applications   335
   Rationale for Recommendation   336

Current State-of-the-Art   336
Baseline Use   338
Trends Impacting Improvement   339
Alternative Approaches and Organisations   339
Leading Aerospace Applications   340
Potential Nonaerospace Applications   340
Drivers for Change   341
Recommended Approach(es)   342
Rationale for Recommendations   342

SECTION II

Nanocomposites and their Applications   *343–519*

21. Nanocomposites   **345–361**

Introduction   345
Applied Condition   346
Processing   347
Properties   347
    Mechanical Properties   347
    Air Permeability   348
    Reduction in Cost   349
Nanocomposites: Novel Polymer/Inorganic Hybrids   349
    Clay Structure   349
    Nanocomposites   350
    Nanocomposite Synthesis   351
    Advantages of Nanocomposites Over Conventional Composites   355
    Industrial Applications   355
Resins for Composites   355
    Main Components of Resin Systems   356
    Advantages of Composites   359

22. Clay Based Nanocomposites   **362-373**

Introduction   362
Flammability Properties and Radiation Resistance Properties   363
Nanosilicates   367
    Nanosilicate Treatments for Fabric   368
    Gas Barrier   369
Ceramic–Metal Nanocomposites   370

23. Processing and Characterisation of Nanoceramic Composites   **374-387**

Introduction   374
Experimental Procedure   376
Results and Discussion   378
    Processing and Microstructure   378

	Mechanical Properties	379
	Electrical Properties	381
	Thermal Properties	382
	Thermoelectric Properties	383
	Ultra-Low-Temperature Superplasticity	384

24. Polyolefin, Polypropylene and Polystyrene Nanocomposites — **388-406**

Introduction	388
Creating Polyolefin Nanocomposites	388
Nanocomposite Performance	388
Properties	389
Gas Barrier Enhancement	391
Preparation and Properties of Polyolefin Nanocomposites	393
Experimental Procedure	393
Results and Discussion	394
Preparation and Characterisation of Polypropylene Nanocomposites Containing	
Polystyrene-grafted Alumina Nanoparticles	397
Experimental Procedure	397
Results and Discussion	399
Optical and Structural Characterisation of Periodic Silver-polystyrene Nanocomposites	402
Sample Preparation	402
Results and Discussion	403

25. Polyester-clay Nanocomposites — **407-416**

Introduction	407
Experimental Details	408
Characterisation	408
Characterisation and Property Evaluation	408
Modal Testing	409
Results and Discussion	409
Dynamic Mechanical Analysis	412
Natural Frequency	413
Damping Factor	415

26. Nylon-6 Nanocomposites — **417–422**

Introduction	417
Clay Platelets	417
Montmorillonite	418
Compatibilising Agents	419
Mechanical Properties	420
Gas Permeability	421
Properties	421
Biomedical Nanocomposites	422

27.   Chemical Synthesis of Inorganic Nanomaterials and Composites   **423–430**

Introduction   423
Synthesis of Nanomaterials and Composites   424
    Physical Methods   424
    Chemical Method   424
Industrial Application of Nanocomposite Fillers Based on Organic Intercalated Bentonites   426

28.   Biologically Derived Synthetic Nanocomposites   **431–469**

Introduction   431
Protein-based Nanostructure Formation   431
DNA-Templated Nanostructure Formation   432
Protein Assembly   435
Biologically Inspired Nanocomposites   436
    Biological Systems for Complex Inorganic Structures   436
    Nanoparticles   437
Lyotropic Liquid-crystal Templating   442
    Amphiphilic Molecules and Segments   443
    Phases Observed in Mixtures of Water and Amphiphile   443
    Direct Templating of Materials   447
    Mineral Growth in Hexagonal and Lamellar Phases   451
    Cubic Phase and Other Phases   452
    Direct Templating of an Inorganic by an Organic Liquid Crystal   454
Liquid-crystal Templating of Thin Films   457
Block Copolymer Templating   458
    Procedure for Block Copolymer   458
Colloidal Templating   459
    Microperiodic Structures and Photonics   459
    Colloidal Crystal Templating   460
    Sol-gel Infilling Colloidal Crystal Templates   460
    Semiconductors for Photonic Crystals   463
    Electrochemical Growth of Conducting Polymers   465

29.   Metal and Oxide Nanocomposites   **470–475**

Introduction   470
Development of Alumina-metal Nanocomposites   471
Improving Physical and Mechanical Properties of Metals by Incorporating Nanoparticles
    of an Oxide   472
Some Other Nanocomposites and Nanoalloys (Besides Al_2O_3)   474

30.   Sustainable Flame Retardant Nanocomposites   **476–480**

Introduction   476
Current Economic and Environmental Climate   476
Studies of Polymers with Layered Silicates   477
Discussion   478

31. Nondestructive Testing of Nanocomposites by Optical Techniques 481–487

Introduction 481
Holographic Interferometry 482
Electronic Speckle Pattern Interferometry 484
Shearography 484
Comparison of Holography, ESPI and Shearography 485
Examples 486

32. Nanoclays for Polymer Nanocomposites, Paints, Cosmetics and Waste-water Treatment 488–502

Introduction 488
Clays and their Modification 489
Polymer Nanocomposites 494
Rheological Modifier 496
Nanoclay as Drug Vehicle 499
Waste-water Treatment 500

33. Hard and Tough Nanocomposite Coatings 503–519

Introduction 503
Design Methods for Hard Yet Tough Nanocomposites 504
  Hardening Mechanisms in Nanocomposite Coatings 504
  Toughening Mechanisms in Nanocomposite Coatings 508
  Toughness Characterisation 511
Superhard Nanocomposite Coating 513
  Generic Design Concept and Thermal Stability of Superhard Nanocomposites 514
  Mechanical Properties of the Superhard Nanocomposites 517
  Industrial Applications 517
Future Trends and Directions 519

References 521

Index 523–529

SECTION I

General Considerations

1. Basic Concepts of Nanotechnology   3
2. Quantum Chemistry   23
3. Solid-state Physics   34
4. Fullerene   50
5. Carbon Nanotubes   59
6. Nanowire   90
7. Nanomechanics   102
8. Micro- and Nano-electromechanical System   113
9. Supramolecular Chemistry   122
10. Microfluidics   158
11. Self-assembled Monolayer   168
12. Nanostructured Advanced Materials   186
13. Gas Phase Cluster   200
14. Nanophotonics   218
15. Nanomaterial Synthesis and Application   232
16. Oxide Nanoprecursors: A Technological Perspective   261
17. Core/Shell Nanoparticles   276
18. Kinetics and Energetics in Nanolubrication   286
19. MEMS Packaging   304
20. Nanosensor   330

Chapter 1

Basic Concepts of Nanotechnology

INTRODUCTION

Nanotechnology, shortened to 'nanotech', is the study of the controlling of matter on an atomic and molecular scale. Generally nanotechnology deals with structures of the size 100 nanometers or smaller in at least one dimension, and involves developing materials or devices within that size. Nanotechnology is very diverse, ranging from extensions of conventional device physics to completely new approaches based upon molecular self-assembly, from developing new materials with dimensions on the nanoscale to investigating whether we can directly control matter on the atomic scale.

There has been much debate on the future implications of nanotechnology. Nanotechnology has the potential to create many new materials and devices with a vast range of applications, such as in medicine, electronics and energy production. On the other hand, nanotechnology raises many of the same issues as with any introduction of new technology, including concerns about the toxicity and environmental impact of nanomaterials, and their potential effects on global economics, as well as speculation about various doomsday scenarios. These concerns have led to a debate among advocacy groups and governments on whether special regulation of nanotechnology is warranted.

ORIGIN

The first use of the concepts found in 'nanotechnology' (but pre-dating use of that name) was in 'There's Plenty of Room at the Bottom,' a talk given by physicist Richard Feynman at an American Physical Society meeting at Caltech on December 29, 1959. Feynman described a process by which the ability to manipulate individual atoms and molecules might be developed, using one set of precise tools to build and operate another proportionally smaller set, and so on down to the needed scale. In the course of this, he noted, scaling issues would arise from the changing magnitude of various physical phenomena: gravity would become less important, surface tension and van der Waals attraction would become increasingly more significant, etc. This basic idea appeared plausible, and exponential assembly enhances it with parallelism to produce a useful quantity of end products. The term 'nanotechnology' was defined by Tokyo Science University Professor Norio Taniguchi in a 1974 paper as follows: 'Nanotechnology' mainly consists of the processing of, separation, consolidation, and deformation of materials by one atom or by one molecule.' In the 1980s the basic idea of this definition was explored in much more depth by Dr. K. Eric Drexler, who promoted the technological significance of nano-scale phenomena and devices through speeches and the books Engines of Creation: The coming era of nanotechnology and nanosystems: molecular machinery, manufacturing, and computation, and so the term acquired its current sense. Nanotechnology and nanoscience got started in the early 1980s with two major developments; the birth of cluster science and the invention of the scanning tunneling microscope (STM). This development led to the discovery of fullerenes in 1985 and carbon nanotubes a few years later. The Buckminsterfullerene C_{60}, also known as the buckyball, is the simplest of the carbon structures known as fullerenes is shown in Fig. 1.1. In another development, the synthesis and properties of semiconductor nanocrystals was studied; this led to a fast increasing number of metal and metal oxide nanoparticles and quantum dots. The atomic force microscope was invented six years after the STM was invented. In 2000, the United States National Nanotechnology Initiative was founded to coordinate Federal nanotechnology research and development.

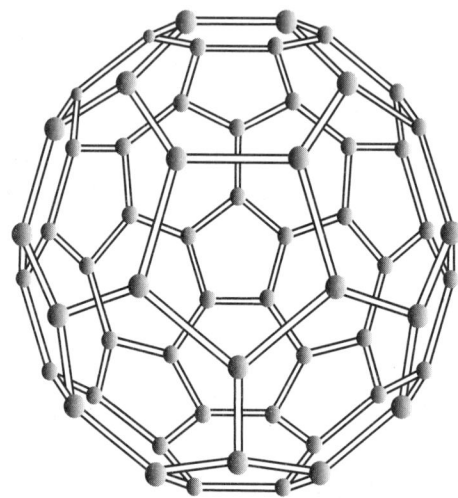

Fig. 1.1. Buckminsterfullerene C_{60}, also known as the buckyball, is the simplest of the carbon structures known as fullerenes. Members of the fullerene family are a major subject of research falling under the nanotechnology umbrella.

FUNDAMENTAL CONCEPTS

One nanometer (nm) is one billionth, or 10^{-9}, of a meter. By comparison, typical carbon-carbon bond lengths, or the spacing between these atoms in a molecule, are in the range 0.12–0.15 nm, and a DNA double-helix has a diameter around 2 nm. On the other hand, the smallest cellular life-forms, the bacteria of the genus *Mycoplasma*, are around 200 nm in length.

To put that scale in another context, the comparative size of a nanometer to a meter is the same as that of a marble to the size of the earth. Or another way of putting it: a nanometer is the amount a man's beard grows in the time it takes him to raise the razor to his face.

Two main approaches are used in nanotechnology. In the 'bottom-up' approach, materials and devices are built from molecular components which assemble themselves chemically by principles of molecular recognition. In the 'top-down' approach, nano-objects are constructed from larger entities without atomic-level control.

Areas of physics such as nanoelectronics, nanomechanics and nanophotonics have evolved during the last few decades to provide a basic scientific foundation of nanotechnology.

Simple to Complex: A Molecular Perspective

Modern synthetic chemistry has reached the point where it is possible to prepare small molecules to almost any structure. These methods are used today to manufacture a wide variety of useful chemicals such as pharmaceuticals or commercial polymers. This ability raises the question of extending this kind of control to the next-larger level, seeking methods to assemble these single molecules into supramolecular assemblies consisting of many molecules arranged in a well defined manner.

These approaches utilise the concepts of molecular self-assembly and/or supramolecular chemistry to automatically arrange themselves into some useful conformation through a bottom-up approach. The concept of molecular recognition is especially important: molecules can be designed so that a specific configuration or arrangement is favoured due to non-covalent intermolecular forces. The Watson–Crick basepairing rules are a direct result of this, as is the specificity of an enzyme being targeted to a single substrate, or the specific folding of the protein itself. Thus, two or more components can be designed to be complementary and mutually attractive so that they make a more complex and useful whole.

Such bottom-up approaches should be capable of producing devices in parallel and be much cheaper than top-down methods, but could potentially be overwhelmed as the size and complexity of the desired assembly increases. Most useful structures require complex and thermodynamically unlikely arrangements of atoms. Nevertheless, there are many examples of self-assembly based on molecular recognition in biology, most notably Watson–Crick basepairing and enzyme-substrate interactions. The challenge for nanotechnology is whether these principles can be used to engineer new constructs in addition to natural ones.

Molecular Nanotechnology

Molecular nanotechnology, sometimes called molecular manufacturing, is a term given to the concept of engineered nanosystems (nanoscale machines) operating on the molecular scale. It is especially associated with the concept of a molecular assembler, a machine that can produce a desired structure or device atom-by-atom using the principles of mechanosynthesis. Manufacturing in the context of productive nanosystems is not related to, and should be clearly distinguished from the conventional technologies used to manufacture nanomaterials such as carbon nanotubes and nanoparticles.

When the term 'nanotechnology' was independently coined and popularised by Eric Drexler (who at the time was unaware of an earlier usage by Norio Taniguchi) it referred to a future manufacturing technology based on molecular machine systems. The premise was that molecular scale biological analogies of traditional machine components demonstrated molecular machines were possible, by the countless examples found in biology, it is known that sophisticated, stochastically optimised biological machines can be produced. It is hoped that developments in nanotechnology will make possible their construction by some other means, perhaps using biomimetic principles. However, Drexler and other researchers have proposed that advanced nanotechnology, although perhaps initially implemented by biomimetic means, ultimately could be based on mechanical engineering principles, namely, a manufacturing technology based on the mechanical functionality of these components (such as gears, bearings, motors, and structural members) that would enable programmable, positional assembly to atomic specification. In general it is very difficult to assemble devices on the atomic scale, as all one has to position atoms are other atoms of comparable size and stickiness. Another view, put forth by Carlo Montemagno, is that future nanosystems will be hybrids of silicon technology and biological molecular machines. Yet another view, put forward by the late Richard Smalley, is that mechanosynthesis is impossible due to the difficulties in mechanically manipulating individual molecules. Though biology clearly demonstrates that molecular machine systems are possible, non-biological molecular machines are today only in their infancy. Leaders in research on non-biological molecular machines Dr. Alex Zettl and his colleagues have constructed at least three distinct molecular devices whose motion is controlled from the desktop with changing voltage: a nanotube nanomotor, a molecular actuator, and a nanoelectromechanical relaxation oscillator (Fig. 1.2).

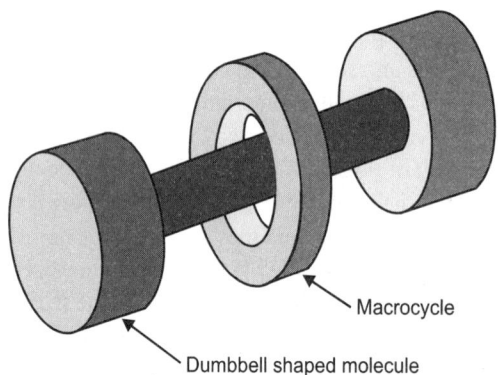

Fig. 1.2. Graphical representation of a rotaxane, useful as a molecular switch.

An experiment indicating that positional molecular assembly is possible was performed by Ho and Lee, who used a scanning tunneling microscope to move an individual carbon monoxide (CO) molecule to an individual iron (Fe) atom sitting on a flat silver crystal, and chemically bound the CO to the Fe by applying a voltage.

CURRENT RESEARCH

Nanomaterials

This includes subfields which develop or study materials having unique properties arising from their nanoscale dimensions:

1. Interface and colloid science has given rise to many materials which may be useful in nanotechnology, such as carbon nanotubes and other fullerenes, and various nanoparticles and nanorods.
2. Nanoscale materials can also be used for bulk applications; most present commercial applications of nanotechnology are of this flavour.
3. Progress has been made in using these materials for medical applications.
4. Nanoscale materials are sometimes used in solar cells which combats the cost of traditional silicon solar cells.
5. Development of applications incorporating semiconductor nanoparticles to be used in the next generation of products, such as display technology, lighting, solar cells and biological imaging.

Bottom-up Approaches

These seek to arrange smaller components into more complex assemblies:

1. DNA nanotechnology utilises the specificity of Watson–Crick basepairing to construct well-defined structures out of DNA and other nucleic acids. Sarfus image of a DNA biochip elaborated by bottom-up approach are shown in Fig. 1.3.

2. Approaches from the field of 'classical' chemical synthesis also aim at designing molecules with well-defined shape (e.g. *bis*-peptides).

3. More generally, molecular self-assembly seeks to use concepts of supramolecular chemistry, and molecular recognition in particular, to cause single-molecule components to automatically arrange themselves into some useful conformation. Device transfers energy from nano-thin layers of quantum wells to nanocrystals above them, causing the nanocrystals to emit visible light is shown in Fig. 1.4.

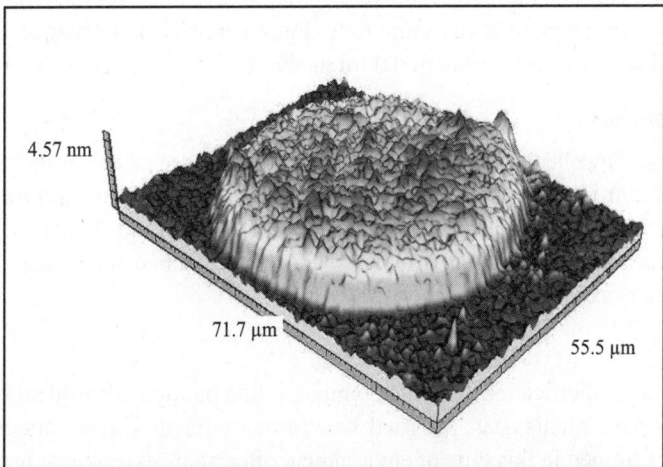

Fig. 1.3. Sarfus image of a DNA biochip elaborated by bottom-up approach.

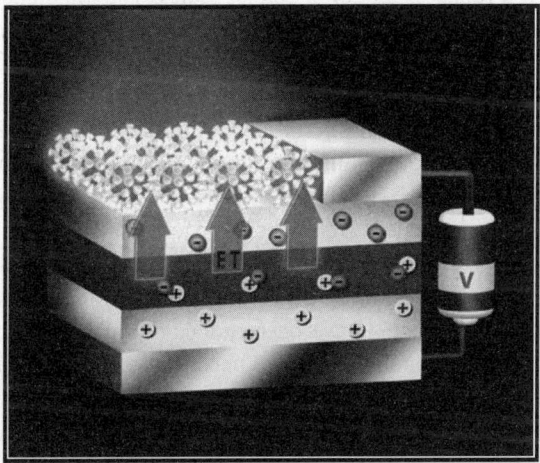

Fig. 1.4. Device transfers energy from nano-thin layers of quantum wells to nanocrystals above them, causing the nanocrystals to emit visible light.

Top-Down Approaches

These seek to create smaller devices by using larger ones to direct their assembly:

1. Many technologies that descended from conventional solid-state silicon methods for fabricating microprocessors are now capable of creating features smaller than 100 nm, falling under the

definition of nanotechnology. Giant magnetoresistance-based hard drives already on the market fit this description, as do atomic layer deposition (ALD) techniques.

2. Solid-state techniques can also be used to create devices known as nanoelectromechanical systems or NEMS, which are related to microelectromechanical systems or MEMS.
3. Atomic force microscope tips can be used as a nanoscale 'write head' to deposit a chemical upon a surface in a desired pattern in a process called dip pen nanolithography. This fits into the larger subfield of nanolithography.
4. Focused ion beams can directly remove material, or even deposit material when suitable precursor gasses are applied at the same time. For example, this technique is used routinely to create sub-100 nm sections of material for analysis in transmission electron microscopy.

Functional Approaches

These seek to develop components of a desired functionality without regard to how they might be assembled.

1. Molecular electronics seeks to develop molecules with useful electronic properties. These could then be used as single-molecule components in a nanoelectronic device.
2. Synthetic chemical methods can also be used to create synthetic molecular motors, such as in a so-called nanocar.

Biomaterials

Biomineralisation (e.g. silicification) is quite common in the biological world and occurs in bacteria, single-celled organisms, plants (e.g. petrified wood), and animals (invertebrates and vertebrates). Crystalline minerals formed in this type of environment often show exceptional mechanical properties (e.g. strength, hardness, fracture toughness) and tend to form hierarchical structures that exhibit microstructural order over a range of length or spatial scales. The minerals are typically crystallised from an environment that is undersaturated with respect to certain metallic elements such as silicon, calcium and phosphorus, which are readily oxidised under conditions of neutral pH and low temperature (0°–40°C). Formation of the mineral may occur either within or outside of the cell wall of an organism, and specific biochemical reactions for mineral deposition exist that include lipids, proteins and carbohydrates. The significance of the cellular machinery cannot be overemphasised, and it is with advances in experimental techniques in cellular biology and the capacity to mimic the biological environment that significant progress is currently being reported.

Examples include silicates in algae and diatoms, carbonates in invertebrates, and calcium phosphates and carbonates in vertebrates. These minerals often form structural features such as sea shells and the bone in mammals and birds. Organisms have been producing mineralised skeletons for nearly 600 million years. The most common biominerals are the phosphate and carbonate salts of calcium that are used in conjunction with organic polymers such as collagen and chitin to give mechanical strength to bones and shells. Other examples include copper, iron and gold deposits involving bacteria.

Thus, most natural (or biological) materials are complex composites whose mechanical properties are often outstanding, considering the weak constituents from which they are assembled. These complex structures, which have risen from hundreds of million years of evolution, are inspiring materials scientists interested primarily in the design of novel materials with exceptional physical properties for high performance in adverse conditions. Their defining characteristics such as hierarchy, multifunctionality, and the capacity for self-healing, are currently being investigated. Sand from Pismo Beach, California including quartz, shell and rock fragments is shown in Fig. 1.5.

Fig. 1.5. Sand from Pismo Beach, California including quartz, shell and rock fragments.

The basic building blocks begin with the 20 amino acids and proceed to polypeptides and polysaccharides. These, in turn, compose the basic proteins, which are the primary constituents of the 'soft tissues' common to most biominerals. With well over 1000 proteins possible, current research emphasises the use of collagen, chitin, keratin, and elastin. The 'hard' phases are often strengthened by crystalline minerals, which nucleate and grow in a biomediated environment that determines the size, shape and distribution of individual crystals. The most important silicate phases have been identified as hydroxyapatite, silica, and aragonite. Using the classification of Wegst and Ashby, the principal mechanical characteristics and structures of a number of biological ceramics, polymer composites, elastomers, and cellular materials have been recently characterised. Selected systems in each class are being investigated with emphasis on the relationship between their microstructure over a range of length scales and their mechanical response. Collagen fibres of woven bone are given in Fig. 1.6.

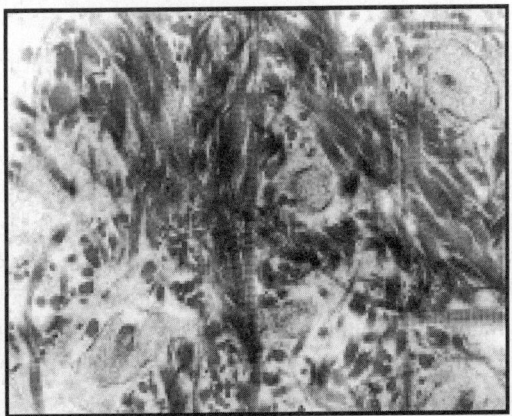

Fig. 1.6. Collagen fibres of woven bone.

Recent joint collaboration at UC Santa Barbara and UC San Diego has produced striking results, including high resolution SEM images of the microstructure of the mother-of-pearl (or nacre) portion of the abalone shell, which exhibits the highest mechanical strength and fracture toughness of any non-metallic substance known.

Clearly visible in these images (http://www.wavesignal.com/Materials/Nacre_Shell.htm) are the neatly stacked (or ordered) mineral tiles separated by thin organic sheets—along with a macrostructure of larger periodic growth bands which collectively form what scientists refer to as a hierarchical composite structure. Iridescent nacre inside a Nautilus shell is given in Fig. 1.7.

Fig. 1.7. Iridescent nacre inside a nautilus shell.

Speculative

These subfields seek to anticipate what inventions nanotechnology might yield, or attempt to propose an agenda along which inquiry might progress. These often take a big-picture view of nanotechnology, with more emphasis on its societal implications than the details of how such inventions could actually be created:

1. Molecular nanotechnology is a proposed approach which involves manipulating single molecules in finely controlled, deterministic ways. This is more theoretical than the other subfields and is beyond current capabilities.

2. Nanorobotics centres on self-sufficient machines of some functionality operating at the nanoscale. There are hopes for applying nanorobots in medicine, but it may not be easy to do such a thing because of several drawbacks of such devices. Nevertheless, progress on innovative materials and methodologies has been demonstrated with some patents granted about new nano-manufacturing devices for future commercial applications, which also progressively helps in the development towards nanorobots with the use of embedded nanobioelectronics concepts.

3. Programmable matter based on artificial atoms seeks to design materials whose properties can be easily, reversibly and externally controlled.

4. Due to the popularity and media exposure of the term nanotechnology, the words picotechnology and femtotechnology have been coined in analogy to it, although these are only used rarely and informally.

TOOLS AND TECHNIQUES

There are several important modern developments. The atomic force microscope (AFM) and the scanning tunneling microscope (STM) are two early versions of scanning probes that launched (Fig. 1.8) nanotechnology. There are other types of scanning probe microscopy, all flowing from the ideas of the scanning confocal microscope developed by Marvin Minsky in 1961 and the scanning acoustic microscope (SAM) developed by Calvin Quate and coworkers in the 1970s, that made it possible to see structures at the nanoscale. The tip of a scanning probe can also be used to manipulate nanostructures (a process called positional assembly). Feature-oriented scanning-positioning methodology suggested by Rostislav Lapshin appears to be a promising way to implement these nanomanipulations in automatic mode. However, this is still a slow process because of low scanning velocity of the microscope. Various techniques of nanolithography such as optical lithography, X-ray lithography dip pen nanolithography, electron beam lithography or nanoimprint lithography were also developed. Lithography is a top-down fabrication technique where a bulk material is reduced in size to nanoscale pattern.

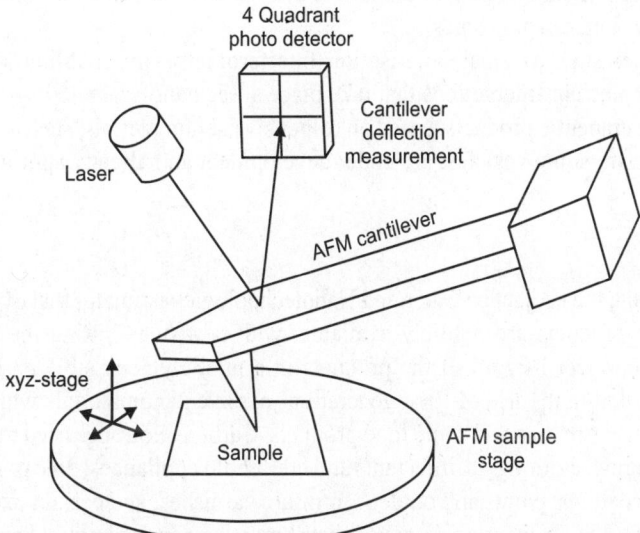

Fig. 1.8. Typical AFM setup. A microfabricated cantilever with a sharp tip is deflected by features on a sample surface, much like in a phonograph but on a much smaller scale. A laser beam reflects off the backside of the cantilever into a set of photodetectors, allowing the deflection to be measured and assembled into an image of the surface.

Another group of nanotechnological techniques include those used for fabrication of nanowires, those used in semiconductor fabrication such as deep ultraviolet lithography, electron beam lithography, focused ion beam machining, nanoimprint lithography, atomic layer deposition, and molecular vapour deposition, and further including molecular self-assembly techniques such as those employing di-block copolymers. However, all of these techniques preceded the nanotech era, and are extensions in the development of scientific advancements rather than techniques which were devised with the sole purpose of creating nanotechnology and which were results of nanotechnology research.

The top-down approach anticipates nanodevices that must be built piece by piece in stages, much as manufactured items are made. Scanning probe microscopy is an important technique both for characterisation and synthesis of nanomaterials. Atomic force microscopes and scanning tunneling microscopes can be used to look at surfaces and to move atoms around. By designing different tips for these microscopes, they can be used for carving out structures on surfaces and to help guide self-assembling structures. By using, for example, feature-oriented scanning-positioning approach, atoms can be moved around on a surface with scanning probe microscopy techniques. At present, it is expensive and time-consuming for mass production but very suitable for laboratory experimentation.

In contrast, bottom-up techniques build or grow larger structures atom by atom or molecule by molecule. These techniques include chemical synthesis, self-assembly and positional assembly. Another variation of the bottom-up approach is molecular beam epitaxy or MBE. Researchers at Bell Telephone Laboratories like John R. Arthur. Alfred Y. Cho, and Art C. Gossard developed and implemented MBE as a research tool in the late 1960s and 1970s. Samples made by MBE were key to the discovery of the fractional quantum Hall effect for which the 1998 Nobel Prize in Physics was awarded. MBE allows scientists to lay down atomically-precise layers of atoms and, in the process, build up complex structures. Important for research on semiconductors, MBE is also widely used to make samples and devices for the newly emerging field of spintronics.

Newer techniques such as Dual Polarisation Interferometry are enabling scientists to measure quantitatively the molecular interactions that take place at the nano-scale.

However, new therapeutic products, based on responsive nanomaterials, such as the ultradeformable, stress-sensitive Transfersome vesicles, are under development and already approved for human use in some countries.

APPLICATIONS

As of August 21, 2008, the Project on Emerging Nanotechnologies estimates that over 800 manufacturer-identified nanotech products are publicly available, with new ones hitting the market at a pace of 3–4 per week. The project lists all of the products in a publicly accessible on-line inventory. Most applications are limited to the use of 'first generation' passive nanomaterials which includes titanium dioxide in sunscreen, cosmetics and some food products. Carbon allotropes used to produce gecko tape; silver in food packaging, clothing, disinfectants and household appliances; zinc oxide in sunscreens and cosmetics, surface coatings, paints and outdoor furniture varnishes; and cerium oxide as a fuel catalyst.

The National Science Foundation (a major distributor for nanotechnology research in the United States) funded researcher David Berube to study the field of nanotechnology. His findings are published in the monograph Nano-Hype (http://www.prometheusbooks.com index.php?main_page= product_info&products_id=1822/): The Truth Behind the Nanotechnology Buzz. This published study (with a foreword by Mikhail Roco, Senior Advisor for Nanotechnology at the National Science Foundation) concludes that much of what is sold as 'nanotechnology' is in fact a recasting of straightforward materials science, which is leading to a 'nanotech industry built solely on selling nanotubes, nanowires, and the like' which will 'end up with a few suppliers selling low margin products in huge volumes.' Further applications which require actual manipulation or arrangement of nanoscale components await further research. Though technologies branded with the term 'nano' are sometimes little related to and fall far short of the most ambitious and transformative technological goals of the sort in molecular manufacturing proposals, the term still connotes such ideas. According to Berube, there may be a danger that a 'nano bubble' will form, or is forming already, from the use of the term by scientists and entrepreneurs to garner funding, regardless of interest in the transformative possibilities of more ambitious and far-sighted work.

Nano-membranes have been produced that are portable and easily-cleaned systems that purify, detoxify and desalinate water meaning that third-world countries could get clean water, solving many water related health issues.

IMPLICATIONS

Due to the far-ranging claims that have been made about potential applications of nanotechnology, a number of serious concerns have been raised about what effects these will have on our society if realised, and what action if any is appropriate to mitigate these risks.

There are possible dangers that arise with the development of nanotechnology. The Centre for Responsible Nanotechnology suggests that new developments could result, among other things, in untraceable weapons of mass destruction, networked cameras for use by the government, and weapons developments fast enough to destabilise arms races (Nanotechnology Basics).

One area of concern is the effect that industrial-scale manufacturing and use of nanomaterials would have on human health and the environment, as suggested by nanotoxicology research. Groups such as the Centre for Responsible Nanotechnology have advocated that nanotechnology should be specially regulated by governments for these reasons. Others counter that overregulation would stifle scientific research and the development of innovations which could greatly benefit mankind.

Other experts, including director of the Woodrow Wilson Centre's Project on Emerging Nano-technologies David Rejeski, have testified that successful commercialisation depends on adequate oversight, risk research strategy, and public engagement. Berkeley, California is currently the only city in the United States to regulate nanotechnology; Cambridge, Massachusetts in 2008 considered enacting a similar law, but ultimately rejected this.

Health and Environmental Concerns

Some of the recently developed nanoparticle products may have unintended consequences. Researchers have discovered that silver nanoparticles used in socks only to reduce foot odour are being released in the wash with possible negative consequences. Silver nanoparticles, which are bacteriostatic, may then destroy beneficial bacteria which are important for breaking down organic matter in waste treatment plants or farms.

A study at the University of Rochester found that when rats breathed in nanoparticles, the particles settled in the brain and lungs, which led to significant increases in biomarkers for inflammation and stress response.

A major study published more recently in Nature Nanotechnology suggests some forms of carbon nanotubes—a poster child for the 'nanotechnology revolution'—could be as harmful as asbestos if inhaled in sufficient quantities. Anthony Seaton of the Institute of Occupational Medicine in Edinburgh, Scotland, who contributed to the article on carbon nanotubes said, 'We know that some of them probably have the potential to cause mesothelioma. So those sorts of materials need to be handled very carefully.' In the absence of specific nano-regulation forthcoming from governments, Paull and Lyons have called for an exclusion of engineered nanoparticles from organic food. A newspaper article reports that workers in a paint factory developed serious lung disease and nanoparticles were found in their lungs.

Regulation

Calls for tighter regulation of nanotechnology have occurred alongside a growing debate related to the human health and safety risks associated with nanotechnology. Furthermore, there is significant debate about who is responsible for the regulation of nanotechnology. While some non-nanotechnology specific regulatory agencies currently cover some products and processes (to varying degrees) — by 'bolting on' nanotechnology to existing regulations — there are clear gaps in these regimes. In 'Nanotechnology Oversight: An Agenda for the Next Administration,' former EPA deputy administrator J. Clarence (Terry) Davies lays out a clear regulatory roadmap for the next presidential administration and describes the immediate and longer term steps necessary to deal with the current shortcomings of nanotechnology oversight.

Stakeholders concerned by the lack of a regulatory framework to assess and control risks associated with the release of nanoparticles and nanotubes have drawn parallels with bovine spongiform encephalopathy (mad cow's disease), thalidomide, genetically modified food, nuclear energy, reproductive technologies, biotechnology, and asbestosis. Dr. Andrew Maynard, chief science advisor to the Woodrow Wilson Centre's Project on Emerging Nanotechnologies, concludes (among others) that there is insufficient funding for human health and safety research, and as a result there is currently limited understanding of the human health and safety risks associated with nanotechnology. As a result, some academics have called for stricter application of the precautionary principle, with delayed marketing approval, enhanced labelling and additional safety data development requirements in relation to certain forms of nanotechnology.

The Royal Society report identified a risk of nanoparticles or nanotubes being released during disposal, destruction and recycling, and recommended that 'manufacturers of products that fall under extended producer responsibility regimes such as end-of-life regulations publish procedures outlining how these materials will be managed to minimise possible human and environmental exposure'. Reflecting the challenges for ensuring responsible life cycle regulation, the Institute for Food and Agricultural Standards (https://www.msu.edu/~ifas/) has proposed standards for nanotechnology research and development should be integrated across consumer, worker and environmental standards. They also propose that NGOs and other citizen groups play a meaningful role in the development of these standards.

In October 2008, the department of toxic substances control (DTSC), within the California Environmental Protection Agency, announced its intent to request information regarding analytical test methods, fate and transport in the environment, and other relevant information from manufacturers of carbon nanotubes. The purpose of this information request will be to identify information gaps and to develop information about carbon nanotubes, an important emerging nanomaterial.

ENERGY APPLICATIONS OF NANOTECHNOLOGY

Over the past few decades, the fields of science and engineering have been seeking to develop new and improved types of energy technologies that have the capability of improving life all over the world. In order to make the next leap forward from the current generation of technology, scientists and engineers have been developing Energy Applications of Nanotechnology. Nanotechnology, a new field in science, is any technology that contains components smaller than 100 nanometers. For scale, a single virus particle is about 100 nanometers in width.

An important subfield of nanotechnology related to energy is nanofabrication. Nanofabrication is the process of designing and creating devices on the nanoscale. Creating devices smaller than 100 nanometers opens many doors for the development of new ways to capture, store, and transfer energy. The inherent level of control that nanofabrication could give scientists and engineers would be critical in providing the capability of solving many of the problems that the world is facing today related to the current generation of energy technologies.

People in the fields of science and engineering have already begun developing ways of utilising nanotechnology for the development of consumer products. Benefits already observed from the design of these products are an increased efficiency of lighting and heating, increased electrical storage capacity, and a decrease in the amount of pollution from the use of energy. Benefits such as these make the investment of capital in the research and development of nanotechnology a top priority.

Consumer Products

Recently, previously established and entirely new companies such as BetaBatt, Inc. and Oxane Materials are focusing on nanomaterials as a way to develop and improve upon older methods for the capture, transfer, and storage of energy for the development of consumer products.

ConsERV, a product developed by the Dais Analytic Corporation, uses nanoscale polymer membranes to increase the efficiency of heating and cooling systems and has already proven to be a lucrative design. The polymer membrane was specifically configured for this application by selectively engineering the size of the pores in the membrane to prevent air from passing, while allowing moisture to pass through the membrane. Polymer membranes can be designed to selectively allow particles of one size and shape to pass through while preventing others of different dimensions. This makes for a powerful tool that can be used in consumer products from biological weapons protection to industrial chemical separations.

A New York based company called Applied Nano Works, Inc. has been developing a consumer product that utilises LED technology to generate light. Light-emitting diodes or LEDs, use only about 10 per cent of the energy that a typical incandescent or fluorescent light bulb use and typically lasts much longer, which makes them a viable alternative to traditional light bulbs. While LEDs have been around for decades, this company and others like it have been developing a special variant of LED called the white LED. White LEDs consist of semi-conducting organic layers that are only about 100 nanometers in distance from each other and are placed between two electrodes, which create an anode, and a cathode. When voltage is applied to the system, light is generated when electricity passes through the two organic layers. This is called electroluminescence. The semiconductor properties of the organic layers are what allow for the minimal amount of energy necessary to generate light. In traditional light bulbs, a metal filament is used to generate light when electricity is run through the filament. Using metal generates a great deal of heat and therefore lowers efficiency.

Research for longer lasting batteries has been an ongoing process for years. Researchers have now begun to utilise nanotechnology for battery technology. mPhase Technologies in conglomeration with Rutgers University and Bell Laboratories have utilised nanomaterials to alter the wetting behaviour of the surface where the liquid in the battery lies to spread the liquid droplets over a greater area on the surface and therefore have greater control over the movement of the droplets. This gives more control to the designer of the battery. This control prevents reactions in the battery by separating the electrolytic liquid from the anode and the cathode when the battery is not in use and joining them when the battery is in need of use.

Thermal applications also are a future applications of nanotechnology creating low cost system of heating, ventilation, and air conditioning, changing molecular structure for better management of temperature.

Economic Benefits

The relatively recent shift toward using nanotechnology with respect to the capture, transfer, and storage of energy has and will continue to have many positive economic impacts on society. The control of materials that nanotechnology offers to scientists and engineers of consumer products is one of the most important aspects of nanotechnology. This allows for an improved efficiency of products across the board.

A major issue with current energy generation is the loss of efficiency from the generation of heat as a by-product of the process. A common example of this is the heat generated by the internal combustion engine. The internal combustion engine loses about 64 per cent of the energy from gasoline as heat and an improvement of this alone could have a significant economic impact. However, improving the internal combustion engine in this respect has proven to be extremely difficult without sacrificing performance. Improving the efficiency of fuel cells through the use of nanotechnology appears to be more plausible by using molecularly tailored catalysts, polymer membranes, and improved fuel storage.

In order for a fuel cell to operate, particularly of the hydrogen variant, a noble-metal catalyst (usually platinum, which is very expensive) is needed to separate the electrons from the protons of the hydrogen atoms. However, catalysts of this type are extremely sensitive to carbon monoxide reactions. In order to combat this, alcohols or hydrocarbons compounds are used to lower the carbon monoxide concentration in the system. This adds an additional cost to the device. Using nanotechnology, catalysts can be designed through nanofabrication that are much more resistant to carbon monoxide reactions, which improves the efficiency of the process and may be designed with cheaper materials to additionally lower costs.

Fuel cells that are currently designed for transportation need rapid start-up periods for the practicality of consumer use. This process puts a lot of strain on the traditional polymer electrolyte membranes, which decreases the life of the membrane requiring frequent replacement. Using nanotechnology, engineers have the ability to create a much more durable polymer membrane, which addresses this problem. Nanoscale polymer membranes are also much more efficient in ionic conductivity. This improves the efficiency of the system and decreases the time between replacements, which lowers costs.

Another problem with contemporary fuel cells is the storage of the fuel. In the case of hydrogen fuel cells, storing the hydrogen in gaseous rather than liquid form improves the efficiency by 5 per cent. However, the materials that we currently have available to us significantly limit fuel storage due to low stress tolerance and costs. Scientists have come up with an answer to this by using a nanoporous styrene material (which is a relatively inexpensive material) that when super-cooled to around $-196°C$, naturally holds on to hydrogen atoms and when heated again releases the hydrogen for use.

APPLICATIONS OF NANOTECHNOLOGY

With nanotechnology, a large set of materials and improved products rely on a change in the physical properties when the feature sizes are shrunk. Nanoparticles for example take advantage of their dramatically increased surface area to volume ratio. Their optical properties, e.g. fluorescence, become a function of the particle diameter. When brought into a bulk material, nanoparticles can strongly influence the mechanical properties of the material, like stiffness or elasticity. For example, traditional polymers can be reinforced by nanoparticles resulting in novel materials which can be used as lightweight replacements for metals. Therefore, an increasing societal benefit of such nanoparticles can be expected. Such nanotechnologically enhanced materials will enable a weight reduction accompanied by an increase in stability and an improved functionality. Overview of many applications of nanotechnology, is discussed below.

Medicine

The biological and medical research communities have exploited the unique properties of nanomaterials for various applications (e.g. contrast agents for cell imaging and therapeutics for treating cancer).

Terms such as biomedical nanotechnology, bionanotechnology, and nanomedicine are used to describe this hybrid field. Functionalities can be added to nanomaterials by interfacing them with biological molecules or structures. The size of nanomaterials is similar to that of most biological molecules and structures; therefore, nanomaterials can be useful for both *in vivo* and *in vitro* biomedical research and applications. Thus far, the integration of nanomaterials with biology has led to the development of diagnostic devices, contrast agents, analytical tools, physical therapy applications, and drug delivery vehicles.

Diagnostics

Nanotechnology-on-a-chip is one more dimension of lab-on-a-chip technology. Magnetic nanoparticles, bound to a suitable antibody, are used to label specific molecules, structures or micro-organisms. Gold nanoparticles tagged with short segments of DNA can be used for detection of genetic sequence in a sample. Multicolour optical coding for biological assays has been achieved by embedding different-sized quantum dots into polymeric microbeads. Nanopore technology for analysis of nucleic acids converts strings of nucleotides directly into electronic signatures.

Drug delivery

The overall drug consumption and side-effects can be lowered significantly by depositing the active agent in the morbid region only and in no higher dose than needed. This highly selective approach reduces costs and human suffering. An example can be found in dendrimers and nanoporous materials. They could hold small drug molecules transporting them to the desired location. Another vision is based on small electromechanical systems; NEMS are being investigated for the active release of drugs. Some potentially important applications include cancer treatment with iron nanoparticles or gold shells. A targeted or personalised medicine reduces the drug consumption and treatment expenses resulting in an overall societal benefit by reducing the costs to the public health system. Nanotechnology is also opening up new opportunities in implantable delivery systems, which are often preferable to the use of injectable drugs, because the latter frequently display first-order kinetics (the blood concentration goes up rapidly, but drops exponentially over time). This rapid rise may cause difficulties with toxicity, and drug efficacy can diminish as the drug concentration falls below the targeted range.

Tissue engineering

Nanotechnology can help to reproduce or to repair damaged tissue. 'Tissue engineering' makes use of artificially stimulated cell proliferation by using suitable nanomaterial-based scaffolds and growth factors. Tissue engineering might replace today's conventional treatments like organ transplants or artificial implants. Advanced forms of tissue engineering may lead to life extension.

For patients with end-state organ failure, there may not be enough healthy cells for expansion and transplantation into the ECM (extracellular matrix). In this case, pluripotent stem cells are needed. One potential source for these cells is iPS (induced pluripontent stem cells); these are ordinary cells from the patients own body that are reprogrammed into a pluripotent state, and has the advantage of avoiding rejection (and the potentially life-threatening complications associated with immunosuppressive treatments). Another potential source of pluripotent cells is from embryos, but this has two disadvantages: (i) it requires that we solve the problem of cloning, which is technically very difficult (especially preventing abnormalities); and (ii) it requires the harvesting of embryos. Given that each one of us was once an embryo, this source is ethically problematic.

Chemistry and Environment

Chemical catalysis and filtration techniques are two prominent examples where nanotechnology already plays a role. The synthesis provides novel materials with tailored features and chemical properties: for example, nanoparticles with a distinct chemical surrounding (ligands), or specific optical properties. In this sense, chemistry is indeed a basic nanoscience. In a short-term perspective, chemistry will provide novel 'nanomaterials' and in the long run, superior processes such as 'self-assembly' will enable energy and time preserving strategies. In a sense, all chemical synthesis can be understood in terms of nanotechnology, because of its ability to manufacture certain molecules. Thus, chemistry forms a base for nanotechnology providing tailor-made molecules, polymers, etc. as well as clusters and nanoparticles.

Catalysis

Chemical catalysis benefits especially from nanoparticles, due to the extremely large surface to volume ratio. The application potential of nanoparticles in catalysis ranges from fuel cell to catalytic converters and photocatalytic devices. Catalysis is also important for the production of chemicals. Platinum nanoparticles are now being considered in the next generation of automotive catalytic converters because the very high surface area of nanoparticles could reduce the amount of platinum required. However, some concerns have been raised due to experiments demonstrating that they will spontaneously combust if methane is mixed with the ambient air. Ongoing research at the Centre National de la Recherche Scientifique (CNRS) in France may resolve their true usefulness for catalytic applications. Nanofiltration may come to be an important application, although future research must be careful to investigate possible toxicity.

Filtration

A strong influence of nanochemistry on waste-water treatment, air purification and energy storage devices is to be expected. Mechanical or chemical methods can be used for effective filtration techniques. One class of filtration techniques is based on the use of membranes with suitable hole sizes, whereby the liquid is pressed through the membrane. Nanoporous membranes are suitable for a mechanical filtration with extremely small pores smaller than 10 nm (nanofiltration) and may be composed of nanotubes. Nanofiltration is mainly used for the removal of ions or the separation of different fluids. On a larger scale, the membrane filtration technique is named ultrafiltration, which works down to between 10 and 100 nm. One important field of application for ultrafiltration is medical purposes as can be found in renal dialysis. Magnetic nanoparticles offer an effective and reliable method to remove heavy metal contaminants from waste-water by making use of magnetic separation techniques. Using nanoscale particles increases the efficiency to absorb the contaminants and is comparatively inexpensive compared to traditional precipitation and filtration methods. Some water-treatment devices incorporating nanotechnology are already on the market, with more in development. Low-cost nanostructured separation membranes methods have been shown to be effective in producing potable water in a recent study.

Energy

The most advanced nanotechnology projects related to energy are: storage, conversion, manufacturing improvements by reducing materials and process rates, energy saving (by better thermal insulation for example), and enhanced renewable energy sources.

Reduction of energy consumption

A reduction of energy consumption can be reached by better insulation systems, by the use of more efficient lighting or combustion systems, and by use of lighter and stronger materials in the transportation sector. Currently used light bulbs only convert approximately 5 per cent of the electrical energy into light. Nanotechnological approaches like light-emitting diodes (LEDs) or quantum caged atoms (QCAs) could lead to a strong reduction of energy consumption for illumination.

Increasing the efficiency of energy production

Today's best solar cells have layers of several different semiconductors stacked together to absorb light at different energies but they still only manage to use 40 per cent of the sun's energy. Commercially available solar cells have much lower efficiencies (15–20 per cent). Nanotechnology could help increase the efficiency of light conversion by using nanostructures with a continuum of bandgaps.

The degree of efficiency of the internal combustion engine is about 30–40 per cent at the moment. Nanotechnology could improve combustion by designing specific catalysts with maximised surface area. In 2005, scientists at the University of Toronto developed a spray-on nanoparticle substance that, when applied to a surface, instantly transforms it into a solar collector.

Use of more environmentally friendly energy systems

An example for an environmentally friendly form of energy is the use of fuel cells powered by hydrogen, which is ideally produced by renewable energies. Probably the most prominent nanostructured material in fuel cells is the catalyst consisting of carbon supported noble metal particles with diameters of 1–5 nm. Suitable materials for hydrogen storage contain a large number of small nanosized pores. Therefore many nanostructured materials like nanotubes, zeolites or alanates are under investigation. Nanotechnology can contribute to the further reduction of combustion engine pollutants by nanoporous filtres, which can clean the exhaust mechanically, by catalytic converters based on nanoscale noble metal particles or by catalytic coatings on cylinder walls and catalytic nanoparticles as additive for fuels.

Recycling of batteries

Because of the relatively low energy density of batteries the operating time is limited and a replacement or recharging is needed. The huge number of spent batteries and accumulators represent a disposal problem. The use of batteries with higher energy content or the use of rechargeable batteries or supercapacitors with higher rate of recharging using nanomaterials could be helpful for the battery disposal problem.

Information and Communication

Current high-technology production processes are based on traditional top down strategies, where nanotechnology has already been introduced silently. The critical length scale of integrated circuits is already at the nanoscale (50 nm and below) regarding the gate length of transistors in CPUs or DRAM devices.

Memory storage

Electronic memory designs in the past have largely relied on the formation of transistors. However, research into crossbar switch based electronics have offered an alternative using reconfigurable interconnections between vertical and horizontal wiring arrays to create ultra high density memories. Two leaders in this area are Nantero which has developed a carbon nanotube based crossbar memory called Nano-RAM and Hewlett-Packard which has proposed the use of memristor material as a future replacement of Flash memory.

Novel semiconductor devices

An example of such novel devices is based on spintronics. The dependence of the resistance of a material (due to the spin of the electrons) on an external field is called magnetoresistance. This effect can be significantly amplified (GMR—Giant Magneto-Resistance) for nanosized objects, for example when two ferromagnetic layers are separated by a nonmagnetic layer, which is several nanometers thick (e.g. Co-Cu-Co). The GMR effect has led to a strong increase in the data storage density of hard disks and made the gigabyte range possible. The so-called tunneling magnetoresistance (TMR) is very similar to GMR and based on the spin dependent tunneling of electrons through adjacent ferromagnetic layers. Both GMR and TMR effects can be used to create a non-volatile main memory for computers, such as the so-called magnetic random access memory or MRAM.

In 1999, the ultimate CMOS transistor developed at the Laboratory for Electronics and Information Technology in Grenoble, France, tested the limits of the principles of the MOSFET transistor with a diameter of 18 nm (approximately 70 atoms placed side by side). This was almost one tenth the size of the smallest industrial transistor in 2003 (130 nm in 2003, 90 nm in 2004, 65 nm in 2005 and 45 nm in 2007). It enabled the theoretical integration of seven billion junctions on a □1 coin. However, the CMOS transistor, which was created in 1999, was not a simple research experiment to study how CMOS technology functions, but rather a demonstration of how this technology functions now that we ourselves are getting ever closer to working on a molecular scale. Today it would be impossible to master the coordinated assembly of a large number of these transistors on a circuit and it would also be impossible to create this on an industrial level.

Novel optoelectronic devices

In the modern communication technology traditional analog electrical devices are increasingly replaced by optical or optoelectronic devices due to their enormous bandwidth and capacity, respectively. Two promising examples are photonic crystals and quantum dots. Photonic crystals are materials with a periodic variation in the refractive index with a lattice constant that is half the wavelength of the light used. They offer a selectable band gap for the propagation of a certain wavelength, thus they resemble a semiconductor, but for light or photons instead of electrons. Quantum dots are nanoscaled objects, which can be used, among many other things, for the construction of lasers. The advantage of a quantum dot laser over the traditional semiconductor laser is that their emitted wavelength depends on the diameter of the dot. Quantum dot lasers are cheaper and offer a higher beam quality than conventional laser diodes.

Displays

The production of displays with low energy consumption could be accomplished using carbon nanotubes (CNT). Carbon nanotubes are electrically conductive and due to their small diameter of several nanometers, they can be used as field emitters with extremely high efficiency for field emission displays (FED). The principle of operation resembles that of the cathode ray tube, but on a much smaller length scale.

Quantum computers

Entirely new approaches for computing exploit the laws of quantum mechanics for novel quantum computers, which enable the use of fast quantum algorithms. The quantum computer has quantum bit memory space termed 'qubit' for several computations at the same time. This facility may improve the performance of the older systems.

Heavy Industry

An inevitable use of nanotechnology will be in heavy industry.

Aerospace

Lighter and stronger materials will be of immense use to aircraft manufacturers, leading to increased performance. Spacecraft will also benefit, where weight is a major factor. Nanotechnology would help to reduce the size of equipment and thereby decrease fuel-consumption required to get it airborne. Hang gliders may be able to halve their weight while increasing their strength and toughness through the use of nanotech materials. Nanotech is lowering the mass of supercapacitors that will increasingly be used to give power to assistive electrical motors for launching hang gliders off flatland to thermal-chasing altitudes.

Construction

Nanotechnology has the potential to make construction faster, cheaper, safer, and more varied. Automation of nanotechnology construction can allow for the creation of structures from advanced homes to massive skyscrapers much more quickly and at much lower cost.

Refineries

Using nanotech applications, refineries producing materials such as steel and aluminium will be able to remove any impurities in the materials they create.

Vehicle manufacturers

Much like aerospace, lighter and stronger materials will be useful for creating vehicles that are both faster and safer. Combustion engines will also benefit from parts that are more hard-wearing and more heat-resistant.

Consumer Goods

Nanotechnology is already impacting the field of consumer goods, providing products with novel functions ranging from easy-to-clean to scratch-resistant. Modern textiles are wrinkle-resistant and stain-repellent; in the mid-term clothes will become 'smart', through embedded 'wearable electronics'. Already in use are different nanoparticle improved products. Especially in the field of cosmetics, such novel products have a promising potential.

Foods

Complex set of engineering and scientific challenges in the food and bioprocessing industry for manufacturing high quality and safe food through efficient and sustainable means can be solved through nanotechnology. Bacteria identification and food quality monitoring using biosensors; intelligent, active, and smart food packaging systems; nanoencapsulation of bioactive food compounds are few examples of emerging applications of nanotechnology for the food industry. Nanotechnology can be applied in the production, processing, safety and packaging of food. A nanocomposite coating process could improve food packaging by placing anti-microbial agents directly on the surface of the coated film. Nanocomposites could increase or decrease gas permeability of different fillers as is needed for different products. They can also improve the mechanical and heat-resistance properties and lower the oxygen transmission rate. Research is being performed to apply nanotechnology to the detection of chemical and biological substances for sensanges in foods.

Nanofoods

New consumer products emerging nanotechnologies (PEN), based on an inventory it has drawn up of 609 known or claimed nano-products. On PEN's list are three foods—a brand of canola cooking oil called canola active oil, a tea called nanotea and a chocolate diet shake called nanoceuticals slim shake chocolate. According to company information posted on PEN's website, the canola oil, by Shemen Industries of Israel, contains an additive called 'nanodrops' designed to carry vitamins, minerals and phytochemicals through the digestive system. The shake, according to US manufacturer RBC Life Sciences Inc., uses cocoa infused 'Nanoclusters' to enhance the taste and health benefits of cocoa without the need for extra sugar.

Household

The most prominent application of nanotechnology in the household is self-cleaning or 'easy-to-clean' surfaces on ceramics or glasses. Nanoceramic particles have improved the smoothness and heat resistance of common household equipment such as the flat iron.

Optics

The first sunglasses using protective and anti-reflective ultrathin polymer coatings are on the market. For optics, nanotechnology also offers scratch resistant surface coatings based on nanocomposites. Nano-optics could allow for an increase in precision of pupil repair and other types of laser eye surgery.

Textiles

The use of engineered nanofibres already makes clothes water- and stain-repellent or wrinkle-free. Textiles with a nanotechnological finish can be washed less frequently and at lower temperatures. Nanotechnology has been used to integrate tiny carbon particles membrane and guarantee full-surface protection from electrostatic charges for the wearer. Many other applications have been developed by research institutions such as the Textiles Nanotechnology Laboratory (http://nanotextiles.human.cornell.edu/) at Cornell University.

Cosmetics

One field of application is in sunscreens. The traditional chemical UV protection approach suffers from its poor long-term stability. A sunscreen based on mineral nanoparticles such as titanium dioxide offer several advantages. Titanium oxide nanoparticles have a comparable UV protection property as the bulk material, but lose the cosmetically undesirable whitening as the particle size is decreased.

Agriculture

Applications of nanotechnology have the potential to change the entire agriculture sector and food industry chain from production to conservation, processing, packaging, transportation, and even waste treatment. Nanoscience concepts and nanotechnology applications have the potential to redesign the production cycle, restructure the processing and conservation processes and redefine the food habits of the people. Major challenges related to agriculture like Low productivity in cultivable areas, large uncultivable areas, shrinkage of cultivable lands, wastage of inputs like water, fertilisers, pesticides, wastage of products and of course food security for growing numbers can be addressed through various applications of nanotechnology.

Chapter 2

Quantum Chemistry

INTRODUCTION

Quantum chemistry is a branch of theoretical chemistry, which applies quantum mechanics and quantum field theory to address problems in chemistry. The description of the electronic behaviour of atoms and molecules as pertaining to their reactivity is one of the applications of quantum chemistry. Quantum chemistry lies on the border between chemistry and physics, and significant contributions have been made by scientists from both fields. It has a strong and active overlap with the field of atomic physics and molecular physics, as well as physical chemistry.

Quantum chemistry mathematically describes the fundamental behaviour of matter at the molecular scale. It is, in principle, possible to describe all chemical systems using this theory. In practice, only the simplest chemical systems may realistically be investigated in purely quantum mechanical terms, and approximations must be made for most practical purposes (e.g. Hartree-Fock, post-Hartree-Fock or density functional theory, see computational chemistry for more details). Hence a detailed understanding of quantum mechanics is not necessary for most chemistry, as the important implications of the theory (principally the orbital approximation) can be understood and applied in simpler terms.

In quantum mechanics the Hamiltonian, or the physical state, of a particle can be expressed as the sum of two operators, one corresponding to kinetic energy and the other to potential energy. The Hamiltonian in the Schrödinger wave equation used in quantum chemistry does not contain terms for the spin of the electron.

Solutions of the Schrödinger equation for the hydrogen atom gives the form of the wave function for atomic orbitals, and the relative energy of the various orbitals. The orbital approximation can be used to understand the other atoms, e.g. helium, lithium and carbon.

ELECTRONIC STRUCTURE

The first step in solving a quantum chemical problem is usually solving the Schrödinger equation (or Dirac equation in relativistic quantum chemistry) with the electronic molecular Hamiltonian. This is called determining the electronic structure of the molecule. It can be said that the electronic structure of a molecule or crystal implies essentially its chemical properties. An exact solution for the Schrödinger equation can only be obtained for the hydrogen atom. Since all other atomic, or molecular systems, involve the motions of three or more 'particles', their Schrödinger equations cannot be solved exactly and so approximate solutions must be sought.

Wave Model

The foundation of quantum mechanics and quantum chemistry is the wave model, in which the atom is a small, dense, positively charged nucleus surrounded by electrons. Unlike the earlier Bohr model of the atom, however, the wave model describes electrons as 'clouds' moving in orbitals, and their positions are represented by probability distributions rather than discrete points. The strength of this model lies in its predictive power. Specifically, it predicts the pattern of chemically similar elements found in the periodic table. The wave model is so named because electrons exhibit properties (such as interference) traditionally associated with waves (see wave-particle duality).

Valence Bond

Although the mathematical basis of quantum chemistry had been laid by Schrödinger in 1926, it is generally accepted that the first true calculation in quantum chemistry was that of the German physicists Walter Heitler and Fritz London on the hydrogen (H_2) molecule in 1927. Heitler and London's method was extended by the American theoretical physicist John C. Slater and the American theoretical chemist Linus Pauling to become the valence-bond (VB) [or Heitler-London-Slater-Pauling (HLSP)] method. In this method, attention is primarily devoted to the pairwise interactions between atoms, and this method therefore correlates closely with classical chemists' drawings of bonds.

Molecular Orbital

An alternative approach was developed in 1929 by Friedrich Hund and Robert S. Mulliken, in which electrons are described by mathematical functions delocalised over an entire molecule. The Hund-Mulliken approach or molecular orbital (MO) method is less intuitive to chemists, but has turned out capable of predicting spectroscopic properties better than the VB method. This approach is the conceptional basis of the Hartree-Fock method and further post-Hartree-Fock methods.

Density Functional Theory

The Thomas-Fermi model was developed independently by Thomas and Fermi in 1927. This was the first attempt to describe many-electron systems on the basis of electronic density instead of wave functions, although it was not very successful in the treatment of entire molecules. The method did provide the basis for what is now known as density functional theory. Though this method is less developed than post Hartree-Fock methods, its lower computational requirements allow it to tackle larger polyatomic molecules and even macromolecules, which has made it the most used method in computational chemistry at present.

CHEMICAL DYNAMICS

A further step can consist of solving the Schrödinger equation with the total molecular Hamiltonian in order to study the motion of molecules. Direct solution of the Schrödinger equation is called quantum molecular dynamics, within the semiclassical approximation semiclassical molecular dynamics, and within the classical mechanics framework molecular dynamics (MD). Statistical approaches, using for example Monte Carlo methods, are also possible.

Adiabatic Chemical Dynamics

In adiabatic dynamics, interatomic interactions are represented by single scalar potentials called potential energy surfaces. This is the Born-Oppenheimer approximation introduced by Born and Oppenheimer in

1927. Pioneering applications of this in chemistry were performed by Rice and Ramsperger in 1927 and Kassel in 1928, and generalised into the RRKM theory in 1952 by Marcus who took the transition state theory developed by Eyring in 1935 into account. These methods enable simple estimates of unimolecular reaction rates from a few characteristics of the potential surface.

Non-adiabatic Chemical Dynamics

Non-adiabatic dynamics consists of taking the interaction between several coupled potential energy surface (corresponding to different electronic quantum states of the molecule). The coupling terms are called vibronic couplings. The pioneering work in this field was done by Stueckelberg, Landau, and Zener in the 1930s, in their work on what is now known as the Landau-Zener transition. Their formula allows the transition probability between two diabatic potential curves in the neighbourhood of an avoided crossing to be calculated.

QUANTUM CHEMISTRY AND QUANTUM FIELD THEORY

The application of quantum field theory (QFT) to chemical systems and theories has become increasingly common in the modern physical sciences. One of the first and most fundamentally explicit appearances of this is seen in the theory of the photomagneton. In this system, plasmas, which are ubiquitous in both physics and chemistry, are studied in order to determine the basic quantisation of the underlying bosonic field. However, quantum field theory is of interest in many fields of chemistry, including: nuclear chemistry, astrochemistry, sonochemistry, and quantum hydrodynamics. Field theoretic methods have also, been critical in developing the *ab initio* effective Hamiltonian theory of semi-empirical pi-electron methods.

COMPUTATIONAL CHEMISTRY

Computational chemistry is a branch of chemistry that uses computers to assist in solving chemical problems. It uses the results of theoretical chemistry, incorporated into efficient computer programs, to calculate the structures and properties of molecules and solids. While its results normally complement the information obtained by chemical experiments, it can in some cases predict hitherto unobserved chemical phenomena. It is widely used in the design of new drugs and materials.

Examples of such properties are structure (i.e. the expected positions of the constituent atoms), absolute and relative (interaction) energies, electronic charge distributions, dipoles and higher multipole moments, vibrational frequencies, reactivity or other spectroscopic quantities, and cross-sections for collision with other particles.

The methods employed cover both static and dynamic situations. In all cases the computer time and other resources (such as memory and disk space) increase rapidly with the size of the system being studied. That system can be a single molecule, a group of molecules, or a solid. Computational chemistry methods range from highly accurate to very approximate; highly accurate methods are typically feasible only for small systems. *Ab initio* methods are based entirely on theory from first principles. Other (typically less accurate) methods are called empirical or semi-empirical because they employ experimental results, often from acceptable models of atoms or related molecules, to approximate some elements of the underlying theory.

Both *ab initio* and semi-empirical approaches involve approximations. These range from simplified forms of the first-principles equations that are easier or faster to solve, to approximations limiting the size of the system (for example, periodic boundary conditions), to fundamental approximations to the underlying equations that are required to achieve any solution to them at all. For example, most *ab initio* calculations make the Born-Oppenheimer approximation, which greatly simplifies the underlying Schrödinger equation by freezing the nuclei in place during the calculation. In principle, *ab initio* methods eventually converge to the exact solution of the underlying equations as the number of approximations is reduced. In practice, however, it is impossible to eliminate all approximations, and residual error inevitably remains. The goal of computational chemistry is to minimise this residual error while keeping the calculations tractable.

In some cases, the details of electronic structure are less important than the long-time phase space behaviour of molecules. This is the case in conformational studies of proteins and protein-ligand binding thermodynamics. Classical approximations to the potential energy surface are employed, as they are computationally less intensive than electronic calculations, to enable longer simulations of molecular dynamics. Furthermore, chemoinformatics uses even more empirical (and computationally cheaper) methods like machine learning based on physicochemical properties. One typical problem in cheminformatics is to predict the binding affinity of drug molecules to a given target.

Concepts

The term *theoretical chemistry* may be defined as a mathematical description of chemistry, whereas *computational chemistry* is usually used when a mathematical method is sufficiently well developed that it can be automated for implementation on a computer. Note that the words *exact* and *perfect* do not appear here, as very few aspects of chemistry can be computed exactly. However, almost every aspect of chemistry can be described in a qualitative or approximate quantitative computational scheme.

Molecules consist of nuclei and electrons, so the methods of quantum mechanics apply. Computational chemists often attempt to solve the non-relativistic Schrödinger equation, with relativistic corrections added, although some progress has been made in solving the fully relativistic Dirac equation. In principle, it is possible to solve the Schrödinger equation in either its time-dependent or time-independent form, as appropriate for the problem in hand; in practice, this is not possible except for very small systems. Therefore, a great number of approximate methods strive to achieve the best trade-off between accuracy and computational cost. Accuracy can always be improved with greater computational cost. Significant errors can present themselves in *ab initio* models comprising many electrons, due to the computational expense of full relativistic-inclusive methods. This complicates the study of molecules interacting with high atomic mass unit atoms, such as transitional metals and their catalytic properties. Present algorithms in computational chemistry can routinely calculate the properties of molecules that contain up to about 40 electrons with sufficient accuracy. Errors for energies can be less than a few kJ/mol. For geometries, bond lengths can be predicted within a few picometers and bond angles within 0.5 degrees. The treatment of larger molecules that contain a few dozen electrons is computationally tractable by approximate methods such as density functional theory (DFT). There is some dispute within the field whether or not the latter methods are sufficient to describe complex chemical reactions, such as those in biochemistry. Large molecules can be studied by semi-empirical approximate methods. Even larger molecules are treated by classical mechanics methods that employ what are called molecular mechanics. In QM/MM methods, small portions of large complexes are treated quantum mechanically (QM), and the remainder is treated approximately (MM).

In theoretical chemistry, chemists, physicists and mathematicians develop algorithms and computer programs to predict atomic and molecular properties and reaction paths for chemical reactions.

Computational chemists, in contrast, may simply apply existing computer programs and methodologies to specific chemical questions. There are two different aspects to computational chemistry:

1. Computational studies can be carried out in order to find a starting point for a laboratory synthesis, or to assist in understanding experimental data, such as the position and source of spectroscopic peaks.
2. Computational studies can be used to predict the possibility of so far entirely unknown molecules or to explore reaction mechanisms that are not readily studied by experimental means.

Thus, computational chemistry can assist the experimental chemist or it can challenge the experimental chemist to find entirely new chemical objects.

Several major areas may be distinguished within computational chemistry:

1. The prediction of the molecular structure of molecules by the use of the simulation of forces, or more accurate quantum chemical methods, to find stationary points on the energy surface as the position of the nuclei is varied.
2. Storing and searching for data on chemical entities (see chemical databases).
3. Identifying correlations between chemical structures and properties (see QSPR and QSAR).
4. Computational approaches to help in the efficient synthesis of compounds.
5. Computational approaches to design molecules that interact in specific ways with other molecules (e.g. drug design and catalysis).

Methods

A single molecular formula can represent a number of molecular isomers. Each isomer is a local minimum on the energy surface (called the potential energy surface) created from the total energy (i.e. the electronic energy, plus the repulsion energy between the nuclei) as a function of the coordinates of all the nuclei. A stationary point is a geometry such that the derivative of the energy with respect to all displacements of the nuclei is zero. A local (energy) minimum is a stationary point where all such displacements lead to an increase in energy. The local minimum that is lowest is called the global minimum and corresponds to the most stable isomer. If there is one particular coordinate change that leads to a decrease in the total energy in both directions, the stationary point is a transition structure and the coordinate is the reaction coordinate. This process of determining stationary points is called geometry optimisation.

The determination of molecular structure by geometry optimisation became routine only after efficient methods for calculating the first derivatives of the energy with respect to all atomic coordinates became available. Evaluation of the related second derivatives allows the prediction of vibrational frequencies if harmonic motion is estimated. More importantly, it allows for the characterisation of stationary points. The frequencies are related to the eigenvalues of the Hessian matrix, which contains second derivatives. If the eigenvalues are all positive, then the frequencies are all real and the stationary point is a local minimum. If one eigenvalue is negative, (i.e. an imaginary frequency), then the stationary point is a transition structure. If more than one eigenvalue is negative, then the stationary point is a more complex one, and is usually of little interest. When one of these is found, it is necessary to move the search away from it if the experimenter is looking solely for local minima and transition structures.

The total energy is determined by approximate solutions of the time-dependent Schrödinger equation, usually with no relativistic terms included, and by making use of the Born-Oppenheimer approximation, which allows for the separation of electronic and nuclear motions, thereby simplifying the Schrödinger equation. This leads to the evaluation of the total energy as a sum of the electronic energy at fixed nuclei positions and the repulsion energy of the nuclei. A notable exception are certain approaches called direct quantum chemistry, which treat electrons and nuclei on a common footing. Density functional methods and semi-empirical methods are variants on the major theme. For very large systems, the relative total energies can be compared using molecular mechanics. The ways of determining the total energy to predict molecular structures are:

Ab initio methods

The programs used in computational chemistry are based on many different quantum-chemical methods that solve the molecular Schrödinger equation associated with the molecular Hamiltonian. Methods that do not include any empirical or semi-empirical parameters in their equations — being derived directly from theoretical principles, with no inclusion of experimental data — are called *ab initio* methods. This does not imply that the solution is an exact one; they are all approximate quantum mechanical calculations. It means that a particular approximation is rigorously defined on first principles (quantum theory) and then solved within an error margin that is qualitatively known beforehand. If numerical iterative methods have to be employed, the aim is to iterate until full machine accuracy is obtained (the best that is possible with a finite word length on the computer, and within the mathematical and/or physical approximations made).

The simplest type of *ab initio* electronic structure calculation is the Hartree-Fock (HF) scheme, an extension of molecular orbital theory, in which the correlated electron-electron repulsion is not specifically taken into account; only its average effect is included in the calculation. As the basis set size is increased, the energy and wave function tend towards a limit called the Hartree-Fock limit. Many types of calculations (known as post-Hartree-Fock methods) begin with a Hartree-Fock calculation and subsequently correct for electron-electron repulsion, referred to also as electronic correlation. As these methods are pushed to the limit, they approach the exact solution of the non-relativistic Schrödinger equation. In order to obtain exact agreement with experiment, it is necessary to include relativistic and spin orbit terms, both of which are only really important for heavy atoms. In all of these approaches, in addition to the choice of method, it is necessary to choose a basis set. This is a set of functions, usually centred on the different atoms in the molecule, which are used to expand the molecular orbitals with the LCAO ansatz. *Ab initio* methods need to define a level of theory (the method) and a basis set. Figure 2.1 illustrating various *ab initio* electronic structure methods in terms of energy. Spacings are not to scale.

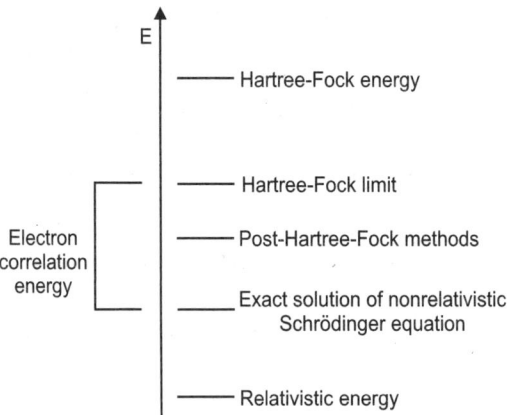

Fig. 2.1. Various *ab initio* electronic structure methods in terms of energy.

The Hartree-Fock wave function is a single configuration or determinant. In some cases, particularly for bond breaking processes, this is quite inadequate, and several configurations need to be used. Here, the coefficients of the configurations and the coefficients of the basis functions are optimised together.

The total molecular energy can be evaluated as a function of the molecular geometry; in other words, the potential energy surface. Such a surface can be used for reaction dynamics. The stationary points of the surface lead to predictions of different isomers and the transition structures for conversion between isomers, but these can be determined without a full knowledge of the complete surface.

A particularly important objective, called computational thermochemistry, is to calculate thermochemical quantities such as the enthalpy of formation to chemical accuracy. Chemical accuracy is the accuracy required to make realistic chemical predictions and is generally considered to be 1 kcal/mol or 4 kJ/mol. To reach that accuracy in an economic way it is necessary to use a series of post-Hartree-Fock methods and combine the results. These methods are called quantum chemistry composite methods.

Density functional methods

Density functional theory (DFT) methods are often considered to be *ab initio* methods for determining the molecular electronic structure, even though many of the most common functionals use parameters derived from empirical data, or from more complex calculations. In DFT, the total energy is expressed in terms of the total one-electron density rather than the wave function. In this type of calculation, there is an approximate Hamiltonian and an approximate expression for the total electron density. DFT methods can be very accurate for little computational cost. Some methods combine the density functional exchange functional with the Hartree-Fock exchange term and are known as hybrid functional methods.

Semi-empirical and empirical methods

Semi-empirical quantum chemistry methods are based on the Hartree-Fock formalism, but make many approximations and obtain some parameters from empirical data. They are very important in computational chemistry for treating large molecules where the full Hartree-Fock method without the approximations is too expensive. The use of empirical parameters appears to allow some inclusion of correlation effects into the methods.

Semi-empirical methods follow what are often called empirical methods, where the two-electron part of the Hamiltonian is not explicitly included. For π-electron systems, this was the Hückel method proposed by Erich Hückel, and for all valence electron systems, the extended Hückel method proposed by Roald Hoffmann.

Molecular mechanics

In many cases, large molecular systems can be modelled successfully while avoiding quantum mechanical calculations entirely. Molecular mechanics simulations, for example, use a single classical expression for the energy of a compound, for instance the harmonic oscillator. All constants appearing in the equations must be obtained beforehand from experimental data or *ab initio* calculations.

The database of compounds used for parameterisation, i.e. the resulting set of parameters and functions is called the force field, is crucial to the success of molecular mechanics calculations. A force field parameterised against a specific class of molecules, for instance proteins, would be expected to only have any relevance when describing other molecules of the same class.

These methods can be applied to proteins and other large biological molecules, and allow studies of the approach and interaction (docking) of potential drug molecules.

Methods for solids

Computational chemical methods can be applied to solid state physics problems. The electronic structure of a crystal is in general described by a band structure, which defines the energies of electron orbitals for each point in the Brillouin zone. *Ab initio* and semi-empirical calculations yield orbital energies;

therefore, they can be applied to band structure calculations. Since it is time-consuming to calculate the energy for a molecule, it is even more time-consuming to calculate them for the entire list of points in the Brillouin zone.

Chemical dynamics

Once the electronic and nuclear variables are separated (within the Born-Oppenheimer representation), in the time-dependent approach, the wave packet corresponding to the nuclear degrees of freedom is propagated via the time evolution operator (physics) associated to the time-dependent Schrödinger equation (for the full molecular Hamiltonian). In the complementary energy-dependent approach, the time-independent Schrödinger equation is solved using the scattering theory formalism. The potential representing the interatomic interaction is given by the potential energy surfaces. In general, the potential energy surfaces are coupled via the vibronic coupling terms.

The most popular methods for propagating the wave packet associated to the molecular geometry are:
1. The split operator technique.
2. The multiconfiguration time-dependent Hartree method (MCTDH).
3. The semiclassical method.

Molecular dynamics

Molecular dynamics (MD) uses Newton's laws of motion to examine the time-dependent behaviour of systems, including vibrations or Brownian motion, using a classical mechanical description. MD combined with density functional theory leads to the Car-Parrinello method.

Interpreting Molecular Wave Functions

The atoms in molecules model developed by Richard Bader was developed in order to effectively link the quantum mechanical picture of a molecule, as an electronic wavefunction, to chemically useful older models such as the theory of Lewis pairs and the valence bond model. Bader has demonstrated that these empirically useful models are connected with the topology of the quantum charge density. This method improves on the use of Mulliken population analysis.

Software Packages

There are many self-sufficient software packages used by computational chemists. Some include many methods covering a wide range, while others concentrating on a very specific range or even a single method. Details of most of them can be found in:
1. Quantum chemistry and solid state physics software supporting several methods.
2. Molecular mechanics programs.
3. Semi-empirical programs.
4. Valence bond programs.
5. Biomolecular modelling programs: proteins, nucleic acid.
6. Molecular design software.

QUANTUM CHEMISTRY COMPUTER PROGRAMS

Quantum chemistry computer programs are used in computational chemistry to implement the methods of quantum chemistry. Most include the Hartree-Fock (HF) and some post-Hartree-Fock methods. They may also include density functional theory (DFT), molecular mechanics or semi-empirical quantum chemistry methods. The programs include both open source and commercial software. Most of them are large, often containing several separate programs, and have developed over many years.

Table 2.1 illustrates the capabilities of the most versatile software packages that show an entry in two or more columns of the table.

Table 2.1. Capabilities of the most versatile software packages that show an entry in two or more columns of the table.

Package	License[†]	Basis	Periodic[‡]	Mol. mech.	Semi- emp.	HF	Post- HF	DFT
ABINIT	GPL	PW	3d	Y	N	N	N	Y
ACES II	acad.	GTO	N	N	N	Y	Y	Y
ACES II MAB	acad.	GTO	N	N	N	Y	Y	N
ADF	comm.	STO	any	Y	Y[4]	Y	N	Y
Atomistix ToolKit	comm.	NAO	3d	Y	N	N	N	Y
BigDFT	GPL	Wavelet	any	Y	N	N	N	Y
CADPAC	acad.	GTO	N	N	N	Y	Y	Y
CASINO (QMC)	acad.	GTO/PW/spline/ grid/STO	any	N	N	Y	Y	N
CASTEP	acad. (UK)/ comm.	PW	3d	Y	N	Y[5]	N	Y
CFOUR	acad.	GTO	N	N	N	Y	Y	N
COLUMBUS	acad.	GTO	N	N	N	Y	Y	N
CONQUEST	acad. (UK)	?	?	?	?	?	?	?
COSMOS	comm.	?	?	Y	Y	N	N	N
CP2K	GPL	Hybrid GTO/PW	any	Y	Y	Y,	N	Y
CPMD	acad.	PW	any	Y	N	Y	N	Y
CRYSTAL	acad. (UK)/ comm.	GTO	any	Y	N	Y	N	Y
DACAPO	GPL[1]	PW	3d	Y	N	N	N	Y
DALTON	acad.	GTO	N	N	N	Y	Y	Y
DFTB+	acad./comm.	NAO	any	Y	Y	N	N	N
DFT++	GPL	PW/Wavelet	3d	Y	N	N	N	Y
DIRAC	acad.	GTO	N	N	N	Y	Y	Y
DMol3	comm.	?	?	Y	N	N	N	Y
FreeON	GPL	GTO	any	Y	N	Y	Y	Y
Firefly/PC GAMESS	acad.	GTO	N	Y[3]	Y	Y	Y	Y
GAMESS (UK)	acad. (UK)/ comm.	GTO	N	N	Y	Y	Y	Y
GAMESS (US)	acad.	GTO	N	Y[2]	Y	Y	Y	Y
GAUSSIAN	comm.	GTO	any	Y	Y	Y	Y	Y
GPAW	GPL	grid/NAO	any	Y	?	Y[5]	N	Y
hBar Lab[7]	comm.	GTO	N	N	N	Y	Y	Y

(Contd ...)

Package	License[†]	Basis	Periodic[‡]	Mol. mech.	Semi- emp.	HF	Post- HF	DFT
HiLAPW	?	FLAPW	3d	N	N	N	N	Y
JAGUAR	comm.	GTO	?	Y	N	Y	Y	Y
Materials Studio	comm.	PW	?	Y	Y	N	N	Y
MedeA	?	?	?	?	N	N	N	Y
MOLCAS	comm.	GTO	N	Y	Y	Y	Y	Y
MOLPRO	comm.	GTO	?	N	N	Y	Y	Y
MOPAC	acad./comm.	?	?	?	Y	N	N	N
MPQC	LGPL	GTO	N	N	N	Y	Y	Y
NWChem	acad.	?	?	Y	N	Y	Y	Y
OCTOPUS	GPL	grid	any	Y	N	N	N	Y
ONETEP	acad. (UK)/ comm.	PW	any	Y	N	Y[5]	N	Y
OpenAtom	acad.	PW	?	Y	N	N	N	Y
OpenMX	GPL	NAO	3d	Y	N	N	N	Y
PLATO	acad.	NAO	any	Y	N	N	N	Y
PQS	comm.	?	?	Y	Y	Y	Y	Y
Priroda 04	?	?	?	N	N	Y	Y	Y
PSI	GPL	GTO	N	N	N	Y	Y	N
PWscf[6]	GPL	PW	3d	N	N	Y	N	Y
PyQuante	BSD	GTO	N	N	Y	Y	Y	Y
Q-Chem	comm.	GTO	N	Y	Y	Y	Y	Y
Quantum ESPRESSO	GPL	PW	3d	N	N	Y	N	Y
SPARTAN 06	comm.	GTO	?	Y	Y	Y	Y	Y
SIESTA	acad.	NAO	?	Y	N	N	N	Y
TURBOMOLE	comm.	GTO	?	Y	N	Y	Y	Y
VASP	acad.	PW	any	N	N	Y	N	Y
WIEN2k	comm.	augPW	?	?	N	N	N	Y

[†] 'acad.': academic (no cost) license possible upon request, 'comm.': commercially distributed.

[‡] Support for periodic systems (3d-crystals, 2d-slabs, 1d-rods and 0d-molecules): 3d-periodic codes always allow the simulation of systems with lower dimensionality within a supercell. Specified here is the capability to actually simulation within lower periodicity.

[1] The CAMPOS project (which includes Dacapo) states that all code is GPL. The Dacapo distribution itself does not contain any license information.

[2] Through interface to TINKER

[3] Through Ascalaph

[4] Through interface to MOPAC

[5] Using exact exchange DFT

[6] Distributed with Quantum ESPRESSO

[7] Web service integrating MPQC.

QUANTUM ELECTROCHEMISTRY

The scientific school of quantum electrochemistry began to form in the 1960s under Revaz Dogonadze. Generally speaking, the field comprises the notions arising in electrodynamics, quantum mechanics, and electrochemistry; and so is studied by a very large array of different professional researchers. The fields they reside in include, chemical, electrical and mechanical engineering, chemistry and physics.

More specifically, quantum electrochemistry is the application of quantum mechanical tools such as density functional theory to the study of electrochemical processes, including electron transfer at electrodes. It also includes models such as Marcus theory.

The first development of 'quantum electrochemistry' is somewhat difficult to pin down. This is not very surprising, since the development of quantum mechanics to chemistry can be summarised as the application of quantum wave theory models to atoms and molecules. This being the case, electrochemistry, which is particularly concerned with the electronic states of some particular system, is already, by its nature, tied into the quantum mechanical model of the electron in quantum chemistry. There were proponents of quantum electrochemistry, who applied quantum mechanics to electrochemistry with unusual zeal, clarity and precision. Among them were Revaz Dogonadze and his co-workers. They developed one of the early quantum mechanical models for proton transfer reactions in chemical systems. Dogonadze is a particularly celebrated promoter of quantum electrochemistry, and is also credited with forming an international summer school of quantum electrochemistry centred in Yugoslavia.

Chapter 3

Solid-state Physics

INTRODUCTION

Solid-state physics, the largest branch of condensed matter physics, is the study of rigid matter, or solids, through methods such as quantum mechanics, crystallography, electromagnetism and metallurgy. Solid-state physics considers how the large-scale properties of solid materials result from their atomic-scale properties. Solid-state physics thus forms the theoretical basis of materials science, as well as having direct applications, for example in the technology of transistors and semiconductors.

Solid materials are formed from densely-packed atoms, with intense interaction forces between them. These interactions are responsible for the mechanical (e.g. hardness and elasticity), thermal, electrical, magnetic and optical properties of solids. Depending on the material involved and the conditions in which it was formed, the atoms may be arranged in a regular, geometric pattern (crystalline solids, which include metals and ordinary water ice) or irregularly (an amorphous solid such as common window glass).

The bulk of solid-state physics theory and research is focused on crystals, largely because the periodicity of atoms in a crystal—its defining characteristic—facilitates mathematical modelling, and also because crystalline materials often have electrical, magnetic, optical, or mechanical properties that can be exploited for engineering purposes.

The forces between the atoms in a crystal can take a variety of forms. For example, in a crystal of sodium chloride (common salt), the crystal is made up of ionic sodium and chlorine, and held together with ionic bonds. In others, the atoms share electrons and form covalent bonds. In metals, electrons are shared amongst the whole crystal in metallic bonding. Finally, the noble gases do not undergo any of these types of bonding. In solid form, the noble gases are held together with van der Waals forces resulting from the polarisation of the electronic charge cloud on each atom. The differences between the types of solid result from the differences between their bonding.

CRYSTAL STRUCTURE AND PROPERTIES

Many properties of materials are affected by their crystal structure. This structure can be investigated using a range of crystallographic techniques, including X-ray crystallography, neutron diffraction and electron diffraction. An example of a closed packed lattice is shown in Fig. 3.1.

The sizes of the individual crystals in a crystalline solid material vary depending on the material involved and the conditions when it was formed. Most crystalline materials encountered in everyday life are polycrystalline, with the individual crystals being microscopic in scale, but macroscopic single crystals can be produced either naturally (e.g. diamonds) or artificially. Real crystals feature defects or irregularities in the ideal arrangements, and it is these defects that critically determine many of the electrical and mechanical properties of real materials.

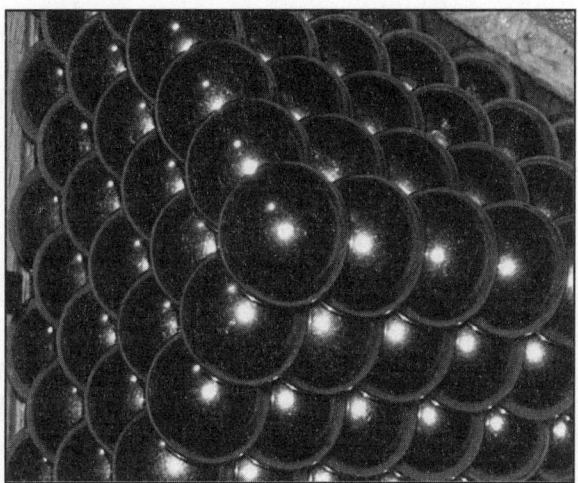

Fig. 3.1. Closed packed lattice.

The crystal lattice can vibrate. These vibrations are found to be quantised, the quantised vibrational modes being known as phonons. Phonons play a major role in many of the physical properties of solids, such as the transmission of sound. In insulating solids, phonons are also the primary mechanism by which heat conduction takes place. Phonons are also necessary for understanding the lattice heat capacity of a solid, as in the Einstein model and the later Debye model.

ELECTRONIC PROPERTIES

Properties of materials such as electrical conduction and heat capacity are investigated by solid state physics. An early model of electrical conduction was the Drude model, which applied kinetic theory to the electrons in a solid. By assuming that the material contains immobile positive ions and an 'electron gas' of classical, non-interacting electrons, the Drude model was able to explain electrical and thermal conductivity and the Hall effect in metals, although it greatly overestimated the electronic heat capacity.

Arnold Sommerfeld combined the classical Drude model with quantum mechanics in the free electron model (or Drude-Sommerfeld model). Here, the electrons are modelled as a Fermi gas, a gas of particles which obey the quantum mechanical Fermi-Dirac statistics. The free electron model gave improved predictions for the heat capacity of metals, however, it was unable to explain the existence of insulators.

The nearly-free electron model is a modification of the free electron model which includes a weak periodic perturbation meant to model the interaction between the conduction electrons and the ions in a crystalline solid. By introducing the idea of electronic bands, the theory explains the existence of conductors, semiconductors and insulators.

The nearly-free electron model rewrites the Schrödinger equation for the case of a periodic potential. The solutions in this case are known as Bloch states. Since Bloch's theorem applies only to periodic potentials, and since unceasing random movements of atoms in a crystal disrupt periodicity, this use of Bloch's theorem is only an approximation, but it has proven to be a tremendously valuable approximation, without which most solid-state physics analysis would be intractable. Deviations from periodicity are treated by quantum mechanical perturbation theory.

QUASICRYSTAL

Quasicrystals are structural forms that are both ordered and nonperiodic. They form patterns that fill all the space but lack translational symmetry. Classical theory of crystals allows only 2, 3, 4, and 6-fold rotational symmetries, but quasicrystals display symmetry of other orders (folds). They can be said to be in a state intermediate between crystal and glass. Just like crystals, quasicrystals produce modified Bragg diffraction, but where crystals have a simple repeating structure, quasicrystals are more complex.

Aperiodic tilings were discovered by mathematicians in the early 1960s, but some twenty years later they were found to apply to the study of quasicrystals. The discovery of these aperiodic forms in nature has produced a paradigm shift in the fields of crystallography and solid state physics. Quasicrystals had been investigated and observed earlier but until the 80s they were disregarded in favour of the prevailing views about the atomic structure of matter.

Roughly, an ordering is non-periodic if it lacks translational symmetry, which means that a shifted copy will never match exactly with its original. The more precise mathematical definition is that there is never translational symmetry in more than $n - 1$ linearly independent directions, where n is the dimension of the space filled, i.e. the three-dimensional tiling displayed in a quasicrystal may have translational symmetry in two dimensions. The ability to diffract comes from the existence of an indefinitely large number of elements with a regular spacing, a property loosely described as long-range order. Experimentally the aperiodicity is revealed in the unusual symmetry of the diffraction pattern, that is, symmetry of orders other than 2, 3, 4, or 6. The first officially reported case of what came to be known as quasicrystals was made by Dan Shechtman and coworkers in 1984. The distinction between quasicrystals and their corresponding mathematical models (e.g. the three-dimensional version of the Penrose tiling) need not be emphasised.

Mathematical Description

One definition, the 'cut and project' construction, is that a quasicrystal consists of a slice (an intersection of one or more hyperplanes) of a higher-dimensional periodic pattern. In order that the quasicrystal itself be aperiodic, this slice must not be a lattice plane of the higher-dimensional lattice. (For example, Penrose tilings can be viewed as two-dimensional slices of five-dimensional hypercubic structures.) Equivalently, the Fourier transform of such a quasicrystal is nonzero only at a dense set of points spanned by integer multiples of a finite set of basis vectors (the projections of the primitive reciprocal lattice vectors of the higher-dimensional lattice).

The intuitive considerations obtained from simple model aperiodic tilings are formally expressed in the concepts of Meyer and Delone sets. The mathematical counterpart of physical diffraction is the Fourier transform and the qualitative description of a diffraction picture as 'clear cut' or 'sharp' means that singularities are present in the Fourier spectrum. There are different methods to construct model quasicrystals. These are the same methods that produce aperiodic tilings with the additional constraint for the diffractive property.

Thus, for a substitution tiling the eigenvalues of the substitution matrix should be Pisot numbers. The aperiodic structures obtained by the cut-and-project method are made diffractive by choosing a suitable orientation for the construction. This is indeed a geometric approach which has also a great appeal for physicists.

Classical theory of crystals reduces crystals to point lattices where each point is the centre of mass of one of the identical units of the crystal. The structure of crystals can by analysed by defining an associated group (mathematics). Quasicrystals, on the other hand, are composed of more than one type of unit, so instead of lattices, quasilattices must be used. Instead of groups, groupoids, the mathematical generalisation of groups in category theory, is the appropriate tool for studying quasicrystals.

Using mathematics for construction and analysis of quasicrystal structures is a difficult task for most experimentalists. Computer modelling, based on the existing theories of quasicrystals, however greatly facilitated this task. Advanced programs have been developed allowing one to construct, visualise and analyse quasicrystal structures and their diffraction patterns.

Physics of Quasicrystals

The first problem in physics is whether the data represent Bragg diffraction. They do not because the order, n, is generally, though not always, logarithmic, instead of linear. The second problem is structural. It is not necessary to employ more than one unit cell; the patterns can be indexed and simulated using a single unit, as is the norm in crystallography and consistent with the driving force. A third problem is dimensional. Because the solids contain multiple interplanar spacings, and because each diffracted beam results from a combination of such spacings, the measured quasilattice spacing is a compromise. It is larger than the Bragg equivalent for crystals, and the dimension is important in model building where atoms must fit allotted spaces.

MATERIAL SCIENCE OF QUASICRYSTALS

Since the original discovery of Shechtman hundreds of quasicrystals have been reported and confirmed. Undoubtedly, the quasicrystals are no longer a unique form of solid; they exist universally in many metallic alloys and some polymers. Quasicrystals are found most often in aluminium alloys (Al-Li-Cu, Al-Mn-Si, Al-Ni-Co, Al-Pd-Mn, Al-Cu-Fe, Al-Cu-V, etc.), but numerous other compositions are also known (Cd-Yb, Ti-Zr-Ni, Zn-Mg-Ho, Zn-Mg-Sc, In-Ag-Yb, Pd-U-Si, etc.).

In theory, there are two types in quasicrystals. One is called polygonal (dihedral) quasicrystals, which have one 8, 10, or 12-fold axis and is periodic along this axis. They are called octagonal, decagonal, and dodecagonal quasicrystals. These structures take an ordered structure (quasiperiodic structure) in a plane normal to such a periodic axis. Another one called an icosahedral quasicrystal has no period along any directions.

Regarding thermal stability, three types of quasicrystals are distinguished:

1. Stable quasicrystals grown by slow cooling or casting with subsequent annealing.
2. Metastable quasicrystals prepared by melt-spinning.
3. Metastable quasicrystals formed by the crystallisation of the amorphous phase.

Except for the Al–Li–Cu system, all the stable quasicrystals are almost free of defects and disorder, as evidenced by X-ray and electron diffraction revealing peak widths as sharp as those of perfect crystals such as Si. Diffraction patterns exhibit fivefold, threefold, and twofold symmetries, and reflections are arranged quasiperiodically in three dimensions.

The origin of the stabilisation mechanism is different for the stable and metastable quasicrystals. Nevertheless, there is a common feature observed in most quasicrystal-forming liquid alloys or their undercooled liquids: a local icosahedral order. The icosahedral order is in equilibrium in the liquid state for the stable quasicrystals, whereas the icosahedral order prevails in the undercooled liquid state for the metastable quasicrystals.

SPIN GLASS

A spin glass is a magnet with frustrated interactions, augmented by stochastic disorder, where usually ferromagnetic and antiferromagnetic bonds are randomly distributed. Its magnetic ordering resembles the positional ordering of a conventional, chemical glass.

Spin glasses display many metastable structures leading to a plenitude of time scales which are difficult to explore experimentally or in simulations.

Magnetic Behaviour

It is the time dependence which distinguishes spin glasses from other magnetic systems. Beginning above the spin glass transition temperature, T_c, where the spin glass exhibits more typical magnetic behaviour (such as paramagnetism as discussed here but other kinds of magnetism are possible), if an external magnetic field is applied and the magnetisation is plotted versus temperature, it follows the typical Curie law (in which magnetisation is inversely proportional to temperature) until T_c is reached, at which point the magnetisation becomes virtually constant (this value is called the field cooled magnetisation). This is the onset of the spin glass phase. When the external field is removed, the spin glass has a rapid decrease of magnetisation to a value called the remanent magnetisation, and then a slow decay as the magnetisation approaches zero (or some small fraction of the original value — this remains unknown). This decay is non-exponential and no single function can fit the curve of magnetisation versus time adequately. This slow decay is particular to spin glasses. Experimental measurements on the order of days have shown continual changes above the noise level of instrumentation.

If a similar test is run on a ferromagnetic substance, when the external field is removed there is a rapid change to a remanent value that then stays constant in time. For a paramagnet, when the external field is removed the magnetisation rapidly goes to zero and stays there. In each case the decay is rapid and exponential.

If instead, the spin glass is cooled below T_c in the absence of an external field and then a field is applied, there is a rapid initial increase to a value called the zero-field-cooled magnetisation followed by a slow upward drift toward the field cooled magnetisation.

Surprisingly, the sum of the two complex functions of time (the zero-field-cooled and remanent magnetisations) is a constant, namely the field-cooled value, and thus both share identical functional forms with time, at least in the limit of very small external fields.

Model of Sherrington and Kirkpatrick

In addition to unusual experimental properties, spin glasses are the subject of extensive theoretical and computational investigations. A substantial part of early theoretical work on spin glasses dealt with a form of mean field theory based on a set of replicas of the partition function of the system.

An important exactly-solvable model of a spin glass was introduced by D. Sherrington and S. Kirkpatrick in 1975. It is an Ising model with long range frustrated ferro- as well as antiferromagnetic couplings. It corresponds to a mean field approximation of spin glasses describing the slow dynamics of the magnetisation and the complex non-ergodic equilibrium state.

The equilibrium solution of the model, after some initial attempts by Sherrington, Kirkpatrick and others, was found by Giorgio Parisi in 1979 within the replica method. The subsequent work of interpretation of the Parisi solution — by M. Mezard, G. Parisi, M.A. Virasoro and many others — revealed the complex nature of a glassy low temperature phase characterised by ergodicity breaking, ultrametricity and non-selfaverageness. Further developments led to the creation of the cavity method, which allowed study of the low temperature phase without replicas. A rigorous proof of the Parisi solution has been provided in the work of Francesco Guerra and Michel Talagrand.

The formalism of replica mean field theory has also been applied in the study of neural networks, where it has enabled calculations of properties such as the storage capacity of simple neural network architectures without requiring a training algorithm (such as backpropagation) to be designed or implemented. More realistic spin glass models with short range frustrated interactions and disorder, like the Gaussian model where the couplings between neighbouring spins follow a Gaussian distribution, have been studied extensively as well, especially using Monte Carlo simulations. These models display spin glass phases bordered by sharp phase transitions. Besides its relevance in condensed matter physics, spin glass theory has acquired a strongly interdisciplinary character, with applications to neural network theory, computer science, theoretical biology, econophysics, etc.

Non-ergodic Behaviour, and Applications

A so-called non-ergodic behaviour happens in spin glasses below the freezing temperature T_f, since below that temperature the system cannot escape from the ultradeep minima of the hierarchically-disordered energy landscape.

Although the freezing temperature is typically as low as 30 Kelvin (= –240 Centigrades) so that the spinglass-magnetism appears to be practically without applications in daily life, there are applications in different contexts, e.g. in the already mentioned theory of neural networks, i.e. in theoretical brain research, and in the mathematical-economical theory of optimisation.

SUPERCONDUCTIVITY

Superconductivity occurs in certain materials at very low temperatures. When superconductive, a material has an electrical resistance of exactly zero and no interior magnetic field (the Meissner effect). It was discovered by Heike Kamerlingh Onnes in 1911. Like ferromagnetism and atomic spectral lines, superconductivity is a quantum mechanical phenomenon. It cannot be understood simply as the idealisation of 'perfect conductivity' in classical physics (Fig. 3.2).

Fig. 3.2. A magnet levitating above a high-temperature superconductor, cooled with liquid nitrogen. Persistent electric current flows on the surface of the superconductor, acting to exclude the magnetic field of the magnet (the Meissner effect). This current effectively forms an electromagnet that repels the magnet.

The electrical resistivity of a metallic conductor decreases gradually as the temperature is lowered. However, in ordinary conductors such as copper and silver, this decrease is limited by impurities and other defects. Even near absolute zero, a real sample of copper shows some resistance. In a superconductor however, despite these imperfections, the resistance drops abruptly to zero when the material is cooled below its critical temperature. An electric current flowing in a loop of superconducting wire can persist indefinitely with no power source.

Superconductivity occurs in many materials: simple elements like tin and aluminium, various metallic alloys and some heavily-doped semiconductors. Superconductivity does not occur in noble metals like gold and silver, nor in pure samples of ferromagnetic metals.

In 1986, it was discovered that some cuprate-perovskite ceramic materials have critical temperatures of more than 90 kelvin. These high-temperature superconductors renewed interest in the topic because the current theory could not explain them. From a practical perspective, 90 kelvin is easy to reach with the readily available liquid nitrogen (boiling point 77 kelvin). This means more experimentation and more commercial applications are feasible, especially if materials with even higher critical temperatures could be discovered.

Elementary Properties of Superconductors

Most of the physical properties of superconductors vary from material to material, such as the heat capacity and the critical temperature, critical field, and critical current density at which superconductivity is destroyed.

On the other hand, there is a class of properties that are independent of the underlying material. For instance, all superconductors have exactly zero resistivity to low applied currents when there is no magnetic field present. The existence of these 'universal' properties implies that superconductivity is a thermodynamic phase, and thus possesses certain distinguishing properties which are largely independent of microscopic details.

Zero electrical 'dc' resistance

The simplest method to measure the electrical resistance of a sample of some material is to place it in an electrical circuit in series with a current source I and measure the resulting voltage V across the sample. The resistance of the sample is given by Ohm's law as $R = V/I$. If the voltage is zero, this means that the resistance is zero and that the sample is in the superconducting state.

Superconductors are also able to maintain a current with no applied voltage whatsoever, a property exploited in superconducting electromagnets such as those found in MRI machines. Experiments have demonstrated that currents in superconducting coils can persist for years without any measurable degradation.

Experimental evidence points to a current lifetime of at least 100,000 years. Theoretical estimates for the lifetime of a persistent current can exceed the estimated lifetime of the universe, depending on the wire geometry and the temperature. Thus, a superconductor does not have exactly zero resistance, however, the resistance is negligibly small.

In a normal conductor, an electrical current may be visualised as a fluid of electrons moving across a heavy ionic lattice. The electrons are constantly colliding with the ions in the lattice, and during each collision some of the energy carried by the current is absorbed by the lattice and converted into heat, which is essentially the vibrational kinetic energy of the lattice ions. As a result, the energy carried by the current is constantly being dissipated. This is the phenomenon of electrical resistance.

The situation is different in a superconductor. In a conventional superconductor, the electronic fluid cannot be resolved into individual electrons. Instead, it consists of bound pairs of electrons known as Cooper pairs. This pairing is caused by an attractive force between electrons from the exchange of phonons. Due to quantum mechanics, the energy spectrum of this Cooper pair fluid possesses an energy gap, meaning there is a minimum amount of energy ΔE that must be supplied in order to excite the fluid. Therefore, if ΔE is larger than the thermal energy of the lattice, given by kT, where k is Boltzmann's constant and T is the temperature, the fluid will not be scattered by the lattice. The Cooper pair fluid is thus a superfluid, meaning it can flow without energy dissipation.

In a class of superconductors known as Type II superconductors, including all known high-temperature superconductors, an extremely small amount of resistivity appears at temperatures not too far below the nominal superconducting transition when an electrical current is applied in conjunction with a strong magnetic field, which may be caused by the electrical current. This is due to the motion of vortices in the electronic superfluid, which dissipates some of the energy carried by the current. If the current is sufficiently small, the vortices are stationary, and the resistivity vanishes. The resistance due to this effect is tiny compared with that of non-superconducting materials, but must be taken into account in sensitive experiments. However, as the temperature decreases far enough below the nominal superconducting transition, these vortices can become frozen into a disordered but stationary phase known as a 'vortex glass'. Below this vortex glass transition temperature, the resistance of the material becomes truly zero.

Superconducting phase transition

In superconducting materials, the characteristics of superconductivity appear when the temperature T is lowered below a critical temperature T_c. The value of this critical temperature varies from material to material. Conventional superconductors usually have critical temperatures ranging from around 20 K to less than 1 K. Solid mercury, for example, has a critical temperature of 4.2 K. As of 2009, the highest critical temperature found for a conventional superconductor is 39 K for magnesium diboride (MgB_2), although this material displays enough exotic properties that there is some doubt about classifying it as a 'conventional' superconductor. Cuprate superconductors can have much higher critical temperatures: $YBa_2Cu_3O_7$, one of the first cuprate superconductors to be discovered, has a critical temperature of 92 K, and mercury-based cuprates have been found with critical temperatures in excess of 130 K. The explanation for these high critical temperatures remains unknown. Electron pairing due to phonon exchanges explains superconductivity in conventional superconductors, but it does not explain superconductivity in the newer superconductors that have a very high critical temperature. Figure 3.3 shows the behaviour of heat capacity and resistivity at the superconducting phase transition.

Similarly, at a fixed temperature below the critical temperature, superconducting materials cease to superconduct when an external magnetic field is applied which is greater than the critical magnetic field. This is because the Gibbs free energy of the superconducting phase increases quadratically with the magnetic field while the free energy of the normal phase is roughly independent of the magnetic field. If the material superconducts in the absence of a field, then the superconducting phase free energy is lower than that of the normal phase and so for some finite value of the magnetic field (proportional to the square root of the difference of the free energies at zero magnetic field) the two free energies will be equal and a phase transition to the normal phase will occur. More generally, a higher temperature and a stronger magnetic field lead to a smaller fraction of the electrons in the superconducting band and consequently a longer London penetration depth of external magnetic fields and currents. The penetration depth becomes infinite at the phase transition.

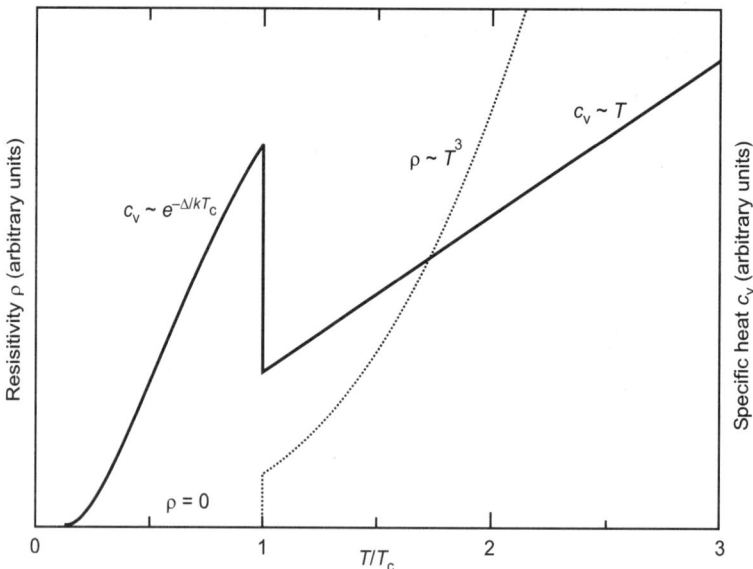

Fig. 3.3. Behaviour of heat capacity (c_v, dark line) and resistivity (ρ, dotted line) at the superconducting phase transition.

The onset of superconductivity is accompanied by abrupt changes in various physical properties, which is the hallmark of a phase transition. For example, the electronic heat capacity is proportional to the temperature in the normal (non-superconducting) regime. At the superconducting transition, it suffers a discontinuous jump and thereafter ceases to be linear. At low temperatures, it varies instead as $e^{-\alpha/T}$ for some constant α. This exponential behaviour is one of the pieces of evidence for the existence of the energy gap.

The order of the superconducting phase transition was long a matter of debate. Experiments indicate that the transition is second-order, meaning there is no latent heat. However in the presence of an external magnetic field there is latent heat, as a result of the fact that the superconducting phase has a lower entropy below the critical temperature than the normal phase. It has been experimentally demonstrated that, as a consequence, when the magnetic field is increased beyond the critical field, the resulting phase transition leads to a decrease in the temperature of the superconducting material.

Calculations in the 1970s suggested that it may actually be weakly first-order due to the effect of long-range fluctuations in the electromagnetic field. In the 1980s it was shown theoretically with the help of a disorder field theory, in which the vortex lines of the superconductor play a major role, that the transition is of second order within the type II regime and of first order (i.e. latent heat) within the type I regime, and that the two regions are separated by a tricritical point. The results were confirmed by Monte Carlo computer simulations.

Meissner effect

When a superconductor is placed in a weak external magnetic field H, the field penetrates the superconductor only a small distance λ, called the London penetration depth, decaying exponentially to zero within the bulk of the material. This is called the Meissner effect, and is a defining characteristic of superconductivity. For most superconductors, the London penetration depth is on the order of 100 nm.

The Meissner effect is sometimes confused with the kind of diamagnetism one would expect in a perfect electrical conductor: according to Lenz's law, when a changing magnetic field is applied to a conductor, it will induce an electrical current in the conductor that creates an opposing magnetic field. In a perfect conductor, an arbitrarily large current can be induced, and the resulting magnetic field exactly cancels the applied field.

The Meissner effect is distinct from this because a superconductor expels all magnetic fields, not just those that are changing. Suppose we have a material in its normal state, containing a constant internal magnetic field. When the material is cooled below the critical temperature, we would observe the abrupt expulsion of the internal magnetic field, which we would not expect based on Lenz's law.

The Meissner effect was explained by the brothers Fritz and Heinz London, who showed that the electromagnetic free energy in a superconductor is minimised provided

$$\nabla^2 H = \lambda^{-2} J$$

where, H is the magnetic field and λ is the London penetration depth.

This equation, which is known as the London equation, predicts that the magnetic field in a superconductor decays exponentially from whatever value it possesses at the surface.

The Meissner effect breaks down when the applied magnetic field is too large. Superconductors can be divided into two classes according to how this breakdown occurs. In Type I superconductors, superconductivity is abruptly destroyed when the strength of the applied field rises above a critical value J_c. Depending on the geometry of the sample, one may obtain an intermediate state consisting of regions of normal material carrying a magnetic field mixed with regions of superconducting material containing no field.

In Type II superconductors, raising the applied field past a critical value J_{c1} leads to a mixed state in which an increasing amount of magnetic flux penetrates the material, but there remains no resistance to the flow of electrical current as long as the current is not too large. At a second critical field strength J_{c2}, superconductivity is destroyed. The mixed state is actually caused by vortices in the electronic superfluid, sometimes called fluxons because the flux carried by these vortices is quantised. Most pure elemental superconductors, except niobium, technetium, vanadium and carbon nanotubes, are Type I, while almost all impure and compound superconductors are Type II.

London moment

Conversely, a spinning superconductor generates a magnetic field, precisely aligned with the spin axis. The effect, the London moment, was put to good use in Gravity Probe B. This experiment measured the magnetic fields of four superconducting gyroscopes to determine their spin axes. This was critical to the experiment since it is one of the few ways to accurately determine the spin axis of an otherwise featureless sphere.

Theories of Superconductivity

Since discovery of superconductivity, great efforts have been devoted to finding out how and why it works. During the 1950s, theoretical condensed matter physicists arrived at a solid understanding of 'conventional' superconductivity, through a pair of remarkable and important theories: the phenomenological Ginzburg-Landau theory and the microscopic BCS theory. Generalisations of these theories form the basis for understanding the closely related phenomenon of superfluidity, because they fall into the Lambda transition universality class, but the extent to which similar generalisations can be applied to unconventional superconductors as well is still controversial. The four-dimensional extension of the Ginzburg-Landau theory, the Coleman-Weinberg model, is important in quantum field theory and cosmology.

High temperature superconductivity

Until 1986, physicists had believed that BCS theory forbade superconductivity at temperatures above about 30 K. In that year, Bednorz and Müller discovered superconductivity in a lanthanum-based cuprate perovskite material, which had a transition temperature of 35 K. It was shortly found that replacing the lanthanum with yttrium, i.e. making YBCO, raised the critical temperature to 92 K, which was important because liquid nitrogen could then be used as a refrigerant (at atmospheric pressure, the boiling point of nitrogen is 77 K). This is important commercially because liquid nitrogen can be produced cheaply on-site from air, and is not prone to some of the problems (for instance solid air plugs) of helium in piping. Many other cuprate superconductors have since been discovered, and the theory of superconductivity in these materials is one of the major outstanding challenges of theoretical condensed matter physics.

From about 1993, the highest temperature superconductor was a ceramic material consisting of thallium, mercury, copper, barium, calcium, and oxygen ($HgBa_2Ca_2Cu_3O_{8+\delta}$) with T_c=138 K.

In February 2008, an iron-based family of high temperature superconductors was discovered. Hideo Hosono, of the Tokyo Institute of Technology, and colleagues found lanthanum oxygen fluorine iron arsenide ($LaO_{1-x}F_xFeAs$), an oxypnictide that superconducts below 26 K. Replacing the lanthanum in $LaO_{1-x}F_xFeAs$ with samarium leads to superconductors that work at 55 K.

Classification

There is not just one criterion to classify superconductors. The most common are:

1. By their physical properties: they can be Type I (if their phase transition is of first order) or Type II (if their phase transition is of second order).
2. By the theory to explain them: they can be conventional (if they are explained by the BCS theory or its derivatives) or unconventional (if not).
3. By their critical temperature: they can be high temperature (generally considered if they reach the superconducting state just cooling them with liquid nitrogen, that is, if T_c > 77 K), or low temperature (generally if they need other techniques to be cooled under their critical temperature).
4. By material: they can be chemical elements (as mercury or lead), alloys (as niobium-titanium or germanium-niobium), ceramics (as YBCO or the magnesium diboride), or organic superconductors (as fullerenes or carbon nanotubes, which technically might be included between the chemical elements as they are made of carbon).

Applications

Superconducting magnets are some of the most powerful electromagnets known. They are used in MRI and NMR machines, mass spectrometers, and the beam-steering magnets used in particle accelerators. They can also be used for magnetic separation, where weakly magnetic particles are extracted from a background of less or non-magnetic particles, as in the pigment industries.

Superconductors have also been used to make digital circuits (e.g. based on the rapid single flux quantum technology) and RF and microwave filters for mobile phone base stations.

Superconductors are used to build Josephson junctions which are the building blocks of SQUIDs (superconducting quantum interference devices), the most sensitive magnetometers known. SQUIDs are used in scanning SQUID microscopes. Series of Josephson devices are used to define the SI volt. Depending on the particular mode of operation, a Josephson junction can be used as a photon detector or as a mixer. The large resistance change at the transition from the normal- to the superconducting state is used to build thermometers in cryogenic micro-calorimeter photon detectors.

Other early markets are arising where the relative efficiency, size and weight advantages of devices based on high-temperature superconductivity outweigh the additional costs involved.

Promising future applications include high-performance smart grid electric power transmission, transformers, power storage devices, electric motors (e.g. for vehicle propulsion, as in vactrains or maglev trains), magnetic levitation devices, fault current limiters, nanoscopic materials such as buckyballs, nanotubes, composite materials, and superconducting magnetic refrigeration. However, superconductivity is sensitive to moving magnetic fields so applications that use alternating current (e.g. transformers) will be more difficult to develop than those that rely upon direct current.

Finding a cost effective room-temperature superconductor has been an elusive dream of superconductivity research scientists for generations. If such materials could be developed in the future, they might revolutionise our understanding and use of nearly everything that is electric.

SEMICONDUCTOR DEVICE FABRICATION

Semiconductor device fabrication is the process used to create the integrated circuits (silicon chips) that are present in everyday electrical and electronic devices. It is a multiple-step sequence of photographic and chemical processing steps during which electronic circuits are gradually created on a wafer made of pure semiconducting material. Silicon is the most commonly used semiconductor material today, along with various compound semiconductors.

The entire manufacturing process from start to packaged chips ready for shipment takes six to eight weeks and is performed in highly specialised facilities referred to as fabs.

Wafers

A typical wafer is made out of extremely pure silicon that is grown into mono-crystalline cylindrical ingots (boules) up to 300 mm (slightly less than 12 inches) in diameter using the Czochralski process. These ingots are then sliced into wafers about 0.75 mm thick and polished to obtain a very regular and flat surface. Once the wafers are prepared, many process steps are necessary to produce the desired semiconductor integrated circuit. In general, the steps can be grouped into two areas:

1. Front-end processing.
2. Back-end processing.

Processing

In semiconductor device fabrication, the various processing steps fall into four general categories: deposition, removal, patterning, and modification of electrical properties.

1. Deposition is any process that grows, coats, or otherwise transfers a material onto the wafer. Available technologies consist of physical vapour deposition (PVD), chemical vapour deposition (CVD), electrochemical deposition (ECD), molecular beam epitaxy (MBE) and more recently, atomic layer deposition (ALD) among others.
2. Removal processes are any that remove material from the wafer either in bulk or selectively and consist primarily of etch processes, either wet etching or dry etching. Chemical-mechanical planarisation (CMP) is also a removal process used between levels.

3. Patterning covers the series of processes that shape or alter the existing shape of the deposited materials and is generally referred to as lithography. For example, in conventional lithography, the wafer is coated with a chemical called a 'photoresist'. The photoresist is exposed by a 'stepper', a machine that focuses, aligns, and moves the mask, exposing select portions of the wafer to short wavelength light. The unexposed regions are washed away by a developer solution. After etching or other processing, the remaining photoresist is removed by plasma ashing.

4. Modification of electrical properties has historically consisted of doping transistor sources and drains originally by diffusion furnaces and later by ion implantation. These doping processes are followed by furnace anneal or in advanced devices, by rapid thermal anneal (RTA) which serve to activate the implanted dopants. Modification of electrical properties now also extends to reduction of dielectric constant in low-K insulating materials via exposure to ultraviolet light in UV processing (UVP).

Many modern chips have eight or more levels produced in over 300 sequenced processing steps.

Front-end processing

'Front-end processing' refers to the formation of the transistors directly on the silicon. The raw wafer is engineered by the growth of an ultrapure, virtually defect-free silicon layer through epitaxy. In the most advanced logic devices, prior to the silicon epitaxy step, tricks are performed to improve the performance of the transistors to be built. One method involves introducing a 'straining step' wherein a silicon variant such as 'silicon-germanium' (SiGe) is deposited.

Once the epitaxial silicon is deposited, the crystal lattice becomes stretched somewhat, resulting in improved electronic mobility.

Another method, called 'silicon on insulator' technology involves the insertion of an insulating layer between the raw silicon wafer and the thin layer of subsequent silicon epitaxy. This method results in the creation of transistors with reduced parasitic effects.

Gate oxide and implants

Front-end surface engineering is followed by: growth of the gate dielectric, traditionally silicon dioxide (SiO_2), patterning of the gate, patterning of the source and drain regions, and subsequent implantation or diffusion of dopants to obtain the desired complementary electrical properties. In memory devices, storage cells, conventionally capacitors, are also fabricated at this time, either into the silicon surface or stacked above the transistor.

Back-end processing

Metal layers

Once the various semiconductor devices have been created they must be interconnected to form the desired electrical circuits. This 'back end of line' (BEOL—the latter portion of the wafer fabrication, not to be confused with 'back end' of chip fabrication which refers to the package and test stages) involves creating metal interconnecting wires that are isolated by insulating dielectrics. The insulating material was traditionally a form of SiO_2 or a silicate glass, but recently new low dielectric constant materials are being used. These dielectrics presently take the form of SiOC and have dielectric constants around 2.7 (compared to 3.9 for SiO_2), although materials with constants as low as 2.2 are being offered to chipmakers.

Interconnect

Historically, the metal wires consisted of aluminium. In this approach to wiring often called 'subtractive aluminium', blanket films of aluminium are deposited first, patterned, and then etched, leaving isolated wires. Dielectric material is then deposited over the exposed wires. The various metal layers are interconnected by etching holes, called 'vias', in the insulating material and depositing tungsten in them with a CVD technique. This approach is still used in the fabrication of many memory chips such as dynamic random access memory (DRAM) as the number of interconnect levels is small, currently no more than four.

More recently, as the number of interconnect levels for logic has substantially increased due to the large number of transistors that are now interconnected in a modern microprocessor, the timing delay in the wiring has become significant prompting a change in wiring material from aluminium to copper and from the silicon dioxides to newer low-K material. This performance enhancement also comes at a reduced cost via damascene processing that eliminates processing steps. In damascene processing, in contrast to subtractive aluminium technology, the dielectric material is deposited first as a blanket film, and is patterned and etched leaving holes or trenches. In 'single damascene' processing, copper is then deposited in the holes or trenches surrounded by a thin barrier film resulting in filled vias or wire 'lines' respectively. In 'dual damascene' technology, both the trench and via are fabricated before the deposition of copper resulting in formation of both the via and line simultaneously, further reducing the number of processing steps. The thin barrier film, called copper barrier seed (CBS), is necessary to prevent copper diffusion into the dielectric. The ideal barrier film is as thin as possible. As the presence of excessive barrier film competes with the available copper wire cross section, formation of the thinnest continuous barrier represents one of the greatest ongoing challenges in copper processing today.

As the number of interconnect levels increases, planarisation of the previous layers is required to ensure a flat surface prior to subsequent lithography. Without it, the levels would become increasingly crooked and extend outside the depth of focus of available lithography, interfering with the ability to pattern. CMP (chemical mechanical planarisation) is the primary processing method to achieve such planarisation although dry 'etch back' is still sometimes employed if the number of interconnect levels is no more than three.

Wafer Test

The highly serialised nature of wafer processing has increased the demand for metrology in between the various processing steps. Wafer test metrology equipment is used to verify that the wafers have not been damaged by previous processing steps up until testing. If the number of dies—the integrated circuits that will eventually become chips—etched on a wafer exceeds a failure threshold (i.e. too many failed dies on one wafer), the wafer is scrapped rather than investing in further processing.

Device Test

Once the front-end process has been completed, the semiconductor devices are subjected to a variety of electrical tests to determine if they function properly. The proportion of devices on the wafer found to perform properly is referred to as the yield.

The fab tests the chips on the wafer with an electronic tester that presses tiny probes against the chip. The machine marks each bad chip with a drop of dye. The fab charges for test time; the prices are on the order of cents per second. Chips are often designed with 'testability features' such as 'built-in self-test' to speed testing, and reduce test costs.

Good designs try to test and statistically manage corners: extremes of silicon behaviour caused by operating temperature combined with the extremes of fab processing steps. Most designs cope with more than 64 corners.

Die Preparation

Once tested, the wafer is scored and then broken into individual die—wafer dicing. Only the good, unmarked chips go on to be packaged.

Packaging

Plastic or ceramic packaging involves mounting the die, connecting the die pads to the pins on the package, and sealing the die. Tiny wires are used to connect pads to the pins. In the old days, wires were attached by hand, but now purpose-built machines perform the task. Traditionally, the wires to the chips were gold, leading to a 'lead frame' (pronounced 'leed frame') of copper, that had been plated with solder, a mixture of tin and lead. Lead is poisonous, so lead-free 'lead frames' are now mandated by ROHS. Chip-scale package (CSP) is another packaging technology. A plastic dual in-line package, like most packages, is many times larger than the actual die hidden inside, whereas CSP chips are nearly the size of the die. CSP can be constructed for each die before the wafer is diced.

The packaged chips are retested to ensure that they were not damaged during packaging and that the die-to-pin interconnect operation was performed correctly. A laser etches the chip's name and numbers on the package.

List of Steps

This is a list of processing techniques that are employed numerous times in a modern electronic device and do not necessarily imply a specific order.
1. Wafer processing:
   (a) Wet cleans.
   (b) Photolithography.
   (c) Ion implantation (in which dopants are embedded in the wafer creating regions of increased [or decreased] conductivity).
   (d) Dry etching.
   (e) Wet etching.
   (f) Plasma ashing.
   (g) Thermal treatments: (i) rapid thermal anneal, (ii) furnace anneals, and (iii) thermal oxidation.
   (h) Chemical vapour deposition (CVD).
   (i) Physical vapour deposition (PVD).
   (j) Molecular beam epitaxy (MBE).
   (k) Electrochemical deposition (ECD).
   (l) Chemical-mechanical planarisation (CMP).
   (m) Wafer testing (where the electrical performance is verified).
   (n) Wafer backgrinding (to reduce the thickness of the wafer so the resulting chip can be put into a thin device like a smartcard or PCMCIA card).
2. Die preparation:
   (a) Wafer mounting.
   (b) Die cutting.

3. IC packaging:
   (a) Die attachment.
   (b) IC bonding: (i) wire bonding, (ii) flip chip, and (iii) tab bonding.
   (c) IC encapsulation: (i) baking, (ii) plating, (iii) lasermarking, and (iv) trim and form.
4. IC testing.

Hazardous Materials

Many toxic materials are used in the fabrication process. These include:

1. Poisonous elemental dopants such as arsenic, antimony and phosphorus.
2. Poisonous compounds like arsine, phosphine and silane.
3. Highly reactive liquids, such as hydrogen peroxide, fuming nitric acid, sulphuric acid and hydrofluoric acid.

It is vital that workers not be directly exposed to these dangerous substances. The high degree of automation common in the IC fabrication industry helps to reduce the risks of exposure of this sort. Most fabrication facilities employ exhaust management systems, such as wet scrubbers, combustors, heated absorber cartridges, etc. to control the risk to workers and also the environment if these toxic materials are released into the atmosphere.

Chapter 4

Fullerene

INTRODUCTION

A fullerene is any molecule composed entirely of carbon, in the form of a hollow sphere, ellipsoid, or tube. Spherical fullerenes are also called buckyballs, and cylindrical ones are called carbon nanotubes or buckytubes. Fullerenes are similar in structure to graphite, which is composed of stacked graphene sheets of linked hexagonal rings; but they may also contain pentagonal (or sometimes heptagonal) rings.

The first fullerene to be discovered, and the family's namesake, was buckminsterfullerene C_{60}, made in 1985 by Robert Curl, Harold Kroto and Richard Smalley. The name was an homage to Richard Buckminster Fuller, whose geodesic domes it resembles. Fullerenes have since been found to occur (if rarely) in nature. The icosahedral fullerene C_{540} is shown in Fig. 4.1.

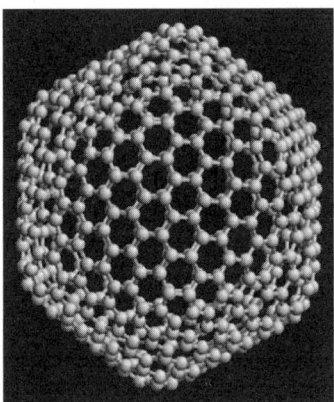

Fig. 4.1. The icosahedral fullerene C_{540}.

The discovery of fullerenes greatly expanded the number of known carbon allotropes, which until recently were limited to graphite, diamond, and amorphous carbon such as soot and charcoal. Buckyballs and buckytubes have been the subject of intense research, both for their unique chemistry and for their technological applications, especially in materials science, electronics, and nanotechnology.

PREDICTION AND DISCOVERY

The existence of C_{60} was predicted by Eiji Osawa of Toyohashi University of Technology in a Japanese magazine in 1970. He noticed that the structure of a corannulene molecule was a subset of a soccer-ball

50

shape, and he made the hypothesis that a full ball shape could also exist. His idea was reported in Japanese magazines, but did not reach Europe or America.

With mass spectrometry, discrete peaks were observed corresponding to molecules with the exact mass of sixty or seventy or more carbon atoms. In 1985, Harold Kroto (then of the University of Sussex), James R. Heath, Sean O'Brien, Robert Curl and Richard Smalley, from Rice University, discovered C_{60}, and shortly thereafter came to discover the fullerenes. Kroto, Curl, and Smalley were awarded the 1996 Nobel Prize in Chemistry for their roles in the discovery of this class of compounds. C_{60} and other fullerenes were later noticed occurring outside the laboratory (e.g. in normal candle soot). By 1991, it was relatively easy to produce gram-sized samples of fullerene powder using the techniques of Donald Huffman and Wolfgang Krätschmer. Fullerene purification remains a challenge to chemists and to a large extent determines fullerene prices. So-called endohedral fullerenes have ions or small molecules incorporated inside the cage atoms. Fullerene is an unusual reactant in many organic reactions such as the Bingel reaction discovered in 1993. The first nanotubes were obtained in 1991.

Minute quantities of the fullerenes, in the form of C_{60}, C_{70}, C_{76}, and C_{84} molecules, are produced in nature, hidden in soot and formed by lightning discharges in the atmosphere. Recently, fullerenes were found in a family of minerals known as Shungites in Karelia, Russia.

Buckminsterfullerene (C_{60}) was named after Richard Buckminster Fuller, a noted architectural modeler who popularised the geodesic dome. Since buckminsterfullerenes have a similar shape to that sort of dome, the name was thought to be appropriate. As the discovery of the fullerene family came after buckminsterfullerene, the shortened name 'fullerene' was used to refer to the family of fullerenes.

VARIATIONS

Since the discovery of fullerenes in 1985, structural variations on fullerenes have evolved well beyond the individual clusters themselves. Examples include:

1. Buckyball clusters: Smallest member is C_{20} (unsaturated version of dodecahedrane) and the most common is C_{60}.
2. Nanotubes: Hollow tubes of very small dimensions, having single or multiple walls; potential applications in electronics industry.
3. Megatubes: Larger in diameter than nanotubes and prepared with walls of different thickness; potentially used for the transport of a variety of molecules of different sizes.
4. Polymers: Chain, two-dimensional and three-dimensional polymers are formed under high pressure high temperature conditions.
5. Nano-onions: Spherical particles based on multiple carbon layers surrounding a buckyball core; proposed for lubricants.
6. Linked 'ball-and-chain' dimers: Two buckyballs linked by a carbon chain.
7. Fullerene rings.

BUCKYBALL

Buckminsterfullerene (IUPAC name (C_{60}–I_h) (fullerene) is the smallest fullerene molecule in which no two pentagons share an edge (which can be destabilising, as in pentalene). It is also the most common in terms of natural occurrence, as it can often be found in soot.

The structure of C_{60} is a truncated (T = 3) icosahedron, which resembles a soccer ball of the type made of twenty hexagons and twelve pentagons, with a carbon atom at the vertices of each polygon and a bond along each polygon edge (Fig. 4.2).

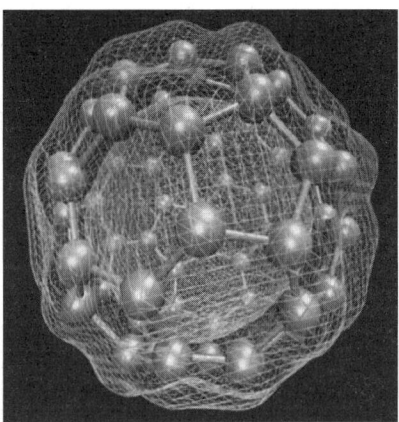

Fig. 4.2. C_{60} with isosurface of ground state electron density as calculated with DFT

The van der Waals diameter of a C_{60} molecule is about 1 nanometer (nm). The nucleus to nucleus diameter of a C_{60} molecule is about 0.7 nm.

The C_{60} molecule has two bond lengths. The 6:6 ring bonds (between two hexagons) can be considered 'double bonds' and are shorter than the 6:5 bonds (between a hexagon and a pentagon). Its average bond length is 1.4 angstroms. Silicon buckyballs have been created around metal ions.

Boron Buckyball

A new type of buckyball utilising boron atoms instead of the usual carbon has been predicted and described by researchers at Rice University. The B-80 structure, with each atom forming 5 or 6 bonds, is predicted to be more stable than the C-60 buckyball. One reason for this given by the researchers is that the B-80 is actually more like the original geodesic dome structure popularised by Buckminster Fuller which utilises triangles rather than hexagons. However, this work has been subject to much criticism by quantum chemists as it was concluded that the predicted Ih symmetric structure was vibrationally unstable and the resulting cage undergoes a spontaneous symmetry break yielding a puckered cage with rare Th symmetry (symmetry of a volleyball). The number of six atom rings in this molecule is 20 and number of five member rings is 12. There is an additional atom in the centre of each six member ring, bonded to each atom surrounding it.

Variations of Buckyballs

Another fairly common buckminsterfullerene is C_{70}, but fullerenes with 72, 76, 84 and even up to 100 carbon atoms are commonly obtained.

In mathematical terms, the structure of a fullerene is a trivalent convex polyhedron with pentagonal and hexagonal faces. In graph theory, the term fullerene refers to any 3-regular, planar graph with all faces of size 5 or 6 (including the external face). It follows from Euler's polyhedron formula, $|V|-|E|+|F| = 2$, (where $|V|$, $|E|$, $|F|$ indicate the number of vertices, edges, and faces), that there are exactly 12 pentagons in a fullerene and $|V|/2-10$ hexagons.

The smallest fullerene is the dodecahedron—the unique C_{20}. There are no fullerenes with 22 vertices. The number of fullerenes C_{2n} grows with increasing $n = 12,13,14...$, roughly in proportion to n^9 (sequence A007894 in OEIS). For instance, there are 1812 non-isomorphic fullerenes C_{60}. Note that only one form of C_{60}, the buckminsterfullerene alias truncated icosahedron, has no pair of adjacent pentagons (the smallest such fullerene). To further illustrate the growth, there are 214,127,713 non-isomorphic fullerenes C_{200}, 15,655,672 of which have no adjacent pentagons.

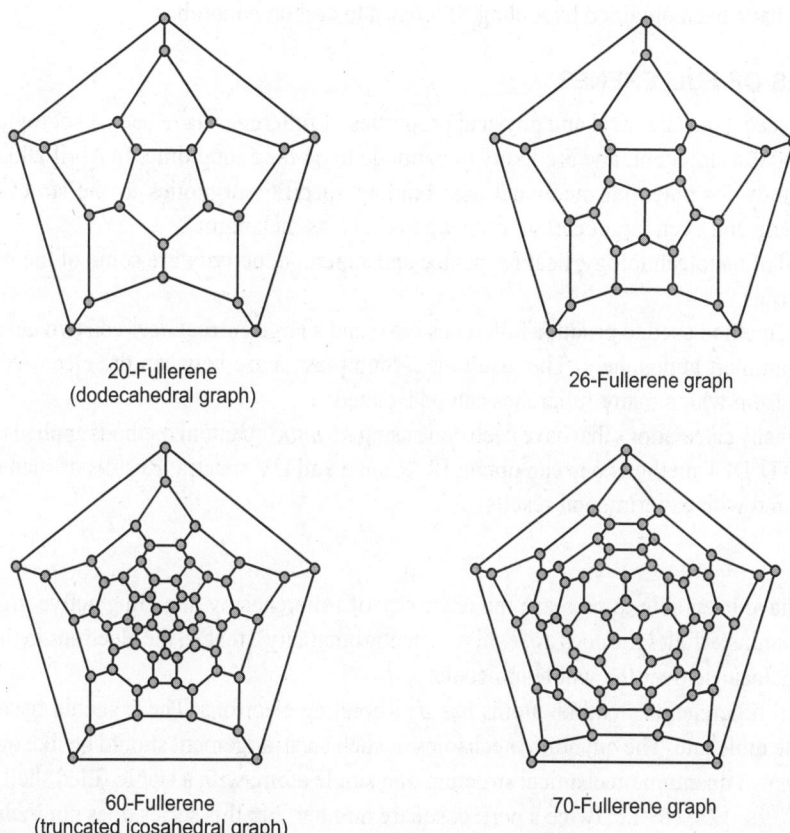

20-Fullerene
(dodecahedral graph)

26-Fullerene graph

60-Fullerene
(truncated icosahedral graph)

70-Fullerene graph

Trimetasphere carbon nanomaterials were discovered by researchers at Virginia Tech and licensed exclusively to Luna Innovations. This class of novel molecules comprises 80 carbon atoms (C_{80}) forming a sphere which encloses a complex of three metal atoms and one nitrogen atom. These fullerenes encapsulate metals which puts them in the subset referred to as metallofullerenes. Trimetaspheres have the potential for use in diagnostics (as safe imaging agents), therapeutics and in organic solar cells.

CARBON NANOTUBES

Nanotubes are cylindrical fullerenes. These tubes of carbon are usually only a few nanometers wide, but they can range from less than a micrometer to several millimeters in length. They often have closed ends, but can be open-ended as well. There are also cases in which the tube reduces in diameter before closing off. Their unique molecular structure results in extraordinary macroscopic properties, including high tensile strength, high electrical conductivity, high ductility, high resistance to heat, and relative chemical inactivity (as it is cylindrical and 'planar'—that is, it has no 'exposed' atoms that can be easily displaced). One proposed use of carbon nanotubes is in paper batteries, developed in 2007 by researchers at Rensselaer Polytechnic Institute. Another proposed use in the field of space technologies and science fiction is to produce high-tensile carbon cables required by a space elevator. Carbon nanotubes are discussed in next Chapter 5.

Nanobuds have been obtained by adding fullerenes to carbon nanotubes.

PROPERTIES OF FULLERENES

For the past decade, the chemical and physical properties of fullerenes have been a hot topic in the field of research and development, and are likely to continue to be for a long time. In April 2003, fullerenes were under study for potential medicinal use: binding specific antibiotics to the structure to target resistant bacteria and even target certain cancer cells such as melanoma.

In the field of nanotechnology, heat resistance and superconductivity are some of the more heavily studied properties.

A common method used to produce fullerenes is to send a large current between two nearby graphite electrodes in an inert atmosphere. The resulting carbon plasma arc between the electrodes cools into sooty residue from which many fullerenes can be isolated.

There are many calculations that have been done using *ab initio* quantum methods applied to fullerenes. By DFT and TD-DFT methods one can obtain IR, Raman and UV spectra. Results of such calculations can be compared with experimental results.

Aromaticity

Researchers have been able to increase the reactivity of fullerenes by attaching active groups to their surfaces. Buckminsterfullerene does not exhibit 'superaromaticity': that is, the electrons in the hexagonal rings do not delocalise over the whole molecule.

A spherical fullerene of n carbon atoms has n pi-bonding electrons. These should try to delocalise over the whole molecule. The quantum mechanics of such an arrangement should be like one shell only of the well-known quantum mechanical structure of a single atom, with a stable filled shell for $n = 2, 8, 18, 32, 50, 72, 98, 128$, etc. i.e. twice a perfect square number; but this series does not include 60. As a result, C_{60} in water tends to pick up two more electrons and become an anion. The nC_{60} described below may be the result of C_{60} trying to form a loose metallic bonding.

Chemistry

Fullerenes are stable, but not totally unreactive. The sp^2-hybridised carbon atoms, which are at their energy minimum in planar graphite, must be bent to form the closed sphere or tube, which produces angle strain. The characteristic reaction of fullerenes is electrophilic addition at 6,6-double bonds, which reduces angle strain by changing sp^2-hybridised carbons into sp^3-hybridised ones. The change in hybridised orbitals causes the bond angles to decrease from about 120 degrees in the sp^2 orbitals to about 109.5 degrees in the sp^3 orbitals. This decrease in bond angles allows for the bonds to bend less when closing the sphere or tube, and thus, the molecule becomes more stable.

Other atoms can be trapped inside fullerenes to form inclusion compounds known as endohedral fullerenes. An unusual example is the egg shaped fullerene $Tb_3N@C_{84}$, which violates the isolated pentagon rule. Recent evidence for a meteor impact at the end of the Permian period was found by analysing noble gases so preserved. Metallofullerene-based inoculates using the rhonditic steel process are beginning production as one of the first commercially-viable uses of buckyballs.

Solubility

Fullerenes are sparingly soluble in many solvents. Common solvents for the fullerenes include aromatics, such as toluene, and others like carbon disulphide. Solutions of pure buckminsterfullerene have a deep purple colour. Solutions of C_{70} are a reddish brown. The higher fullerenes C_{76} to C_{84} have a variety of colours. C_{76} has two optical forms, while other higher fullerenes have several structural isomers. Fullerenes are the only known allotrope of carbon that can be dissolved in common solvents at room temperature.

Some fullerene structures are not soluble because they have a small band gap between the ground and excited states. These include the small fullerenes C_{28}, C_{36} and C_{50}. The C_{72} structure is also in this class, but the endohedral version with a trapped lanthanide-group atom is soluble due to the interaction of the metal atom and the electronic states of the fullerene. Researchers had originally been puzzled by C_{72} being absent in fullerene plasma-generated soot extract, but found in endohedral samples. Small band gap fullerenes are highly reactive and bind to other fullerenes or to soot particles.

Solvents that are able to dissolve buckminsterfullerene (C_{60}) are listed below in order from highest solubility. The value in parentheses is the approximate saturated concentration:

1. 1-Chloronaphthalene (51 mg/ml).
2. 1-Methylnaphthalene (33 mg/ml).
3. 1,2-Dichlorobenzene (24 mg/ml).
4. 1,2,4-Trimethylbenzene (18 mg/ml).
5. Tetrahydronaphthalene (16 mg/ml).
6. Carbon disulphide (8 mg/ml).
7. 1,2,3-Tribromopropane (8 mg/ml).
8. Bromoform (5 mg/ml).
9. Cumene (4 mg/ml).
10. Toluene (3 mg/ml).
11. Benzene (1.5 mg/ml).
12. Cyclohexane (1.2 mg/ml).
13. Carbon tetrachloride (0.4 mg/ml).
14. Chloroform (0.25 mg/ml).
15. n-Hexane (0.046 mg/ml).
16. Tetrahydrofuran (0.006 mg/ml).
17. Acetonitrile (0.004 mg/ml).
18. Methanol (0.00004 mg/ml).
19. Water (1.3×10^{-11} mg/ml).

Solubility of C_{60} in some solvents shows unusual behaviour due to existence of solvate phases (analogues of crystallohydrates). For example, solubility of C_{60} in benzene solution shows maximum at about 313 K. Crystallisation from benzene solution at temperatures below maximum results in formation of triclinic solid solvate with four benzene molecules $C_{60} \cdot 4C_6H_6$ which is rather unstable in air. Out of solution, this structure decomposes into usual fcc C_{60} in few minutes' time. At temperatures above solubility maximum the solvate is not stable even when immersed in saturated solution and melts with formation of fcc C_{60}. Crystallisation at temperatures above the solubility maximum results in formation of pure fcc C_{60}. Large millimeter size crystals of C_{60} and C_{70} can be grown from solution both for solvates and for pure fullerenes.

Quantum Mechanics

In 1999, researchers from the University of Vienna demonstrated that wave-particle duality applied to molecules such as fullerene. One of the co-authors of this research, Julian Voss-Andreae, has since created several sculptures symbolising wave-particle duality in fullerenes.

Science writer Marcus Chown stated on the CBC radio show *Quirks and Quarks* in May 2006 that scientists are trying to make buckyballs exhibit the quantum behaviour of existing in two places at once (quantum superposition).

Safety and Toxicity

When considering toxicological data, care must be taken to distinguish as necessary between what are normally referred to as fullerenes: (C_{60}, C_{70},...); fullerene derivatives: C_{60} or other fullerenes with covalently bonded chemical groups; fullerene complexes (e.g. water-solubilised with surfactants, such as C_{60}-PVP; host-guest complexes, such as with cyclodextrin), where the fullerene is physically bound to another molecule; C_{60} nanoparticles, which are extended solid-phase aggregates of C_{60} crystallites; and nanotubes, which are generally much larger (in terms of molecular weight and size) compounds, and are different in shape to the spheroidal fullerenes C_{60} and C_{70}, as well as having different chemical and physical properties.

The above different compounds span the range from insoluble materials in either hydrophilic or lipophilic media, to hydrophilic, lipophilic, or even amphiphilic compounds, and with other varying physical and chemical properties. Therefore any broad generalisation extrapolating for example results from C_{60} to nanotubes or vice versa is not possible, though technically all are fullerenes, as the term is defined as a close-caged all-carbon molecule. Any extrapolation of results from one compound to other compounds must take into account considerations based on a quantitative structural analysis relationship study (QSARS), which mostly depends on how close the compounds under consideration are in physical and chemical properties.

In 1996 and 1997, Moussa studied the *in vivo* toxicity of C_{60} after intra-peritoneal administration of large doses. No evidence of toxicity was found and the mice tolerated a dose of 5000 mg/kg of body weight (BW). Mori could not find toxicity in rodents for C_{60} and C_{70} mixtures after oral administration of a dose of 2000 mg/kg BW and did not observe evidence of genotoxic or mutagenic potential *in vitro*. Other studies could not establish the toxicity of fullerenes: on the contrary, the work of Gharbi suggested that aqueous C_{60} suspensions failing to produce acute or subacute toxicity in rodents could also protect their livers in a dose-dependent manner against free-radical damage.

A comprehensive and recent review on fullerene toxicity is given by Kolosnjaj and Smith. These authors review the works on fullerene toxicity beginning in the early 1990s to present, and conclude that very little evidence gathered since the discovery of fullerenes indicate that C_{60} is toxic.

With reference to nanotubes, a recent study of Poland on carbon nanotubes introduced into the abdominal cavity of mice led the authors to suggest comparisons to 'asbestos-like pathogenicity'. It should be noted that this was not an inhalation study, though there have been several performed in the past, therefore it is premature to conclude that nanotubes should be considered to have a toxicological profile similar to asbestos. Conversely, and perhaps illustrative of how the various classes of compounds which fall under the general term fullerene cover a wide range of properties. Sayes, found that *in vivo* inhalation of $C_{60}(OH)_{24}$ and nano-C_{60} in rats gave no effect, whereas in comparison quartz particles produced an inflammatory response under the same conditions. As stated above, nanotubes are quite different in chemical and physical properties to C_{60}, i.e. molecular weight, shape, size, physical properties (such as solubility) all are very different, so from a toxicological standpoint, different results for C_{60} and nanotubes are not suggestive of any discrepancy in the findings.

Superconductivity

After the synthesis of macroscopic amounts of fullerenes, their physical properties could be investigated. Very soon Haddon found that intercalation of alkali-metal atoms in solid C_{60} leads to metallic behaviour. In 1991, it was revealed that potassium-doped C_{60} becomes superconducting at 18 K. This was the highest transition temperature for a molecular superconductor. Since then, superconductivity has been reported in fullerene doped with various other alkali metals.

It has been shown that the superconducting transition temperature in alkaline-metal-doped fullerene increases with the unit-cell volume V. As caesium forms the largest alkali ion, caesium-doped fullerene is an important material in this family. Recently, superconductivity at 38 K has been reported in bulk Cs_3C_{60}, but only under applied pressure. The highest superconducting transition temperature of 33 K at ambient pressure is reported for Cs_2RbC_{60}.

The increase of transition temperature with the unit-cell volume had been believed to be evidence for the BCS mechanism of C_{60} solid superconductivity, because inter C_{60} separation can be related to an increase in the density of states on the Fermi level, $N(\varepsilon_F)$. Therefore, there have been many efforts to increase the interfullerene separation, in particular, intercalating neutral molecules into the $A_3 C_{60}$ lattice to increase the interfullerene spacing while the valence of C_{60} is kept unchanged. However, this ammoniation technique has revealed a new aspect of fullerene intercalation compounds: the Mott-Hubbard transition and the correlation between the orientation/orbital order of C_{60} molecules and the magnetic structure.

The C_{60} molecules compose a solid of weakly bound molecules. The fullerites are therefore molecular solids, in which the molecular properties still survive. The discrete levels of a free C_{60} molecule are only weakly broadened in the solid, which leads to a set of essentially non-overlapping bands with a narrow width of about 0.5 eV. For an undoped C_{60} solid, the 5-fold h_u band is the HOMO level, and the 3-fold t_{1u} band is the empty LUMO level, and this system is a band insulator. But when the C_{60} solid is doped with metal atoms, the metal atoms give electrons to the t_{1u} band or the upper 3-fold t_{1g} band. This partial electron occupation of the band leads to sometimes metallic behaviour. However, A_4C_{60} is an insulator, although the t_{1u} band is only partially filled and it should be a metal according to band theory. This unpredicted behaviour may be explained by the Jahn-Teller effect, where spontaneous deformations of high-symmetry molecules induce the splitting of degenerate levels to gain the electronic energy. The Jahn-Teller type electron-phonon interaction is strong enough in C_{60} solids to destroy the band picture for particular valence states.

A narrow band or strongly correlated electronic system and degenerated ground states are important points to understand in explaining superconductivity in fullerene solids. When the inter-electron repulsion U is greater than the bandwidth, an insulating localised electron ground state is produced in the simple Mott-Hubbard model. This explains the absence of superconductivity at ambient pressure in caesium-doped C_{60} solids.

Electron-correlation-driven localisation of the t_{1u} electrons exceeds the critical value, leading to the Mott insulator. The application of high pressure decreases the interfullerene spacing, therefore caesium-doped C_{60} solids turn to metallic and superconducting.

A fully developed theory of C_{60} solids superconductivity is still lacking, but it has been widely accepted that strong electronic correlations and the Jahn-Teller electron-phonon coupling produce local electron-pairings that show a high transition temperature close to the insulator-metal transition.

Chirality: Few fullerenes (e.g. C_{76}, C_{78}, C_{80}, and C_{84}) are inherently chiral because they are D_2-symmetric and have been successfully resolved. Research efforts are ongoing to develop specific sensors for their enantiomers.

FULLERITE (SOLID STATE)

Fullerites are the solid-state manifestation of fullerenes and related compounds and materials. Ultrahard fullerite, buckyball: 'Ultrahard fullerite' is a coined term frequently used to describe material produced by high-pressure high-temperature (HPHT) processing of fullerite. Such treatment converts fullerite into a nanocrystalline form of diamond which exhibits remarkable mechanical properties.

Chapter 5

Carbon Nanotubes

INTRODUCTION

Carbon nanotubes (CNTs) are allotropes of carbon with a cylindrical nanostructure. Nanotubes have been constructed with length-to-diameter ratio of up to 28,000,000:1, which is significantly larger than any other material. These cylindrical carbon molecules have novel properties that make them potentially useful in many applications in nanotechnology, electronics, optics and other fields of materials science, as well as potential uses in architectural fields. They exhibit extraordinary strength and unique electrical properties, and are efficient thermal conductors. Their final usage, however, may be limited by their potential toxicity and controlling their property changes in response to chemical treatment.

Nanotubes are members of the fullerene structural family, which also includes the spherical buckyballs. The ends of a nanotube might be capped with a hemisphere of the buckyball structure. Their name is derived from their size, since the diameter of a nanotube is on the order of a few nanometers (approximately 1/50,000th of the width of a human hair), while they can be up to several millimetres in length (as of 2008). Nanotubes are categorised as single-walled nanotubes (SWNTs) and multiwalled nanotubes (MWNTs). The nature of the bonding of a nanotube is described by applied quantum chemistry, specifically, orbital hybridisation. The chemical bonding of nanotubes is composed entirely of sp^2 bonds, similar to those of graphite. This bonding structure, which is stronger than the sp^3 bonds found in diamonds, provides the molecules with their unique strength. Nanotubes naturally align themselves into 'ropes' held together by van der Waals forces.

TYPES OF CARBON NANOTUBES AND RELATED STRUCTURES

Single-walled

Most single-walled nanotubes (SWNT) have a diameter of close to 1 nanometer, with a tube length that can be many millions of times longer. The structure of a SWNT can be conceptualised by wrapping a one-atom-thick layer of graphite called graphene into a seamless cylinder. The way the graphene sheet is wrapped is represented by a pair of indices (n,m) called the chiral vector. The integers n and m denote the number of unit vectors along two directions in the honeycomb crystal lattice of graphene. If $m = 0$, the nanotubes are called 'zigzag'. If $n = m$, the nanotubes are called 'armchair'. Otherwise, they are called 'chiral'. Figure 5.1 shows the different types of single-walled nanotubes (SWNT).

Single-walled nanotubes are an important variety of carbon nanotube because they exhibit electric properties that are not shared by the multiwalled carbon nanotube (MWNT) variants. Single-walled nanotubes are the most likely candidate for miniaturising electronics beyond the micro electromechanical scale currently used in electronics.

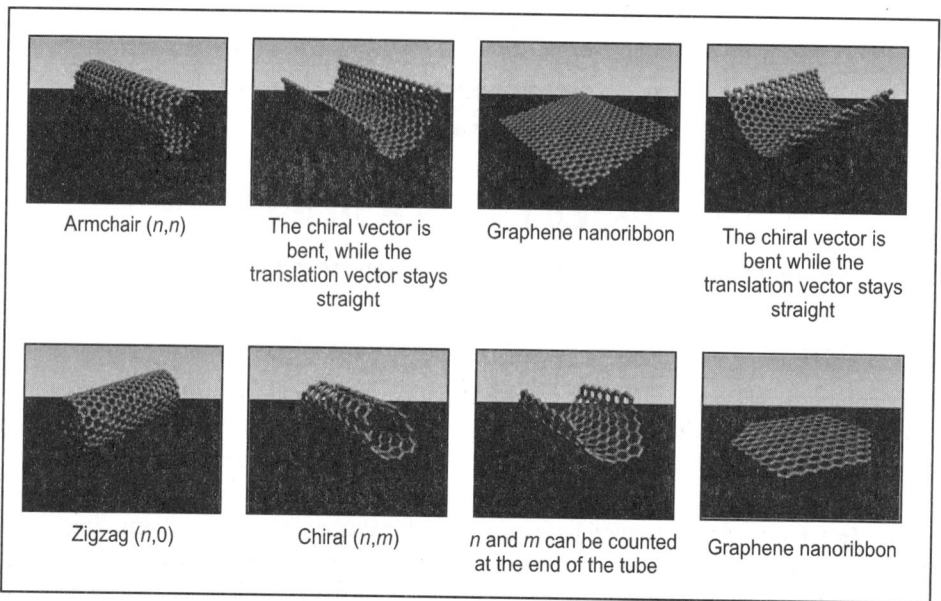

Fig. 5.1. Different types of single-walled nanotubes (SWNT).

Figure 5.2 shows the (n,m) nanotube naming scheme can be thought of as a vector (C_h) in an infinite graphene sheet that describes how to 'roll up' the graphene sheet to make the nanotube. T denotes the tube axis, and a_1 and a_2 are the unit vectors of graphene in real space.

The most basic building block of these systems is the electric wire, and SWNTs can be excellent conductors. One useful application of SWNTs is in the development of the first intramolecular field effect transistors (FET). Production of the first intramolecular logic gate using SWNT FETs has recently become possible as well. To create a logic gate you must have both a p-FET and an n-FET. Because SWNTs are p-FETs when exposed to oxygen and n-FETs otherwise, it is possible to protect half of an SWNT from oxygen exposure, while exposing the other half to oxygen. This results in a single SWNT that acts as a NOT logic gate with both p and n-type FETs within the same molecule. Figure 5.3 shows electron micrograph showing a single-walled nanotube. Single-walled nanotubes are still very expensive to produce, around $1500 per gram as of 2000, and the development of more affordable synthesis techniques is vital to the future of carbon nanotechnology. If cheaper means of synthesis cannot be discovered, it would make it financially impossible to apply this technology to commercial-scale applications. Several suppliers offer as-produced arc discharge SWNTs for ~$50–100 per gram as of 2007. Figure 5.4 shows the SEM image of carbon nanotubes bundles.

Multiwalled

Multiwalled nanotubes (MWNT) consist of multiple rolled layers (concentric tubes) of graphite. There are two models which can be used to describe the structures of multiwalled nanotubes. In the Russian Doll model, sheets of graphite are arranged in concentric cylinders, e.g. a single-walled nanotube (SWNT) within a larger single-walled nanotube. In the Parchment model, a single sheet of graphite is rolled in around itself, resembling a scroll of parchment or a rolled newspaper. The interlayer distance in multiwalled nanotubes is close to the distance between graphene layers in graphite, approximately 3.3 Å (330 pm).

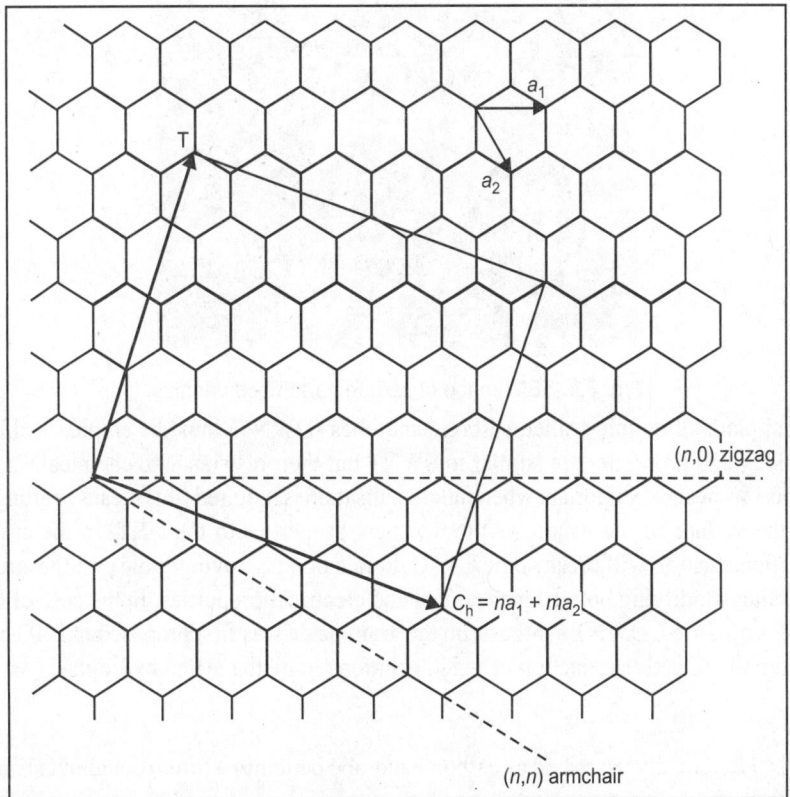

Fig. 5.2. The (n,m) nanotube naming scheme can be thought of as a vector (C_h) in an infinite graphene sheet that describes how to 'roll up' the graphene sheet to make the nanotube. T denotes the tube axis, and a_1 and a_2 are the unit vectors of graphene in real space.

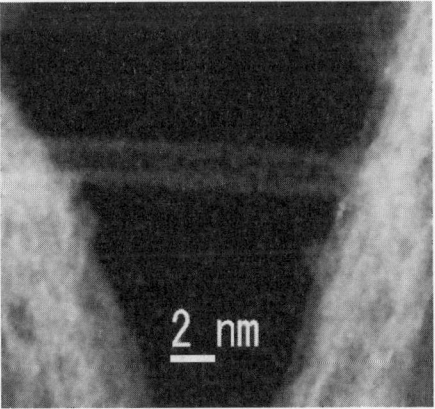

Fig. 5.3. Electron micrograph showing a single-walled nanotube.

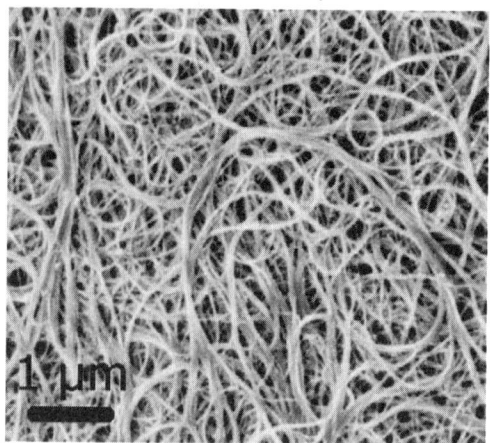

Fig. 5.4. SEM image of carbon nanotubes bundles.

The special place of double-walled carbon nanotubes (DWNT) must be emphasised here because their morphology and properties are similar to SWNT but their resistance to chemicals is significantly improved. This is especially important when functionalisation is required (this means grafting of chemical functions at the surface of the nanotubes) to add new properties to the CNT. In the case of SWNT, covalent functionalisation will break some $C=C$ double bonds, leaving 'holes' in the structure on the nanotube and thus modifying both its mechanical and electrical properties. In the case of DWNT, only the outer wall is modified. DWNT synthesis on the gram-scale was first proposed in 2003 by the CCVD technique, from the selective reduction of oxide solutions in methane and hydrogen.

Torus

A nanotorus is theoretically described as carbon nanotube bent into a torus (doughnut shape). Nanotori are predicted to have many unique properties, such as magnetic moments 1000 times larger than previously expected for certain specific radii. Properties such as magnetic moment, thermal stability, etc. vary widely depending on radius of the torus and radius of the tube.

Nanobud

Carbon nanobuds are a newly created material combining two previously discovered allotropes of carbon: carbon nanotubes and fullerenes. In this new material fullerene-like 'buds' are covalently bonded to the outer sidewalls of the underlying carbon nanotube. This hybrid material has useful properties of both fullerenes and carbon nanotubes (Fig. 5.5).

Fig. 5.5. A stable nanobud structure.

In particular, they have been found to be exceptionally good field emitters. In composite materials, the attached fullerene molecules may function as molecular anchors preventing slipping of the nanotubes, thus improving the composite's mechanical properties.

PROPERTIES

Strength

Carbon nanotubes are the strongest and stiffest materials yet discovered in terms of tensile strength and elastic modulus respectively. This strength results from the covalent sp^2 bonds formed between the individual carbon atoms. In 2000, a multiwalled carbon nanotube was tested to have a tensile strength of 63 gigapascals (GPa). (This, for illustration, translates into the ability to endure tension of 6300 kg on a cable with cross-section of 1 mm^2.) Since carbon nanotubes have a low density for a solid of 1.3 to 1.4 g·cm^{-3}, its specific strength of up to 48,000 kN·m·kg^{-1} is the best of known materials, compared to high-carbon steel's 154 kN·m·kg^{-1}.

Under excessive tensile strain, the tubes will undergo plastic deformation, which means the deformation is permanent. This deformation begins at strains of approximately 5 per cent and can increase the maximum strain the tubes undergo before fracture by releasing strain energy.

CNTs are not nearly as strong under compression. Because of their hollow structure and high aspect ratio, they tend to undergo buckling when placed under compressive, torsional or bending stress. Table 5.1 listed the comparison of mechanical properties.

Table 5.1. Comparison of mechanical properties.

Material	Young's modulus (TPa)	Tensile strength (GPa)	Elongation at break (%)
SWNT	~1 (from 1 to 5)	13–53^E	16
Armchair SWNT	0.94^T	126.2^T	23.1
Zigzag SWNT	0.94^T	94.5^T	15.6–17.5
Chiral SWNT	0.92	–	–
MWNT	0.8–0.9^E	150	–
Stainless steel	~0.2	~0.65–1	15–50
Kevlar	~0.15	~3.5	~2
Kevlar^T	0.25	29.6	–

^{E}Experimental observation; ^{T}Theoretical prediction.

The above discussion referred to axial properties of the nanotube, whereas simple geometrical considerations suggest that carbon nanotubes should be much softer in the radial direction than along the tube axis. Indeed, TEM observation of radial elasticity suggested that even the van der Waals forces can deform two adjacent nanotubes. Nanoindentation experiments, performed by several groups on multiwalled carbon nanotubes, indicated Young's modulus of the order of several GPa confirming that CNTs are indeed rather soft in the radial direction.

Kinetic

Multiwalled nanotubes, multiple concentric nanotubes precisely nested within one another, exhibit a striking telescoping property whereby an inner nanotube core may slide, almost without friction, within its outer nanotube shell thus creating an atomically perfect linear or rotational bearing. This is one of the first true examples of molecular nanotechnology, the precise positioning of atoms to create useful machines. Already this property has been utilised to create the world's smallest rotational motor. Future applications such as a gigahertz mechanical oscillator are also envisaged.

Electrical

Because of the symmetry and unique electronic structure of graphene, the structure of a nanotube strongly affects its electrical properties. For a given (n,m) nanotube, if $n = m$, the nanotube is metallic; if $n - m$ is a multiple of 3, then the nanotube is semiconducting with a very small band gap, otherwise the nanotube is a moderate semiconductor. Thus all armchair $(n = m)$ nanotubes are metallic, and nanotubes, etc. are semiconducting. In theory, metallic nanotubes can carry an electrical current density of 4×10^9 A/cm^2 which is more than 1000 times greater than metals such as copper.

Optical

Optical properties of carbon nanotubes are discussed latter in this section.

Thermal

All nanotubes are expected to be very good thermal conductors along the tube, exhibiting a property known as 'ballistic conduction', but good insulators laterally to the tube axis. It is predicted that carbon nanotubes will be able to transmit up to 6000 W·m^{-1}·K^{-1} at room temperature; compare this to copper, a metal well-known for its good thermal conductivity, which transmits 385 W·m^{-1}·K^{-1}. The temperature stability of carbon nanotubes is estimated to be up to 2800°C in vacuum and about 750°C in air.

Defects

As with any material, the existence of a crystallographic defect affects the material properties. Defects can occur in the form of atomic vacancies. High levels of such defects can lower the tensile strength by up to 85 per cent. Another form of carbon nanotube defect is the Stone Wales defect, which creates a pentagon and heptagon pair by rearrangement of the bonds. Because of the very small structure of CNTs, the tensile strength of the tube is dependent on its weakest segment in a similar manner to a chain, where the strength of the weakest link becomes the maximum strength of the chain.

Crystallographic defects also affect the tube's electrical properties. A common result is lowered conductivity through the defective region of the tube. A defect in armchair-type tubes (which can conduct electricity) can cause the surrounding region to become semiconducting, and single monoatomic vacancies induce magnetic properties.

Crystallographic defects strongly affect the tube's thermal properties. Such defects lead to phonon scattering, which in turn increases the relaxation rate of the phonons. This reduces the mean free path and reduces the thermal conductivity of nanotube structures. Phonon transport simulations indicate that substitutional defects such as nitrogen or boron will primarily lead to scattering of high-frequency optical phonons. However, larger-scale defects such as Stone Wales defects cause phonon scattering over a wide range of frequencies, leading to a greater reduction in thermal conductivity.

One-dimensional Transport

Due to their nanoscale dimensions, electron transport in carbon nanotubes will take place through quantum effects and will only propagate along the axis of the tube. Because of this special transport property, carbon nanotubes are frequently referred to as 'one-dimensional' in scientific articles.

Toxicity

Determining the toxicity of carbon nanotubes has been one of the most pressing questions in nanotechnology. Unfortunately such research has only just begun and the data is still fragmentary and subject to criticism. Preliminary results highlight the difficulties in evaluating the toxicity of this heterogeneous material. Parameters such as structure, size distribution, surface area, surface chemistry, surface charge, and agglomeration state as well as purity of the samples, have considerable impact on the reactivity of carbon nanotubes. However, available data clearly show that, under some conditions, nanotubes can cross membrane barriers, which suggests that if raw materials reach the organs they can induce harmful effects such as inflammatory and fibrotic reactions.

A study led by Alexandra Porter from the University of Cambridge shows that CNTs can enter human cells and accumulate in the cytoplasm, causing cell death.

Results of rodent studies collectively show that regardless of the process by which CNTs were synthesised and the types and amounts of metals they contained, CNTs were capable of producing inflammation, epithelioid granulomas (microscopic nodules), fibrosis, and biochemical/toxicological changes in the lungs. Comparative toxicity studies in which mice were given equal weights of test materials showed that SWCNTs were more toxic than quartz, which is considered a serious occupational health hazard when chronically inhaled. As a control, ultrafine carbon black was shown to produce minimal lung responses.

The needle-like fibre shape of CNTs, similar to asbestos fibres, raises fears that widespread use of carbon nanotubes may lead to mesothelioma, cancer of the lining of the lungs often caused by exposure to asbestos. A recently-published pilot study supports this prediction. Scientists exposed the mesothelial lining of the body cavity of mice, as a surrogate for the mesothelial lining of the chest cavity, to long multiwalled carbon nanotubes and observed asbestos-like, length-dependent, pathogenic behaviour which included inflammation and formation of lesions known as granulomas. Authors of the study conclude:

'This is of considerable importance, because research and business communities continue to invest heavily in carbon nanotubes for a wide range of products under the assumption that they are no more hazardous than graphite. Our results suggest the need for further research and great caution before introducing such products into the market if long-term harm is to be avoided.'

According to co-author Dr. Andrew Maynard:

'This study is exactly the kind of strategic, highly focused research needed to ensure the safe and responsible development of nanotechnology. It looks at a specific nanoscale material expected to have widespread commercial applications and asks specific questions about a specific health hazard. Even though scientists have been raising concerns about the safety of long, thin carbon nanotubes for over a decade, none of the research needs in the current US federal nanotechnology environment, health and safety risk research strategy address this question.'

Although further research is required, results presented today clearly demonstrate that, under certain conditions, especially those involving chronic exposure, carbon nanotubes can pose a serious risk to human health.

SYNTHESIS

Techniques have been developed to produce nanotubes in sizeable quantities, including arc discharge, laser ablation, high pressure carbon monoxide (HiPCO), and chemical vapour deposition (CVD). Most of these processes take place in vacuum or with process gases. CVD growth of CNTs can occur in vacuum or at atmospheric pressure. Large quantities of nanotubes can be synthesised by these methods; advances in catalysis and continuous growth processes are making CNTs more commercially viable.

Arc Discharge

Nanotubes were observed in 1991 in the carbon soot of graphite electrodes during an arc discharge, by using a current of 100 amps, that was intended to produce fullerenes. However, the first macroscopic production of carbon nanotubes was made in 1992 by two researchers at NEC's Fundamental Research Laboratory. The method used was the same as in 1991. During this process, the carbon contained in the negative electrode sublimates because of the high discharge temperatures. Because nanotubes were initially discovered using this technique, it has been the most widely-used method of nanotube synthesis.

The yield for this method is up to 30 per cent by weight and it produces both single- and multiwalled nanotubes with lengths of up to 50 micrometres with few structural defects.

Laser Ablation

In the laser ablation process, a pulsed laser vapourises a graphite target in a high-temperature reactor while an inert gas is bled into the chamber. Nanotubes develop on the cooler surfaces of the reactor as the vapourised carbon condenses. A water-cooled surface may be included in the system to collect the nanotubes.

This process was developed by Dr. Richard Smalley and co-workers at Rice University, who at the time of the discovery of carbon nanotubes, were blasting metals with a laser to produce various metal molecules. When they heard of the existence of nanotubes they replaced the metals with graphite to create multiwalled carbon nanotubes. Later that year the team used a composite of graphite and metal catalyst particles (the best yield was from a cobalt and nickel mixture) to synthesise single-walled carbon nanotubes. The laser ablation method yields around 70 per cent and produces primarily single-walled carbon nanotubes with a controllable diameter determined by the reaction temperature. However, it is more expensive than either arc discharge or chemical vapour deposition.

Chemical Vapour Deposition (CVD)

The catalytic vapour phase deposition of carbon was first reported in 1959, but it was not until 1993 that carbon nanotubes were formed by this process. In 2007, researchers at the University of Cincinnati (UC) developed a process to grow aligned carbon nanotube arrays of 18 mm length on a FirstNano ET3000 carbon nanotube growth system.

During CVD, a substrate is prepared with a layer of metal catalyst particles, most commonly nickel, cobalt, iron, or a combination. The metal nanoparticles can also be produced by other ways, including reduction of oxides or oxides solid solutions. The diameters of the nanotubes that are to be grown are related to the size of the metal particles. This can be controlled by patterned (or masked) deposition of the metal, annealing, or by plasma etching of a metal layer. The substrate is heated to approximately 700°C. To initiate the growth of nanotubes, two gases are bled into the reactor: a process gas (such as ammonia, nitrogen or hydrogen) and a carbon-containing gas (such as acetylene, ethylene, ethanol or methane). Nanotubes grow at the sites of the metal catalyst; the carbon-containing gas is broken apart at the surface of the catalyst particle, and the carbon is transported to the edges of the particle, where it forms the nanotubes. This mechanism is still being studied. The catalyst particles can stay at the tips of the growing nanotube during the growth process, or remain at the nanotube base, depending on the adhesion between the catalyst particle and the substrate.

CVD is a common method for the commercial production of carbon nanotubes. For this purpose, the metal nanoparticles are mixed with a catalyst support such as MgO or Al_2O_3 to increase the surface area for higher yield of the catalytic reaction of the carbon feedstock with the metal particles. One issue in this synthesis route is the removal of the catalyst support via an acid treatment, which sometimes could destroy the original structure of the carbon nanotubes. However, alternative catalyst supports that are soluble in water have proven effective for nanotube growth.

If a plasma is generated by the application of a strong electric field during the growth process (plasma enhanced chemical vapour deposition), then the nanotube growth will follow the direction of the electric field. By adjusting the geometry of the reactor it is possible to synthesise vertically aligned carbon nanotubes (i.e. perpendicular to the substrate), a morphology that has been of interest to researchers interested in the electron emission from nanotubes. Without the plasma, the resulting nanotubes are often randomly oriented. Under certain reaction conditions, even in the absence of a plasma, closely spaced nanotubes will maintain a vertical growth direction resulting in a dense array of tubes resembling a carpet or forest.

Of the various means for nanotube synthesis, CVD shows the most promise for industrial-scale deposition, because of its price/unit ratio, and because CVD is capable of growing nanotubes directly on a desired substrate, whereas the nanotubes must be collected in the other growth techniques. The growth sites are controllable by careful deposition of the catalyst. In 2007, a team from Meijo University demonstrated a high-efficiency CVD technique for growing carbon nanotubes from camphor. Researchers at Rice University, until recently led by the late Dr. Richard Smalley, have concentrated upon finding methods to produce large, pure amounts of particular types of nanotubes. Their approach grows long fibres from many small seeds cut from a single nanotube; all of the resulting fibres were found to be of the same diameter as the original nanotube and are expected to be of the same type as the original nanotube. Further characterisation of the resulting nanotubes and improvements in yield and length of grown tubes are needed.

CVD growth of multiwalled nanotubes is used by several companies to produce materials on the ton scale, including NanoLab, Bayer, Arkema, Nanocyl, Nanothinx, Hyperion Catalysis, Mitsui, and Showa Denko.

Natural, Incidental, and Controlled Flame Environments

Fullerenes and carbon nanotubes are not necessarily products of high-tech laboratories; they are commonly formed in such mundane places as ordinary flames, produced by burning methane, ethylene, and benzene, and they have been found in soot from both indoor and outdoor air. However, these naturally occurring varieties can be highly irregular in size and quality because the environment in which they are produced is often highly uncontrolled. Thus, although they can be used in some applications, they can lack in the high degree of uniformity necessary to meet many needs of both research and industry. Recent efforts have focused on producing more uniform carbon nanotubes in controlled flame environments. Nano-C, Inc of Westwood, Massachusetts, is producing flame synthesised single-walled carbon nanotubes. This method has promise for large-scale, low-cost nanotube synthesis, though it must compete with rapidly developing large scale CVD production.

POTENTIAL AND CURRENT APPLICATIONS

The strength and flexibility of carbon nanotubes makes them of potential use in controlling other nanoscale structures, which suggests they will have an important role in nanotechnology engineering. The highest tensile strength an individual multiwalled carbon nanotube has been tested to be is 63 GPa. Carbon nanotubes were found in Damascus steel, possibly helping to account for the legendary strength of the (almost ancient) swords made of it.

Structural

Because of the carbon nanotube's superior mechanical properties, many structures have been proposed ranging from everyday items like clothes and sports gear to combat jackets and space elevators. However, the space elevator will require further efforts in refining carbon nanotube technology, as the practical tensile strength of carbon nanotubes can still be greatly improved. Figure 5.6 shows the joining of two carbon nanotubes with different electrical properties to form a diode has been proposed.

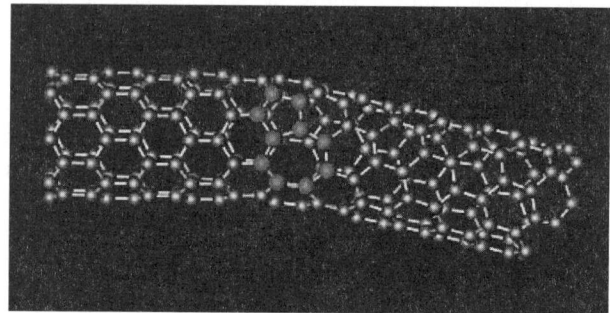

Fig. 5.6. The joining of two carbon nanotubes with different electrical properties to form a diode has been proposed.

For perspective, outstanding breakthroughs have already been made. Pioneering work led by Ray H. Baughman at the NanoTech Institute has shown that single and multiwalled nanotubes can produce materials with toughness unmatched in the man-made and natural worlds.

In Electrical Circuits

Carbon nanotubes have many properties—from their unique dimensions to an unusual current conduction mechanism—that make them ideal components of electrical circuits. For example, they have shown to exhibit strong electron-phonon resonances, which indicate that under certain direct current (DC) bias and doping conditions their current and the average electron velocity, as well as the electron concentration on the tube oscillate at terahertz frequencies. These resonances could potentially be used to make terahertz sources or sensors.

Nanotube based transistors have been made that operate at room temperature and that are capable of digital switching using a single electron.

One major obstacle to realisation of nanotubes has been the lack of technology for mass production. However, in 2001 IBM researchers demonstrated how nanotube transistors can be grown in bulk, somewhat like silicon transistors. Their process is called 'constructive destruction' which includes the automatic destruction of defective nanotubes on the wafer.

The IBM process has been developed further and single-chip wafers with over ten billion correctly aligned nanotube junctions have been created. In addition it has been demonstrated that incorrectly aligned nanotubes can be removed automatically using standard photolithography equipment.

The first nanotube integrated memory circuit was made in 2004. One of the main challenges has been regulating the conductivity of nanotubes. Depending on subtle surface features a nanotube may act as a plain conductor or as a semiconductor. A fully automated method has, however, been developed to remove non-semiconductor tubes.

Most recently, collaborating American and Chinese researchers at Duke University and Peking University announced a new CVD recipe involving a combination of ethanol and methanol gases and quartz substrates resulting in horizontally aligned arrays of 95–98 per cent semiconducting nanotubes. This is considered a large step towards the ultimate goal of producing perfectly aligned, 100 per cent semiconducting carbon nanotubes for mass production of electronic devices.

Another way to make carbon nanotube transistors has been to use random networks of them. By doing so one averages all of their electrical differences and one can produce devices in large scale at the wafer level. This approach was first patented by Nanomix Inc. It was first published in the academic literature by the United States Naval Research Laboratory in 2003 through independent research work. This approach also enabled Nanomix to make the first transistor on a flexible and transparent substrate.

Nanotubes are usually grown on nanoparticles of magnetic metal (Fe, Co), which facilitates production of electronic (spintronic) devices. In particular control of current through a field-effect transistor by magnetic field has been demonstrated in such a single-tube nanostructure.

Large structures of carbon nanotubes can be used for thermal management of electronic circuits. An approximately 1 mm thick carbon nanotube layer was used as a special material to fabricate coolers, this materials has very low density, ~20 times lower weight than a similar copper structure, while the cooling properties are similar for the two materials.

As Paper Batteries

A paper battery is a battery engineered to use a paper-thin sheet of cellulose (which is the major constituent of regular paper, among other things) infused with aligned carbon nanotubes. The nanotubes act as electrodes; allowing the storage devices to conduct electricity. The battery, which functions as both a lithium-ion battery and a supercapacitor, can provide a long, steady power output comparable to a conventional battery, as well as a supercapacitor's quick burst of high energy — and while a conventional battery contains a number of separate components, the paper battery integrates all of the battery components in a single structure, making it more energy efficient.

As a Vessel for Drug Delivery

The nanotube's versatile structure allows it to be used for a variety of tasks in and around the body. Although often seen especially in cancer-related incidents, the carbon nanotube is often used as a vessel for transporting drugs into the body. The nanotube application potentially allows for the drug dosage to be lowered by localising its distribution.

The nanotube commonly carries the drug one of two ways: the drug can be attached to the side or trailed behind, or the drug can actually be placed inside the nanotube. Both of these methods are effective for the delivery and distribution of drugs inside the body.

Current Applications

Current use and application of nanotubes has mostly been limited to the use of bulk nanotubes, which is a mass of rather unorganised fragments of nanotubes. Bulk nanotube materials may never achieve a tensile strength similar to that of individual tubes, but such composites may nevertheless yield strengths sufficient for many applications. Bulk carbon nanotubes have already been used as composite fibres in polymers to improve the mechanical, thermal and electrical properties of the bulk product.

Easton-Bell Sports, Inc. have been in partnership with Zyvex, using CNT technology in a number of their bicycle components — including flat and riser handlebars, cranks, forks, seatposts, stems and aero bars.

Solar cells

Solar cells developed at the New Jersey Institute of Technology use a carbon nanotube complex, formed by a mixture of carbon nanotubes and carbon buckyballs (known as fullerenes) to form snake-like structures. Buckyballs trap electrons, although they can't make electrons flow. Add sunlight to excite the polymers, and the buckyballs will grab the electrons. Nanotubes, behaving like copper wires, will then be able to make the electrons or current flow.

Ultracapacitors

MIT Laboratory for Electromagnetic and Electronic Systems uses nanotubes to improve ultracapacitors. The activated charcoal used in conventional ultracapacitors has many small hollow spaces of various size, which create together a large surface to store electric charge. But as charge is quantised into elementary charges, i.e. electrons, and each such elementary charge needs a minimum space, a significant fraction of the electrode surface is not available for storage because the hollow spaces are not compatible with the charge's requirements. With a nanotube electrode the spaces may be tailored to size — few too large or too small — and consequently the capacity should be increased considerably.

Other applications

Carbon nanotubes have been implemented in nanoelectromechanical systems, including mechanical memory elements (NRAM being developed by Nantero Inc.) and nanoscale electric motors.

Carbon nanotubes have been proposed as a possible gene delivery vehicle and for use in combination with radiofrequency fields to destroy cancer cells.

In May 2005, Nanomix Inc placed on the market a hydrogen sensor which integrated carbon nanotubes on a silicon platform. Since then Nanomix has been patenting many such sensor applications such as in the field of carbon dioxide, nitrous oxide, glucose, DNA detection, etc.

Eikos Inc of Franklin, Massachusetts and Unidym Inc. of Silicon Valley, California are developing transparent, electrically conductive films of carbon nanotubes to replace indium tin oxide (ITO). Carbon nanotube films are substantially more mechanically robust than ITO films, making them ideal for high-reliability touchscreens and flexible displays. Printable water-based inks of carbon nanotubes are desired to enable the production of these films to replace ITO. Nanotube films show promise for use in displays for computers, cell phones, PDAs, and ATMs.

A nanoradio, a radio receiver consisting of a single nanotube, was demonstrated in 2007. In 2008 it was shown that a sheet of nanotubes can operate as a loudspeaker if an alternating current is applied. The sound is not produced through vibration but thermoacoustically.

Due to the high mechanical strength of carbon nanotubes, research is being made into weaving them into clothes to create stab-proof and bulletproof clothing. The nanotubes would effectively stop the bullet from penetrating the body, although the bullet's kinetic energy would likely cause broken bones and internal bleeding.

A flywheel made of carbon nanotubes could be spun at extremely high velocity on a floating magnetic axis, and potentially store energy at a density approaching that of conventional fossil fuels. Since energy can be added to and removed from flywheels very efficiently in the form of electricity, this might offer a way of storing electricity, making the electrical grid more efficient and variable power suppliers (like wind turbines) more useful in meeting energy needs.

The practicality of this depends heavily upon the cost of making massive, unbroken nanotube structures, and their failure rate under stress. Rheological properties can also be shown very effectively by carbon nanotubes.

Nitrogen-doped carbon nanotubes may replace platinum catalysts used to reduce oxygen in fuel cells. A forest of vertically-aligned nanotubes can reduce oxygen in alkaline solution more effectively than platinum, which has been used in such applications since the 1960s. The nanotubes have the added benefit of not being subject to carbon monoxide poisoning.

BORON NITRIDE NANOTUBES

A boron nitride nanotube can be imagined as a rolled on itself graphite-like sheet, where carbon atoms are alternately substituted by nitrogen and boron atoms. Structurally, it is a close analog of the carbon nanotube, namely a long cylinder with diameter of several to hundred nanometers and length of many microns. However, the properties of BN nanotubes are very different: whereas carbon nanotubes can be metallic or semiconducting depending on the rolling direction and radius, a BN nanotube is an electrical insulator with a bandgap of ~5.5 eV, basically independent of tube chirality and morphology. In addition, a layered BN structure is much more thermally and chemically stable than a graphitic carbon structure.

All reliable and well-established techniques of carbon nanotube growth, e.g. arc-discharge, laser ablation, and chemical vapour deposition, do not effectively work for BN nanotubes. However, the latter can be produced by ball milling of amorphous boron, mixed with a catalyst–iron powder, under NH_3 atmosphere. Subsequent annealing at ~1100°C in nitrogen flow transforms most of the product into BN nanotubes.

Electrical and field emission properties of the thus prepared nanotubes can be tuned by doping with gold atoms via sputtering of gold on the nanotubes. Doping rare-earth atoms of europium turns a BN nanotube into a phosphor material emitting visible light under electron excitation.

Like BN fibres, boron nitride nanotubes show promise for aerospace applications where integration of boron and in particular the light isotope of boron (^{10}B) into structural materials improves their radiation-shielding properties; the improvement is due to strong neutron absorption by ^{10}B. Such ^{10}BN materials are of particular theoretical value as composite structural materials in future manned interplanetary spacecraft, where absorption-shielding from cosmic ray spallation neutrons is expected to be a particular asset in light construction materials.

Composites Containing BN

Addition of boron nitride to silicon nitride ceramics improves the thermal shock resistance of the resulting material. For the same purpose, BN is added also to silicon nitride-alumina and titanium nitride-alumina ceramics. Other materials being reinforced with BN are, e.g. alumina and zirconia, borosilicate glasses, glass ceramics, enamels, and composite ceramics with titanium boride-boron nitride and titanium boride-aluminium nitride-boron nitride and silicon carbide-boron nitride composition.

Health issues: Boron nitride (along with Si_3N_4, NbN, and BNC) is reported to show weak fibrigenic activity and cause pneumoconiosis. The maximum concentration recommended for nitrides of nonmetals is 10 mg/m^3 for BN and 4 for AlN or ZrN.

Carbon Nanotubes in Photovoltaics

Organic photovoltaic devices (OPVs) are fabricated from thin films of organic semiconductors, such as polymers and small-molecule compounds, and are typically on the order of 100 nm thick. Because polymer based OPVs can be made using a coating process such as spin coating or inkjet printing, they are an attractive option for inexpensively covering large areas as well as flexible plastic surfaces. A promising low cost alternative to silicon solar cells, there is a large amount of research being dedicated throughout industry and academia towards developing OPVs and increasing their power conversion efficiency.

Carbon Nanotube Composites in the Photoactive Layer

Combining the physical and chemical characteristics of conjugated polymers with the high conductivity along the tube axis of carbon nanotubes (CNTs) provides a great deal of incentive to disperse CNTs into the photoactive layer in order to obtain more efficient OPV devices. The interpenetrating bulk donor–acceptor heterojunction in these devices can achieve charge separation and collection because of the existence of a bicontinuous network. Along this network, electrons and holes can travel toward their respective contacts through the electron acceptor and the polymer hole donor. Photovoltaic efficiency enhancement is proposed to be due to the introduction of internal polymer/nanotube junctions within the polymer matrix. The high electric field at these junctions can split up the excitons, while the single-walled carbon nanotube (SWCNT) can act as a pathway for the electrons.

The dispersion of CNTs in a solution of an electron donating conjugated polymer is perhaps the most common strategy to implement CNT materials into OPVs. Generally poly(3-hexylthiophene) (P3HT) or poly(3-octylthiophene) (P3OT) are used for this purpose. These blends are then spin coated onto a transparent conductive electrode with thicknesses that vary from 60 to 120 nm. These conductive electrodes are usually glass covered with indium tin oxide (ITO) and a 40 nm sublayer of poly(3,4-ethylenedioxythiophene) (PEDOT) and poly(styrenesulphonate) (PSS). PEDOT and PSS help to smooth the ITO surface, decreasing the density of pinholes and stifling current leakage that occurs along shunting paths. Through thermal evaporation or sputter coating, a 20 to 70 nm thick layer of aluminium and sometimes an intermediate layer of lithium fluoride are then applied onto the photoactive material. Multiple research investigations with both multiwalled carbon nanotubes (MWCNTs) and single-walled carbon nanotubes (SWCNTs) integrated into the photoactive material have been completed.

Enhancements of more than two orders of magnitude have been observed in the photocurrent from adding SWCNTs to the P3OT matrix. Improvements were speculated to be due to charge separation at polymer-SWCNT connections and more efficient electron transport through the SWCNTs. However, a rather low power conversion efficiency of 0.04 per cent under 100 mW/cm^2 white illumination was observed for the device suggesting incomplete exciton dissociation at low CNT concentrations of 1.0 per cent wt. Because the lengths of the SWCNTs were similar to the thickness of photovoltaic films, doping a higher percentage of SWCNTs into the polymer matrix was believed to cause short circuits. To supply additional dissociation sites, other researchers have physically blended functionalised MWCNTs into P3HT polymer to create a P3HT-MWCNT with fullerene C_{60} double-layered device. However, the power efficiency was still relatively low at 0.01 per cent under 100 mW/cm^2 white illumination. Weak exciton diffusion toward the donor-acceptor interface in the bilayer structure may have been the cause in addition to the fullerene C_{60} layer possibly experiencing poor electron transport.

More recently, a polymer photovoltaic device from C_{60}-modified SWCNTs and P3HT has been fabricated. Microwave irradiating a mixture of aqueous SWCNT solution and C_{60} solution in toluene was the first step in making these polymer-SWCNT composites. Conjugated polymer P3HT was then added resulting in a power conversion efficiency of 0.57 per cent under simulated solar irradiation (95 mW/cm^2). It was concluded that improved short circuit current density was a direct result of the addition of SWCNTs into the composite causing faster electron transport via the network of SWCNTs. It was also concluded that the morphology change led to an improved the fill factor. Overall, the main result was improved power conversion efficiency with the addition of SWCNTs, compared to cells without SWCNTs; however, further optimisation was thought to be possible.

Additionally, it has been found that heating to the point beyond the glass transition temperature of either P3HT or P3OT after construction can be beneficial for manipulating the phase separation of the blend. This heating also affects the ordering of the polymeric chains because the polymers are microcrystalline systems and it improves charge transfer, charge transport, and charge collection throughout the OPV device. The hole mobility and power efficiency of the polymer-CNT device also increased significantly as a result of this ordering.

Emerging as another valuable approach for deposition, the use of tetraoctylammonium bromide in tetrahydrofuran has also been the subject of investigation to assist in suspension by exposing SWCNTs to an electrophoretic field. In fact, photoconversion efficiencies of 1.5 and 1.3 per cent were achieved when SWCNTs were deposited in combination with light harvesting cadmium sulphide (CdS) quantum dots and porphyrins, respectively.

Among the best power conversions achieved to date using CNTs were obtained by depositing a SWCNT layer between the ITO and the PEDOT : PSS or between the PEDOT : PSS and the photoactive blend in a modified ITO/PEDOT : PSS/P3HT : (6,6)-phenyl-C$_{61}$-butyric acid methyl ester (PCBM)/Al solar cell. By dip-coating from a hydrophilic suspension, SWCNT were deposited after an initially exposing the surface to an argon plasma to achieve a power conversion efficiency of 4.9 per cent, compared to 4 per cent without CNTs. However, even though CNTs have shown potential in the photoactive layer, they have not resulted in a solar cell with a power conversion efficiency greater than the best tandem organic cells (6.5 per cent efficiency). But, it has been shown in most of the previous investigations that the control over a uniform blending of the electron donating conjugated polymer and the electron accepting CNT is one of the most difficult as well as crucial aspects in creating efficient photocurrent collection in CNT-based OPV devices. Therefore, using CNTs in the photoactive layer of OPV devices is still in the initial research stages and there is still room for novel methods to better take advantage of the beneficial properties of CNTs.

Carbon Nanotubes as a Transparent Electrode

ITO is currently the most popular material used for the transparent electrodes in OPV devices; however, it has a number of deficiencies. For one, it is not very compatible with polymeric substrates due to its high deposition temperature of around 600°C. Traditional ITO also has unfavourable mechanical properties such as being relatively fragile. In addition, the combination of costly layer deposition in vacuum and a limited supply of indium results in high quality ITO transparent electrodes being very expensive. Therefore, developing and commercialising a replacement for ITO is a major focus of OPV research and development.

Conductive CNT coatings have recently become a prospective substitute based on wide range of methods including spraying, spin coating, casting, layer-by-layer, and Langmuir-Blodgett deposition. The transfer from a filter membrane to the transparent support using a solvent or in the form of an adhesive film is another method for attaining flexible and optically transparent CNT films. Other research efforts have shown that films made of arc-discharge CNT can result in a high conductivity and transparency. Furthermore, the work function of SWCNT networks is in the 4.8 to 4.9 eV range (compared to ITO which has a lower work function of 4.7 eV) leading to the expectation that the SWCNT work function should be high enough to assure efficient hole collection. Another benefit is that SWCNT films exhibit a high optical transparency in a broad spectral range from the UV-visible to the near-infrared range. Only a few materials retain reasonable transparency in the infrared spectrum while maintaining transparency in the visible part of the spectrum as well as acceptable overall electrical conductivity. SWCNT films are highly flexible, do not creep, do not crack after bending, theoretically have high thermal conductivities to tolerate heat dissipation, and have high radiation resistance. However, the electrical sheet resistance of ITO is an order of magnitude less than the sheet resistance measured for SWCNT films. Nonetheless, initial research studies demonstrate SWCNT thin films can be used as conducting, transparent electrodes for hole collection in OPV devices with efficiencies between 1 per cent and 2.5 per cent confirming that they are comparable to devices fabricated using ITO. Thus, possibilities exist for advancing this research to develop CNT-based transparent electrodes that exceed the performance of traditional ITO materials.

CNTs in Dye-Sensitised Solar Cells

Due to the simple fabrication process, low production cost, and high efficiency, there is significant interest in dye-sensitised solar cells (DSSCs). Thus, improving DSSC efficiency has been the subject of a variety of research investigations because it has the potential to be manufactured economically enough to compete with other solar cell technologies. Titanium dioxide nanoparticles have been widely used as a working electrode for DSSCs because they provide a high efficiency, more than any other metal oxide semiconductor investigated. Yet the highest conversion efficiency under air mass (AM) 1.5 (100 mW/cm^2) irradiation reported for this device to date is about 11 per cent. Despite this initial success, the effort to further enhance efficiency has not produced any major results. The transport of electrons across the particle network has been a key problem in achieving higher photoconversion efficiency in nanostructured electrodes. Because electrons encounter many grain boundaries during the transit and experience a random path, the probability of their recombination with oxidised sensitiser is increased. Therefore, it is not adequate to enlarge the oxide electrode surface area to increase efficiency because photo-generated charge recombination should be prevented. Promoting electron transfer through film electrodes and blocking interface states lying below the edge of the conduction band are some of the non-CNT based strategies to enhance efficiency that have been employed.

With recent progress in CNT development and fabrication, there is promise to use various CNT based nanocomposites and nanostructures to direct the flow of photogenerated electrons and assist in charge injection and extraction. To assist the electron transport to the collecting electrode surface in a DSSC, a popular concept is to utilise CNT networks as support to anchor light harvesting semiconductor particles. Research efforts along these lines include organising CdS quantum dots on SWCNTs. Charge injection from excited CdS into SWCNTs was documented upon excitation of CdS nanoparticles. Other varieties of semiconductor particles including CdSe and CdTe can induce charge-transfer processes under visible light irradiation when attached to CNTs. Including porphyrin and C$_{60}$ fullerene, organisation of photoactive donor polymer and acceptor fullerene on electrode surfaces has also been shown to offer considerable improvement in the photoconversion efficiency of solar cells. Therefore, there is an opportunity to facilitate electron transport and increase the photoconversion efficiency of DSSCs utilising the electron-accepting ability of semiconducting SWCNTs.

Other researchers fabricated DSSCs using the sol-gel method to obtain titanium dioxide coated MWCNTs for use as an electrode. Because pristine MWCNTs have a hydrophobic surface and poor dispersion stability, pretreatment was necessary for this application. A relatively low-destruction method for removing impurities, H_2O_2 treatment was used to generate carboxylic acid groups by oxidation of MWCNTs. Another positive aspect was the fact that the reaction gases including CO_2 and H_2O were non-toxic and could be released safely during the oxidation process. As a result of treatment, H_2O_2 exposed MWCNTs have a hydrophilic surface and the carboxylic acid groups on the surface have polar covalent bonding. Also, the negatively charged surface of the MWCNTs improved the stability of dispersion. By then entirely surrounding the MWCNTs with titanium dioxide nanoparticles using the sol-gel method, an increase in the conversion efficiency of about 50 per cent compared to a conventional titanium dioxide cell was achieved. The enhanced interconnectivity between the titanium dioxide particles and the MWCNTs in the porous titanium dioxide film was concluded to be the cause of the improvement in short circuit current density. Here again, the addition of MWCNTs was thought to provide more efficient electron transfer through film in the DSSC.

OPTICAL PROPERTIES OF CARBON NANOTUBES

Within materials science, the optical properties of carbon nanotubes refer specifically to the absorption, photoluminescence, and Raman spectroscopy of carbon nanotubes. Spectroscopic methods offer the possibility of quick and non-destructive characterisation of relatively large amounts of carbon nanotubes. There is a strong demand for such characterisation from the industrial point of view: numerous parameters of the nanotube synthesis can be changed, intentionally or unintentionally, to alter the nanotube quality. As shown below, optical absorption, photoluminescence and Raman spectroscopies allow quick and reliable characterisation of this 'nanotube quality' in terms of non-tubular carbon content, structure (chirality) of the produced nanotubes, and structural defects. Those features determine nearly any other properties such as optical, mechanical, and electrical properties.

Carbon nanotubes are unique 'one dimensional systems' which can be envisioned as rolled single sheets of graphite (or more precisely graphene). This rolling can be done at different angles and curvatures resulting in different nanotube properties. The diameter typically varies in the range 0.4–40 nanometers (i.e. only ~100 times), but the length can vary ~10,000 times reaching 4 cm. Thus the nanotube aspect ratio, or the length-to-diameter ratio, can be as high as 28,000,000:1, which is unequalled by any other material. Consequently, all the properties of the carbon nanotubes relative to those of typical semiconductors are extremely anisotropic (directionally dependent) and tunable.

Whereas mechanical, electrical and electrochemical (supercapacitor) properties of the carbon nanotubes are well established and have immediate applications, the practical use of optical properties is yet unclear. The aforementioned tunability of properties is potentially useful in optics and photonics. In particular, light-emitting diodes (LEDs) and photo-detectors based on a single nanotube have been produced in the lab. Their unique feature is not the efficiency, which is yet relatively low, but the narrow selectivity in the wavelength of emission and detection of light and the possibility of its fine tuning through the nanotube structure. In addition, bolometer and optoelectronic memory devices have been realised on ensembles of single-walled carbon nanotubes.

Electronic Structure of Carbon Nanotube

A single-wall carbon nanotube can be imagined as graphene sheet rolled at a certain 'chiral' angle with respect to a plane perpendicular to the tube's long axis. Consequently, SWCNT can be defined by its diameter and chiral angle. The chiral angle can range from 0 to 30 degrees.

However, more conveniently, a pair of indexes (n, m) is used instead. The indexes refer to equally long unit vectors at 60° angles to each other across a single 6-member carbon ring. Taking the origin as carbon number 1, the a_1 unit vector may be considered the line drawn from carbon 1 to carbon 3, and the a_2 unit vector is then the line drawn from carbon 1 to carbon 5. To visualise a CNT with indexes (n, m), draw n a_1 unit vectors across the graphene sheet, then draw m a_2 unit vectors at a 60° angle to the a_1 vectors, then add the vectors together. The line representing the sum of the vectors will define the circumference of the CNT along the plane perpendicular to its long axis, connecting one end to the other. In the Fig. 5.7, C_h is a (4,2) vector: the sum of 4 unit vectors from the origin directly to the right, then 2 unit vectors at a 60° angle down and to the right. Figure 5.7 shows the armchair and zigzag nanotube.

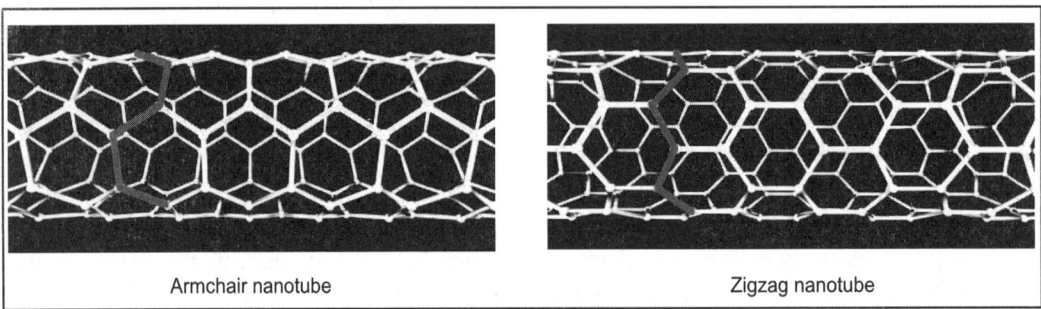

| Armchair nanotube | Zigzag nanotube |

Fig. 5.7. The armchair and zigzag nanotube structures.

Tubes having $n = m$ (chiral angle = 0°) are called 'armchair' and those with $m = 0$ (chiral angle = 30°) 'zigzag'. Those indexes uniquely determine whether CNT is a metal, semimetal or semiconductor, as well as its band gap: when $|m - n| = 3k$ (k is integer) the tube is metallic; but if $|m - n| = 3k + 1$ or $3k - 1$ the tube is semiconducting. The nanotube diameter d is related to m and n as:

$$d = \frac{a}{\pi}\sqrt{(n^2 + nm + m^2)}$$

In this equation, a is the magnitude of either unit vector a_1 or a_2.

The situation in multiwall CNTs is complicated as their properties are determined by contribution of all individual shells; those shells have different structures, and, because of the synthesis, are usually more defective than SWCNTs. Therefore, optical properties of MWCNTs will not be considered here.

van Hove Singularities

Optical properties of carbon nanotubes derive from electronic transitions within one-dimensional density of states (DOS). A typical feature of one-dimensional nanomaterials is that their DOS is not a continuous function of energy, but it descends gradually and then increases in a discontinuous spike. In contrast, three-dimensional materials have continuous DOS. The sharp peaks found in one-dimensional materials are called van Hove singularities (Fig. 5.8).

van Hove singularities result in the following remarkable optical properties of carbon nanotubes:

1. Optical transitions occur between the $v_1 - c_1$, $v_2 - c_2$, etc. states of semiconducting or metallic nanotubes and are traditionally labelled as S_{11}, S_{22}, M_{11}, etc. or, if the 'conductivity' of the tube is unknown or unimportant, as E_{11}, E_{22}, etc. Crossover transitions $c_1 - v_2$, $c_2 - v_1$, etc. are dipole-

forbidden and thus are extremely weak, but they were possibly observed using cross-polarised optical geometry.

2. The energies between the van Hove singularities depend on the nanotube structure. Thus by varying this structure, one can tune the optoelectronic properties of carbon nanotube. Such fine tuning has been experimentally demonstrated using UV illumination of polymer-dispersed CNTs.

3. Optical transitions are rather sharp (~10 meV) and strong. Consequently, it is relatively easy to selective excite nanotubes having certain (n, m) indexes, as well as to detect optical signals from individual nanotubes.

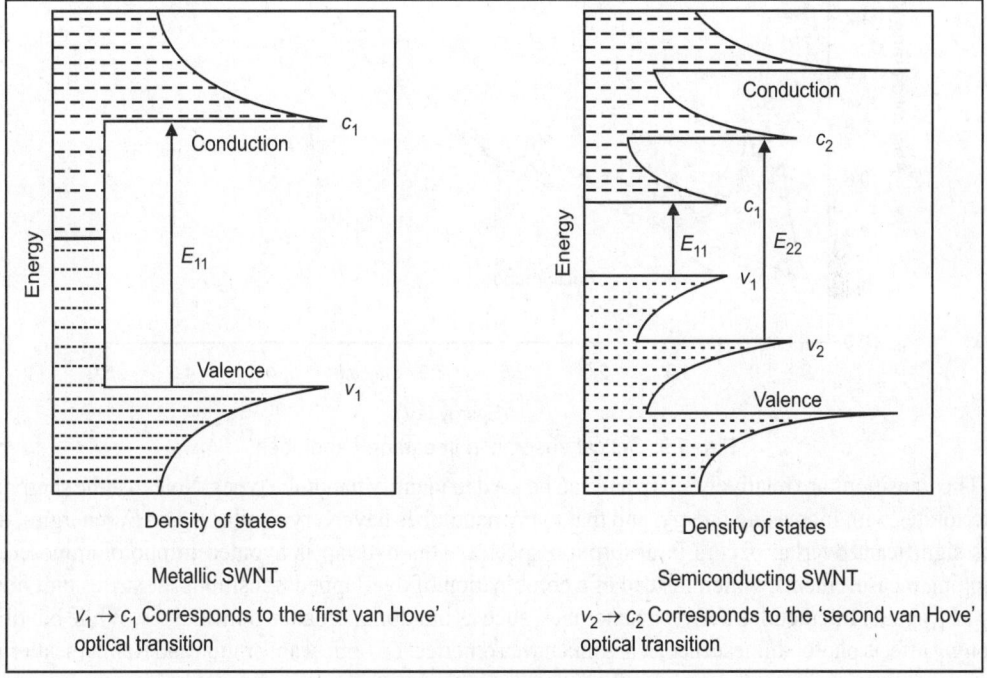

Fig. 5.8. van Hove singularities.

Kataura Plot

The band structure of carbon nanotubes having certain (n, m) indexes can be calculated. However, the calculation is complex and not accurate enough. To provide the needed information without calculations, an experimental graph named 'Kataura plot', named after Hiromichi Kataura, was designed in 1999. A Kataura plot relates the nanotube diameter and its bandgap energies. The oscillating shape of every branch of the Kataura plot reflects the intrinsic strong dependence of the SWCNT properties on the (n, m) index rather than on its diameter. For example, (10,0) and (8,3) tubes have almost the same diameter, but very different properties: the former is a metal, but the latter is semiconductor.

Optical Absorption

Optical absorption in carbon nanotubes differs from absorption in conventional 3D materials by presence of sharp peaks (1D nanotubes) instead of an absorption threshold followed by an absorption increase (most 3D solids). Absorption in nanotubes originates from electronic transitions from the v_2 to c_2 (energy E_{22}) or v_1 to c_1 (E_{11}) levels, etc. (Fig. 5.9).

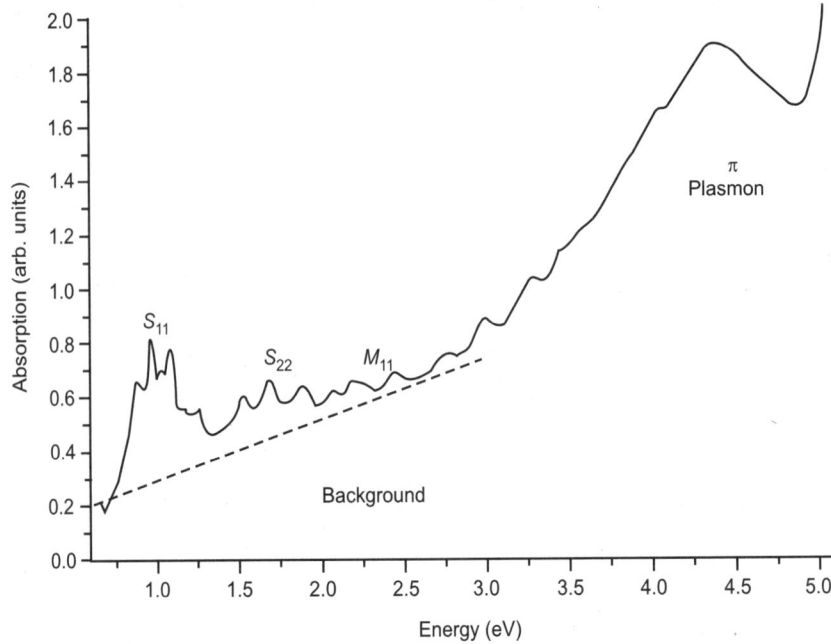

Fig. 5.9. Optical absorption in carbon nanotubes.

The transitions are relatively sharp and can be used to identify nanotube types. Note that the sharpness deteriorates with increasing energy, and that many nanotubes have very similar E_{22} or E_{11} energies, and thus significant overlap occurs in absorption spectra. This overlap is avoided in photoluminescence mapping measurements, which instead of a combination of overlapped transitions identifies individual (E_{22}, E_{11}) pairs. Interactions between nanotubes, such as bundling, broaden optical lines. While bundling strongly affects photoluminescence, it has much weaker effect on optical absorption and Raman scattering. Consequently, sample preparation for the latter two techniques is relatively simple.

Optical absorption is routinely used to quantify quality of the carbon nanotube powders. The spectrum is analysed in terms of intensities of nanotube-related peaks, background and pi-carbon peak; the latter two mostly originate from non-nanotube carbon.

Luminescence

Excitation mechanism

Photoluminescence (PL) is one of the important tools for nanotube characterisation. The excitation of PL usually occurs as follows: an electron in a nanotube absorbs excitation light via S_{22} transition, creating an electron-hole pair (exciton). Both electron and hole rapidly relax (via phonon-assisted processes) from c_2 to c_1 and from v_2 to v_1 states, respectively. Then they recombine through a $c_1 - v_1$ transition resulting in light emission.

No excitonic luminescence can be produced in metallic tubes—electron can be excited, thus resulting in optical absorption, but the hole is immediately filled by another electron out of many available in metal. Therefore, no exciton is produced.

Salient properties

1. Photoluminescence from SWCNT, as well as optical absorption and Raman scattering, is linearly polarised along the tube axis. This allows monitoring of the SWCNTs orientation without direct microscopic observation.

2. PL is quick: relaxation typically occurs within 100 picoseconds.

3. PL efficiency is usually low (~0.01 per cent), however, there is a room for enhancement, e.g. through improving the structural quality of the nanotubes and clever nanotube isolation strategies. For example, the efficiency of 3 per cent has been reported in nanotubes sorted by diameter and length through gradient centrifugation, and it has been further increased to 20 per cent by optimising the procedure of isolating individual nanotubes in solution.

4. The spectral range of PL is rather wide. Emission wavelength can vary between 0.8 and 2.1 micrometers depending on the nanotube structure.

5. Interaction between nanotubes or between nanotube and another material (e.g. substrate) quenches PL. For this reason, no PL is observed in multiwall carbon nanotubes. PL from double-wall carbon nanotubes strongly depends on how they were prepared: CVD grown DWCNTs show emission both from inner and outer shells. However, DWCNTs produced by encapsulating fullerenes into SWCNTs and annealing show PL only from the outer shells. Isolated SWCNTs lying on the substrate show extremely weak PL which has been detected in few studies only. Detachment of the tubes from the substrate drastically increases PL.

6. Position of the (S_{22}, S_{11}) PL peaks depends slightly (within 2 per cent) on the nanotube environment (air, dispersant, etc.). However, the shift depends on the (n, m) index, and thus the whole PL map not only shifts, but also warps upon changing the CNT medium.

Applications

Because of low efficiency, no commercial application of PL from pure carbon nanotubes is viable yet. However, PL is widely used to deduce (n, m) indexes: first nanotubes are isolated (dispersed) using an appropriate chemical agent (dispersant) to reduce the intertube quenching. Then PL is measured, scanning both the excitation and emission energies and thereby producing a PL map. The ovals in the map define (S_{22}, S_{11}) pairs, which unique identify (n, m) index of a tube. The data of Weisman and Bachillo are conventionally used for the identification.

Sensitisation

Optical properties, including the PL efficiency, can be modified by encapsulating organic dyes (carotene, lycopene, etc.) inside the tubes. Efficient energy transfer occurs between the encapsulated dye and nanotube — light is efficiently absorbed by the dye and without significant loss is transferred to the SWCNT. Thus potentially, optical properties of a carbon nanotube can be controlled by encapsulating certain molecule inside it.

Besides, encapsulation allows isolation and characterisation of organic molecules which are unstable under ambient conditions. For example, Raman spectra are extremely difficult to measure from dyes because of their strong PL (efficiency close to 100 per cent). However, encapsulation of dye molecules inside SWCNTs completely quenches dye PL, thus allowing measurement and analysis of their Raman spectra.

Cathodoluminescence

Cathodoluminescence (CL)—light emission excited by electron beam—is a process commonly observed in TV screens. An electron beam can be finely focused and scanned across the studied material. This technique is widely used to study defects in semiconductors and nanostructures with nanometer-scale spatial resolution. It would be beneficial to apply this technique to carbon nanotubes.

However, no reliable CL, i.e. sharp peaks assignable to certain (n, m) indexes, has been detected from carbon nanotubes yet.

Electroluminescence

If appropriate electrical contacts are attached to a nanotube, electron-hole pairs (excitons) can be generated by injecting electrons and holes from the contacts. Subsequent exciton recombination results in electroluminescence (EL). Electroluminescent devices have been produced from single nanotubes.

Raman Scattering

Raman spectroscopy has good spatial resolution (~0.5 micrometers) and sensitivity (single nanotubes); it requires only minimal sample preparation and is rather informative. Consequently, Raman spectroscopy is probably the most popular technique of carbon nanotube characterisation. Raman scattering in SWCNTs is resonant, i.e. only those tubes are probed which have one of the bandgaps equal to the exciting laser energy. Several scattering modes dominate the SWCNT spectrum, as discussed below.

Similar to photoluminescence mapping, the energy of the excitation light can be scanned in Raman measurements, thus producing Raman maps. Those maps also contain oval-shaped features uniquely identifying (n, m) indexes. Contrary to PL, Raman mapping detects not only semiconducting but also metallic tubes, and it is less sensitive to nanotube bundling than PL. However, requirement of a tunable laser and a dedicated spectrometer is a strong technical impediment.

Radial breathing mode

Radial breathing mode (RBM) corresponds to radial expansion-contraction of the nanotube. Therefore, its frequency v_{RBM} (in cm^{-1}) depends on the nanotube diameter d (in nanometers) and can be estimated as $v_{RBM} = 223/d + 10$, which is very useful in deducing the CNT diameter from the RBM position. Typical RBM range is 100–350 cm^{-1}. If RBM intensity is particularly strong, its weak second overtone can be observed at double frequency (Fig. 5.10).

Bundling mode

The bundling mode is a special form of RBM supposedly originating from collective vibration in a bundle of SWCNTs.

G mode

Another very important mode is the G mode (G from graphite). This mode corresponds to planar vibrations of carbon atoms and is present in most graphite-like materials. G band in SWCNT is shifted to lower frequencies relative to graphite (1580 cm^{-1}) and is split into several peaks. The splitting pattern and intensity depend on the tube structure and excitation energy; they can be used, though with much lower accuracy compared to RBM mode, to estimate the tube diameter and whether the tube is metallic or semiconducting.

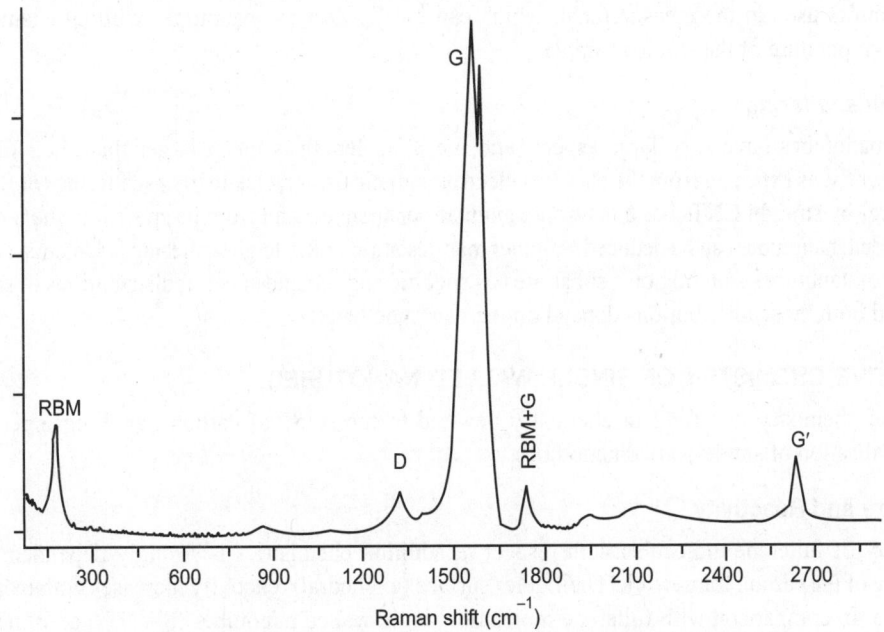

Fig. 5.10. Raman spectrum of single-wall carbon nanotubes.

D mode

D mode is present in all graphite-like carbons and originates from structural defects. Therefore, the ratio of the *G/D* modes is conventionally used to quantify the structural quality of carbon nanotubes. High-quality nanotubes have this ratio significantly higher than 100.

G' mode

The name of this mode is misleading: it is given because in graphite, this mode is usually the second strongest after the G mode. However, it is actually the second overtone of the defect-induced D mode (and thus should logically be named D'). Its intensity is stronger than that of the D mode due to different selection rules. In particular, D mode is forbidden in the ideal nanotube and requires a structural defect, providing a phonon of certain angular momentum, to be induced. In contrast, G' mode involves a 'self-annihilating' pair of phonons and thus does not require defects. The spectral position of G' mode depends on diameter, so it can be used roughly to estimate the SWCNT diameter. In particular, G' mode is a doublet in double-wall carbon nanotubes, but the doublet is often unresolved due to line broadening.

Other overtones, such as a combination of RBM+G mode at ~1750 cm^{-1}, are frequently seen in CNT Raman spectra. However, they are less important and are not considered here.

Anti-stokes scattering

All the above Raman modes can be observed both as Stokes and anti-Stokes scattering. As mentioned above, Raman scattering from CNTs is resonant in nature, i.e. only tubes whose band gap energy is similar to the laser energy are excited. The difference between those two energies, and thus the band gap of individual tubes, can be estimated from the intensity ratio of the Stokes/anti-Stokes lines. This estimate, however, relies on the temperature factor (Boltzmann factor), which is often miscalculated—focused laser beam is used in the measurement, which can locally heat the nanotubes without changing the overall temperature of the studied sample.

Rayleigh scattering

Carbon nanotubes have very large aspect ratio, i.e. their length is much larger than their diameter. Consequently, as expected from the classical electromagnetic theory, elastic light scattering (or Rayleigh scattering) by straight CNTs has anisotropic angular dependence, and from its spectrum, the band gaps of individual nanotubes can be deduced. Another manifestation of Rayleigh scattering is 'antenna effect'— an array of nanotubes standing on a substrate has specific angular and spectral distributions of reflected light, and both those distributions depend on the nanotube length.

SELECTIVE CHEMISTRY OF SINGLE-WALLED NANOTUBES

Nanotube chemistry is a field in chemistry devoted to the study of carbon nanotubes specifically functionalisation of single-walled nanotubes.

Structure and Reactivity

Reactivity of fullerene molecules with respect to addition chemistries is strongly dependent on the curvature of the carbon framework. Their outer surface (exohedral) reactivity increases with increase in curvature. In comparison with fullerene molecules single-walled nanotubes (SWNTs) are moderately curved. Consequently nanotubes are expected to be less reactive than most fullerene molecules due to their smaller curvature, but more reactive than a graphene sheet due to pyramidalisation and misalignment of *pi*-orbitals. The strain of a carbon framework is also reflected in the pyramidalisation angle (θ_p) of the carbon constituents. Trigonal carbon atoms (sp^2 hybridised) prefer a planar orientation with $\theta_p=0°$ (i.e. graphene) and fullerene molecules have $\theta_p= 11.6°$. The (5,5) SWNT has $\theta_p\sim6°$ for the sidewall. Values for other (n, n) nanotubes show a trend of increasing θ_p (sidewall) with decrease in n. Therefore, generally the chemical reactivity of SWNT increases with decrease in diameter (or n, diameter increases with n). Apart from the curvature SWNT reactivity is also highly sensitive to chiral wrapping (n,m) which determine its electronic structure. Nanotubes with $n-m = 3i$ (i is an integer) are all metals and rest are all semiconducting (SC).

Sidewall Functionalisation

Carbon nanotubes are metallic or semiconducting, based upon delocalised electrons occupying a 1-D density of states. However, any covalent bond on SWNT sidewall causes localisation of these electrons. In the vicinity of localised electrons, the SWNT can no longer be described using a band model that assumes delocalised electrons moving in a periodic potential. Two important addition reactions of SWNT sidewall are: (i) fluorination, and (ii) aryl diazonium salt addition. These functional groups on SWNT improve solubility and processibility. Moreover, these reactions allow for combining unique properties of SWNTs with those of other compounds. Above all, the selective diazonium chemistry can be used to separate the semiconducting and metallic nanotubes (Fig. 5.11).

Fluorination

The first extensive SWNT sidewall reaction was fluorination in 1998 by Mickleson. These fluorine moieties can be removed from the nanotube by treatment in hydrazine and the spectroscopic properties of the SWNT can be restored completely.

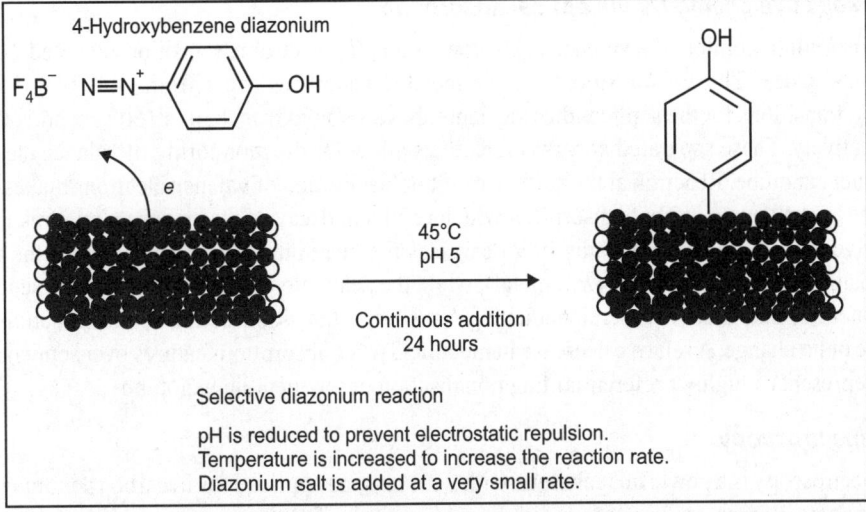

Fig. 5.11. Selective diazonium reaction.

Diazonium chemistry

One of the most important SWNT sidewall reaction is that with diazonium reagent which if done under controlled conditions can be used to do selective covalent chemistry. Water-soluble diazonium salts react with carbon nanotubes via charge transfer in which they extract electrons from SWNT and form a stable covalent aryl bond. This covalent aryl bond forms with extremely high affinity for electrons with energies near the Fermi level, E_f of the nanotube. Metallic SWNT have a greater electron density near E_f resulting in their higher reactivity over semiconducting nanotubes. The reactant forms a charge-transfer complex at the nanotube surface, where electron donation from the latter stabilises the transition state and accelerates the forward rate. Once the bond symmetry of the nanotube is disrupted by the formation of this defect, adjacent carbons increase in reactivity and initial selectivity for metallic SWNT is amplified. Under carefully controlled conditions this behaviour can be exploited to obtain highly selective functionalisation of metallic nanotubes to the near exclusion of the semiconductors.

Selective reaction conditions

Primary condition is addition of reactant molecules at a very small rate to SWNT solution for a sufficient long time. This ensures reaction with only metallic SWNTs and with no semiconducting SWNTs as all the reactant molecules are taken up by the metallic SWNTs. Long time injection ensures that all metallic tubes are reacted. For example one highly selective condition is: addition of 500 µl of 4-hydroxybenzene diazonium tetrafluoroborate solution in water (0.245 mM) at an injection rate of 20.83 µl/hr into 5 ml of SWNT solution (1 wt % sodium dodecyl sulphate (SDS)) over 24 hours. However, if the entire diazonium solution is added all instantaneously then semiconducting SWNTs will also react due to presence of excess reactant.

Spectroscopy and functionalisation

SWNTs have unique optical and spectroscopic properties largely due to one-dimensional confinement of electronic and phonon states, resulting in so-called van Hove singularities in the nanotube density of states (DOS).

Probing selective chemistry via optical absorption

Optical absorption monitors the valence (v) to conduction (c) electronic transitions denoted E_{nn} where, n is the band index. The E_{11} transitions for the metallic nanotubes occur from ~440 to 645 nm. The E_{11} and E_{22} transitions for the semiconducting nanotubes are found from 830 to 1600 nm and 600 to 800 nm, respectively. These separated absorption features allow for the monitoring of valence electrons in each distinct nanotube. Reaction at the surface result in localisation of valence electrons makes them no longer free to participate in photoabsorption which results in decay of the spectrum features.

Selective diazonium chemistry abruptly decreases the peak intensities that represent the first van Hove transition of metallic species (E_{11}, metal), while the peak intensities representing the second (E_{22}, semiconducting) and first (E_{11}, semiconducting) van Hove transition of the semiconducting species show little or no change. A relative decrease in metallic SWNT absorption features over semiconducting features represents a highly preferential functionalisation of the metallic nanotubes.

Raman spectroscopy

Raman spectroscopy is a powerful technique with wide ranging applications in carbon nanotube studies. Some important Raman features are radial breathing mode (RBM), tangential mode (G-band), and disorder-related mode.

RBM features correspond to the coherent vibration of the C atoms in the radial direction of the nanotube. These features are unique in carbon nanotubes and occur with the frequencies ω_{RBM} between 120 and 350 cm^{-1} for SWNT in the diameter range (0.7 nm–2 nm). They can be used to probe the SWNT diameter, electronic structure through their frequency and intensity (I_{RBM}) respectively and hence perform an (n,m) assignment to their peaks. The addition of the moiety to the sidewall of the nanotube disrupts the oscillator strength that gives rise to RBM feature and hence causes decay of these features. These features are distinct for species of a particular nanotube (n,m) and hence enables to probe which SWNTs are functionalised and to what extent.

Two main components of the tangential mode include G$^+$ at 1590 cm^{-1} and G$^-$ at 1570 cm^{-1}. G$^+$ feature is associated with carbon atom vibrations along the nanotube axis. The G$^-$ feature is associated with vibrations of carbon atoms along the circumferential direction. The G-band frequency can be used: (i) to distinguish between metallic and semiconducting SWNTs, and (ii) to probe charge transfer arising from doping a SWNT. Frequency of G$^+$ is sensitive to charge transfer. It upshifts for acceptors and downshifts for donors. Lineshape of G$^-$ is highly sensitive to be whether SWNT is metallic (Breit-Wigner-Fano lineshape) or semiconducting (Lorentzian lineshape).

The disorder-related mode (D peak) is a phonon mode at 1300 cm^{-1} and involves the resonantly enhanced scattering of an electron via phonon emission by a defect that breaks the basic symmetry of the graphene plane. This mode corresponds to the conversion of a sp^2-hybridised carbon to a sp^3-hybridised on the surface. Intensity of D peak measures covalent bond made with the nanotube surface. This feature does not increase as a result of surfactant or hydronium ion adsorption on the nanotube surface. Figure 5.12 shows the density of states control the selectivity of covalent bond formation.

Selective reaction and Raman features

Selective functionalisation increases the intensity of the D peak due to formation of aryl-nanotube bond and decreases the tangential mode due loss of electronic resonance. These two effects are generally summed together as increase in their peaks ratio (D/G). RBM peaks of metallic nanotubes decay and the peaks corresponding to those of semiconducting nanotubes remain almost unchanged (Fig. 5.13).

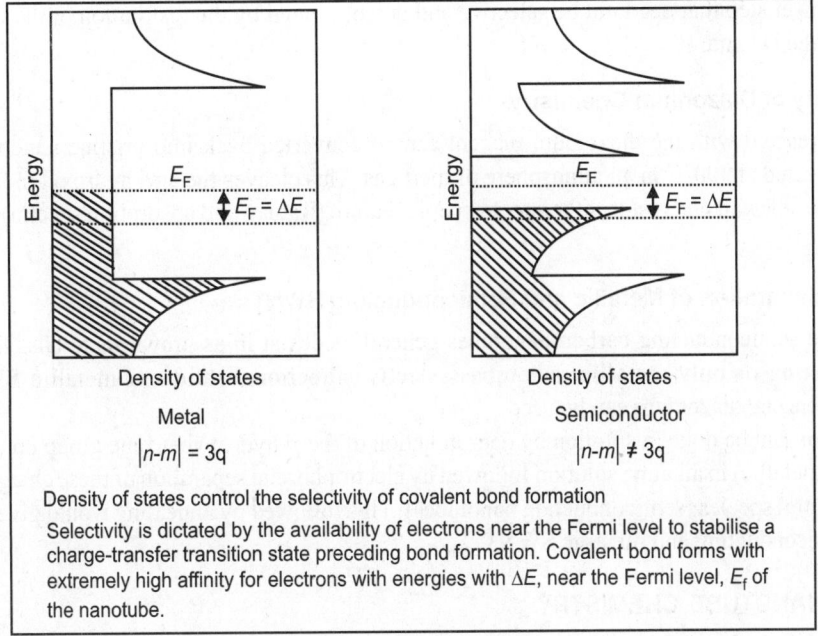

Density of states control the selectivity of covalent bond formation

Selectivity is dictated by the availability of electrons near the Fermi level to stabilise a charge-transfer transition state preceding bond formation. Covalent bond forms with extremely high affinity for electrons with energies with ΔE, near the Fermi level, E_f of the nanotube.

Fig. 5.12. The density of states control the selectivity of covalent bond formation.

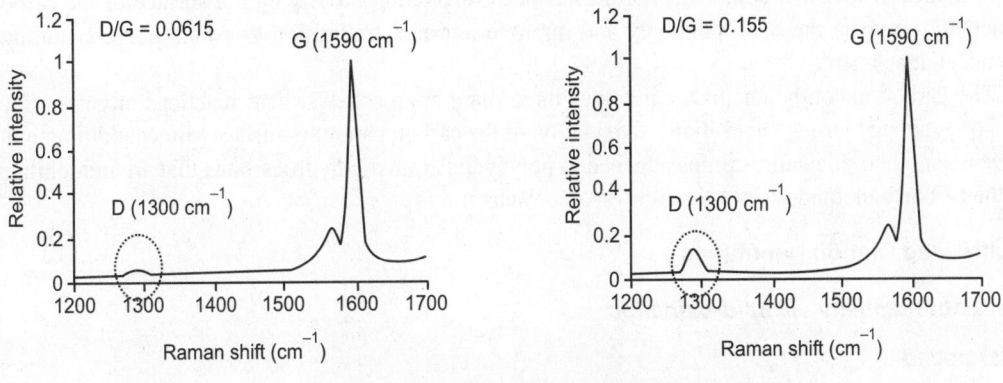

Raman spectrum and covalent reaction

Disorder mode grows with increase in functionalisation.
(Raman spectra taken at 785-nm excitation)

Fig. 5.13. Raman spectrum and covalent reaction.

Reaction Mechanism

Diazonium reagent and SWNT reaction has a two step mechanism. First, the diazonium reagent adsorbs noncovalently to an empty site on the nanotube surface, forming a charge-transfer complex. This is a fast, selective noncovalent adsorption and diazonium group in this complex partially dopes the nanotube, diminishing the tangential mode in the Raman spectrum. Desorption of A from nanotube is negligible $(k_{-1} \sim 0)$. In second step complex B then decomposes to form a covalent bond with the nanotube surface.

This is a slower step that need not be selective and is represented by the restoration of the G peak and increase in the D band.

Reversibility of Diazonium Chemistry

Nanotubes reacted with the diazonium reagent can be converted back into pristine nanotubes when thermally treated at 300°C in an atmosphere of inert gas. This cleaves the aryl hydroxyl moieties from the nanotube sidewall and restores the spectroscopic feature (Raman and absorption spectra) of pristine nanotube.

Chemical Separation of Metallic and Semiconducting SWNTs

Metallic and semiconducting carbon nanotubes generally coexist in as-grown materials. To get only semiconducting or only metallic nanotubes selective functionalisation of metallic SWNTs via 4-hydroxybenzene diazonium can be used.

Separation can be done in solution by deprotonation of the p-hydroxybenzene group on the reacted nanotubes (metallic) in alkaline solution followed by electrophoretic separation of these charged species from the neutral species (semiconducting nanotubes). This followed by annealing would give separated pristine semiconducting and metallic SWNT.

CARBON NANOTUBE CHEMISTRY

Because of their hydrophobic nature, carbon nanotubes tend to agglomerate hindering their dispersion is solvents or viscous polymer melts. Unsurprisingly, the resulting aggregates reduce the mechanical performance of the final composite. Effort has been directed at modifying the surface of the carbon nanotube to reduce the hydrophobicity and improve interfacial adhesion to a bulk polymer through chemical attachment.

The carbon nanotube chemistry involves three main approaches, where reactions target: surface groups generated through acid-induced oxidation of the carbon nanotube surface; direct addition to the carbon nanotube sidewalls; groups attached to polycyclic aromatic hydrocarbons that are immobilised to the carbon nanotube surface through van der Waals forces.

Multiwalled Carbon Nanotubes

Covalent reactivity via acid-oxidation

Background

The purification and oxidation of carbon nanotubes (CNTs) has been well represented in literature. These processes were essential for low yield production of carbon nanotubes where carbon particles, amorphous carbon particles and coatings comprised a significant percentage of the overall material and are still important for the introduction of surface functional groups. During acid oxidation, the carbon-carbon bonded network of the graphitic layers is broken allowing the introduction of oxygen units in the form of carboxyl, phenolic and lactone groups, which have been extensively exploited for further chemical functionalisation.

The problem

In the mass production of single-walled carbon nanotubes in 1998, treatment with an aqueous base liberated yellow solutions, which were recognised as small polycyclic aromatic structures that contained oxygen groups. However, this purification step has not appeared in a large number of papers for the preparation and subsequent reaction of carboxylated CNTs. Studies in 2007 indicated that the liberated yellow alkaline solutions contained carbonaceous fragments from the oxidation of multiwalled carbon nanotubes (MWCNTs).

These smaller fragments of the MWCNT lattice were determined to be humic acids and therefore are strongly immobilised on the MWCNT surface in acidic solution, becoming detached in basic solution, and account for up to 45 per cent of the total number of oxygen containing groups. The problem, therefore, is that the majority of research publications over the last 15 years have not used an alkaline purification procedure to remove these lattice fragments and it is highly likely that subsequent chemical modification of their acid-oxidised carbon nanotubes occurs through carboxylic groups on these detachable fragments rather than the carboxylic groups covalently attached to the carbon nanotube lattice. Moreover, the carboxylic group readily form carboxylates with amines and are quite stable to various treatments.

So the problem with existing publications is compounded by the fact that such reports did not use an acid wash after coupling amines to the carbon nanotube carboxylic groups to ensure complete removal of the carboxylates.

This renders ca. 98 per cent of the existing research publications, which involve chemistry of acid-oxidised carbon nanotubes, potentially spurious.

The solution

The Whitby group at the University of Brighton used acid-base (or Boehm) titrations with careful purification procedures for carbon nanotubes at each stage of the relevant process, it was demonstrated that after removal of acid-generated humic acids from the surface of multiwalled carbon nanotubes, the remaining carboxylic groups undergo covalent amidation reactions, through carbodiimide coupling, with around 50–75 per cent conversion depending on the length of acid-oxidation initially used.

Moreover, the use of glutamic acid in the amidation step reveals that the intermediate complex of carbon nanotubes modified with 1-ethyl-3-(3-dimethylaminopropyl) carbodiimide hydrochloride (EDAC) and N-hydroxysuccinimide (NHS) do react to form the final amide.

This work highlights the importance of using alkaline solutions to remove humic acids generated and immobilised *in situ* to the carbon nanotube surface when subjected to nitric acid reflux and the use of acids at key stages to remove carboxylates.

Figure 5.14 shows the complete reaction scheme of carbon nanotubes for the direct attestation of their covalent carboxylic group reactivity.

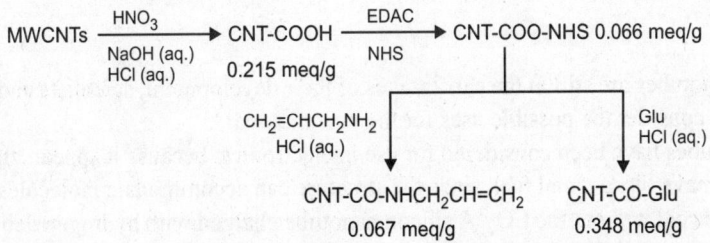

Fig. 5.14. Complete reaction scheme of carbon nanotubes for the direct attestation of their covalent carboxylic group reactivity.

Single-walled Carbon Nanotubes

Covalent reactivity via acid-oxidation

Studies in 2007 revealed that the acid-oxidation of SWCNTs generated carbonaceous fragments, resulting from the acid-oxidation of the SWCNT structure, which are immobilised to the outer surface of the SWCNT. After purification and removal of these fragments indicated that the final SWCNT structure bore no acidic groups and that these carbonaceous fragments may be the sole carrier of the carboxylic groups.

COLOSSAL CARBON TUBE

Colossal carbon tubes (CCTs) are a tubular form of carbon. In contrast to the carbon nanotubes (CNTs), colossal carbon tubes have much larger diameters ranging between 40 and 100 μm. Their walls have a corrugated structure with abundant pores, like in corrugated fibreboard, where the solid membranes have a graphite-like layered structure.

CCTs have technologically attractive properties such as ultralight weight, extremely high strength, excellent ductility and high conductivity — which make them possibly suitable for clothing. They are excellent conductors, are 15 times stronger than the strongest carbon fibre (T1000), have 30 times the tenacity of Kevlar and are 224 times stronger than individual cotton fibres. The tubes exhibit an ultra low density comparable to that of carbon nanofoams.

CCTs have a tensile strength of 7 GPa, and a high specific strength, which is tensile strength per density. It exceeds the specific strength of the strongest carbon nanotube, the strongest material known. The strength as published is sufficient to support a space elevator if fabricated in very long fibres with strengths similar to the regular scale, which is already beyond the nanoscale.

SILICON NANOTUBES

Silicon nanotubes are nanoparticles which create a tube-like structure from silicon atoms. The nanotubes' discovery has many significant implications for electronics development, as silicon is already a vastly important material in the semiconductor industry. Only recently has it been possible to prepare these nanotubes which are similar to carbon nanotubes. Nano-materials are complex, and understanding how the behaviour of silicon materials differs from their carbon-based cousins is still under research.

Synthesis

The nanotubes are created using a reactor employing an electric arc without the use of any catalyst. To ensure purity, the reactor was evacuated and filled with the nonreactive noble gas argon. The actual formation of the nanotubes relies on the process of chemical vapour deposition.

Applications

While silicon nanotubes are still in the early stages of their development, scientists and engineers have already begun to consider the possible uses for the new material.

Silicon nanotubes have been considered for use in electronics, because it appears that silicon nano-materials may behave like a metal fuel, since the structure can accommodate molecules of hydrogen so it might resemble coal without the CO_2. A silicon nanotube charged with hydrogen delivers energy and in the process leaves residual water, ethanol, silicon and sand.

However, as the hydrogen production requires considerable energy, so this only a proposed method of storing energy, not producing it.

Lithium Ion Batteries

The most profound application of silicon nanotubes has been their possible use in lithium-ion batteries. Conventional Li-ion batteries use graphitic carbon as the anode, but replacing this with silicon nanotubes experimentally increases the specific (by mass) anode capacity by a factor of 10 (but the battery capacity improvement is less due to far lower specific cathode capacities).

Researchers from LG Chem (the makers of the battery in the Chevrolet Volt), Stanford University and Hanyang University have been researching ways in which this technology could improve electric cars by making the battery smaller. However, the costs currently associated with the nanotubes is high and therefore, the research is currently theoretical—any employment of the silicon nanotubes will have to wait until the technology becomes cost effective.

Chapter 6

Nanowire

INTRODUCTION

Nanowire is a nanostructure, with the diameter of the order of a nanometer (10^{-9} metres). Alternatively, nanowires can be defined as structures that have a thickness or diameter constrained to tens of nanometers or less and an unconstrained length. At these scales, quantum mechanical effects are important—which coined the term 'quantum wires'.

Many different types of nanowires exist, including metallic (e.g. Ni, Pt, Au), semiconducting (e.g. Si, InP, GaN, etc.), and insulating (e.g. SiO_2, TiO_2). Molecular nanowires are composed of repeating molecular units either organic (e.g. DNA) or inorganic (e.g. $Mo_6S_{9-x}I_x$).

The nanowires could be used, in the near future, to link tiny components into extremely small circuits. Using nanotechnology, such components could be created out of chemical compounds.

Typical nanowires exhibit aspect ratios (length-to-width ratio) of 1000 or more. As such they are often referred to as one-dimensional (1-D) materials. Nanowires have many interesting properties that are not seen in bulk or 3-D materials. This is because electrons in nanowires are quantum confined laterally and thus occupy energy levels that are different from the traditional continuum of energy levels or bands found in bulk materials.

Peculiar features of this quantum confinement exhibited by certain nanowires manifest themselves in discrete values of the electrical conductance. Such discrete values arise from a quantum mechanical restraint on the number of electrons that can travel through the wire at the nanometer scale. These discrete values are often referred to as the quantum of conductance and are integer values of:

$$\frac{2e^2}{h} \simeq 12.9 \ k\Omega^{-1}$$

They are inverse of the well-known resistance unit h/e^2, which is roughly equal to 25812.8 ohms, and referred to as the von Klitzing constant R_K (after Klaus von Klitzing, the discoverer of exact quantisation). Since 1990, a fixed conventional value R_{K-90} is accepted.

Examples of nanowires include inorganic molecular nanowires ($Mo_6S_{9-x}I_x$, $Li_2Mo_6Se_6$), which can have a diameter of 0.9 nm be hundreds of micrometers long. Other important examples are based on semiconductors such as InP, Si, GaN, etc. dielectrics (e.g. SiO_2, TiO_2), or metals (e.g. Ni, Pt).

There are many applications where nanowires may become important in electronic, opto-electronic and nanoelectromechanical devices, as additives in advanced composites, for metallic interconnects in nanoscale quantum devices, as field-emitters and as leads for biomolecular nanosensors.

SYNTHESIS OF NANOWIRES

There are two basic approaches of synthesising nanowires: top-down and bottom-up approach. In a top-down approach a large piece of material is cut down to small pieces through different means such as lithography and electrophoresis. Whereas in a bottom-up approach the nanowire is synthesised by the combination of constituents ad-atoms. Most of the synthesis techniques are based on bottom-up approach.

Nanowire structures are grown through several common laboratory techniques including suspension, deposition (electrochemical or otherwise), and VLS growth. Figure 6.1 shows an SEM image of epitaxial nanowire heterostructures grown from catalytic gold nanoparticles.

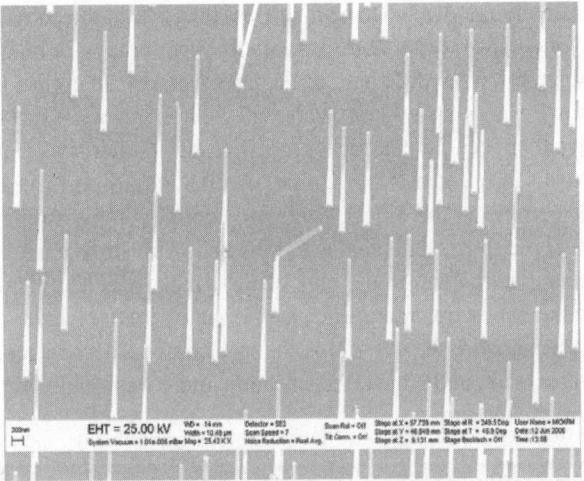

Fig. 6.1. An SEM image of epitaxial nanowire heterostructures grown from catalytic gold nanoparticles.

Suspension

A suspended nanowire is a wire produced in a high-vacuum chamber held at the longitudinal extremities. Suspended nanowires can be produced by:

1. The chemical etching, or bombardment (typically with highly energetic ions) of a larger wire.
2. Indenting the tip of a STM in the surface of a metal near its melting point, and then retracting it.

VLS Growth

A common technique for creating a nanowire is the vapour–liquid–solid (VLS) synthesis method. This technique uses as source material either laser ablated particles or a feed gas (such as silane). The source is then exposed to a catalyst. For nanowires, the best catalysts are liquid metal (such as gold) nanoclusters, which can either be purchased in colloidal form and deposited on a substrate or self-assembled from a thin film by dewetting. This process can often produce crystalline nanowires in the case of semiconductor materials. The source enters these nanoclusters and begins to saturate it. Once supersaturation is reached, the source solidifies and grows outward from the nanocluster. The final product's length can be adjusted by simply turning off the source. Compound nanowires with super-lattices of alternating materials can be created by switching sources while still in the growth phase.

Inorganic nanowires such as $Mo_6S_{9-x}I_x$ (which are alternatively viewed as cluster polymers) are synthesised in a single-step vapour phase reaction at elevated temperature.

Solution-phase Synthesis

Nanowires of many types of materials can grow in solution. Solution-phase synthesis has the advantage that it can be scaled-up to produce very large quantities of nanowires as compared to methods that produce nanowires on a surface. The polyol synthesis, in which ethylene glycol is both solvent and reducing agent, has proven particularly versatile at producing nanowires of Pb, Pt, and silver.

PHYSICS OF NANOWIRES

Conductivity of Nanowires

The conductivity of a nanowire is expected to be much less than that of the corresponding bulk material. This is due to a variety of reasons. First, there is scattering from the wire boundaries, when the wire width is below the free electron mean free path of the bulk material. In copper, for example, the mean free path is 40 nm. Nanowires less than 40 nm wide will shorten the mean free path to the wire width.

Nanowires also show other peculiar electrical properties due to their size. Unlike carbon nanotubes, whose motion of electrons can fall under the regime of ballistic transport (meaning the electrons can travel freely from one electrode to the other), nanowire conductivity is strongly influenced by edge effects. The edge effects come from atoms that lay at the nanowire surface and are not fully bonded to neighbouring atoms like the atoms within the bulk of the nanowire. The unbonded atoms are often a source of defects within the nanowire, and may cause the nanowire to conduct electricity more poorly than the bulk material. As a nanowire shrinks in size, the surface atoms become more numerous compared to the atoms within the nanowire, and edge effects become more important. Figure 6.2 shows an SEM image of a 15 micrometre nickel wire.

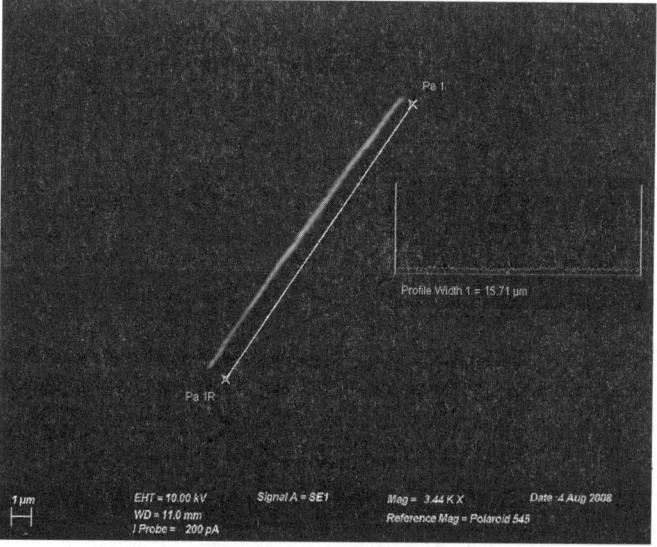

Fig. 6.2. An SEM image of a 15 micrometre nickel wire.

Furthermore the conductivity can undergo a quantisation in energy: i.e. the energy of the electrons going through a nanowire can assume only discrete values, multiple of the Von Klitzing constant $G = 2e^2/h$ (where, e is the charge of the electron and h is the Planck constant).

The conductivity is hence described as the sum of the transport by separate channels of different quantised energy levels. The thinner the wire is, the smaller the number of channels available to the transport of electrons. The conductivity of a nanowire can be studied suspending it between two electrodes. This has been proven by measuring the conductivity of a nanowire while pulling it: as its diameter is reduced, its conductivity decreases in a stepwise fashion and the plateaus correspond to multiples of G.

The quantised conductivity is more pronounced in semiconductors like Si or GaAs than in metals, due to lower electron density and lower effective mass. Quantised conductance can be observed in 25 nm wide silicon fins, resulting in increased threshold voltage.

Welding Nanowires

To incorporate nanowire technology into industrial applications, researchers recently developed a method of welding nanowires together: a sacrificial metal nanowire is placed adjacent to the ends of the pieces to be joined (using the manipulators of a scanning electron microscope); then an electric current is applied, which fuses the wire ends. The technique fuses wires as small as 10 nm.

Uses of Nanowires

Nanowires still belong to the experimental world of laboratories. However, they may complement or replace carbon nanotubes in some applications. Some early experiments have shown how they can be used to build the next generation of computing devices.

To create active electronic elements, the first key step was to chemically dope a semiconductor nanowire. This has already been done to individual nanowires to create *p*-type and *n*-type semiconductors.

The next step was to find a way to create a *p-n* junction, one of the simplest electronic devices. This was achieved in two ways. The first way was to physically cross a *p*-type wire over an *n*-type wire. The second method involved chemically doping a single wire with different dopants along the length. This method created a *p-n* junction with only one wire.

After *p-n* junctions were built with nanowires, the next logical step was to build logic gates. By connecting several *p-n* junctions together, researchers have been able to create the basis of all logic circuits: the AND, OR, and NOT gates have all been built from semiconductor nanowire crossings.

It is possible that semiconductor nanowire crossings will be important to the future of digital computing. Though there are other uses for nanowires beyond these, the only ones that actually take advantage of physics in the nanometer regime are electronic. Nanowires are being studied for use as photon ballistic waveguides as interconnects in quantum dot/quantum effect well photon logic arrays. Photons travel inside the tube, electrons travel on the outside shell. When two nanowires acting as photon waveguides cross each other the juncture acts as a quantum dot. Conducting nanowires offer the possibility of connecting molecular-scale entities in a molecular computer. Dispersions of conducting nanowires in different polymers are being investigated for use as transparent electrodes for flexible flat-screen displays. Because of their high Young's moduli, their use in mechanically enhancing composites is being investigated. Because nanowires appear in bundles, they may be used as tribological additives to improve friction characteristics and reliability of electronic transducers and actuators.

Because of their high aspect ratio, nanowires are also uniquely suited to dielectrophoretic manipulation.

MOLECULAR WIRES

Molecular wires (or sometimes called molecular nanowires) are molecular-scale objects which conduct electrical current. They are the fundamental building blocks for molecular electronic devices. Their typical diameters are less than three nanometers, while their bulk lengths may be macroscopic, extending to centimetres or more.

Materials

Most work thus far has consisted of organic molecules. Higher conductivities originate from highly conjugated systems, while alkane chains are important in understanding basic charge transport and tunnelling. A molecular wire occurring in nature is DNA. Prominent inorganic examples include polymeric materials such as $Li_2Mo_6Se_6$ and $Mo_6S_{9-x}I_x$, and single-molecule extended metal atom chains (EMACs) which comprise strings of late transition metal atoms directly bonded to each other. Molecular wires containing paramagnetic inorganic moieties are interesting, in particular, because they can lead to observations of Kondo peaks.

Structure

Unlike the more usual nanowires (which are very thin crystals), molecular nanowires are composed of repeating molecular units, which may be organic (e.g. DNA) or inorganic (e.g. $Mo_6S_{9-x}I_x$). In the case of DNA, the repeat units are the nucleotides with a backbone made of sugars and phosphate groups joined by ester bonds. Attached to each sugar is one of four types of bases. In case of $Mo_6S_{9-x}I_x$, the repeat units are $Mo_6S_{9-x}I_x$ clusters, which are joined together by sulphur or iodine bridges. Molecular nanowires often aggregate in solution into swatches or bundles. In the case of the Mo chalcogenide-halides, they grow in the form of ordered strands, in which the individual strands are linked by very weak van der Waals forces. By contrast, EMAC molecular wires consist of distinct molecules that do not aggregate and hence enable control of the exact length of the wire at the atomic scale (Fig. 6.3).

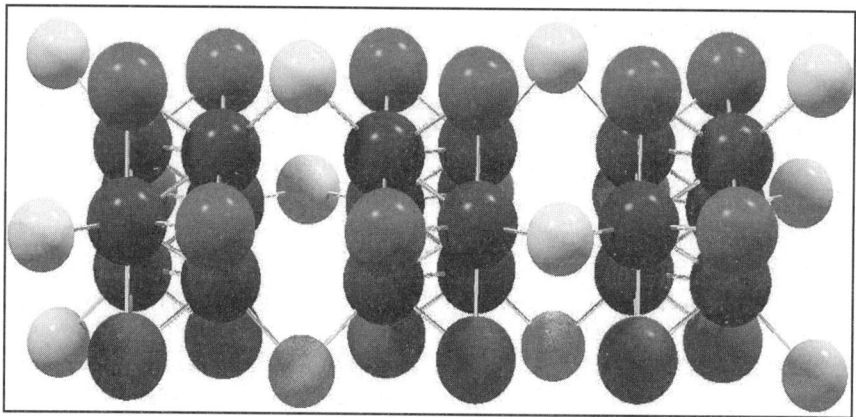

Fig. 6.3. The structure of a $Mo_6S_{9-x}I_x$ molecular wire.

Conduction of Electrons

Molecular wires conduct electricity. They typically have non-linear current-voltage characteristics, and do not behave as simple ohmic conductors. The conductance follows typical power law behaviour as a function of temperature or electric field, whichever is the greater, arising from their strong one-dimensional character. Numerous theoretical ideas have been used in an attempt to understand the conductivity of one-dimensional systems, where strong interactions between electrons lead to departures from normal metallic (Fermi liquid) behaviour. Important concepts are those introduced by Tomonaga,

Luttinger and Wigner. Effects caused by classical Coulomb repulsion (called Coulomb blockade Coulomb blockade) have also been found to be important in determining the properties of molecular wires.

Use of Nanowires in Molecular Electronics

To be of use for connecting molecules together, MWs need to display some very important characteristics. The connectors between elements need to be able to self-assemble following well-defined routes and form reliable electrical contacts between them. To reproducibly self-assemble a complex circuit based on single molecules, it is essential that the connectors which join them have recognitive ability. They should be able to connect to diverse materials, such as gold metal surfaces (for connections to outside world), biomolecules (for nanosensors, nanoelectrodes, molecular switches) and most importantly, they must allow branching. The connectors should also be available of predetermined diameter and length. They should also have covalent bonding to ensure reproducible transport and contact properties. DNA-like molecules have specific molecular-scale recognition and can be used very effectively in molecular scaffold fabrication. Very complex shapes have recently been demonstrated, but unfortunately metal coated DNA which is electrically conducting is much too thick to connect to individual molecules. Thinner coated DNA lacks electronic connectivity, and are not suitable for connecting molecular electronics elements. Some varieties of carbon nanotubes (CNTs) are conducting, and connectivity at their ends can be achieved by attachment of connecting groups. Unfortunately manufacturing CNTs with predetermined properties is impossible at present, and the functionalised ends are typically not conducting, limiting their usefulness as molecular connectors. Individual CNTs can be soldered in an electron microscope, but the contact is not covalent and cannot be self-assembled. Recently possible routes for the construction of larger functional circuits using $Mo_6S_{9-x}I_x$ MWs have been demonstrated, either via gold nanoparticles as linkers, or by direct connection to thiolated molecules. The two approaches may lead to different possible applications. The use of GNPs offers the possibility of branching and construction of larger circuits.

FABRICATION OF NANOWIRES AT SURFACES

The goal of this project is the design of artificial materials that consist of ultrafine wires or linear arrays of dots, ten to hundred times finer than those produced with commercial micro-structure fabrication techniques. In fact, we have gone all the way down to atom chains which may be viewed as the ultimate nanowires. These patterns are formed by self-assembly, where atoms arrange themselves naturally at stepped silicon surfaces.

An important aspect in fabricating nanowires is the ability to prepare wires of an any material on any substrate with any thickness. In particular, using silicon wafers as substrate is highly-desirable. To achieve this goal we suggest the following 'universal' process. First, a silicon substrate with a regular array of steps is prepared (A). Then, stripes (B) or dots (C) of a passivating material are attached to the step edges. This part is analogous to creating a photoresist mask in traditional lithography. As mask material we use calcium fluoride, which is lattice-matched to silicon and chemically inert. Eventually, the desired material is deposited on the remaining silicon, for example by substrate-selective chemical vapour deposition (CVD) or electroplating. Alternatively, calcium fluoride could become useful as an etch mask for producing trenches in the silicon that can be filled with new materials to achieve a planar structure.

Figure 6.4 shows the preparation of calcium fluoride masks in schematic form (top), together with actual data (bottom).

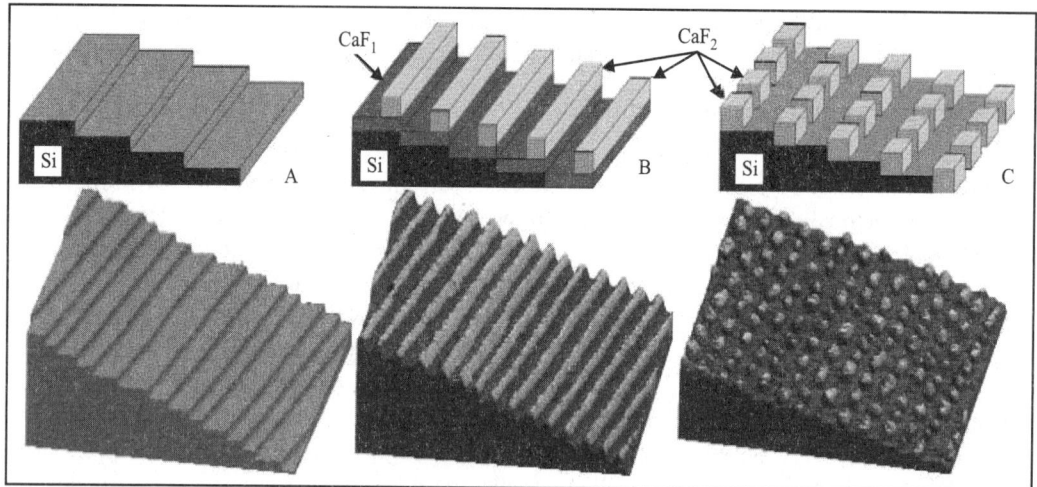

Fig. 6.4. The preparation of calcium fluoride masks in schematic form (top), together with actual data (bottom).

To start along this pathway, we determine the conditions for obtaining highly-regular step structures on silicon. The images below demonstrate the range of step arrays that can be formed on silicon surfaces by self-assembly. Typically, the step spacing is comparable to the size of a virus. These images are taken with a scanning tunnelling microscope (STM). They show the derivative of the tip height. That gives the impression of a surface illuminated from the left, with the steps casting dark shadows to the right (Fig. 6.5).

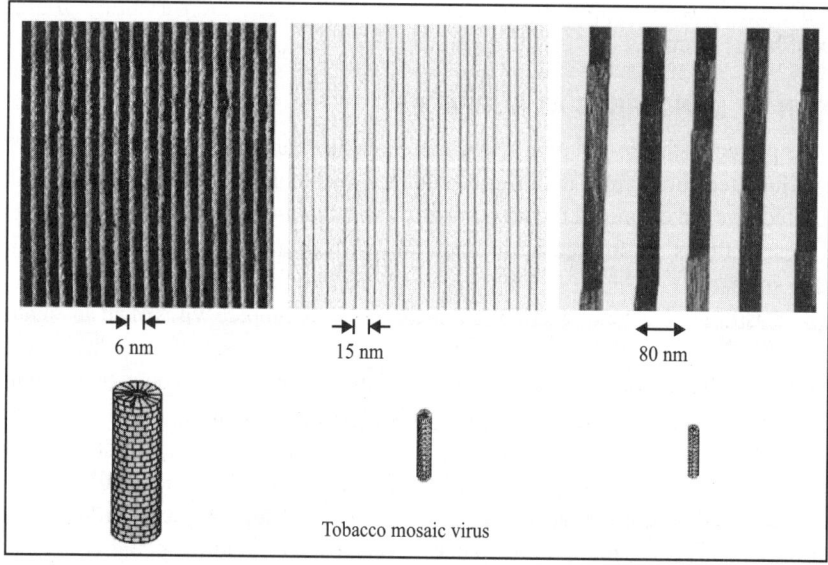

Fig. 6.5. Step structures on silicon.

Particularly perfect step arrays could be achieved on the Si 7×7 surface. The 7×7 structure causes steps running along the [011]-direction to become extremely straight, because each kink requires generating 14 new rows of silicon atoms (7 rows, two layers deep). The step edges are atomically-straight with a kink spacing as low as a single kink in 20,000 atoms, as seen in the Fig. 6.6. These are taken with a scanning tunnelling microscope (STM). The x-derivative of the topography is displayed, which makes steps appear as dark lines. The image on the right zooms in on the terrace between two steps (heavy dark lines). The atomic pattern of the 7×7 structure is resolved, which has fine grooves built in that run parallel to the step edges. These grooves are 2.3 nm apart and determine the location of possible step edges. They may be viewed as the LEGO blocks of the aspiring nano-engineer.

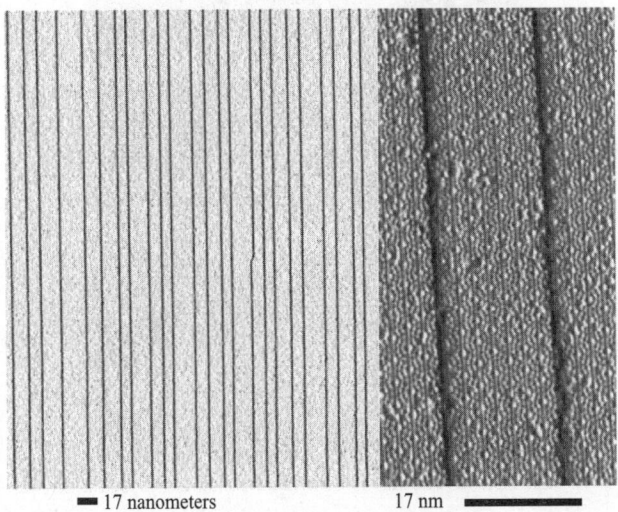

17 nanometers   17 nm

Fig. 6.6. Single kink in 20,000 atoms.

Our 'universal process' can be carried further by producing a calcium fluoride mask, as shown in the Fig. 6.7. A stepped silicon surface is coated with a layer of CaF_1 and CaF_2 stripes are formed on top of that layer. These stripes are continuous and do not touch each other, because adjacent stripes cannot bond to each other.

7 nanometers

Fig. 6.7. Calcium fluoride mask.

The stripe width of 7 nm achieved here is well below the resolution of 180 nm achieved in commercial lithography for chip fabrication.

The third step of the 'universal process' involves selective deposition or etching between the masked areas. The picture below shows that molecules can be deposited selectively in the CaF_1 grooves between CaF_2 stripes. Using organometallic molecules, such as ferrocene, it is possible to fabricate iron wires 3 nanometers wide (Fig. 6.8).

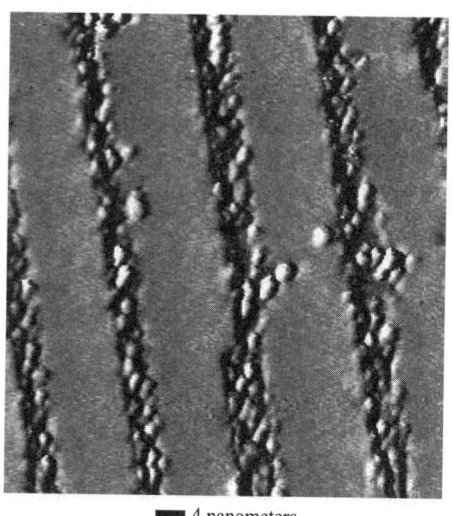

■ 4 nanometers

Fig. 6.8. Selective deposition.

ATOM CHAINS, THE ULTIMATE NANOWIRES

Self-assembly can reach atomic precision for very small structures (up to 10 nm in size). It is possible to go all the way to the ultimate limit for nanowires, i.e. chainse of single atoms with a single set of orbitals connecting them.

Such atomic wires are obtained by depositing a fraction of a monolayer of metal atoms onto a stepped silicon surface. An example is the Si (557)-Au surface shown in Fig. 6.9. It contains a step every five silicon atom rows and a row of gold atoms in the middle of the terrace. The STM image below shows two rows of fine white dots, which are magnified in the inset. They correspond to silicon atoms with dangling bonds.

Being able to fabricate atomic wires opens up the possibility of studying one-dimensional electrons moving along the atom chains. The properties of electrons become more and more exotic when progressing from the three-dimensional world into lower dimensions. In a two-dimensional electron gas one observes surprising phenomena already, such as fractional charge and statistics with the fractional quantum Hall effect. Predictions for one-dimensional electrons are even farther out. The concept of a single electron fails because electrons constantly penetrate each other when moving along the same one-dimensional line.

They lose their identity and separate into two quasiparticles, a spinon that carries spin without charge, and a holon that carries the positive charge of a hole without its spin. That is the theoretical prediction, and the hunt is on whether such strange behaviour can be observed experimentally.

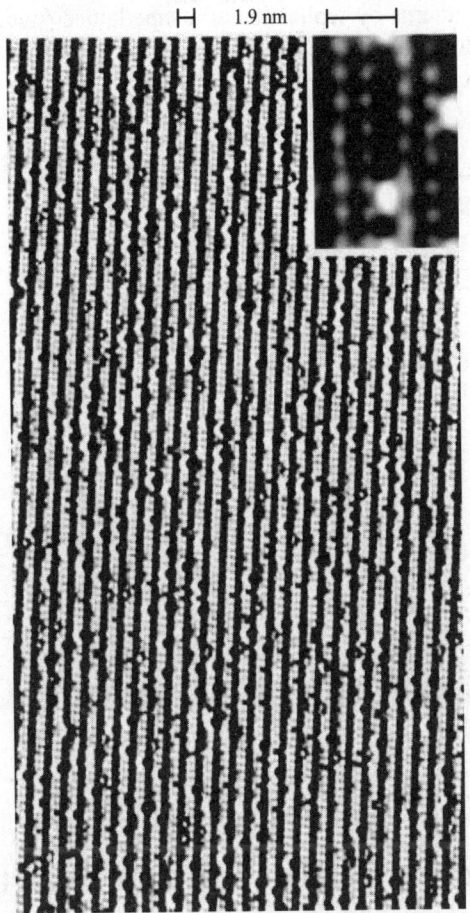

H⊢ 1.9 nm ⊢H

Fig. 6.9. Si(557)-Au surface.

The complete set of quantum numbers for electrons in a solid is revealed by angle-resolved photoemission with tunable synchrotron radiation. The two principal quantum numbers energy and momentum are combined in a band structure plot. The dark bands represent regions with high photoemission intensity. With multitedection techniques for energy and angle it is now possible to see the band structure directly on a television screen.

Some of the atomic wires are found to be metallic. The metallic behaviour is deduced from the observation that the bands extend all the way up to the Fermi level E_F for Si(557)-Au, as in a metal. By way of contrast, the flat Si(111)-Au surface in the panel on the right shows a band that does not reach E_F. Even though the metal atoms are strongly coupled to the substrate, metallic electrons do not interact with the silicon substrate because their energy lies in the band gap of silicon (Fig. 6.10).

Electrons at the Fermi level are particularly interesting because they assemble themselves into exotic phases in low dimensions. These electrons are best characterised by the Fermi surface, which describes their momentum distribution. Such Fermi surfaces are shown below for two- and one-dimensional electrons at silicon surfaces with extra metal atoms (typically a monolayer for a 2D structure and 1/5 of a monolayer for a 1D structure). Two-dimensional Fermi surfaces typically consist of circles (top), which may form intricate patterns by replication at a superlattice (middle). One-dimensional Fermi surfaces are straight lines, which become modulated by small wiggles if a weak two-dimensional coupling is present between the atom chains (bottom).

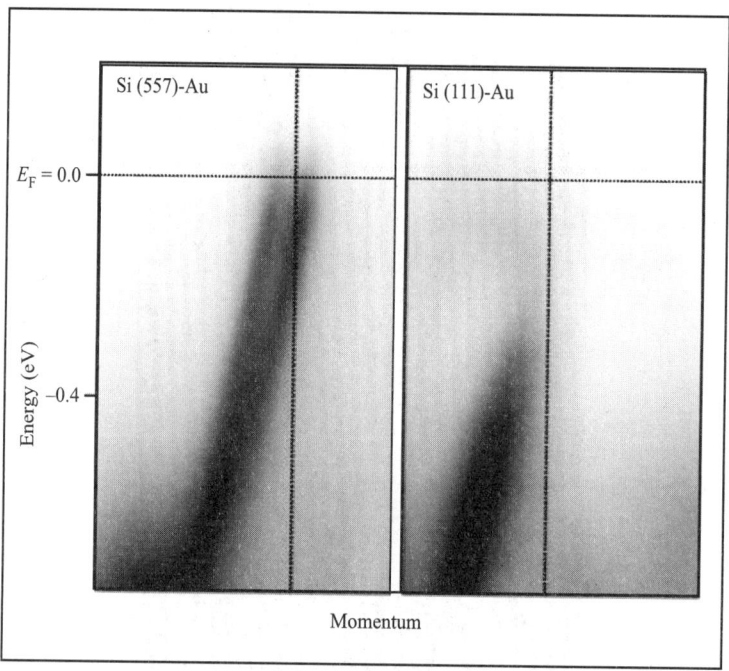

Fig. 6.10. Si (557)-Au and Si(111)-Au surface.

SOLAR NANOWIRES PROMISE EFFICIENT, LOW-COST SOLAR POWER

University researchers claim to have found a way of channelling solar energy directly into electrical appliances without the need for conventional solar panels.

Researchers at the Georgia Institute of Technology say they have efficiently turned light into electricity using fibre-optic cables that have been attached to solar cells.

The system uses fibre optics of the type used by the telecommunications industry and covers them with zinc oxide nanowires to increase their surface area. The nanowires are then coated with dye-sensitised solar cells that convert light to electricity.

The researchers said that sunlight entering the optical fibre passes into the nanowires, where it interacts with the dye molecules in the cells to produce an electrical current.

They added that the cells are inexpensive to manufacture, flexible and mechanically robust. While they are currently less efficient than the silicon-based cells used in traditional solar panels, this can be countered by using the nanostructure arrays to increase the available surface area, making them in practice six times more efficient than conventional solar panels.

'You have multiple light reflections within the fibre, and multiple reflections within the nanostructures', explained Zhong Lin Wang, professor at the Georgia Tech School of Materials Science and Engineering, who led the research. 'These interactions increase the likelihood that the light will interact with the dye molecules, and that increases the efficiency.'

Wang and his research team hope to double the efficiency of the system again by further improving the surface area. The amount of light entering the optical fibre could also be increased by using lenses to focus the incoming light.

Wang added that the breakthrough had the potential to revolutionise the way solar energy is harvested. 'Using this technology, we can make photovoltaic generators that are foldable, concealed and mobile', he said. 'Optical fibre could conduct sunlight into a building's walls where the nanostructures would convert it to electricity. This is truly a three-dimensional solar cell.'

He admitted that the new solar cells are unlikely to replace silicon-based panels within the next few years, but argued that they will broaden the potential applications for photovoltaic energy.

'This is a different way to gather power from the sun', he said. 'To meet our energy needs, we need all the approaches we can get.'

NANOWIRE BATTERY

A nanowire battery is a lithium-ion battery invented by a team led by Dr. Yi Cui at Stanford University in 2007. The team's invention consists of a stainless steel anode covered in silicon nanowires, to replace the traditional graphite anode. Silicon, which stores ten times more lithium than graphite, allows a far greater energy density on the anode, thus reducing the mass of the battery. The high surface area further allows for fast charging and discharging.

Traditional silicon anodes were researched and dismissed due to the tendency of silicon to crack and become useless as it swelled with lithium during operation. The nanowires, on the other hand, do not suffer from this flaw. According to Dr. Cui, the battery only reached 10x density on the first charge and levelled out at 8x density on subsequent charges. Since this is only an anode advancement, an equivalent cathode advancement would be needed to get the full energy storage density improvements.

Commercialisation is expected to take approximately five years (as of 2007), with the batteries costing similar or less per watt hour than conventional lithium-ion. The next milestone, life-cycle testing, should be completed, and the team expects to get at least a thousand cycles out of the battery.

Chapter 7

Nanomechanics

INTRODUCTION

Nanomechanics is a branch of nanoscience studying fundamental mechanical (elastic, thermal and kinetic) properties of physical systems at the nanometer scale. Nanomechanics has emerged on the cross-road of classical mechanics, solid-state physics, statistical mechanics, materials science, and quantum chemistry. As an area of nanoscience, nanomechanics provides a scientific foundation of nanotechnology.

Nanomechanics is that branch of nanoscience, which deals with the study and application of fundamental mechanical properties of physical systems at the nanoscale, like elastic, thermal, and kinetic.

Often, nanomechanics is viewed as a branch of nanotechnology, i.e. an applied area with a focus on the mechanical properties of engineered nanostructures and nanosystems (systems with nanoscale components of importance).

Examples of the latter include nanoparticles, nanopowders, nanowires, nanorods, nanoribbons, nanotubes, including carbon nanotubes (CNT) and boron nitride nanotubes (BNNTs); nanoshells, nanomembranes, nanocoatings, nanocomposite/nanostructured materials, (fluids with dispersed nanoparticles), nanomotors, etc.

Some of the well-established fields of nanomechanics are: nanomaterials, nanotribology (friction, wear and contact mechanics at the nanoscale), nanoelectromechanical systems (NEMS), and nanofluidics.

As a fundamental science, nanomechanics is based on some empirical principles (basic observations): (i) general mechanics principles, and (ii) specific principles arising from the smallness of physical sizes of the object of study or research.

General mechanics principles include:

1. Energy and momentum conservation principles.
2. Variational Hamilton's principle.
3. Symmetry principles.

Due to smallness of the studied object, nanomechanics also accounts for:

1. Discreteness of the object, whose size is comparable with the interatomic distances.
2. Plurality, but finiteness, of degrees of freedom in the object.
3. Importance of thermal fluctuations.
4. Importance of entropic effects.
5. Importance of quantum effects.

These principles serve to provides a basic insight into novel mechanical properties of nanometer objects. Novelty is understood in the sense that these properties are not present in similar macroscale objects or much different from the properties of those (e.g. nanorods vs. usual macroscopic beam structures). In particular, smallness of the subject itself gives rise to various surface effects determined by higher surface-to-volume ratio of nanostructures, and thus affects mechanoenergetic and thermal properties (melting point, heat capacitance, etc.) of nanostructures. Discreteness serves a fundamental reason, for instance, for the dispersion of mechanical waves in solids, and some special behaviour of basic elastomechanics solutions at small scales. Plurality of degrees of freedom and the rise of thermal fluctuations are the reasons for thermal tunnelling of nanoparticles through potential barriers, as well as for the cross-diffusion of liquids and solids. Smallness and thermal fluctuations provide the basic reasons of the Brownian motion of nanoparticles. Increased importance of thermal fluctuations and configuration entropy at the nanoscale give rise to superelasticity, entropic elasticity (entropic forces), and other exotic types of elasticity of nanostructures. Aspects of configuration entropy are also of great interest in the context self-organisation and cooperative behaviour of open nanosystems.

Quantum effects determine forces of interaction between individual atoms in physical objects, which are introduced in nanomechanics by means of some averaged mathematical models called interatomic potentials. Subsequent utilisation of the interatomic potentials within the classical multibody dynamics provide deterministic mechanical models of nanostructures and systems at the atomic scale/resolution. Numerical methods of solution of these models are called molecular dynamics (MD), and sometimes molecular mechanics (especially, in relation to statically equilibrated [still] models). Nondeterministic numerical approaches include Monte-Carlo, Kinetic More-Carlo (KMC), and other methods. Contemporary numerical tools include also hybrid multiscale approaches allowing concurrent or sequential utilisation of the atomistic scale methods (usually, MD) with the continuum (macro) scale methods (usually, FEM) within a single mathematical model. Development of these complex methods is a separate subject of applied mechanics research.

Quantum effects also determine novel electrical, optical and chemical properties of nanostructures, and therefore they find even greater attention in adjacent areas of nanoscience and nanotechnology, such as nanoelectronics, advanced energy systems, and nanobiotechnology.

MOLECULAR MACHINE

A molecular machine, or nanomachine, has been defined as a discrete number of molecular components that perform mechanical-like movements (output) in response to specific stimuli (input). It is often applied more generally to molecules that simply mimic functions at the macroscopic level. The term is also common in nanotechnology, and a number of highly complex molecular machines have been proposed towards the goal of constructing a molecular assembler. Molecular machines can be divided into two broad categories: synthetic and biological.

Molecular systems that are able to shift a chemical or mechanical process away from equilibrium represent a potentially important branch of chemistry and nanotechnology. By definition, these types of systems are examples of molecular machinery, as the gradient generated from this process is able to perform useful work.

Historical Insight and Studies

There are two thought experiments that form the historical basis for molecular machines: Maxwell's demon and Feynman's ratchet (or Brownian ratchet).

Imagine a very small system of two paddles or gears connected by a rigid axle and that it is possible to keep these two paddles at two different temperatures. One of the gears (at T2) has a pawl that is rectifying the system motion, and therefore, the axle can only move in a clockwise rotation, and in doing so, it could lift a weight (m) upward upon ratcheting. Now imagine if the paddle in box T1 was in a much hotter environment than the gear in box T2; it would be expected that the kinetic energy of the gas molecules (dark circles) hitting the paddle in T1 would be much higher than the gas molecules hitting the gear at T2. Therefore, with lower kinetic energy of the gases in T2, there would be very little resistance from the molecules on colliding with the gear in the statistically opposite direction. Further, the ratcheting would allow for directionality, and slowly over time, the axle would rotate and ratchet, lifting the weight (m).

As described, this system may seem like a perpetual motion machine; however, the key ingredient is the heat gradient within the system. This ratchet does not threaten the second law of thermodynamics, because this temperature gradient must be maintained by some external means. Brownian motion of the gas particles provides the power to the machine, and the temperature gradient allows the machine to drive the system cyclically away from equilibrium. An interesting design concept in Feynman's ratchet is that random Brownian motion is not fought against, but instead, harnessed and rectified. Unfortunately, temperature gradients cannot be maintained over molecular scale distances because of molecular vibration redistributing the energy to other parts of the molecule. Furthermore, despite Feynman's machine doing useful work in lifting the mass, using Brownian motion to power a molecular level machine does not provides any insight on how that power (or potential energy of the lifted weight, m) can be used to perform nanoscale tasks (Fig. 7.1).

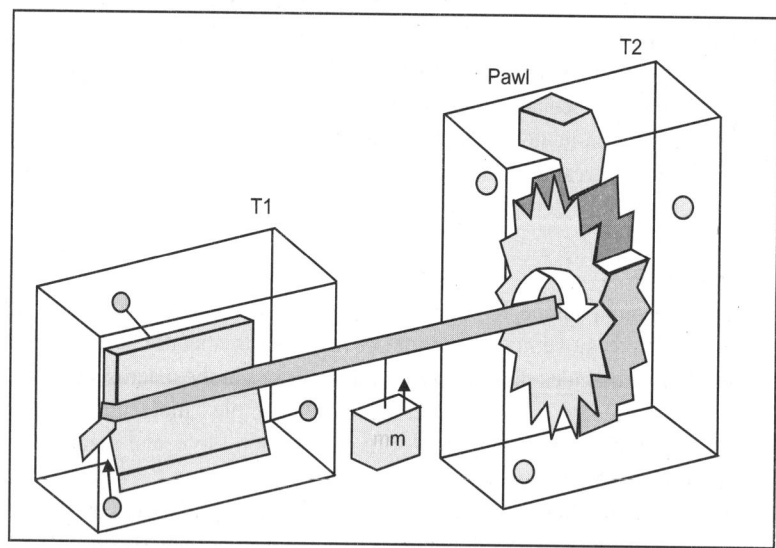

Fig. 7.1. Schematic diagram of Feynman's ratchet

Modern Insights and Studies

Unlike macroscopic motion, molecular systems are constantly undergoing significant dynamic motions subject to the laws of Brownian mechanics (or Brownian motion), and as such, harnessing molecular motion is a far more difficult process. At the macroscopic level, many machines operate in the gas phase, and often, air resistance is neglected, as it is insignificant, but analogously for a molecular system in a Brownian environment, molecular motion is similar 'to walking in a hurricane, or swimming in molasses'. The phenomenon of Brownian motion (observed by Robert Brown [botanist], 1827) was later explained by Albert Einstein in 1905. Einstein found that Brownian motion is a consequence of scale and not the nature of the surroundings. As long as thermal energy is applied to a molecule, it will undergo Brownian motion with the kinetic energy appropriate to that temperature. Therefore, like Feynman's strategy, when designing a molecular machine, it seems sensible to utilise Brownian motion rather than attempt to fight against it.

Like macroscopic machines, molecular machines typically have movable parts. However, while the macroscopic machines we encounter in everyday life may provide inspiration for molecular machines, it is misleading to draw analogies between their design strategy; the dynamics of large and small length scales are simply too different. Harnessing Brownian motion and making molecular level machines is regulated by the second law of thermodynamics, with its often counter-intuitive consequences, and as such, we need another inspiration. Although it is a challenging process to harness Brownian motion, nature has provided us with several blueprints for molecular motion performing useful work. Nature has created many useful structures for compartmentalising molecular systems, hence creating distinct non-equilibrium distributions; the cell membrane is an excellent example. Lipophilic barriers make use of a number of different mechanisms to power motion from one compartment to another.

Examples of molecular machines

From a synthetic perspective, there are two important types of molecular machines: molecular switches (or shuttles) and molecular motors. The major difference between the two systems is that a switch influences a system as a function of state, whereas a motor influences a system as function of trajectory. A switch (or shuttle) may appear to undergo translational motion, but returning a switch to its original position undoes any mechanical effect and liberates energy to the system. Furthermore, switches cannot use chemical energy to repetitively and progressively drive a system away from equilibrium where a motor can.

Synthetic

A wide variety of rather simple molecular machines have been synthesised by chemists. They can consist of a single molecule; however, they are often constructed for mechanically-interlocked molecular architectures, such as rotaxanes and catenanes.

1. Molecular motors are molecules that are capable of unidirectional rotation motion powered by external energy input. A number of molecular machines have been synthesised powered by light or reaction with other molecules.
2. A molecular propeller is a molecule that can propel fluids when rotated, due to its special shape that is designed in analogy to macroscopic propellers. It has several molecular-scale blades attached at a certain pitch angle around the circumference of a nanoscale shaft.
3. A molecular switch is a molecule that can be reversibly shifted between two or more stable states. The molecules may be shifted between the states in response to changes in e.g. pH, light, temperature, an electrical current, microenvironment, or the presence of a ligand.
4. A molecular shuttle is a molecule capable of shuttling molecules or ions from one location to another. A common molecular shuttle consists of a rotaxane where the macrocycle can move between two sites or stations along the dumbbell backbone.

5. Molecular tweezers are host molecules capable of holding items between its two arms. The open cavity of the molecular tweezers binds items using non-covalent bonding including hydrogen bonding, metal coordination, hydrophobic forces, van der Waals forces, π-π interactions, and/or electrostatic effects. Examples of molecular tweezers have been reported that are constructed from DNA and are considered DNA machines.

6. A molecular sensor is a molecule that interacts with an analyte to produce a detectable change. Molecular sensors combine molecular recognition with some form of reporter, so the presence of the item can be observed.

7. A molecular logic gate is a molecule that performs a logical operation on one or more logic inputs and produces a single logic output. Unlike a molecular sensor, the molecular logic gate will only output when a particular combination of inputs are present.

Biological

The most complex molecular machines are found within cells. These include motor proteins, such as myosin, which is responsible for muscle contraction, kinesin, which moves cargo inside cells away from the nucleus along microtubules, and dynein, which produces the axonemal beating of cilia and flagella. These proteins and their nanoscale dynamics are far more complex than any molecular machines that have yet been artificially constructed.

A high-level-abstraction summary is that, 'in effect, the motile cilium is a nanomachine composed of perhaps over 600 proteins in molecular complexes, many of which also function independently as nanomachines'.

Theoretical

The construction of more complex molecular machines is an active area of theoretical research. A number of molecules, such as molecular propellers, have been designed, although experimental studies of these molecules are inhibited by the lack of methods to construct these molecules. These complex molecular machines form the basis of areas of nanotechnology, including molecular assembler.

DNA MACHINE

A DNA machine is a molecular machine constructed from DNA. Research into DNA machines was pioneered in the late 1980s by Nadrian Seeman and co-workers from New York University. DNA is used because of the numerous biological tools already found in nature that can affect DNA, and the immense knowledge of how DNA works previously researched by biochemists.

DNA machines can be logically designed since DNA assembly of the double helix is based on strict rules of base pairing that allow portions of the strand to be predictably connected based on their sequence. The 'selective stickiness' is a key advantage in the construction DNA machines.

An example of a DNA machine was reported by Bernard Yurke and co-workers at Lucent Technologies in the year 2000, who constructed molecular tweezers out of DNA. The DNA tweezers contains three strands: A, B and C. Strand A latches onto half of strand B and half of strand C, and so it joins them all together. Strand A acts as a hinge so that the two 'arms'—AB and AC—can move. The structure floats with its arms open wide. They can be pulled shut by adding a fourth strand of DNA (D) 'programmed' to stick to both of the dangling, unpaired sections of strands B and C. The closing of the tweezers was proven by tagging strand A at either end with light-emitting molecules that do not emit light when they are close together. To reopen the tweezers add a further strand (E) with the right sequence to pair up with strand D. Once paired up, they have no connection to the machine BAC, so float away. The DNA machine can be opened and closed repeatedly by cycling between strands D and E. These tweezers can be used for removing drugs from inside fullerenes as well as from a self-assembled DNA tetrahedron. The state of the device can be determined by measuring the separation between donor and acceptor fluorophores using FRET.

DNA NANOTECHNOLOGY

DNA nanotechnology is a subfield of nanotechnology which seeks to use the unique molecular recognition properties of DNA and other nucleic acids to create novel, controllable structures out of DNA. The DNA is thus used as a structural material rather than as a carrier of genetic information, making it an example of bionanotechnology. This has possible applications in molecular self-assembly and in DNA computing.

DNA nanotechnology is that branch, which deals with the study and application of molecular recognition properties of DNA and other nucleic acids to create controllable structures out of them for computing and assembly.

DNA nanotechnology makes use of branched DNA structures to create DNA complexes with useful properties. DNA is normally a linear molecule, in that its axis is unbranched. However, DNA molecules containing junctions can also be made. For example, a four-arm junction can be made using four individual DNA strands which are complementary to each other in the correct pattern. Due to Watson-Crick base pairing, only portions of the strands which are complementary to each other will attach to each other to form duplex DNA. This four-arm junction is an immobile form of a Holliday junction (Fig. 7.2).

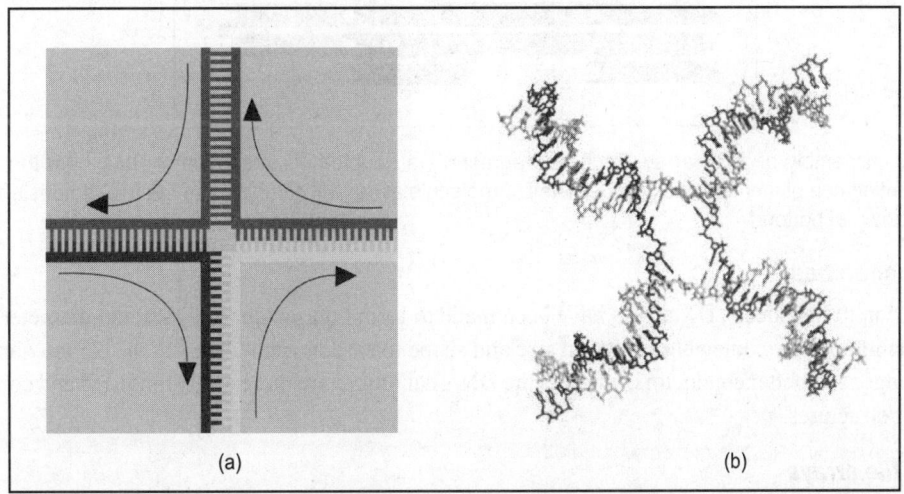

Fig. 7.2. Structure of the 4-arm junction: (a) a schematic, and (b) a more realistic model.

Junctions can be used in more complex molecules. The most important of these is the 'double-crossover' or DX motif. Here, two DNA duplexes lie next to each other, and share two junction points where strands cross from one duplex into the other.

This molecule has the advantage that the junction points are now constrained to a single orientation as opposed to being flexible as in the four-arm junction. This makes the DX motif suitable as a structural building block for larger DNA complexes.

Tile-based Arrays

DX arrays

DX, Double Crossover, molecules can be equipped with sticky ends in order to combine them into a two-dimensional periodic lattice. Each DX molecule has four termini, one at each end of the two double-helical domains, and these can be equipped with sticky ends that program them to combine into a specific pattern. More than one type of DX can be used which can be made to arrange in rows or any other tessellated pattern. They thus form extended flat sheets which are essentially two-dimensional crystals of DNA (Fig. 7.3).

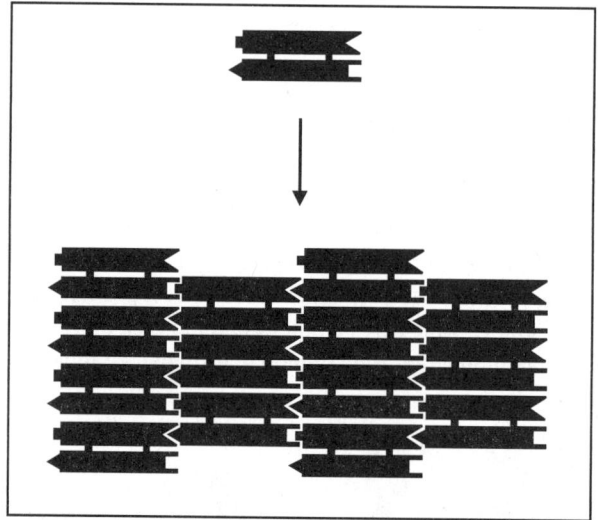

Fig. 7.3. Assembly of a DX array. Each bar represents a double-helical domain of DNA, with the shapes representing complimentary sticky ends. The DX molecule at top will combine into the two-dimensional DNA array shown at bottom.

DNA nanotubes

In addition to flat sheets, DX arrays have been made to form hollow tubes of 4–20 nm diameter. These DNA nanotubes are somewhat similar in size and shape to carbon nanotubes, but the carbon nanotubes are stronger and better conductors, whereas the DNA nanotubes are more easily modified and connected to other structures.

Other tile arrays

Two-dimensional arrays have been made out of other motifs as well, including the Holliday junction rhombus array as well as various DX-based arrays in the shapes of triangles and hexagons. Another motif, the six-helix bundle, has the ability to form three-dimensional DNA arrays as well.

DNA Origami

As an alternative to the tile-based approach, two-dimensional DNA structures can be made from a single, long DNA strand of arbitrary sequence which is folded into the desired shape by using shorter, 'staple' strands. This allows the creation of two-dimensional shapes at the nanoscale using DNA.

DNA Polyhedra

A number of three-dimensional DNA molecules have been made which have the connectivity of a polyhedron such as an octahedron or cube. In other words, the DNA duplexes trace the edges of a polyhedron with a DNA junction at each vertex. The earliest demonstrations of DNA polyhedra involved multiple ligations and solid-phase synthesis steps to create catenated polyhedra. More recently, there have been demonstrations of a DNA truncated octahedron made from a long single strand designed to fold into the correct conformation, as well as a tetrahedron which can be produced from four DNA strands in a single step.

DNA Nanomechanical Devices

DNA complexes have been made which change their conformation upon some stimulus. These are intended to have applications in nanorobotics. One of the first such devices, called 'molecular tweezers', changes from an open to a closed state based upon the presence of control strands.

DNA machines have also been made which show a twisting motion. One of these makes use of the transition between the B-DNA and Z-DNA forms to respond to a change in buffer conditions. Another relies on the presence of control strands to switch from a paranemic-crossover (PX) conformation to a double-junction (JX2) conformation.

Stem Loop Controllers

A design called a stem loop, consisting of a single strand of DNA which has a loop at an end, are a dynamic structure that opens and closes when a piece of DNA bonds to the loop part. This effect has been exploited to create several logic gates. These logic gates have been used to create the computers MAYA I and MAYA II which can play tick-tac-toe to some extent.

Applications

Algorithmic self-assembly

DNA nanotechnology has been applied to the related field of DNA computing. A DX array has been demonstrated whose assembly encodes an XOR operation, which allows the DNA array to implement a cellular automaton which generates a fractal called the Sierpinski gasket. This shows that computation can be incorporated into the assembly of DNA arrays, increasing its scope beyond simple periodic arrays (Fig. 7.4).

Note that DNA computing overlaps with, but is distinct from, DNA nanotechnology. The latter uses the specificity of Watson-Crick basepairing to make novel structures out of DNA. These structures can be used for DNA computing, but they do not have to be. Additionally, DNA computing can be done without using the types of molecules made possible by DNA nanotechnology (Fig. 7.5).

Nanoarchitecture

The idea of using DNA arrays to template the assembly of other functional molecules has been around for a while, but only recently has progress been made in reducing these kinds of schemes to practice. In 2006, researchers covalently attached gold nanoparticles to a DX-based tile and showed that self-assembly of the DNA structures also assembled the nanoparticles hosted on them. A non-covalent hosting scheme was shown in 2007, using Dervan polyamides on a DX array to arrange streptavidin proteins on specific kinds of tiles on the DNA array. Previously in 2006 LaBean demonstrated the letters 'D', 'N' and 'A' created on a 4×4 DX array using streptavidin. DNA has also been used to assemble a single walled carbon nanotube Field-effect transistor.

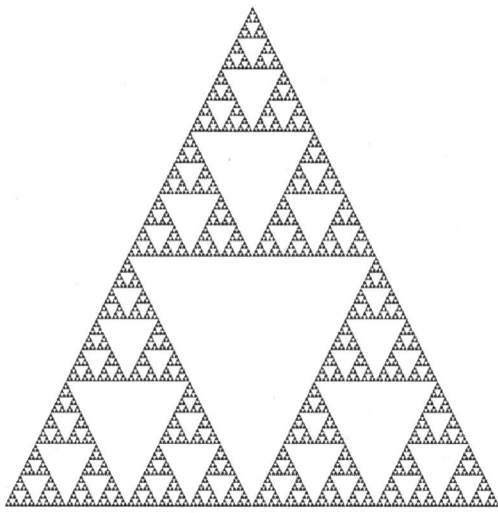

Fig. 7.4. The Sierpinski gasket.

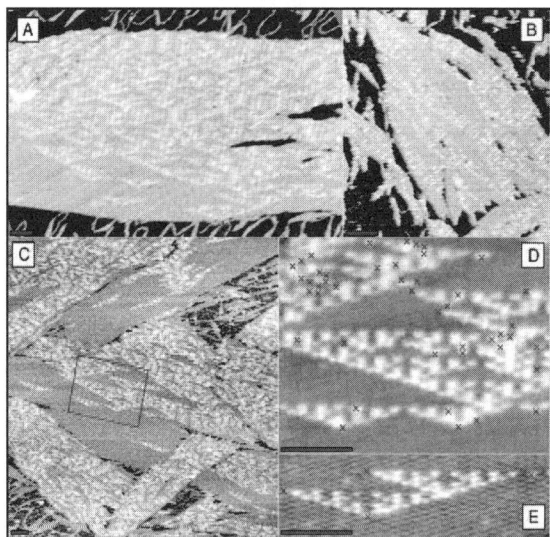

Fig. 7.5. DNA arrays that display a representation of the Sierpinski gasket on their surfaces..

GEOMETRIC PHASE

In mechanics (including classical mechanics as well as quantum mechanics), the Geometric phase, or the Pancharatnam-Berry phase (named after S. Pancharatnam and Sir Michael Berry), also known as the Pancharatnam phase or Berry phase, is a phase acquired over the course of a cycle, when the system is subjected to cyclic adiabatic processes, resulting from the geometrical properties of the parameter space of the Hamiltonian. The phenomenon was first discovered in 1956, and rediscovered in 1984. It can be seen in the Aharonov-Bohm effect and in the conical intersection of potential energy surfaces. In the case of the Aharonov-Bohm effect, the adiabatic parameter is the magnetic field inside the solenoid, and cyclic means that the difference involved in measuring the effect by interference corresponds to a closed loop, in the usual way. In the case of the conical intersection, the adiabatic parameters are the molecular coordinates. Apart from quantum mechanics, it arises in a variety of other wave systems, such as classical optics. As a rule of thumb, it occurs whenever there are at least two parameters affecting a wave, in the vicinity of some sort of singularity or some sort of hole in the topology.

Waves are characterised by amplitude and phase, and both may vary as a function of those parameters. The Berry phase occurs when both parameters are changed simultaneously but very slowly (adiabatically), and eventually brought back to the initial configuration. In quantum mechanics, this could, e.g. involve rotations but also translations of particles, which are apparently undone at the end. Intuitively one expects that the waves in the system return to the initial state, as characterised by the amplitudes and phases (and accounting for the passage of time). However, if the parameter excursions correspond to a cyclic loop instead of a self-retracing back-and-forth variation, then it is possible that the initial and final states differ in their phases. This phase difference is the Berry phase, and its occurrence typically indicates that the system's parameter dependence is singular (undefined) for some combination of parameters.

To measure the Berry phase in a wave system, an interference experiment is required. The Foucault pendulum is an example from classical mechanics that is sometimes used to illustrate the Berry phase. This mechanics analogue of the Berry phase is known as the Hannay angle.

Examples of Geometric Phases

Foucault pendulum

One of the easiest examples is the Foucault pendulum. An easy explanation in terms of geometric phases is given by Frank Wilczek:

> How does the pendulum process when it is taken around a general path C? For transport along the equator, the pendulum will not precess. Now if C is made up of geodesic segments, the precession will all come from the angles where the segments of the geodesics meet; the total precession is equal to the net deficit angle which in turn equals the solid angle enclosed by C modulo 2π. Finally, we can approximate any loop by a sequence of geodesic segments, so the most general result (on or off the surface of the sphere) is that the net precession is equal to the enclosed solid angle.

In summary, there are no inertial forces that could make the pendulum precess. Thus the orientation of the pendulum undergoes parallel transport along the path of fixed latitude. By the Gauss-Bonnet theorem the phase shift is given by the enclosed solid angle.

Polarised light in an optical fibre

Imagine linearly polarised light entering a single-mode optical fibre. Suppose the fibre traces out some path in space and the light exits the fibre in the same direction as it entered. Then compare the initial and final polarisations. In semiclassical approximation the fibre functions like a waveguide and the momentum of the light is at all times tangent to the fibre. The polarisation can be thought of as an orientation perpendicular to the momentum. As the fibre traces out its path, the momentum vector of the light traces out a path on the sphere in momentum space. The path is closed since initial and final directions of the light coincide, and the polarisation is a vector tangent to the sphere. Going to momentum space is equivalent to taking the Gauss map. There are no forces that could make the polarisation turn, just the constraint to remain tangent to the sphere. Thus the polarisation undergoes parallel transport and the phase shift is given by the enclosed solid angle (times the spin, which in case of light is 1).

Stochastic pump effect

A stochastic pump is a classical stochastic system that responds with nonzero, on average, currents to periodic changes of parameters. The stochastic pump effect can be interpreted in terms of a geometric phase in evolution of the moment generating function of stochastic currents.

Chapter 8

Micro- and Nano-electromechanical System

INTRODUCTION

Micro-electromechanical systems (MEMS) (also written as micro-electro-mechanical, or micro-electro mechanical) is the technology of the very small, and merges at the nano-scale into nanoelectromechanical systems (NEMS) and nanotechnology. MEMS are also referred to as micromachines (in Japan), or micro systems technology—MST (in Europe). MEMS are separate and distinct from the hypothetical vision of molecular nanotechnology or molecular electronics. MEMS are made up of components between 1 to 100 micrometres in size (i.e. 0.001 to 0.1 mm) and MEMS devices generally range in size from 20 micrometres (20 millionths of a metre) to a millimetre. They usually consist of a central unit that processes data, the microprocessor and several components that interact with the outside such as microsensors. At these size scales, the standard constructs of classical physics are not always useful. Due to MEMS' large surface area to volume ratio, surface effects such as electrostatics and wetting dominate volume effects such as inertia or thermal mass.

The potential of very small machines was appreciated long before the technology existed that could make them. MEMS became practical once they could be fabricated using modified semiconductor device fabrication technologies, normally used to make electronics. These include moulding and plating, wet etching (KOH, TMAH) and dry etching (RIE and DRIE), electro discharge machining (EDM), and other technologies capable of manufacturing very small devices.

MEMS technology can be implemented using a number of different materials and manufacturing techniques; the choice of which will depend on the device being created and the market sector in which it has to operate.

MATERIALS FOR MEMS MANUFACTURING

Silicon

Silicon is the material used to create most integrated circuits used in consumer electronics in the modern world. The economies of scale, ready availability of cheap high-quality materials and ability to incorporate electronic functionality make silicon attractive for a wide variety of MEMS applications. Silicon also has significant advantages engendered through its material properties. In single crystal form, silicon is an almost perfect Hookean material, meaning that when it is flexed there is virtually no hysteresis and hence almost no energy dissipation. As well as making for highly repeatable motion, this also makes silicon very reliable as it suffers very little fatigue and can have service lifetimes in the range of billions to trillions of cycles without breaking. The basic techniques for producing all silicon based MEMS devices are deposition of material layers, patterning of these layers by photolithography and then etching to produce the required shapes.

Polymers

Even though the electronics industry provides an economy of scale for the silicon industry, crystalline silicon is still a complex and relatively expensive material to produce. Polymers on the other hand can be produced in huge volumes, with a great variety of material characteristics. MEMS devices can be made from polymers by processes such as injection moulding, embossing or stereolithography and are especially well suited to microfluidic applications such as disposable blood testing cartridges.

Metals

Metals can also be used to create MEMS elements. While metals do not have some of the advantages displayed by silicon in terms of mechanical properties, when used within their limitations, metals can exhibit very high degrees of reliability.

Metals can be deposited by electroplating, evaporation, and sputtering processes. Commonly used metals include gold, nickel, aluminium, chromium, titanium, tungsten, platinum, and silver.

MEMS BASIC PROCESSES

Figure 8.1 gives brief idea of MEMS basic processes.

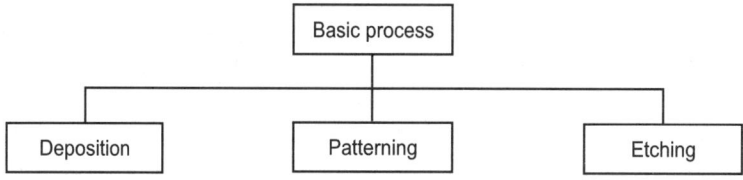

Fig. 8.1. MEMS basic processes.

Deposition Processes

One of the basic building blocks in MEMS processing is the ability to deposit thin films of material with a thickness anywhere between a few nanometers to about 100 micrometres.

Physical Deposition

There is a type of physical deposition.

Physical vapour deposition (PVD):

1. Sputtering.
2. Evaporation.

Chemical Deposition

There are two types of chemical deposition:

1. Chemical vapour deposition.
2. LPCVD: Low pressure CVD PECVD: Plasma enhanced CVD.
3. Thermal oxidation.

Patterning

Patterning in MEMS is the transfer of a pattern into a material.

Lithography

Lithography in MEMS context is typically the transfer of a pattern into a photosensitive material by selective exposure to a radiation source such as light. A photosensitive material is a material that experiences a change in its physical properties when exposed to a radiation source. If a photosensitive material is selectively exposed to radiation (e.g. by masking some of the radiation) the pattern of the radiation on the material is transferred to the material exposed, as the properties of the exposed and unexposed regions differs.

This exposed region can then be removed or treated providing a mask for the underlying substrate. Photolithography is typically used with metal or other thin film deposition, wet and dry etching.

1. Photolithography.
2. Electron beam lithography.
3. Ion beam lithography.
4. X-ray lithography.

Diamond Patterning

Etching processes

There are two basic categories of etching processes: wet and dry etching. In the former, the material is dissolved when immersed in a chemical solution. In the latter, the material is sputtered or dissolved using reactive ions or a vapour phase etchant.

Wet etching

Wet chemical etching consists in a selective removal of material by dipping a substrate into a solution that can dissolve it. Due to the chemical nature of this etching process, a good selectivity can often be obtained, which means that the etching rate of the target material is considerably higher than that of the mask material if selected carefully.

Etchant etching

Isotropic etching

Etching progresses at the same speed in all directions. Long and narrow holes in a mask will produce V-shaped grooves in the silicon. The surface of these grooves can be atomically smooth if the etch is carried out correctly, with dimensions and angles being extremely accurate.

Anisotropic etching

Some single crystal materials, such as silicon, will have different etching rates depending on the crystallographic orientation of the substrate. This is known as anisotropic etching and one of the most common examples is the etching of silicon in KOH (potassium hydroxide), where Si <111> planes etch approximately 100 times slower than other planes (crystallographic orientations). Therefore, etching a rectangular hole in a (100)-Si wafer will result in a pyramid shaped etch pit with 54.7° walls, instead of a hole with curved sidewalls as it would be the case for isotropic etching.

Electrochemical etching

Electrochemical etching (ECE) for dopant-selective removal of silicon is a common method to automate and to selectively control etching. An active *p-n* diode junction is required, and either type of dopant can be the etch-resistant ('etch-stop') material. Boron is the most common etch-stop dopant. In combination with wet anisotropic etching as described above, ECE has been used successfully for controlling silicon diaphragm thickness in commercial piezoresistive silicon pressure sensors. Selectively doped regions can be created either by implantation, diffusion, or epitaxial deposition of silicon.

Dry etching

Vapour etching

Xenon difluoride etching

Xenon difluoride (XeF_2) is a dry vapour phase isotropic etch for silicon originally applied for MEMS in 1995 at University of California, Los Angeles. Primarily used for releasing metal and dielectric structures by undercutting silicon, XeF_2 has the advantage of a stiction-free release unlike wet etchants. Its etch selectivity to silicon is very high, allowing it to work with photoresist, SiO_2, silicon nitride, and various metals for masking. Its reaction to silicon is 'plasmaless', is purely chemical and spontaneous and is often operated in pulsed mode. Models of the etching action are available, and university laboratories and various commercial tools offer solutions using this approach.

HF etching

Hydrofluoric acid is commonly used as an aqueous etchant for silicon dioxide (SiO_2, aka BOX for SOI). HF are usually in 49 per cent concentrated form, 5:1, 10:1 or 20:1 BOE (Buffered Oxide Etchant) or BHF (Buffered HF). They were first used in medieval times for glass etching. It was used in IC fabrication for patterning the gate oxide until the process step was replaced by RIE.

HF is considered one of the more dangerous acids in the cleanroom. It penetrates the skin upon contact and it diffuses straight to the bone. Therefore, the damage will not be felt until it is too late.

Plasma etching

Sputtering

Reactive ion etching (RIE)

In reactive ion etching (RIE), the substrate is placed inside a reactor in which several gases are introduced. A plasma is struck in the gas mixture using an RF power source, breaking the gas molecules into ions. The ions are accelerated towards, and react with, the surface of the material being etched, forming another gaseous material. This is known as the chemical part of reactive ion etching. There is also a physical part which is similar in nature to the sputtering deposition process. If the ions have high enough energy, they can knock atoms out of the material to be etched without a chemical reaction. It is a very complex task to develop dry etch processes that balance chemical and physical etching, since there are many parameters to adjust. By changing the balance it is possible to influence the anisotropy of the etching, since the chemical part is isotropic and the physical part highly anisotropic the combination can form sidewalls that have shapes from rounded to vertical. RIE can be deep and its name will be Deep RIE or DRIE Deep reactive ion etching (DRIE).

A special subclass of RIE which continues to grow rapidly in popularity is deep RIE (DRIE). In this process, etch depths of hundreds of micrometres can be achieved with almost vertical sidewalls. The primary technology is based on the so-called 'Bosch process', named after the German company Robert Bosch which filed the original patent, where two different gas compositions are alternated in the reactor. Currently there are two variations of the DRIE. The first variation consists of three distinct steps (the Bosch Process as used in the UNAXIS tool) while the second variation only consists of two steps (ASE used in the STS tool). In the 1st variation, the etch cycle is as follows: (i) SF_6 isotropic etch; (ii) C_4F_8 passivation, and (iii) SF_6 anisoptropic etch for floor cleaning. In the 2nd variation, steps (i) and (iii) are combined.

Both variations operate similarly. The C_4F_8 creates a polymer on the surface of the substrate, and the second gas composition (SF_6 and O_2) etches the substrate. The polymer is immediately sputtered away by the physical part of the etching, but only on the horizontal surfaces and not the sidewalls. Since the polymer only dissolves very slowly in the chemical part of the etching, it builds up on the sidewalls and protects them from etching. As a result, etching aspect ratios of 50 to 1 can be achieved. The process can easily be used to etch completely through a silicon substrate, and etch rates are 3–6 times higher than wet etching.

MEMS MANUFACTURING TECHNOLOGIES

Bulk Micromachining

Bulk micromachining is the oldest paradigm of silicon based MEMS. The whole thickness of a silicon wafer is used for building the micro-mechanical structures. Silicon is machined using various etching processes. Anodic bonding of glass plates or additional silicon wafers is used for adding features in the third dimension and for hermetic encapsulation. Bulk micromachining has been essential in enabling high performance pressure sensors and accelerometers that have changed the shape of the sensor industry in the 80's and 90's.

Surface Micromachining

Surface micromachining uses layers deposited on the surface of a substrate as the structural materials, rather than using the substrate itself. Surface micromachining was created in the late 1980s to render micromachining of silicon more compatible with planar integrated circuit technology, with the goal of combining MEMS and integrated circuits on the same silicon wafer. The original surface micromachining concept was based on thin polycrystalline silicon layers patterned as movable mechanical structures and released by sacrificial etching of the underlying oxide layer. Interdigital comb electrodes were used to produce in-plane forces and to detect in-plane movement capacitively. This MEMS paradigm has enabled the manufacturing of low cost accelerometers for e.g. automotive air-bag systems and other applications where low performance and/or high g-ranges are sufficient. Analog devices have pioneered the industrialisation of surface micromachining and have realised the co-integration of MEMS and integrated circuits.

High Aspect Ratio (HAR) Silicon Micromachining

Both bulk and surface silicon micromachining are used in the industrial production of sensors, ink-jet nozzles, and other devices. But in many cases the distinction between these two has diminished. A new etching technology, deep reactive-ion etching, has made it possible to combine good performance typical of bulk micromachining with comb structures and in-plane operation typical of surface micromachining. While it is common in surface micromachining to have structural layer thickness in the range of 2 μm, in HAR silicon micromachining the thickness can be from 10 to 100 μm. The materials commonly used in HAR silicon micromachining are thick polycrystalline silicon, known as epi-poly, and bonded silicon-on-insulator (SOI) wafers although processes for bulk silicon wafer also have been created (SCREAM). Bonding a second wafer by glass frit bonding, anodic bonding or alloy bonding is used to protect the MEMS structures. Integrated circuits are typically not combined with HAR silicon micromachining. The consensus of the industry at the moment seems to be that the flexibility and reduced process complexity obtained by having the two functions separated far outweighs the small penalty in packaging.

A forgotten history regarding surface micromachining revolved around the choice of polysilicon. Coarse grain polysilicon was advocated by Prof Henry Guckel (U. Wisconsin) while fine grain polysilicon was advocated by the UC Berkeley group.

Applications

In one viewpoint MEMS application is categorised by type of use: (i) in sensor, (ii) in actuator, and (iii) in structure.

In another viewpoint MEMS applications are categorised by the field of application (Commercial applications include):

1. Inkjet printers, which use piezoelectrics or thermal bubble ejection to deposit ink on paper.
2. Accelerometers in modern cars for a large number of purposes including airbag deployment in collisions.
3. Accelerometers in consumer electronics devices such as game controllers (Nintendo Wii), personal media players/cell phones (Apple iPhone, various Nokia mobile phone models, various HTC PDA models) and a number of Digital Cameras (various Canon Digital IXUS models). Also used in PCs to park the hard disk head when free-fall is detected, to prevent damage and data loss.
4. MEMS gyroscopes used in modern cars and other applications to detect yaw; e.g. to deploy a roll over bar or trigger dynamic stability control.
5. Silicon pressure sensors e.g. cartyre pressure sensors, and disposable blood pressure sensors.
6. Displays e.g. the DMD chip in a projector based on DLP technology has on its surface several hundred thousand micromirrors.
7. Optical switching technology which is used for switching technology and alignment for data communications.
8. Bio-MEMS applications in medical and health related technologies from Lab-On-Chip to MicroTotalAnalysis (biosensor, chemosensor).
9. Interferometric modulator display (IMOD) applications in consumer electronics (primarily displays for mobile devices). Used to create interferometric modulation — reflective display technology as found in mirasol displays.

Companies with strong MEMS programs come in many sizes. The larger firms specialise in manufacturing high volume inexpensive components or packaged solutions for end markets such as automobiles, biomedical, and electronics. The successful small firms provide value in innovative solutions and absorb the expense of custom fabrication with high sales margins. In addition, both large and small companies work in R & D to explore MEMS technology.

Research and Development

Researchers in MEMS use various engineering software tools to take a design from concept to simulation, prototyping and testing. Finite element analysis is often used in MEMS design. Simulation of dynamics, heat, and electrical domains, among others, can be performed by ANSYS, COMSOL and CoventorWare-ANALYZER. Other software, such as CoventorWare-ARCHITECT and MEMS-PRO, is used to produce a design layout suitable for delivery to a fabrication firm and even simulate the MEMS embedded in a system. Once prototypes are on-hand, researchers can test the specimens using various instruments, including laser doppler scanning vibrometers, microscopes, and stroboscopes.

Industry Structure

The global market for micro-electromechanical systems, which includes products such as automobile airbag systems, display systems and inkjet cartridges totalled $40 billion in 2006 according to Global MEMS/Microsystems Markets and Opportunities, a comprehensive new market research report from SEMI and Yole Developpement (http://www.yole.fr/). (htp://www.azonano.com/news.asp?newsID=4479). A 2009 report from 'The Information Network (http://www.theinformationnet.com/) points out that the market in 2008 was $6.9 billion'.

MEMS devices are defined as die-level components of first-level packaging, and include pressure sensors, accelerometers, gyroscopes, microphones, digital mirror displays, micro fluidic devices, etc. The materials and equipment used to manufacture MEMS devices topped $1 billion worldwide in 2006. Materials demand is driven by substrates, making up over 70 per cent of the market, packaging coatings and increasing use of chemical mechanical planarisation (CMP). While MEMS manufacturing continues to be dominated by used semiconductor equipment, there is a migration to 200 mm lines and select new tools, including etch and bonding for certain MEMS applications.

MEMS THERMAL ACTUATOR

A MEMS thermal actuator is a micromechanical device that typically generates motion by thermal expansion amplification. A small amount of thermal expansion of one part of the device translates to a large amount of deflection of the overall device. Usually fabricated out of doped Single Crystal Silicon or Polysilicon as a complex compliant member, the increase in temperature can be achieved internally by electrical resistive heating or externally by a heat source capable of locally introducing heat.

Some of the important thermal actuator are: (i) asymmetric (bimorph), (ii) symmetric (bent beam, chevron), (iii) electrostatic—parallel plate or comb drive, (iv) magnetic, and (v) piezoelectric.

MICRO-OPTO-ELECTROMECHANICAL SYSTEMS

Micro-opto-electromechanical systems (MOEMS) are a special class of Micro-electro-mechanical systems (MEMS) which involves sensing or manipulating optical signals on a very small size scale using integrated mechanical and electrical systems. MOEMS includes a wide variety of devices including optical switch, optical cross-connect, tunable VCSEL, microbolometers amongst others. These devices are usually fabricated using standard micromachining technologies using materials like silicon, silicon dioxide, silicon nitride and gallium arsenide.

NANOELECTROMECHANICAL SYSTEMS

The term Nanoelectromechanical systems or NEMS is used to describe devices integrating electrical and mechanical functionality on the nanoscale. NEMS form the logical next miniaturisation step from so-called microelectromechanical systems, or MEMS devices. NEMS typically integrate transistor-like nanoelectronics with mechanical actuators, pumps, or motors, and may thereby form physical, biological, and chemical sensors. The name derives from typical device dimensions in the nanometer range, leading to low mass, high mechanical resonance frequencies, potentially large quantum mechanical effects such as zero point motion, and a high surface to volume ratio useful for surface-based sensing mechanisms. Uses include accelerometers, or detectors of chemical substances in the air.

Because of the scale on which they can function, NEMS are expected to significantly impact many areas of technology and science and eventually replace MEMS. As noted by Richard Feynman's in his famous talk in the 60s, There's Plenty of Room at the Bottom, there are a lot of potential applications of machines at smaller and smaller sizes; by building and controlling devices at smaller scales, all technology benefits. Among the expected benefits include greater efficiencies and reduced size, decreased power consumption and lower costs of production in electromechanical systems.

In 2000, the first very large scale integration (VLSI) NEMS device was demonstrated by researchers from IBM. Its premise was an array of AFM tips which can heat/sense a deformable substrate in order to function as a memory device. In 2007, the International Technical Roadmap for Semiconductors (ITRS) contains NEMS Memory as a new entry for the Emerging Research Devices section.

Importance for AFM

A key application of NEMS is Atomic force microscope tips. The increased sensitivity achieved by NEMS leads to smaller and more efficient sensors to detect stresses, vibrations, forces at the atomic level, and chemical signals. AFM tips and other detection at the nanoscale rely heavily on NEMS. If implementation of better scanning devices becomes available, all of nanoscience could be better understood by AFM tips.

Approaches to Miniaturisation

Two complementary approaches to fabrication of NEMS systems can be found. The top-down approach uses the traditional microfabrication methods, i.e. optical and electron beam lithography, to manufacture devices. While being limited by the resolution of these methods, it allows a large degree of control over the resulting structures. Typically, devices are fabricated from metallic thin films or etched semiconductor layers.

Bottom-up approaches, in contrast, use the chemical properties of single molecules to cause single-molecule components to (i) self-organise or self-assemble into some useful conformation, and (ii) rely on positional assembly. These approaches utilise the concepts of molecular self-assembly and/or molecular recognition. This allows fabrication of much smaller structures, albeit often at the cost of limited control of the fabrication process.

A combination of these approaches may also be used, in which nanoscale molecules are integrated into a top-down framework. One such example is the carbon nanotube nanomotor.

Materials

Carbon allotropes

Many of the commonly used materials for NEMS technology have been carbon based, specifically carbon nanotubes and graphene. This is mainly because of the useful properties of carbon based materials which directly meet the needs of NEMS. The mechanical properties of carbon (such as large Young's modulus) are fundamental to the stability of NEMS while the metallic and semiconductor conductivities of carbon based materials allow them to function as transistors.

Both graphene and carbon exhibit high Young's modulus, excessively low density, low friction and large surface area. The low friction of CNTs, allow practically frictionless bearings and has thus been a huge motivation towards practical applications of CNTs as constitutive elements in NEMS, such as nanomotors, switches, and high-frequency oscillators. Carbon nanotubes and graphene's physical strength allows carbon based materials to meet higher stress demands, when common materials would normally fail and thus further support their use as a major materials in NEMS technological development.

Along with the mechanical benefits of carbon based materials, the electrical properties of carbon nanotubes and graphene allow it to be used in many electrical components of NEMS. Nanotransistors have been developed for both carbon nanotubes as well as graphene. Transistors are one of the basic building blocks for all electronic devices, so by effectively developing usable transistors, carbon nanotubes and graphene are both very crucial to NEMS. Metallic carbon nanotubes have also been proposed for nanoelectronic interconnects since they can carry high current densities. This is a very useful property as wires to transfer current are another basic building block of any electrical system. Carbon nanotubes have specifically found so much use in NEMS that methods have already been discovered to connect suspended carbon nanotubes to other nanostructures. This allows carbon nanotubes to be structurally set up to make complicated nanoelectric systems. Because carbon based products can be properly controlled and act as interconnects as well as transistors, they serve as a fundamental material in the electrical components of NEMS.

Difficulties

Despite all of the useful properties of carbon nanotubes and graphene for NEMS technology, both of these products face several hindrances to their implementation. One of the main problems is carbon's response to real life environments. Carbon nanotubes exhibit a large change in electronic properties when it is exposed to oxygen. Similarly, other changes to the electronic and mechanical attributes of carbon based materials must fully be explored before their implementation, especially because of their high surface area which can easily react with surrounding environments. Carbon nanotubes were also found to have varying conductivities, being either metallic or semiconducting depending on their helicity when processed. Because of this, very special treatment must be given to the nanotubes during processing, in order to assure that all of the nanotubes have appropriate conductivities. Graphene also has very complicated electric conductivity properties compared to traditional semiconductors as it lacks a energy band gap and essentially changes all the rules for how electrons move through a graphene based device. This means that traditional constructions of electronic devices will likely not work and completely new architectures must be designed for these new electronic devices.

Future of NEMS

Before NEMS devices can actually be implemented, reasonable integrations of carbon based products must be created. The focus is currently shifting from experimental work towards practical applications and device structures that will implement and profit from the use of carbon nanotubes. At this point in NEMS research, there is a general understanding of the properties of carbon nanotubes and graphene. The next challenge to overcome involves understanding all of the properties of these carbon based tools, and using the properties to make efficient and durable NEMS.

NEMS devices, if implemented into everyday technologies, could further reduce the size of modern devices and allow for better performing sensors. Carbon based materials have served as prime materials for NEMS use, because of their highlighted mechanical and electrical properties. Once NEMS interactions with outside environments are integrated with effective designs, they will likely become useful products to everyday technologies.

Chapter 9

Supramolecular Chemistry

INTRODUCTION

Supramolecular chemistry refers to the area of chemistry beyond the molecules and focuses on the chemical systems made up of a discrete number of assembled molecular subunits or components (Fig. 9.1). The forces responsible for the spatial organisation may vary from weak (intermolecular forces, electrostatic or hydrogen bonding) to strong (covalent bonding), provided that the degree of electronic coupling between the molecular component remains small with respect to relevant energy parameters of the component. While traditional chemistry focuses on the covalent bond, supramolecular chemistry examines the weaker and reversible noncovalent interactions between molecules. These forces include hydrogen bonding, metal coordination, hydrophobic forces, van der Waals forces, pi-pi interactions and electrostatic effects. Supramolecular complex of a chlorine ion, cucurbituril, and cucurbituril are shown in Fig. 9.2. Important concepts that have been demonstrated by supramolecular chemistry include molecular self-assembly, folding, molecular recognition, host-guest chemistry, mechanically-interlocked molecular architectures, and dynamic covalent chemistry. The study of non-covalent interactions is crucial to understanding many biological processes from cell structure to vision that rely on these forces for structure and function. Biological systems are often the inspiration for supramolecular research. An example of mechanically-interlocked molecular architecture in this case a rotaxane which is shown in Fig. 9.3.

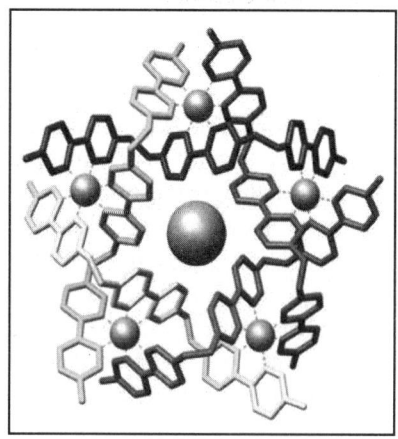

Fig. 9.1. An example of a supramolecular assembly.

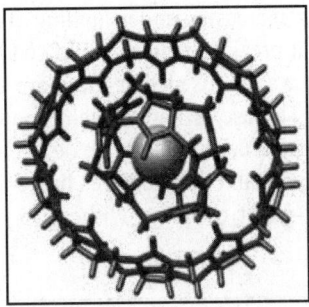

Fig. 9.2. Supramolecular complex of a chlorine ion, cucurbituril, and cucurbituril.

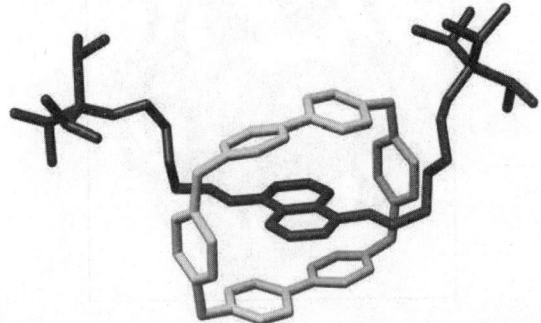

Fig. 9.3. An example of a mechanically-interlocked molecular architecture in this case a rotaxane.

The existence of intermolecular forces was first postulated by Johannes Diderik van der Waals in 1873. However, it is with Nobel laureate Hermann Emil Fischer that supramolecular chemistry has its philosophical roots. In 1890, Fischer suggested that enzyme-substrate interactions take the form of a 'lock and key', pre-empting the concepts of molecular recognition and host-guest chemistry. In the early twentieth century noncovalent bonds were understood in gradually more detail, with the hydrogen bond being described by Latimer and Rodebush in 1920. An example of a host-guest chemistry is shown in Fig. 9.4.

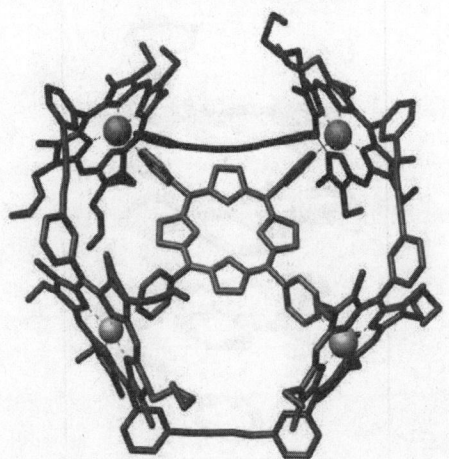

Fig. 9.4. An example of a host-guest chemistry.

The use of these principles led to an increasing understanding of protein structure and other biological processes. For instance, the important breakthrough that allowed the elucidation of the double helical structure of DNA occurred when it was realised that there are two separate strands of nucleotides connected through hydrogen bonds. The use of noncovalent bonds is essential to replication because they allow the strands to be separated and used to template new double stranded DNA. Concomitantly, chemists began to recognise and study synthetic structures based on noncovalent interactions, such as micelles and microemulsions. Host-guest complex with a *p*-xylylenedimmonium bound within a cucurbituril is shown in Fig. 9.5.

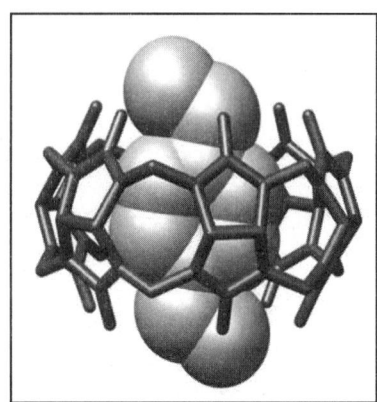

Fig. 9.5. Host-guest complex with a *p*-xylylenediammonium bound within a cucurbituril.

Eventually, chemists were able to take these concepts and apply them to synthetic systems. The breakthrough came in the 1960s with the synthesis of the crown ethers by Charles J. Pedersen. Following this work, other researchers such as Donald J. Cram, Jean-Marie Lehn and Fritz Vogtle became active in synthesising shape- and ion-selective receptors, and throughout the 1980s research in the area gathered a rapid pace with concepts such as mechanically-interlocked molecular architectures emerging. Intramolecular self-assembly of foldamer is shown in Fig. 9.6.

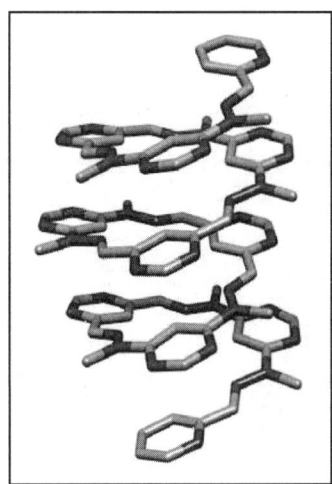

Fig. 9.6. Intramolecular self-assembly of a foldamer.

The importance of supramolecular chemistry was established by the 1987 Nobel Prize for Chemistry which was awarded to Donald J. Cram, Jean-Marie Lehn, and Charles J. Pedersen in recognition of their work in this area. The development of selective 'host-guest' complexes in particular, in which a host molecule recognises and selectively binds a certain guest, was cited as an important contribution.

In the 1990s, supramolecular chemistry became even more sophisticated, with researchers such as James Fraser Stoddart developing molecular machinery and highly complex self-assembled structures, and Itamar Willner developing sensors and methods of electronic and biological interfacing. During this period, electrochemical and photochemical motifs became integrated into supramolecular systems in order to increase functionality, research into synthetic self-replicating system began, and work on molecular information processing devices began. The emerging science of nanotechnology also had a strong influence on the subject, with building blocks such as fullerenes, nanoparticles, and dendrimers becoming involved in synthetic systems.

CONTROL OF SUPRAMOLECULAR CHEMISTRY

Thermodynamics

Supramolecular chemistry deals with subtle interactions, and consequently control over the processes involved can require great precision. In particular, noncovalent bonds have low energies and often no activation energy for formation. As demonstrated by the Arrhenius equation, this means that, unlike in covalent bond-forming chemistry, the rate of bond formation is not increased at higher temperatures. In fact, chemical equilibrium equations show that the low bond energy results in a shift towards the breaking of supramolecular complexes at higher temperatures.

However, low temperatures can also be problematic to supramolecular processes. Supramolecular chemistry can require molecules to distort into thermodynamically disfavoured conformations (e.g. during the 'slipping' synthesis of rotaxanes), and may include some covalent chemistry that goes along with the supramolecular. In addition, the dynamic nature of supramolecular chemistry is utilised in many systems (e.g. molecular mechanics), and cooling the system would slow these processes.

Thus, thermodynamics is an important tool to design, control, and study supramolecular chemistry. Perhaps the most striking example is that of warm-blooded biological systems, which cease to operate entirely outside a very narrow temperature range.

Environment

The molecular environment around a supramolecular system is also of prime importance to its operation and stability. Many solvents have strong hydrogen bonding, electrostatic, and charge-transfer capabilities, and are therefore able to become involved in complex equilibria with the system, even breaking complexes completely. For this reason, the choice of solvent can be critical.

CONCEPTS IN SUPRAMOLECULAR CHEMISTRY

Molecular Self-assembly

Molecular self-assembly is the construction of systems without guidance or management from an outside source (other than to provide a suitable environment). The molecules are directed to assemble through noncovalent interactions. Self-assembly may be subdivided into intermolecular self-assembly (to form a supramolecular assembly), and intramolecular self-assembly (or folding as demonstrated by foldamers and polypeptides). Molecular self-assembly also allows the construction of larger structures such as micelles, membranes, vesicles, liquid crystals, and is important to crystal engineering.

Molecular Recognition and Complexation

Molecular recognition is the specific binding of a guest molecule to a complementary host molecule to form a host-guest complex. Often, the definition of which species is the 'host' and which is the 'guest' is arbitrary. The molecules are able to identify each other using noncovalent interactions. Key applications of this field are the construction of molecular sensors and catalysis.

Template-directed Synthesis

Molecular recognition and self-assembly may be used with reactive species in order to preorganise a system for a chemical reaction (to form one or more covalent bonds). It may be considered a special case of supramolecular catalysis. Noncovalent bonds between the reactants and a 'template' hold the reactive sites of the reactants close together, facilitating the desired chemistry. This technique is particularly useful for situations where the desired reaction conformation is thermodynamically or kinetically unlikely, such as in the preparation of large macrocycles. This pre-organisation also serves purposes such as minimising side reactions, lowering the activation energy of the reaction, and producing desired stereochemistry. After the reaction has taken place, the template may remain in place, be forcibly removed, or may be 'automatically' decomplexed on account of the different recognition properties of the reaction product. The template may be as simple as a single metal ion or may be extremely complex.

Mechanically-interlocked Molecular Architectures

Mechanically-interlocked molecular architectures consist of molecules that are linked only as a consequence of their topology. Some noncovalent interactions may exist between the different components (often those that were utilised in the construction of the system), but covalent bonds do not. Supramolecular chemistry, and template-directed synthesis in particular, is key to the efficient synthesis of the compounds. Examples of mechanically-interlocked molecular architectures include catenanes, rotaxanes, molecular knots, and molecular Borromean rings.

Dynamic Covalent Chemistry

In dynamic covalent chemistry covalent bonds are broken and formed in a reversible reaction under thermodynamic control. While covalent bonds are key to the process, the system is directed by noncovalent forces to form the lowest energy structures.

Biomimetics

Many synthetic supramolecular systems are designed to copy functions of biological systems. These biomimetic architectures can be used to learn about both the biological model and the synthetic implementation. Examples include photoelectrochemical systems, catalytic systems, protein design and self-replication.

Imprinting

Molecular imprinting describes a process by which a host is constructed from small molecules using a suitable molecular species as a template. After construction, the template is removed leaving only the host. The template for host construction may be subtly different from the guest that the finished host bind. In its simplest form, imprinting utilises only steric interactions, but more complex systems also incorporate hydrogen bonding and other interactions to improve binding strength and specificity.

Molecular Machinery

Molecular machines are molecules or molecular assemblies that can perform functions such as linear or rotational movement, switching, and entrapment. These devices exist at the boundary between supramolecular chemistry and nanotechnology, and prototypes have been demonstrated using supramolecular concepts.

BUILDING BLOCKS OF SUPRAMOLECULAR CHEMISTRY

Supramolecular systems are rarely designed from first principles. Rather, chemists have a range of well-studied structural and functional building blocks that they are able to use to build up larger functional architectures. Many of these exist as whole families of similar units, from which the analog with the exact desired properties can be chosen.

Synthetic Recognition Motifs

1. The pi-pi charge-transfer interactions of bipyridinium with dioxyarenes or diaminoarenes have been used extensively for the construction of mechanically interlocked systems and in crystal engineering.
2. The use of crown ether binding with metal or ammonium cations is ubiquitous in supramolecular chemistry.
3. The formation of carboxylic acid dimers and other simple hydrogen bonding interactions.
4. The complexation of bipyridines or tripyridines with ruthenium, silver or other metal ions is of great utility in the construction of complex architectures of many individual molecules.
5. The complexation of porphyrins or phthalocyanines around metal ions gives access to catalytic, photochemical and electrochemical properties as well as complexation. These units are used a great deal by nature.

Macrocycles

Macrocycles are very useful in supramolecular chemistry, as they provide whole cavities that can completely surround guest molecules and may be chemically modified to fine-tune their properties.

1. Cyclodextrins, calixarenes, cucurbiturils and crown ethers are readily synthesised in large quantities, and are therefore convenient for use in supramolecular systems.
2. More complex cyclophanes, and cryptands can be synthesised to provide more taliored recognition properties.

Structural Units

Many supramolecular systems require their components to have suitable spacing and conformations relative to each other, and therefore easily-employed structural units are required.

1. Commonly used spacers and connecting groups include polyether chains, biphenyls and triphenyls, and simple alkyl chains. The chemistry for creating and connecting these units is very well understood.
2. Nanoparticles, nanorods, fullerenes and dendrimers offer nanometer-sized structure and encapsulation units.

3. Surfaces can be used as scaffolds for the construction of complex systems and also for interfacing electrochemical systems with electrodes. Regular surfaces can be used for the construction of self-assembled monolayers and multilayers.

Photo-/Electro-chemically Active Units

1. Porphyrins, and phthalocyanines have highly tunable photochemical and electrochemical activity as well as the potential for forming complexes.
2. Photochromic and photoisomerisable groups have the ability to change their shapes and properties (including binding properties) upon exposure to light.
3. TTF and quinones have more than one stable oxidation state, and therefore can be switched with redox chemistry or electrochemistry. Other units such as benzidine derivatives, viologens groups and fullerenes, have also been utilised in supramolecular electrochemical devices.

Biologically-derived Units

1. The extremely strong complexation between avidin and biotin is instrumental in blood clotting, and has been used as the recognition motif to construct synthetic systems.
2. The binding of enzymes with their cofactors has been used as a route to produce modified enzymes, electrically contacted enzymes, and even photoswitchable enzymes.
3. DNA has been used both as a structural and as a functional unit in synthetic supramolecular systems.

APPLICATIONS

Material Technology

Supramolecular chemistry and molecular self-assembly processes in particular have been applied to the development of new materials. Large structures can be readily accessed using bottom-up synthesis as they are composed of small molecules requiring fewer steps to synthesise. Thus most of the bottom-up approaches to nanotechnology are based on supramolecular chemistry.

Catalysis

A major application of supramolecular chemistry is the design and understanding of catalysts and catalysis. Noncovalent interactions are extremely important in catalysis, binding reactants into conformations suitable for reaction and lowering the transition state energy of reaction. Template-directed synthesis is a special case of supramolecular catalysis. Encapsulation systems such as micelles and dendrimers are also used in catalysis to create microenvironments suitable for reactions (or steps in reactions) to progress that is not possible to use on a macroscopic scale.

Medicine

Supramolecular chemistry is also important to the development of new pharmaceutical therapies by understanding the interactions at a drug binding site. The area of drug delivery has also made critical advances as a result of supramolecular chemistry providing encapsulation and targeted release mechanisms. In addition, supramolecular systems have been designed to disrupt protein-protein interactions that are important to cellular function.

Data Storage and Processing

Supramolecular chemistry has been used to demonstrate computation functions on a molecular scale. In many cases, photonic or chemical signals have been used in these components, but electrical interfacing of these units has also been shown by supramolecular signal transduction devices. Data storage has been accomplished by the use of molecular switches with photochromic and photoisomerisable units, by electrochromic and redox-switchable units, and even by molecular motion. Synthetic molecular logic gates have been demonstrated on a conceptual level. Even full-scale computations have been achieved by semi-synthetic DNA computers.

Green Chemistry

Research in supramolecular chemistry also has application in green chemistry where reactions have been developed which proceed in the solid state directed by non-covalent bonding. Such procedures are highly desirable since they reduce the need for solvents during the production of chemicals.

Other Devices and Functions

Supramolecular chemistry is often pursued to develop new functions that cannot appear from a single molecule. These functions also include magnetic properties, light responsiveness, self-healing polymers, molecular sensors, etc. Supramolecular research has been applied to develop high-tech sensors, processes to treat radioactive waste, and contrast agents for CAT scans.

CATEGORIES AND SUB-CATEGORIES OF SUPRAMOLECULAR CHEMISTRY

Some of the categories and sub-categories of supramolecular chemistry are discussed below.

Calixarene

A calixarene is a macrocycle or cyclic oligomer based on a hydroxyalkylation product of a phenol and an aldehyde. The word calixarene is derived from calix or chalice because this type of molecule resembles a vase and from the word arene that refers to the aromatic building block. Calixarenes have hydrophobic cavities that can hold smaller molecules or ions and belong to the class of cavitands known in Host-guest chemistry. Calixarene nomenclature is straightforward and involves counting the number of repeating units in the ring and include it in the name. A calix[4]arene has 4 units in the ring and a calix[6]arene has 6. A substituent in the meso position R_b is added to the name with a prefix C- as in C-methylcalix[6] arene.

Host guest interactions

Calixarenes are efficient sodium ionophores and are applied as such in chemical sensors. With the right chemistry these molecules exhibit great selectivity towards other cations. Calixarenes are used in commercial applications as sodium selective electrodes for the measurement of sodium levels in blood. Calixarenes also form complexes with cadmium, lead, lanthanides and actinides. Calix[5]arene and the C_{70} fullerene in *p*-xylene form a ball-and-socket supramolecular complex. Calixarenes also form exo-calix ammonium salts with aliphatic amines such as piperidine.

Molecular self-assembly

Molecular self-assembly of resorcinarenes and pyrogallolarenes lead to larger supramolecular assemblies. Both in the crystalline state and in solution, they are known to form hexamers that are akin to certain Archimedean solids with an internal volume of around one cubic nanometer (nanocapsules). (Isobutylpyrogallol[4]arene)$_6$ is held together by 48 intermolecular hydrogen bonds. The remaining 24 hydrogen bonds are intramolecular. The cavity is filled by a number of solvent molecules.

Applications

Calixarenes are applied in enzyme mimetics, ion sensitive electrodes or sensors, selective membranes, non-linear optics and in HPLC stationary phase. In addition, in nanotechnology calixarenes are used as negative resist for high-resolution electron beam lithography.

A tetrathia[4] arene is found to mimic aquaporin proteins. This calixarene adopts a 1,3-alternate conformation (methoxy groups populate the lower ring) and water is not contained in the basket but grabbed by two opposing tert-butyl groups on the outer rim in a pincer. The nonporous and hydrophobic crystals are soaked in water for 8 hours in which time the calixarene:water ratio nevertheless acquires the value of one.

Calixarenes are able to accelerate reactions taking place inside the concavity by a combination of local concentration effect and polar stabilisation of the transition state. An extended resorcin[4]arene cavitand is found to accelerate the reaction rate of a Menshutkin reaction between quinuclidine and butylbromide by a factor of 1600.

In heterocalixarenes the phenolic units are replaced by heterocycles, for instance by furans in calix[n]furanes and by pyridines in calix[n]pyridines. Calixarenes have been used as the macrocycle portion of a rotaxane and two calixarene molecules covalently joined together by the lower rims form carcerands.

Calix[4]arene was used as scaffold to assemble a construct bearing four Tn-antigen unit, at upper rim, and the immunoadjuvant P3CS, at lower rim. The construct showed a cluster effect in the production of Tn specific IgG antibodies in mice when compared to an analogous monovalent construct. This reveals perspectives for potential applications in cancer immunotherapy.

Carcerand

A carcerand is a host molecule that completely entraps its guest so that it will not escape even at high temperatures. This type of molecule was first described by Donald J. Cram in 1985 and is derived from the Latin *carcer*, or prison. The complexes formed by a carcerand with permanently imprisoned guests are called carceplexes.

In contrast hemicarcerands allow guests to enter and exit the cavity at high temperatures but will form stable complexes at ambient temperatures. The complexes formed by a hemicarcerand and a guests are called hemicarceplexes.

Reactivity of bound guests

Cram described the interior of the container compound as the inner phase in which radically different reactivity was observed. He used a hemicarcerand to isolate highly unstable, antiaromatic cylobutadiene at room temperature. The hemicarcerand stabilises guests within its cavity by preventing their reaction with other molecules.

Large carcerands

The internal cavity of a carcerand can be as large as 1700 Å^3 (1.7 nm^3) when six hemicarcerands form a single octahedral compound. This is accomplished by dynamic covalent chemistry in a one-pot condensation of 6 equivalents of a tetraformyl calixarene and 12 equivalents of ethylene diamine with trifluoroacetic acid catalyst in chloroform at room temperature followed by reduction of the imine bonds with sodium borohydride.

Catenane

A catenane is a mechanically-interlocked molecular architecture consisting of two or more interlocked macrocycles. The interlocked rings cannot be separated without breaking the covalent bonds of the macrocycles. Catenane is derived from the Latin *catena* meaning 'chain'. They are conceptually related to other mechanically-interlocked molecular architectures, such as rotaxanes, molecular knots or molecular Borromean rings. Recently the terminology 'mechanical bond' has been coined that describes the connection between the macrocycles of a catenane.

Properties and applications

A particularly interesting property of many catenanes is the ability of the rings to rotate through each other. This motion can often be detected and measured by NMR spectroscopy, among other methods. When molecular recognition motifs exist in the finished catenane (usually those that were used to synthesise the catenane), the catenane can have one or more thermodynamically preferred positions of the rings with respect to each other. In the case where one recognition site is a switchable moiety, a mechanical molecular switch results. When a catenane is synthesised by coordination of the macrocycles around a metal ion, then removal and re-insertion of the metal ion can switch the free motion of the rings on and off.

Families of catenanes

There are a number of distinct methods of holding the precursors together prior to the ultimate ring-closing reaction in a template-directed catenane synthesis. Each noncovalent approach to catenane formation results in what can be considered different families of catenanes.

Another family of catenanes are called pretzelanes or bridged catenanes after their likeness to pretzels with a spacer linking the two macrocycles.

In catenane nomenclature, a number in square brackets precedes the word 'catenane' in order to indicate how many rings are involved. Discrete catenanes up to a catenane have been synthesised and isolated.

Cavitand

A cavitand is a container shaped molecule. The cavity of the cavitand allows it to engage in host-guest chemistry with guest molecules of a complementary shape and size. Examples include cyclodextrins, calixarenes, and cucurbiturils.

Crown Ether

Crown ethers are heterocyclic chemical compounds that consist of a ring containing several ether groups. The most common crown ethers are oligomers of ethylene oxide, the repeating unit being ethyleneoxy, i.e. $-CH_2CH_2O-$. Important members of this series are the tetramer ($n = 4$), the pentamer ($n = 5$), and the hexamer ($n = 6$). The term 'crown' refers to the resemblance between the structure of a crown ether bound to a cation, and a crown sitting on a head. The first number in a crown ether's name refers to the number of atoms in the cycle, and the second number refers to the number of those atoms that are oxygen. Crown ethers are much broader than the oligomers of ethylene oxide; an important group are derived from catechol.

Crown ethers strongly bind certain cations, forming complexes. The oxygen atoms are well situated to coordinate with a cation located at the interior of the ring, whereas the exterior of the ring is hydrophobic. The resulting cations often form salts that are soluble in nonpolar solvents, and for this reason crown ethers are useful in phase transfer catalysis. The denticity of the polyether influences the affinity of the crown ether for various cations. For example, 18-crown-6 has high affinity for potassium cation, 15-crown-5 for sodium cation, and 12-crown-4 for lithium cation. The high affinity of 18-crown-6 for potassium ions contributes towards its toxicity.

Crown ethers are not the only macrocyclic ligands that have affinity for the potassium cation. Ionophores such as valinomycin also display a marked preference for the potassium cation over other cations.

Affinity for cations

Apart from its high affinity for potassium cations, 18-crown-6 can also bind to protonated amines and form very stable complexes in both solution and the gas phase. Some amino acids, such as lysine, contain a primary amine on their side chains. Those protonated amino groups can bind to the cavity of 18-crown-6 and form stable complexes in the gas phase. Hydrogen-bonds are formed between the three hydrogen atoms of protonated amines and three oxygen atoms of 18-crown-6. These hydrogen-bonds make the complex a stable adduct.

Aza-crowns

Some or all of the oxygen atoms in crown ethers can be replaced by nitrogens to form cryptands. A well-known tetrazacrown is cyclen in which there are no oxygens.

Cryptand

Cryptands are a family of synthetic bi- and polycyclic multidentate ligands for a variety of cations. The term cryptand implies that this ligand binds substrates in a crypt, interring the guest as in a burial. These molecules are three dimensional analogues of crown ethers but are more selective and complex the guest ions more strongly. The resulting complexes are lipophilic.

Properties

The three-dimensional interior cavity of a cryptand provides a binding site or nook for 'guest' ions. The complex between the cationic guest and the cryptand is called a cryptate. Cryptands form complexes with many 'hard cations' including NH_4^+, lanthanoids, alkali metals, and alkaline earth metals. In contrast to crown ethers, cryptands bind the guest ions using both nitrogen and oxygen donors. This three-dimensional encapsulation mode confers some size-selectivity, enabling discrimination among alkali metal cations (e.g. Na^+ vs. K^+).

Uses

Cryptands are more expensive and difficult to prepare, but offer much better selectivity and strength of binding than other complexants for alkali metals, such as crown ethers. They are able to bind otherwise insoluble salts into organic solvents. They can also be used as phase transfer catalysts by transferring ions from one phase to another. Cryptands enabled the synthesis of the alkalides and electrides. They have also been used in the crystallisation of Zintl ions such as Sn_9^{4-}.

Cryptophane

Cryptophanes are a class of organic supramolecular compounds studied and synthesised primarily for molecular encapsulation and recognition. One possible noteworthy application of cryptophanes is encapsulation and storage of hydrogen gas for potential use in fuel cell automobiles. Cryptophanes can also serve as containers in which organic chemists can carry out reactions that would otherwise be difficult to run under normal conditions. Due to their unique molecular recognition properties, cryptophanes also hold great promise as a potentially new way to study the binding of organic molecules with substrates, particularly as pertaining to biological and biochemical applications.

Crystal Engineering

Crystal engineering is the design and synthesis of molecular solid-state structures with desired properties, based on an understanding and exploitation of intermolecular interactions. The two main strategies currently in use for crystal engineering are based on hydrogen bonding and coordination complexation. These may be understood with key concepts such as the supramolecular synthon and the secondary building unit.

Crystal engineering relies on noncovalent bonding to achieve the organisation of molecules and ions in the solid state. Much of the initial work on purely organic systems focused on the use of hydrogen bonds, though with the more recent extension to inorganic systems, the coordination bond has also emerged as a powerful tool. Other intermolecular forces such as $\pi...\pi$, halogen...halogen, and Au...Au interactions have all been exploited in crystal engineering studies, and ionic interactions can also be important. However, the two most commonly used strategies in crystal engineering exploit hydrogen bonds and coordination bonds.

Molecular self-assembly is at the heart of crystal engineering, and it typically involves an interaction between complementary hydrogen-bonding faces or a metal and a ligand. By analogy with the retrosynthetic approach to organic synthesis, Desiraju coined the term 'supramolecular synthon' to describe building blocks that are common to many structures and hence can be used to order specific groups in the solid state. The carboxylic acid dimer represents a simple supramolecular synthon, though in practice this is only observed in approximately 30 per cent of crystal structures in which it is theoretically possible. The Cambridge structural database (CSD) provides an excellent tool for assessing the efficiency of particular synthons. The supramolecular synthon approach has been successfully applied in the synthesis of one-dimensional tapes, two-dimensional sheets and three-dimensional structures. The CSD today contains atomic positional parameters for nearly 3,00,000 crystal structures, and this forms the basis for heuristic or synthon-based or 'experimental' crystal engineering.

Polymorphism

Polymorphism is the phenomenon wherein the same chemical compound exists in different crystal forms. In the initial days of crystal engineering, polymorphism was not properly understood and incompletely studied. Today, it is one of the most exciting branches of the subject partly because polymorphic forms of drugs may be entitled to independent patent protection if they show new and improved properties over the known crystal forms. With the growing importance of generic drugs, the importance of crystal engineering to the pharmaceutical industry is expected to grow exponentially.

Cucurbituril

A cucurbituril is a macrocyclic molecule consisting of several glycoluril [$=C_4H_2N_4O_2=$] repeat units, each joined to the next one by two methylene [$-CH_2-$] bridges to form a closed band. The oxygen atoms are located along the edges of the band and are tilted inwards, forming a partly enclosed cavity. The name is derived from the resemblance of this molecule with a pumpkin of the family of Cucurbitaceae.

Cucurbiturils are commonly written as cucurbit[n]uril, where n is the number of repeat units. A common abbreviation is CB[n].

These compounds are particularly interesting to chemists because they are capable of binding other molecules within their cavities. The cavity of cucurbit[6]uril has nanoscale dimensions with an approximate height of 9.1 Å, outer diameter 5.8 Å and inner diameter 3.9 Å.

Cucurbiturils were first synthesised in 1905 by Behrend, by condensing glycoluril with formaldehyde, but their structure was not elucidated until 1981. To date cucurbiturils composed of 5, 6, 7, 8, and 10 repeat units have all been isolated, which have internal cavity volumes of 82, 164, 279, 479, and 870 Å respectively. A cucurbituril composed of 9 repeat units has yet to be isolated (as of 2009). Other common molecular capsules that share a similar molecular shape with cucurbiturils include cyclodextrins and calixarenes.

Applications

Supramolecular host molecules

Cucurbiturils are efficient host molecules in molecular recognition and have a particularly high affinity for positively charged or cationic compounds. High association constants with positively charged molecules are attributed to the carbonyl groups that line each end of the cavity and can interact with cations is a similar fashion to crown ethers. The affinity of cucurbiturils can be very high. For example the affinity equilibrium constant of cucurbit[7]uril with the positively charged 1-aminoadamantane hydrochloride is experimentally determined at 4.23×10^{12}. Host guest interactions also significantly influence solubility behaviour of cucurbiturils. Cucurbit[6]uril dissolves poorly in just about any solvent but solubility is greatly improved in a solution of potassium hydroxide or in an acidic solution. The cavitand forms a positively charged inclusion compound with a potassium ion or a hydronium ion respectively which have much greater solubility than the uncomplexed neutral molecule.

CB is large enough to hold other molecular hosts such as a calixarene molecule. With a calixarene guest different chemical conformations (cone, 1,2-alternate, 1,3-alternate) are in rapid equilibrium. Allosteric control is provided when an adamantane molecule forces a cone conformation with a calixarene-adamantane inclusion complex within a CB molecule.

Rotaxane macrocycles

Given their high affinities to form inclusion complexes cucurbiturils have been employed as the macrocycles component of a rotaxane. After formation of the supramolecular assembly or threaded complex with a guest molecule such as hexamethylene diamine the two ends of the guest can be reacted with bulky groups that will then act as a stoppers preventing the two separate molecules from dissociating.

In another rotaxane system with a CB wheel, the axle is a 4,4'-bipyridinium or viologen subunit with two carboxylic acid terminated aliphatic N-substituents at both ends. In water at concentration higher than 0.5 M complexation is quantitative without need of stoppers. At pH = 2 the carboxylic end-groups are protonated and the wheel shuttles back and forth between them as evidenced by the presence of just two aromatic viologen protons in the proton NMR spectrum. At pH = 9 the wheel is locked around the viologen centre.

Drug delivery vehicles

Cucurbituril's host-guest properties have been explored for drug delivery vehicles. The potential of this application has been explored with cucurbit[7]uril that forms an inclusion compound with the important cancer fighting drug oxaliplatin. CB was employed despite the fact that it is more difficult to isolate since it has much greater solubility in water and its larger cavity size can accommodate the drug molecule.

The resulting complex was found to have increased stability and greater selectivity that may lead to less side effects.

Supramolecular catalysts

Cucurbiturils have also been explored as supramolecular catalysts. Larger cucurbiturils, such as cucurbit[8]uril can bind multiple guest molecules. CB forms a complex 2:1 (guest:host) with (E)-diaminostilbene dihydrochloride which is accommodated by CB's larger internal diameter of 8.8 angstrom and height 9.1 angstrom. The close proximity and optimal orientation of the guest molecules within the cavity enhances the rate of the photochemical cyclisation to give cyclobutane dimer with a 19:1 stereoselectivity for the syn configuration when bound to CB. In the absence of CB the cyclisation reaction does not occur, but only the isomerisation of the *trans*-isomer to the *cis*-isomer is observed.

Cyclodextrin

Cyclodextrins (sometimes called cycloamyloses) make up a family of cyclic oligosaccharides, composed of 5 or more α-D-glucopyranoside units linked 1->4, as in amylose (a fragment of starch). The 5-membered macrocycle is not natural. Recently, the largest well-characterised cyclodextrin contains 32 1,4-anhydroglucopyranoside units, while as a poorly characterised mixture, even at least 150-membered cyclic oligosaccharides are also known. Typical cyclodextrins contain a number of glucose monomers ranging from six to eight units in a ring, creating a cone shape.

Cyclodextrins are produced from starch by means of enzymatic conversion. Over the last few years they have found a wide range of applications in food, pharmaceutical and chemical industries as well as agriculture and environmental engineering. It is also the chief active compound found in Procter and Gamble's deodourising product 'Febreze' under the brand name 'Clenzaire'.

Uses

Cyclodextrins are able to form host-guest complexes with hydrophobic molecules given the unique nature imparted by their structure. As a result, these molecules have found a number of applications in a wide range of fields. Other than the above mentioned pharmaceutical applications for drug release, cyclodextrins can be employed in environmental protection: these molecules can effectively immobilise inside their rings toxic compounds, like trichloroethane or heavy metals, or can form complexes with stable substances, like trichlorfon (an organophosphorus insecticide) or sewage sludge, enhancing their decomposition.

In the food industry cyclodextrins are employed for the preparation of cholesterol free products: the bulky and hydrophobic cholesterol molecule is easily lodged inside cyclodextrin rings that are then removed.

Weight loss supplements are marketed from alpha-cyclodextrin which claim to bind to fat and be an alternative to other anti-obesity medications.

Other food applications further include the ability to stabilise volatile or unstable compounds and the reduction of unwanted tastes and odour. Reportedly cyclodextrins are used in alcohol powder, a powder for mixing alcoholic drinks.

The strong ability of complexing fragrances can also be used for another purpose: first dry, solid cyclodextrin microparticles are exposed to a controlled contact with fumes of active compounds, then they are added to fabric or paper products. Such devices are capable of releasing fragrances during ironing or when heated by human body. Such a device commonly used is a typical 'dryer sheet'. The heat from a clothe's dryer releases the fragrance into the clothing.

The ability of cyclodextrins to form complexes with hydrophobic molecules has led to their usage in supramolecular chemistry. In particular they have been used to synthesise certain mechanically-interlocked molecular architectures, such as rotaxanes and catenanes, by reacting the ends of the threaded guest.

The application of cyclodextrin as supramolecular carrier is also possible in organometallic reactions. The mechanism of action probably takes place in the interfacial region. Wipff also demonstrated by computational study that the reaction occurs in the interfacial layer. The application of cyclodextrins as supramolecular carrier is possible in various organometallic catalysis.

Dendrimer

Dendrimers are repeatedly branched molecules. The huge number of papers on dendritic architectures such as dendrimers, dendronised, hyperbranched and brush-polymers has generated a vast variety of inconsistent terms and definitions making a clear and concise unfolding of this topic highly difficult. The purpose of this section is to provide the vocabulary required for the description of chemical and physical phenomena as well as application aspects associated with the research in the area of dendritic molecules.

Dendritic molecules are repeatedly branched species that are characterised by structural perfection. This is based on the evaluation of both symmetry and polydispersity. The field of dendritic molecules can roughly be divided into low-molecular weight and high-molecular weight species. The first category includes dendrimers and dendrons, and the second includes dendronised polymers, hyperbranched polymers and brush-polymers (called also bottle-brushes).

Properties and applications

The properties of dendrimers are dominated by the functional groups on the molecular surface, however, there are examples of dendrimers with internal functionality. Dendritic encapsulation of functional molecules allows for the isolation of the active site, a structure that mimics the structure of active sites in biomaterials because dendritic scaffolds separate internal and external functions. For example, a dendrimer can be water-soluble when its end-group is a hydrophilic group, like a carboxyl group. It is theoretically possible to design a water-soluble dendrimer with internal hydrophobicity, which would allow it to carry a hydrophobic drug in its interior. Recently it has been shown that redox-active nanoparticles can be synthesised, placing the redox molecules between the nanoparticle core and the dendritic wedges; despite their isolation, some of the redox molecules (COOH in this case) remained uncoupled, and thus still reactive.

Another property is that the volume of a dendrimer increases when it has a positive charge. If this property can be applied, dendrimers can be used for drug delivery systems (DDS) that can give medication to the affected part inside a patient's body directly.

Dynamic Covalent Chemistry

In supramolecular chemistry, dynamic covalent chemistry is a strategy that aims at synthesising large complex molecules. In it a reversible reaction is under thermodynamic reaction control and a specific reaction product out of many is captured. Because all the components in the reaction mixture are able to equilibrate quickly, (according to its advocates) some degree of error checking and proof reading is enabled. The concept of dynamic covalent chemistry was demonstrated in the development of specific molecular Borromean rings.

The underlying idea is that rapid equilibration allows the coexistence of a huge variety of different species among which one can select molecules with desired chemical, pharmaceutical and biological properties. For instance, the addition of a proper template will shift the equilibrium toward the component that forms the complex of higher stability (thermodynamic template effect). After the new equilibrium is established, the researcher modifies the reaction conditions so as to stop equilibration. The optimal binder for the template is then extracted from the reactional mixture by the usual laboratory procedures.

Endohedral Fullerene

Endohedral fullerenes are fullerenes that have additional atoms, ions, or clusters enclosed within their inner spheres. Two types of endohedral complexes exist: endohedral metallofullerenes and non-metal doped fullerenes.

Endohedral metallofullerenes

Doping fullerenes with electropositive metals takes place in an arc reactor or via laser evaporation. The metals can be transition metals like scandium, yttrium as well as lanthanides like lanthanum and cerium. Also possible are endohedral complexes with elements of the alkaline earth metals like barium and strontium, alkali metals like potassium and tetravalent metals like uranium, zirconium and hafnium. The synthesis in the arc reactor is however unspecific. Besides unfilled fullerenes, endohedral metallofullerenes develop with different cage sizes like La@C_{60} or La@C_{82} and as different isomer cages. Aside from the dominant presence of mono-metal cages, numerous di-metal endohedral complexes and the tri-metal carbide fullerenes like Sc_3C_2@C_{80} were also isolated.

In 1998 a discovery drew large attention. With the synthesis of the Sc_3N@C_{80}, the inclusion of a molecule fragment in a fullerene cage had succeeded for the first time. This compound can be prepared by arc-vapourisation at temperatures up to 1100°C of graphite rods packed with scandium(III) oxide iron nitride and graphite powder in a K-H generator in a nitrogen atmosphere at 300 Torr.

Endohedral metallofullerenes are characterised by the fact that electrons will transfer from the metal atom to the fullerene cage and that the metal atom takes a position off-centre in the cage.

Non-metal doped fullerenes

Saunders in 1993 showed the formation of endohedral complexes He@C_{60} and Ne@C_{60} when C_{60} is exposed to a pressure of around 3 bar of the noble gases. Under these conditions about one out of every 6,50,000 C_{60} cages was doped with a helium atom. The formation of endohedral complexes with helium, neon, argon, krypton and xenon as well as numerous adducts of the He@C_{60} compound was also demonstrated with pressures of 3 kbars and incorporation of up to 0.1 per cent of the noble gases.

While noble gases are chemically very inert and commonly exist as individual atoms, this is not the case for nitrogen and phosphorus and so the formation of the endohedral complexes N@C_{60}, N@C_{70} and P@C_{60} is more surprising. The nitrogen atom is in its electronic initial state ($^4S_{3/2}$) and is therefore to be highly reactive. Nevertheless N@C_{60} is sufficiently stable that exohedral derivatisation from the mono- to the hexa adduct of the malonic acid ethyl ester is possible. In these compounds no charge transfer of the nitrogen atom in the centre to the carbon atoms of the cage takes place.

Therefore ^{13}C-couplings, which are observed very easily with the endohedral metallofullerenes, could only be observed in the case of the N@C_{60} in a high resolution spectrum as shoulders of the central line. The central atom in these endohedral complexes is located in the centre of the cage. While other atomic traps require complex equipment, e.g. laser cooling or magnetic traps, endohedral fullerenes represent an atomic trap that is stable at room temperature and for an arbitrarily long time. Atomic or ion traps are of great interest since particles are present free from (significant) interaction with their environment, allowing unique quantum mechanical phenomena to be explored. For example, the compression of the atomic wave function as a consequence of the packing in the cage could be observed with ENDOR spectroscopy. The nitrogen atom can be used as a probe, in order to detect the smallest changes of the electronic structure of its environment.

Foldamer

A foldamer is a discrete chain molecule or oligomer that adopts a secondary structure stabilised by non-covalent interactions. They are artificial molecules that mimic the ability of proteins, nucleic acids, and polysaccharides to fold into well-defined conformations, such as helices and β-sheets. Foldamers have been demonstrated to display a number of interesting supramolecular properties including molecular self-assembly, molecular recognition, and host-guest chemistry. They are studied as models of biological molecules and have been shown to display antimicrobial activity. They also have great potential application to the development of new functional materials.

Examples

1. *m*-Phenylene ethynylene oligomers are driven to fold into a helical conformation by solvophobic forces and aromatic stacking interactions.
2. β-peptides are composed of amino acids containing an additional methylene unit between the amine and carboxylic acid. They are more stable to enzymatic degradation and have been demonstrated to have antimicrobial activity.
3. Aedamers that fold in aqueous solutions driven by hydrophobic and aromatic stacking interactions.

Folding (Chemistry)

In chemistry folding is the process by which a molecule assumes its shape or conformation. The process can also be described as intramolecular self-assembly where the molecule is directed to form a specific shape through noncovalent interactions, such as hydrogen bonding, metal coordination, hydrophobic forces, van der Waals forces, pi-pi interactions, and/or electrostatic effects.

The most active area of interest in the folding of molecules is the process of protein folding, which is the shape that is assumed by a specific sequence of amino acids in a protein. The shape of the folded protein can be used to understand its function and design drugs to influence the processes that it is involved in.

There is also a great deal of interest in the construction of artificial folding molecules or foldamers. They are studied as models of biological molecules and potential application to the development of new functional materials.

Host-Guest Chemistry

In supramolecular chemistry, host-guest chemistry describes complexes that are composed of two or more molecules or ions held together in unique structural relationships by hydrogen bonding or by ion pairing or by van der Waals force other than those of full covalent bonds. The host component is defined as an organic molecule or ion whose binding sites converge in the complex and the guest component is defined as any molecule or ion whose binding sites diverge in the complex.

Common host molecules

Common host molecules: (i) cyclodextrins, (ii) calixarenes, (iii) cucurbiturils, (iv) porphyrins, (v) crown ethers, (vi) zeolites, (vii) cryptophanes, and (viii) carcerands.

Host-guest chemistry is observed in: (i) inclusion compounds, (ii) intercalation compounds, (iii) clathrates (iv) cryptands, and (v) molecular tweezers.

Hydrogen Bond

A hydrogen bond is the attractive interaction of a hydrogen atom with an electronegative atom, like nitrogen, oxygen or fluorine (thus the name 'hydrogen bond', which must not be confused with a covalent bond to hydrogen). The hydrogen must be covalently bonded to another electronegative atom to create the bond. These bonds can occur between molecules (intermolecularly), or within different parts of a single molecule (intramolecularly). The hydrogen bond (5 to 30 kJ/mole) is stronger than a van der Waals interaction, but weaker than covalent or ionic bonds. This type of bond occurs in both inorganic molecules such as water and organic molecules such as DNA.

Intermolecular hydrogen bonding is responsible for the high boiling point of water (100°C). This is because of the strong hydrogen bond, as opposed to other group 16 hydrides. Intramolecular hydrogen bonding is partly responsible for the secondary, tertiary, and quaternary structures of proteins and nucleic acids.

Hydrogen bonding also plays an important role in determining the three-dimensional structures adopted by proteins and nucleic bases. In these macromolecules, bonding between parts of the same macromolecule cause it to fold into a specific shape, which helps determine the molecule's physiological or biochemical role. The double helical structure of DNA, for example, is due largely to hydrogen bonding between the base pairs, which link one complementary strand to the other and enable replication.

A symmetric hydrogen bond is a special type of hydrogen bond in which the proton is spaced exactly halfway between two identical atoms. The strength of the bond to each of those atoms is equal.

The hydrogen bond can be compared with the closely related dihydrogen bond, which is also an intermolecular bonding interaction involving hydrogen atoms.

Hydrophobic Effect

The hydrophobic effect is the property that non-polar molecules tend to form aggregates of like molecules in water and analogous intramolecular interactions. The name arises from the combination of water in Attic Greek hydro- and for fear phobos, which describes the apparent repulsion between water and hydrocarbons. At the macroscopic level, the hydrophobic effect is apparent when oil and water are mixed together and form separate layers or the beading of water on hydrophobic surfaces such as waxy leafs. At the molecular level, the hydrophobic effect is an important driving force for biological structures and responsible for protein folding, protein-protein interactions, formation of lipid bilayer membranes, nucleic acid structures, and protein-small molecule interactions.

According to the solvophobic theory of Reversed Phase Chromatography (RPC), the hydrophobic effect is driven by the loss of hydrogen bonding and the higher entropic cost of forming a cavity around nonpolar molecules. These losses can be minimised by forcing nonpolar molecules together. The effect does not involve forces of repulsion between the components; hydration of hydrophobic substances is enthalpically favourable.

Amphiphiles

Amphiphiles are molecules that have both hydrophobic and hydrophilic domains. Detergents are composed of amphiphiles that allow hydrophobic molecules to be solubilised in water by forming micelles and bilayers (as in soap bubbles). They are also important to cell membranes composed of amphiphilic phospholipids that prevent the internal aqueous environment of a cell from mixing with external water.

Biological folding

In the case of protein folding, the hydrophobic effect is important to understand the structure of proteins that have hydrophobic amino acids, such as alanine, valine, leucine, isoleucine, phenylalanine, and methionine grouped together with the protein. Most folded proteins have a hydrophobic core in which side chain packing stabilises the folded state, and charged or polar side chains on the solvent-exposed surface where they interact with surrounding water molecules. It is generally accepted that minimising the number of hydrophobic side chains exposed to water is the principal driving force behind the folding process, although a recent theory has been proposed which reassesses the contributions made by hydrogen bonding.

The energetics of DNA tertiary structure assembly were determined to be primarily driven by the hydrophobic effect, as opposed to Watson-Crick base pairing (which is responsible for sequence selectivity), although there is also a significant contribution from stacking interactions between the aromatic bases.

Protein purification

In biochemistry, the hydrophobic effect can be used to separate mixtures of proteins based on their hydrophobicity. Column chromatography with a hydrophobic stationary phase such as phenyl-sepharose will cause more hydrophobic proteins to travel more slowly, while less hydrophobic ones elute from the column sooner. To achieve better separation, a salt may be added (higher concentrations of salt increase the hydrophobic effect) and its concentration decreased as the separation goes on.

Inclusion Compound

In host-guest chemistry an inclusion compound is a complex in which one chemical compound (the 'host') forms a cavity in which molecules of a second 'guest' compound are located. The definition of inclusion compounds is very broad, extending to channels formed between molecules in a crystal lattice in which guest molecules can fit. If the spaces in the host lattice are enclosed on all sides so that the guest species is 'trapped' as in a cage, the compound is known as a clathrate. In molecular encapsulation a guest molecule is actually trapped inside another molecule.

Cyclodextrin inclusion compounds

Inclusion complexes are formed between cyclodextrins and ferrocene. When a solution of both compounds in a 2:1 ratio in water is boiled for 2 days and then allowed to rest for 10 hours at room temperature orange-yellow crystals form. X-ray diffraction analysis of these crystals reveals a 4:5 inclusion complex with 4 molecules of ferrocene included in the cavity of 4 cyclodextrine molecules and with the fifth ferrocene molecule sandwiched between two stacks of ferrocene-cyclodextrine dimers.

Cyclodextrin also forms inclusion compounds with fragrance molecules. As a result the fragrance molecules have a reduced vapour pressure and are more stable towards exposure to light and air. When incorporated into textiles the fragrance lasts much longer due to the slow-release action.

Intercalation (Chemistry)

In chemistry, intercalation is the reversible inclusion of a molecule (or group) between two other molecules (or groups). Examples include DNA intercalation and in graphite intercalation compounds.

DNA intercalation

There are several ways molecules—in this case, also known as 'ligands'—can interact with DNA. Ligands may interact with DNA by covalently binding, electrostatically binding, or intercalating. Intercalation occurs when ligands of an appropriate size and chemical nature fit themselves in between base pairs of DNA. These ligands are mostly polycyclic, aromatic, and planar, and therefore often make good nucleic acid stains. Intensively studied DNA intercalators include ethidium bromide, proflavine, daunomycin, doxorubicin, and thalidomide. DNA intercalators are used in chemotherapeutic treatment to inhibit DNA replication in rapidly growing cancer cells. Examples include, doxorubicin (adriamycin) and daunorubicin (both of which are used in treatment of Hodgkin's lymphoma), and dactinomycin (used in Wilm's tumour, Ewing's Sarcoma, rhabdomyosarcoma).

Intercalation as a mechanism of interaction between cationic, planar, polycyclic aromatic systems of the correct size (on the order of a base pair) was first proposed by Leonard Lerman in 1961. One proposed mechanism of intercalation is as follows: in aqueous isotonic solution, the cationic intercalator is attracted electrostatically to the polyanionic DNA. The ligand displaces a sodium and/or magnesium cation that always surround DNA (to balance its charge) and forms a weak electrostatic bond with the outer surface of DNA. From this position, the ligand may then slide into the hydrophobic environment found between the base pairs and away from the hydrophilic outer environment surrounding the DNA. The base pairs transiently form such openings due to energy absorbed during collisions with solvent molecules.

Macrocycle

A macrocycle is, as defined by IUPAC, 'a cyclic macromolecule or a macromolecular cyclic portion of a molecule'. In the chemical literature, organic chemists may consider any molecule containing a ring of seven or more atoms to be a macrocycle. Coordination chemists generally define a macrocycle more narrowly as a cyclic molecule with three or more potential donor atoms that can coordinate to a metal centre.

Macrocycle effect

The macrocyclic effect was discovered in 1969. Coordination chemists study macrocycles with three or more potential donor atoms in rings of greater than nine atoms as these compounds often have strong and specific binding with metals. This property of coordinating macrocyclic molecules is the macrocycle effect. It is in essence a specific case of the chelation effect: complexes of bidentate and polydentate ligands are more stable than those with unidentate ligands of similar strength (or similar donor atoms). A macrocycle has donor atoms arranged in more fixed positions and thus there is less of an entropic effect in the binding energy of macrocycles than monodentate or bidentate ligands with an equal number of donor atoms. Thus the macrocycle effect states that complexes of macrocyclic ligands are more stable than those with linear polydentate ligands of similar strength (or similar donor atoms). The same can be said for multicyclic macrocycles, or cryptates, being stronger complexing agents (a cryptate effect).

Applications

1. Removal of heavy metals from aqueous solution for water purification.
   (a) Chelation therapy whereby the use of chelating agents such as EDTA to remove heavy metals from the body.
2. Molecular switches and linear motors for constructing artificial nanoscale machinery (rotaxanes).
3. Chemical sensors.
4. Mimicry of cellular receptors.

5. Molecular recognition.
   (a) Recognition of peptides.
   (b) Small molecules.
6. Organic light-emitting diodes (OLEDs).

Biological macrocycles

1. Heme, the active site in the haemoglobin (the protein in blood that transports oxygen) is a porphyrin containing iron.
2. Chlorophyll, the green photosynthetic pigment found in plants contains a chlorin ring.
3. Vitamin B12, contains a corrin ring.

Mechanical Bond

The mechanical bond is a type of chemical bond found in mechanically-interlocked molecular architectures such as catenanes and rotaxanes. Unlike classical molecular structures, interlocked molecules consist of two or more separate components which are not connected by chemical (i.e. covalent) bonds. These structures are true molecules and not a supramolecular species, as each component is intrinsically linked to the other—resulting in a mechanical bond which prevents dissociation without cleavage of one or more covalent bonds. 'Mechanical bond' is a relatively new term and at this point has limited usage in chemical literature relative to more well established bonds, such as covalent, hydrogen, or ionic bonds.

Mechanically-interlocked Molecular Architectures

Mechanically-interlocked molecular architectures are connections of molecules not through traditional bonds, but instead as a consequence of their topology. This connection of molecules is analogous to keys on a key chain loop. The keys are not directly connected to the key chain loop but they cannot be separated without breaking the loop. On the molecular level the interlocked molecules cannot be separated without significant distortion of the covalent bonds that make up the conjoined molecules. Examples of mechanically-interlocked molecular architectures include catenanes, rotaxanes, molecular knots, and molecular Borromean rings.

The synthesis of such entangled architectures has been made efficient through the combination of supramolecular chemistry with traditional covalent synthesis, however mechanically-interlocked molecular architectures have properties that differ from both 'supramolecular assemblies' and 'covalently-bonded molecules'. Recently the terminology 'mechanical bond' has been coined to describe the connection between the components of mechanically-interlocked molecular architectures. Although research into mechanically-interlocked molecular architectures is primarily focused on artificial compounds many examples have been found in biological systems including: cystine knots, cyclotides or lasso-peptides such as microcin J25 are protein, and a variety of peptides. There is a great deal of interest in mechanically-interlocked molecular architectures to develop molecular machines by manipulating the relative position of the components.

Examples of mechanically-interlocked molecular architectures are (i) rotaxane, (ii) catenane, (iii) molecular knot, and (iv) molecular borromean rings.

Metallacrown

Metallacrowns are inorganic analogues of crown ethers, discovered by Pecoraro and Lah in 1989. The crown analogy best describes the similarity between these two classes of macrocycle compounds. Since then their study has enormously increased in many fields of chemistry. In the literature metallacrowns are being currently studied as potential MRI contrast agents, as SMM (single molecule magnets) and anion sensors both in aqueous solution and in the solid state, not to mention the extensive solid state study to use them as building blocks for one-, two- and three-dimensional solids.

Micelle

A micelle is an aggregate of surfactant molecules dispersed in a liquid colloid. A typical micelle in aqueous solution forms an aggregate with the hydrophilic 'head' regions in contact with surrounding solvent, sequestering the hydrophobic single tail regions in the micelle centre. This phase is caused by the insufficient packing issues of single tailed lipids in a bilayer. The difficulty filling all the volume of the interior of a bilayer, while accommodating the area per head group forced on the molecule by the hydration of the lipid head group leads to the formation of the micelle. This type of micelle is known as a normal phase micelle (oil-in-water micelle). Inverse micelles have the headgroups at the centre with the tails extending out (water-in-oil micelle). Micelles are approximately spherical in shape. Other phases, including shapes such as ellipsoids, cylinders, and bilayers are also possible. The shape and size of a micelle is a function of the molecular geometry of its surfactant molecules and solution conditions such as surfactant concentration, temperature, pH, and ionic strength. The process of forming micellae is known as micellisation and forms part of the phase behaviour of many lipids according to their polymorphism.

Solvation

Individual surfactant molecules that are in the system but are not part of a micelle are called 'monomers'. Lipid micelles represent a molecular assembly in which the individual components are thermodynamically in equilibrium with monomers of the same species in the surrounding medium. In water, the hydrophilic 'heads' of surfactant molecules are always in contact with the solvent, regardless of whether the surfactants exist as monomers or as part of a micelle. However, the lipophilic 'tails' of surfactant molecules have less contact with water when they are part of a micelle — this being the basis for the energetic drive for micelle formation. In a micelle, the hydrophobic tails of several surfactant molecules assemble into an oil-like core the most stable form of which has no contact with water. By contrast, surfactant monomers are surrounded by water molecules that create a 'cage' of molecules connected by hydrogen bonds. This water cage is similar to a clathrate and has an ice-like crystal structure and can be characterised according to the hydrophobic effect. The extent of lipid solubility is determined by the unfavourable entropy contribution due to the ordering of the water structure according to the hydrophobic effect.

Micelles composed of ionic surfactants have an electrostatic attraction to the ions that surround them in solution, the latter known as counterions. Although the closest counterions partially mask a charged micelle (by up to 90 per cent), the effects of micelle charge affect the structure of the surrounding solvent at appreciable distances from the micelle. Ionic micelles influence many properties of the mixture, including its electrical conductivity. Adding salts to a colloid containing micelles can decrease the strength of electrostatic interactions and lead to the formation of larger ionic micelles. This is more accurately seen from the point of view of an effective change in hydration of the system.

Energy of formation

Micelles only form when the concentration of surfactant is greater than the critical micelle concentration (CMC), and the temperature of the system is greater than the critical micelle temperature, or Krafft temperature. The formation of micelles can be understood using thermodynamics: micelles can form spontaneously because of a balance between entropy and enthalpy. In water, the hydrophobic effect is the driving force for micelle formation, despite the fact that assembling surfactant molecules together reduces their entropy. At very low concentrations of the lipid, only monomers are present in true solution. As the concentration of the lipid is increased, a point is reached at which the unfavourable entropy considerations, derived from the hydrophobic end of the molecule, become dominant. At this point, the lipid hydrocarbon chains of a portion of the lipids must be sequestered away from the water. Therefore, the lipid starts to form micelles. Broadly speaking, above the CMC, the entropic penalty of assembling the surfactant molecules is less than the entropic penalty of caging the surfactant monomers with water molecules. Also important are enthalpic considerations, such as the electrostatic interactions that occur between the charged parts surfactants.

Inverse/reverse micelles

In a non-polar solvent, it is the exposure of the hydrophilic head groups to the surrounding solvent that is energetically unfavourable, giving rise to a water-in-oil system. In this case the hydrophilic groups are sequestered in the micelle core and the hydrophobic groups extend away from the centre. These inverse micelles are proportionally less likely to form on increasing headgroup charge, since hydrophilic sequestration would create highly unfavourable electrostatic interactions.

Uses

When surfactants are present above the CMC (critical micelle concentration), they can act as emulsifiers that will allow a compound that is normally insoluble (in the solvent being used) to dissolve. This occurs because the insoluble species can be incorporated into the micelle core, which is itself solubilised in the bulk solvent by virtue of the head groups' favourable interactions with solvent species. The most common example of this phenomenon is detergents, which clean poorly soluble lipophilic material (such as oils and waxes) that cannot be removed by water alone. Detergents also clean by lowering the surface tension of water, making it easier to remove material from a surface. The emulsifying property of surfactants is also the basis for emulsion polymerisation.

Micelle formation is essential for the absorption of fat-soluble vitamins and complicated lipids within the human body. Bile salts formed in the liver and secreted by the gall bladder allow micelles of fatty acids to form. This allows the absorption of complicated lipids (e.g. lecithin) and lipid soluble vitamins (A, D, E and K) within the micelle by the small intestine.

Molecular Borromean Rings

Molecular Borromean rings are an example of a mechanically-interlocked molecular architecture in which three macrocycles are interlocked in such a way that breaking any macrocycle allows the others to disassociate. They are the smallest examples of Borromean rings. The synthesis of molecular Borromean rings was reported in 2004 by the group of J. Fraser Stoddart. The so-called Borromeate is made up of three interpenetrated macrocycles formed from the reaction between 2,6-diformylpyridine and diamine compounds, complexed with zinc.

This compound was synthesised from two building blocks: 2,6-diformylpyridine (a pyridine with two aldehyde groups) and a diamine containing a 2,2′-bipyridine group. Zinc acetate is added as the template for the reaction, resulting in one zinc atom in each of a total of 6 pentacoordinate complexation sites. Trifluoroacetic acid (TFA) is added to catalyse the imine bond-forming reactions. The preparation of the tri-ring Borromeate involves a total of 18 precursor molecules and is only possible because the building blocks self-assemble through 12 aromatic pi-pi interactions and 30 zinc to nitrogen dative bonds. Because of these interactions, the Borromeate is thermodynamically the most stable reaction product out of potentially many others. As a consequence of all the reactions taking place being equilibria, the Borromeate is the predominant reaction product.

Molecular Encapsulation

Molecular encapsulation in supramolecular chemistry is the confinement of a guest molecule inside the cavity of a supramolecular host molecule (molecular capsule, molecular container or cage compounds). Examples of supramolecular host molecule include carcerands and endohedral fullerenes.

Reactivity of guests

An important implication of encapsulating a molecule at this level is that the guest is prevented from contacting other molecules that it might otherwise react with. Thus the encapsulated molecule behaves very differently from the way it would when in solution. The guest molecule tends to be extremely unreactive and often has much different spectroscopic signatures. Compounds normally highly unstable in solution, such as arynes or cycloheptatetraene have been successfully isolated at room temperature when molecularly encapsulated.

Examples: One of the first examples of encapsulating a structure at the molecular level was demonstrated by Cram and coworkers in which they were able to isolate highly unstable, antiaromatic cylobutadiene at room temperature by encapsulating it within a hemicarcerand. Isolation of cyclobutadiene allowed chemists to experimentally confirm one of the most fundamental predictions of the rules of aromaticity.

Molecular Tweezers

Molecular tweezers, sometimes termed molecular clips, are noncyclic macrocyclic molecular complexes with open cavities capable of binding guests. The open cavity of the molecular tweezer may bind guests using non-covalent bonding which includes hydrogen bonding, metal coordination, hydrophobic forces, van der Waals forces, $\pi-\pi$ interactions, and/or electrostatic effects. These complexes are a subset of macrocyclic molecular receptors and their structure is that the two 'arms' that bind the guest molecule between them are only connected at one end.

Examples: One example of molecular tweezers has been reported by Lehn and coworkers. This molecule is capable of binding aromatic guests. The molecular tweezers are composed of two anthracene arms held at a distance that allows aromatic guests to gain $\pi-\pi$ interactions from both.

Another class of molecular tweezers is composed of two substituted porphyrin macrocycles tethered by a amide linker with a variable length. This example of a molecular tweezer shows the potential mobility of this class of molecules, as the orientation of the porphyrin planes which comprise the tweezer can be altered by the guest which is bound.

The above examples show the potential reactivity and specificity of these molecules. The binding site between the planes of the tweezer can be designed to bind to an appropriate guest with resulting high association constants and consequent stability, depending on the design of the tweezer. That makes this overall class of macromolecule truly a synthetic molecular receptor.

Molecular Knot

In chemistry, a molecular knot, or knotane, is a mechanically-interlocked molecular architecture that is analogous to a macroscopic knot. A molecular knot in a trefoil knot configuration is chiral, having at least two enantiomers. Examples of naturally formed knotanes are DNA and certain proteins. Lactoferrin has an unusual biochemical reactivity compared to its linear analogue. Other synthetic molecular knots have a distinct globular shape and nanometer sized dimensions that make them potential building blocks in nanotechnology. Molecular knots are also referred to by some chemists as 'knotanes', a term coined by Fritz Vögtle in Angewandte Chemie International Edition in 2000 by analogy with rotaxane and catenane. The term however has yet to be adopted by IUPAC.

Molecular Imprinting

In chemistry, molecular imprinting is a technique to create template-shaped cavities in polymer matrices with memory of the template molecules to be used in molecular recognition. This technique is based on the system used by enzymes for substrate recognition, which is called the 'lock and key' model. The active binding site of an enzyme has a unique geometric structure that is particularly suitable for a substrate.

A substrate that has a corresponding shape to the site is recognised by selectively binding to the enzyme, while an incorrectly shaped molecule that does not fit the binding site is not recognised.

In a similar way, molecularly imprinted materials are prepared using a template molecule and functional monomers that assemble around the template and subsequently get crosslinked to each other. The functional monomers, which are self-assembled around the template molecule by interaction between functional groups on both the template and monomers, are polymerised to form an imprinted matrix (commonly known in the scientific community as a molecularly imprinted polymer, i.e. MIP). Then the template molecule is removed from the matrix under certain conditions, leaving behind a cavity complementary in size and shape to the template. The obtained cavity can work as a selective binding site for a specific template molecule (Fig. 9.7).

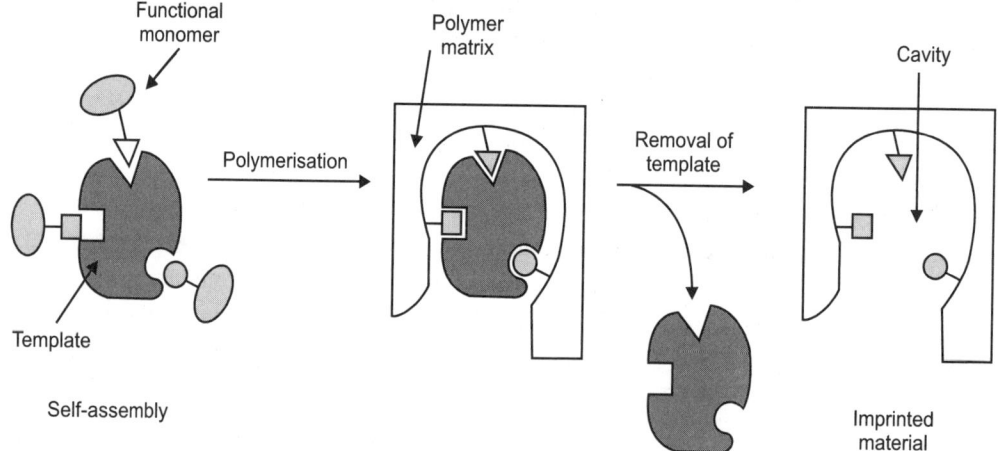

Fig. 9.7. Molecular imprinting.

In recent decades, the molecular imprinting technique has been developed for use in receptors, chromatographic separations, fine chemical sensing, etc. Taking advantage of the shape selectivity of the cavity, use in catalysis for certain reactions has also been facilitated.

Molecular Machine

A molecular machine, or nanomachine, has been defined as a discrete number of molecular components that perform mechanical-like movements (output) in response to specific stimuli (input). It is often applied more generally to molecules that simply mimic functions at the macroscopic level. The term is also common in nanotechnology, and a number of highly complex molecular machines have been proposed towards the goal of constructing a molecular assembler. Molecular machines can be divided into two broad categories: synthetic and biological.

Molecular systems that are able to shift a chemical or mechanical process away from equilibrium represent a potentially important branch of chemistry and nanotechnology. By definition, these types of systems are examples of molecular machinery, as the gradient generated from this process is able to perform useful work.

As described, this system may seem like a perpetual motion machine; however, the key ingredient is the heat gradient within the system. This ratchet does not threaten the second law of thermodynamics, because this temperature gradient must be maintained by some external means. Brownian motion of the gas particles provides the power to the machine, and the temperature gradient allows the machine to drive the system cyclically away from equilibrium. An interesting design concept in Feynman's ratchet is that random Brownian motion is not fought against, but instead, harnessed and rectified. Unfortunately, temperature gradients cannot be maintained over molecular scale distances because of molecular vibration redistributing the energy to other parts of the molecule. Furthermore, despite Feynman's machine doing useful work in lifting the mass, using Brownian motion to power a molecular level machine does not provides any insight on how that power (or potential energy of the lifted weight, m) can be used to perform nanoscale tasks (Fig. 9.8).

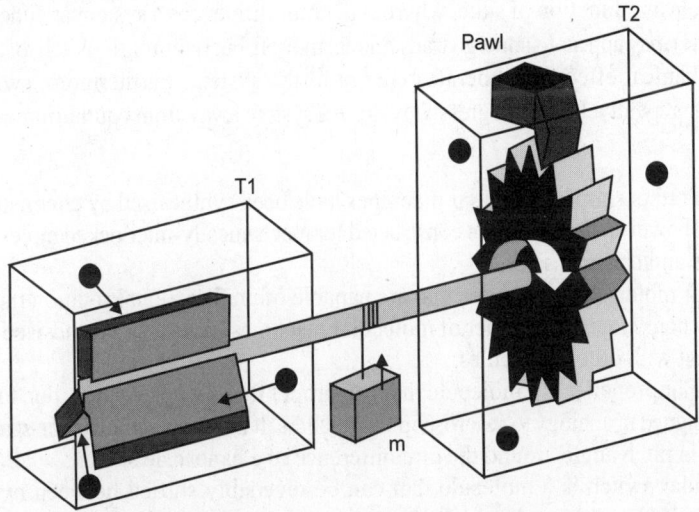

Fig. 9.8. Schematic figure of Feynman's ratchet.

Modern insights and studies

Unlike macroscopic motion, molecular systems are constantly undergoing significant dynamic motions subject to the laws of Brownian mechanics (or Brownian motion), and as such, harnessing molecular motion is a far more difficult process. At the macroscopic level, many machines operate in the gas phase, and often, air resistance is neglected, as it is insignificant, but analogously for a molecular system in a Brownian environment, molecular motion is similar 'to walking in a hurricane, or swimming in molasses'. The phenomenon of Brownian motion (observed by Robert Brown) was later explained by Albert Einstein in 1905. Einstein found that Brownian motion is a consequence of scale and not the nature of the surroundings. As long as thermal energy is applied to a molecule, it will undergo Brownian motion with the kinetic energy appropriate to that temperature. Therefore, like Feynman's strategy, when designing a molecular machine, it seems sensible to utilise Brownian motion rather than attempt to fight against it.

Like macroscopic machines, molecular machines typically have movable parts. However, while the macroscopic machines we encounter in everyday life may provide inspiration for molecular machines, it is misleading to draw analogies between their design strategy; the dynamics of large and small length scales are simply too different. Harnessing Brownian motion and making molecular level machines is regulated by the second law of thermodynamics, with its often counter-intuitive consequences, and as such, we need another inspiration.

Although it is a challenging process to harness Brownian motion, nature has provided us with several blueprints for molecular motion performing useful work. Nature has created many useful structures for compartmentalising molecular systems, hence creating distinct non-equilibrium distributions; the cell membrane is an excellent example. Lipophilic barriers make use of a number of different mechanisms to power motion from one compartment to another.

Examples of molecular machines

From a synthetic perspective, there are two important types of molecular machines: molecular switches (or shuttles) and molecular motors. The major difference between the two systems is that a switch influences a system as a function of state, whereas a motor influences a system as function of trajectory. A switch (or shuttle) may appear to undergo translational motion, but returning a switch to its original position undoes any mechanical effect and liberates energy to the system. Furthermore, switches cannot use chemical energy to repetitively and progressively drive a system away from equilibrium where a motor can.

Synthetic

A wide variety of rather simple molecular machines have been synthesised by chemists. They consist of a single molecule; however, they are often constructed for mechanically-interlocked molecular architectures, such as rotaxanes and catenanes.

1. Molecular motors are molecules that are capable of unidirectional rotation motion powered by external energy input. A number of molecular machines have been synthesised powered by light or reaction with other molecules.
2. A molecular propeller is a molecule that can propel fluids when rotated, due to its special shape that is designed in analogy to macroscopic propellers. It has several molecular-scale blades attached at a certain pitch angle around the circumference of a nanoscale shaft.
3. A molecular switch is a molecule that can be reversibly shifted between two or more stable states. The molecules may be shifted between the states in response to changes in, e.g. pH, light, temperature, an electrical current, microenvironment, or the presence of a ligand.
4. A molecular shuttle is a molecule capable of shuttling molecules or ions from one location to another. A common molecular shuttle consists of a rotaxane where the macrocycle can move between two sites or stations along the dumbbell backbone.

5. Molecular tweezers are host molecules capable of holding items between its two arms. The open cavity of the molecular tweezers binds items using non-covalent bonding including hydrogen bonding, metal coordination, hydrophobic forces, van der Waals forces, $\pi-\pi$ interactions, and/or electrostatic effects. Examples of molecular tweezers have been reported that are constructed from DNA and are considered DNA machines.

6. A molecular sensor is a molecule that interacts with an analyte to produce a detectable change. Molecular sensors combine molecular recognition with some form of reporter, so the presence of the item can be observed.

7. A molecular logic gate is a molecule that performs a logical operation on one or more logic inputs and produces a single logic output. Unlike a molecular sensor, the molecular logic gate will only output when a particular combination of inputs are present.

Biological

The most complex molecular machines are found within cells. These include motor proteins, such as myosin, which is responsible for muscle contraction, kinesin, which moves cargo inside cells away from the nucleus along microtubules, and dynein, which produces the axonemal beating of cilia and flagella. These proteins and their nanoscale dynamics are far more complex than any molecular machines that have yet been artificially constructed.

The detailed mechanism of ciliary motility has been described by Satir in a 2008 review article. A high-level-abstraction summary is that, '[i]n effect, the [motile cilium] is a nanomachine composed of perhaps over 600 proteins in molecular complexes, many of which also function independently as nanomachines'.

Theoretical

The construction of more complex molecular machines is an active area of theoretical research. A number of molecules, such as molecular propellers, have been designed, although experimental studies of these molecules are inhibited by the lack of methods to construct these molecules. These complex molecular machines form the basis of areas of nanotechnology, including molecular assembler.

Molecular Self-Assembly

Molecular self-assembly is the process by which molecules adopt a defined arrangement without guidance or management from an outside source. There are two types of self-assembly, intramolecular self-assembly and intermolecular self-assembly. Most often the term molecular self-assembly refers to intermolecular self-assembly, while the intramolecular analogue is more commonly called folding.

Supramolecular systems

Molecular self-assembly is a key concept in supramolecular chemistry since assembly of the molecules is directed through noncovalent interactions (e.g. hydrogen bonding, metal coordination, hydrophobic forces, van der Waals forces, $\pi-\pi$ interactions, and/or electrostatic) as well as electromagnetic interactions. Common examples include the formation of micelles, vesicles, liquid crystal phases, and Langmuir monolayers by surfactant molecules. Further examples of supramolecular assemblies demonstrate that a variety of different shapes and sizes can be obtained using molecular self-assembly.

Molecular self-assembly has allowed the construction of challenging molecular topologies. An example are Borromean rings, interlocking rings wherein removal of one ring unlocks each of the other rings. DNA has been used to prepare a molecular analogue of Borromean rings. More recently, a similar structure has been prepared using non-biological building blocks.

Biological systems

Molecular self-assembly is crucial to the function of cells. It is exhibited in the self-assembly of lipids to form the membrane, the formation of double helical DNA through hydrogen bonding of the individual strands, and the assembly of proteins to form quaternary structures. Molecular self-assembly of incorrectly folded proteins into insoluble amyloid fibres is responsible for infectious prion-related neurodegenerative diseases.

Nanotechnology

Molecular self-assembly is an important aspect of bottom-up approaches to nanotechnology. Using molecular self-assembly the final (desired) structure is programmed in the shape and functional groups of the molecules. Self-assembly is referred to as a 'bottom-up' manufacturing technique in contrast to a 'top-down' technique such as lithography where the desired final structure is carved from a larger block of matter. In the speculative vision of molecular nanotechnology, microchips of the future might be made by molecular self-assembly. An advantage to constructing nanostructure using molecular self-assembly for biological materials is that they will degrade back into individual molecules that can be broken down by the body.

DNA nanotechnology

DNA nanotechnology is an area of current research that uses the bottom-up, self-assembly approach for nanotechnological goals. DNA nanotechnology uses the unique molecular recognition properties of DNA and other nucleic acids to create self-assembling branched DNA complexes with useful properties. DNA is thus used as a structural material rather than as a carrier of biological information, to make structures such as two-dimensional periodic lattices (both tile-based as well as using the 'DNA origami' method) and three-dimensional structures in the shapes of polyhedra. These DNA structures have also been used to template the assembly of other molecules such as gold nanoparticles and streptavidin proteins.

Molecular Sensor

A molecular sensor or chemosensor is a molecule that interacts with an analyte to produce a detectable change. Molecular sensors combine molecular recognition with some form of reporter so the presence of the guest can be observed. The term supramolecular analytical chemistry has recently been coined to describe the application of molecular sensors to analytical chemistry.

Early examples of molecular sensors are crown ethers with large affinity for sodium ions but not for potassium and forms of metal detection by so-called complexones which are traditional pH indicators retrofitted with molecular groups sensitive to metals. This receptor-spacer-reporter concept is a recurring theme often with the reporter displaying photoinduced electron transfer. One example is a sensor sensitive to heparin.

Other receptors are sensitive not to a specific molecule but to a molecular compound class. One example is the grouped analysis of several tannic acids that accumulate in ageing scotch whiskey in oak barrels. The grouped results demonstrated a correlation with the age but the individual components did not. A similar receptor can be used to analyse tartrates in wine.

The compound saxitoxin is a neurotoxin found in shellfish and a chemical weapon. An experimental sensor for this compound is again based on PET. Interaction of saxitoxin with the sensor's crown ether moiety kills its PET process towards the fluorophore and fluorescence is switched from off to on. The unusual boron moiety makes sure the fluorescence takes place in the visible light part of the electromagnetic spectrum. In another strategy called indicator-displacement assay (IDA) an analyte such as citrate or phosphate ions displace a fluorescent indicator in an indicator-host complex. The so-called UT taste chip is a prototype electronic tongue and combines supramolecular chemistry with charge-coupled devices based on silicon wafers and immobilised receptor molecules.

Molecular Shuttle

A molecular shuttle in supramolecular chemistry is a special type of molecular machine capable of shuttling molecules or ions from one location to another. This field is of relevance to nanotechnology in its quest for nanoscale electronic components and also to biology where many biochemical functions are based on molecular shuttles. Academic interest also exists for synthetic molecular shuttles, the first prototype reported in 1991 based on a rotaxane (Fig. 9.9).

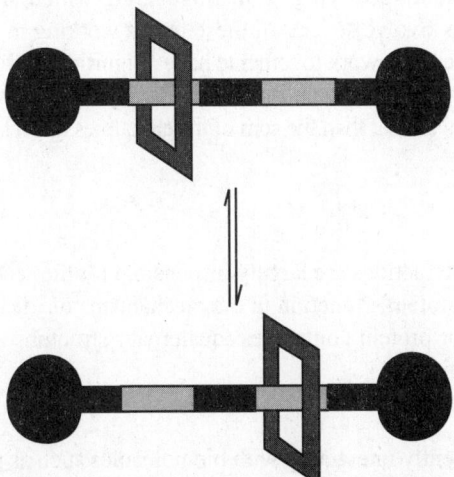

Fig. 9.9. An example of a molecular shuttle where the macrocycle moves between two stations.

This device is based on a molecular thread composed of a ethyleneglycol chain interrupted by two arene groups acting as so-called stations. The terminal units (or stoppers) on this wire are bulky triisopropylsilyl groups. The bead is a tetracationic cyclophane based on two bipyridine groups and two para-phenylene groups. The bead is locked to one of the stations by pi-pi interactions but since the activation energy for migration from one station to the other station is only 13 kcal/mol (54 kJ/mol) the bead shuttles between them. The stoppers prevent the bead from slipping from the thread. Chemical synthesis of this device is based on molecular self-assembly from a preformed thread and two bead fragments (32 per cent chemical yield).

Noncovalent Bonding

A noncovalent bond is a type of chemical bond, typically between macromolecules, that does not involve the sharing of pairs of electrons, but rather involves more dispersed variations of electromagnetic interactions. The noncovalent bond is the dominant type of bond between supermolecules in supermolecular chemistry. Noncovalent bonds are critical in maintaining the three-dimensional structure of large molecules, such as proteins and nucleic acids, and are involved in many biological processes in which large molecules bind specifically but transiently to one another. The energy released in the formation of noncovalent bonds is of the order of 1–5 kcal per mol. There are four commonly mentioned types of non-covalent interactions: hydrogen bonds, ionic bonds, van der Waals forces, and hydrophobic interactions. The noncovalent interactions hold together the two strands DNA in the double helix, stabilise secondary and tertiary structures of proteins, and enable enzyme-substrate binding and antibody-antigen association.

In general, noncovalent bonding refers to a variety of interactions that are not covalent in nature between molecules or parts of molecules that provide force to hold the molecules or parts of molecules together, usually in a specific orientation or conformation. Noncovalent bonding is the dominant type of bonding in supramolecular chemistry. These noncovalent interactions include: ionic bonds, hydrophobic interactions, hydrogen bonds, van der Waals forces, i.e. 'London dispersion forces', and Dipole-dipole bonds.

The terms 'noncovalent bonding', 'noncovalent interactions', and 'noncovalent forces' all refer to these forces as a whole without specifying or distinguishing which specific forces are involved: noncovalent interactions often involve several of these forces working in concert. Noncovalent bonds are weak by nature and must therefore work together to have a significant effect. In addition, the combined bond strength is greater than the sum of the individual bonds. This is because the free energy of multiple bonds between two molecules is greater than the sum of the enthalpies of each bond due to entropic effects.

Examples

Protein structure

Intramolecular noncovalent interactions are largely responsible for the secondary and tertiary structure of proteins and therefore the protein's function in the mechanisms of life. Intermolecular noncovalent interactions are responsible for protein complexes (quaternary structure) where two or more proteins function in a coherent mechanism.

Pharmaceuticals

Most drugs work by noncovalently interacting with biomolecules such as proteins or RNA. Relatively few drugs actually form covalent bonds with the biomolecules they interact with; instead, they interfere with or activate some biological mechanism through noncovalently interacting in very specific locations on specific biomolecules which present the perfect combination of noncovalent binding partners in just the right geometry.

Porphyrin

Porphyrins are a group of organic compounds of which many occur in nature, most well-known as the pigment in red blood cells. They are heterocyclic macrocycles characterised by the presence of four modified pyrrole subunits interconnected at their α carbon atoms via methine bridges (=CH–). Porphyrins are aromatic, and they obey Hückel's rule for aromaticity in that they possess $4n + 2$ π electrons ($n = 4$ for the shortest cyclic path) that are delocalised over the macrocycle. The macrocycles, therefore, are highly-conjugated systems and, as a consequence, have very intense absorption in the visible region and therefore are deeply coloured; the name porphyrin comes from a Greek word for purple. The macrocycle has 26 pi electrons in total. The parent porphyrin is porphine, and substituted porphines are called porphyrins.

Applications

Although natural porphyrin complexes are essential for life, synthetic porphyrins and their complexes have limited utility. Complexes of meso-tetraphenylporphyrin, e.g. the iron(III) chloride complex

(TPPFeCl) catalyses a variety of reactions in organic synthesis, but none is of practical value. Porphyrin-based compounds are of interest in molecular electronics and supramolecular building blocks. Phthalocyanines, which are structurally related to porphyrins, are used in commerce as dyes and catalysts. Synthetic porphyrin dyes that are incorporated in the design of solar cells are the subject of ongoing research.

In 2008 the corporation Destiny Pharma reported successful clinical trials of an intra-nasally applied porphyrin XF-73 against methicillin-resistant *Staphylococcus aureus*.

Supramolecular chemistry

Porphyrins are often used to construct structures in supramolecular chemistry. These systems take advantage of the Lewis acidity of the metal, typically zinc. An example of a host-guest complex that was constructed from a macrocycle composed of four porphyrins. A guest-free base porphyrin is bound to the centre by coordination with its four pyridine substituents.

Organic geochemistry

The field of organic geochemistry, the study of the impacts and processes that organisms have had on the earth, had its origins in the isolation of porphyrins from petroleum. This finding helped establish the biological origins of petroleum. Petroleum is sometimes 'fingerprinted' by analysis of trace amounts of nickel and vanadyl porphyrins.

Resorcinarene

A resorcinarene (also resorcarene or calix resorcinarene) is a macrocycle, or a cyclic oligomer, based on the condensation of resorcinol (1,3-dihydroxybenzene) and an aldehyde. Resorcinarenes are a type of calixarene.

The resorcinarene macrocycle is typically prepared by condensation of resorcinol and an aldehyde in concentrated acid solution. Recrystallisation typically gives the desired isomer in quite pure form. However, for certain aldehydes, the reaction conditions lead to significant by-products. Therefore, alternative condensation conditions have been developed, including the use of Lewis acid catalysts.

Supramolecular chemistry

Yasuhiro Aoyama's research group at Kyoto University first noted the potential for resorcinarenes to interact with other molecules as a host-guest complex. It was later found that resorcinarenes and pyrogallolarenes self-assemble into to larger supramolecular structures. Both in the crystalline state and in organic solvents, six resorcinarene molecules are known to form hexamers with an internal volume of around one cubic nanometer (nanocapsules) and shapes similar to the Archimedean solids. Hydrogen bonds appear to hold the assembly together. A number of solvent or other molecules reside inside. The resorcinarene is also the basic structural unit for other molecular recognition scaffolds. A number of chemists, including Nobel-laureate Donald J. Cram, constructed novel molecular structures based on this macrocycle, namely cavitands and carcerands.

Rotaxane

A rotaxane is a mechanically-interlocked molecular architecture consisting of a 'dumbbell shaped molecule' which is threaded through a 'macrocycle'. The name is derived from the Latin for wheel (rota) and axle (axis). The two components of a rotaxane are kinetically trapped since the ends of the dumbbell (often called stoppers) are larger than the internal diameter of the ring and prevent disassociation (unthreading) of the components since this would require significant distortion of the covalent bonds.

Much of the research concerning rotaxanes and other mechanically-interlocked molecular architectures, such as catenanes, has been focused on their efficient synthesis. However, examples of rotaxane have been found in biological systems including: cystine knot peptides, cyclotides or lasso-peptides such as microcin J25 are protein, and a variety of peptides with rotaxane substructure.

Synthesis

To obtain a reasonable quantity of rotaxane the macrocycle was attached to a solid phase support and treated with both halves of the dumbbell 70 times and then severed from the support to give a 6 per cent yield. However, the synthesis of rotaxanes has advanced significantly and efficient yields can be obtained by preorganising the components utilising hydrogen bonding, metal coordination, hydrophobic forces, covalent bonds, or coulombic interactions. The three most common strategies to synthesise rotaxane are capping, clipping, and slipping, though others do exist.

Capping

Synthesis via the capping method relies strongly upon a thermodynamically driven template effect; that is the 'thread' is held within the 'macrocycle' by non-covalent interactions. This dynamic complex or pseudorotaxane is then converted to the rotaxane by reacting the ends of the threaded guest with large groups preventing disassociation.

Clipping

The clipping method is similar to the capping reaction except that in this case the dumbbell shaped molecule is complete and is bound to a partial macrocycle. The partial macrocycle then undergoes a ring closing reaction around the dumbbell shaped molecule forming the rotaxane.

Slipping

The method of slipping is one which exploits the kinetic stability of the rotaxane. If the end groups of the dumbbell are an appropriate size it will be able to reversibly thread through the macrocycle at higher temperatures. By cooling the dynamic complex it becomes kinetically trapped as a rotaxane at the lower temperature.

Potential applications

Molecular machines

Rotaxane-based molecular machines have been of initial interest for their potential use in molecular electronics as logic molecular switching elements and as molecular shuttles. These molecular machines are usually based on the movement of macrocycle on the dumbbell. The macrocycle can rotate around the axis of the dumbbell like a wheel and axle or it can slide along its axis from one site to another. Controlling the position of the macrocycle allows the rotaxane to function as molecular switch with each possible location of the macrocycle corresponding to a different state. These rotaxane machines can be manipulated both by chemical and photochemical inputs. Rotaxane based systems have also been demonstrated as molecular muscles. Very recently, it has been reported a domino effect from one extremity to the other in a Glycorotaxane Molecular Machine. In this case, the $^4C_1$ or $^1C_4$ chair-like conformation of the mannopyranoside stopper can be controlled, depending on the localisation of the macrocycle.

Ultrastable dyes

Potential application as long lasting dyes is based on the enhanced stability of the inner portion of the dumbbell shaped molecule. Studies with cyclodextrin protected rotaxane azo dyes established this characteristic. More reactive squaraine dyes have also been shown to have enhanced stability by preventing nucleophilic attack of the inner squaraine moiety. The enhanced stabilities of rotaxane dyes is attributed to the insulating effect of the macrocycle which is able to block interactions with other molecules.

Nanorecording

In a nanorecording application a certain rotaxane is deposited as a Langmuir-Blodgett film on ITO coated glass. When a positive voltage is applied with the tip of a scanning tunnelling microscope probe, the rotaxane rings in the tip area switch to a different part of the dumbbell and the resulting new conformation makes the molecules stick out from the surface by 0.3 nanometer and this height difference turns out to be sufficient for a memory dot. It is not yet possible to erase such a nanorecording film.

Stacking (Chemistry)

Stacking in supramolecular chemistry refers to a stacked arrangement of often aromatic molecules, which is adopted due to interatomic interactions. The most common example of a stacked system is found for consecutive base pairs in DNA. Stacking also frequently occurs in proteins where two relatively non-polar rings overlap. Which intermolecular forces contribute to stacking is a matter of debate.

Controlling forces

Stacking is often referred to as $\pi-\pi$ interaction, though effects due to the presence of a π-orbital are only one source of such interactions, and in many common cases appear not to be the dominant contributors.

In ring-systems with fewer than three rings, *ab initio* calculations suggest that aromaticity contributes little to stacking forces, and that the strength of these forces, which stabilise the stacked conformation, does not differ significantly from the van der Waals forces also experienced by similarly-sized saturated molecules when stacked. Therefore, DNA nucleobases (having one or two rings) probably do not significantly stabilise DNA's stacked structure as a result of their aromaticity, but do so by those intermolecular forces experienced by all closed-shell neutral molecules.

For larger ring-systems (perhaps including twelve or more atoms), there does appear to be such a $\pi-\pi$ effect, caused by a larger orbital-dependent (i.e. non-atom-pairwise) contribution to the dispersion component of the van der Waals force than in an equivalent saturated molecule. This contribution is coupled with the optimal stacking of π-orbitals minimising exchange 'repulsion' for geometric reasons. Electrostatic forces actually considerably weaken this effect in aromatics, though do not entirely cancel it, whereas the induction component of van der Waals forces makes no significant contribution.

Stacking within supramolecular chemistry

In supramolecular chemistry, an aromatic interaction (or $\pi-\pi$ interaction) is a noncovalent interaction between organic compounds containing aromatic moieties. $\pi-\pi$ interactions are caused by intermolecular overlapping of π-orbitals in π-conjugated systems, so they become stronger as the number of π-electrons increases. Other noncovalent interactions include hydrogen bonds, van der Waals forces, charge-transfer interactions, and dipole-dipole interactions.

$\pi-\pi$ interactions act strongly on flat polycyclic aromatic hydrocarbons such as anthracene, triphenylene, and coronene because of the many delocalised π-electrons. This interaction, which is a bit stronger than other noncovalent interactions, plays an important role in various parts of supramolecular chemistry. For example, $\pi-\pi$ interactions have a large influence on molecule-based crystal structures of aromatic compounds.

A powerful demonstration of stacking is found in the buckycatcher depicted below. This molecular tweezer is based on two concave buckybowls with a perfect fit for one convex fullerene molecule. Complexation takes place simply by evaporating a toluene solution containing both compounds. In solution an association constant of $8600 \ M^{-1}$ is measured based on changes in NMR chemical shifts.

Stacking in biology

In DNA, pi stacking occurs between adjacent nucleotides and adds to the stability of the molecular structure. The nitrogenous bases of the nucleotides are made from either purine or pyrimidine rings, consisting of aromatic rings. Within the DNA molecule, the aromatic rings are positioned nearly perpendicular to the length of the DNA strands. Thus, the faces of the aromatic rings are arranged parallel to each other, allowing the bases to participate in aromatic interactions. Through aromatic interactions, the pi bonds, extending from atoms participating in double bonds, overlap with pi bonds of adjacent bases.

This is a type of non-covalent chemical bond. Though a non-covalent bond is weaker than a covalent bond, the sum of all pi stacking interactions within the double-stranded DNA molecule creates a large net stabilising energy.

Uses in materials

Many discotic liquid crystals can form columnar structures by $\pi-\pi$ interactions. In addition, $\pi-\pi$ interactions are an important factor in molecular self-assembly techniques in bottom-up nanotechnology.

Supermolecule

The term supermolecule or supramolecule, was introduced by K.L. Wolf (*Übermoleküle*) in 1937 to describe hydrogen bonded acetic acid dimers. The study of non-covalent association of complexes of molecules has since developed into the field of supramolecular chemistry. The term supermolecule is sometimes used to describe supramolecular assemblies, which are complexes of two or more molecules often macromolecules that are not covalently bonded. The term supermolecule is also used in biochemistry to describe complexes of biomolecules, such as peptides and oligonucleotides composed of multiple strands.

Supramolecular Assembly

A supramolecular assembly or 'supermolecule' is a well defined complex of molecules held together by noncovalent bonds. While a supramolecular assembly can be simply composed of two molecules (e.g. a DNA double helix or an inclusion compound), it is more often used to denote larger complexes of molecules that form sphere-, rod-, or sheet-like species. The dimensions of supramolecular assemblies can range from nanometers to micrometers. Thus they allow access to nanoscale objects using a bottom-up approach in far fewer steps than a single molecule of similar dimensions.

The process by which a supramolecular assembly forms is called molecular self-assembly. Some try to distinguish self-assembly as the process by which individual molecules form the defined aggregate.

Self-organisation, then, is the process by which those aggregates create higher-order structures. This can become useful when talking about liquid crystals and block copolymers.

Applications

Supramolecular assemblies are being investigated as new materials in a variety of contexts. For instance, Samuel Stupp and coworkers at Northwestern University showed that a supramolecular assembly of peptide amphiphiles in the form of nanofibres could be used to promote the growth of neurons. A great advantage to this supramolecular approach is that the nanofibres will degrade back into the individual peptide molecules that can be broken down by the body.

Another example with implications at the biology/materials science interface is of self-assembling dendritic dipeptides, which form hollow cylindrical supramolecular assemblies in solution and in bulk. The cylindrical assemblies possess internal helical order and self-organise into columnar liquid crystalline lattices. When inserted into vesicular membranes, the porous cylindrical assemblies mediate transport of protons across the membrane.

Self-assembling dendrons have also been used to generate arrays of nanowires. Electron donor-acceptor complexes comprise the core of the cylindrical supramolecular assemblies, which further self-organise into two-dimensional columnar liquid crystaline lattices. Each cylindrical supramolecular assembly functions as an individual wire. High charge carrier mobilities for holes and electrons were obtained.

Supramolecular Polymers

A supramolecular polymer is a polymer whose monomer repeat units are held together by noncovalent bonds. Noncovalent forces that hold supramolecular polymers together include coordination, π-π interactions, and hydrogen bonding. One system that has been demonstrated uses quadruple hydrogen bonds to form supramolecular polymers. Functionalisation of polymers with the quadruple hydrogen bonding unit from the Meijer group introduces reversible cross-links and a virtual increase in the polymer's molecular weight. In general, polymers with a higher molecular weight possess better material properties.

Topoisomer

Topoisomers or topological isomers are molecules with the same chemical formula and stereochemical bond connectivities but different topologies. Examples of molecules for which there exist topoisomers include DNA, which can form knots, and catenanes. DNA topoisomers can be interchanged by enzymes called topoisomerases.

Chapter 10

Microfluidics

INTRODUCTION

Microfluidics is rapidly emerging as an enabling technology, having applications ranging from unmanned aerial vehicles to inkjet printing to biochemical sensing, filtration and purification processes, among many other current and potential applications. Given the emerging importance of micro- and nanoscale transport phenomena, this chapter will provide working level engineers, faculty and managers with an overview and understanding of the fundamental fluid mechanics, heat and mass transfer, and chemistry involved in such devices, as well as the biochemistry and engineering principles governing the design of micro- and nanofluidic devices.

Microfluidics deals with the behaviour, precise control and manipulation of fluids that are geometrically constrained to a small, typically sub-millimetre, scale. Typically, micro means one of the following features:

1. Small volumes (nl, pl, fl).
2. Small size.
3. Low energy consumption.
4. Effects of the micro domain.

It is a multidisciplinary field intersecting engineering, physics, chemistry, microtechnology and biotechnology, with practical applications to the design of systems in which such small volumes of fluids will be used. Microfluidics has emerged in the beginning of the 1980s and is used in the development of inkjet printheads, DNA chips, lab-on-a-chip technology, micro-propulsion, and micro-thermal technologies.

MICROSCALE BEHAVIOUR OF FLUIDS

The behaviour of fluids at the microscale can differ from 'macrofluidic' behaviour in that factors such as surface tension, energy dissipation, and fluidic resistance start to dominate the system. Microfluidics studies show these behaviours change, and how they can be worked around, or exploited for new uses.

At small scales (channel diameters of around 100 nanometers to several hundred micrometers) some interesting and sometimes unintuitive properties appear. In particular, the Reynolds number (which compares the effect of momentum of a fluid to the effect of viscosity) can become very low. A key consequence of this is that fluids, when side-by-side, do not necessarily mix in the traditional sense; molecular transport between them must often be through diffusion. This property is important in many microfluidic devices (Fig. 10.1).

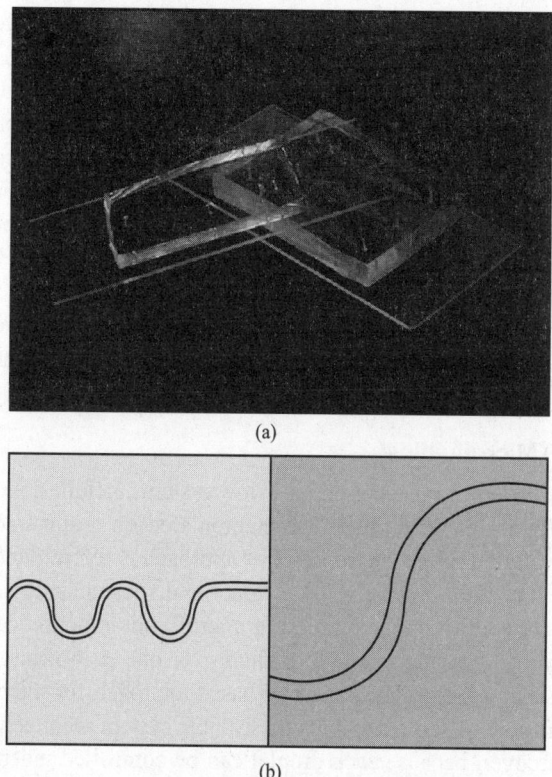

(a)

(b)

Fig. 10.1. Silicone rubber and glass microfluidic devices. (a) A photograph of the devices, and (b) DIC micrographs of a serpentine channel ~15 μm wide.

Effects of micro domain is related to: (i) laminar flow, (ii) surface tension, (iii) electrowetting, (iv) fast thermal relaxation, (v) electrical surface charges, and (vi) diffusion.

KEY APPLICATION AREAS

Microfluidic structures include micropneumatic systems, i.e. microsystems for the handling of off-chip fluids (liquid pumps, gas valves, etc.), and microfluidic structures for the on-chip handling of nano- and picolitre volumes. The commercially most successful application today is the inkjet printhead.

Advances in microfluidics technology are revolutionising molecular biology procedures for enzymatic analysis (e.g. glucose and lactate assays), DNA analysis (e.g. polymerase chain reaction and high-throughput sequencing), and proteomics. The basic idea of microfluidic biochips is to integrate assay operations such as detection, as well as sample pretreatment and sample preparation on one chip.

An emerging application area for biochips is clinical pathology, especially the immediate point-of-care diagnosis of diseases. In addition, microfluidics-based devices, capable of continuous sampling and real-time testing of air/water samples for biochemical toxins and other dangerous pathogens, can serve as an always-on 'bio-smoke alarm' for early warning.

Continuous-flow Microfluidics

These technologies are based on the manipulation of continuous liquid flow through microfabricated channels. Actuation of liquid flow is implemented either by external pressure sources, external mechanical pumps, integrated mechanical micropumps, or by combinations of capillary forces and electrokinetic mechanisms. Continuous-flow microfluidic operation is the mainstream approach because it is easy to implement and less sensitive to protein fouling problems. Continuous-flow devices are adequate for many well-defined and simple biochemical applications, and for certain tasks such as chemical separation, but they are less suitable for tasks requiring a high degree of flexibility or complicated fluid manipulations. These closed-channel systems are inherently difficult to integrate and scale because the parameters that govern flow field vary along the flow path making the fluid flow at any one location dependent on the properties of the entire system. Permanently-etched microstructures also lead to limited reconfigurability and poor fault tolerance capability.

Process monitoring capabilities in continuous-flow systems can be achieved with highly sensitive microfluidic flow sensors based on MEMS technology which offer resolutions down to the nanolitre range.

Digital (Droplet-based) Microfluidics

Alternatives to the above closed-channel continuous-flow systems include novel open structures, where discrete, independently controllable droplets are manipulated on a substrate using electrowetting. Following the analogy of digital microelectronics, this approach is referred to as digital microfluidics, which was pioneered as the 'fluid transistor' by Cytonix and subsequently commercialised by Duke University. By using discrete unit-volume droplets, a microfluidic function can be reduced to a set of repeated basic operations, i.e. moving one unit of fluid over one unit of distance. This 'digitisation' method facilitates the use of a hierarchical and cell-based approach for microfluidic biochip design. Therefore, digital microfluidics offers a flexible and scalable system architecture as well as high fault-tolerance capability. Moreover, because each droplet can be controlled independently, these systems also have dynamic reconfigurability, whereby groups of unit cells in a microfluidic array can be reconfigured to change their functionality during the concurrent execution of a set of bioassays. Although droplets are manipulated in confined microfluidic channels, since the control on droplets is not independent, it should not be confused as 'digital microfluidics'. One common actuation method for digital microfluidics is electrowetting-on-dielectric (EWOD). Many lab-on-a-chip applications have been demonstrated within the digital microfluidics paradigm using electrowetting. However, recently other techniques for droplet manipulation have also been demonstrated using surface acoustic waves, opto-electrowetting, etc.

DNA Chips (Microarrays)

Early biochips were based on the concept of a DNA microarray, e.g. the GeneChip DNAarray from Affymetrix, which is a piece of glass, plastic or silicon substrate on which pieces of DNA (probes) are affixed in a microscopic array. Similar to a DNA microarray, a protein array is a miniature array where a multitude of different capture agents, most frequently monoclonal antibodies, are deposited on a chip surface; they are used to determine the presence and/or amount of proteins in biological samples, e.g. blood. A drawback of DNA and protein arrays is that they are neither reconfigurable nor scalable after manufacture.

Molecular Biology

In addition to microarrays biochips have been designed for two-dimensional electrophoresis, transcriptome analysis, and PCR amplification. Other applications include various electrophoresis and liquid chromatography applications for proteins and DNA, cell separation, in particular blood cell separation, protein analysis, cell manipulation and analysis including cell viability analysis and micro-organism capturing.

Acoustic Droplet Ejection (ADE)

Acoustic droplet ejection uses a pulse of ultrasound to move low volumes of fluids (typically nanolitres or picolitres) without any physical contact. This technology focuses acoustic energy into a fluid sample in order to eject droplets as small as a millionth of a millionth of a litre (picolitre = 10^{-12} litre). ADE technology is a very gentle process, and it can be used to transfer proteins, high molecular weight DNA and live cells without damage or loss of viability. This feature makes the technology suitable for a wide variety of applications including proteomics and cell-based assays.

Fuel Cells

Microfluidic fuel cells can use laminar flow to separate the fuel and its oxidant to control the interaction of the two fluids without a physical barrier as would be required in conventional fuel cells.

DIGITAL MICROFLUIDICS

Digital microfluidics is an alternative technology for lab-on-a-chip systems based upon micromanipulation of discrete droplets. Microfluidic processing is performed on unit-sized packets of fluid which are transported, stored, mixed, reacted, or analysed in a discrete manner using a standard set of basic instructions. In analogy to digital microelectronics, these basic instructions can be combined and reused within hierarchical design structures so that complex procedures (e.g. chemical synthesis or biological assays) can be built up step-by-step. And in contrast to continuous-flow microfluidics, digital microfluidics works much the same way as traditional bench-top protocols, only with much smaller volumes and much higher automation. Thus a wide range of established chemistries and protocols can be seamlessly transferred to a nanolitre droplet format. Electrowetting, dielectrophoresis, and immiscible-fluid flows are the three most commonly used principles, which have been used to generate and manipulate microdroplets in a digital microfluidic device.

Working Principle

Droplets are formed using the surface tension properties of liquid. For example, water placed on a hydrophobic surface will lower its contact with the surface by creating drops whose contact angle with the substrate will increase as the hydrophobicity increases. However, in some cases it is possible to control the hydrophobicity of the substrate by using electrical fields. This is referred to as electrowetting on dielectric or EWOD. In thin layers of Teflon AF, FluoroPel V-polymer or CYTOP, for example, while no field is applied the surface will be extremely hydrophobic and a droplet of water will try to 'stay away' from the surface, resulting in a droplet with steep walls. When a field is applied, a polarised hydrophilic surface is created, and the water droplet tries to 'get closer' to the surface, resulting in much more spread out droplet. By controlling the localisation of this polarisation it is possible to control the displacement of the droplet.

Implementation

In one of various embodiments of EWOD-based microfluidic biochips, investigated first by Cytonix in 1987 and subsequently commercialised by Advanced Liquid Logic, there are two parallel glass plates, and the bottom plate contains a patterned array of individually controllable electrodes, and the top plate is coated with a continuous grounding electrode. A dielectric insulator coated with a hydrophobic is added to the plates to decrease the wettability of the surface and to add capacitance between the droplet and the control electrode. The droplet containing biochemical samples and the filler medium, such as the silicone oil, a fluorinated oil or air are sandwiched between the plates; the droplets travel inside the filler medium. In order to move a droplet, a control voltage is applied to an electrode adjacent to the droplet, and at the same time, the electrode just under the droplet is deactivated. By varying the electric potential along a linear array of electrodes, electrowetting can be used to move droplets along this line of electrodes.

LAB-ON-A-CHIP

A lab-on-a-chip (LOC) is a device that integrates one or several laboratory functions on a single chip of only millimetres to a few square centimetres in size. LOCs deal with the handling of extremely small fluid volumes down to less than picolitres. Lab-on-a-chip devices are a subset of MEMS devices and often indicated by 'micro total analysis systems' (μTAS) as well. Microfluidics is a broader term that describes also mechanical flow control devices like pumps and valves or sensors like flowmeters and viscometers. However, strictly regarded 'lab-on-a-chip' indicates generally the scaling of single or multiple lab processes down to chip-format, whereas 'μTAS' is dedicated to the integration of the total sequence of lab processes to perform chemical analysis. The term 'lab-on-a-chip' was introduced later on when it turned out that μTAS technologies were more widely applicable than only for analysis purposes (Fig. 10.2).

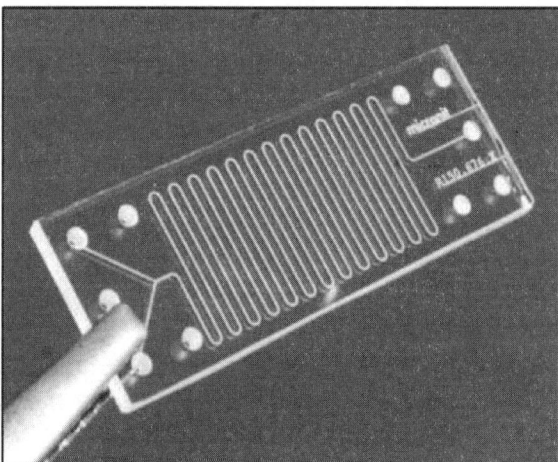

Fig. 10.2. Lab-on-a-chip made of glass.

After the invention of microtechnology for realising integrated semiconductor structures for microelectronic chips, these lithography-based technologies were soon applied in pressure sensor manufacturing as well. Due to further development of these usually CMOS-compatibility limited processes, a tool box became available to create micrometer or sub-micrometer sized mechanical structures in silicon wafers as well: the microelectromechanical systems (MEMS) era [also indicated with micro system technology (MST)] had started (Fig. 10.3).

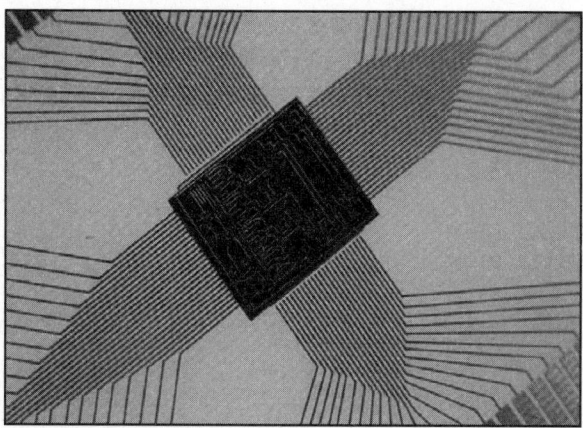

Fig. 10.3. Microelectromechanical systems chip, sometimes called 'lab-on-a-chip'.

Next to pressure sensors, airbag sensors and other mechanically movable structures, fluid handling devices were developed. Examples are: channels (capillary connections), mixers, valves, pumps and dosing devices. The first LOC analysis system was a gas chromatograph, developed in 1975 by S.C. Terry, Stanford University. However, only at the end of the 1980's, and beginning of the 1990's, the LOC research started to seriously grow as a few research groups in Europe developed micropumps, flowsensors and the concepts for integrated fluid treatments for analysis systems. These µTAS concepts demonstrated that integration of pretreatment steps, usually done at lab-scale, could extend the simple sensor functionality towards a complete laboratory analysis, including e.g. additional cleaning and separation steps.

A big boost in research and commercial interest came in the mid 1990's, when µTAS technologies turned out to provide interesting tooling for genomics applications, like capillary electrophoresis and DNA microarrays. A big boost in research support also came from the military, especially from Defense Advanced Research Projects Agency (DARPA), for their interest in portable bio- or chemical warfare agent detection systems. The added value was not only limited to integration of lab processes for analysis but also the characteristic possibilities of individual components and the application to other, non-analysis, lab processes. Hence the term 'lab-on-a-chip' was introduced.

Although the application of LOCs is still novel and modest, a growing interest of companies and applied research groups is observed in different fields such as analysis (e.g. chemical analysis, environmental monitoring, medical diagnostics and cellomics) but also in synthetic chemistry (e.g. rapid screening and microreactors for pharmaceutics). Besides further application developments, research in LOC systems is expected to extend towards down scaling of fluid handling structures as well, by using nanotechnology. Sub-micrometer and nano-sized channels, DNA labyrinths, single cell detection an analysis and nano-sensors might become feasible that allow new ways of interaction with biological species and large molecules.

Chip Materials and Fabrication Technologies

The basis for most LOC fabrication processes is photolithography. Initially most processes were in silicon, as these well-developed technologies were directly derived from semiconductor fabrication. Because of demands for, e.g. specific optical characteristics, bio- or chemical compatibility, lower production costs and faster prototyping, new processes have been developed such as glass, ceramics and metal etching, deposition and bonding, PDMS processing (e.g. soft lithography), thick-film- and stereolithography as well as fast replication methods via electroplating, injection moulding and embossing. Furthermore the LOC field more and more exceeds the borders between lithography-based microsystem technology, nanotechnology and precision engineering.

Advantages of LOCs

LOCs may provide advantages, which are specific to their application. Typical advantages are:
1. Low fluid volumes consumption (less waste, lower reagents costs and less required sample volumes for diagnostics).
2. Faster analysis and response times due to short diffusion distances, fast heating, high surface to volume ratios, small heat capacities.
3. Better process control because of a faster response of the system (e.g. thermal control for exothermic chemical reactions).
4. Compactness of the systems due to integration of much functionality and small volumes.
5. Massive parallelisation due to compactness, which allows high-throughput analysis.
6. Lower fabrication costs, allowing cost-effective disposable chips, fabricated in mass production.
7. Safer platform for chemical, radioactive or biological studies because of integration of functionality, smaller fluid volumes and stored energies.

Disadvantages of LOCs

1. Novel technology and therefore not yet fully developed.
2. Physical and chemical effects—like capillary forces, surface roughness, chemical interactions of construction materials on reaction processes—become more dominant on small-scale. This can sometimes make processes in LOCs more complex than in conventional lab equipment.
3. Detection principles may not always scale down in a positive way, leading to low signal-to-noise ratios.
4. Although the absolute geometric accuracies and precision in microfabrication are high, they are often rather poor in a relative way, compared to precision engineering for instance.

Examples of LOC applications

1. Real-time PCR detect bacteria, viruses and cancers.
2. Biochemical assays.
3. Immunoassay detect bacteria, viruses and cancers based on antigen-antibody reactions.
4. Dielectrophoresis detecting cancer cells and bacteria.
5. Blood sample preparation can crack cells to extract DNA.
6. Cellular lab-on-a-chip for single-cell analysis.
7. Ion channel screening.

LOCs and Global Health

Lab-on-a-chip technology may soon become an important part of efforts to improve global health, particularly through the development of point-of-care testing devices. In countries with few healthcare resources, infectious diseases that would be treatable in a developed nation are often deadly. In some cases, poor healthcare clinics have the drugs to treat a certain illness but lack the diagnostic tools to identify patients who should receive the drugs. Many researchers believe that LOC technology may be the key to powerful new diagnostic instruments. The goal of these researchers is to create microfluidic chips that will allow healthcare providers in poorly equipped clinics to perform diagnostic tests such as immunoassays and nucleic acid assays with no laboratory support.

Global challenges

For the chips to be used in areas with limited resources, many challenges must be overcome. In developed nations, the most highly valued traits for diagnostic tools include speed, sensitivity, and specificity; but in countries where the healthcare infrastructure is less well developed, attributes such ease of use and shelf life must also be considered. The reagents that come with the chip, for example, must be designed so that they remain effective for months even if the chip is not kept in a climate-controlled environment. Chip designers must also keep cost, scalability, and recyclability in mind as they choose what materials and fabrication techniques to use.

Examples of global LOC application

One active area of LOC research involves ways to diagnose and manage HIV infections. Around 40 million people are infected with HIV in the world today, yet only 1.3 million of these people receive anti-retroviral treatment. Around 90 per cent of people with HIV have never been tested for the disease. Measuring the number of CD4+ T lymphocytes in a person's blood is an accurate way to determine if a person has HIV and to track the progress of an HIV infection. At the moment, flow cytometry is the gold standard for obtaining CD4 counts, but flow cytometry is a complicated technique that is not available in most developing areas because it requires trained technicians and expensive equipment.

NANOFLUID: ENGINEERING THE FLUID

Nanofluids, a suspension of nanoparticles in a base fluid, have grabbed attention of scientists all over the world in the last decade. Nanofluids show an incredible increase in the thermal properties in heat transfer of the fluid. The following review is aimed to bring up some recent advances in nanofluid technology and identifies opportunities for further research.

In today's world miniaturisation has become a trend, as we see smaller and more efficient devices emerge in the market. This miniaturisation has also led to the development and demand for high performance in cooling.

Cooling is one of the most important technical challenges facing many diverse industries, including microelectronics, transportation, defense, solid-state lighting, power generation, etc. We are familiar with the fact that thermal conductivity of solids is more than that of liquids, as seen in Table 10.1. Thus, fluids containing suspended solid metallic particles are expected to display significantly enhanced thermal conductivities relative to conventional heat transfer fluids.

The use of suspensions is not new, but was developed long ago. People have worked with particle size in millimetres to micrometres, but faced various drawbacks. The abrasive action of particles causing erosion of components; clogging in small flow passages; increases in the pressure drop; and sedimentation are just few of the drawbacks.

As nanotechnology gave the tool to produce nanoparticles, nanofluids, i.e. fluid suspension of nanometer sized solid particles and fibres, have been proposed as a route for surpassing the performance of heat transfer liquids currently available.

The term nanofluid is envisioned to describe a solid-liquid mixture consisting of nanoparticles and a base fluid. Choi was the first to coin the term nanofluid. It originates from nanoparticle and base fluid.

Table 10.1. Thermal conductivities of solids and liquids.

	Material	*Thermal conductivity (W/mK)*
Metallic solids	Silver	429
	Copper	401
	Gold	318
	Aluminium	237
Nonmetallic solids	Silicon	148
	Copper oxide	76.5
	Alumina	40
	Silica	1.38
Metallic liquid	Sodium @ 644K	72.3
Nonmetallic liquids	Water	0.613
	Ethylene glycol	0.253
	Engine oil	0.145

Preparation

Preparation of a nanofluid is the first step in understanding it. Nanostructured or nano-phase materials are made of nanometer sized substances engineered on the atomic or molecular scale to produce either new or enhanced physical properties not exhibited by conventional bulk solids. The nanofluid simply does not refer to a liquid-solid mixture, but various parameters like stable suspension, uniform particle distribution, low agglomeration and no chemical change of the fluid. Depending on the application, many particle and fluid combinations can be used; for example nanoparticles of oxides, metal nitrides, metal carbides, carbon nanotubes, diamond, polymers, and nonmetals dispersed in liquids like water, ethylene glycol and oil. There are two methods for nanofluid production. The first and most commonly used is the two step method where nanoparticles are prepared by a process like chemical reduction, microemulsion, UV light irradiation, etc. and then mixed into the base fluid. Xuan and Li used dispersants like laurate salts and oleic acid for the stability of the solution whereas Smith used ultrasonic vibration to keep the particles from agglomerating.

The second method is the direct one-step method developed at Argonne National Laboratory (ANL), USA. In this method, the nanoparticles when formed are directly dispersed in a base fluid. An advantage of this method is that agglomeration is minimised. This method is only applicable for the production of oxide nanofluids. Zhu developed a one-step chemical method for the production of nanofluids, by reduction under microwave irradiation. Non-agglomerated and stably suspended nanofluid was produced. The particle size was 5 nm but few particles showed sizes as high as 20–40 nm also. Phuoc produced Ag-deionised water nanofluid using multi-beam laser ablation in a liquid. The particle size ranged from 20–30 nm. These samples were stable for many months, without the use of any dispersant or surfactant.

Experimental Investigation on Thermal Conductivity of Nanofluids

Thermal conductivity is an important parameter in enhancing the thermal conductivity of the heat transfer fluid. Since the thermal conductivity of metals is far more than that of the liquid, the thermal conductivity of suspensions would lie in between these two extremes. Many researchers have reported experimental studies on the thermal conductivity of nanofluids. Al_2O_3 and CuO are most commonly used nanoparticles by researchers, followed by copper, TiO_2, carbon nanotubes. Eastman measured thermal conductivity of CuO-water nanofluid and found an increase in thermal conductivity by 60 per cent with 5 vol% particle concentration. They used the transient hot-wire method for the measurement of thermal conductivity of nanofluid. Xuan and Li worked and presented data on the thermal conductivity of Cu nanoparticles in different base fluids. The ratio of thermal conductivity of nanofluid to the base fluid varied from 1.24 to 1.75, if the volume fraction of the ultra-fine particles increased from 2.5 to 7.5 per cent.

Smith measured temperature effect of thermal conductivity enhancement in nanofluids. It was observed that a dramatic increase in the enhancement of conductivity takes place with temperature. The observations showed that a two- to four-fold increase in the thermal conductivity of a nanofluid can take place over a temperature range of $21°$ to $5°C$.

Yoo compared the thermal conductivity of TiO_2, Al_2O_3, Fe, and WO_3 in water nanofluid. They found that the surface to volume ratio of nanoparticles is a principle factor in determining the thermal conductivities of nanofluids. Even though there are various parameters for thermal conductivity dependence, suspended particles do not give primary effect on the thermal conductivity of nanofluids.

Tyler and others reported thermal transport of diamond-midel oil nanofluid. They found the effect of sedimentation on the thermal transport in nanofluids. They also found negligible chance in viscosity of nanofluid for small particle concentration, which increased for higher particle concentration. Smith and Chopkar studied the thermal conductivity of alloy nanofluid. They showed that the conductivity of this nanofluid is significantly greater (1.2–2.4 times) than that of the respective base fluid. The increase in thermal conductivity ratio is a function of identity/composition, size, volume fraction and thermal properties of solid suspension. Collin and Allen used a combination of carbon nanotubes with gold and copper nanoparticles (hybrid nanofluids) in deionised water. They found no considerable increase in the thermal conductivity of the combination as compared to the single component nanofluid.

Flow, Convection and Boiling

Nanofluids, as stated above, are a suspension of nanoparticles in a base fluid, but nanoparticles act as fluid in the bulk. The nanofluid does not need any pumping power compensation nor do they block any of the micro flow channels, but show a considerable increase in heat transfer capacity.

Recently, the heat transfer coefficients of nanofluids in natural and forced convection have been studied by few scientists. In the heat transfer of nanofluids, the heat transfer coefficient depends not only on the thermal conductivity, but also on various other properties such as specific heat, density and dynamic viscosity. The viscosity of nanofluid depends on the methods used to disperse and stabilise the nanoparticle suspension. As mentioned before, the enhancements of thermal conductivity of nanofluids make them attractive for nanofluid cooling application. While using nanofluids for cooling in high heat flux applications, heat transfer process follows the boiling regime. As nanoparticles increase the thermal conductivity of conventional fluids, many researchers expected that nanoparticles would also have a reasonable potential to enhance the boiling heat transfer.

Thus, this chapter presented a glimpse of the potential that nanofluids show. There are opportunities associated with nanofluids. With better heat/thermal management the load on environment and fuel consumption is lowered. Future application of nanofluids is yet to be brought out to various other areas of science and engineering. Few industries have shown interest and involved themselves in future use of these novel materials at an industrial scale. The theory of nanofluids and the explanation of its behaviour and mechanism, remains an enigma and its understanding will help understand transport properties better.

Chapter 11

Self-assembled Monolayer

INTRODUCTION

A self-assembled monolayer (SAM) is an organised layer of amphiphilic molecules in which one end of the molecule, the 'head group' shows a special affinity for a substrate. SAMs also consist of a tail with a functional group at the terminal end as seen in Fig. 11.1.

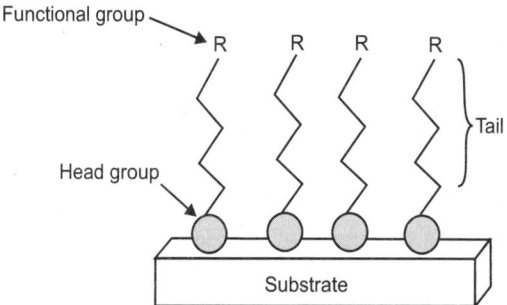

Fig. 11.1. Representation of a SAM structure.

SAMs are created by the chemisorption of hydrophilic 'head groups' onto a substrate from either the vapour or liquid phase followed by a slow two-dimensional organisation of hydrophobic 'tail groups'. Initially, adsorbate molecules form either a disordered mass of molecules or form a 'lying down phase' and over a period of hours, begin to form crystalline or semicrystalline structures on the substrate surface. The hydrophilic 'head groups' assemble together on the substrate, while the hydrophobic tail groups assemble far from the substrate. Areas of close-packed molecules nucleate and grow until the surface of the substrate is covered in a single monolayer.

Adsorbate molecules adsorb readily because they lower the surface energy of the substrate and are stable due to the strong chemisorption of the 'head groups'. These bonds create monolayers that are more stable than the physisorbed bonds of Langmuir-Blodgett films. Thiol-metal bonds, for example, are on the order of 100 kJ/mol, making the bond stable in a wide variety of temperature, solvents, and potentials. The monolayer packs tightly due to van der Waals interactions, thereby reducing its own free energy. The adsorption can be described by the Langmuir adsorption isotherm if lateral interactions are neglected. If they cannot be neglected, the adsorption is better described by the Frumkin isotherm.

Self-assembly is a new paradigm for micro- and nanofabrication. This chapter has focused on self-assembled monolayers (SAMs) as model systems for self-assembled materials. The successful development of SAMs from an intellectual curiosity to a material used in several applications illustrates the promise of self-assembly in fabrication.

Self-assembly a concept for fabrication that warrants the substantial development effort that will be required to convert laboratory processes into manufacturing practices. New long-range strategies for microfabrication are certainly needed. The methods for micro fabrication currently used may be reaching their limits in terms of scale; other considerations (capital and processing costs, waste management/ environmental concerns, degree of perfection) are also becoming increasingly important. Self-assembly is one option that addresses some of these problems. It offers an efficient route to complex structures ranging in size from a few nanometers to hundreds of microns, with a relatively low level of defects and with molecular control over structure and composition; it requires only simple equipment and facilities, has low capital costs and, in some cases, generates only small quantities of wastes.

Several applications for self-assembly in the short-term are possible. The use of SAMs for passivation of surfaces (protection from corrosion or contamination) is a plausible point of entry into several manufacturing processes; the use of patterned SAM as resists against wet and dry chemical etches or as materials to control the surface free energy of materials may also find applications. At present, patterned SAMs are exceptionally convenient for fabrication at the micron-scale and larger- a scale that is well-suited to applications in optics and biotechnology. Patterned SAMs with dimensions as small as 200 nm have been prepared but are not yet routine or reliable; techniques aiming for still smaller scales are being developed.

Whether self-assembly will be effective in high-end microelectronics fabrication remains to be determined. At this early stage of development, patterned SAMs may compete effectively as low-cost alternative techniques for microscale printing such as silk-screening. At present, however, no method of self-assembly can compete with photolithography for smaller (sub-micrometer) scale microelectronics fabrication.

TYPES OF SAMS

Selecting the type of head group depends on the application of the SAM. Typically, head groups are connected to an alkyl chain in which the terminal end can be functionalised (i.e. adding $-OH$, $-NH_3$, or $-COOH$ groups) to vary the wetting and interfacial properties. An appropriate substrate is chosen to react with the head group. Substrates can be planar surfaces, such as silicon and metals, or curved surfaces, such as nanoparticles. Thiols and disulphides are the most commonly used molecules for SAMs on noble metal substrates because of the strong affinity of sulphur for these metals. In addition, gold is an inert and biocompatible material that is easy to acquire. It is also easy to pattern via lithography, a useful feature for applications in nanoelectromechanical systems (NEMS). Additionally, it can withstand harsh chemical cleaning treatments. Silanes are generally used on nonmetallic oxide surfaces.

PREPARATION OF SAMS

Metal substrates for use in SAMs can be produced through physical vapour deposition techniques, electrodeposition or electroless deposition. Alkanethiol SAMs produced by adsorption from solution are made by immersing a substrate into a dilute solution of alkanethiol in ethanol for 12 to 72 hours at room temperature and dried with nitrogen. SAMs can also be adsorbed from the vapour phase. For example, chlorosilane SAMs (which can also be adsorbed from the liquid phase), are often created in a reaction chamber by silanisation in which silane vapour flows over the substrate to form the monolayer.

CHARACTERISATION OF SAMs

The structures of SAMs are most commonly determined using scanning probe microscopy techniques such as atomic force microscopy (AFM) and scanning tunnelling microscopy (STM). More recently, however, diffractive methods have also been used.

The structure can be used to characterise the kinetics and defects found on the monolayer surface. These techniques have also shown physical differences between SAMs with planar substrates and nanoparticle substrates.

Kinetics

There is evidence that SAM formation occurs in two steps, an initial fast step of adsorption and a second slower step of monolayer organisation. Many of the SAM properties, such as thickness, are determined in the first few minutes. However, it may take hours for defects to be eliminated via annealing and for final SAM properties to be determined. The exact kinetics of SAM formation depends on the adsorbate, solvent and substrate properties.

In general, however, the kinetics are dependent on both preparations conditions and material properties of the solvent, adsorbate and substrate. Specifically, kinetics for adsorption from a liquid solution are dependent on:

1. Temperature — room temperature preparation improves kinetics and reduces defects.
2. Concentration of adsorbate in the solution — low concentrations require longer immersion times and often create highly crystalline domains.
3. Purity of the adsorbate — impurities can affect the final physical properties of the SAM.
4. Dirt or contamination on the substrate — imperfections can cause defects in the SAM.

The final structure of the SAM is also dependent on the chain length and the structure of both the adsorbate and the substrate. Steric hindrance and metal substrate properties, for example, can affect the packing density of the film, while chain length affects SAM thickness.

Defects

Though the slow step in SAM formation often removes defects from the film, defects are included in the final SAM structure. Defects can be caused by both external and intrinsic factors. External factors include the cleanliness of the substrate, method of preparation, and purity of the adsorbates. SAMs intrinsically form defects due to the thermodynamics of formation. The high coverage of the adsorbate present in the SAM is, in fact, thermodynamically unstable.

Nanoparticle Properties

The structure of SAMs is also dependent on the curvature of the substrate. SAMs on nanoparticles including colloids and nanocrystals, 'stabilise the reactive surface of the particle and present organic functional groups at the particle-solvent interface'. These organic functional groups are useful for applications, such as immunoassays, that are dependent on chemical composition of the surface.

SELF-ASSEMBLED MONOLAYERS IN ORGANIC CHEMISTRY

Self-assembled monolayers (SAMs) are formed when surfactant molecules spontaneously adsorb in a monomolecular layer on surfaces. Two of the most widely studied systems of SAMs are gold-alkylthiolate monolayers and alkylsilane monolayers (Fig. 11.2).

The first gold-alkylthiolate monolayer was produced by Allara and Nuzzo at Bell laboratories in 1983. They realised the utility of combining a relatively inert gold surface with a bifunctional organic molecule in well-ordered, regularly oriented array.

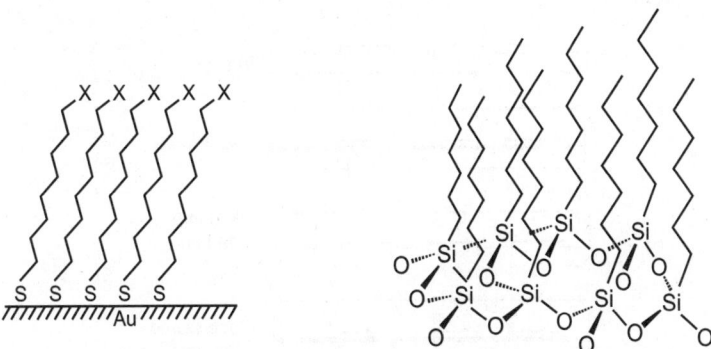

Fig. 11.2. Gold thiolate monolayer and alkylsilane monolayer.

SAMs offer a unique combination of physical properties that allow fundamental studies of interfacial chemistry, solvent-molecule interactions and self-organisation. Their well-ordered arrays and ease of functionalisation make them ideal model systems in many fields. SAMs are invaluable substrates in bioanalytical, organometallic, physical organic, bioorganic and electrochemistry.

Preparation

Gold-thiol monolayers are stable when exposed to air, and aqueous or ethanolic solutions for several months. They are also quite facile to produce. A 1–5 nm film of titanium is evaporated onto a glass coverslip or silicon wafer to promote adhesion of gold to the surface. A 10–200 nm film of gold is then evaporated onto the surface. The resulting gold surface is then immersed into a 2 mm solution of ethanolic alkylthiol, (disulphide solutions may also be used). Mixed monolayers may be formed if the ethanolic solution of ω-functionalised alkylthiols contains two or more different thiols.

Several procedures exist for producing patterned mixed monolayers. Lithiography lies at the heart of all of these techniques. One popular method is microcontact printing, μCP. Stamps with patterned reliefs are formed from elastomers, such as poly(dimethylsiloxane), PDMS, that have been poured over a master, cured and then peeled. The masters are manufactured from photolithography, e-beam writing, micromachining or relief structures etched into metals. Each master may be used to produce up to 50 stamps, and each stamp may be used multiple times. The stamp is inked with an ethanolic solution of ω-functionalised thiol and brought into contact with the gold surface for 10–20 seconds resulting in a gold thiolate monolayer at the areas of contact (Fig. 11.3).

Characterisation

Methods for studying monolayers can differ greatly from characterisation techniques of solution chemistry. Ellipsometry measures the change in amplitude and phase of light upon reflection. Using these values, the thickness and refractive index of a film can be calculated. This procedure can be used to determine the thickness of a monolayer before and after reactions to detect adsorption of molecules onto the SAM. Since the commercialisation of surface plasmon resonance spectrometers, surface plasmon resonance spectroscopy has become a vital tool in for imaging reactions on SAMs. Surface plasmon resonance is an *in situ* technique that measures changes in the refractive index of a monolayer attached to a metal surface. This allows for collection of both kinetic and thermodynamic information about a system in real time. X-ray photoelectron spectroscopy (XPS) is a technique which quantifies the elemental composition of monolayers.

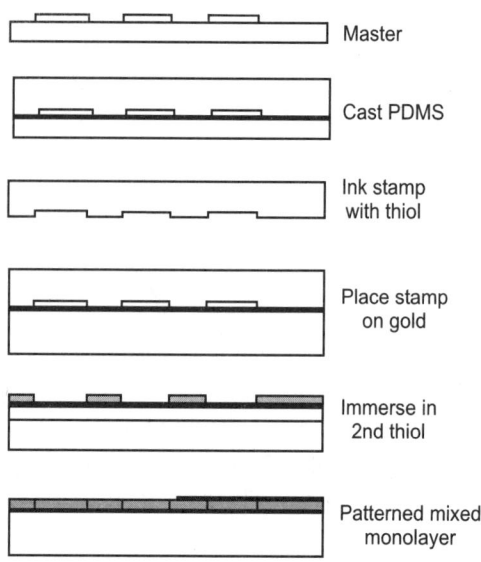

Master

Cast PDMS

Ink stamp with thiol

Place stamp on gold

Immerse in 2nd thiol

Patterned mixed monolayer

Fig. 11.3. Microcontact printing.

Reactivity

Reactivity of monolayers varies substantially from analogous solution reactions. Due to the close proximity of the alkyl chains, interchain reactions, also known as intrafilm reactions, can occur. Interchain reactions can lead to stabilising hydrogen bonding, dimerisations or chain polymerisations.

Reaction kinetics can vary widely from those observed in bulk solution. Often, heterogeneous kinetics are observed for interfacial chemistry. This arises from the rate of the reaction changing as the surface is modified.

Using cyclic voltammetry, Mrksich and co-workers investigated the kinetics of the Diels-Alder reaction of cyclopentadiene with monolayer-bound benzoquinone. In this system, kinetic rates stayed constant over the course of the reaction. The kinetics were dependent on the environment of the immobilised dienophile. With alkyl hydroxy chains co-adsorbed onto the monolayer, the reaction had a second-order rate constant consistent with the bimolecular solution reaction. With methyl terminated mixed monolayer, the reaction did not follow second-order kinetics, but was consistent with the cyclopentadiene adsorbing on the monolayer and then following a first order Diels-Alder reaction. This demonstrates the importance of solution-monolayer dynamics and how the environment of the monolayer can effect the kinetics of a reaction.

Acidity and basicity can be effected by the presence of a monolayer. Ionisation of the monolayer leads to an accumulation of charge across the surface. The formation of a double-layer changes the pH in the vicinity of the monolayer from that of the bulk solution. In general, acids become less acidic and bases become less basic by 2–5 pK_a units (Fig. 11.4).

Sterics play a large role in the reactivity of monolayers. Tightly packed monolayers can inhibit cis-trans photoisomerism of azobenzene molecules. When spacer molecules are integrated into the monolayer surface, photoisomerism is observed. S_N2 reactions on planar SAMs were investigated by Fryxell and coworkers with ω-bromoalkylsilane monolayers. They noted that only NaN_3 completely displaced bromide after 48 hours of reaction. Other nucleophiles, such as thiocyanate and cysteine thiolate, only reached 75 per cent completion, even after 48 hours. These reactions are significantly slower than the analogous solution reactions. The bromide reactivity was probed with hexamethylditin, and the radical reaction was quite facile. Since the bromide was shown to be reactive, the authors postulated that steric crowding around the monolayer causes the lack of reactivity in S_N2 reactions. The incoming nucleophile has to approach the electron-deficient carbon-bromine antibonding (σ*) orbital, which is located under the rigid surface of the monolayer. This demand gives rise to a substantial kinetic barrier.

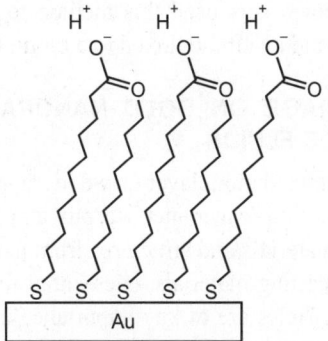

Fig. 11.4. Doublelayer.

One way to lower this barrier is to create a more flexible monolayer. Alkyl thiolates can adsorb onto gold nanoparticles and form monolayer-protected gold clusters, MPCs. The reactivity of these 3D molecules is quite different than that of planar SAMs. Due to the curvature of the gold core, the chain density of the monolayer decreases as the chain radiates from the core. ω-Bromoalkylthiolates adsorbed onto gold can undergo S_N2 reactions with primary amines with rates that are similar to the solution state. Submerging the bromide under the monolayer, by forming mixed monolayers with longer alkyl chains than bromoalkyl chains, caused a reduction in the rate of reaction, but the reaction still went to completion (Fig. 11.5).

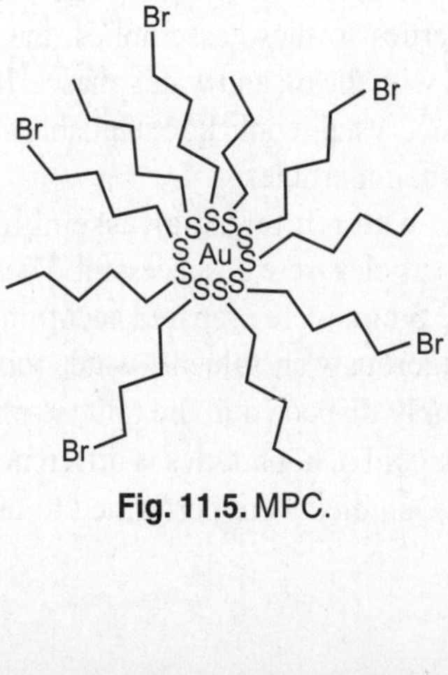

Fig. 11.5. MPC.

Mrksich and Houseman noticed an interesting steric effect in the glycosylation reaction of N-acetylglucosamine. They observed that the enzymatic activity of bovine β-1,4-galactosyltransferase (GalTase) is dependent on the density of adsorbed ligand. As the density of immobilised N-acetylglucosamine increased, the amount of ^{14}C-labelled galactose increased linearly. When the monolayer had a ligand density of 70 per cent, incorporation of ^{14}C-labelled galactose reached a maximum, then decreased with increasing ligand density. This is explained by the steric crowding of the ligands inhibiting substrate-enzyme interactions or the steric crowding of the disaccharide.

Common functionalisations of monolayers are similar to solid phase peptide synthesis. Carbonyl group activation, with coupling reagents such as DDC, EDC or $SOCl_2$, followed by reaction with alcohols or amines yield esters or amides, respectively. Microcontact printing can also be used to selectively functionalise patterned monolayers. Once a monolayer has activated carbonyl groups, the PDMS stamp can be inked with an amine or alcohol. Then amide or ester formation occurs only in the areas of contact with the stamp. Whitesides and coworkers used this method to create an amide bond between an immobilised activated ester group and an amine linked to a biotin molecule.

MIXED MONOLAYER COVERAGE ON GOLD NANOPARTICLES FOR INTERFACIAL STABILISATION OF IMMISCIBLE FLUIDS

Gold nanoparticles covered with a mixed monolayer of n-dodecanethiol and 11-mercapto-1-undecanol were prepared and found to mediate the oil–water interface, providing access to stable water droplets in oil.

The synthesis of well-defined materials and structures from nanoscopic components, using simple approaches and readily available starting materials, is essential for the use of these new materials in technological applications. Nanoparticles are of key importance in this regard for the properties they impart when integrated effectively into polymer materials, fluids, or biological systems. The self-assembly of particles at the fluid–fluid interface can be applied to nanoparticles as a simple method to achieve microscale assemblies, capsules, and ultra-thin sheets. Such assemblies retain the characteristic properties of both the nanoparticles (i.e. conductivity, fluorescence, magnetism, etc.) and the ligands attached to the nanoparticles, and are uniquely suited as functional materials for encapsulation and controlled release.

Smith recently reported the interfacial assembly of tri-n-octylphosphine oxide (TOPO)-covered cadmium selenide (CdSe) nanocrystals, or quantum dots. These nanocrystals were synthesised by high-temperature methods from CdO and Se(O) precursors, precipitated to remove excess surfactant, and found to assemble at the oil–water interface to encapsulate water droplets in a continuous oil phase, or oil droplets in a continuous water phase. The interfacial activity observed for the CdSe nanoparticles reduces interfacial tension. While quantum dots (i.e. the CdSe core nanocrystalline material) impart electronic and photophysical properties to these assemblies, the ligands define the solubility of the nanoparticles, and their interactions with the oil and water phases. In this sense, the oil–water interfacial assembly of TOPOcovered CdSe nanocrystals could be extendable to other ligandstabilised nanoparticles, such as n-alkanethiol-covered gold nanoparticles.

Initial efforts to achieve oil–water interfacial assemblies with the commonly prepared n-dodecanethiol-covered gold nanoparticles were unsuccessful. These alkane-covered gold nanoparticles, with an average diameter of about 2.5 nm, were prepared according to the methods of Brust. When the gold nanoparticles were shaken vigorously in toluene–water mixtures, no interfacial assembly was observed, and the nanoparticles simply dispersed in the toluene phase. This suggests that the extent of the hydrocarbon ligand coverage on gold nanoparticles is different, and likely denser, than on the CdSe nanocrystalline quantum dots. Thus, studies were performed to tailor the ligand coverage of the gold nanoparticles to enable their use in interfacial assemblies. To this end, gold nanoparticles with a more hydrophilic periphery were prepared by ligand exchange chemistry, where the n-dodecanethiol monolayer on the gold nanoparticle surface was replaced entirely with 11-mercapto-1-undecanol, to give gold nanoparticles with a hydroxyl rich periphery. These hydroxyl-functionalised gold nanoparticles also failed to stabilise the oil–water interface; they did not disperse in either the water or toluene phases, but instead precipitated when shaken in the presence of these two solvents.

Gold nanoparticles are distinct from many other types of nanoparticles in that their coverage with thiol-based ligands provides a robust periphery with strong interactions (near covalent bonding) of the thiol ligands to the gold surface. Yet, the thiol ligands are mobile, and can move laterally along the monolayer surface. Furthermore, gold nanoparticles are amenable to 'mixed monolayer' coverages, where different ligands (i.e. alkane thiols with different terminal groups) can be affixed to the particle surface in well-defined ratios. Given our observations with n-dodecanethiol and 11-mercapto-1-undecanol coverages, a mixed monolayer of these two ligands was considered as an approach to achieving interfacial assembly. A range of gold nanoparticle samples was thus prepared with different relative ratios of n-dodecanethiol and 11-mercapto-1-undecanol ligands in the mixed monolayer surface coverage. These samples were obtained by partial ligand exchange of 11-mercapto-1-undecanol on n-dodecanethiol functionalised gold nanoparticles.

The mixed monolayer coverage was characterised by ^{1}H-NMR spectroscopy in CDCl$_3$–CD$_3$OD solution to determine the relative amounts of the two ligands on the gold surface by integration of the methylene proton resonance of the undecanol ligand (R–CH$_2$OH at δ 3.4 ppm) against the methyl proton resonance of the dodecanethiol (R–CH$_3$ at δ 0.70 ppm).

Several gold nanoparticle samples with mixed monolayer coverages were tested for their ability to stabilise the toluene–water interface. These experiments showed that gold nanoparticles with less than 40 mole per cent undecanol embedded within the mixed monolayer dispersed in toluene, and those with more than 50 per cent undecanol coverage did not disperse in either solvent. However, as shown in Fig. 11.6, gold nanoparticles with an undecanol-to-dodecane ligand ratio of approximately 1 to 1.1 (determined by ^{1}H NMR spectroscopy) were found to stabilise the toluene–water interface, and, consequently, water droplets in a continuous organic phase. When these partially hydroxylfunctionalised gold nanoparticles were sonicated in toluene–water mixtures (MilliQ-purified water), water droplets formed that proved stable against coalescence over considerable time periods (weeks or longer). Performing the same experiment in the presence of the aqueous soluble sulphorhodamine B confirmed the aqueous nature of the encapsulated phase. As shown in Fig. 11.7a, fluorescence confocal microscopy revealed the characteristic dye fluorescence (λ_{max} = 554 nm) in the encapsulated aqueous region. For this confocal characterisation, the droplets were transferred to curved-well glass slides with flat coverslips, and diluted with toluene. The gold nanoparticle-covered droplets were drop-cast onto carbon-coated copper grids and allowed to dry. Transmission electron microscopy (TEM) revealed individual gold nanoparticles (Fig.11.7c). The nanoparticle assemblies were stabilised further by the presence of a hydrogel in the capsule interior. An aqueous solution of poly(vinyl alcohol) was added to the mixed monolayer protected gold nanoparticles in toluene, where the nanoparticles were labelled with rhodamine B as part of the ligand structure. Methanol was then used to remove toluene and water from the system, whereupon the assembly was maintained as shown by the presence of the fluorescently labelled particles in the cross-sectional image of Fig. 11.7d. Droplet size could be controlled roughly by agitation— sonication gave droplets of less than 50 mm diameter, while vigorous shaking resulted in droplets in the 100–200 mm range.

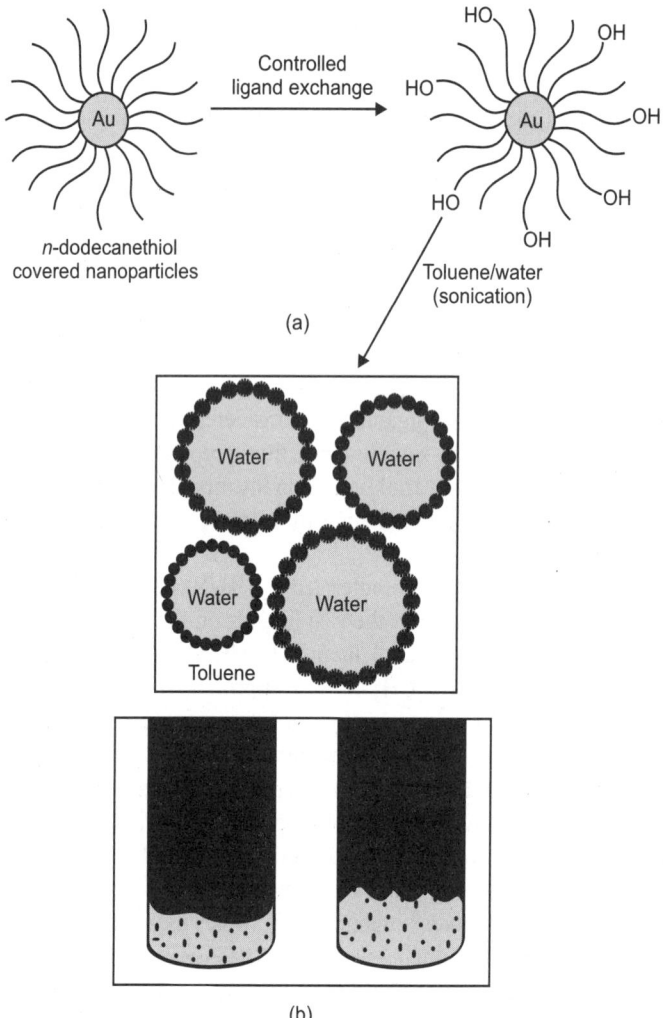

(a)

(b)

Fig. 11.6. (a) Schematic of controlled ligand exchange to provide functional gold nanoparticles for interfacial assemblies and the stabilisation of water droplets in toluene; (b) left-hand side: photograph of dodecanethiolcovered gold nanoparticles in toluene as a dark dispersion resting on top of an aqueous solution of sulphorhodamine B, where no droplet stabilisation is observed; right-hand side: photograph of water droplets stabilised by the functionalised gold nanoparticles resting on the bottom of the vial below a dispersion of excess nanoparticles.

The mixed monolayer-covered gold nanoparticles that are found to assemble at the oil–water interface are not amphiphilic in the traditional sense, in that they are not soluble in either water or oil. Nonetheless, the ligand coverage is balanced such that interfacial mediation and droplet stabilisation is enabled. The nearly equivalent molar ratio of the dodecane and undecanol ligands used on the particles that gave successful assemblies, combined with the surface mobility of alkane thiols on gold, suggests the possibility that a fluid-induced phase separation of ligands on the gold surface occurred, yielding distinct polar and apolar hemispheres. Related studies using similar ligand coverages on gold nanoparticles have been conducted at the air-water interface.

This would give the nanoparticles a surfactant-like structure. Future studies will examine the ligand distribution on the surface of the assembled particles to probe this possibility.

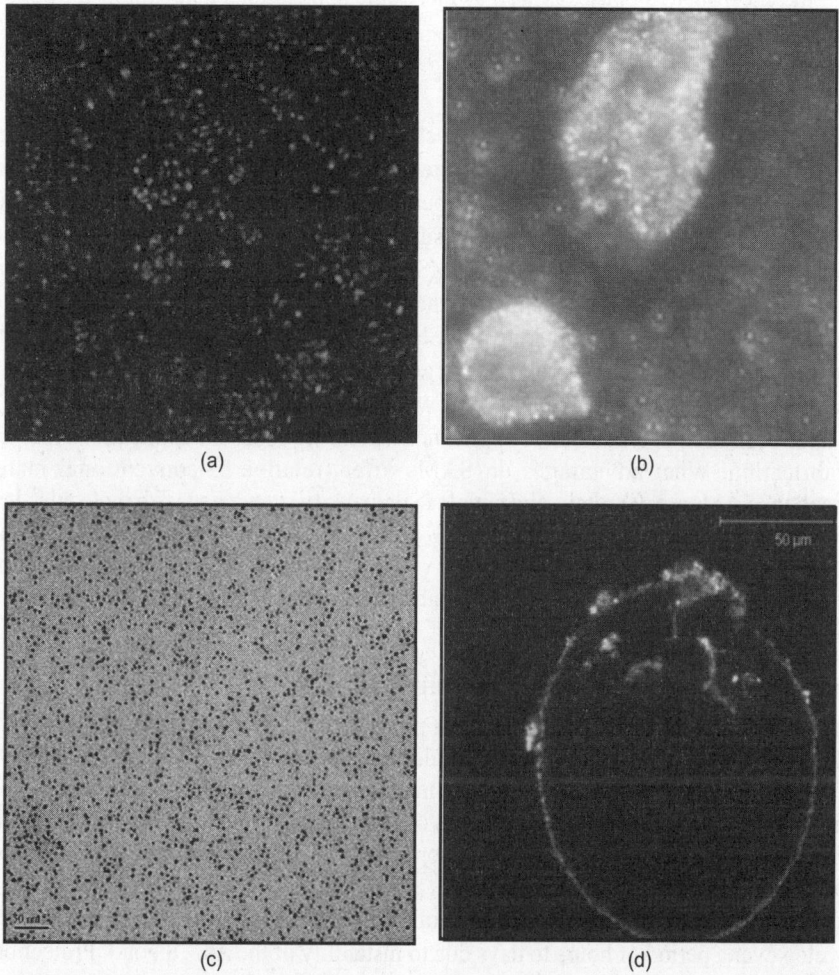

Fig. 11.7. (a) Fluorescence confocal microscope image, and (b) optical microscope image of gold nanoparticle-stabilised water droplets with sulphorhodamine B in the encapsulated phase (scale bar = 40 microns), (c) TEM image taken on part of a dried droplet showing individual nanoparticles (scale bar = 25 nm), (d) Fluorescence confocal image of nanoparticle-stabilised water droplets with a poly(vinyl alcohol) interior, and rhodamine B labelled gold nanoparticles after removal of the oil–water interface by washing with methanol (scale bar = 80 mm).

Thus, the ability to generate capsular assemblies of gold nanoparticles at the oil–water interface has been demonstrated through the use of functional ligands and mixed monolayers on the nanoparticle surface. While there have been several recent reports on the mediation of fluid–fluid interfaces by gold nanoparticles these have typically utilised citrate-stabilised gold nanoparticles, co-solvent mediation, or homogenous coverages with specially prepared ligands, containing, for example, bromopropionate groups, that serve to balance the water contact angle of the nanoparticles. The ability to use mixed monolayers to tune interfacial interactions for mediation of the oil–water interface provides a very simple approach to impart precise levels of surface functionality to the monolayer. The current studies include the use of chemistries that exploit such surface functionality in postassembly chemistry with materials in the encapsulated phase, as well as electronic applications that exploit the conductivity of the gold nanoparticles used in the assemblies.

APPLICATIONS OF SAMS

Areas of application for SAMs include biology, electrochemistry and electronics, nanoelectromechanical systems (NEMS) and microelectromechanical systems (MEMS), and everyday household goods. SAMs can serve as models for studying membrane properties of cells and organelles and cell attachment on surfaces. SAMs can also be used to modify the surface properties of electrodes for electrochemistry, general electronics, and various NEMS and MEMS. For example, the properties of SAMs can be used to control electron transfer in electrochemistry. They can serve to protect metals from harsh chemicals and etchants. SAMs can also reduce sticking of NEMS and MEMS components in humid environments. In the same way, SAMs can alter the properties of glass. A common household product, Rain-X, utilises SAMs to create a hydrophobic monolayer on car windshields to keep them clear of rain.

SAMs have been used in a number of applications. Here we focus on applications that are relevant to microfabrication. What advantages do SAMs offer (relative to conventional materials) for microfabrication? SAMs are: (i) robust and simple to prepare; (ii) generated with molecular-level control over their size and composition; (iii) ultra-thin (~1–2 nm) and therefore plausible resists for nanometer-scale, high-resolution lithography; (iv) compatible with a wide-variety of functional groups which allows control of surface properties such as wettability and surface free energy; and (v) can be patterned by several techniques.

Passivation of Surfaces: Protection from Corrosion

Self-assembled monolayers can provide excellent protection of an underlying metal against corrosion. A study of SAMs of alkanethiols on copper concluded that: (i) SAMs protected the metal from oxidation; (ii) increasing the length of the alkyl chain in the monolayer decreased the rate of oxidation (increased protection), and (iii) the kinetics of oxidation was consistent with a model in which the SAM protected the copper by inhibiting transport of O_2 to the copper surface. SAMs of alkanethiolates also provide protection for GaAs against oxidation.

SAMs of octadecanethiol retarded the growth of oxide on GaAs significantly, although the oxide did slowly develop over a period of hours to days due to instability of the As – S bond. Protection provided by other types of thin films against corrosion for Fe and Ni suggests that SAMs will probably be effective in providing corrosion resistance to many other materials.

Fabrication Using Patterned SAMs as Resists

SAMs of alkanethiolates protect gold from corrosion for a variety of wet, chemical etchants, including basic solutions of CN^- and ferricyanide. Alkanethiolate SAMs terminated by CH_3 provide the best protection: SAMs with terminal groups such as COOH, OH or CN are less effective. The extent of protection depends on the length of the alkyl chain. For a C_{16} SAM of HDT on gold, the protection is sufficient for fabrication of gold structures using patterned SAMs as resists: in the time required to remove a 100 nm thick film of bare gold fully with an etchant solution, gold protected by SAM formed from HDT on Au showed only pinhole defects.

Thinner SAMs are much less effective. SAMs of hexanethiol (C_6) on gold, for example, offer only marginal protection from the same etchant relative to bare gold.

Fabrication on Curved Surfaces

Microcontact printing can also be used to fabricate micron-scale features on curved surfaces. Microcontact printing on curved substrates uses stamps fabricated from PDMS. Because they are elastomeric, these stamps can conform to a nonplanar substrate without distorting the pattern on their surface. Figure 11.8 outlines a procedure of structures formed. At present, there is no comparable technique for microfabrication on curved surfaces. This work opens the door immediately to new optical and perhaps optoelectronic structures.

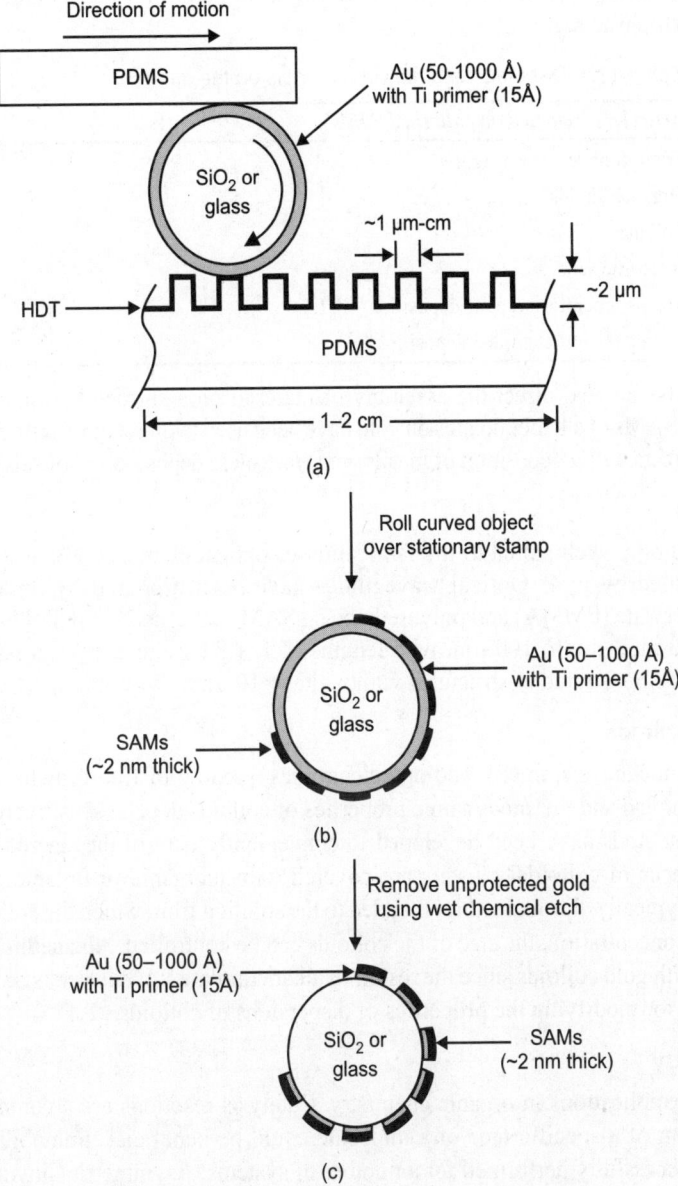

Fig. 11.8. Scheme for μCP on curved surfaces.

Directed Assembly of Materials on the Surface of a Patterned SAM

Patterned SAMs can be used to extend the principle of molecular self-assembly fabrication in the plane perpendicular to the plane of the monolayer (Table 11.1) by using a SAM in which different regions of the surface are patterned in different organic functional groups. These functional groups make different regions of the surface have different values of surface free energy and different wettabilities. These differences in surface free energy and wettability can then be used to direct the assembly of certain material to specific regions of the surface. In one example, when patterned SAMs are exposed to water vapour (in air) at low temperatures (or high humidity), droplets of water condense on the surface selectively on hydrophilic regions.

Table 11.1. Directed assembly of materials on the surface.

Material assembled on patterned SAMs
Water (condensation figures)
Organic liquids
Polymers
Inorganic salts
Metals (chemical vapour deposition, CVD)
Metals (electrochemical deposition)

Patterned SAMs can also direct the assembly of material on a surface by inhibiting a deposition process. Patterned SAMs of alkanethiolates on gold have been used to direct the electrochemical deposition of metals and polymers, CVD deposition of metals, and electroless deposition of metals in selected areas.

Optical Systems

Microcontact printing is well suited to the fabrication of optical elements. Metal diffraction gratings have been fabricated by μCP. Optical waveguides have been prepared by directed assembly of polymethylmethacrylate (PMMA) and polyurethane on SAMs patterned by μCP (Fig. 11.9). Structures ranging in width from 3 μm to 100 μm, with lengths of 1–1.5 cm are easily accessible. Multi-mode waveguiding was demonstrated in structures with widths >10 μm.

Fabrication of Colloids

Colloids are used in catalysis, micro- and optoelectronics, seeding of film growth, and spectroscopy. The electrical, chemical and thermodynamic properties of colloids depend sensitively on their size and shape. Recently, methods have been developed to conveniently control the size of colloids. In these methods, the surfaces of colloidal clusters are covered with inorganic or organic functional groups during synthesis. Typically, the alkanethiol is added to the solution from which the colloids are prepared; by controlling its concentration, the size of the colloids can be controlled. Alkanethiols are particularly effective for use with gold colloids since the resulting alkanethiolate SAM is very stable. SAMs provide a useful technique for modifying the properties of dispersions of colloids.

Organic Chemistry

SAMs have many applications in organic chemistry. Catalysis reactions are a common area where a defined presentation of a specific face of a molecule could be beneficial. Immobilising catalysts on SAMs has been successfully performed for a number of systems. Asymmetric dihydroxylation can be carried out on MPCs. If the monolayer immobilises the catalyst in a beneficial orientation, desired reaction rates can increase dramatically. Tremel and coworkers attached a ruthenium ROMP catalyst to MPCs and found that the catalyst not only catalysed ring-opening metathesis of norborene, its turnover frequency increased from 3000 to 16,000 h^{-1}. When the catalyst was immobilised to a planar monolayer, the turnover frequency increased to 80,000 h^{-1}.

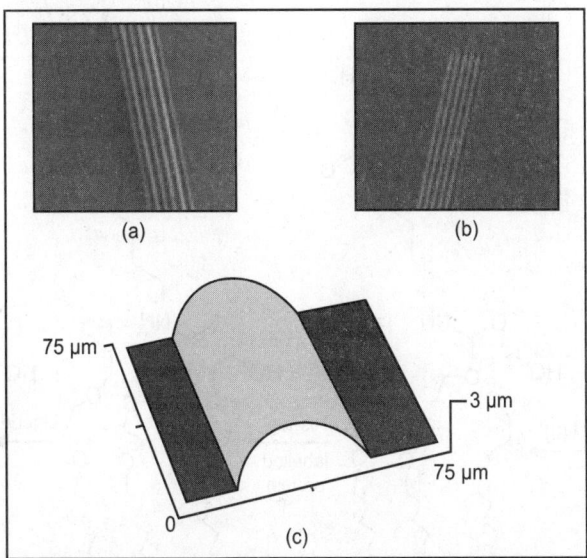

Fig. 11.9. Scanning electron micrographs (a–b) and an AFM image (c) of polymeric waveguides prepared by directed assembly of polymethylmethacrylate on a patterned SAM.

Molecular recognition is another important application for SAMs. Macromolecules, including DNA, proteins, and even cells can be immobilised onto monolayers. A major breakthrough came when Whitesides and Prime realised that oligio(ethylene glycol) chains prevent non-specific binding of proteins. This allows for adherence of specific biomolecules without unspecific binding events. Utilising techniques, such as SPR and ellipsometry, binding and recognition of biomolecules can be measured in real time to allow for calculation of association and dissociation constants. Competitive binding experiments and many other physical properties of biological systems can be assayed using SAMs (Fig. 11.10).

Applications in Other Areas

Self-assembled monolayers on a gold surface constitute an ideal system for practising fundamental studies related to the electron transfer process. The electrode surface used for this purpose is modified with electoactive monolayers through self-assembly. The attachment of thiols with the active terminal group allows further derivatisation through classical organic reactions. In one attempt, cycloaddition was also used to derivatise a surface modified with thiol containing azide at the termini.

Chemical force microscopy (CFM) combined with chiral discrimination by a molecule can be used to distinguish different chiral forms of the same molecule. In this technique, an AFM tip is functionalised with a chiral molecule. This chiral probe is then used to discriminate between the two chiral forms of the same molecule on a surface. The gold-coated AFM tip is functionalised with a chiral probe by using acylated phenyl glycine modified with alkanethiol. The changes in the friction or adhesion forces are used to distinguish between the two enantiomers of mandelic acid.

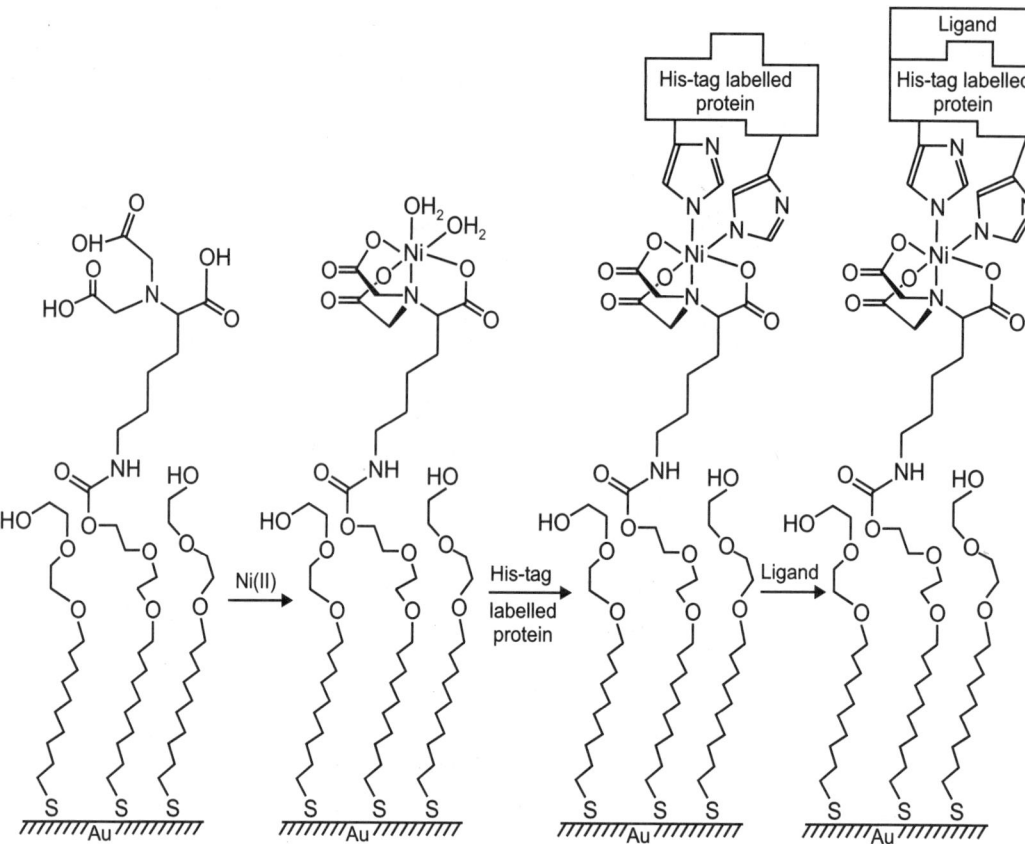

Fig. 11.10. Protein immobilisation of His-tag labelled proteins.

Wetting control

The wetting properties of a surface can be modified, to a great extent, by coating it with a monolayer of molecules. This is one of the important applications of SAMs. A low coverage surface is made intentionally by the self-assembly of alkanethiol with a bulky end group on an Au (111) surface. The end group is then hydrolysed to form a carboxy terminated monolayer with low surface coverage. The carboxylate groups generated by the hydrolysis can be attracted to the gold surface by applying an electric field. Thus the surface can now be made hydrophilic or hydrophobic by changing the electrochemical potential.

Molecular electronics

SAMs are also used to make electrical contacts. By using a self-assembled monolayer of dithiol on gold, one can make a surface with pendent thiol groups and can also attach a gold nanoparticle. This attachment with a covalent linkage showed four orders of magnitude higher current than when the nanoparticle was physisorbed on the SAM surface at the same tip bias voltage. This showed that monolayer-based electrical contacts are feasible. In various studies, different kinds of interactions such as van der Waals, hydrogen bonding and covalent interactions have been made between monolayers attached to metallic surfaces such as gold and mercury. These studies have shown that the electron transfer rates increased in the order, van der Waals > hydrogen bonding < covalent. In all these experiments, it is necessary to make a contact. This is done through monolayers. The experimental protocol is illustrated. In the experiment, a nanoparticle solution is exposed to a monolayer making nanoparticles sit on the monolayer. The current flowing between the tip and the surface is measured at a bias voltage.

Templates

SAMs are excellent templates on which nanofabrication can be done. A given structure can be produced on the surface by using a number of methodologies. Various nanoscopic objects can be used for producing a structure, several of these objects are detailed below.

Nanoparticles

Monolayer-protected nanoparticles are new kinds of materials. In this class of systems, monolayers are grown on the surface of nanoparticles of metals, semiconductors and insulators. The monolayer protection allows the nanoparticle to preserve its size and shape. Through a judicious choice of the functionality on the nanoparticle surface, one can bind the material on to an SAM in a well-defined fashion. The nanoparticle surface can be coated with various molecules having additional properties. The chemistry discussed on the planar monolayer surface can be eminently done on the nanoparticle surface as well. One of the important aspects in the use of nanoparticles is that with monolayer functionalisation, nanoparticles can be incorporated in any matrix such as polymers. It is also possible to attach nanoparticles on solid surfaces through a monolayer anchor. This allows one to produce device structures with nanoparticles.

Nanotubes

By functionalising carbon nanotubes, one can attach them to SAMs through specific chemistry. This facilitates the arrangement of aligned nanotubes on monolayer surfaces. In this way, nanotubes can be made available on stable supports. They can be used as devices for applications such as field emission.

Crystal growth

As mentioned earlier, SAMs can nucleate crystal growth and such low temperature chemical routes are important for making ultra-thin layers of materials. By creating a functional molecular surface at specific locations, it is possible to grow another material at these locations. Several examples of these are known. The formation of tetragonal zirconia on SAMs has been reported. In Fig. 11.11 a SEM image of ZrO_2 crystals grown on a SAM is shown. The crystals are highly faceted, and signify one of the very early examples of low temperature growth of ordered materials. The SAMs act as templates on which initial nucleation occurs. There are also examples of this kind of assisted growth on metal nanoparticles, wherein a ZrO_2 shell is grown on silver nanoparticles by solution chemistry.

By patterning an SAM with mercaptohexanoic acid (MHA) and the balance with mercaptohexanol (MCH) it has been shown that calcite crystals are grown selectively at locations of MHA instead of MCH. Several other materials have also been grown in this way.

Polymers

One can build polymers on an organised structure. The variety in polymeric structures is diverse so that almost anything can be grown on a SAM surface through an appropriate choice of precursors.

Fig. 11.11. Scanning electron micrograph of tetragonal ZrO_2 grown on self-assembled monolayers.

Complex molecules

One can also think of placing biological molecules at specific locations just as in materials chemistry. The most common biomaterials are proteins and placing them on metal surfaces is important. However, proteins placed directly on surfaces get denatured, which is why placing them on SAMs is a feasible alternative. The problem, however, is that SAMs have a non-specific affinity for proteins, which needs to be controlled. Ethylene glycol units on the surface prevent non-specific protein adsorption and SAMs with 4 to 7 poly(ethyleneglycol) units are used for this purpose. Specific thiolated monolayers can be grown on surfaces and the protein can have a biorecognition function. Such an approach can be used to adhere cells onto monolayers. This is mediated by proteins of the extracellular matrix (ECM) such as fibronectin and collagen.

These proteins can be anchored to surfaces by using patterned monolayers. All the unpatterned region of the SAM is covered with poly(ethyleneglycol) monolayer so that no protein adsorption occurs. Cells can then be attached to the protein modified sites. The size of the islands affects the cell growth. This kind of capability, along with the spatial control that is possible in monolayer growth, will make it possible to have lab on a chip.

Layer-by-layer-structure

SAMs involve growing layers. If such monolayers can be grown one over the other, step by step, microscopic thickness can be generated. For achieving this, it is important to have the link the monolayers through covalent chemistry. In one such approach, a benzenedimethane thiol (BDMT) monolayer was linked to another thiol through the covalent chemistry of the pendent thiol group of the BDMT monolayer.

CURRENT ISSUES

A number of issues will need to be addressed as self-assembly is developed for applications in the fabrication of micro- and nanosystems. These issues are discussed below.

Developing new systems of SAMs and optimising existing systems: SAMs of alkanethiolates on gold are presently the best developed system. Gold, however, is not a good substrate for many applications, particularly in silicon microelectronics. SAMs of alkanethiolates on gold are also too unstable thermally for many processing applications. Other systems of SAMs exist, but these systems need further development to be used generally for fabrication. Iron, nickel, chromium, titanium, aluminum, and stainless steel are possible targets for new substrates.

Decreasing the density of defects in the monolayer: At present, the density of defects in metal structures fabricated by etching using patterned SAMs as resists is too high for applications in high-density microelectronics. Improved methods for evaluating and reducing the density of defects are needed.

Using SAMs to augment current technologies for microelectronics fabrication: The most plausible application is surface passivation. Initial results that use SAMs to direct CVD processes are promising.

Developing SAMs as components of optical systems: Self-assembly at the micron-scale seems particularly well-suited for use in optical systems. Initial work with waveguides and microlens arrays suggests wide application, but both types of systems are in very early stages of development.

Techniques for registration of multiple patterning steps: Most complex systems involve multilayer processing and structures. For techniques that rely on elastomeric stamps (such as microcontact printing), developing methods for registration will be crucial.

Exploring techniques for self-assembly in three dimensions: Most of the work to date using SAMs for fabrication has generated patterns in the plane of the monolayer. Directed assembly of materials demonstrates the practicality of the fabrication of mesoscopic, 3-dimensional structures by self-assembly.

Extending current techniques to structures with nanometer-scale lateral resolution: At present, only electron beam and STM writing have been used to fabricate patterned SAMs with features smaller than 100 nm. Many important phenomena involving electrons appear only in structures with lateral dimensions less than 50 nm. To make an impact in nano-electronics, techniques that use self-assembly must provide reliable and convenient access to this regime of size.

Exploring non-organic self-assembled structures: There is no intrinsic reason that self-assembly should be restricted to organic structures. Self-assembled inorganic multilayers have been reported. Extension of this work to other materials such as ceramics is an important goal.

Developing SAMs showing useful electrical function: Most SAMs are based on organic molecules, which are electrical insulators. Developing methods of building organic conductivity into organic self-assembled systems is an important issue. Several attempts have been made to enhance the electrical properties of SAMs, including the incorporation of ferrocenes, the synthesis of thiol-terminated polythiophene and polyphenylene chains, and the synthesis of polypyrrole-terminated alkanethiolate SAMs.

Chapter 12

Nanostructured Advanced Materials

INTRODUCTION

A focus of frontline interdisciplinary research today is the development of the conceptual framework and the experimental background of the science of nanostructured materials and the perspectives of its technological applications. We consider some current directions in the preparation, characterisation, manipulation, and interrogation of nanomaterials, in conjunction with the modelling of the unique structure -dynamics function relations of nanostructures and their assemblies. The implications of quantum size and shape effects on the energetics, nuclear-electronic level structure, electric optical response and dynamics, reveal new unique physical phenomena that qualitatively differ from those of the bulk matter and provide avenues for the control of the function of nanostructures. Current applications in the realm of nanoelectronics, nanooptoelectronics, and information nanoprocessing are addressed, and other directions highlighted. Chemical sciences make a central contribution to this novel and exciting scientific technological area.

PROLOGUE

Nanoscience and nanotechnology pertain to the synthesis, characterisation, exploration, interrogation, exploitation, and utilisation of nanostructured materials, which are characterised by at least one dimension in the nanometer (1 nm = 10^{-9} m) range. Such nanostructured systems constitute a bridge between single molecules and infinite bulk systems. Individual nanostructures involve clusters, nanoparticles, nanocrystals, quantum dots, nanowires, and nanotubes, while collections of nanostructures involve arrays, assemblies, and superlattices of individual nanostructures. Table 12.1 lists some typical dimensions of nanomaterials. The chemical and physical properties of nanomaterials can significantly differ from those of the atomic-molecular or the bulk materials of the same chemical composition. The uniqueness of the structural characteristics, energetics, response, dynamics, and chemistry of nanostructures is novel and constitutes the experimental and conceptual background for the novel field of nanoscience. Suitable control of the properties and response of nanostructures can lead to new devices and technologies. The underlying themes of nanoscience and nanotechnology are dual: first, the bottom-up approach of miniaturisation of the components, as advanced by Richard Feynman in his often-cited lecture stating that 'there is plenty of room at the bottom'; second, the top-down approach of the self-assembly of molecular components, where each molecular or nanostructured component plugs itself into a suprastructure. This approach was pioneered by Jean-Marie Lehn, revealing that 'there is plenty of room at the top'.

Table 12.1. Nanostructures and their assemblies.

Nanostructure	Size	Material
Clusters Nanocrystals Quantum dots	Radius: 1–10 nm	Insulators, semiconductors, metals, magnetic materials
Other nanoparticles	Radius: 1–100 nm	Ceramic oxides
Nanobiomaterials Photosynthetic reaction center	Radius: 5–10 nm	Membrane protein
Nanowires	Diameter: 1–100 nm	Metals, semiconductors, oxides, sulphides, nitrides
Nanotubes	Diameter: 1–100 nm	Carbon, layered chalcogenides
Nanobiorods	Diameter: 5 nm	DNA
2D arrays of nanoparticles	Area: several nm^2–μm^2	Metals, semiconductors, magnetic materials
Surfaces and thin films	Thickness: 1–1000 nm	Insulators, semiconductors, metals, DNA
3D superlattices of nanoparticles	Radius: several nm	Metals, semiconductors, magnetic materials

The IUPAC Conference on New Directions in Chemistry, the Workshop on Nanostructured Advanced Materials (WAM II), addressed the recent developments in the broad, interdisciplinary research field of nanoscience and nanotechnology, focusing on:

1. Quantum structures, that is, nanoparticles and nanocrystals of metals and of semiconductors, nanostructures, nanowires, and nanobiological systems.
2. Assemblies of nanostructures (e.g. nanoparticles and nanowires) and the use of biological systems (e.g. DNA) as molecular nanowires, as well as templates for metallic or semiconducting nanostructures.
3. Theoretical and computational studies that provided the conceptual framework for structure, dynamics, response, and transport in nanostructures. Theory and simulations in chemical sciences are unique in the building of conceptual bridges with experiment.

This chapter will highlight some significant aspects of the characterisation, interrogation, and response of nanostructures, in conjunction with theoretical modelling of the unique structure, dynamics, and function of quantum structures and their assemblies. We focus on the current state of the art for the development of the conceptual framework and experimental background of nanoscience and nanotechnology.

QUANTUM STRUCTURES: SYNTHESIS, CHARACTERISATION, MANIPULATION, AND ASSEMBLY

Impressive growth in nanoscience and technology in the last decade stems from new methods for the synthesis of nanomaterials in conjunction with the advent of tools for characterisation and manipulation. The synthesis of nanomaterials spans inorganic, organic, and biological systems and manipulation with control of structure, size, and shape (Table 12.2). The subsequent assembling of the individual nanostructures into ordered arrays is often imperative for their functions. Uniqueness of properties of nanostructures originates from the combination of the individual nanosystems and their assembly methods. Notable examples for the synthesis of novel nanobuilding units are:

1. Nanocrystals of metals, semiconductors and magnetic materials, involving the utilisation of colloid chemistry methods.
2. The use of physical and chemical methods for synthesis of nanoparticles of ceramic materials.
3. Surface deposition of clusters and nanocrystals on graphite and other metallic or semiconducting surfaces provides novel 3-dimensional or 2-dimensional nanosytems. Deposition of semiconducting clusters from a cluster beam provides novel and interesting systems of deposited fractal clusters, whose dimensionality is characterised by the Hausdorf fractal dimensionality, which is of considerable interest in the context of their response and chemistry.
4. Single- and multi-walled carbon nanotubes, as well as nanotubes of layered metal chalcogenides (e.g. MoS_2, WS_2, $MoSe_2$, NbS_2, V_2O_5).
5. Nanowires of metals, semiconductors, oxides, nitrides, sulphides, and other materials.
6. Utilisation of quantum well superlattices for growing nanowires.
7. New polymeric structures involving dendrimers and block copolymers.
8. Nanobiological structures (e.g. bacterial and plant photosynthetic reaction centers and segments of DNA. Mutagenesis and chemical engineering of the protein structure and of the prosthetic groups in the reaction center, as well as chemical modifications of the DNA double strand, provide means of control of the response of these systems.

Table 12.2. Synthesis, characterisation and description of nanomaterials.

Scale (approx.)	Synthetic methods	Structural tools	Theory and simulation
0.1–10 nm	Covalent synthesis	Vibrational spectroscopy	Electronic structure
		NMR	Molecular dynamics
		Diffraction methods	Transport
		Scanning probe microscopies (SPM)	
<1–100 nm	Self-assembly techniques	SEM, TEM, SPM	Molecular dynamics and mechanics
100 nm–1 μm	Processing, modifications	SEM, TEM	Coarse-grained models for electronic interactions, vibronic effects, and transport

We now briefly address the characterisation of nanostructures. In addition to the well-established techniques of crystallography and transmission electron microscopy and spectroscopy, the modern methods of scanning probe and tunnelling spectroscopies and of extended X-ray absorption fine structure (EXAFS) synchrotron radiation spectroscopy have provided powerful tools for the characterisation of nanostructures. Unique quantum nanostructures involve the familiar examples of scanning tunnelling spectroscopy (STM) imaging of quantum dots (e.g. a germanium pyramid on a silicon surface) and of a quantum corral of 48 Fe atoms placed in a circle of 7.3 nm.

Perspectives of advances in the area of characterisation and dynamic interrogation may involve time-resolved structural studies of nanostructures using synchrotron radiation sources. Temporal resolution on the ns time-scale was already achieved with the Grenoble ESRF synchrotron source for time-resolved structures of biomolecules (e.g. myoglobin). The interrogation of time-resolved structures for the dynamics and response of nanomaterials will be of considerable interest.

Self-assembly provides new ways to pattern a huge number (in the range of 10^6) of nanocrystals, nanowires, etc. to produce complex circuits for optical, magnetic, or electrical response. Examples for the self-assembly of individual nanosystems are:

1. 2-dimensional and 3-dimensional structures of nanocrystals of semiconductors, metals, and magnetic materials being self-assembled using suitable organic solvents.
2. Polymer-coated nanocrystals assembled to form giant nanoparticles.
3. Self-assembled carbon nanotubes forming single crystals.
4. Self-assembly of colloid nanostructures.
5. Lithography-induced self-assembly.
6. The utilisation of the unique features of recognition, assembly, and specific binding of nucleobases in DNA duplexes for the construction of conductive blocks or as insulating/conducting templates for the assembly of other nanoelements.
7. In the realm of biosystems, viral particles are decorated with metal nanoparticles, with the aim of allowing the viruses to assemble themselves into arrays to create networks of the nanoparticles.

NANOWORLD IS DIFFERENT

The conceptual framework and practice of nanoscience encompasses both nanostructures and their ensembles. In this broad context, the physical and chemical properties of nanostructures are distinct from both the single atom or the molecule and from the bulk matter of the same chemical composition. These fundamental differences between the nanoworld on the one hand, and the molecular and condensed phase worlds on the other hand, pertain to the spatial structures and shapes, phase changes, energetics, electronic–nuclear level structure, spectroscopy, response, dynamics, chemical reactivity, and catalytic properties of large, finite systems and their assemblies. Central issues in this broad, interdisciplinary research area of nanoscience pertain to size effects, shape phenomena, confinement of elementary excitations, level structure of elementary excitations, and the response to external electric and optical excitations of individual finite systems and of coupled finite systems. The ubiquity of these phenomena reflects on quantum effects in finite nanostructures.

Size Effects

A key concept for the quantification of the unique characteristics of individual nanostructures pertains to size effects. These involve the evolution of structural, thermodynamic, electronic, energetic, spectroscopic, electromagnetic, dynamic, and chemical features of finite systems with increasing size (Fig. 12.1).

This concept emerged from cluster chemical physics, but is applicable to other nanostructures (e.g. nanocrystals or nanowires). Size effects fall into two categories (Fig. 12.1):

1. Specific size effects: These involve self-selection and existence of 'magic numbers' for small and moderately sized clusters and nanostructures. An irregular variation of the relevant property $\chi(n)$ (where n is the number of constitutents), with increasing the size of the nanostructure, is manifested. Accordingly, $\chi(n)$ is not amenable to size-scaling.
2. Smooth size effects for 'large' nanostructures: In this size domain, a quantitative description (Fig. 12.1) was advanced for the 'transition' of the physical and chemical attributes of clusters

to the infinite bulk system in terms of the size Equation $\chi(n) = \chi(\infty) + An^{-\beta}$, where A is the constant and $\beta(\beta \geq 0)$ is a positive exponent. For a spherical nanoparticle of radius R, the size Equation is $\chi(n) = \chi(\infty) + An^{-3\beta}$. These size equations can be traced to two distinct physical origins: cluster packing and excluded volume effects. Size equations constitute scaling laws for the nuclear-electronic level structure, energetics, and dynamics, providing the quantitative basis for the description of optical and electrical response of nanostructures.

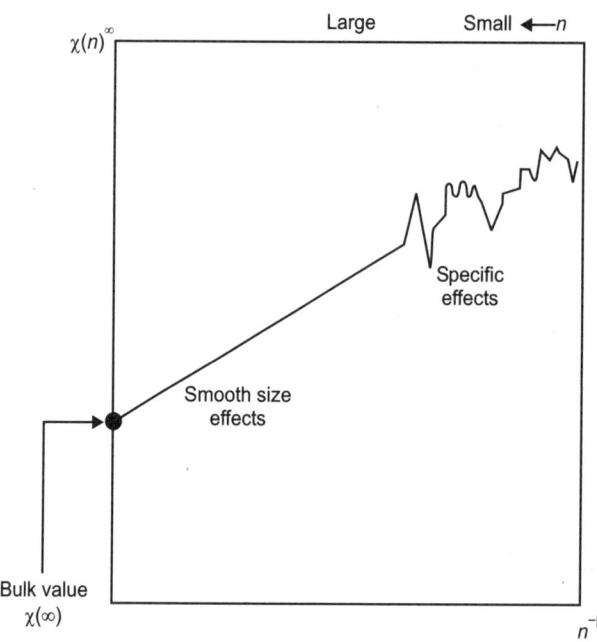

Fig. 12.1. The cluster size dependence of a cluster property $\chi(n)$ on the number n of cluster constituents. The data are plotted vs. $n^{-\beta}$, where $\beta \geq 0$. 'Small' clusters reveal specific size effects, while 'large' clusters are expected to exhibit for many properties a smooth size dependence of $\chi(n)$, which converges for $n \to \infty$ to the bulk value $\chi(\infty)$.

When is such size-scaling partial and incomplete? Several examples come to mind in the context of energetics, nuclear dynamics, and cooperative effects. First, the structural characterisation and specification of phase-like forms, for example, solid (rigid) and liquid (nonrigid) or solid (rigid) and solid (rigid) configurations and 'smeared' (rounded-off) first-order or second-order structural-phase changes between them in nanoparticles, as well as for second-order phase transitions (e.g. superfluidity in finite Boson systems) or onset of superconductivity in finite Fermion systems, may differ from the corresponding feature in bulk matter.

Second, nuclear adiabatic dynamics of clusters manifests new collective excitations, (e.g. compression modes, which do not have an analog in the bulk. Third, finite systems exhibit novel fragmentation patterns, such as cluster fission and Coulomb explosion, which are unique for finite systems and do not have an analogue in the dynamics of the corresponding bulk matter. A striking example constitutes the dynamics of Coulomb explosion of multicharged single clusters, which may also prevail in nanostructures, whose energetics is characterised by a divergent scaling size equation.

Threshold Size Effects from a Single Particle to Collective Phenomena

An important variant of specific size effects pertains to the onset of a qualitatively new feature of the electronic or nuclear-level structure. For excess electron binding to atomic $(Xe)_n$ or polar molecular $(H_2O)_m$, $(NH_3)_n$ clusters the issue is: what is the minimal cluster size required for the localisation of an excess electron? Experimental photoelectron spectroscopy, in conjunction with quantum mechanical calculations, established the threshold size of Xe_n^- for electron localisation in the range $n_c = 4$–6. The threshold size for excess electron localisation in polar clusters manifests a marked system specificity. For water clusters, both experiment and theory reveal that an excess electron is bound by the dimer $(H_2O)_2$ forming a diffuse ($<r> \simeq 20$ Å) weakly bound ($E_b = -50$ meV), dipole bound state. In contrast, for ammonia clusters the 'critical' cluster size is $n_c \simeq 35$, with the formation of a localised ($E_b = -0.5$ eV) solvated electron in the $(NH_3)_{35}^-$ cluster at the onset. Another significant area of many-particle threshold size effects pertains to the important problem of metal–nonmetal transition (MNMT) in clusters (in the section 'Metal–nonmetal transition in finite systems').

Some fascinating novel collective phenomena in large finite systems pertain to the nuclear dynamics and phase changes of finite ultracold gases, which, in the temperature domain of $T = 100$ nk–100 µK, involve gases in magnetooptical traps, optical molasses, and Bose–Einstein condensates. While ultracold finite gases are macroscopic, some features of their optical response were already related to the dynamics of nanostructures. These ultracold, finite systems may be of considerable interest in the context of superfluidity and superconductivity in nanostructures. An interesting example for the collective nuclear dynamics of an ultracold finite system pertains to the Bose–Einstein condensation of atoms with a negative scattering length (e.g. ^7Li atoms confined in a harmonic trap at $T \cong 100$ nK). A molecular description of the nuclear dynamics of this system showed that the effective potential contains only metastable states subjected to macroscopic tunnelling. For an assembly of ^7Li atoms, practical stability is insured for a critical size of $n_{CR} \simeq 1500$, manifesting some novel features of the collective nuclear tunnelling of finite ultracold systems. Of course, such instability, induced by macroscopic tunnelling, cannot be described in terms of an algebraic size-scaling relation. In contrast, finite size effects on the critical temperature for Bose–Einstein condensation of a noninteracting Bose gas in a harmonic trap manifest the reduction of the condensate fraction and the lowering of the transition temperature, as compared to the infinite system. In an n particles condensate, the shift of the critical temperature T_c, relative to that for the $n \rightarrow \infty$ limit, T_c^0, is given by the size-scaling relation $(T_c^0 - T_c)/T_c^0 \propto n^{-1/3}$. Alternatively, one can express the lowering of the critical temperature as $(T_c^0 - T_c) \propto L^{-1}$, where L is the characteristic length of the three-dimensional system. The perspectives of exploring the response (e.g. optical excitations) of ultracold low-density nanosystems is of considerable future interest.

Metal–Nonmetal Transition in Finite Systems

The MNMT in bulk systems constitutes a quantum transformation driven by the effects of electron interactions, disorder, and temperature. While the MNMT in the bulk is manifested by the combination of electron delocalisation, localisation, and electron-scattering phenomena, the corresponding effects in an atomic cluster (e.g. Na_n or Hg_n) pertain to the implications of the discrete electronic-level structure in a finite system. One approach to the visualisation of the relevant electronic-level structure in a cluster or nanoparticle containing one valence electron per atom is to start with a bulk metal and gradually decrease its size to form smaller 'metallic' particles. As the size of the particles is reduced, the electronic energy levels become discrete, with the energy spacing between the lowest unoccupied molecular orbital (LUMO) and highest occupied molecular orbital (HOMO) levels being given by $\Delta \cong E_F/n$, where E_F is the Fermi energy of the bulk metal. Kubo proposed that the formation of a spatially filled 'band' in the finite system can be realised at a finite temperature, when the spacing Δ is lower than the thermal energy. Under these circumstances, the electrical response of the nanostructure (e.g. held between nanoelectrodes) will result in gapless 'metallic' conduction. Thus, the condition for an MNMT in a cluster of monovalent atoms is given by the Kubo relation $\Delta \leq k_B T$. For $(Na)_n$ sodium clusters ($E_F = 3.2$ eV) the Kubo condition for the MNMT is realised at $T > 5$ K for $n = 16000$ ($d = 10$ nm), at $T > 50$ K for $n = 2000$ ($d = 5$ nm), and at $T > 300$ K for $n = 125$ ($d = 2$ nm). For smaller Na cluster sizes ($n \lesssim 100$), this simple description breaks down, and a molecular description of these clusters is adequate.

For clusters containing two or more valence electrons per metallic atom, band-overlap effects induce the formation of the metallic state, for example, s–p band overlap in Hg_n clusters. While early ionisation potential studies of $(Hg)_n$ clusters predicted the MNMT and $n \simeq 10$, subsequent photoelectron spectroscopy data for Hg_n^- clusters demonstrated s–p band closure for considerably larger clusters of $n = 400$. There is currently a discrepancy between the experimental result for s–p closure and theory for the MNMT in these finite systems. One expects that distinct relevant electrical or optical response properties, manifesting an MNMT in clusters, will be characterised by a different 'critical size'. This property dependence of the 'critical' size reveals the generality, but not the universality of the MNMT in finite systems. The description of the MNMT in finite systems is general, but not universal.

Up to this point, we focused on clusters and nanocrystals of metallic elements. An interesting issue is the realisation of metal clusters of nonmetallic elements, where metallic properties are induced by a melting-phase transition. Some semiconducting solids (e.g. silicon or germanium) become metallic in the liquid state, owing to the change in the coordination number. There is a distinct possibility that 'melting' (rigid–nonrigid transition) of Si or Ge clusters, which is exhibited at lower temperatures than that of the bulk, will result in metal clusters. Finally, we move from individual nanostructures to their assemblies. The phenomena of the change of electronic structure and transport in an assembly of nanostructures (e.g. semiconducting or metal clusters) provides a means for controlling functionality of nanomaterials. A 2-dimensional array of Pd nanocrystals was induced to undergo a macroscopic MNMT by varying the spacer length or the cluster size. The Mott–Hubbard model is applicable to describe the MNMT induced in the nanomaterial by correlation effects between the nanoparticles, where the condition for the metal-insulator transition is $B/U \simeq 1.15$, where B is the bandwidth (without correlation) and U is the interparticle energy between nanoparticles.

Coulomb Blockade

An important feature of electrical properties of finite systems refers to scaling relations for the charging energy of nanostructures, as manifested by the Coulomb blockade and the Coulomb staircase. In a double-junction nanodevice, consisting of a cluster of nanocrystals confined between two metallic electrodes, which is subjected to an external voltage V, excess charges can accumulate on the nanoparticle, giving rise to a Coulomb blockade. The addition of a single electron to the nanoparticle of radius R results in the changing energy $W = W(\infty) + b/R$, where $W(\infty)$ can be identified with the changing energy of the bulk. The minimal voltage $V_{MIN} = W/e$ required to inject an extra electron into the cluster, gives rise to the Coulomb blockade staircase, with the separation voltage steps exhibiting a size equation, i.e. $V_{MIN} = W(\infty)/e + b/eR$. The observation of the Coulomb staircase in the current voltage response is facilitated in nanocrystals (Figs. 12.2 and 12.3). With a higher resolution, the observation of the eigenvalue staircase is facilitated, monitoring one-particle electronic excitations of the nanostructure. The observation of the Coulomb staircase and the eigenvalue staircase in the electronic response of nanostructures provides a direct demonstration of the discreteness of the electronic-level structure in finite systems.

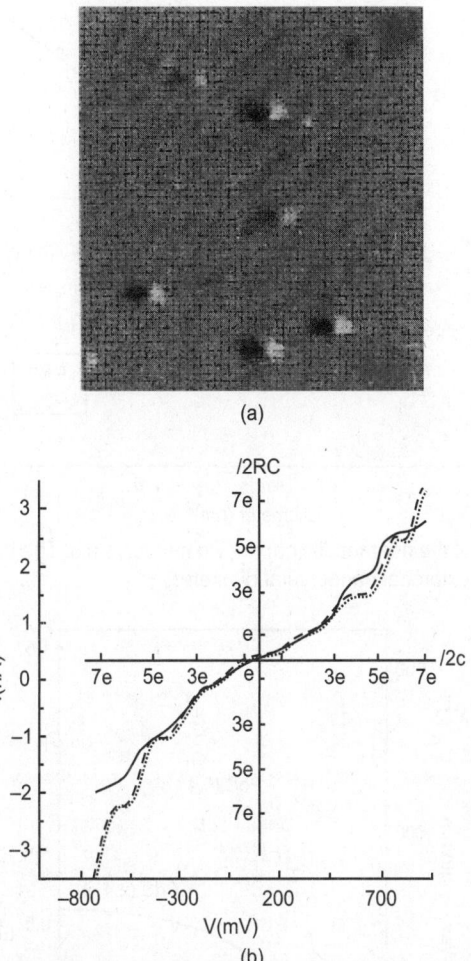

(a)

(b)

Fig. 12.2. Coulomb blockade for Pd nanoparticles. (a) STM imaging of the nanoparticles on a graphite substrate. (b) The Coulomb staircase.

Confinement

Boundary effects on electrons or excitons in nanostructures result in the localisation of these elementary excitations, provided that this interaction is repulsive. Characteristic examples are confinement of large-radius Wannier–Mott excitons (i.e. electron-hole pairs) in nanostructures (Fig. 12.4). In this context, one encounters the interesting problem of the interplay between the attractive electron-hole interaction and the repulsive boundary repulsion. In the weak confinement limit, when the cluster size exceeds the exciton radius, the exciton energy is given by the size equation $E = E(\infty) + C/R^2$, where $E(\infty)$ is the bulk exciton energy, and $C = \pi^2 \hbar^2/2m^*$, with m^* being the exciton mass. The size effects describe confinement of an exciton in a spherical box. This scaling relation accounts well for the optical properties of clusters and nanostructures, providing a central mechanism for the tuning of the optical response of nanostructures.

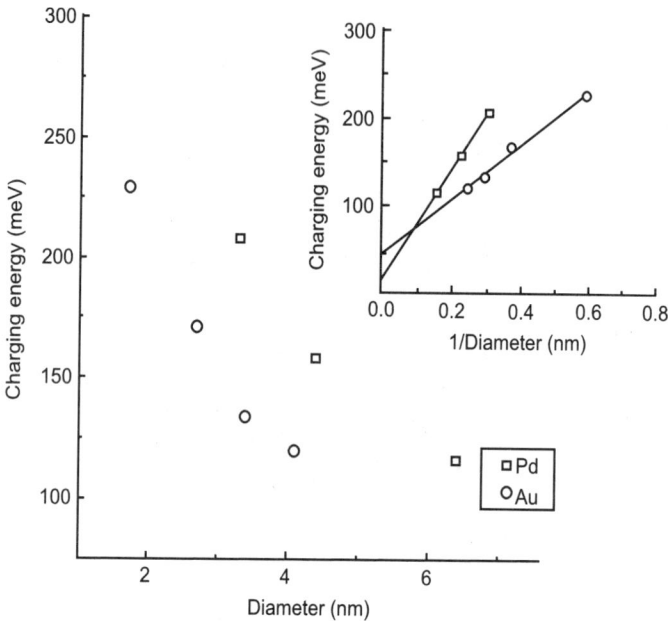

Fig. 12.3. Size dependence of the nonmetallic gap for Pd nanocrystals. The insert shows the dependence of the charging energy on the reciprocal nanocrystal diameter.

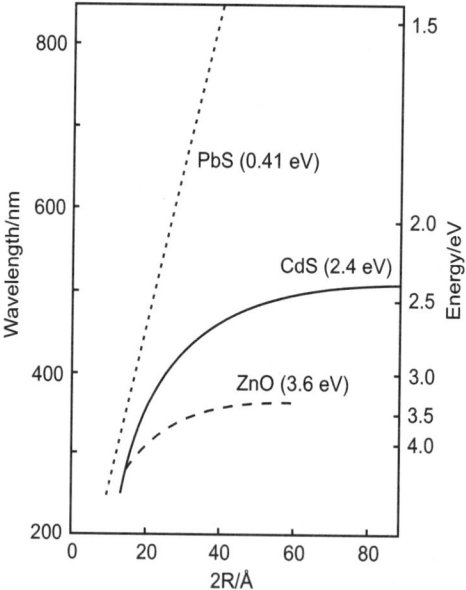

Fig. 12.4. Size dependence of the wavelength of the absorption threshold in semiconductors manifesting exciton confinement.

For the case of a single charge carrier, boundary scattering affects the electronic-level structure. For a semiconducting nanowire of width (diameter) d, one expects that the band-gap, E_G, is given by energetics of a particle confined in a 2-dimensional box, $E_G = E_G(\infty) + B/d^2$, where $E_G(\infty)$ is the bulk gap value. Such a 2-dimensional confinement provides a control mechanism for the electrical properties of conducting nanowires.

Size Effects for Transport in Nanostructures

The ubiquity of quantum effects in finite systems reflects on unique transport properties of nanostructures. We have already considered the phenomenon of Coulomb blockade in nanoparticles. The electronic quantum size effects in nanowires are manifested when $d/\lambda_F \lesssim 1$, where λ_F is the electronic wavelength at the Fermi energy E_F. Accordingly, novel quantum effects in nanowires are expected to be exhibited when $d \lesssim 2\pi \; \hbar^2/(2mE_F)^{1/2}$, where $E_F \sim 1$ eV, i.e. when $d \lesssim 1$ nm. Under these conditions, quantised conductance is exhibited in nanowires, being given by the Landauer expression $g = \left(2e^2/h\right)\sum T_i$ where T_i is the scattering matrix in channel i. When $T_i = 1$ for all the channels, the conductance is an integer multiple of the conductance quantum $2e^2/h = (12.9 \text{ k}\Omega)^{-1}$. This quantised conductivity (with $T_i = 1$) corresponds to ballistic transport in nanostructures. This transport quantisation was experimentally demonstrated in carbon nanowires (Fig. 12.5) and in metal nanowires. In longer nanowires (>10 nm), heat dissipation effects may lead to instability.

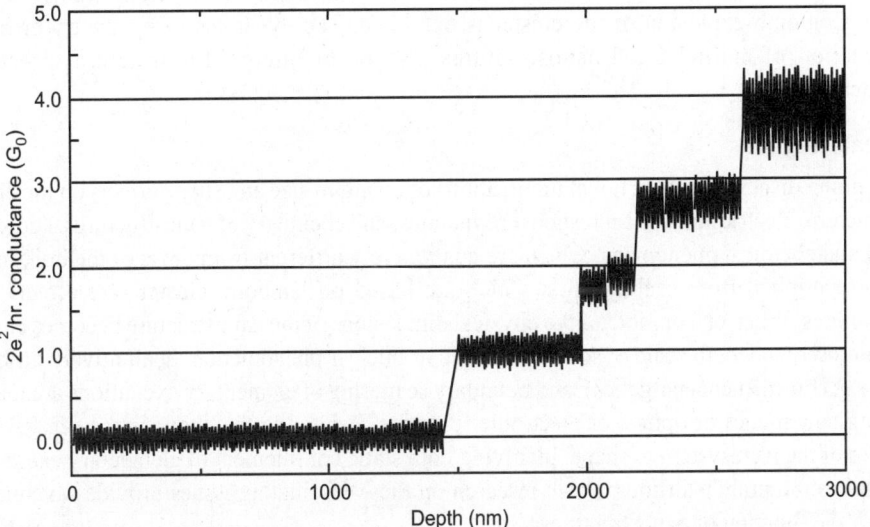

Fig. 12.5. Conductance of carbon nanotubes observed using the set-up in an AFM, showing the independence of the conductance of the depth into the liquid mercury in which the tip is being inserted. The second step is introduced as another nanotube touches the mercury.

Shape Effects

The central question of the nanostructure shape analysis is: How do structural deformations affect the electronic–nuclear level structure, energetics, dynamics, and transport? The concept of spontaneous symmetry-breaking advanced by Anderson in 1984 pertains to the reduction of symmetry of the cluster or nanostructure driven by the instability thus acquired. Interesting implications to cluster and nanosystems involve spontaneous symmetry-breaking at zero and low magnetic fields in single quantum dots and quantum dot molecules in zero and low magnetic fields, which are reflected in the electronic charge density and spin densities. These may involve Wigner crystallisation, localisation on individual electron puddles, as well as the formation of spin density waves without localisation.

Another class of shape effects involves nuclear collective excitations revealing shape deformation of spherical clusters. Such compressive modes are treated by the liquid drop model with the lowest excitation energy being given by $\hbar\omega_b = \pi\hbar u/R$, where u is the velocity of sound and R the cluster radius. The size equation $\hbar\omega_B \propto n^{-1/3}$ was experimentally verified by He scattering from Ar_n clusters. Coherent acoustic mode oscillations in supported silver nanoparticles ($R = 3$–20 nm) were experimentally demonstrated, and evidence for plasmon–collective vibration excitations was provided.

Another class of interesting, symmetry-breaking, dynamic–effects, involves fission of multicharged clusters. The ubiquity of fission phenomena of charged droplets and clusters was described in terms of the celebrated liquid drop model, where a classically charged drop deforms through elongated shapes to form separate droplets. The fissibility parameter, advanced by Lord Rayleigh in 1882, is $X = E(\text{Coulomb})/2E(\text{surface})$ and characterises the relative contribution of repulsive (Coulomb) and cohesive (surface) energies to the fission barrier. For $X < 1$, the spatially unisotropic fission limit prevails, with the formation of a small number of large nanostructures, as is the case for the Coulomb instability of nuclei, droplets, and metal clusters. The Rayleigh limit is overcome by a marked enhancement of the Coulomb energy, which can be accomplished by multielectron ionisation in intense laser fields. When $X > 1$, spatially isotropic Coulomb explosion of the cluster is exhibited. The dynamics of symmetry-breaking and fragmentation of multicharged nanostructures may be of interest for structural processing of nanomaterials.

Nanoscale Regime

The foregoing discussion of the novel implications of quantum size and shape effects on the energetics, level structure, electric and optical response, dynamics, and chemistry of nanostructures, addresses new physical and chemical phenomena, which are qualitatively different from those of the bulk matter and from those implied from scaling laws, which are based on 'smooth' cluster size effects for large nanostructures. When one or more of the physical dimensions of the nanostructure becomes comparable to the microscopic length scale of some electronic or nuclear phenomenon, qualitatively different new effects (e.g. confinement energetics), and boundary scattering of elementary excitations are manifested, exhibiting new modes of optical or electronic response. Related to the above is the sensitivity of the properties of the nanosystem to shape, involving both static confinement of elementary excitations and of dynamic structural distortions. Basic research on these fascinating issues provides avenues for the control of the function of nanostructures.

Nanoelectronics, Nano-optoelectronics, and Information Nanoprocessing

One of the most important and far-reaching potential applications of nanomaterials will be in the field of nanoelectronics. While the field of molecular electronics was fraught with some conceptual—practical difficulties in the context of connecting molecular devices to the 'outside world', these issues were solved by nanodevice fabrication, the design of surface-nanodevice chemical contacts, and chemical engineering of molecular-nanoparticles or biomolecular-nanoparticle hybridisation. This multidisciplinary research technology area of nanoelectronics has dual goals:

1. The utilisation of a single, individual nanostructure (e.g. cluster, nanoparticle, nanocrystal, quantum dot, nanowire, or nanotube) for the processing of optical, electrical, magnetic, chemical, or biological signals.

2. Providing nanostructured materials, consisting of assemblies of nanostructures, for electronic, optoelectronic, chemical-catalytic, or biological-diagnostic applications.

The distinction between classes (I) and (II) is always practical and sometimes also conceptual. While class (II) is aimed toward the miniaturisation of electronic circuitry and of catalytic and biological templates, class (I) is aimed toward the realisation of single-electron nanodevices. There are already significant advances in the utilisation of single nanostructures for single-electron memory devices based on Coulomb blockade and on a single-electron transistor (Fig. 12.6). Progress for the class (II) system involves scanning probe tips in arrays, LED and laser diodes of semiconductor nanostructures, arrays of semiconductor quantum dots, and nanowires. Nanocircuits making use of carbon nanotubes were described. Metallic and semiconducting properties of multiwalled nanotubes have been constructed by the stepwise burning of layers and by chirality control. These approaches allow for the use of nanotubes in nanocircuitry, with special potential advances in the use of Y junction nanotubes. Another significant area involves nanomaterials for optoelectronics, where functional devices, based on confinement, show potential for photonic switching and optical communication.

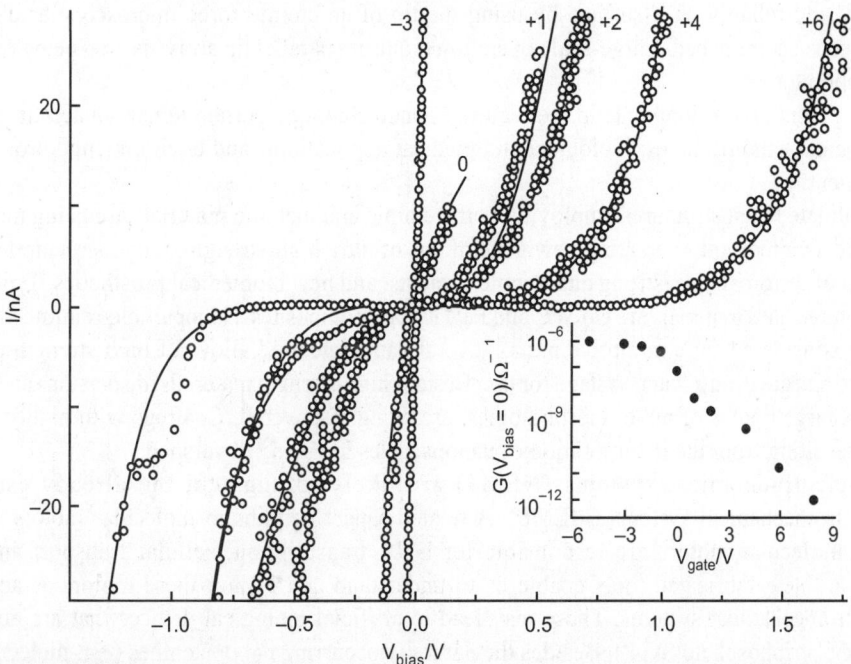

Fig. 12.6. Current-potential characteristics at different gate voltages of a field-effect transistor based on a single-walled nanotube.

The miniaturisation of electronic circuitry will allow for the advent of novel information storage, reading, retrieval, and programming systems. Both classes (I) and (II) of nanoelectronics show great potential for nanoscale information processing. The information paradigm in nanostructures may involve two alternative routes. First, the bottom-up approach, starting from a single nanostructure being based on nanofabrication, miniaturisation, and assembly of nanostructures to produce a nanostructured computer. Resonant tunnelling devices deserve special mention in this context, since they have already demonstrated success in multivalued logic and memory circuits. Second, the top-down approach will utilise and apply the conceptual framework of supramolecular chemistry and self-assembly of nanostructures to produce organised suprastructures for information processes. Spintropic memory based on magnetic, semiconducting nanoparticles, provides a promising direction.

Perspectives in Nanoscience and Nanotechnology

The emerging nanoworld encompasses entirely new and novel means of investigating structures and systems. Species as small as single atoms and molecules will be manipulated and even exploited as atomic switches. Computer-controlled scanning probe microscopy enables real-time, hands-on nanostructures manipulation. Nanomanipulators have been designed to operate in scanning and transmission electron microscopes as well. A nanomanipulator gives virtual telepresence on the surface, with a scale factor of a million to one. Optical tweezers provide another approach to hold and move nanometer structures, a capability specially useful in investigating dynamics of molecules and particles. Questions such as 'How does a polymer move, generate force, respond to an applied force and unfold?' can be answered by the use of optical tweezers. It is noteworthy that the positioning of nanoparticles accurately and reliably on a surface by using the tip of an atomic force microscope as a robot has already been accomplished. Large-scale operation requiring parallel tip arrays is now being explored in several laboratories.

Novel potential developments in the realm of nanotechnology pertain to nanomaterials, molecular and biological nanomachines, biological and medical applications, and environmental protection and improvement.

Consolidated nanostructures employing both ceramic and metallic materials are being increasingly recognised as important in creating new generations of ultra high-strength, tough structural materials, new types of ferromagnets, strong and ductile cements, and new biomedical prosthetics. Typical of the nanostructured hard materials are Co/WC and Fe/TiC nanocomposites. Nanoparticle-reinforced polymers are being considered for automotive parts. Several nanostructured alloys of high strength have been discovered and are in an advanced stage for use. Besides high-strength materials, dispersions and powders, as well as large bodies of novel morphologies, are being discovered. Coatings with highly improved features resulting from the incorporation of nanoparticles are being developed.

Nanoelectrochemical systems (NEMS) are likely to augment the already established microelectromechanical systems (MEMS). A related aspect pertains to molecular motors. In wiring systems, molecular motors are responsible for DNA transcription, cellular transport and muscle contraction. New fabricated tools enable us to understand and exploit these motors as actuators in nanoelectromechanical systems. These may lead to artificial biological devices that are powered by adenosine 5'-triphosphate (ATP). Besides the naturally occurring nanomachines (e.g. molecular motor protein F1-ATPase), organic chemists are synthesising molecules (e.g. rotaxanes) capable of various kinds of motions at the nanolevel. Using molecular motors as nanomachines and interfacing them with inorganic energy sources and other nanodevices is an aspect of great interest.

DNA chips and microarrays represent a technology with immediate applications in diagnostics and genetic research. DNA chips and arrays are devices wherein different DNA sequences are arrayed on a solid support, the arrays generally having 100 to 1,00,000 different pixels (DNA sites) on the chip surface. The chip devices will be useful in genomic research, drug discovery, gene expression analysis, forensics, and various types of detection and diagnostics. Electronically active DNA microarrays and electronically directed DNA self-assembly technology hold promise in various areas including photonic and electronic devices. Since genetic medicine appears to be emerging as a part of future health care, gene and drug delivery assume importance. Appropriate nanoparticles containing DNA may indeed provide viable means of delivery in the near future. The gene gun is already being used to deliver genetic materials to transfect plant and animal cells.

Semiconductor nanocrystals have been used as fluorescent biological labels. It is likely that sensors based on nanotechnology will revolutionise health care, climate control, and detection of toxic substances. Integrated nanoscale sensors could monitor the condition of a living organism, the environment, or nutrient supply. It is altogether possible that we will have nanochips to carry out complete chemical analysis. Such nano-total analysis systems will have to employ new approaches to valves, pipes, pumps, separations, and detection.

A knowledge of the processes related to nanoscale structures, natural as well as man-made, is not only useful for understanding transport and other aspects of those materials, but also helps in developing technologies for preventing or minimising harm to the environment. The use of homogeneous and heterogeneous catalysis (including nanocatalysis) for improving energy efficiency and reducing waste is fairly well documented. The design of environmentally benign nanocomposites, the use of nanomaterials (e.g. nanoparticles of TiO_2) for various environmental cleansing processes and of nanoporous solids (e.g. MCM41) for sorption, are typical examples of the applications of nanotechnology for the protection and improvement of the environment. The use of nanoporous polymers for water purification and purification of fluids by photocatalysis using nanostructural TiO_2 are two other examples.

Thus, the preceding overview serves to provide a glimpse of the current status and future prospects of nanostructured materials. Clearly, there is a great vitality in this area of immense opportunities. While it is truly an interdisciplinary area covering physics, chemistry, biology, materials, and engineering, chemists have a major role to play. Synthesis, self-assembly, manipulation, simulation, and theory require dedicated efforts of chemists, and there is every reason to believe that nanoscience will become an integral part of chemical science. Interaction among scientists with different backgrounds will undoubtedly create new materials and new science with novel technological possibilities.

Nanoscience and technology are likely to benefit various industrial sectors, including chemical and electronic industries, as well as manufacturing. Health care and medical practice will undoubtedly benefit from nanoscience, as will environmental protection. One of the difficult problems facing the design of nanostructures-based systems is in understanding how they are to be interconnected and addressed. There is much to be learnt about the preferred architectures of systems and the design and fabrication of devices. Developing techniques for fabricating a large number of nanostructures also requires substantial effort. The eventual success of nanoscience will depend on the development of new manufacturing technologies. There is every reason to believe that there will be much progress in the coming decade.

There is considerable effort in nanotechnology in a large number of academic and industrial laboratories all over the world, and it is becoming imperative that we establish dedicated centers with the required infrastructure and experimental facilities for nanomaterials research. The subject has caused excitement not only in the advanced countries but also in the developing nations, and international cooperation would certainly be most beneficial.

Chapter 13

Gas Phase Cluster

INTRODUCTION

The theoretical investigations of gas phase clusters enable the evaluation of intrinsic molecular properties and intermolecular interactions, one can predict the macroscopic properties of bulk matter, from a microscopic determination of the properties of individual atoms, molecules, or clusters. Based on the insights obtained from theoretical investigations of the properties of a large number of cluster systems (ranging from simple water clusters to large π-systems), we have investigated the properties of various novel molecular systems including endo/exohedral fullerenes, nanotori, nonlinear optical materials, ionophores/receptors, polypeptides, enzymes, organic nanotubes, nanowires, and electronic and nano-mechanical molecular devices. The present minireview highlights some of the interesting results obtained in the course of extensive theoretical investigations of clusters and nanomaterials.

During the last few decades, conventional host-guest molecular systems including supra-molecules with novel and exotic structures have been the focus of extensive investigations. However, the practical exploitation of most of these systems requires a more detailed insight of the structure function relationships. In this context, rigorous quantum chemical investigations have proven to be of value because in addition to aiding the understanding of such structure function relationships, they can also be used in the *de novo* design of novel molecular systems and functional materials.

Given the ability of nano-chemistry to provide functional materials of practical utility in the near future, the predictive power of quantum chemical calculations are of significant value because interesting individual molecules or small clusters can be developed as viable functional materials and devices. Interest in molecular level devices is also high because they aid the development of extremely small, fast, and powerful computers. Another area of interest has been the use of molecular recognition events as microscopic chemical/biochemical sensors/monitors such as DNA chips. It should be emphasised that most of the above research thrives on ideas gleaned from condensed matter physics, chemistry, computational science and biology.

Consequently, the prediction of properties of nanomaterials from a nanoscopic determination of the properties of individual atoms, molecules, or clusters, has begun to attract a lot of attention. Much of this interest stems from the utility of such studies in the design and development of novel functional nanosystems with potential applications in electronics, photonics, chemistry, biology, neuroscience and medicine, based on molecular interactions (with ion/atom/molecule/electron/photon), molecular recognition, nanorecognition, molecular clustering/aggregation, self-assembly, and self-synthesis. In this context, detailed theoretical investigations of the properties of a large number of cluster systems, ranging from simple water clusters to large π-systems has been carried out.

The intrinsic molecular properties and intermolecular interactions obtained from the gas phase clusters enable us to predict structures and properties of nanomaterials. Therefore, an effective design strategy would require a thorough understanding of various interaction forces and mechanisms. In this connection, we have theoretically characterised novel interaction forces. We have employed diverse theoretical methods ranging from traditional *ab initio*, density functional theory, tight binding, path-integral, Monte Carlo to molecular dynamics simulations. In the course of these investigations, we have been successful in elucidating the properties of a diverse range of novel molecular systems. These include molecular clusters (water clusters, solvated cations, solvated anions, solvated electrons, solvated chemical compounds, solvated biomolecules, solvated surfaces, inorganic/metal clusters), endo/exhohedral fullerenes/nanotori, nonlinear optical materials, ionophores/receptors, polypeptides/membranes, enzymes, organic species, organic nanotubes/nanowires, photo/electro-nanodevices, and nano-mechanical molecular devices (Fig. 13.1).

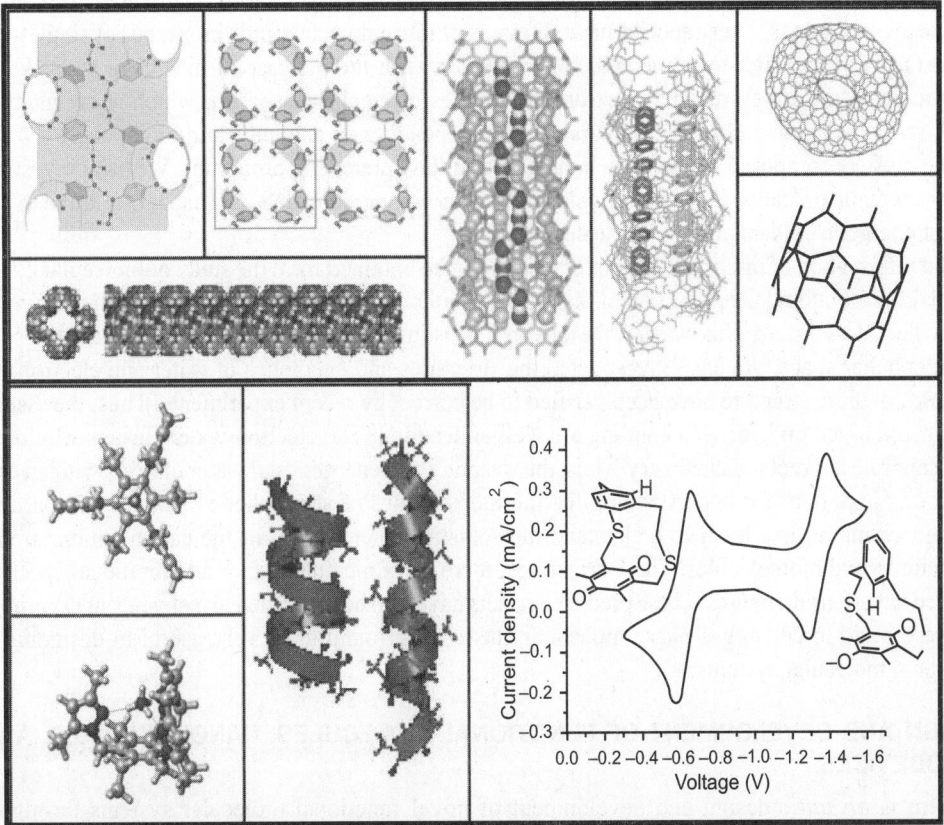

Fig. 13.1. Organic nanotubes (one-dimensional H-bonds relays and displaced π-π stacks); nanotorus, beltenes, ionophores/receptors, and left-handed lambda *vs* right-handed alpha helices, and molecular flipper.

The success of design strategy is validated by experimental characterisation of novel ionophores, organic nanotubes, and molecular flippers as well as many other experimental demonstration by other groups. For example, we have designed receptors with high affinity and selectivity for specific cations (acetylcholine, NH_4^+) or anions (F^-, Cl^-) which are biologically important. Quantum nanostructures have been synthesised. Our designed photo-electronic and nano-mechanical molecular devices would be useful for computer memory with nonlinear optical switch phenomena and molecular vehicles/tweezers for drug delivery and nano-surgery, respectively. In the course of this review, we elaborate on the theoretical accomplishments which eventually led to the *de novo* design of these novel functional molecular systems.

MOLECULAR CLUSTERS AND INTERACTION FORCES

Studies of carbon clusters, oxygen/sulphur clusters, metal clusters would provide the intrinsic properties of these clusters. We find that endohedral fullerenes (by the presence of N, P, As, O, S) would be useful for quantum computing due to the presence of isolated and controllable spins. We have demonstrated the unusual magnetic phase transition in exohedral fullerenes. Structures, energetics, and electronic structures of carbon nanotubes have been investigated, and the carbon nanotubes have recently been characterised. We also investigated the smallest building blocks of carbon nanotubes: trannulenes, cyclacenes, collarenes. These species have shown very intriguing electronic properties of singlet-triplet degeneracy and insulator-to-semiconductor transitions with the increase of molecular size. We have also predicted the structures and electronic properties of smallest nanotori, which show interesting metal, semiconductor, and insulator characteristics depending on nanotube building blocks.

The clusters composed of inorganic atoms often show interesting properties. We have investigated highly energetic oxidants, O_n clusters, which have been compared with S_n clusters. O_4 is found to be the highest energetic oxidant among O_n clusters.

The information of intermolecular interactions can be obtained from the study of molecular clusters. The hydrogen bonding would be one of the most important interactions in biosystems. The most typical case is the water clusters, since water is the universal constituent in biosystems, environments, atmosphere, and interstellar space. We have investigated the structures and energetics of water and electron-water clusters. All these structure have been verified to be correct by recent experiments. Thus, the assembly phenomena based on hydrogen bonding are well understood. The electron-water clusters would be an important role in precloud chemistry. Thus, the structure and energetics of water clusters with an excess electron is important for understanding the thermodynamics of atmospheric chemistry. In particular, this understanding is critical to understand the solvation phenomena for the cation, anion, aromatic compounds, and biomolecules. These structures, thermodyamic quantities, and vibrational spectra are reported in various literatures. These predicted results have also been verified in many recent experiments. All these basic forces in gas phase molecular clusters are found to be very useful in designing new functional molecular systems.

DESIGN AND DEVELOPMENT OF FUNCTIONAL MOLECULES, NANOMATERIALS, AND NANODEVICES

Our aim is *ab initio* design and development of novel functional molecular systems through our understanding of molecular functional mechanisms and assemblies. In sharp contrast to conventional host-guest systems, the guests in our case include photons, electrons, and protons. To this end, we investigate the capabilities of manipulating individual photons/electrons/protons/atoms/molecules (for understanding the function in molecular devices/sensors) and the principles governing biomolecular recognition, solvation mechanisms including proton exchange, signal transduction, knowledge recognition, and memory retention in living systems. For practical utility, we design novel molecular electronic/photonic devices and molecular sensors. These include single electron/photon devices and molecular tools to manipulate a single electron/photon. Some of predicted devices/sensors have been synthesised and characterised. We include the study of electron tunnelling, proton tunnelling, and current flow at the molecular level. Our planned approaches are as follows: (i) designing novel superfunctional molecular systems through elucidation of host-guest interactions and molecular self-assembly/self-synthesis, (ii) designing electron/photon/proton-host systems and studying its capture/release/transfer mechanisms and dynamics, and (iii) designing nanodevices. All these aims have been successfully carried out. Firstly, novel types of interaction forces have been elucidated. Secondly, the catalytic role of enzymes (in particular, the importance of electron dissipation and charge buffering) has been further clarified. Thirdly, novel functional organic nanotubes and nanowires have been characterised. Finally, novel ionophores and nanomechanical devices have been designed, synthesised, and characterised.

Ionophores/Receptors

Molecular level studies of intermolecular interaction forces and microscopic structures have been accomplished to understand molecular aggregates and self-assembly phenomena. The design and synthesis of receptors capable of binding anionic or cationic guests is of crucial importance due to its potential applications in environmental and biological processes. Based on this, we have been successful in designing and synthesising various types of novel ionophores. Utilising our earlier work of the cation-π interactions, we investigated a new type of ionophore family of [n]beltenes. We also recently designed, synthesised, and characterised tripodal cation ionophores/receptors. In particular, acetylcholine receptors [1,3,5-*tris*(pyrrolylmethyl) benzene] can aid the discoveries of novel drugs with understanding the binding mechanism. Novel amphi-ionophores in gas phase and in aqueous solution have been designed for the first time. Tripodal receptors for halide anion binding have been designed and synthesised. These receptors show extremely high affinity and selectivity for F^- and Cl^-. A tripodal receptor shows strong affinity and high selectivity for F^- through $(C–H)^+$—X^- hydrogen bonds. Another type of tripodal receptors for halide anion binding has been designed and synthesised. These show extremely high affinity and selectivity for Cl^-. In this case, we employed both charge-charge interactions and charge-dipole interactions. This approach would also aid design of novel functional molecular systems and biologically important chemosensors based on ion selective electrode.

Organic Nanotubes and Nanowires

Using the computerised molecular design approach, we recently reported the synthesis of calix hydroquinone (CHQ) nanotube arrays self-assembled with infinitely long one-dimensional (1-D) short hydrogen bonds (H-bonds) and aromatic-aromatic interactions. For the design of the CHQ nanotubes and the study of their assembly process, we employed both chemistry approach (based on *ab initio* and density functional theories) and physics approach (based on first principles calculations using ultrasoft pseudopotential plane wave methods). Since X-ray structures do not contain the positions of H atoms, it is necessary to analyse the system using quantum theoretical calculations. The competition between H-bonding and displaced π-π stacking in the assembling process has been clarified. The IR spectroscopic features and NMR chemical shifts of 1-D short H-bonds have been investigated both experimentally and theoretically. The dissection of the two most important interaction components leading to self-assembly processes would help design new functional materials and nano-materials.

For the computer-aided molecular design strategy, on the basis of intermolecular interaction forces (i.e. hydrogen bonding and aromatic-aromatic interactions), we investigated the assembling phenomena of CHQs with density functional calculations of various possible combinations of assembled structures derived from previously reported calixarene-based dimers, trimers, tertramers, hexamers, and polymers. The results suggest that in the presence of bridging water molecules, a linear tubular polymeric structure is highly stabilised by the formation of H-bonded bridges between repeating tubular octamer units. These predicted organic nanotubes have been successfully synthesised. After the synthesis and characterisation of the nanotubes, we have further investigated the details of predicted assembly phenomena as well as the refined structure and electronic properties of the CHQ nanotubes. Based on our calculation results, we have clarified the origins of 1-D H-bonding and displaced π-π stacking in the assembling process and these competitions. The spectroscopic features of 1-D short H-bonds have been elucidated. These interesting structures and functions of the organic nanotubes would find various applications in chemistry, physics, and biology. The crystal structures shows that the nanotubes have infinitely long 1-D H-bond arrays. The crystals can be grown into thick and long multi-channel bundles (up to 0.5 mm wide and 5 mm long). The crystal structure shows that the bundles of CHQ nanotubes form novel chessboard-like rectangular structures. Each nanotube has 17×17 Å^2 cross section with 8×8 Å^2 square pore (with the van der Waals volume excluded). Since the nanotubes are electrochemically and photochemically active, they should find numerous applications such as a model for selective water/ion channels in biological systems and a nano-host to include size specific guest molecules.

Using these organic nanotubes as templates, we synthesised ultrathin silver nanowire arrays. These could be applied as nanoconnectors for nanodevices. First principles calculations suggest the existence of three conducting channels for electron transport as a quantum wire. The ultrathin silver nanowires with infinitely high aspect-ratio would serve as model systems for investigating many exciting 1-D physical phenomena as well as nano-electronic devices. The nanotubes arrays can be utilised in many interesting nanosystems. The redox reaction of the nanotube with novel metal ions allows to form silver nanowire arrays. The wires exist as uniformly oriented three-dimensional arrays of ultrahigh density, and thus could be employed as model systems for investigating one-dimensional phenomena and as nanoconnectors for designing nanoelectronic devices.

Nanowires have attracted extensive interests in recent years because of their unusual quantum properties and potential use as nanoconnectors and nanoscale devices. To obtain enhanced physical properties, the wires need to be of small diameter and high aspect ratio, and to be uniformly oriented. The characterisation of metal nanowires has been extended from silver to gold, platinum, palladium and mercury.

Catalytic Residues in Enzymatic Reaction

We have investigated the role of catalytic residues in enzymatic mechanisms. Our previous research on enzymatic reaction mechanism has been expanded to have more generalised concept. In particular, the charge buffering/dissipation role has been clarified. We have detailed the structures of the active site of ketosteroid isomerase and the role of various catalytic residues, and have explained the origin of its fast reactivity by carrying out a detailed investigation of the enzymatic reaction mechanism. The catalytic residues, through short strong hydrogen bonds, play the role of charge buffer to stabilise the negative charge built up on the intermediates in the course of the reaction.

Left-handed Helix of Polypeptides

In polypeptides, we have first noted that the left-handed helix can exist, in contrast to the conventional wisdom of right-handed helix. The diameter of the left-handed helix is larger than the conventional right-handed helix. The left-handed helix is stabilised when the terminal residues are charged, because the dipole moments of carbonyl groups for the former are aligned opposite to those of the latter. This indicates that the protein folding can be controlled by the charged moieties of either the polypeptides or the residues of other molecules around the terminal ends of the polypeptides.

Nanomechanical Device

A molecular flipper has been designed, synthesised, and characterised. The flipping/flapping motion, which is due to the changes of edge-to-face and face-to-face aromatic interactions, can be electrochemically controlled by reduction/oxidation of the quinone moiety in the molecular system. We believe that the present investigation would spur the development of novel nanomechanical devices whose motion can be controlled through electrochemical means. The conformational change of the upper benzene ring in the normal state reduces to the dianionic state, upon accepting two electrons by voltage change. Thus, the conformational changes between reduced and oxidised states can be made very fast by electrochemical process.

Utilising the knowledge of molecular interactions obtained from the gas phase clusters, we have tried to design and synthesise functional molecular systems: nanowires/nanotubes, receptors/sensors, and molecular robotics. These would be eventually useful for molecular nanoelectronic/mechanical devices, quantum computing devices, biomolecular sensors, and nano-surgery. Indeed, we have succeeded in synthesising functional organic nanotubes which are composed of electrochemically and photochemically reactive functional groups. Using these organic nanotubes, we have made very long and thin silver nanowire arrays. Subnanowires can form super-crystalline structure with 3-dimensional arrays which are all coherently oriented atomwise. Other interesting metal wires are also formed inside the organic nanotubes in ambient experimental condition. For practical utility for nanodevices, we have characterised nanowires of gold, platinum, palladium, and mercury. Since these metals have their own characteristics, their utility would be interesting. In addition, we are also interested in designing ferromagnetic metal nanowires and superconducting nanowires.

Based on novel molecular interaction forces, we have also been successful in designing various receptors and ionophores, and in understanding the reaction mechanisms of enzymes. This research will be further expanded to include drug design. The knowledge of the novel interaction forces in functional molecular systems would be utilised for the development of bio-nano-robotics, such as molecular vehicles and electron/proton/molecular tweezers. Novel functional molecular systems could be designed so as to have the capacity of controlled assembling. Artificial receptors, bioinformatics, nano-sensors (including DNA chips), and bio-nano-robotics would be utilised to design novel drugs, to obtain genetic information, to examine illness, and to carry out local-surgery. These fields are still in embryo. We have been working on this project, based on computer-aided molecular design strategy for developing functional organic/bio-organic molecular systems. It is encouraging to note the worldwide progress of various kinds of research from molecular interactions to material/device design.

GAS CLUSTER ION BEAM

Gas cluster ion beams (GCIB) is a new technology for nanoscale modification of surfaces. It can smooth a wide variety of surface material types to within an angstrom of roughness without subsurface damage. It is also used to chemically alter surfaces through infusion or deposition.

Process

Using GCIB a surface is bombarded by a beam of high energy nanoscale cluster ions. The clusters are formed when a high pressure gas (approximately 10 atmospheres pressure) expands into a vacuum

(1e-5 atmospheres). The gas expands adiabatically and cools then condenses into clusters. The clusters are nano sized bits of crystalline matter with unique properties intermediate between the realms of atomic physics and those of solid state physics. The expansion takes place inside of a nozzle that shapes the gas flow and facilitates the formation of a jet of clusters. The jet of clusters passes through differential pumping apertures into a region of high vacuum (1e-8 atmospheres) where the clusters are ionised by collisions with energetic electrons. The ionised clusters are accelerated electrostatically to very high velocities, and are focused into a tight beam.

The GCIB beam is then used to treat a surface—typically the treated substrate is mechanically scanned in the beam to allow uniform irradiation of the surface. Argon is a commonly used gas in GCIB treatments because it is chemically inert and inexpensive. Argon forms clusters readily, the atoms in the cluster are bound together with van der Waals forces. Typical parameters for a high energy Argon GCIB are: average cluster size 10,000 atoms, average cluster charge +3, average cluster energy 65 keV, average cluster velocity 6.5 km/s, with a total electrical current of 200 µA or more. When an Argon cluster with these parameters strikes a surface, a shallow crater is formed with a diameter of approximately 20 nm and a depth of 10 nm. When imaged using Atomic Force Microscopy (AFM) the craters have an appearance much like craters on planetary bodies. A typical GCIB surface treatment allows every point on the surface to be struck by many cluster ions, resulting in smoothing of surface irregularities.

Lower energy GCIB treatments can be used to further smooth the surface, and GCIB can be used to produce an atomic level smoothness on both planar and nonplanar surfaces. Almost any gas can be used for GCIB, and there are many more uses for chemically reactive clusters such as for doping semiconductors (using B_2H_6 gas), cleaning and etching (using NF_3 gas), and for depositing chemical layers.

Industrial Applications

In industry, GCIB has been used for the manufacture of semiconductor devices, optical thin films, trimming SAW and FBAR filter devices, fixed disk memory systems and for other uses. GCIB smoothing of high voltage electrodes has been shown to reduce field electron emission, and GCIB treated RF cavities are being studied for use in future high energy particle accelerators.

CLUSTER CHEMISTRY

In chemistry, a cluster is an ensemble of bound atoms intermediate in size between a molecule and a bulk solid. Clusters exist of diverse stoichiometries and nuclearities. For example, carbon and boron atoms form fullerene and borane clusters, respectively. Transition metals and main group elements form especially robust clusters.

The phrase cluster was coined by F.A. Cotton in the early 1960s to refer to compounds containing metal–metal bonds. In another definition a cluster compound contains a group of two or more metal atoms where direct and substantial metal bonding is present.

The main cluster types are 'naked' clusters (without stabilising ligands) and those with ligands. Typical ligands that stabilise clusters include carbon monoxide, halides, isocyanides, alkenes, and hydrides.

Applications of Clusters in Catalysis

Synthetic metal carbonyl cluster compounds have been evaluated as catalysts for a wide range of industrial reactions, especially related to carbon monoxide utilisation, but no industrial applications exist. The clusters $Ru_3(CO)_{12}$ and $Ir_4(CO)_{12}$ catalyse the Water gas shift reaction, also catalysed by iron oxide, and $Rh_6(CO)_{16}$ catalyses the conversion of carbon monoxide into hydrocarbons, reminiscent of the Fischer-Tropsch process, although again iron-oxide based heterogeneous catalysts are used industrially.

Although discrete clusters have no well-defined role in industrial catalysis, they are widespread in nature. Most prevalent are the iron-sulphur proteins, which are involved with electron-transfer but also catalyse certain transformations. Nitrogen is reduced to ammonia at an Fe-Mo-S cluster at the heart of the enzyme nitrogenase. CO is oxidised to CO_2 by the Fe-Ni-S cluster carbon monoxide dehydrogenase. Hydrogenases rely on Fe_2 and NiFe clusters.

The term cluster should be pertinent to assembly of more than two metal atoms bound together in a planar or polyhedron arrangements such as Re_3Cl_9 and Mo_6Cl_8 units. Metal-Metal cluster could be classified as cages compounds or not when it is planar.

Electronic Structure

Metal clusters are frequently composed of refractory metal atoms. In general metal centres with large d-orbitals form stable clusters because of favourable overlap of valence orbitals. Thus, metals with a low oxidation state for the later metals and mid-oxidation states for the early metals tend to form stable clusters. Polynuclear metal carbonyls are generally found in late transition metals with low formal oxidation states. The polyhedral skeletal electron pair theory or Wade's electron counting rules predict trends in the stability and structures of many metal clusters.

The development of cluster chemistry occurred contemporaneously along several independent lines, which are roughly classified in the following sections. The first synthetic metal cluster was probably calomel, which was known in India already in the 12th century. The existence of a mercury to mercury bond in this compound was established in beginning of the 20th century.

Transition metal carbonyl clusters: The development of metal carbonyl compounds such as $Ni(CO)_4$ and $Fe(CO)_5$ led quickly to the isolation of $Fe_2(CO)_9$ and $Fe_3(CO)_{12}$. Rundle and Dahl discovered that $Mn_2(CO)_{10}$ featured an 'unsupported' Mn-Mn bond, thereby verifying the ability of metals to bond to one another in molecules. In the 1970's, Paolo Chini demonstrated that very large clusters could be prepared from the platinum metals, one example being $[Rh_{13}(CO)_{24}H_3]_{2-}$.

Transition metal halide clusters: Linus Pauling showed that '$MoCl_2$' consisted of Mo_6 octahedra. F. Albert Cotton established that '$ReCl_3$' in fact features subunits of the cluster Re_3Cl_9, which could be converted to a host of adducts without breaking the Re-Re bonds. Because this compound is diamagnetic and not paramagnetic the rhenium bonds are double bonds and not single bonds. In the solid state further bridging occurs between neighbours and when this compound is dissolved in hydrochloric acid a $Re_3Cl_{12}^{3-}$ complex forms. An example of a tetranuclear complex is hexadecamethoxytetratungsten $W_4(OCH_3)_{12}$ with tungsten single bonds and molybdenum chloride $(Mo_6Cl_8)Cl_4$ is a hexanuclear molybdenum compound and an example of an octahedral cluster. A related group of clusters with the general formula $M_xMo_6X_8$ such as $PbMo_6S_8$ form a Chevrel phase, which exhibit superconductivity at low temperatures. The eclipsed structure of potassium octachlorodirhenate(III), $K_2Re_2Cl_8$ was explained by invoking Quadruple bonding. This discovery led to a broad range of derivatives including di-tungsten tetra(hpp), the current record holder low ionisation energy.

Boron hydrides: Contemporaneously with the development of metal cluster compounds, numerous boron hydrides were discovered by Alfred Stock and his successors who popularised the use of vacuum-lines for the manipulation of these often volatile, air-sensitive materials. Clusters of boron are boranes such as pentaborane and decaborane. Composite clusters containing CH and BH vertices are carboranes.

Fe-S clusters in biology. In the 1970s, ferredoxin was demonstrated to contain Fe_4S_4 clusters and later nitrogenase was shown to contain a distinctive $MoFe_7S_9$ active site. With the development of bioinorganic chemistry, a variety of synthetic analogues of these clusters have been described.

Zintl clusters. Zintl compounds feature naked anionic clusters that are generated by reduction of heavy main group *p* elements, mostly metals or semimetals, with alkali metals, often as a solution in anhydrous liquid ammonia or ethylenediamine. Examples of Zintl anions are $[Bi_3]^{3-}$, $[Sn_9]^{4-}$, $[Pb_7]^{4-}$, and $[Sb_7]^{3-}$. Although these species are called 'naked clusters', they are usually strongly associated with alkali metal cations. Some examples have been isolated using cryptate complexes of the alkali metal cation, e.g. $[Pb_{10}]^{2-}$ anion, which features a capped square antiprismatic shape. According to Wade's rules ($2n+2$) the number of cluster electrons is 22 and therefore a closo cluster. The compound is prepared from oxidation of K_4Pb_9 by Au^+ in PPh_3AuCl (by reaction of tetrachloroauric acid and triphenylphosphine) in ethylene diamine with 2.2.2-crypt. This type of cluster was already known as is the endohedral $Ni@Pb_{10}^{2-}$ (the cage contains one nickel atom). The icosahedral tin cluster Sn_{12}^{2-} or stannaspherene anion is another closed shell structure observed (but not isolated) with photoelectron spectroscopy. With an internal diameter of 6.1 Angstrom it is of comparable size to fullerene and should be capable of containing small atoms as in endohedral fullerenes.

Gas-phase clusters and fullerenes: Unstable clusters can also be observed in the gas-phase by means of mass spectrometry even though they may be thermodynamically unstable and aggregate easily upon condensation. Such naked clusters, i.e. those that are not stabilised by ligands, are often produced by laser induced evaporation—or ablation—of a bulk metal or metal-containing compound. Typically, this approach produces a broad distribution of size distributions. Their electronic structures can be interrogated by techniques such as photoelectron spectroscopy, while infrared multiphoton dissociation spectroscopy is more probing the clusters geometry. Their properties (Reactivity, ionisation potential, HOMO-LUMO-gap) often show a pronounced size dependence. Examples of such clusters are certain aluminium clusters as superatoms and certain gold clusters. Certain metal clusters are considered to exhibit metal aromaticity. In some cases, the results of laser ablation experiments are translated to isolated compounds, and the premier cases are the clusters of carbon called the fullerenes, notably clusters with the formula C_{60}, C_{70}, and C_{84}. The fullerene sphere can be filled with small molecules in Endohedral fullerenes.

Extended metal atom chains: Extended metal atom chain complexes (EMAC) are a novel topic in academic research. They are comprised of linear chains of metal atoms stabilised with ligands. EMACS are known based on nickel (with 9 atoms), chromium and cobalt (7 atoms) and ruthenium (5 atoms). In theory it should be possible to obtain infinite one-dimensional molecules and research is oriented towards this goal. In one study an EMAC was obtained that consisted of 9 chromium atoms in a linear array with 4 ligands (based on an oligo pyridine) wrapped around it. In it the chromium chain contains 4 quadruple bonds.

EXCURSIONS IN CLUSTER SCIENCE

During the last two decades, the chemical physics group of Tel-Aviv University explored the structure, energetics, spectroscopy and dynamics of clusters, focusing on the energy landscapes, spatial structures and shapes, phase changes, superfluidity, energetics, level structure, electronic-vibrational spectroscopy, size effects, response and nuclear-electronic dynamics of large finite systems. Recently, our dynamic studies were extended for the adiabatic nuclear dynamics of multicharged atomic and molecular clusters, which manifest unique fragmentation patterns, such as cluster fission and Coulomb explosion.

Concurrently, a fascinating analogy was established between Coulomb explosion of multicharged clusters and nuclear dynamics of finite, ultracold gases, i.e. optical molasses, in the temperature domain of $T = 10\ \mu K$–$100\ \mu K$. Cluster science constitutes the art of building bridges, i.e. bridging between the structure, energetics, thermodynamics, response and dynamics of molecular and condensed phase systems in terms of size scaling laws, bridging between the electron-nuclear dynamics and response of clusters and of nanostructures, and bridging between nuclear dynamics of clusters and of ultracold, large, finite, quantum systems.

From Fission to Coulomb Explosion

The fragmentation of multiply charged finite systems driven by long-range Coulomb (or pseudo-Coulomb) forces, i.e. nuclei, clusters, droplets, and optical molasses, raises the following interesting questions regarding the energetics and dynamics of dissociation:

1. How does a finite system respond to a large excess charge or to an effective charge?
2. What are the topography and topology of the multidimensional energy landscape that guide the system's shape evolution and fragmentation?
3. What are the fragmentation channels and under what conditions are they realised?
4. What is the interplay between fission, i.e. instability towards dissociation of the finite system into two (or a small number of) fragments and Coulomb explosion into a large number ($\sim n$, where n is the number of constituents) of ionic species?

The ubiquity of fission phenomena of droplets, nuclei, and clusters was traditionally described by the classically liquid drop model (LDM), where a classically charged drop deforms through elongated shapes to form separate droplets. The fissibility parameter $X = E(Coulomb)/2E(surface)$ characterises the relative contribution of repulsive (Coulomb) and cohesive (surface) energies to the fission barrier, separating between the bound initial states and the fission products. For $X < 1$, thermally activated fission over the barrier prevails. At the Rayleigh instability limit of $X = 1$ the barrier height is zero. Many features of nuclear and metal cluster fission require to account for quantum shell effects. Nevertheless, the simple LDM expression $X = Z^2 e^2 / 16\pi R_0^3 = (Z^2/n) / (Z^2/n)_{cr}$, with the proportionality factor $(Z^2/n)_{cr} = 16\pi R_0^3/e^2$ (where γ is the surface tension, Z the total charge, R_0 the system's radius and r_0 the constituent radius), provided the conceptual framework for the fission of charged finite systems. All the diverse phenomena of fission were realised for fissibility parameters below the Rayleigh instability limit of $X = 1$, i.e. nuclear fission, the fission of metal clusters, and of hydrogen-bonded clusters. Beyond the fissibility limit ($X > 1$) barrierless fission and other dissociative channels open up. We have transcended the Rayleigh instability limit ($X = 1$) for Coulomb instability of large finite systems, demonstrating the prevalence of a qualitatively different fragmentation pattern of Coulomb explosion beyond the Rayleigh instability limit. We studied the fragmentation patterns and dynamics of highly charged Morse clusters by varying the range of the pair potential and of the fissibility parameters. The instability of multicharged Morse clusters directly reflects on covalently or dispersion-bound chemical and biophysical finite systems. The Rayleigh instability limit separates between nearly binary or tertiary spatially unisotropic fission for $X < 1$ and spatially isotropic Coulomb explosion into a large number of ionic fragments for $X > 1$ (Fig. 13.2).

We explored the Coulomb instability of multicharged, or effectively charged, finite systems (Fig. 13.3). The majority of the currently available experimental information on the Coulomb instability of nuclei (i.e. $X = 0.7$ for 235 U and $X = 0.9$ for the recently discovered $Z = 114$ element), of charged droplets (i.e. $X = 0.7 - 1.0$ for hydrogen bonded systems), and of multiply charged metal clusters ($X = 0.85 \pm 0.07$ for Na_n^{+z} clusters) pertains to the fission limit, i.e. $X < 1$ (Fig. 13.3).

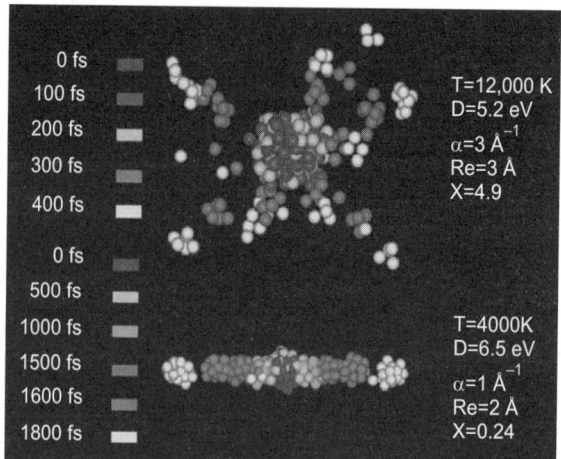

Fig. 13.2. Time resolved nuclear dynamics of the fragmentation of highly charged $(A^+)_{55}$ Morse clusters (mass of A is 100 amu). The two panels show superimposed temporal patterns of the fragmentation, where each colour corresponds to a different time for a one-colour snapshot, as marked on the two panels. The Morse potential parameters and the fissibility parameter X are marked on the panels. The time $t = 0$ corresponds to the T jump to the final temperature (see text). Note the dramatic difference between the spatially isotropic Coulomb explosion (for $X = 4.9$) on the upper panel, and cluster fusion (for $X = 0.24$) on the lower panel.

How can the Rayleigh limit for the Coulomb instability of a finite system be overcome? The $X \gg 1$ domain can be accomplished either by a marked enhancement of the repulsive Coulomb energy, or by a dramatic reduction of the cohesive surface energy (Fig. 13.3).

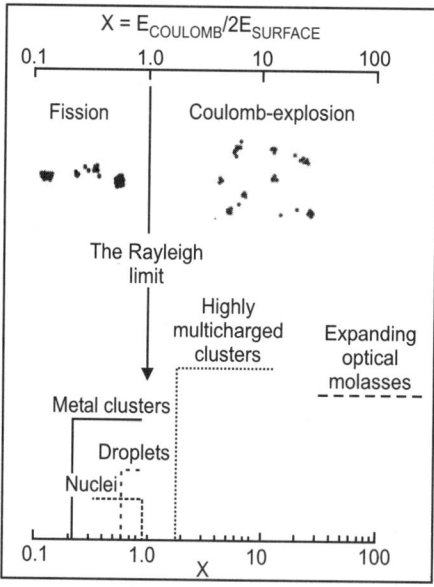

Fig. 13.3. A classification of fragmentation patterns of multicharged and effectively charged large, finite systems.

The increase of E (Coulomb) can be attained by cluster multielectron ionisation in ultraintense (peak intensity $I = 10^{15} - 10^{20}$ Wcm^{-2}) laser fields, while the dramatic decrease of E(surface) can be accomplished in three-dimensional, ultracold optical molasses, where pseudo-Coulomb forces result in isotropic cloud expansion, in analogy with Coulomb explosion.

The traditional view of Coulomb explosion involves uniform ion expansion. Such is the case for the explosion of multicharged homonuclear clusters (e.g. $(D_2)_{n/2}$ or $(Xe)_n$ with the expansion of (e.g. D^+ or Xe^{q+}) ions retaining a uniform spatial distribution (as is the case for X > 1 in Fig. 13.2), with an energy distribution being proportional to the square root of the energy, up light-heavy heteroclusters consisting of light and heavy ions, e.g. vertically ionised heteroclusters of hydrogen iodide, $(H^+I^{q+})n$ or $(D^+I^{q+})n$ ($q = 7$–35). In this case, kinematic overrun effects of the light ions (e.g. H^+ or D^+) will result in thin, two-dimensional shells of these light ions, with the monolayer expansion occurring on the femtosecond time scale (Fig. 13.4).

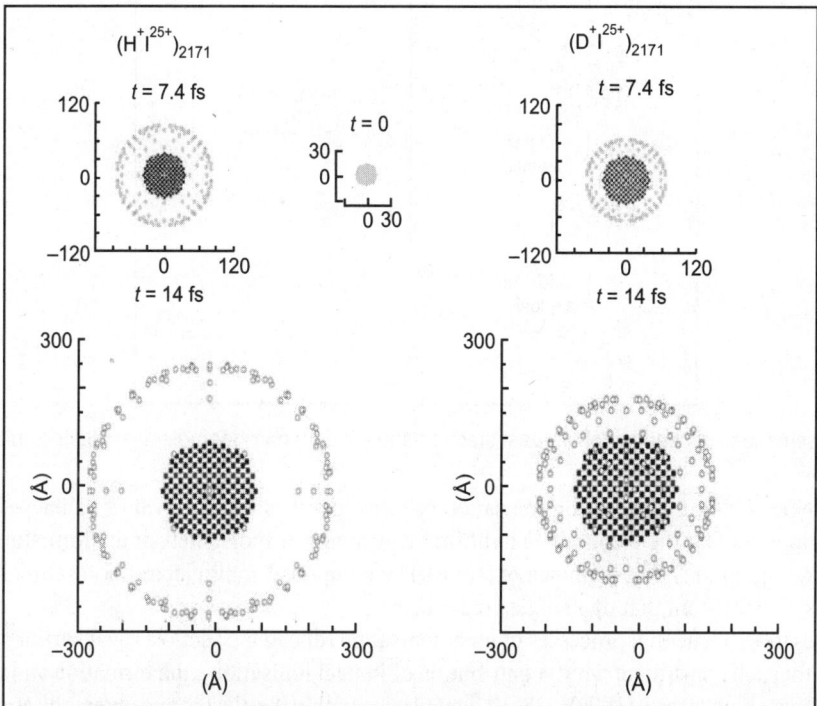

Fig. 13.4. A two-dimensional picture of the spatial structure of Coulomb expanding $(H^+I^{25+})_{2171}$ light-heavy heteroclusters at $t = 0$, 7.4 fs and 14 fs, obtained from molecular dynamics simulations. Black squares (■) represent I^{25+} ions, while circles (O) represent H^+ ions. This pictorial representation reveals the formation of narrow expanding shells of the light ions.

Such an expanding nanoshell of light ions, corresponding to transient soft matter, is analogous to a 'soap bubble' characterised by negative surface tension and is being driven by Coulomb pressure. This transient halo of an expanding, regular monoionic spherical nanointerface manifests transient self-organisation on the molecular level in complex systems. Future experimental interrogations of these novel phenomena will emerge from the exploration of the energetics of the light ions in the Coulomb explosion of multicharged light-heavy heteroclusters, involving a narrow energy distribution with a low-energy cutoff. An exciting experimental approach pertains to the application of ultrafast electron diffraction methods for the exploration of the transient structure of the exploding clusters.

Ultraintense Laser—Cluster Interactions

Table top lasers delivering an energy of 1 Joule per pulse on the time scale of ~100 fs, can deliver a power of ~ 1020 Wcm^{-2}, constituting the highest light intensity on earth. Highly charged molecular clusters can be prepared by the irradiation of a cluster beam by ultrashort (tens of fs) and ultraintense (intensity $I = 10^{15}$–10^{20} Wcm^{-2}) laser pulses (Fig. 13.5).

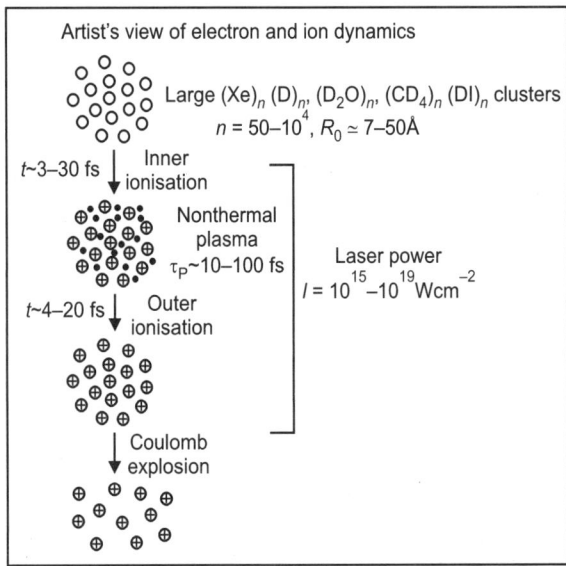

Fig. 13.5. A schematic description of ultrafast electron and ion dynamics for molecular clusters in ultraintense laser pulses.

The extreme cluster multielectron ionisation process involves the removal of valence electrons or complete stripping of all the electrons in light first-row atoms or molecules, or the formation of highly charged ions, e.g. up to Xe^{36+}, from heavy atoms. The compound multielectron ionisation mechanism of clusters is distinct from that of a single constituent.

It involves three sequential processes of inner ionisation (due to the semiclassical barrier suppression mechanism for each constituent with a contribution of impact ionisation), the formation and response of a nonequilibrium, high energy (100eV–3keV) nanoplasma within the cluster, and outer ionisation (induced by barrier suppression for the entire cluster and by quasiresonance effects). Femtosecond electron dynamics of inner ionisation on the time scale of ~1–5 fs and of outer ionisation on the time scale of ~5–20 fs results in multielectron ionisation.

For the intensity domain of $I = 10^{16}$–10^{17} Wcm^{-2} the cluster molecules loose all their valence electrons, with the nanoplasma being persistent, while for the highest intensity range of 10^{18}–10^{19} Wcm^{-2} both valence and inner shell electrons can be stripped off, with the nanoplasma being completely depleted. The Coulomb instability of a highly charged finite cluster triggers simultaneous and concurrent ultrafast Coulomb explosion on the time scales of 10–200 fs (Fig. 13.5).

Analytical expressions for the fs time scales of Coulomb explosion and of (divergent) scaling laws for the energetics of the highly charged ions were derived and were confirmed by molecular dynamics simulations with attosecond time steps describing fs dynamics. Ultrahigh ion energies in the range of 1keV–1MeV are released by cluster Coulomb explosion, as portrayed in Fig. 13.6 for deuterium containing homonuclear and heteronuclear clusters, where deuteron energies in the range of 1–100 keV can be obtained.

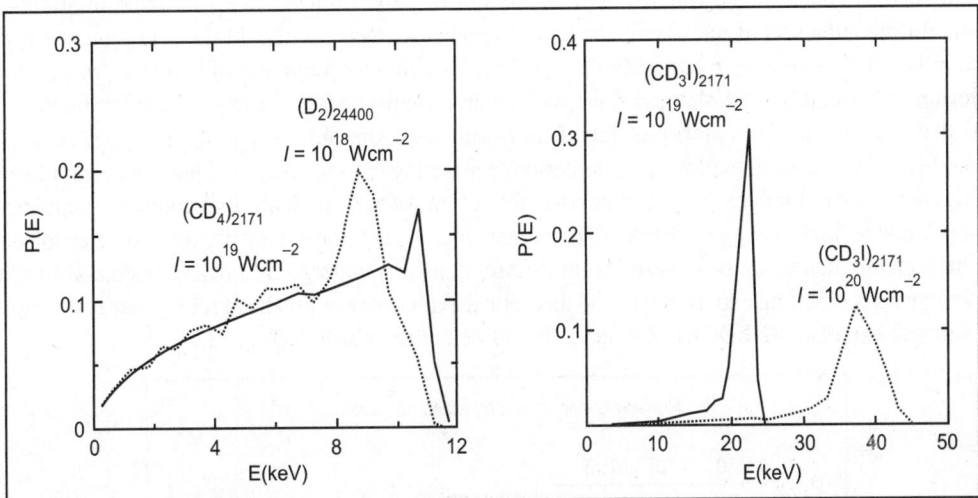

Fig. 13.6. Energy distributions of D^+ ions from Coulomb explosion of $(D^+)_n$ homonuclear clusters ($n = 2.44 \cdot 10^4$) and deuterium containing heteronuclear clusters $(C^6 + D_4^+)n$ ($n = 2171$) and $(C^{+4} + H_3^+ I^{q+})_n$ ($n = 2171$, $q = 25$ at $I = 10^{19}$ Wcm^{-2}, and $q = 35$ at $I = 10^{20}$ Wcm^{-2}).

A significant implication of these high ion energies pertains to nuclear fusion reactions of highly energetic D^+ (as well as T^+ or H^+) ions produced by Coulomb explosion of multicharged clusters in extreme multielectron ionisation in ultraintense laser fields, which will be addressed later in this chapter. Cluster dynamics is moving from ultrafast femtosecond to picosecond nuclear dynamics, towards ultrafast attosecond to femtosecond electron dynamics, and towards electron-nuclear dynamics in ultraintense laser fields. 'Pure' electron dynamics constitutes new dynamic processes in chemistry and physics. Ultrafast cluster dynamics is not limited to the dynamics of ions on the time scale of nuclear motion, but is extended to the realm of electron dynamics, which by-passes the constraints imposed by the Franck-Condon principle.

Nuclear Fusion Driven by Cluster Coulomb Explosion

Eighty years of search for table-top nuclear fusion, driven by bulk or surface chemical reactions, which involved catalytic dissociation or electrochemical productions of deuterium, reflects on a multitude of experimental and conceptual failures. In 1926 the German physicist Fritz Paneth reported on the apparent observation of helium from hydrogen absorbed on powdered palladium, which might have originated from nuclear fusion. A year later this claim was retracted. In 1935 Adalbert Farkas and Ladislaus Farkas, the founders of physical chemistry in Israel, worked on ortho- and parahydrogen and on deuterium chemistry in the Department of Colloid Science at Cambridge University, England, where they found shelter as refugees from Germany. When passing deuterium gas through a palladium tube, they seemed to observe traces of helium, which might have originated from dd ($D^+ + D^+$) nuclear fusion. However, a search for neutron emission in this system, conducted by Lord Rutherford at the request of the Farkas brothers, was negative and eliminated any possibility of nuclear fusion. In this category of negative results for nuclear fusion belongs the widely publicised 1989 'cold fusion' controversy, which did not provide any acceptable scientific information.

These spectacular failures are not surprising as, to the best of our knowledge, no theoretical evidence is available to support any valid mechanism of nuclear fusion driven by chemical reactions in infinite bulk or surface systems. The fragmentation dynamics of large finite systems involves an alternative avenue for the induction of nuclear fusion by chemical reactions, e.g. the dd $(D^+ + D^+)$ nuclear fusion reaction $D^+ + D^+ \rightarrow {}^3He^{2+} + n$ (2.45 MeV) + 3.27 MeV, with the production of neutrons (n). Coulomb explosion of extremely multicharged finite molecular systems strives towards the exploration of new areas that are alien to the majority of the chemical physics community. These areas involve nuclear fusion driven by Coulomb explosion of deuterium containing homonuclear and heteronuclear clusters.

High-energy Coulomb explosion of an assembly of multicharged, deuterium containing, molecular clusters produces high-energy (1–100 keV) deuterons (Fig. 13.6) in the energy domain of nuclear physics. The high energy deuterons originating from different clusters undergo dd nuclear fusion. During the last four years compelling experimental and theoretical evidence was advanced for nuclear fusion driven by Coulomb explosion (NFDCE) in an assembly of deuterium clusters (Fig. 13.7).

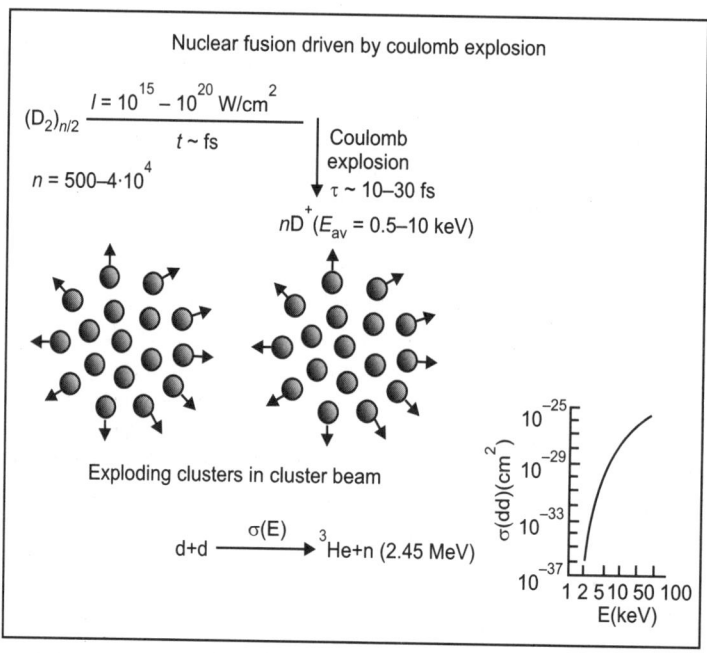

Fig. 13.7. dd Nuclear fusion driven by Coulomb explosion of deuterium clusters. Top: Multielectron ionisation in ultraintense laser fields $(I > 10^{17}$ Wcm$^{-2})$ strips the $(D_2)_{n/2}$ clusters of all their valence electrons via consecutive inner and outer ionisation. Parallel and concurrent with outer ionisation, cluster Coulomb explosion of $(D^+)_n$ clusters occurs. In the size domain of $n = 459$–$7.6 \cdot 10^4$ the D^+ average energy increases from 0.3 keV to 9.0 keV. Energetic deuterons (D^+ ions) emerging from different clusters in the cluster beam undergo dd nuclear fusion. Bottom: Energy dependence of the cross sections $\sigma(dd)$ for dd fusion adopted from the data of reference.

Completely ionised $(D^+)_n$ clusters are produced by multielectron ionisation of homonuclear $(D_2)_{n/2}$ $(n = 500 - 4 \cdot 10^4$, $R_0 = 10$–75 Å$)$ clusters in ultraintense laser fields $(I > 10^{17}$ Wcm$^{-2})$, stripping the clusters from all their electrons. For Coulomb explosion of very large homonuclear deuterium $(D^+)_n$ clusters $(n = 3.8 \cdot 10^4$ and cluster radius $R_0 = 72$Å$)$, the average deuteron (D^+) energy is $E_{av} = 9$ keV and the maximal energy is $E_M = 13$ keV. For these deuteron energies the cross section for dd $(D^+ + D^+)$

nuclear fusion is $\sigma(dd) \simeq 10^{-28}$ cm^2 (Fig. 13.7), being sufficiently high to induce the dd fusion reaction. Collisions between energetic deuterons, which originate from Coulomb explosion of different deuterium clusters (Fig. 13.7), result in NFDCE, which was experimentally observed in the Lawrence-Livermore laboratory. Our theoretical and computational work proposed and demonstrated that an effective way to produce highly energetic d nuclei (D$^+$ ions) for nuclear fusion involves multielectron ionisation and Coulomb explosion of molecular heteroclusters of deuterium bound to heavy atoms. Highly ionised heteroclusters for high-energy Coulomb explosion involve heavy water clusters $(D^+D^+O^{q+})_n$ $(q = 6-8)$, heavy methane clusters $(C^{q+}(D^+)_4)_n$ $(q = 4-6)$, or deuterated hydroiodic clusters $(D^+I^{q+})_n$ $(q = 7-35)$ in the size domain of $n = 55-4 \cdot 10^3$ $(R_0 = 10-40$Å$)$. A dramatic energy enhancement of deuteron energy from these heteroclusters, as compared to deuterons from homonuclear deuterium clusters, is manifested (Fig. 13.6). For Coulomb explosion of heteroclusters the heavy multicharged ions (e.g. C^{4+}, C^{6+}, O^{6+}, O^{8+}) act as energetic triggers driving the light D$^+$ clusters to considerably higher kinetic energies than for totally ionised deuterium clusters of the same size. In addition, kinematic effects, which manifest a sharp energy maximum in the vicinity of E_M, in the energy spectra of the D$^+$ ions from heteronuclear clusters, provide a supplementary contribution to the efficiency of the NFDCE. The effects of energetic and kinematic triggering on the energetics of the D$^+$ ions in Coulomb explosion of multicharged deuterium containing homonuclear and heteronuclear clusters are manifested by the neutron yields for NFDCE (Fig. 13.8) calculated under the conditions of the Lawrence-Livermore experiment.

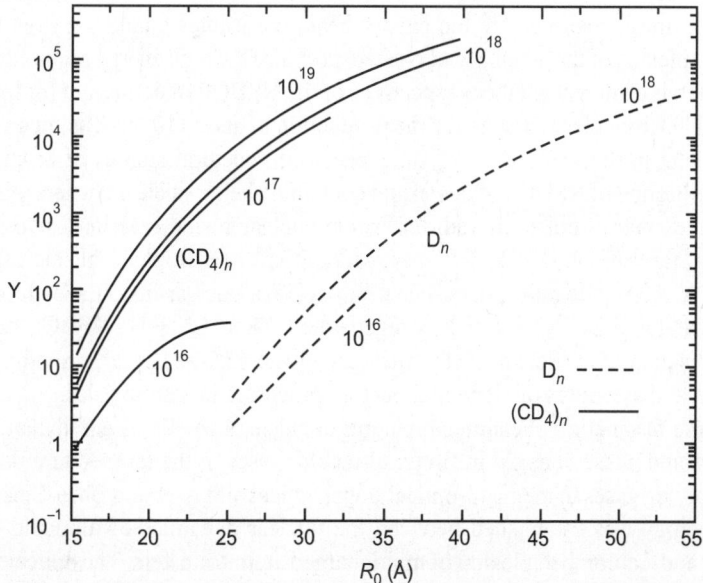

Fig. 13.8. Cluster size dependence of neutron yields per laser pulse for dd NFDCE in an assembly of $(CD_4)_n$ ($n = 55-4213$) heteroclusters (solid curves), and of $(D_2)_{n/2}$ ($n = 55-33573$) homonuclear clusters (dashed curves) in the laser intensity range $I = 10^{16}-10^{19}$ Wcm^{-2}. The NFDE for heteronuclear clusters manifests a considerably larger neutron yield than for homonuclear clusters of the same size, exhibiting energetic and kinematic effects, as discussed in the text.

The neutron yields per laser pulse (Fig. 13.8) for laser intensities of $I > 10^{17}$ Wcm^{-2} are higher by 2–3 orders of magnitude for Coulomb explosion of $(CD_4')_n$ clusters than for $(D_2)_{n/2}$ clusters of the same size. This theoretical prediction was confirmed by the experiments of the Saclay group, which demonstrated a marked enhancement of neutron yields from dd fusion in an assembly of Coulomb exploding $(CD_4)_n$ clusters, as compared to $(D_2)_{n/2}$ clusters.

An extreme way to attain highly effective energetic and kinematic triggering for driving deuterons to very high energies can be achieved for Coulomb explosion of deuterated methyl iodide $(C^{6+}D_3^+I^{q+})_n$ and hydroiodic acid $(D^+I^{q+})_n$, which produces heavily charged I^{q+} ions in very intense laser fields, i.e. $q = 25$ at $I = 10^{19}$ Wcm^{-2} and $q = 35$ at $I = 10^{20}$ Wcm^{-2}. For such iodine containing heteroclusters (with $n \simeq 4213$, $q = 25$ and $R_0 \simeq 40$ Å) the deuteron energies are $E_{av} = 40$ keV, being considerably higher than those for homonuclear $(D_2)_{15000}$ ($E_{av} \simeq 2.8$ keV) and heteronuclear $(CD_4)_{4213}$ clusters ($E_{av} = 10$ keV) of the same size. When the D^+ energy increases by a numerical factor of 3, the cross-section for dd fusion and the neutron yield increase by 2–3 orders of magnitude (Fig. 13.7). We infer from the data of Fig. 13.8 and from the foregoing discussion that the neutron yields for Coulomb explosion of deuterium containing homonuclear and heteronuclear clusters with $R_0 = 40$ Å are predicted to be $Y = 10^3$ neutrons/laser pulse for $(D_2)_{15000}$, $Y = 10^5$ neutrons/laser pulse for $(CD_4)_{4213}$, and $Y = 10^8$ neutrons/laser pulse for $(DI)_{4213}$. A semi-quantitative confirmation of these predictions was provided from experiments for Coulomb exploding $(D_2)_n$, and $(CD_4)_n$ clusters. The dream of table-top nuclear fusion in the chemical physics laboratory came true.

Perspectives

The NFDCE of molecular clusters induced by multielectron ionisation and Coulomb explosion involves a 'cold-hot' fusion mechanism, where the cluster beam constitutes a cold (or even ultracold) target, while Coulomb explosion of the assembly of clusters provides the high energy required to induce nuclear fusion. Of considerable interest are the perspectives of the NFDCE of deuterium (or tritium) containing homonuclear and heteronuclear clusters for the production of short (100ps–1ns) neutron pulses, which may be instrumental in the exploration of time–resolved structural studies of biomolecules or large molecules. The utilisation of NFDCE of deuterium containing heteronuclear clusters will greatly enhance the intensity of the neutron pulse. In addition, some nuclear fusion reactions involving protons and heavy nuclei, e.g. the $^{12}C^{6+} + H^+ \rightarrow ^{13}N^{7+} + \gamma$ reaction, are of astrophysical interest for the carbon-nitrogen-oxygen (CNO) cycle in hot stars. The CNO cycle of nuclear fusion, which supplies energy to the hot stars, is catalysed by $^{12}C^{6+}$, which is regenerated. The $^{12}C^{6+} + H^+$ NFDCE can be induced by multielectron ionisation of sufficiently large methane clusters ($R_0 = 120$ Å), providing information on the cross sections and dynamics of elemental nuclear processes in astrophysics.

Some novel and fascinating phenomena relating to cluster size effects and dynamics pertain to the nuclear dynamics and phase changes in finite, ultracold, gases in the temperature domain of $T = 100$ nK–100 µK, involving gases in magneto-optical traps, optical molasses and Bose-Einstein condensates.

A striking analogy was established between the nuclear dynamics of ultracold optical molasses ($T = 10$–100 µK) and Coulomb explosion of multicharged atomic clusters. The optical molasses involve a cloud of trapped, laser irradiated, neutral atoms, e.g. Rb, in a magnetic trap (Fig. 13.9) which is characterised by a density of $\rho = 10^{11}$–10^{13} atoms/cm^3 and by an interatomic distance of $r_0 \simeq 10^4$ Å. When the magnetic trap is suppressed, the cloud expands via the radiative trapping force, which prevails between radiation–emitting and reabsorbing atoms.

An isomorphism was established between the radiative trapping force and the electrostatic Coulomb force, with the effective charge characterising the radiative trapping force being $\sim 4 \cdot 10^{-5} e$, with e being the electron charge. The theory of the dynamics of cluster Coulomb explosion of multicharged molecular clusters was applied for the expansion of optical molasses. While the Coulomb explosion time of $(Xe^+)_n$

clusters is 10^{-13} s, the expansion time of optical molasses of Cs atoms was predicted to be 10 orders of magnitude longer, i.e. ~10^{-3}s. This estimate is in accord with experiments (Fig. 13.9). These studies, together with the exploration of superfluidity of helium–4 clusters, bridge between the dynamics of clusters and ultracold, large finite quantum systems.

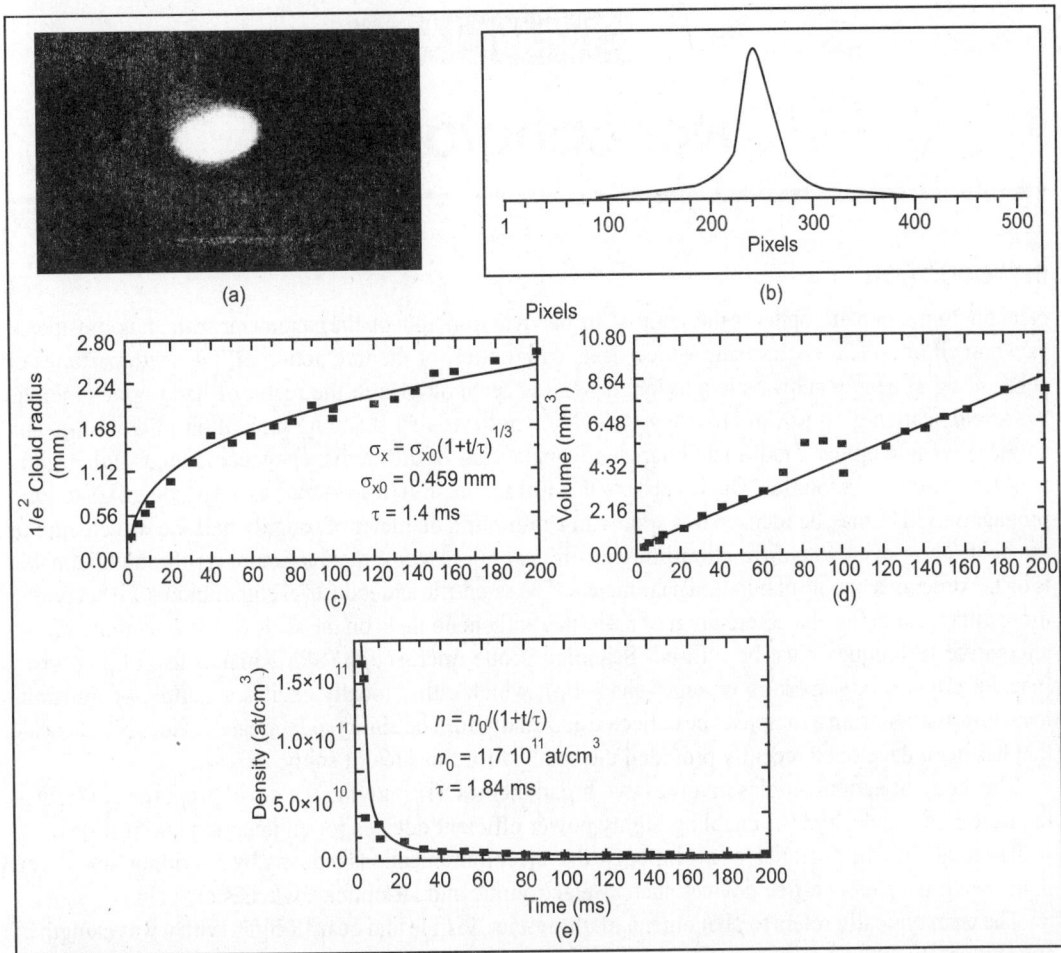

Fig. 13.9. Nuclear dynamics of the spatial extension of optical molasses of Rb. (a) A photograph of the irradiated cloud at $t = 0$, (b) distribution of excited atoms in an irradiated cloud at $t = 0$, (c) time dependence of the radius of the irradiated cloud. Characteristic expansion time $\tau = 1.4$ ms, (d) time dependence of the volume of the irradiated cloud, and (e) time dependence of the density of the irradiated expanding cloud. Characteristic expansion time $\tau = 1.8$ ms.

During the last decade, cluster science explored new fascinating scientific territories, bridging between cluster electron-nuclear dynamics and nuclear dd fusion, and bridging between cluster dynamics and ultracold quantum clouds.

Chapter 14

Nanophotonics

INTRODUCTION

Nanophotonics or nano-optics is the study of the behaviour of light on the nanometer scale. It is considered as a branch of optical engineering which deals with optics, or the interaction of light with particles or substances, at deeply subwavelength length scales. Technologies in the realm of nano-optics include near-field scanning optical microscopy (NSOM), photoassisted scanning tunnelling microscopy, and surface plasmon optics. Traditional microscopy makes use of diffractive elements to focus light tightly in order to increase resolution. But because of the diffraction limit (also known as the Rayleigh Criterion), propagating light may be focused to a spot with a minimum diameter of roughly half the wavelength of the light. Thus, even with diffraction-limited confocal microscopy, the maximum resolution obtainable is on the order of a couple of hundred nanometers. The scientific and industrial communities are becoming more interested in the characterisation of materials and phenomena on the scale of a few nanometers, so alternative techniques must be utilised. Scanning probe microscopy (SPM) makes use of a 'probe', (usually either a tiny aperture or super-sharp tip), which either locally excites a sample or transmits local information from a sample to be collected and analysed. The ability to fabricate devices in nanoscale that has been developed recently provided the catalyst for this area of study.

The study of nanophotonics involves two broad themes: (i) studying the novel properties of light at the nanometer scale, and (ii) enabling highly power efficient devices for engineering applications.

The study has the potential to revolutionise the telecommunications industry by providing low power, high speed, interference-free devices such as electrooptic and all-optical switches on a chip.

The term typically refers to phenomena of ultraviolet, visible and near IR light, with a wavelength of approximately 300 nm to 1.2 micrometres.

The interaction of light with these nanoscale features leads to confinement of the electromagnetic field to the surface or tip of the nanostructure resulting in a region referred to as the optical near field. This effect is to some extent analogous to a lightning rod, where the field concentrates at the tip. In this region, the field may need to adjust to the topography of the nanostructure. This means that the electromagnetic field will be dependent on the size and shape of the nanostructure that the light is interacting with.

This optical near field can also be described as a surface bound optical oscillation which can vary on length scale of tens or hundreds of nanometers—a length scale smaller than the wavelength of the incoming light. This can provide higher spatial resolution beyond the limitations imposed by the law of diffraction in conventional far-field microscopy. The technique derived from this effect is known as near-field microscopy, and opens up many new possibilities for imaging and spectroscopy on the nanoscale. A novel embodiment which has picometer resolution in the vertical plane above the waveguide surface is dual polarisation interferometry. Novel optical properties of materials can result from their extremely small size. A typical example of this type of effect is the colour change associated with colloidal gold. In contrast to bulk gold, known for its yellow colour, gold particles of 10 to 100 nm in size exhibit a rich red colour. The critical size where these and related effects take place are correlated with the mean free path of the conduction electrons of the metal.

In addition to these extrinsic size effects that determine a material's optical response to incoming light, the intrinsic properties of the material can change. These size effects occur as particles become even smaller. At this stage some of the intrinsic electronic properties of the medium itself change. One example of this phenomenon is in semiconductor nanostructures where the extremely small particle size confines the quantum mechanical wavefunction, leading to discrete optical transitions, e.g. fluorescence colours that depend on the size of the particle. The changing bandgap of the semiconductor is the reason for this colour change. This effect, however, since not directly correlated with optical wavelength, is not unanimously included when referring to nano-optics.

COMPONENTS OF A NANOPHOTONIC SYSTEM

1. Waveguides.
2. Couplers.
3. Optical switch.
4. Photo detectors.
5. Electro-optic modulators.
6. Wavelength division multiplexors.
7. Amplifiers.
8. Lasers.
9. Optical circulators.

Waveguide

A waveguide is a structure which guides waves, such as electromagnetic waves or sound waves. There are different types of waveguide for each type of wave. Waveguides differ in their geometry which can confine energy in one-dimension such as in slab waveguides or two-dimensions as in fibre or channel waveguides.

Electromagnetic waveguides

Waveguides can be constructed to carry waves over a wide portion of the electromagnetic spectrum, but are especially useful in the microwave and optical frequency ranges. Depending on the frequency, they can be constructed from either conductive or dielectric materials. Waveguides are used for transferring both power and communication signals.

Optical waveguides

Waveguides used at optical frequencies are typically dielectric waveguides, structures in which a dielectric material with high permittivity, and thus high index of refraction, is surrounded by a material with lower permittivity. The structure guides optical waves by total internal reflection. The most common optical waveguide is optical fibre.

Other types of optical waveguide are also used, including photonic-crystal fibre, which guides waves by any of several distinct mechanisms. Guides in the form of a hollow tube with a highly reflective inner surface have also been used as light pipes for illumination applications. The inner surfaces may be polished metal, or may be covered with a multilayer film that guides light by Bragg reflection (this is a special case of a photonic-crystal fibre). One can also use small prisms around the pipe which reflect light via total internal reflection — such confinement is necessarily imperfect, however, since total internal reflection can never truly guide light within a lower-index core (in the prism case, some light leaks out at the prism corners).

Acoustic waveguides

An acoustic waveguide is a physical structure for guiding sound waves. A duct for sound propagation also behaves like a transmission line. The duct contains some medium, such as air, that supports sound propagation.

Sound synthesis

Uses digital delay lines as computational elements to simulate wave propagation in tubes of wind instruments and the vibrating strings of string instruments.

Coupler

In telecommunications, the term coupler has the following meanings:

1. An interface device for coupling electrical signals by acoustical means — usually into and out of a telephone instrument.
2. A terminal device used to link data terminals and radio sets with the telephone network.

The link is achieved through acoustic (sound) signals rather than through direct electrical connection (Fig. 14.1).

Fig. 14.1. The novation CAT coupled modem.

Miniaturisation and integration are key drivers for future optical communication networks. Nanophotonic components are very interesting for ultra-dense photonic circuits, but the coupling with the outside world remains an important problem. An attractive solution is provided by grating couplers. In this chapter, we present the design and fabrication of compact and efficient grating couplers in InP-membrane, for coupling between nanophotonic waveguides and single mode fibre. A high vertical index contrast is achieved by wafer bonding. First components show a coupling efficiency of 30 per cent.

Today, a large breakthrough of optical communications is compromised by high coupling losses between the chip and the outside world (optical fibre), caused by the large deviation in dimensions between an optical fibre mode and an optical waveguide mode. Several solutions have been proposed to address this issue. By using an inverse taper approach, low loss and broadband operation was demonstrated, but these structures require lensed or special fibres with high numerical aperture.

Optical Switch

In telecommunication, an optical switch is a switch that enables signals in optical fibres or integrated optical circuits (IOCs) to be selectively switched from one circuit to another.

The word is used on several levels. In commercial terms (such as 'the telecom optical switch market size') it refers to any piece of circuit switching equipment between fibres. The majority of installed systems in this category actually use electronic switching between fibre transponders. Systems that perform this function by physically switching light are often referred to as 'photonic' switches, independent of how the light itself is switched. Away from the world of telecom systems, an optical switch is the unit that actually switches light between fibres, and a photonic switch is one that does this by exploiting nonlinear material properties to steer light (i.e. to switch wavelengths or signals within a given fibre). Hence a certain portion of the optical switch market is made up of photonic switches. These will contain within them an optical switch, which will, in a small number of cases, be a photonic switch.

An optical switch may operate by mechanical means, such as physically shifting an optical fibre to drive one or more alternative fibres, or by electro-optic effects, magneto-optic effects, or other methods. Slow optical switches, such as those using moving fibres, may be used for alternate routing of an optical transmission path, such as routing around a fault.

Fast optical switches, such as those using electro-optic or magneto-optic effects, may be used to perform logic operations; also included in this category are the semiconductor optical amplifiers, which are optoelectronic devices that can be used as optical switches and be integrated with discrete or integrated microelectronic circuits.

Photodetector

Photosensors or photodetectors are sensors of light or other electromagnetic energy. There are several varieties:

1. Optical detectors, which are mostly quantum devices in which an individual photon produces a discrete effect.
2. Chemical detectors, such as photographic plates, in which a silver halide molecule is split into an atom of metallic silver and a halogen atom. The photographic developer causes adjacent molecules to split similarly.
3. Photoresistors or light dependent resistors (LDR) which change resistance according to light intensity.
4. Photovoltaic cells or solar cells which produce a voltage and supply an electric current when illuminated.
5. Photodiodes which can operate in photovoltaic mode or photoconductive mode.

6. Photomultiplier tubes containing a photocathode which emits electrons when illuminated, the electrons are then amplified by a chain of dynodes.

7. Phototubes containing a photocathode which emits electrons when illuminated, such that the tube conducts a current proportional to the light intensity.

8. Phototransistors, which act like amplifying photodiodes.

9. Optical detectors that are effectively thermometers, responding purely to the heating effect of the incoming radiation, such as pyroelectric detectors, Golay cells, thermocouples and thermistors, but the latter two are much less sensitive.

10. Cryogenic detectors are sufficiently sensitive to measure the energy of single X-ray, visible and near infrared photons.

11. Charge-coupled devices (CCD), which are used to record images in astronomy, digital photography, and digital cinematography. Although before the 1990s photographic plates were the most common in astronomy. Glass-backed plates were used rather than film, because they do not shrink or deform in going between wet and dry condition, or under other disturbances. Unfortunately, Kodak discontinued producing several kinds of plates between 1980 and 2000, terminating the production of important sky surveys. The next generation of astronomical instruments, such as the Astro-E2, include cryogenic detectors. In experimental particle physics, a particle detector is a device used to track and identify elementary particles.

12. LEDs reverse-biased to act as photodiodes.

Electro-Optic Modulator

Electro-optic modulator (EOM) is an optical device in which a signal-controlled element displaying electro-optic effect is used to modulate a beam of light. The modulation may be imposed on the phase, frequency, amplitude, or direction of the modulated beam. Modulation bandwidths extending into the gigahertz range are possible with the use of laser-controlled modulators.

Generally a nonlinear optical material (organic polymers have the fastest response rates, and thus are best for this application) with an incident static or low frequency optical field will see a modulation of its refractive index.

Types of EOMs

Phase modulation

The simplest kind of EOM consists of a crystal, such as Lithium niobate, whose refractive index is a function of the strength of the local electric field. That means that if lithium niobate is exposed to an electric field, light will travel more slowly through it.

But the phase of the light leaving the crystal is directly proportional to the length of time it took that light to pass through it. Therefore, the phase of the laser light exiting an EOM can be controlled by changing the electric field in the crystal.

Note that the electric field can be created placing a parallel plate capacitor across the crystal. Since the field inside a parallel plate capacitor depends linearly on the potential, the index of refraction depends linearly on the field (for crystals where Pockel's effect dominates), and the phase depends linearly on the index of refraction, the phase modulation must depend linearly on the potential applied to the EOM.

Liquid crystal devices are electro-optical phase modulators if no polarisers are used.

Amplitude modulation

A phase modulating EOM can be also be used as an amplitude modulator by using a Mach-Zehnder interferometer. A beam splitter divides the laser light into two paths, one of which has a phase modulator as described above. The beams are then recombined. Changing the electric field on the phase modulating path will then determine whether the two beams interfere constructively or destructively at the output, and thereby control the amplitude or intensity of the exiting light. This device is called a Mach-Zehnder modulator. A very common application of EOMs is for creating sidebands in a monochromatic laser beam.

Wavelength-Division Multiplexing

In fibre-optic communications, wavelength-division multiplexing (WDM) is a technology which multiplexes multiple optical carrier signals on a single optical fibre by using different wavelengths (colours) of laser light to carry different signals. This allows for a multiplication in capacity, in addition to enabling bidirectional communications over one strand of fibre. This is a form of frequency division multiplexing (FDM) but is commonly called wavelength division multiplexing.

The term wavelength-division multiplexing is commonly applied to an optical carrier (which is typically described by its wavelength), whereas frequency-division multiplexing typically applies to a radio carrier (which is more often described by frequency). However, since wavelength and frequency are inversely proportional, and since radio and light are both forms of electromagnetic radiation, the two terms are equivalent in this context.

WDM systems

A WDM system uses a multiplexer at the transmitter to join the signals together, and a demultiplexer at the receiver to split them apart. With the right type of fibre it is possible to have a device that does both simultaneously, and can function as an optical add-drop multiplexer. The optical filtering devices used have traditionally been etalons, stable solid-state single-frequency Fabry-Perot interferometers in the form of thin-film-coated optical glass (Fig. 14.2).

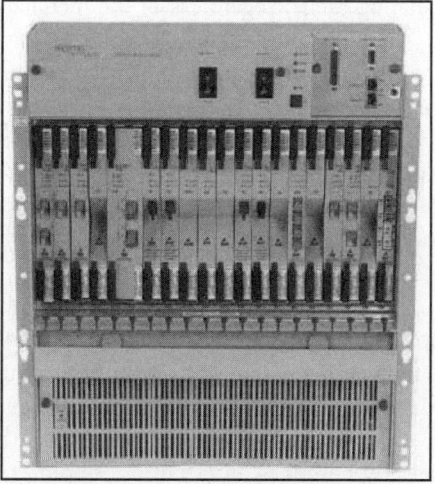

Fig. 14.2. Nortel's WDM system.

WDM systems are popular with telecommunications companies because they allow them to expand the capacity of the network without laying more fibre. By using WDM and optical amplifiers, they can accommodate several generations of technology development in their optical infrastructure without having to overhaul the backbone network. Capacity of a given link can be expanded by simply upgrading the multiplexers and demultiplexers at each end.

This is often done by using optical-to-electrical-to-optical (O/E/O) translation at the very edge of the transport network, thus permitting interoperation with existing equipment with optical interfaces.

Coarse WDM

Originally, the term 'coarse wavelength division multiplexing' was fairly generic, and meant a number of different things. In general, these things shared the fact that the choice of channel spacings and frequency stability was such that erbium doped fibre amplifiers (EDFAs) could not be utilised. Prior to the relatively recent ITU standardisation of the term, one common meaning for coarse WDM meant two (or possibly more) signals multiplexed onto a single fibre, where one signal was in the 1550 nm band, and the other in the 1310 nm band.

CWDM is also being used in cable television networks, where different wavelengths are used for the downstream and upstream signals. In these systems, the wavelengths used are often widely separated, for example the downstream signal might be at 1310 nm while the upstream signal is at 1550 nm.

Passive CWDM is an implementation of CWDM that uses no electrical power. It separates the wavelengths using passive optical components such as bandpass filters and prisms. Many manufacturers are promoting passive CWDM to deploy fibre to the home.

Dense WDM

Dense wavelength division multiplexing, or DWDM for short, refers originally to optical signals multiplexed within the 1550 nm band so as to leverage the capabilities (and cost) of erbium doped fibre amplifiers (EDFAs), which are effective for wavelengths between approximately 1525-1565 nm (C band), or 1570-1610 nm (L band). EDFAs were originally developed to replace SONET/SDH optical-electrical-optical (OEO) regenerators, which they have made practically obsolete. EDFAs can amplify any optical signal in their operating range, regardless of the modulated bit rate. In terms of multi-wavelength signals, so long as the EDFA has enough pump energy available to it, it can amplify as many optical signals as can be multiplexed into its amplification band (though signal densities are limited by choice of modulation format). EDFAs therefore allow a single-channel optical link to be upgraded in bit rate by replacing only equipment at the ends of the link, while retaining the existing EDFA or series of EDFAs through a long haul route. Furthermore, single-wavelength links using EDFAs can similarly be upgraded to WDM links at reasonable cost. The EDFAs cost is thus leveraged across as many channels as can be multiplexed into the 1550 nm band.

Amplifier

Generally, an amplifier or simply amp, is any device that changes, usually increases, the amplitude of a signal. The relationship of the input to the output of an amplifier—usually expressed as a function of the input frequency—is called the transfer function of the amplifier, and the magnitude of the transfer function is termed the gain.

In popular use, the term usually describes an electronic amplifier, in which the input 'signal' is usually voltage or current. In audio applications, amplifiers operate loudspeakers used in PA systems to make the human voice louder or play recorded music. Amplifiers may be classified according to the input (source) they are designed to amplify (such as a guitar amplifier, to perform with an electric guitar), the device they are intended to drive (such as a headphone amplifier), the frequency range of the signals (Audio, IF, RF, and VHF amplifiers, for example), whether they invert the signal (inverting amplifiers and non-inverting amplifiers), or the type of device used in the amplification (valve or tube amplifiers, FET amplifiers, etc.). A related device that emphasises conversion of signals of one type to another (for example, a light signal in photons to a DC signal in amperes) is a transducer, a transformer, or a sensor. However, none of these amplify power.

Laser

Light amplification by stimulated emission of radiation, LASER (laser), is a mechanism for emitting light within the electromagnetic radiation region of the spectrum, via the process of stimulated emission. The emitted laser light is (usually) a spatially coherent, narrow low-divergence beam, that can be manipulated with lenses. In laser technology, 'coherent light' denotes a light source that produces (emits) light of in-step waves of identical frequency and phase.

The laser's beam of coherent light differentiates it from light sources that emit incoherent light beams, of random phase varying with time and position; whereas the laser light is a narrow-wavelength electromagnetic spectrum monochromatic light; yet, there are lasers that emit a broad spectrum light, or simultaneously, at different wavelengths. Figure 14.3 shows the laser beams in fog and on a car windshield.

Fig. 14.3. Laser beams in fog and on a car windshield.

The word laser originally was the upper-case LASER, the acronym from light amplification by stimulated emission of radiation, wherein light broadly denotes electromagnetic radiation of any frequency, not only the visible spectrum; hence infrared laser, ultraviolet laser, X-ray laser, etc. (Fig. 14.4). Because the microwave predecessor of the laser, the maser, was developed first, devices that emit microwave and radio frequencies are denoted 'masers'. In the early technical literature, especially in that of the Bell Telephone Laboratories researchers, the laser was also called optical maser, a currently uncommon term, moreover, since 1998, Bell Laboratories adopted the laser usage. Linguistically, the back-formation verb to lase means 'to produce laser light' and 'to apply laser light to'. The word laser sometimes is inaccurately used to describe a non-laser-light technology, e.g. a coherent-state atom source is an atom laser.

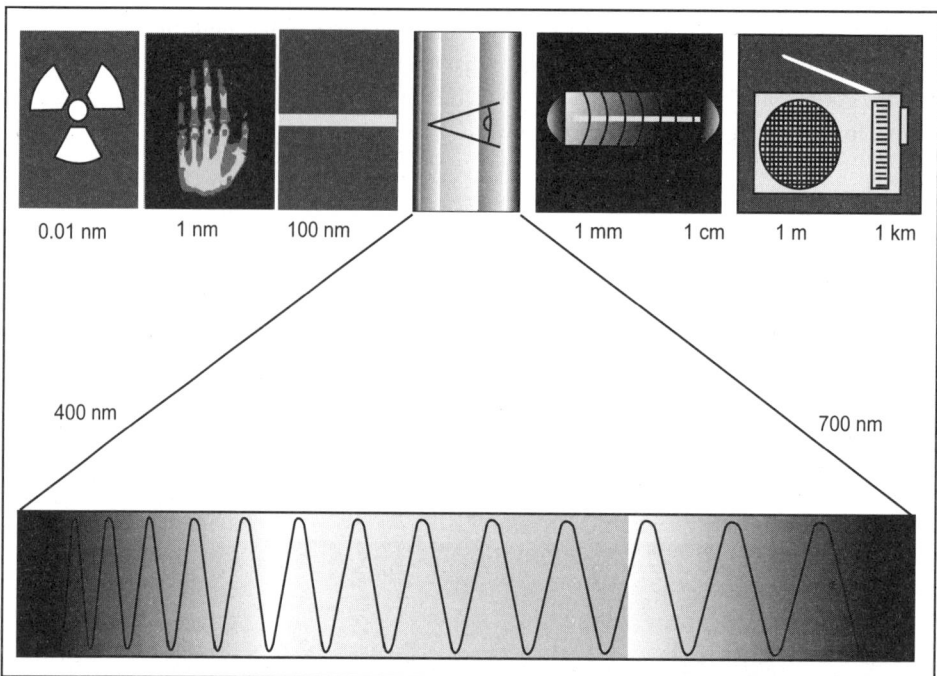

Fig. 14.4. From left to right: gamma rays, X-rays, ultraviolet rays, visible spectrum, infrared, microwaves, radio waves.

A laser consists of a gain medium inside a highly reflective optical cavity, as well as a means to supply energy to the gain medium. The gain medium is a material with properties that allow it to amplify light by stimulated emission. In its simplest form, a cavity consists of two mirrors arranged such that light bounces back and forth, each time passing through the gain medium. Typically one of the two mirrors, the output coupler, is partially transparent. The output laser beam is emitted through this mirror.

Light of a specific wavelength that passes through the gain medium is amplified (increases in power); the surrounding mirrors ensure that most of the light makes many passes through the gain medium, being amplified repeatedly.

Part of the light that is between the mirrors (that is, within the cavity) passes through the partially transparent mirror and escapes as a beam of light.

The process of supplying the energy required for the amplification is called pumping. The energy is typically supplied as an electrical current or as light at a different wavelength. Such light may be provided by a flash lamp or perhaps another laser.

Most practical lasers contain additional elements that affect properties such as the wavelength of the emitted light and the shape of the beam. Figure 14.5 shows the principal components of laser.

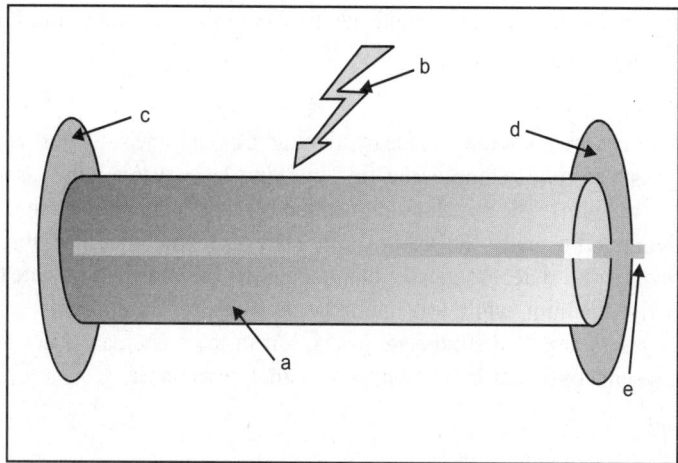

Fig. 14.5. Principal components: (a) gain medium, (b) laser pumping energy, (c) high reflector, (d) output coupler, and (e) laser beam.

Types and operating principles

Gas lasers

Gas lasers using many gases have been built and used for many purposes. The helium-neon laser (HeNe) emits at a variety of wavelengths and units operating at 633 nm are very common in education because of its low cost. Carbon dioxide lasers can emit hundreds of kilowatts at 9.6 μm and 10.6 μm, and are often used in industry for cutting and welding. The efficiency of a CO_2 laser is over 10 per cent.

Argon-ion lasers emit light in the range 351–528.7 nm. Depending on the optics and the laser tube a different number of lines is usable but the most commonly used lines are 458 nm, 488 nm and 514.5 nm.

A nitrogen transverse electrical discharge in gas at atmospheric pressure (TEA) laser is an inexpensive gas laser producing UV light at 337.1 nm.

Metal ion lasers are gas lasers that generate deep ultraviolet wavelengths. Helium-silver (HeAg) 224 nm and neon-copper (NeCu) 248 nm are two examples. These lasers have particularly narrow oscillation line widths of less than 3 GHz (0.5 picometers), making them candidates for use in fluorescence suppressed Raman spectroscopy.

Chemical lasers

Chemical lasers are powered by a chemical reaction, and can achieve high powers in continuous operation. For example, in the Hydrogen fluoride laser (2700–2900 nm) and the Deuterium fluoride laser (3800 nm) the reaction is the combination of hydrogen or deuterium gas with combustion products of ethylene in nitrogen trifluoride.

Excimer lasers

Excimer lasers are powered by a chemical reaction involving an excited dimer, or excimer, which is a short-lived dimeric or heterodimeric molecule formed from two species (atoms), at least one of which is in an excited electronic state. They typically produce ultraviolet light, and are used in semiconductor photolithography and in LASIK eye surgery. Commonly used excimer molecules include F_2 (fluorine, emitting at 157 nm), and noble gas compounds (ArF [193 nm], KrCl [222 nm], KrF [248 nm], XeCl [308 nm], and XeF [351 nm]).

Solid-state lasers

Solid-state laser materials are commonly made by 'doping' a crystalline solid host with ions that provide the required energy states. For example, the first working laser was a ruby laser, made from ruby (chromium-doped corundum). The population inversion is actually maintained in the 'dopant', such as chromium or neodymium. Formally, the class of solid-state lasers includes also fibre laser, as the active medium (fibre) is in the solid state. Practically, in the scientific literature, solid-state laser usually means a laser with bulk active medium, while waveguide lasers are caller fibre lasers.

'Semiconductor lasers' are also solid-state lasers, but in the customary laser terminology, 'solid-state laser' excludes semiconductor lasers, which have their own name.

Fibre-hosted lasers

Solid-state lasers where the light is guided due to the total internal reflection in an optical fibre are called fibre lasers. Guiding of light allows extremely long gain regions providing good cooling conditions; fibres have high surface area to volume ratio which allows efficient cooling. In addition, the fibres waveguiding properties tend to reduce thermal distortion of the beam. Erbium and ytterbium ions are common active species in such lasers.

Quite often, the fibre laser is designed as a double-clad fibre. This type of fibre consists of a fibre core, an inner cladding and an outer cladding. The index of the three concentric layers is chosen so that the fibre core acts as a single-mode fibre for the laser emission while the outer cladding acts as a highly multimode core for the pump laser. This lets the pump propagate a large amount of power into and through the active inner core region, while still having a high numerical aperture (NA) to have easy launching conditions.

Photonic crystal lasers

Photonic crystal lasers are lasers based on nanostructures that provide the mode confinement and the density of optical states (DOS) structure required for the feedback to take place. They are typical micrometer-sized and tunable on the bands of the photonic crystals.

Semiconductor lasers

Semiconductor lasers are also solid-state lasers but have a different mode of laser operation. Commercial laser diodes emit at wavelengths from 375 nm to 1800 nm, and wavelengths of over 3 µm have been demonstrated.

Low power laser diodes are used in laser printers and CD/DVD players. More powerful laser diodes are frequently used to optically pump other lasers with high efficiency. The highest power industrial laser diodes, with power up to 10 kW (70 dBm), are used in industry for cutting and welding. External-cavity semiconductor lasers have a semiconductor active medium in a larger cavity. These devices can generate high power outputs with good beam quality, wavelength-tunable narrow-linewidth radiation, or ultrashort laser pulses.

Dye lasers

Dye lasers use an organic dye as the gain medium. The wide gain spectrum of available dyes allows these lasers to be highly tunable, or to produce very short-duration pulses (on the order of a few femtoseconds).

Free electron lasers

Free electron lasers, or FELs, generate coherent, high power radiation, that is widely tunable, currently ranging in wavelength from microwaves, through terahertz radiation and infrared, to the visible spectrum, to soft X-rays. They have the widest frequency range of any laser type. While FEL beams share the same optical traits as other lasers, such as coherent radiation, FEL operation is quite different. Unlike gas, liquid, or solid-state lasers, which rely on bound atomic or molecular states, FELs use a relativistic electron beam as the lasing medium, hence the term free electron.

Uses

Some of the uses of laser are given below:

1. Medicine: Bloodless surgery, laser healing, surgical treatment, kidney stone treatment, eye treatment, dentistry.
2. Industry: Cutting, welding, material heat treatment, marking parts.
3. Defense: Marking targets, guiding munitions, missile defence, electro-optical countermeasures (EOCM), alternative to radar, blinding enemy troops.
4. Research: Spectroscopy, laser ablation, laser annealing, laser scattering, laser interferometry, LIDAR, laser capture microdissection.
5. Product development/commercial: laser printers, CDs, barcode scanners, thermometers, laser pointers, holograms, bubblegrams.
6. Laser lighting displays: Laser light shows.
7. Laser skin procedures such as acne treatment, cellulite reduction, and hair removal.

Examples by power

Different applications need lasers with different output powers. Lasers that produce a continuous beam or a series of short pulses can be compared on the basis of their average power. Lasers that produce pulses can also be characterised based on the peak power of each pulse. The peak power of a pulsed laser is many orders of magnitude greater than its average power. The average output power is always less than the power consumed.

The continuous or average power required for some uses:

1. Less than 1 mW – laser pointers.
2. 5 mW – CD-ROM drive.
3. 5–10 mW – DVD player or DVD-ROM drive.
4. 100 mW – high-speed CD-RW burner.
5. 250 mW – consumer DVD-R burner.
6. 1 W – green laser in current Holographic Versatile Disc prototype development.
7. 1–20 W – output of the majority of commercially available solid-state lasers used for micro-machining.
8. 30–100 W – typical sealed CO_2 surgical lasers.
9. 100–3000 W (peak output 1.5 kW) – typical sealed CO_2 lasers used in industrial laser cutting.
10. 1 kW – output power expected to be achieved by a prototype 1 cm diode laser bar.

Examples of pulsed systems with high peak power:

1. 700 TW (700×10^{12} W) – National Ignition Facility, a 192-beam, 1.8-megajoule laser system adjoining a 10-metre-diameter target chamber.

2. 1.3 PW (1.3 × 10^{15} W) – world's most powerful laser as of 1998, located at the Lawrence Livermore Laboratory.

Laser safety

Even the first laser was recognised as being potentially dangerous. Theodore Maiman characterised the first laser as having a power of one 'Gillette' as it could burn through one Gillette razor blade. Today, it is accepted that even low-power lasers with only a few milliwatts of output power can be hazardous to human eyesight, when the beam from such a laser hits the eye directly or after reflection from a shiny surface. At wavelengths which the cornea and the lens can focus well, the coherence and low divergence of laser light means that it can be focused by the eye into an extremely small spot on the retina, resulting in localised burning and permanent damage in seconds or even less time (Fig. 14.6).

Fig. 14.6. Warning system for lasers.

Lasers are usually labelled with a safety class number, which identifies how dangerous the laser is:

1. Class I/1 is inherently safe, usually because the light is contained in an enclosure, for example in CD players.
2. Class II/2 is safe during normal use; the blink reflex of the eye will prevent damage. Usually up to 1 mW power, for example laser pointers.
3. Class IIIa/3R lasers are usually up to 5 mW and involve a small risk of eye damage within the time of the blink reflex. Staring into such a beam for several seconds is likely to cause (minor) eye damage.
4. Class IIIb/3B can cause immediate severe eye damage upon exposure. Usually lasers up to 500 mW, such as those in CD and DVD writers.
5. Class IV/4 lasers can burn skin, and in some cases, even scattered light can cause eye and/or skin damage. Many industrial and scientific lasers are in this class.

The indicated powers are for visible-light, continuous-wave lasers. For pulsed lasers and invisible wavelengths, other power limits apply. People working with class 3B and class 4 lasers can protect their eyes with safety goggles which are designed to absorb light of a particular wavelength. Certain infrared lasers with wavelengths beyond about 1.4 micrometres are often referred to as being 'eye-safe'. This is because the intrinsic molecular vibrations of water molecules very strongly absorb light in this part of the spectrum, and thus a laser beam at these wavelengths is attenuated so completely as it passes through the eye's cornea that no light remains to be focused by the lens onto the retina.

The label 'eye-safe' can be misleading, however, as it only applies to relatively low power continuous wave beams and any high power or Q-switched laser at these wavelengths can burn the cornea, causing severe eye damage.

Lasers as weapons

Laser beams are famously employed as weapon systems in science fiction, but actual laser weapons are only beginning to enter the market. The general idea of laser-beam weaponry is to hit a target with a train of brief pulses of light. The rapid evaporation and expansion of the surface causes shockwaves that damage the target.

The power needed to project a high-powered laser beam of this kind is difficult for current mobile power technology. Public prototypes are chemically-powered gas dynamic lasers.

Lasers of all but the lowest powers can potentially be used as incapacitating weapons, through their ability to produce temporary or permanent vision loss in varying degrees when aimed at the eyes. The degree, character, and duration of vision impairment caused by eye exposure to laser light varies with the power of the laser, the wavelength(s), the collimation of the beam, the exact orientation of the beam, and the duration of exposure.

Lasers of even a fraction of a watt in power can produce immediate, permanent vision loss under certain conditions, making such lasers potential non-lethal but incapacitating weapons. The extreme handicap that laser-induced blindness represents makes the use of lasers even as non-lethal weapons morally controversial, and weapons designed to cause blindness have been banned by the Protocol on Blinding Laser Weapons.

In the field of aviation, the hazards of exposure to ground-based lasers deliberately aimed at pilots have grown to the extent that aviation authorities have special procedures to deal with such hazards.

Optical Circulator

An optical circulator is a special fibre-optic component that can be used to separate optical powers that travel in opposite directions in one single optical fibre, analogous to the operation of an electronic circulator. An optical circulator is a three-port device that allows light to travel in only one direction— from port 1 to port 2, then from port 2 to port 3. This means that if some of the light emitted from port 2 is reflected back to the circulator, it is directed not back to port 1, but on to port 3. Circulators can also be used to achieve bi-directional transmission over a single fibre. Because of its high isolation of the input and reflected optical powers and its low insertion loss, optical circulators are widely used in advanced communication systems and fibre-optic sensor applications.

Optical circulators are non-reciprocal optics, which means that changes in the properties of light passing through the device are not reversed when the light passes through in the opposite direction. This can only happen when the symmetry of the system is broken, for example by an external magnetic field. A Faraday rotator is another example of a non-reciprocal optical device.

Chapter 15

Nanomaterial Synthesis and Application

INTRODUCTION

In nanotechnology, a particle is defined as a small object that behaves as a whole unit in terms of its transport and properties. It is further classified according to size: in terms of diameter, fine particles cover a range between 100 and 2500 nanometers, while ultrafine particles, on the other hand, are sized between 1 and 100 nanometers. Similar to ultrafine particles, nanoparticles are sized between 1 and 100 nanometers. Nanoparticles may or may not exhibit size-related properties that differ significantly from those observed in fine particles or bulk materials. Although the size of most molecules would fit into the above outline, individual molecules are usually not referred to as nanoparticles (Fig. 15.1).

Nanoclusters have at least one dimension between 1 and 10 nanometers and a narrow size distribution. Nanopowders are agglomerates of ultrafine particles, nanoparticles, or nanoclusters. Nanometer-sized single crystals, or single-domain ultrafine particles, are often referred to as nanocrystals. Nanoparticle research is currently an area of intense scientific interest due to a wide variety of potential applications in biomedical, optical and electronic fields. The National Nanotechnology Initiative has led to generous public funding for nanoparticle research in the United States.

Although nanoparticles are generally considered an invention of modern science, they actually have a very long history. Nanoparticles were used by artisans as far back as the 9th century in Mesopotamia for generating a glittering effect on the surface of pots.

Even these days, pottery from the Middle Ages and Renaissance often retain a distinct gold or copper coloured metallic glitter. This so-called lustre is caused by a metallic film that was applied to the transparent surface of a glazing. The lustre can still be visible if the film has resisted atmospheric oxidation and other weathering.

The lustre originated within the film itself, which contained silver and copper nanoparticles dispersed homogeneously in the glassy matrix of the ceramic glaze. These nanoparticles were created by the artisans by adding copper and silver salts and oxides together with vinegar, ochre and clay, on the surface of previously-glazed pottery. The object was then placed into a kiln and heated to about 600°C in a reducing atmosphere.

In the heat the glaze would soften, causing the copper and silver ions to migrate into the outer layers of the glaze. There the reducing atmosphere reduced the ions back to metals, which then came together forming the nanoparticles that give the colour and optical effects.

Lustre technique showed that ancient craftsmen had a rather sophisticated empirical knowledge of materials. The technique originated in the islamic world. As Muslims were not allowed to use gold in artistic representations, they had to find a way to create a similar effect without using real gold. The solution they found was using lustre.

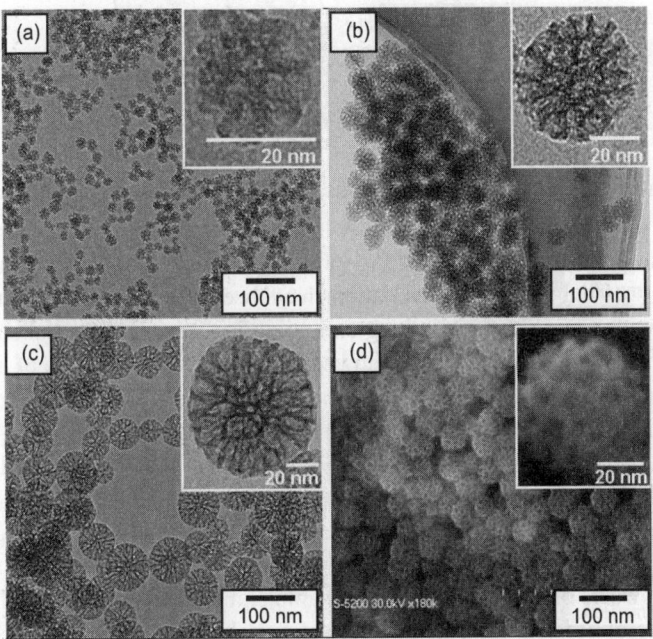

Fig. 15.1. TEM (a, b, and c) images of prepared mesoporous silica nanoparticles with mean outer diameter: (a) 20 nm, (b) 45 nm, and (c) 80 nm. SEM (d) image corresponding to (b). The insets are a high magnification of mesoporous silica particle.

Michael Faraday provided the first description, in scientific terms, of the optical properties of nanometer-scale metals in his classic 1857 paper. In a subsequent paper, the author (Turner) points out that: 'It is well known that when thin leaves of gold or silver are mounted upon glass and heated to a temperature which is well below a red heat (~500°C), a remarkable change of properties takes place, whereby the continuity of the metallic film is destroyed. The result is that white light is now freely transmitted, reflection is correspondingly diminished, while the electrical resistivity is enormously increased.'

UNIFORMITY

The chemical processing and synthesis of high performance technological components for the private, industrial and military sectors requires the use of high purity ceramics, polymers, glass-ceramics and material composites. In condensed bodies formed from fine powders, the irregular particle sizes and shapes in a typical powder often lead to non-uniform packing morphologies that result in packing density variations in the powder compact (Fig. 15.2).

Uncontrolled agglomeration of powders due to attractive van der Waals forces can also give rise to in microstructural inhomogeneities. Differential stresses that develop as a result of non-uniform drying shrinkage are directly related to the rate at which the solvent can be removed, and thus highly dependent upon the distribution of porosity. Such stresses have been associated with a plastic-to-brittle transition in consolidated bodies, and can yield to crack propagation in the unfired body if not relieved.

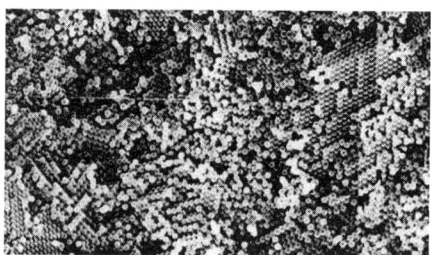

Fig. 15.2. Colloidal crystal composed of amorphous hydrated colloidal silica (particle diameter 600 nm).

In addition, any fluctuations in packing density in the compact as it is prepared for the kiln are often amplified during the sintering process, yielding inhomogeneous densification. Some pores and other structural defects associated with density variations have been shown to play a detrimental role in the sintering process by growing and thus limiting end-point densities. Differential stresses arising from inhomogeneous densification have also been shown to result in the propagation of internal cracks, thus becoming the strength-controlling flaws.

It would, therefore, appear desirable to process a material in such a way that it is physically uniform with regard to the distribution of components and porosity, rather than using particle size distributions which will maximise the green density. The containment of a uniformly dispersed assembly of strongly interacting particles in suspension requires total control over interparticle forces. Monodisperse nanoparticles and colloids provide this potential.

Monodisperse powders of colloidal silica, for example, may therefore, be stabilised sufficiently to ensure a high degree of order in the colloidal crystal or polycrystalline colloidal solid which results from aggregation. The degree of order appears to be limited by the time and space allowed for longer-range correlations to be established. Such defective polycrystalline colloidal structures would appear to be the basic elements of submicrometre colloidal materials science, and, therefore, provide the first step in developing a more rigorous understanding of the mechanisms involved in microstructural evolution in high performance materials and components.

Properties

Nanoparticles are of great scientific interest as they are effectively a bridge between bulk materials and atomic or molecular structures. A bulk material should have constant physical properties regardless of its size, but at the nano-scale this is often not the case where size-dependent properties are often observed. Thus, the properties of materials change as their size approaches the nanoscale and as the percentage of atoms at the surface of a material becomes significant. For bulk materials larger than one micrometre (or micron), the percentage of atoms at the surface is insignificant in relation to the number of atoms in the bulk of the material. The interesting and sometimes unexpected properties of nanoparticles are therefore largely due to the large surface area of the material, which dominates the contributions made by the small bulk of the material. Figure 15.3 shows silicon nanopowder.

An excellent example of this is the absorption of solar radiation in photovoltaic cells, which is much higher in materials composed of nanoparticles than it is in thin films of continuous sheets of material. In this case, the smaller the particles, the greater the solar absorption. Another good example is the bending of bulk copper (wire, ribbon, etc.) occurs with movement of copper atoms/clusters at about the 50 nm scale. Copper nanoparticles smaller than 50 nm, on the other hand, are considered super hard materials that do not exhibit the same malleability and ductility as bulk copper.

Fig. 15.3. Silicon nanopowder.

Other size-dependent property changes include quantum confinement in semiconductor particles, surface plasmon resonance in some metal particles and superparamagnetism in magnetic materials. Ironically, the changes in physical properties are not always desirable. Ferroelectric materials smaller than 10 nm can switch their magnetisation direction using room temperature thermal energy, thus making them unsuitable for memory storage.

Suspensions of nanoparticles are possible since the interaction of the particle surface with the solvent is strong enough to overcome density differences, which otherwise usually result in a material either sinking or floating in a liquid. Nanoparticles also often possess unexpected optical properties as they are small enough to confine their electrons and produce quantum effects. For example gold nanoparticles appear deep red to black in solution.

Nanoparticles have a very high surface area to volume ratio, which provides a tremendous driving force for diffusion, especially at elevated temperatures. Sintering can take place at lower temperatures, over shorter time scales than for larger particles. This theoretically does not affect the density of the final product, though flow difficulties and the tendency of nanoparticles to agglomerate complicates matters. The large surface area to volume ratio also reduces the incipient melting temperature of nanoparticles.

Moreover nanoparticles have been found to impart some extra properties to various day-to-day products. For example the presence of titanium dioxide nanoparticles imparts what we call the self-cleaning effect, and the size being nanorange, the particles cannot be observed. Zinc oxide particles have been found to have superior UV blocking properties compared to its bulk substitute. This is one of the reasons why it is often used in the preparation of sunscreen lotions.

Clay nanoparticles when incorporated into polymer matrices increase reinforcement, leading to stronger plastics, verifiable by a higher glass transition temperature and other mechanical property tests. These nanoparticles are hard, and impart their properties to the polymer (plastic). Nanoparticles have also been attached to textile fibres in order to create smart and functional clothing.

Metal, dielectric, and semiconductor nanoparticles have been formed, as well as hybrid structures (e.g., core-shell nanoparticles). Nanoparticles made of semiconducting material may also be labelled quantum dots if they are small enough (typically sub 10 nm) that quantisation of electronic energy levels occurs. Such nanoscale particles are used in biomedical applications as drug carriers or imaging agents.

Semi-solid and soft nanoparticles have been manufactured. A prototype nanoparticle of semi-solid nature is the liposome. Various types of liposome nanoparticles are currently used clinically as delivery systems for anticancer drugs and vaccines.

Synthesis

There are several methods for creating nanoparticles, including both attrition and pyrolysis. In attrition, macro or microscale particles are ground in a ball mill, a planetary ball mill, or other size reducing mechanism. The resulting particles are air classified to recover nanoparticles. In pyrolysis, a vapourous precursor (liquid or gas) is forced through an orifice at high pressure and burned. The resulting solid (a version of soot) is air classified to recover oxide particles from by-product gases. Pyrolysis often results in aggregates and agglomerates rather than singleton primary particles.

A thermal plasma can also deliver the energy necessary to cause evaporation of small micrometre size particles. The thermal plasma temperatures are in the order of 10,000 K, so that solid powder easily evaporates. Nanoparticles are formed upon cooling while exiting the plasma region. The main types of the thermal plasma torches used to produce nanoparticles are DC plasma jet, DC arc plasma and radio frequency (RF) induction plasmas. In the arc plasma reactors, the energy necessary for evaporation and reaction is provided by an electric arc which is formed between the anode and the cathode. For example, silica sand can be vapourised with an arc plasma at atmospheric pressure. The resulting mixture of plasma gas and silica vapour can be rapidly cooled by quenching with oxygen, thus ensuring the quality of the fumed silica produced. In RF induction plasma torches, energy coupling to the plasma is accomplished through the electromagnetic field generated by the induction coil. The plasma gas does not come in contact with electrodes, thus eliminating possible sources of contamination and allowing the operation of such plasma torches with a wide range of gases including inert, reducing, oxidising and other corrosive atmospheres.

The working frequency is typically between 200 kHz and 40 MHz. Laboratory units run at power levels in the order of 30-50 kW while the large scale industrial units have been tested at power levels up to 1 MW. As the residence time of the injected feed droplets in the plasma is very short it is important that the droplet sizes are small enough in order to obtain complete evaporation. The RF plasma method has been used to synthesise different nanoparticle materials, for example synthesis of various ceramic nanoparticles such as oxides, carbours/carbides and nitrides of Ti and Si.

Inert-gas condensation is frequently used to make nanoparticles from metals with low melting points. The metal is vapourised in a vacuum chamber and then supercooled with an inert gas stream. The supercooled metal vapour condenses in to nanometer-sized particles, which can be entrained in the inert gas stream and deposited on a substrate or studied *in situ*.

Sol-gel

The sol-gel process is a wet-chemical technique (also known as chemical solution deposition) widely used recently in the fields of materials science and ceramic engineering. Such methods are used primarily for the fabrication of materials (typically a metal oxide) starting from a chemical solution (*sol*, short for solution) which acts as the precursor for an integrated network (or *gel*) of either discrete particles or network polymers.

Typical precursors are metal alkoxides and metal chlorides, which undergo hydrolysis and polycondensation reactions to form either a network 'elastic solid' or a colloidal suspension (or dispersion) — a system composed of discrete (often amorphous) submicrometre particles dispersed to various degrees in a host fluid. Formation of a metal oxide involves connecting the metal centres with oxo (M-O-M) or hydroxo (M-OH-M) bridges, therefore generating metal-oxo or metal-hydroxo polymers in solution. Thus, the sol evolves towards the formation of a gel-like diphasic system containing both a liquid phase and solid phase whose morphologies range from discrete particles to continuous polymer networks.

In the case of the colloid, the volume fraction of particles (or particle density) may be so low that a significant amount of fluid may need to be removed initially for the gel-like properties to be recognised. This can be accomplished in any number of ways. The most simple method is to allow time for sedimentation to occur, and then pour off the remaining liquid. Centrifugation can also be used to accelerate the process of phase separation.

Removal of the remaining liquid (solvent) phase requires a drying process, which is typically accompanied by a significant amount of shrinkage and densification. The rate at which the solvent can be removed is ultimately determined by the distribution of porosity in the gel. The ultimate microstructure of the final component will clearly be strongly influenced by changes implemented during this phase of processing. Afterwards, a thermal treatment, or firing process, is often necessary in order to favour further polycondensation and enhance mechanical properties and structural stability via final sintering, densification and grain growth. One of the distinct advantages of using this methodology as opposed to the more traditional processing techniques is that densification is often achieved at a much lower temperature.

The precursor sol can be either deposited on a substrate to form a film (e.g. by dip-coating or spin-coating), cast into a suitable container with the desired shape (e.g. to obtain a monolithic ceramics, glasses, fibres, membranes, aerogels), or used to synthesise powders (e.g. microspheres, nanospheres). The sol-gel approach is a cheap and low-temperature technique that allows for the fine control of the product's chemical composition. Even small quantities of dopants, such as organic dyes and rare earth metals, can be introduced in the sol and end up in uniformly dispersed in the final product. It can be used in ceramics processing and manufacturing as an investment casting material, or as a means of producing very thin films of metal oxides for various purposes. Sol-gel derived materials have diverse applications in optics, electronics, energy, space, (bio)sensors, medicine (e.g. controlled drug release) and separation (e.g. chromatography) technology.

The interest in sol-gel processing can be traced back in the mid-1880s with the observation that the hydrolysis of tetraethyl-orthosilicate (TEOS) under acidic conditions led to the formation of SiO_2 in the form of fibres and monoliths. Sol-gel research grew to be so important that in the 1996s more than 50,000 papers were published worldwide on the process.

Colloids

The term colloid is used primarily to describe a broad range of solid-liquid (and/or liquid-liquid) mixtures, all of which contain distinct solid (and/or liquid) particles which are dispersed to various degrees in a liquid medium. The term is specific to the size of the individual particles, which are larger than atomic dimensions but small enough to exhibit Brownian motion. If the particles are large enough, then their dynamic behaviour in any given period of time in suspension would be governed by forces of gravity and sedimentation. But if they are small enough to be colloids, then their irregular motion in suspension can be attributed to the collective bombardment of a myriad of thermally agitated molecules in the liquid suspending medium, as described originally by Albert Einstein in his dissertation. Einstein proved the existence of water molecules by concluding that this erratic particle behaviour could adequately be described using the theory of Brownian motion, with sedimentation being a possible long-term result. This critical size range (or particle diameter) typically ranges from nanometers (10^{-9} m) to micrometres (10^{-6} m). Figure 15.4 shows nanostars of vanadium (IV) oxide.

5 μm

Fig. 15.4. Nanostars of vanadium (IV) oxide.

Morphology

Scientists have taken to naming their particles after the real world shapes that they might represent. Nanospheres, nanoreefs, nanoboxes and more have appeared in the literature. These morphologies sometimes arise spontaneously as an effect of a templating or directing agent present in the synthesis such as miscellar emulsions or anodised alumina pores, or from the innate crystallographic growth patterns of the materials themselves. Some of these morphologies may serve a purpose, such as long carbon nanotubes being used to bridge an electrical junction, or just a scientific curiosity like the stars shown at right.

Generally speaking, amorphous particles will adopt a spherical shape (due to their microstructural isotropy), whereas anisotropic microcrystalline whiskers will adopt the geometrical form corresponding to their particular crystal habit. At the small end of the size range, nanoparticles are often referred to as clusters. Spheres, rods, fibres, and cups are just a few of the shapes that have been grown. The study of fine particles is called micromeritics.

Characterisation

Nanoparticle characterisation is necessary to establish understanding and control of nanoparticle synthesis and applications. Characterisation is done by using a variety of different techniques, mainly drawn from materials science. Common techniques are electron microscopy (TEM, SEM), atomic force microscopy (AFM), dynamic light scattering (DLS), X-ray photoelectron spectroscopy (XPS), powder X-ray diffraction (XRD), Fourier transform infrared spectroscopy (FTIR), matrix-assisted laser desorption/

ionisation time-of-flight mass spectrometry (MALDI-TOF), ultraviolet-visible spectroscopy, dual polarisation interferometry and nuclear magnetic resonance (NMR).

Whilst the theory has been known for over a century, the technology for nanoparticle tracking analysis (NTA) allows direct tracking of the Brownian motion and this method therefore allows the sizing of individual nanoparticles in solution.

NANOSCALE IRON PARTICLES

Nanoscale iron particles are sub-micrometre particles of iron metal. They are highly reactive because of their large surface area. In the presence of oxygen and water, they rapidly oxidise to form free iron ions. They are widely used in medical and laboratory applications and have also been studied for remediation of industrial sites contaminated with chlorinated organic compounds.

Chemistry

Synthesis

When exposed to oxygen and water, iron oxidises. This redox process can occur under either acidic or neutral/basic conditions:

$$2 \ Fe^0(s) + 4 \ H^+(aq) + O_2(aq) \rightarrow 2 \ Fe^{2+}(aq) + 2 \ H_2O(l)$$
$$Fe^0(s) + 2 \ H_2O \ (aq) \rightarrow Fe^{2+}(aq) + H_2(g) + 2 \ OH^-(aq)$$

Research

Research has shown that nanoscale iron particles can be effectively used to treat several forms of ground contamination, including grounds contaminated by polychlorinated biphenyls (PCBs), chlorinated organic solvents, and organochlorine pesticides. Nanoscale iron particle are easily transportable through ground water, allowing for *in situ* treatment. Additionally, the nanoparticle-water slurry can be injected into the contaminated area and stay there for long periods of time. These factors combine to make this method of cheaper than most currently used alternative. Researchers have found that although metallic iron nanoparticles remediate contaminants well, they tend to agglomerate on the soil surfaces. In response, carbon nanoparticles and water-soluble polyelectrolytes have been used as supports to the metallic iron nanoparticles. The hydrophobic contaminants adsorb to these supports, improving permeability in sand and soil. In field tests have generally confirmed lab findings. However, research is still ongoing and nanoscale iron particles are not yet commonly used for treating ground contamination.

MAGNETIC NANOPARTICLES

Magnetic nanoparticles are a class of nanoparticle which can be manipulated using magnetic field. Such particles commonly consist of magnetic elements such as iron, nickel and cobalt and their chemical compounds. These particles have been the focus of much research recently because they possess attractive properties which could see potential use in catalysis, biomedicine, magnetic resonance imaging, data storage and environmental remediation (Fig. 15.5).

Properties

The physical and chemical properties of magnetic nanoparticles largely depend on the synthesis method and chemical structure. In most cases, the particles range from 1 to 100 nm in size and may display superparamagnetism.

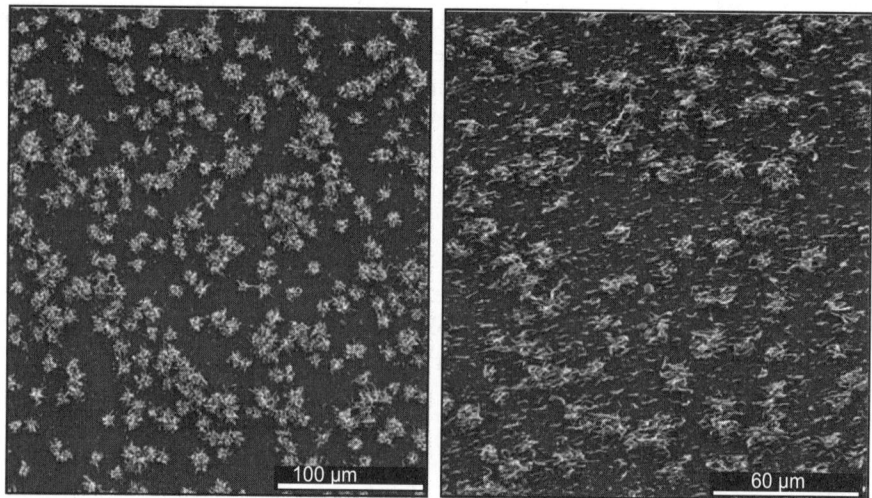

Fig. 15.5. Magnetite: an example of magnetic nanoparticles.

Synthesis

The established methods of magnetic nanoparticle synthesis include.

Co-precipitation

Co-precipitation is a facile and convenient way to synthesise iron oxides (either Fe_3O_4 or γ-Fe_2O_3) from aqueous Fe^{2+}/Fe^{3+} salt solutions by the addition of a base under inert atmosphere at room temperature or at elevated temperature. The size, shape, and composition of the magnetic nanoparticles very much depends on the type of salts used (e.g. chlorides, sulphates, nitrates), the Fe^{2+}/Fe^{3+} ratio, the reaction temperature, the pH value and ionic strength of the media.

Thermal decomposition

Monodisperse magnetic nanocrystals with smaller size can essentially be synthesised through the thermal decomposition of organometallic compounds in high-boiling organic solvents containing stabilising surfactants.

Microemulsion

Using the microemulsion technique, metallic cobalt, cobalt/platinum alloys, and gold-coated cobalt/platinum nanoparticles have been synthesised in reverse micelles of cetyltrimethylammonium bromide, using 1-butanol as the cosurfactant and octane as the oil phase.

Applications

A wide variety of applications have been envisaged for this class of particles include:

Medical diagnostics and treatments

Magnetic nanoparticles are used in an experimental cancer treatment called magnetic hyperthermia in which the fact that nanoparticles heat when they are placed in an alternative magnetic field is used.

Another potential treatment of cancer includes attaching magnetic nanoparticles to free-floating cancer cells, allowing them to be captured and carried out of the body. The treatment has been tested in the laboratory on mice and will be looked at in survival studies.

Magnetic immunoassay

Magnetic immunoassay (MIA) is a novel type of diagnostic immunoassay utilising magnetic beads as labels in lieu of conventional, enzymes, radioisotopes or fluorescent moieties.

NANOSHELL PARTICLES: SYNTHESIS, PROPERTIES AND APPLICATIONS

Nanoshells, which are thin coatings deposited on core particles of different material have gained considerable attention. These materials show novel properties which are different from their single component counterpart. By simply tuning the core to shell ratio, the properties can be altered. With emerging new techniques it is now possible to synthesise these nanostructures in desired shape, size and morphology. Various synthesis routes have been developed over the past few years to prepare these nanostructures. They can be prepared with customised properties such as increased stability, surface area, magnetic, optical and catalytic properties. A review of their synthesis techniques, properties and applications is given here.

Monodispersed colloids have attracted a lot of attention since a long time because of their novel properties and potential applications. Efforts had been limited earlier to produce them in uniform sizes. Now, with the emerging new synthesis techniques, it is possible to synthesise them not only in uniform sizes but also in desired shapes such as rods, tubes, cubes, prisms, etc. Advances in synthesis techniques have allowed creation of other novel structures such as nanoshell or core shell particles, hollow particles, colloidal crystals, etc.

Nanoshell particles constitute a special class of nanocomposite materials. They consist of concentric particles, in which particles of one material are coated with a thin layer of another material using specialised procedures. Nanoshell particles are highly functional materials with tailored properties, which are quite different than either of the core or of the shell material. Indeed, they show modified and improved properties than their singlecomponent counterparts or nanoparticles of the same size.

Therefore, nanoshell particles are preferred over nanoparticles. Their properties can be modified by changing either the constituting materials or core-to-shell ratio. The term nanoshell is used specifically because thickness of the shell is ca 1–20 nm. Properties of shell materials (metal or semiconductor) having thickness in nanometers, become important when they are coated on dielectric cores to achieve higher surface area. Sometimes they are referred as core shell or core@shell particles also. Thicker shells can also be prepared, but their synthesis is restricted mainly to achieve some specific goal, such as providing thermal stability to core particles. Synthesis of nanoshells can be useful for creating novel materials with different morphologies, as it is not possible to synthesise all the materials in desired morphologies. Core particles of different morphologies such as rods, wires, tubes, rings, cubes, etc. can be coated with thin shell to get desired morphology in core shell structures. These materials can be of economic interest also, as precious materials can be deposited on inexpensive cores. By doing so, expensive material is required in lesser amount than usual. These particles are synthesised for a variety of purposes like providing chemical stability to colloids, enhancing luminescence properties, engineering band structures, biosensors, drug delivery, etc.

Nanoshell materials can be synthesised practically using any material, like semiconductors, metals and insulators. Usually dielectric materials such as silica and polystyrene are commonly used as core because they are highly stable. They are chemically inert and water-soluble; therefore they can be useful in biological applications. Nanoshell particles can be synthesised in a variety of combinations such as (core-shell) dielectric–metal, dielectric–semiconductor, dielectric–dielectric, semiconductor–metal, metal–metal, semiconductor–semiconductor, semiconductor–dielectric, metal–dielectric, dye–

dielectric, etc. Core shell particles can be assembled and further utilised for creation of another class of novel materials like colloidal crystal or quantum bubbles. It is indeed possible to create unique core shell structures having multishells. Multishell particles can be visualised as core particles having a number of shells around them. Core particles can be coated with a shell to obtain a single nanoshell. Further, these combinations of core and shell can be repeated again to get multishells. These structures show tunable optical properties from the visible to infrared region of the electromagnetic spectrum by choosing different combinations of core and shell.

This review includes synthesis aspects of nanoshells, their properties and applications.

A schematic diagram showing a variety of core shell particles is depicted in Fig. 15.6. Surface of the core particle can be modified using bifunctional molecules and then small particles can be anchored on it (Fig. 15.6a). Nanoparticles grow around the core particle and form a complete shell (Fig. 15.6b). In some cases, a smooth layer of shell material can be deposited directly on the core by co-precipitation method (Fig. 15.6c). Small core particles such as gold or silver (10–50 nm) can be uniformly encapsulated with silica (Fig. 15.6d). Also a number of colloidal particles can be encapsulated inside a single particle (Fig. 15.6e). Core particles can be removed either by calcination or by dissolving them in a proper solvent. This gives rise to hollow particles also known as quantum bubbles (Fig. 15.6f). Concentric shells also can be grown on core particles to form a novel structure known as multishell or nanomatryushka (named after the Russian doll; Fig. 15.6g).

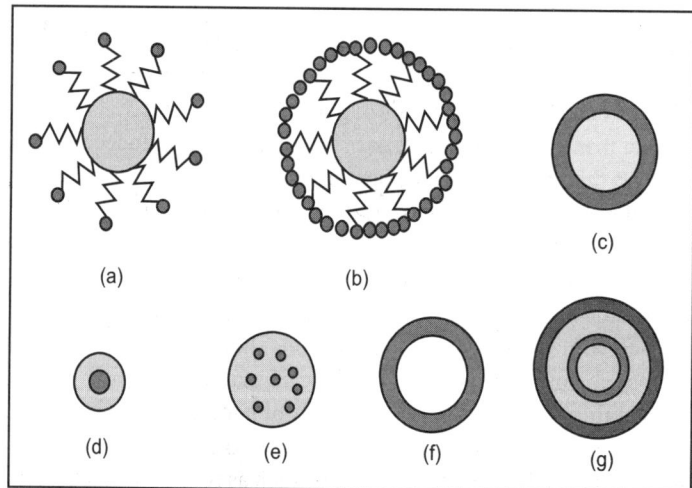

Fig. 15.6. Variety of core shell particles: (a) surface-modified core particles anchored with shell particles, (b) more shell particles reduced onto core to form a complete shell, (c) smooth coating of dielectric core with shell, (d) encapsulation of very small particles with dielectric material, (e) embedding number of small particles inside a single dielectric particle, (f) quantum bubble, and (g) multishell particle.

Synthesis of Nanoshell Particles

Numerous techniques have been developed to synthesise nanoshell particles. Preparation of nanoshell particles involves multistep synthesis procedure. It requires highly controlled and sensitive synthesis protocols to ensure complete coverage of core particles with the shell material. There are various methods to fabricate core shell structures, e.g. precipitation, grafted polymerisation, microemulsion, reverse micelle sol-gel condensation, layer-by-layer adsorption technique, etc. Although several methods have been established, it is still difficult to control the thickness and homogeneity of the coating. If the reaction is not controlled properly, it eventually leads to aggregation of core particles, formation of separate particles of shell material or incomplete coverage. However, with the new emerging techniques it is now possible to deposit homogeneous coatings. One such method is to coat the surface of colloids with appropriate primer (coupling agent) to enhance coupling of the shell material with the core. Also, the surface of the core material can be charged, and shell material can be adsorbed on its surface by electrostatic attraction.

Synthesis procedures for preparation of core particles of dielectric, metal and semiconductor materials and various ways of incorporating them to form core shell particles are discussed in the next section.

Synthesis of dielectric cores

Silica (SiO_2) (dielectric constant ~4.5) is a popular material to form core shell particles because of its extraordinary stability against coagulation. Its non-coagulating nature is due to very low value of Hamaker constant, which defines the van der Waal forces of attraction among the particles and the medium. It is also chemically inert, optically transparent and does not affect redox reactions at core surfaces. For various purposes it is desirable that particles remain well dispersed in the medium, which can be achieved by coating silica on them to form an encapsulating shell.

Silica particles with narrow size distribution can be synthesised following the procedure developed by Stöber. This method involves hydrolysis and successive condensation of TEOS (tetraethylorthosilicate $Si(C_2H_5O)$) in alcoholic medium in the presence of ammonium hydroxide (NH_4OH) as catalyst. The reaction mechanism is explained in Eq. 15.1. The first step is hydrolysis, in which ethoxy groups are replaced by OH groups. In the second step, silicon hydroxides undergo polycondensation process to form SiO_2.

$$Si(OC_2H_5)_4 + 4H_2O \rightarrow Si(OH)_4 + 4C_2H_5OH$$

$$Si(OH)_4 \rightarrow SiO_2 + 2H_2O \qquad \qquad ...(15.1)$$

By varying relative ratio of TEOS to solvent (dilution) and amount of catalyst, one can synthesise these particles in various sizes ranging from ~ 50 nm to 1 mm. Reduction in TEOS concentration leads to the formation of smaller particles. Silica particles synthesised by this procedure are amorphous and porous. Variation of NH_4OH (catalyst) concentration changes the size, porosity as well as morphology of these particles. Figure 15.7(a) shows SEM image of silica particles prepared by the synthesis procedure described above. It can be seen that particles are quite uniform in size (~ 420 nm). They are noncrystalline, but produce a single broad diffraction peak in powder diffraction [Fig. 15.7(b)]. The other important core material is polystyrene (PS).

PS particles along with high stability offer one added advantage. They decompose at relatively lower temperature, i.e. 450°C and are soluble in commonly available solvent such as toluene. Therefore, they are frequently used for the preparation of hollow structures (also known as quantum bubbles). Nanoshells can be synthesised on PS core and the core can be removed easily either by calcinations or by dissolution. Use of other cores requires strong acids (HF or HNO_3) or harsh conditions, which may affect the shell material. PS can be prepared by seed mediated emulsion polymerisation method. There are several other methods like batch, semicontinuous and dispersion emulsion polymerisation methods. However, the main disadvantage of these techniques is the lack of control over the size of the particles. A broad or manytimes bimodal size distribution is obtained. Seed-mediated methods offer good control over the size of the particles. Preparation of PS spheres using seed mediated method involves dissolution of a monomer in suitable medium, in the presence of emulsifier and initiator. Polymerisation of monomer starts when the initiator decomposes to form nuclei. The nuclei grow to form polymer spheres. PS spheres are synthesised using the procedures described above, in which styrene is used as a monomer, sodium styrene sulphonate as an emulsifier and potassium persulphate as initiator. Sodium bisulphite and sodium bicarbonate act as buffer and reaction is carried out in inert atmosphere. Using this method, monodispersed spherical particles of size from 1 to 5 mm can be prepared. Preparation of small PS spheres (~ 20 nm) is also possible. In this procedure, small amount of PMMA (poly(methyl methacrylate)) is used as a seed. Sodium dodecyl sulphate is used as a surfactant and ammonium persulphate acts as an initiator.

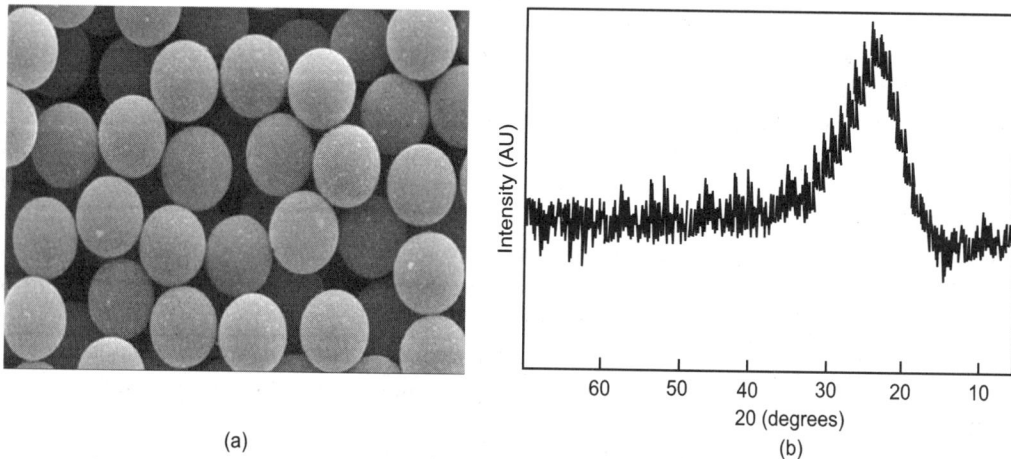

(a)   (b)

Fig. 15.7. (a) SEM image of colloidal particles of silica, and (b) XRD pattern of the same.

Synthesis of metal nanoparticles

Various reviews have appeared on metal nanoparticles, owing to their fascinating optical properties. Gold and silver nanoparticles can be easily prepared by reducing their salts (such as chloroauric acid and silver nitrate respectively) in aqueous solutions. Usually trisodium citrate and sodium borohydride have been used as reducing agents. Sodium borohydride is a strong reducing agent. The reaction takes place almost instantly and small particles are generated (size ~ 10 nm). Reduction using sodium borohydride is done at room temperature, while trisodium citrate needs higher temperature (around 80°C). Other reducing agents such as ascorbic acid, alcohols, hydrazine and ethylene glycol also have been used. Although aqueous synthesis is well established, it is also possible to synthesise metal nanoparticles in other solvents such as alcohol, THF (tetrahydrofuran) and DMF, ethylene glycol, etc. Alcohols and ethylene glycol play a dual role in the reactions. They act as reducing agents as well as solvents. A comparison of various reducing agents is given in Table 15.1.

Synthesis of semiconductor nanoparticles

Semiconductor nanoparticles having sizes in the range of 1–10 nm display novel optical, electronic and physical properties. In these materials, when the size of the nanoparticle becomes less or comparable to the Bohr diameter of the exciton, various size-quantisation effects such as widening of band gap and formation of discrete orbitals come into the picture. In a semiconductor, when an electron is excited from the valence band to the conduction band, a hole is left behind in valence band.

Table 15.1. Comparison of various reducing agents along with reaction conditions for synthesis of metal nanoparticles.

Metal species	Redox potential (V)	Reducing agent	Condition	Rate
Cu^{2+}, Ru^{3+}, Re^{3+}	< 0.7 and ≥ 0	$NaBH_4$,	Ambient	Fast
		Hydrazine, hydrogen	< 70°C	Moderate
		Aldehydes, sugars	70°–100°C	Slow
		Polyols	>120°C	Slow
Rh^{3+}, Pd^{2+}, Ag^+, Ir^{3+}, $Pt^{4+,2+}$, $Au^{3+,+}$, Hg^{2+}	≥ 0.7	Hydrazine, H_2SO_3, H_3PO_2, $NaBH_4$, boranes, hydrated e^-	Ambient	Very fast
		Aldehydes, sugars	Ambient	Fast
		Polyols	< 50°C	Moderate
		Organic acids, alcohols	≥ 70°C	Slow
Cr^{3+}, Mn^{2+}, V^{2+}, Ta^{5+}	< – 0.6	Hydrated e^-, radicals	Ambient	Fast
		$NaBH_4$, boranes	Temperature and Pressure > Ambient	Slow
Fe^{2+}, Co^{2+}, Ni^{2+}, Mo^{3+}, Cd^{2+}, In^{3+}, Sn^{2+}, W^{6+}	< 0 and ≥ – 0.5	Hydrated e^-, radicals	Ambient	Very fast
		$NaBH_4$, boranes	Ambient	Fast
		Hydrazine, hydroxylamine	70°–100°C	Slow
		Polyols	>180°C	Slow

The electron and hole can form a bound state which is known as exciton. When the dimension of the nanoparticle approaches the Bohr diameter of the exciton, the electron and hole pair is localised inside the nanoparticle. Bulk semiconductor band gap opens up in the nanoparticle, like HOMO (highest occupied molecular orbital) and LUMO (lowest unoccupied molecular orbital) gap for molecules (Fig. 15.8). Reduction in the size of nanoparticles leads to widening of the band gap. This widening can be witnessed by optical absorption spectroscopy, in which a blue shift is observed with reduction in particle size. Brus has developed a theory to explain the observed spectral shifts known as effective mass approximation (EMA), according to which size-dependent band gap of nanoparticles is given as:

$$E' = E_g + \Delta E \qquad \qquad ... (15.2)$$

$$E' = E_g + \frac{h^2 \pi^2}{2R^2} \left[\frac{1}{m_e} + \frac{1}{m_h} \right] - \frac{1.8 e^2}{\varepsilon_2 R} + \text{polarisation term} \qquad ... (15.3)$$

where, E_g is the bulk semiconductor band gap; e^2 the dielectric constant of the semiconductor; R the radius of the particle and m_e and m_h are the effective masses of the electron and hole respectively.

Although EMA was useful to understand variation of band gap with size, it quantitatively failed to explain the results for very small particles (< 5 nm). This is because the values for effective mass of electrons and holes are those of the bulk material. Therefore, other approaches like tight-binding approximation, empirical pseudo-potential method and effective bond orbital method have been developed.

Nanoparticles have high free energy because of their large surface area. In order to minimise it, they agglomerate to form bigger particles. They are stabilised by various procedures. Reviews are available on synthesis of these particles. They have been synthesised by methods such as arrested precipitation in aqueous as well as nonaqueous solutions, reverse micelle, using organomettalic precursors and by chemical capping.

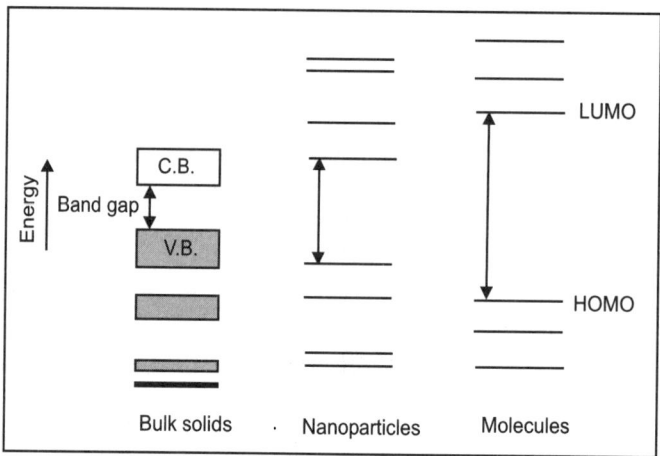

Fig. 15.8. Comparison of energy levels in bulk material, nanoparticles and molecules.

Preparation of core shell assemblies

Nanoshells can be prepared by a variety of approaches as explained earlier. One approach involves synthesis of core and shell particles separately. Later shell particles can be anchored on cores by specialised procedures. In the first method, the surface of the core particles is often modified with bifunctional molecules to enhance coverage of shell material on their surfaces. Surface of core particles such as silica can be modified using bifunctional organic molecules such as APS (3-aminopropyltriethoxysilane). The APS molecule has an ethoxy group at one end, and NH group at the other end. APS forms a covalent bond with silica particles through the OH group and their surface becomes NH-terminated (Fig. 15.9). Now metal and semiconductor nanoparticles can be attached through the NH group on the silica core.

There are several other modifiers such as APTMS (3-amino-propyltrimethoxysilane) and AEAPTMS (*N*-(2-aminoethyl)-3-aminopropyltrimethoxysilane) which produce surface terminated with amine group, MPTMS (3-mercaptopropyltrimethoxysilane) which makes surface terminated with thiols, DPPETES (2-(diphenylphosphino)ethyltriethoxysilane) leaves the surface terminated by diphenylphosphine group and PTMS (propyltrimethoxysilane) giving surface terminated with methyl group. Thus using a variety of linker molecules with different functional groups, a number of materials can be anchored on dielectric cores. Linkage can be verified using techniques like FTIR.

Graf have demonstrated a general method to coat various colloidal particles (gold colloids, silver colloids, boehmite rods, gibbsite platelets positively or negatively charged polystyrene, etc.) with silica. They have functionalised the surface of colloids using PVP (polyvinylpyrrolidone), which is an amphiphilic nonionic polymer. This method can be used for a variety of colloids, it is fast and does not require the use of silane coupling agents or precoating step with sodium silicate. Using this polymer it is possible to alter the interaction potential on the surface of the colloids. Hence it is possible to disperse them in a wide range of solvents.

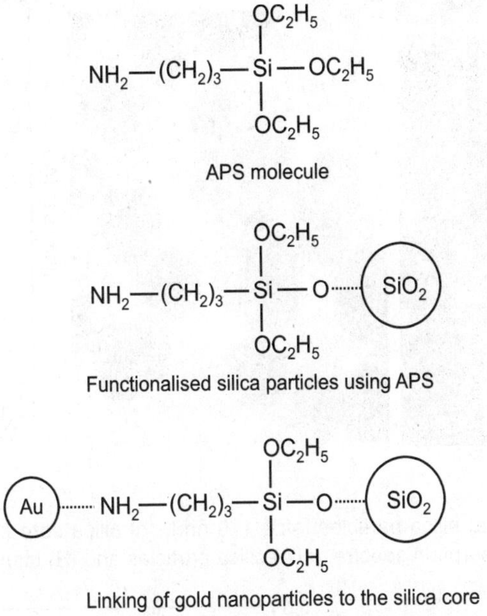

$$OC_2H_5$$
$$|$$
$$NH_2-(CH_2)_3-Si-OC_2H_5$$
$$|$$
$$OC_2H_5$$

APS molecule

$$OC_2H_5$$
$$|$$
$$NH_2-(CH_2)_3-Si-O\cdots SiO_2$$
$$|$$
$$OC_2H_5$$

Functionalised silica particles using APS

$$Au\cdots NH_2-(CH_2)_3-Si-O\cdots SiO_2$$

Linking of gold nanoparticles to the silica core

Fig. 15.9. Linking of bifunctional molecule (APS) to core and shell particles to form nanoshells.

It is also possible to make certain nanoshell particles without functionalising core particles. In this case opposite charges can be developed on the core and shell materials to couple them together by electrostatic attraction.

In the second approach, known as controlled precipitation, synthesis of shell particles can be carried out in the presence of cores. The core particles act as nuclei and hydrolysed shell material gets condensed on these cores forming nanoshells. Reactant concentrations and the amount of added core particles play an important role in deciding the shell thickness. Figure 15.10 illustrates images of titania-coated silica particles prepared using this technique. Silica particles [Fig. 15.10(a)] have been coated with a thin layer of titania [Fig. 15.10(b)].

Silica particles do not show any absorption peak in the range 300–800 nm. When these particles are coated with a thin layer (~ 23 nm) of titania, an absorption band at 320 nm is observed [Fig. 15.10(c)]. Hence UV absorption can be an indirect technique to know the presence of coating on dielectric cores. Metallic nanoshells such as gold and silver are usually synthesised by functionalising the core particle with a linking molecule. It is also possible to deposit metal nanoparticles on the dielectric core using controlled precipitation. Reduction of metallic species can be carried out by commonly used reducing agents such as trisodium citrate or sodium borohydride. Reduction of metal ions on the core, using trisodium citrate is found to be more favourable than reduction using sodium borohydride. Coverage may not be complete in one coating step, but the procedure can be repeated several times till a homogeneous shell layer is obtained.

Using similar techniques other nanoshells such as silica core silver shell can also be synthesised. The TEM image (Fig. 15.11) shows a particle after a single coating step. Optical absorption spectra for the same are also shown. For silver nanoshells a broad band at 446 nm is observed, while silver nanoparticles show absorption at 431 nm.

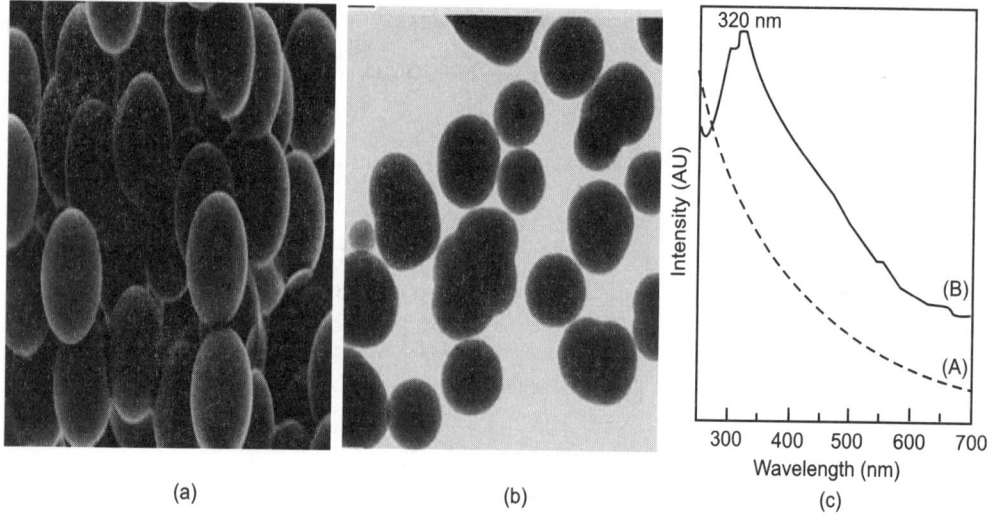

Fig. 15.10. TEM image of (a) silica particles (size 170 nm), (b) silica core (170 nm) titania shell (23 nm) particles, and (c) UV-Vis absorption spectra of (A) silica particles and (B) titania-coated silica particles.

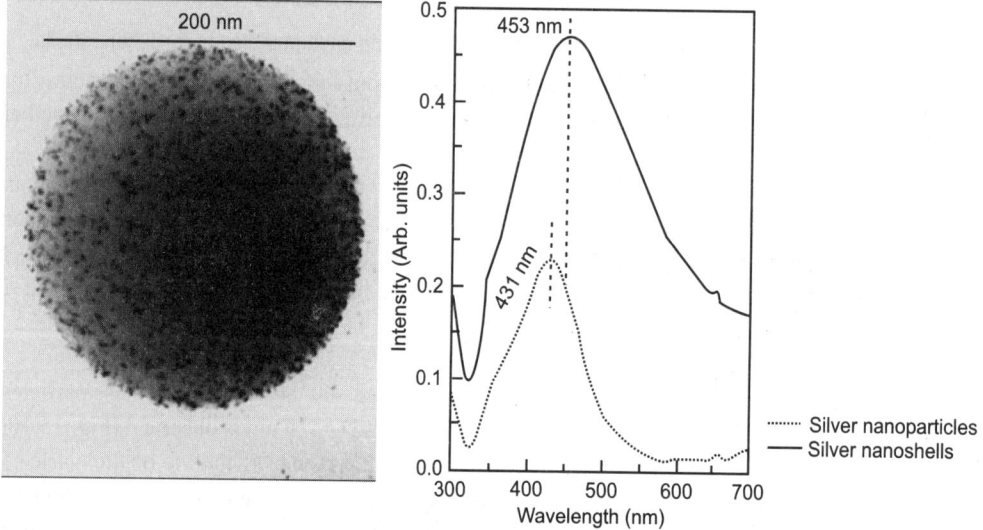

Fig. 15.11. TEM image of silica silver core shell particle and corresponding UV-Vis absorption spectrum.

In another technique, known as layer-by-layer technique, alternating layers of anionic particles and cationic polymer are deposited on surface-modified template molecule by heterocoagulation. Both these methods lead to the formation of homogeneous and dense coatings. Successive removal of core material (either by calcination or dissolution in suitable solvents such as toluene or HF) yields hollow particles consisting of shell material. Selection of a suitable pair for the core and shell assemblies requires understanding of individual properties of core and shell materials. The core particle should withstand the process used for coating of the shell material. Core and shell particles should not interdiffuse and surface energies of the core and shell particles must be similar, so that the probability of heterogeneous nucleation is more than that of homogeneous nucleation.

Surface Plasmon Resonance of Metal Nanoparticles and Metallic Nanoshells

Colloidal dispersions of metal nanoparticles having a size much smaller than the wavelength of visible radiation exhibit intense colours. The art of making coloured glass is known over thousand of years. Windows of churches, palaces, etc. are often decorated with such glasses. Small amount of metal nanoparticles such as gold, silver, copper, etc. doped in glass gives rise to beautiful colours. Although the technique of making coloured glass had been known since a long time, it was not known that this is because of presence of metal nanoparticles in glass, until in 1857, Michael Faraday synthesised gold nanoparticles by reducing chloroauric acid. He obtained ruby-red and pink colour dispersion of gold particles. These were probably the first colloidal particles reported in the literature. The nanoparticles synthesised by Faraday are still stable and can be viewed on visiting the website of the London museum. Thus attempts were made to explain the observed intense colour using classical electromagnetic theory. In 1908, Mie explained this phenomenon using Maxwell's equations, which is qualitatively explained as follows:

Metals can be considered as confined plasma of positive ions (consisting of nuclei and core electrons which are fixed) and conduction electrons (free and mobile). In neutral case, the positive charge cloud of ions and negative charge cloud of electrons overlap with each other. By some external disturbance, i.e. irradiation by electrons or electromagnetic radiation, the charge cloud is disturbed and electrons are moved away from the equilibrium position. If the density of electrons in one region increases, they repel each other and tend to return to their original equilibrium position. As electrons move towards their original positions, they pick up kinetic energy and instead of coming to rest in equilibrium configuration, they overshoot. They oscillate back and forth. The collective oscillations of conduction electrons in metals upon excitation with electromagnetic radiation are known as plasmons. These oscillations give rise to a strong absorption band in the visible range of the electromagnetic spectra. In short, the origin of this band (known as plasmon resonance band) is attributed to resonance between the collective oscillations of the conduction electrons and the incident electromagnetic radiation. For bulk metals, the plasmon frequency (ω_p) can be shown to be:

$$\omega_p = \sqrt{\frac{4\pi n e^2}{m}} \qquad \qquad ... (15.4)$$

Plasmon energy for bulk gold is 9.0 eV and for silver it is 8.9 eV, and it lies in the UV range. However, for nanoparticles the plasmon energy is small and lies in the UV-Vis range.

The electromagnetic wave striking the metal surface has smaller penetration depth. Hence the electrons on the surface are most significant and their collective oscillations are known as surface plasmons. The extinction cross-section C_{ext}, which is composed of absorption and scattering, is given by:

$$C_{ext} = \frac{24\pi^2 R^3 \varepsilon_m^{3/2}}{\lambda} \frac{\varepsilon''}{(\varepsilon' + 2\varepsilon_m)^2 + \varepsilon''^2} \qquad \qquad ... (15.5)$$

where, R is the radius of the particle, l the wavelength of the incident electromagnetic radiation, ε' and ε'' are the real and imaginary parts of the dielectric constant of the particles respectively, and ε_m is the dielectric constant of the embedding medium.

Extinction would be maximum when $\varepsilon_1 + 2\varepsilon_m = 0$, which gives rise to a surface plasmon resonance (SPR) band. The position of the SPR band is dependent on the size and shape of the particle and dielectric constant of the medium in which the particles are dispersed. Silver nanoparticles of size around 10 nm show an intense SPR band at 420 nm. Figure 15.12 shows the absorption spectra of silver nanoparticles of various sizes. The inset shows these silver nanoparticles dispersed in water. Synthesis was carried out at room temperature using sodium borohydride as a reducing agent and trisodium citrate as a capping agent.

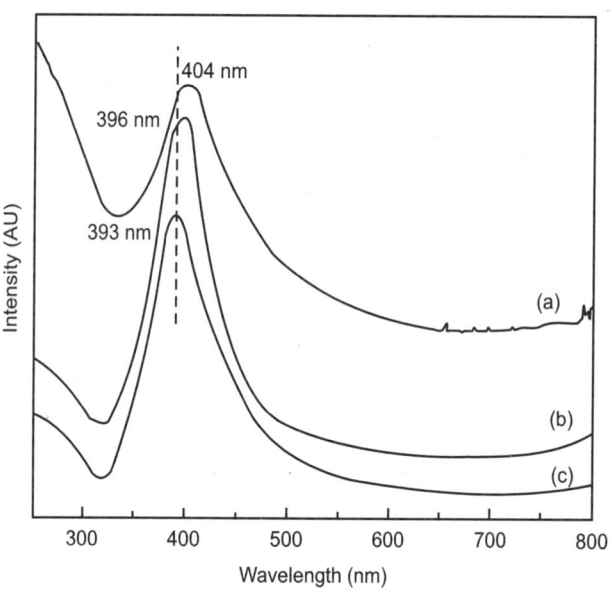

Fig. 15.12. UV-Vis absorption spectra of nanosized silver particles.

Similar to metal nanoparticles, metallic nanoshells also show the SPR band. Intensity of this SPR is higher than metal nanoparticles of the same size. It is possible to tune the position of the SPR band from visible to IR range of the electromagnetic spectrum by changing the core-to-shell ratio. Several models and theories have been put forward for calculating the SPR arising from the nanoshell assemblies. The Mie theory has been extended for nanoshell particles by considering them as concentric spheres of different materials. Recently, Prodan have put forward another model, known as the hybridisation model, to explain plasmon response from nanoshells. This model is in good agreement with the Mie theory within dipole limits. Plasmon excitation from nanoshell particles can be viewed as an interaction between plasmon response from a nanosphere and a nanocavity as shown in Fig. 15.13.

Electromagnetic excitations in the nanosphere and nanocavity induce charges on the inner and outer interfaces of the metal shell. Strength of interaction between the nanosphere and nanocavity plasmons depends on the thickness of the shell layer. Interaction is weak in the case of a thick layer, while it is strong when the layer is thin. Because of the interaction between two plasmon resonances, hybridisation of plasmon resonance band occurs analogous to the interaction between two atoms. They hybridise into lower energy symmetric or bonding plasmon resonance and higher energy antisymmetric or antibonding plasmon resonance. The frequencies of these modes (bonding and antibonding) can be expressed as:

$$\omega_{n\pm}^2 = \frac{\omega_B^2}{2}\left[1 \pm \frac{1}{2n+1}\sqrt{1+4n(n+1)\left(\frac{r_1}{r_2}\right)^{2n+1}}\right] \qquad \dots (15.6)$$

where, r_1 is the inner radius of the shell, r_2 the outer radius of the shell, n the order of spherical harmonics, ω_B the bulk plasmon frequency, ω_{n+} the antisymmetric plasmon and ω_{n-} the symmetric plasmon.

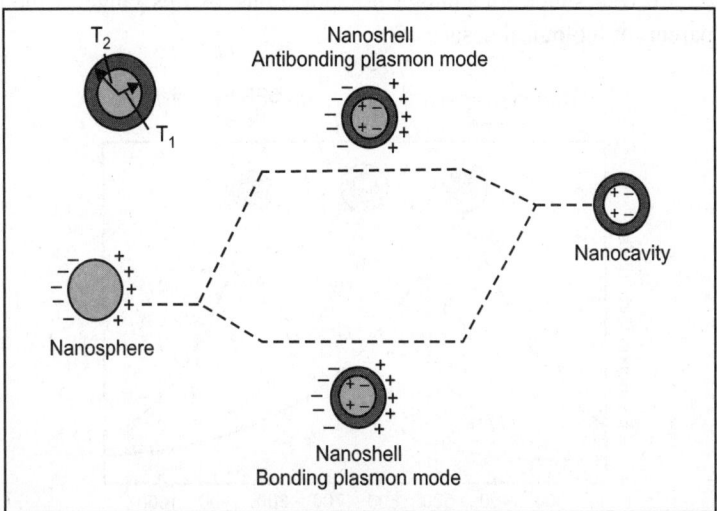

Fig. 15.13. Hybridisation model describing interaction between sphere and cavity plasmons to give rise to nanoshell plasmon.

This model is in full agreement with the Mie theory, expecting a blue shift in energy as the shell thickness is reduced. Plasmon hybridisation model suggests that reduction in the shell layer thickness results in increased interaction between the two plasmon resonances. This causes higher hybridisation between the two and a blue shift in energy is observed.

Properties of Nanoshell Particles

Coating of colloidal particles with shells offers the most simple and versatile way of modifying their surface chemical, reactive, optical, magnetic and catalytic properties. Silica particles coated with gold shell have been studied for fascinating optical properties. CdSe nanoparticles coated with CdS or ZnTe and CdTe nanoparticles coated with CdSe have been studied for enhancement in their luminescent properties. Similarly, semiconductor particles such as ZnS doped with Mn can be embedded inside a single silica particle. Such doped semiconductor nanoparticles are known to be highly efficient fluorescent materials. Coating of dielectric materials such as silica on them can enhance the luminescence properties. Magnetic particles of iron oxide have been coated with dyeincorporated silica shell. Such particles show magnetic properties arising from the core as well as luminescent optical properties arising from the shell. Thus functional materials with novel properties can be synthesised using various combinations of core shell materials and by varying shell thickness. Here is a brief review of properties of these materials.

Optical properties

Metal nanoparticles show optical absorption in the visible range of the electromagnetic spectrum. Position of absorption band shows small variations with particle size. Oldenburg have done pioneering work on optical properties of gold nanoshells. Coating of metallic shells on silica allows one to tune the absorption band from visible to infrared region. Relative thickness of core-to-shell layer is sensitive towards the position of the SPR band. Thus by changing the shell thickness, one can tune the SPR band position in the desired wavelength range (as shown in Fig. 15.14). Metal nanoshells having plasmon resonance in the infrared region are well suited for biological applications, as this range of the electromagnetic spectrum is transparent for biological tissues.

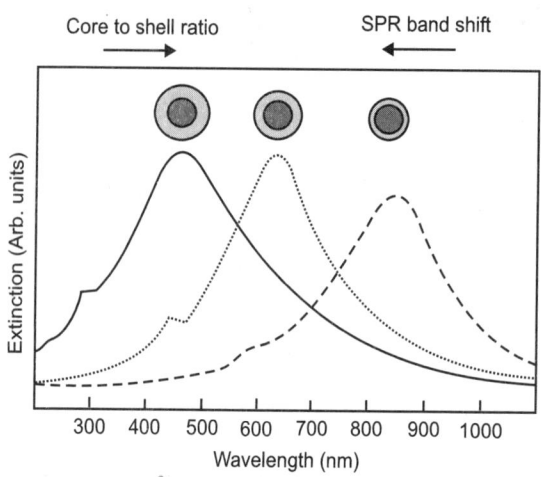

Fig. 15.14. Variation in SPR band with shell thickness.

Interaction between nanoparticles depends upon the separation between neighbouring particles. Thick coating leads to larger separation of the metal particles, whereas thin coating leads to lesser separation. Dipole–dipole coupling between the particles is responsible for red shift of the plasmon band. If the particles are well separated (thick coating), the dipole–dipole coupling is fully suppressed and the plasmon band is located nearly at the same position as the individual metal particle.

By varying the thickness of the shell by a small value, the colour of the core shell particles can be tuned from red, pink, purple to blue and at the same time changes can be monitored spectroscopically by monitoring the SPR bands.

Enhancement of luminescence

Semiconductor nanoparticles are well-known fluorescent materials. Coating of silica is often done on them to reduce photobleaching. Semiconductor nanoparticles coated with another layer of semiconductor have proved to be of great importance to enhance the luminescence from these core shell assemblies. The choice of shell material is important for localisation of the electron–hole pair. There are type-I nanostructures such as CdSe@CdS or CdSe@ZnS in which the conduction band of the shell material (which is a higher band-gap material) is at higher energy than the core and valence band of the shell is at lower energy than that of the core (as shown in Fig. 15.15). In these materials, electrons and holes are confined in the core.

In type-II nanostructures such as (CdSe@ZnTe or CdTe@CdSe), both valence and conduction bands of the core material are at higher (or lower) energy than in the shell (Fig. 15.15). In this case one carrier is confined in the core and the other in the shell. Type-I and type-II nanostructures have properties different from each other because of the spatial separation of carriers. It has been noticed that lifetime decay of exciton and quantum yield of core shell nanoparticles is much higher than individual semiconductor nanoparticles.

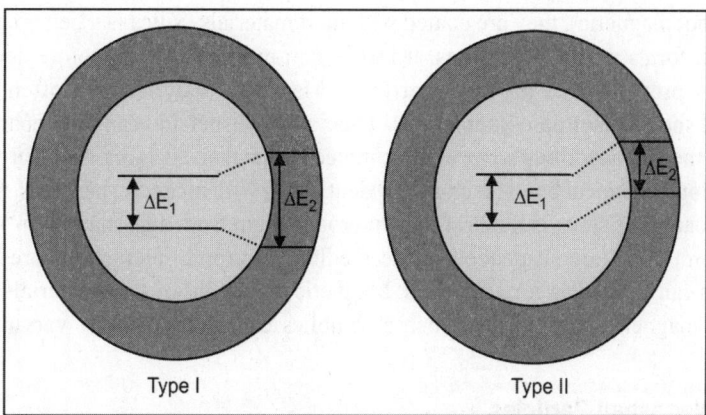

Fig. 15.15. Type-I and type-II semiconductor core shell structures.

Similar to semiconductor nanoparticles, organic dyes are also well-known phosphor materials. They are used as fluorescent biological labels. These dyes are however, not photostable and bleach out quite fast. Some of the dyes cannot be dispersed homogeneously in water. When these dye molecules were entrapped in silica shell, enhancement of luminescence was observed. Silica coating on these dye molecules makes them disperse uniformly in water.

Enhancement of thermal stability

Depression of melting point in nanoparticles compared to bulk is observed. This has been attributed to large surface tension in the case of nanoparticles. In order to release this tension, they melt fast compared to bulk. Encapsulation of silica on these nanoparticles greatly improves the thermal stability of these particles. By changing the thickness of the shell, variation in melting point is observed. In some nanoshell assemblies (metallic shells on dielectric cores) thermal instabilities are observed. Complete distortion of the shell was observed when silica gold nanoshell particles were heated at 325°C. Bulk gold has a melting temperature of ~1064°C. Melting of nanoshells was observed at significantly lower temperature. Due to higher surface area of core shell particle, more number of particles are exposed to the surface and are affected by faster melting. When these particles were encapsulated with silica (to form a multishell having silica–gold–silica layers), enhancement in thermal stability was observed. It has been observed that 60–70 nm thick coating of silica greatly improves (about 300°C higher) the thermal stability of gold nanoshells. Coating of silica on such shells is a way of preserving the identity of individual core particles because of high temperature stability of silica.

Surface chemical and catalytic properties

Core shell particles offer high surface area and can be used as efficient catalysts. Titania is an important photocatalytic material. It has been found that nanoshells and nanoparticles show different catalytic behaviour from bulk titania. It is thermally unstable and loses its surface area readily. Coating a thin layer of some other stable oxide (such as silica) on titania can greatly improve its catalytic activity.

Magnetic properties

Stability of magnetic materials is important when studying their magnetic properties. To improve the surface characteristics and protect them from reacting with various species (to form oxides, e.g. all the oxides of iron are not magnetic), they are coated with inert materials. Silica is a better choice for such a purpose, because it forms stable dispersions. It is also non-magnetic and therefore does not interfere with the magnetic properties of the core particles. Magnetic materials are often susceptible to agglomeration and show anisotropic interactions. Their stable dispersion can be prepared by inducing surface charges on them or adsorbing some organic molecules on their surfaces. Since organic molecules do not form any strong chemical bond (such as covalent bond) with magnetic particles, they can be also desorbed. A thin coating of silica is the best way to protect them from agglomeration. When magnetic particles coated with silica are suspended in the medium, isotropic interactions are observed. Two magnetic materials can be used as core and shell. Magnetic transitions of such materials can be studied. It was found that magnetic properties of such assemblies can be tailored by varying core to shell dimensions.

Applications of Nanoshell Particles

Nanoshell materials have received considerable attention in recent years because of potential applications associated with them. These materials have been synthesised for a variety of applications like fluorescent diagnostic labels, catalysis, avoiding photo degradation, enhancing photoluminescence, creating photonic crystals, preparation of bioconjugates, chemical and colloidal stability, etc. A few applications are discussed here in detail.

Colloidal stability

Nanoparticles are susceptible to coalescence and oxidation. Their surface is unstable because of the presence of dangling bonds, surface strains, etc. Often their surface is passivated by coating another stable material on them. Although stable nanoparticles can be synthesised by capping with organic molecules, these capping agents decompose at elevated temperature (300°C), resulting in agglomeration of nanoparticles. Stability of colloids is greatly enhanced by coating them with stable material. Silica, because of its unusual stability, is a popular choice for shell material as has been discussed earlier.

Photonic band gap materials

Photonic band gap materials are optical equivalents of semiconductors. Behaviour of photons in such materials can be analogically correlated to behaviour of electrons in semiconductors. Colloidal particles can self-assemble to form a three-dimensional crystal having long-range periodicity. Air gaps between the particles form a region of low refractive index, while the particles form a region of high refractive index. When photons are incident on these materials, they pass through regions of the high and low refractive indices. For photons, this contrast in refractive index is similar to the periodic potential that an electron experiences while passing through a semiconductor. If the contrast in refractive index is large, then the photons are either totally reflected or confined inside the dielectric material. The colloidal crystal blocks wavelengths in the photonic band gap, while allowing other wavelengths to pass through. The photonic band gap can be tuned by changing the size of the particles. Similar material can be prepared using core shell particles. Core shell particles are better suited for this application, as relative refractive index contrast in core and shell particles is more. Band gap can be tuned from visible to IR range by changing index contrast.

Chemical libraries

Chemical library has applications in drug delivery, gene screening, barcoding and biological imaging. Semiconductor nanoparticles are photostable and have continuous excitation spectra above the threshold of absorption. They have narrow emission spectra, the position of which is dependent on the size of the particles. These particles are, therefore, well suited for the preparation of chemical libraries. However, dye molecules also have been used for this purpose. Synthesis of chemical library mainly involves combinations of several functional molecules on a single template. Various fluorescent particles (dye or semiconductor nanoparticles) having emission at different energies can be immobilised on a single particle. Excitation of this particle with a single narrow band source produces emission at different wavelengths.

Nanoshells which comprise of template particles (core) and immobilised fluorescent particles (shell) can be used as bar codes. The bar code generated on template particles can be easily decoded by fluorescence microscopy. Similar libraries can be synthesised using encapsulation of nanoparticles inside silica or polymer spheres.

Colorimetry and biosensing

Colorimetric sensing is monitoring changes in the colour of the nanoparticles which act as sensors. Usually gold nanoparticles are used for this purpose and polynucleotides, oligonucleotides and DNA have been detected successfully. Mirkin have used gold nanoparticles for detection of DNA. Single strand of DNA was immobilised on gold nanoparticles and used for detection of complementary DNA strand. It is shown that an intense ruby red-colour changes to blue upon agglomeration when complementary DNA was added. Core shell particles can be a better choice for this purpose because of enhanced sensitivity.

Core shell particles with highly controlled optical properties are used in several biotechnological applications. It is possible to modify these particles to enhance their integration with biomolecules. This can be achieved by modifying their surface layer for enhanced aqueous solubility, biocompatibility and biorecognition. For such applications, mostly silica core and gold or silver shells have been used because they offer highly favourable optical and chemical properties for biomedical imaging and therapeutic applications. It is possible to attach biomolecules to these core shell assembly and form an immunoassay to detect analytes, cancer cells, tumours, antibodies and micro-organisms.

These particles offer a sensitive, reliable and rapid detection of biomolecules. For such kind of detection, it is necessary that nanoshells have an affinity towards the specific analyte which is to be detected. This can be achieved by immobilising specific antibodies onto the nanoshells. Antibodies can be conjugated on nanoshells via the amino group present in them. Since these antibodies are raised against a particular analyte, they have an affinity for it and bind to it. It is like a lock and key arrangement, which fits perfectly. The detection is carried out by monitoring changes in the UV-Vis extinction spectra. Nanoshells show extinction maxima at a particular wavelength, which can be taken as reference. When analytes are added to it, because of agglomeration or charge transfer interactions or changes in refractive index (or all), shifts are produced in the spectra.

These changes to the nanoshell spectra are an indication of the presence of certain biomolecules. A simple detection test has been developed for checking the potability of water. Silver nanoshells can also be used for detection of toxic ions such as Cd, Hg and Pb present in water. The extinction spectrum of silver nanoshells changes when it is mixed with a solution containing these ions. Figure 15.16(a) shows the absorption spectra of silver nanoshells with different amount of Hg ions. This test is sensitive and can detect even small quantities of Hg ions. However, this test lacks specificity. Figure 15.16(b) shows variation of intensity with the amount of added Hg ions.

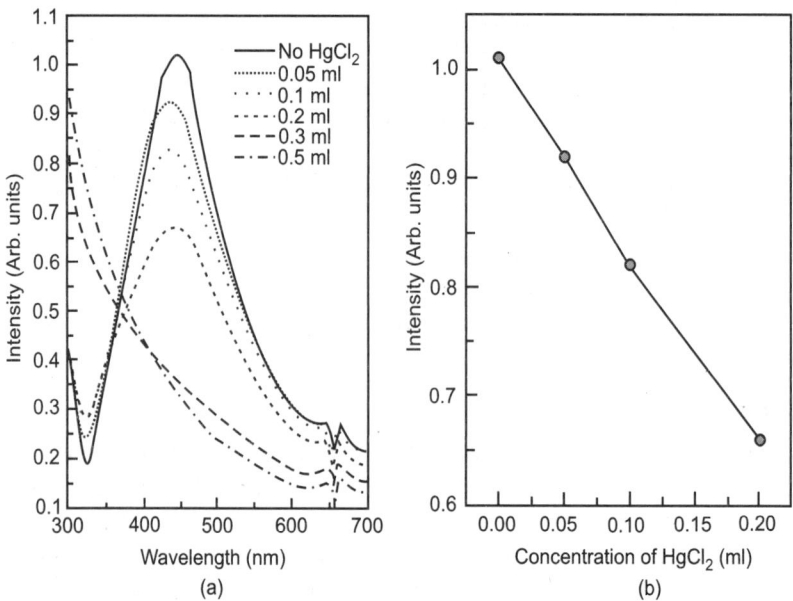

Fig. 15.16. (a) Effect of addition of HgCl$_2$ on SPR of silica silver core shell particles; (b) plot of intensity of SPR band as a function of amount of HgCl$_2$.

Therapeutic applications and drug delivery

Nanoshells have gained considerable attention in clinical and therapeutic applications. By carefully choosing the core-to-shell ratio, it is possible to design novel nanoshell structures, which either absorb light or scatter it effectively. Strong absorbers can be used in photothermal therapy, while efficient scatterers can be used in imaging applications (Fig. 15.17).

Core shell (mostly gold nanoshells) particles conjugated with enzymes and antibodies can be embedded in a matrix of the polymer. These polymers, such as Nisopropylacrylamide (NIPAAm), and acrylamide (AAm), have a melting temperature which is slightly above body temperature. When such a nanoshell and polymer matrix is illuminated with resonant wavelength, nanoshells absorb heat and transfer to the local environment. This causes collapse of the network and release of the drug, as shown in Fig. 15.17(a). In core shell particles-based drug delivery systems either the drug can be encapsulated or adsorbed onto the shell surface. The shell interacts with the drug via a specific functional group or by electrostatic stabilisation method. When it comes in contact with the biological system, it directs the drug.

In imaging applications, nanoshells can be tagged with specific antibodies for diseased tissues or tumours. When these nanoshells are inserted in the body, they get attached to diseased cells and can be imaged [Fig. 15.17(b)]. Once the tumour has been located, it is irradiated with resonance wavelength of the nanoshells. This leads to localised heating of the tumour and it is destroyed. The power required for destroying diseased cells is almost half that required to kill healthy cells.

The usual methods of tumour treatment, such as chemotherapy or radiotherapy have various side effects like substantial loss of hair, lack of appetite, diarrhoea, etc. The process of attacking the tumour, also leads to the loss of many healthy cells. Nanoshells offer an effective and relatively safer strategy to cure these ailments.

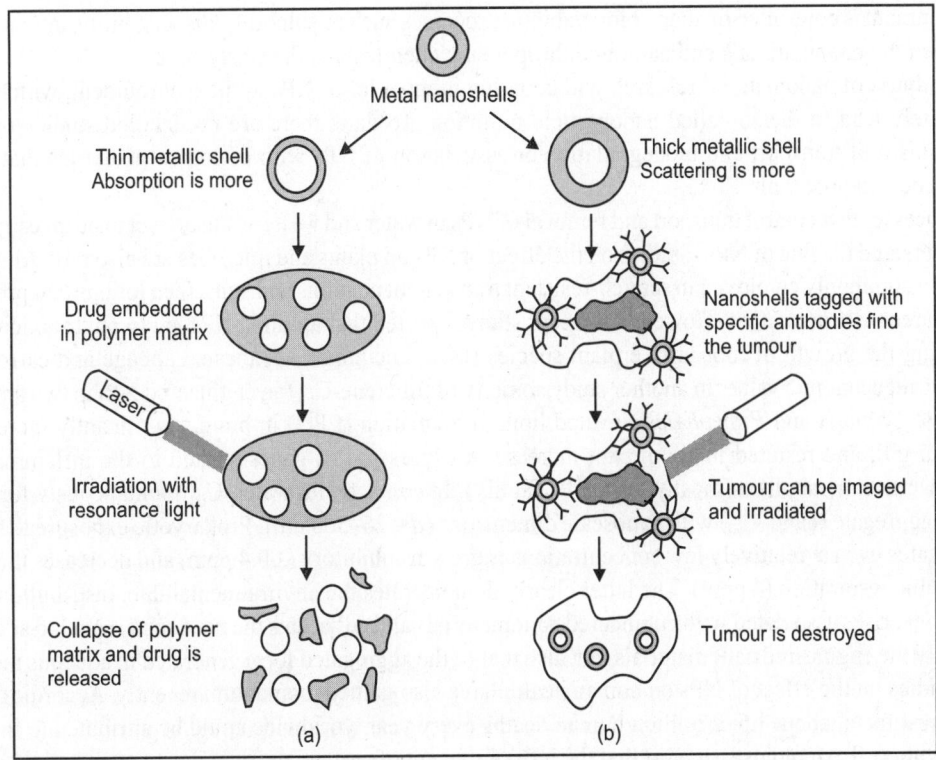

Fig. 15.17(a), (b). Drug delivery and imaging system based on nanoshells.

NANOTOXICITY: THREAT POSED BY NANOPARTICLES

Nanotechnology is the convergence of engineering and molecular biology, leading to the development of structures, devices and systems that have novel functional properties with size ranging between 1 and 100 nm. Nanotechnology has tremendous potential to change and improve many sectors of the economy, including consumer products, healthcare, transportation, energy and agriculture. It is estimated that over 200 consumer goods that are already available consist of nanomaterials. Nanoparticles (NPs) are present in some sunscreens, cosmetics, toothpastes, sanitary-ware coatings, silicon chips and even in food products. Worldwide investment on nanotechnology is on the rise and the trend is expected to continue over the next decade.

While the tremendous positive impacts of nanotechnology are widely publicised, potential threats or risks to human health and the environment are just beginning to emerge. With limited information available for support, critics are presenting a number of concerns on the devastating effects of nanotoxicity on human health and the environment. Several aspects of societal implications of nanoscience and nanotechnology in developing countries have been discussed in various literatures. Detailed studies on the long-term effects of NPs are the need of the day to overcome or reduce possible threats. Simultaneous agglomeration, sedimentation and diffusion at physiologically relevant concentrations should be taken into account while conducting quantitative studies on the uptake of NPs into biological systems, to assess the corresponding risks on human health. In addition, ecological risk assessment is essential to understand environmental implications of nanomaterials. The fate of nanomaterials in aqueous environment is controlled by many biotic/abiotic processes such as solubility/dispersability, interactions between the nanomaterials and natural/anthropogenic chemicals in the ecosystem.

Outburst of nanomaterial research will certainly pump a lot of NPs to the environment, which will ultimately lead to the so-called nanoparticle pollution. To date, there are no detailed studies on the mechanism of transport and biodegradation or association of NPs with biological materials that may eliminate nanomaterials.

Processes that control transport and removal of NPs in water and waste-water are yet to be investigated to understand the fate of NPs. Studies on the effect of NPs on plants and microbes are also rare. Alumina NPs are commonly employed in scratch-resistant transparent coatings and sunscreen lotions that provide transparent-UV protection. However, a recent study reported that alumina NPs led to phytotoxicity by retarding the growth of root in five plant species (corn, cucumber, soyabean, cabbage and carrot) of significant economic value. In another study, toxicity of fullerene-C_{60} (an engineered NP) in two aquatic species, *Daphnia* and *Pimephales* elevated lipid peroxidation (LPO) in brain, significantly increased LPO in gill, and resulted in significant increase in expression of genes related to the inflammatory response and metabolism (mostly the CYP2 family). In contact with water, C_{60} spontaneously forms a stable aggregate (nano-C_{60}) with nanoscale dimensions ($d = 25$–500 nm). Prokaryotic exposure to these aggregates even at relatively low concentrations is growth-inhibitory (≥ 0.4 ppm) and decreases the rate of aerobic respiration (4 ppm). The latter clearly demonstrates the environmental fate, distribution and biological risks associated with engineered nanomaterials and advocates the need for a model to address not only the engineered nanomaterials, but also that of the aggregated form generated in aqueous media.

Studies on the effect of NPs on human health have also gained momentum recently. According to a recent estimate, about 1.5–2 million human deaths every year worldwide could be attributed to indoor air pollution. Toxicologists suggest that the NP component of particulate air is most potent and likely to be responsible for adverse health effects. Whether the inhaled carbon NPs translocate directly into the circulation in humans or not, is a conflicting issue. A study suggests that inhaled carbon NPs are capable of rapid translocation into the circulation. On the contrary, a conflicting report suggests that inhaled carbon NPs remain within the lung up to 6 hours after inhalation, without passing to systemic circulation. In a recent study, suitability of mouse spermatogonial stem cell line as a model system to assess nanotoxicity was evaluated in the male germline *in vitro*. Light microscopy, cell proliferation and standard cytotoxicity assays revealed the effect of different types of NPs on these cells. The results of these experiments revealed concentration dependent toxicity for all types of NPs tested, whereas the corresponding soluble salts had no significant effect. In the assay conditions, silver NPs were the most toxic, while molybdenum trioxide (MoO_3) was least toxic. The toxicity of copper NPs (23.5 nm) was assessed *in vivo* based on LD_{50}, morphological changes, pathological examinations and blood biochemical indices of experimental mice. Grave nanotoxicity was attributed to several factors; for instance, huge specific surface area, ultrahigh reactivity and exceeding consumption of H^+.

Not all NPs are dangerous. The toxicology and biodynamics of silica NPs investigated in a mice model revealed that silica NPs were not toxic and can be used *in vivo*. Results generated from the protocol developed by an insurance company for the purpose of calculating insurance premium for chemical manufacturers revealed that the relative environmental risk from manufacturing five nanomaterials (single-walled carbon nanotubes, bucky balls (C_{60}), one variety of quantum dots, alumoxane NPs and nano-titanium dioxide) was comparatively lower than common industrial manufacturing processes. This study should not be misunderstood to promote the manufacture of these nanomaterials without detailed assessment of environmental and human risks. Besides the workers in the manufacturing wing, others who get exposed (e.g. occupational health nurses) to NPs should be aware of the potential risks and possible means to avoid health risks. There is a need to identify specific regulatory regimes to protect personnel involved in the production and use of NPs for cosmetic, medical and agricultural purposes.

In nanotechnology, we have a unique opportunity to test hazards and control risks as and when the technology develops. At present, safety assessment protocols for products possessing engineered NPs are poorly structured. However, at the current pace of research, we may need several years or decades to clearly establish the health and environmental risks from engineered nano-scale particles. Although research on the adverse effects of NPs on human health is progressing rapidly, environmental fate of NPs is still in its infancy.

Before unknowingly dumping a huge amount of dangerous nanomaterials into the environment, we need to investigate the solubility and degradability of engineered NPs in soils and waters, to establish baseline information on their safety, toxicity and adaptation of soil and aquatic life. Development of novel NPs must be followed by the assessment of their potential risks on life and environment, and possible remedial measures.

Safety

Nanoparticles present possible dangers, both medically and environmentally. Most of these are due to the high surface to volume ratio, which can make the particles very reactive or catalytic. They are also able to pass through cell membranes in organisms, and their interactions with biological systems are relatively unknown. However, free nanoparticles in the environment quickly tend to agglomerate and thus leave the nano-regime, and nature itself presents many nanoparticles to which organisms on earth may have evolved immunity (such as salt particulates from ocean aerosols, terpenes from plants, or dust from volcanic eruptions).

According to the *San Francisco Chronicle*, 'Animal studies have shown that some nanoparticles can penetrate cells and tissues, move through the body and brain and cause biochemical damage they also have shown to cause a risk factor in men for testicular cancer. But whether cosmetics and sunscreens containing nanomaterials pose health risks remains largely unknown, pending completion of long-range studies recently begun by the FDA and other agencies.' Diesel nanoparticles have been found to damage the cardiovascular system in a mouse model.

In October 2008, the Department of Toxic Substances Control (DTSC), within the California Environmental Protection Agency, announced its intent to request information regarding analytical test methods, fate and transport in the environment, and other relevant information from manufacturers of carbon nanotubes. The term 'manufacturers' includes persons and businesses that produce nanotubes in California, or import carbon nanotubes into California for sale. The purpose of this information request will be to identify information gaps and to develop information about carbon nanotubes, an important emerging nanomaterial.

On January 22, 2009, a formal information request letter was sent to manufacturers who produce or import carbon nanotubes in California, or who may export carbon nanotubes into the State. This letter constitutes the first formal implementation of the authorities placed into statute by AB 289 (2006) and is directed to manufacturers of carbon nanotubes, both industry and academia within the State, and to manufacturers outside California who export carbon nanotubes to California. This request for information must be met by the manufacturers within one year.

CONCLUSION

Various methods of synthesis of core shell particles are reported in the literature. Also there are new techniques emerging with which one can ensure complete coverage of core particle and homogeneous coating. However, synthesis of core shell particles is a big challenge. It needs skillful monitoring and highly controlled reaction parameters to produces functional materials. With variations in core and shell materials, one can tune their properties, which make them highly commendable. Nanoshells have a plethora of applications associated with them. These particles are used in imaging cancer cells and other therapeutic applications. Although such applications make them highly fascinating materials, one has to look at the ill-effect of these particles as well. The effect of these particles when inserted inside a human body is not known. For using these particles, intelligent and meticulous studies are needed.

Chapter 16

Oxide Nanoprecursors: A Technological Perspective

INTRODUCTION

Way back in 1959, famous Scientist Richard Feynman was able to visualise the properties of nanoparticles. He had predicted that the reduction in size of a substance to atomic scale would lead to materials with extremely new properties. Scientists experienced the truth in his prediction when they could actually bring the size of the material down to nanoscale and develop different techniques for characterisation. Research activities in nanomaterials are two-fold. Scientists have studied technological advantages of using precursors in finite size range. At the same time rich and fascinating physics of fine particles has attracted the research workers.

After studying the physics of different oxide fine particles, we have established a general rule. According to researchers, with the reduction in size to nanometer scale, the lattice of oxide material tends to acquire higher structural symmetry. As a result, the substance exhibits altogether new properties since there is always a structure-property correlationship. However, this chapter deals with the technological advantages of nanoparticles, which we have come across during our study in past few years. With the reduction in particle size obviously there is increase in specific surface area of the sample. Reduction in size also enhances the ability of particles to cover large surface more uniformly. Fine particles therefore find applications in catalysis, paints and pigments, magnetic recording, etc. Rate of sintering is inversely proportional to initial particle size. As a result, fine particles are more effective when used as precursors to obtain highly dense products, which can even replace single crystal.

SYNTHESIS OF OXIDE NANOPARTICLES

Methods used for the synthesis can be categorised into three types:

1. Mechanical: Ball milling.
2. Physical: Solid-solid phase transition, spray drying, liquid drying.
3. Chemical: Coprecipitation, sol-gel, and micro-emulsion.

Each technique has its own figure of merit and the choice of the technique depends upon the material of interest. However, it has been observed that the particles derived through chemical route have controlled chemical homogeneity, stoichiometry and size distribution. In case of multi-cation systems, it is possible to avoid solid state diffusion step since the complex formed in wet chemical route directly leads to reacted phase on calcination. It helps to reduce the calcination temperature and hence to control the particle size.

Mechanical Method

Ball milling

It is the simplest and easiest method to achieve particles at sub-micron level. High purity oxide materials, in the proportion required in the final product, are mixed together. Mixing is often performed by wet milling in a rubber-lined pot using stainless steel or alumina balls. The mixture is then dried and calcined. The hard cake obtained during calcination process is ball milled once again for a long time to get fine particles. Though process is simple it has some serious limitations such as, lack of control on chemical purity, homogeneity and particle size distribution. The process is lengthy and takes about a month to bring the particle size to sub-micron level.

Physical Methods

Solid-solid phase transition

Palkar were the first to show that the process of phase transformation could be effectively used to synthesise nanoparticles. It was observed, that, irreversible phase transformation is accompanied by marked reduction in particle size. The reduction in particle size is proportional to change in unit cell volume at transition. In order to control the particle size, the calcination temperature needs to be restricted just above the phase transformation. It helps to avoid the growth of the freshly nucleated phase. The method is applicable if suitable metastable polymorph is available. In certain cases it could be an alternative to wet chemical route.

Spray drying

Figure 16.1 shows the set-up used for spray drying. In this method, the solution to be dried is fed to the eye of the centrifugal atomiser disc. Due to rotation of the disc at high speed the solution is thrown out in the form of a fine mist. Hot dried air is simultaneously introduced through an annular opening into the drying chamber. This hot air mixes continuously with the mist of atomised liquid and an instantaneous evaporation of the liquid takes place. The non-volatile part is left in the form of dry powder. The powder falls down towards the outlet aperture in the conical bottom of the chamber and from there it is carried into the dynamic cyclone separator by the air current. The powder falls down into a container and the drying air is discharged from the top of the cyclone. The dry powder is then calcined to get the required phase. In this technique, particle size is controlled by speed of the disc, solution concentration and calcination temperature.

Special care is required to preserve the particle size if the material is hygroscopic. In case of multi-cation systems, to obtain reacted phase, the reactants have to undergo solid-state diffusion since there is no complex formation.

Liquid drying

Liquid drying of an aqueous solution of metal salt requires in general, a drying liquid having a high degree of solubility for water but not for metal ions. Since the interest is in getting oxide phase at final stage, the soluble salts like acetate, citrate, etc. are commonly used. In this method, the stoichiometric aqueous solution is sprayed as a broken stream of droplets into the vortex of swirling bath of hygroscopic liquid. Acetone is used as a hygroscopic liquid in most of the experiments. The ratio of the total volume of acetone to the salt concentration is ~10:1. Under these conditions nucleation of particles is essentially instantaneous. Acetone-water mixture is removed from the dry powder by vacuum nitration. Since no formation of complex occurs, in multi-cation systems solid-state diffusion process is unavoidable to achieve the reacted phase. Particle size is decided by the solution concentration, and calcination temperature.

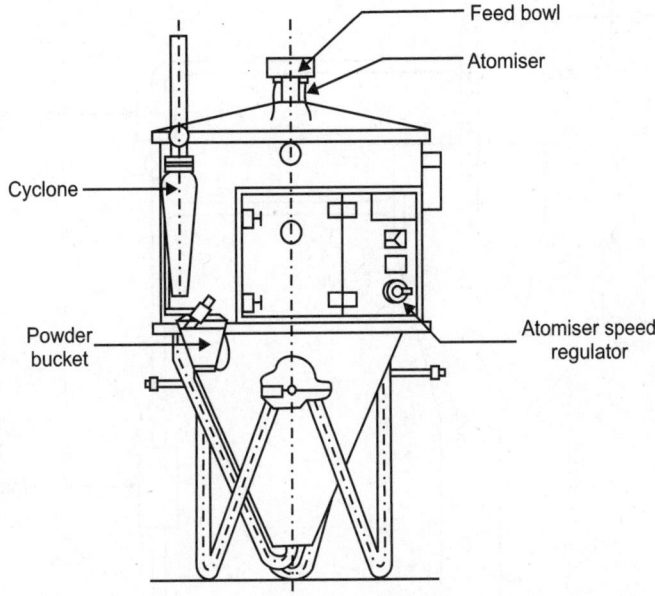

Fig. 16.1. Spray dryer.

Chemical Methods

Co-precipitation

The lengthy ball-milling process involved in mixing the constituents of multi-cation system could be avoided in co-precipitation route. The constituents are weighed in stoichiometric proportion and dissolved to get the stock solution. All the cations are then co-precipitated to achieve the complex formation by selecting suitable precipitating agent and pH.

In case of oxides the precipitate could be in the chemical form of hydroxide, carbonate, oxalate, etc. so that after decomposition it converts to oxide. In this process the particle size is determined by the solution concentration, rate of precipitation and calcination temperature. However, during the drying of the precipitate, there is a possibility of agglomeration of the particles and hard cake obtained in this step needs ball milling.

Sol-gel

This is one of the well-proven methods to produce fine particles with narrow size distribution and controlled chemical composition at comparatively low temperatures. In this process, dispersion of the particles of the metal compound (sol) usually in an aqueous phase is first prepared and then converted to gel particles. Gelation is achieved by means of chemical dehydrating agent with surfactant. 2-ethylhexanol (dehydrating agent) and SPAN-80 (surfactant) is commonly used in the process. In principle, particle size achieved at the precipitation state is maintained during sol and gel formation. Since agglomeration is avoided the process gives better control on particle size and size distribution.

Column technology is used, if geleted material is required in spherical shape (Fig. 16.2). The gel is then dried and calcined to get reacted product.

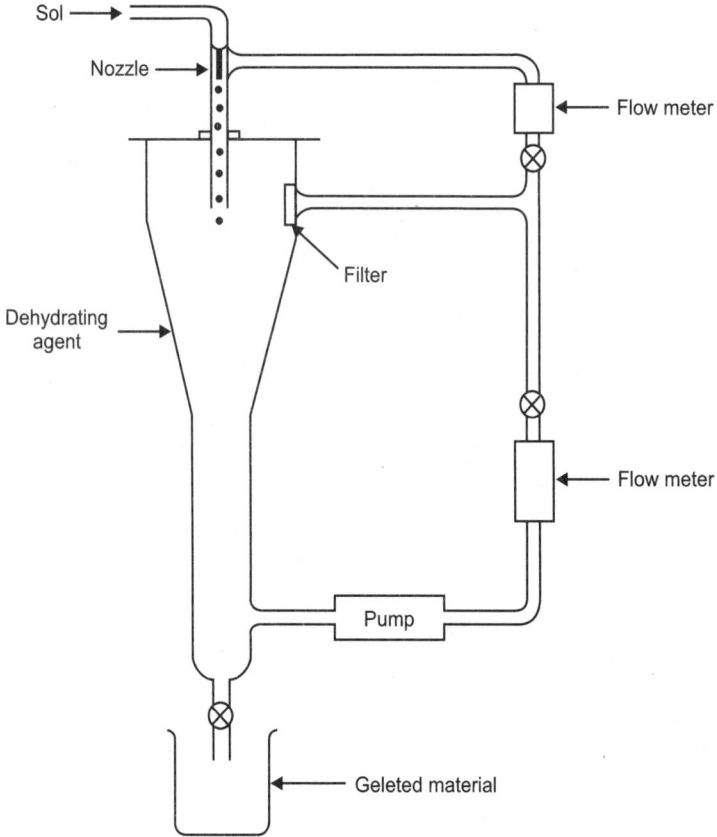

Fig. 16.2. Glass column used in sol-gel process to produce geleted material in spherical shape.

Micro-emulsion

When two immiscible liquids are mixed together, emulsion is formed. However, to get micro-emulsion (stable and transparent system) the globule size of the dispersed phase needs to be less than 1000 Å. Micro-emulsion is basically a three phase system, i.e. oil (organic phase), water (inorganic phase) and surfactant (cationic, anionic or non-ionic). Surfactant remains at the interface and helps to prevent coagulation of dispersed globules. Different types of micro-emulsions [water dispersed in oil (W/O << 1), oil dispersed in water (W/O >> 1) or bicontinuous structure, (W/O ~ 1)] are shown in Fig. 16.3. The method mainly consists of the following steps:

1. Formation of micro-emulsion with proper choice of three phases.
2. Precipitation of inorganic phase.
3. Separation of precipitate.
4. Drying and calcination.

The particle size is mainly controlled by the solution concentration and calcination temperature. Palker were the first to show that micro-emulsion could be used to synthesise oxide fine particles.

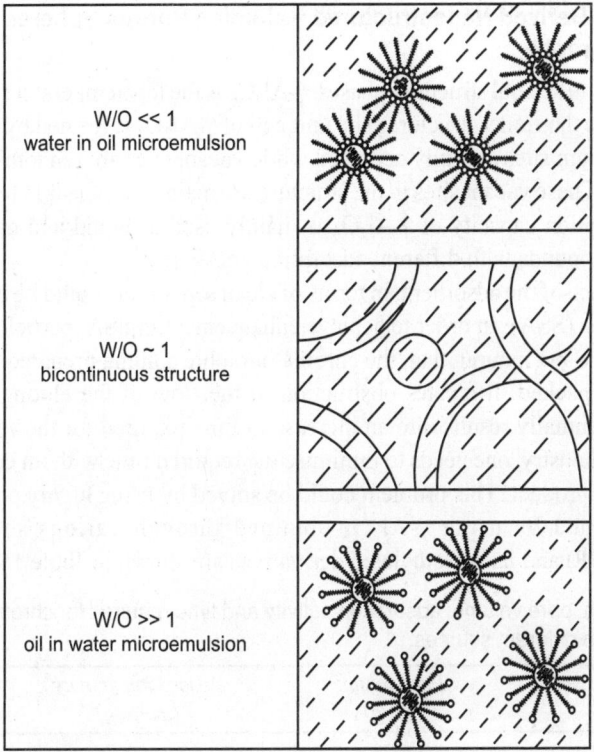

Fig. 16.3. Water in oil, bicontinuous structures and oil in water microemulsion.

SIZE DETERMINATION OF NANOPARTICLES

After the synthesis of phase pure sample, it is important to determine particle size and size distribution before carrying out further studies. X-ray line broadening, scanning electron microscopy (SEM), transmission electron microscopy (TEM) and surface area analysis are some of the techniques that are commonly used for the purpose. Though there are various techniques available for size determination, no technique is fool proof. Size calculated from X-ray line broadening using Scherrer's formula, is actually the size of the coherently diffracting domains. Unless the particles are single domain, the size determined by X-ray is always smaller than that obtained by other techniques. It is not possible to find out the size distribution using X-ray line broadening. SEM and TEM could be used to determine size as well as size distribution. However, during sample preparation bigger particles tend to settle at the bottom of the container. Hence, the sample picked up for the measurement, does not always give true representation of the particle size. This can lead to error in measurements. Size calculated from surface area actually gives equivalent spherical diameter. It is assumed, that, all particles are spherical and there is only monolayer adsorption of gas during surface area measurement. Under the circumstances, there are chances of error in size determination. Moreover, the technique cannot be used to estimate the size distribution.

RESULTS

In order to reveal the technological ability of nanoparticles, results obtained on different oxide (structural ceramic, high T_c oxide superconductor, electronic ceramics) samples are presented in this section.

Efficiency of Sol-gel Derived Nanostructured γ-alumina Porous Spheres as an Adsorbent in Liquid Chromatography

Alumina (Al_2O_3) exists in several structural phases. γ-Al_2O_3 is the low temperature phase with crystalline lattice closely related to the spinel structure. The unit cell of γ-Al_2O_3 is formed by cubic closed packing. The octahedral aluminium sites are fully occupied while vacant sites are randomly distributed over the tetrahedral interstices. These vacant sites in the structure are mainly responsible for the adsorbent nature of γ-Al_2O_3. The adsorption capacity of γ-Al_2O_3 is mainly used in liquid and gas chromatography to separate chemical compounds with different adsorption coefficients.

Efficient performance of the adsorbent, in terms of clean separation, could be obtained by increasing the specific surface area (SSA). In order to achieve enhancement in SSA, particle size of the adsorbent needs to be reduced. It is but natural, that, the chromatographic column prepared by making use of fine particles gets densely packed. It creates obstruction in the flow of the eluting agent used in liquid chromatography. It eventually results into an increase in time required for the separation process. For on line separations in industry, one needs to minimise the required time without compromising with the purity of the separated product. This problem could be solved by using highly porous fine particle.

The results obtained by using γ-Al_2O_3 obtained through various sources for purifying tetraphylporphyrin (TPP) and its n-methylated derivatives are shown in Table 16.1 and Fig. 16.4.

Table 16.1. Surface area, pore volume, adsorption activity and time required for chromatographic separation for γ-Al_2O_3 obtained from different sources.

Sample	Surface area (m^2/g)	Pore volume (cc/g)	Adsorption activity (moles/g)	Time required (hr)
Sol-gel	300	1.28	2.9	9
Fluka	135.8	0.386	0.9	22
BDH	80	0.28	0.1	27

Table 16.1 reveals that sol-gel derived alumina has very high specific surface area and hence a very high adsorption capacity among all tested samples. High surface area coupled with high porosity helps to achieve improved performance and desired economy in processing time when used as an adsorbent in liquid chromatography (Fig. 16.4). This superior behaviour could be highly useful during on line separation in chemical industry.

Efficiency of Pd Impregnated Sol-gel Derived γ-alumina Porous Spheres as Catalyst

The study of catalysts is essentially a study of the surface chemistry. The efficiency of the catalyst is in its interaction with the environment. In surface assembled metal catalyst, metal is distributed in the form of fine particles on support material. The classification of support material is done in terms of porosity and surface area. An enormous increase in activity could be obtained if an appropriate support material could be synthesised.

Generally, the metals like ruthenium, rhodium, palladium, silver, platinum and gold are used as catalysts while alumina, silica, magnetia, etc. are used as supports. The supported catalysts have several advantages over simple catalyst like, large area of metal is exposed, sintering of metal is reduced, and the surface accessibility of the catalyst to reactant is improved. The heat produced in an exothermic reaction is dissipated. The approach allows the controlled use of expensive materials, since they are more or less confined to the surface of the support.

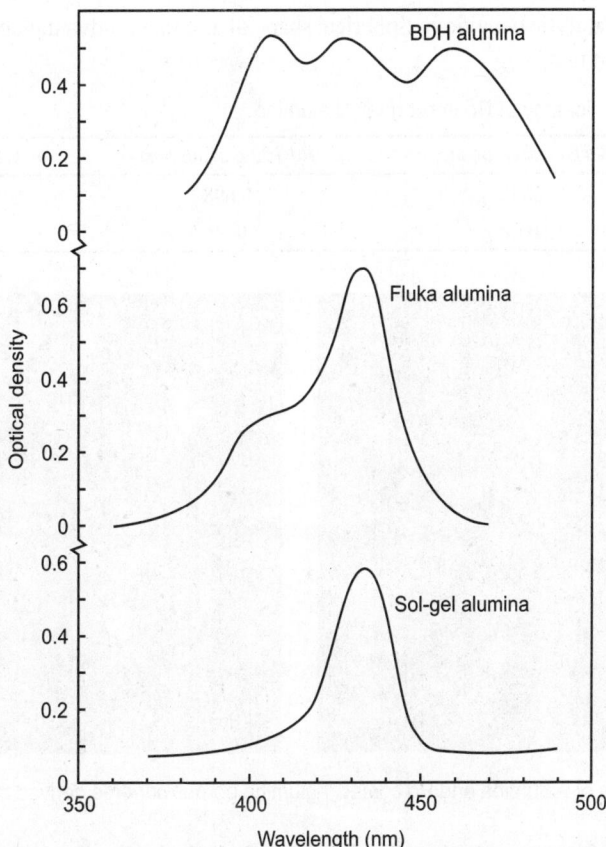

Fig. 16.4. Absorption spectra of TPP and its *n*-methylated derivatives.

The activity of a supported catalyst for a given reaction should be referred to as the unit area of metal surface exposed to reactants. The activity per unit area of accessible surface is called the surface activity. The rapid ability of the catalyst to interact with a reacting system to give a certain chemical composition and structure is desirable. Porous γ-alumina spheres synthesised through novel sol-gel process, were tested as catalyst support after impregnating with palladium. The results were compared with the commercially available support material but impregnated with Pd in laboratory, under identical conditions. Hydrogen-oxygen combination reaction was used to determine the catalytic behaviour. Catalytic efficiency was calculated from the percentage of the unreacted gases. Table 16.2 clearly indicates, that, though both alumina samples are soaked under identical conditions, the amount of Pd impregnated on sol-gel alumina is more. This could be attributed to higher adsorption activity of the material. The specific surface area of the sol-gel alumina is almost four times to that of commercially obtained alumina. Hence, Pd on sol-gel alumina is distributed over larger surface. With the result, that, larger surface of Pd is available for the catalytic reaction. γ-Alumina is a known adsorbent material. Hence, it is more likely that during soaking Pd remains adsorbed on the surface rather than getting absorbed. SEM pictures also indicate uniform coating of Pd on the surface (Fig. 16.5).

Equivalent spherical diameter as calculated from specific surface area indicated that Pd particle size could be of the order of 5 nm. Enhanced surface of nanostructured Pd, available for catalytic action, helps to improve the catalytic behaviour. Spherical shape of alumina is advantageous in forming uniform packing of the catalyst bed.

Table 16.2. Catalytic efficiency of Pd impregnated alumina.

Catalyst support	Surface area of support	Pd/100 g of support	Catalytic efficiency
Sol-gel alumina	300 m^2 g^{-1}	0.698	100%
Commercial alumina	80 m^2 g^{-1}	0.50	93%

 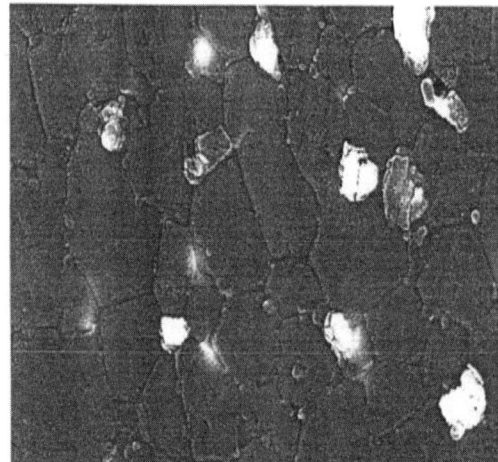

Fig. 16.5. SEM pictures of γ-alumina and Pd coated γ-alumina porous spheres derived through sol-gel process.

Synthesis of High T_c Superconductor YBa$_2$Cu$_4$O$_8$ (1–2–4) Under the Condition of Normal Oxygen Pressure

The superconductor (1–2–4) was originally observed as a lattice defect structure and then as an ordered defect structure in microstructural studies of YBa$_2$Cu$_3$O$_7$. Synthesis of (1–2–4) in bulk form was first reported by Karpinski, using high oxygen pressure (400 bar) and high temperature (1040°C). Cava reported a two-step method, using alkali carbonate catalyst. However, no technique could produce phase pure (1–2–4) sample. Studies on (1–2–4) system had indicated, that, it might turn out to have several advantages over other ceramic superconductors discovered till then. It was, therefore, interesting to find out a simplified method that could produce single phase (1–2–4) material.

It was seen that the sol-gel route could be used to produce single-phase (1–2–4) sample without any requirement of high oxygen pressure. The hydroxy carbonate complex was obtained from stoichiometric solution of the constituents by using sodium carbonate as the precipitating agent. The particle size of the order of 4–5 nm obtained in precipitation step was maintained during the sol and gel formation. The geletion was carried out using chemical dehydrating agent (2-ethyl hexanol) along with surfactant (SPAN – 80). Single step decomposition of the complex (900°C/30 min.) coupled with an enhanced oxygen uptake due to increased specific surface area, facilitated the formation of the (1–2–4) phase during calcination from the oxygen rich atmosphere. For the first time synthesis of phase pure (1–2–4) has been reported.

Densification of Lead Zirconium Titanate (PZT)

PZT ceramic is well known for its best piezoelectric properties. It is, therefore, used in a variety of sensors and actuators, MEMs etc. The piezoelectric properties of PZT are very sensitive to composition, grain size and density of the sample. If the grains are too small the ferroelectric polarisation is 'locked-in' and is difficult to realign by an electric field. To get optimum properties the ceramic should be highly dense (+95 per cent of theoretical value) with grain size of the order of 4 micron. The difficulty in sintering of PZT with desired density and grain size arises due to volatile constituents like PbO. If the temperature is raised during sintering, there are PbO losses. PbO losses affect the stoichiometry of PZT, which ultimately impairs piezoelectric properties. Highly sintered and stoichiometric PZT in pellet form is also required as a target for pulsed laser deposition to get good quality PZT thin films. PZT thin films have variety of important applications including non-volatile memories. It was observed, that, by using sol-gel derived PZT precursor (~16 nm in size), it was possible to bring down the sintering temperature to 1200°C from 1400°C (required for PZT obtained by conventional route). The reduction in sintering temperature helped to minimise PbO losses and maintain the stoichiometry. The highly dense product (+99 per cent) could withstand poling voltage of the order of 70 kV/cm. Ultimately it could give d_{33} coefficient (piezoelectric coupling factor) of the order of 293×10^{-12} Coul/Newton^{-1} (highest reported value) as seen in Fig. 16.6.

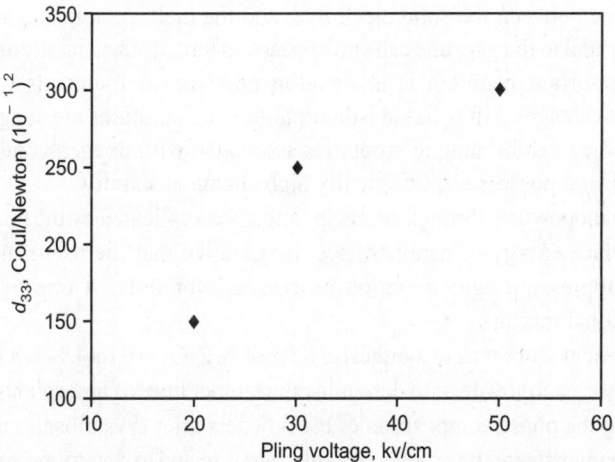

Fig. 16.6. Effect of poling voltage on piezoelectric coupling factor, d_{33} of PZT.

Aqueous-sol Derived Thin Films

Growth of thin films, using physical preparation methods such as, evaporation, sputtering and laser ablation has difficulty in achieving large area depositions. Moreover, it requires costly equipment. Sol-gel processing, which is a solution based method, is attractive due to easier fabrication of large area thin films at low cost. However, commonly used alkoxide precursors for producing films by sol-gel method, are highly reactive, and require careful control of the hydrolysis-condensation reaction. The compatibility between the starting reagents constitutes the main problem. In addition, methoxyethanol used as a solvent, is a hazardous material. The problem has been solved in aqueous sol route to grow thin films. In this technique, nanoparticles are suspended in aqueous medium using suitable electrolyte in desired quantity. It helps to set-up an electrical double layer between the particle and the medium and prevents coagulation. The stable suspension (sol) is then deposited on substrate by spin or dip coating followed by annealing. Ferroelectric $PbTiO_3$, and antiferroelectric $PbZrO_3$ thin films have been successfully grown by novel aqueous sol route. The quality of the films obtained by using this simple technique is comparable to that obtained by other techniques. In short, the capacity of the nanoparticles to produce continuous and smooth coating on effectively large surface has been successfully used in realisation of thin films. The technique could be extended to obtain large area thin films of a variety of materials, provided stable suspension (sol) in the required chemical composition could be prepared. The possibility to avoid an exposure to hazardous chemical reagent is an added advantage of the process.

SYNTHESIS OF LOW-AGGLOMERATED NANOPRECURSORS IN THE ZrO_2-HfO_2-Y_2O_3 SYSTEM

Yttria-stabilised ZrO_2-based solid solutions with a fluorite-type cubic structure are widely used as a solid electrolyte in high-temperature oxygen sensors. Such sensors are essential for analysis of gaseous and liquid media in glassmaking and blast furnaces, in cement production, in energetic plants to control a completeness of a fuel burning-away and so on. The efficient and long-time use of the sensors is hampered by a number of physico-chemical processes; the main of them are: (i) the possible formation of the $Zr_4Y_3O_{12}$ rhombohedral compound, which leads to a drastic change in the unit cell volume and, as a consequence, the destruction of a solid electrolyte material, and (ii) the interaction of these composite compounds with oxide melts. A transition layer containing the products of ceramic decomposition is formed at the interface between the solid electrolyte and the melt. As a result, a variable contribution from the junction potential to the galvanic cell emf appears; in turn, it substantially affects the measurement accuracy. So, an important problem is to develop new sensor materials which are free of the aforementioned disadvantages. HfO_2-based isomorphous solid solutions are suggested as a perspective materials here since they exhibit unique properties associated with oxygen conduction, does not form $Hf_4Y_3O_{12}$ compound, and possesses a sufficiently high chemical stability.

The synthesis of nanopowders through coprecipitation is complicated by the process of agglomeration due to excessive surface energy of nanoparticles. It is known that the use of ultrasound in a sol-gel synthesis permits suppressing agglomeration processes. Moreover, it improves the mixing of the components in reactional mixture.

The aim of the present work was to synthesise 82 mol.% ZrO_2–10 mol.% HfO_2–8 mol.% Y_2O_3 low-agglomerated precursor nanopowders; to determine the temperature of amorphous-crystalline transition in these powders, and the phase composition of the powders after crystallisation; to evaluate the mean size of crystallites; to investigate the process of agglomeration and to determine conditions for preparing precursor powders with the lowest degree of agglomeration.

Theory

The method of amorphous hydroxide co-precipitation from salt solutions of $ZrO(NO_3)_2 \cdot 2H_2O$, $HfOCl_2 \cdot 6H_2O$, and $Y(NO_3)_3 \cdot 6H_2O$ by ammonium aqueous solution was chosen to obtain precursor powders in the ZrO_2-HfO_2-Y_2O_3 system. Figure 16.7 presents pH-metric curves for precipitation of hydroxides $Zr(OH)_4$, $Hf(OH)_4$, and $Y(OH)_3$. It is seen from the figure that pH values for the precipitation of $Zr(OH)_4$ and $Hf(OH)_4$ are close and equal to 2.3 and 2.5, respectively, whereas the pH value for the precipitation of $Y(OH)_3 \approx 7$. This made it possible to use the back-precipitation method with the aim of attaining sufficiently high homogeneity of the powders prepared. The salt solution was added into the precipitant solution. A constant pH equal to 9.5 was maintained at the expense of excess NH_4OH.

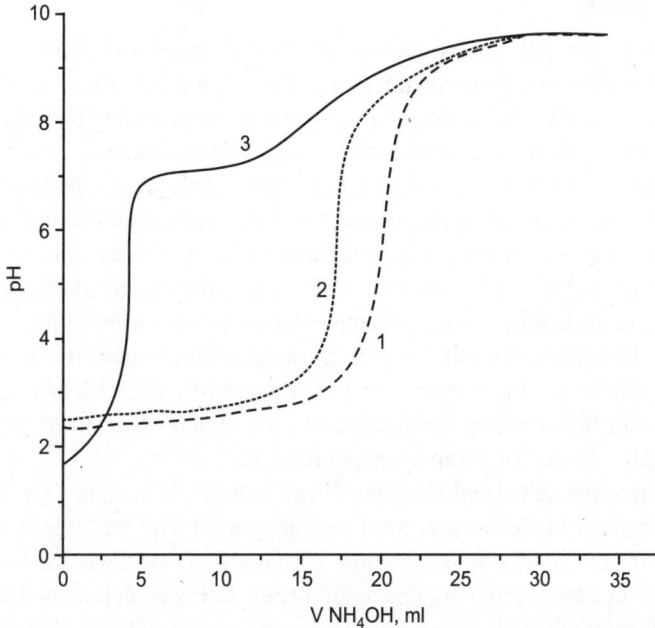

Fig. 16.7. pH-metric curves for precipitation of amorphous hydroxides: $Zr(OH)_2$ (1), $Hf(OH)_2$ (2), and $Y(OH)_3$ (3).

The co-precipitation products are X-ray amorphous gels containing nanoparticles of $Zr(OH)_4$, $Hf(OH)_4$, and $Y(OH)_3$ hydroxides. Upon further thermal decomposition, they are transformed into nanocrystallites of ZrO_2-HfO_2-Y_2O_3 system. The structure and size of agglomerates of crystalline oxide particles completely inherit the structure and size of the agglomerates of amorphous hydroxide particles. Therefore, it is very important to obtain particles with a low degree of agglomeration at the co-precipitation stage. The size, density, and strength of agglomerates can be controlled by varying the reaction conditions of hydroxide preparing.

Numerous investigations carried out by Smith of the present work on sol-gel synthesis of nanosized materials made it possible to find optimal conditions of the synthesis. In order to obtain finely disperse precipitates with a low degree of agglomeration, the following conditions should be satisfied.

1. The precipitation should be performed from a diluted salt solution with a diluted precipitant solution.
2. The precipitation rate should be minimal.
3. Upon pouring, the solutions should be mixed thoroughly.
4. The reaction should be carried out at room or lower temperature, because a decrease in the precipitation temperature makes it possible to control the amount of nuclei of gel-forming particles and their size.
5. Surface-active substances, like NH_4Cl or TSA, should be introduced into the reaction mixture in order to blockade agglomeration.
6. Co-precipitation should be carried out under ultrasound so that an ultrasonic wave may directly act on the entire volume of reaction mixture. An ultrasonic wave gives rise to intensive tensions in gel precipitates and favours a decrease in the size of the agglomerates.
7. The time of keeping the precipitate in a mother solution upon completion of the reaction should be sufficiently short to exclude the agglomerate growth in the course of ageing.

Experimental Procedure

The precipitation of amorphous hydroxides was performed from 0.1M diluted salt solution. A 1M ammonium aqueous solution was used as a precipitant. The salt solution was added dropwise at a rate of 10 ml/min into the precipitant solution upon vigorous stirring with a many-paddled glass agitator. The precipitation was carried out at three temperatures—at room temperature 20°C, 0°C, and –5°C. At room temperature the precipitation was also performed with the use of an ultrasonic bath. The bath has an output of 50 watts and an operating frequency of 35 kHz. Immediately after the completion of the reaction, the precipitate was separated from the solution and washed using a Buchner filter. The amorphous hydroxides obtained were dried and heat treated at 550°C for 30 min. and 1000°C for 10 min.

After the calcination and heat treatment, the samples were examined using differential thermal analysis (DTA) by a MOM-3 derivatograph on air. The powder samples were heated up to 1000°C in an electrical furnace at a rate of 10°C/min. The temperature was measured with Pt/Pt-Rh thermocouple.

X-ray powder diffraction analysis was performed on a D-500/HS Siemens diffractometer (CuK$_a$ radiation, Ni filter, 2 θ = 15–65°) at room temperature on air.

The crystallite size was calculated from the X-ray powder diffraction data. The X-ray powder diffraction curves obtained in the electron form were processed with the 'Winfit version 1.2.1 @ St. Krumm' program using the Fourier transformation. The Gaussian model was used as an approximation function of the Fourier transformation. The agglomerate size was determined using an HORIBA LA-920 laser sedimentograph.

RESULTS AND DISCUSSION

The DTA curves for amorphous hydroxides are shown in Fig. 16.8. As can be seen from the Fig. 16.8, the DTA curves for sample 1 (precipitation at 20°C) and for sample 2 (precipitation at 0°C) exhibit a pronounced endothermal effect in the temperature range 160°–165°C (Table 16.3) due to a considerable loss of structural water by crystal hydrates of amorphous hydroxides. This process is accompanied by a considerable weight loss which is observed in the thermal gravimetric (TG) curves. For sample 4 (precipitation at 20°C in an ultrasonic bath), the above-mentioned effect is shifted to the range of lower temperatures and equals 120°C.

In the DTA curves for sample 3 (precipitation at –5°C) the endothermal effect due to dehydration (220°C) is preceded by the exothermal effect at 120°C. According to TG curve, this effect is not accompanied by the weight loss, and, most likely, is associated with the formation of crystal hydrates of amorphous hydroxides (with a low water content) from hydrated adducts of variable composition which were obtained through the precipitation at –5°C.

Table 16.3. Temperatures of endothermal and exothermal effects observed during heating the synthesised powders (DTA).

Effect	Sample			
T, °C	1	2	3	4
Exo	–	–	–	–
Endo	165	165	165	160
Exo	480	480	500	460

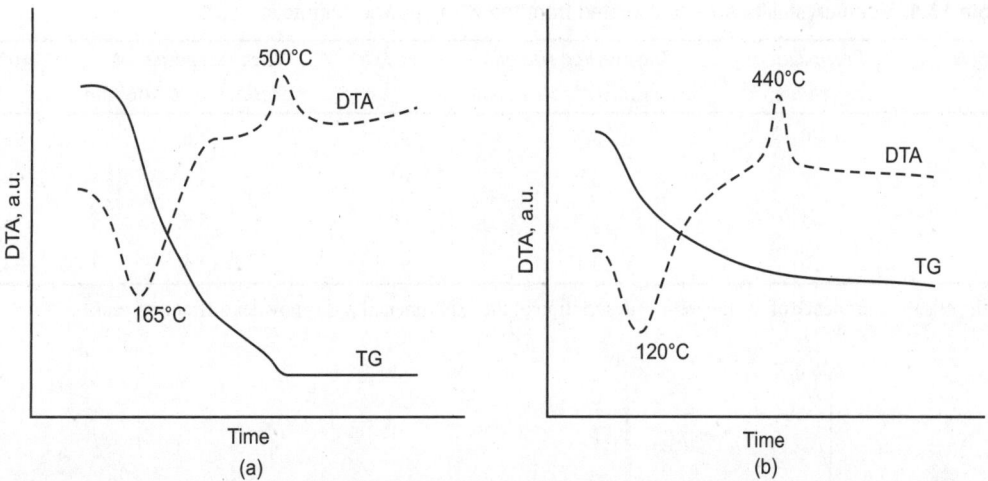

Fig. 16.8. DTA curves for the samples of amorphous hydroxides: (a) sample 1 (precipitation temperature 20°C), (b) sample 4 (precipitation temperature 20°C, in an ultrasonic bath).

The DTA curves for all the samples exhibit an exothermal effect in the temperature range 460–500°C (Table 16.3) due to crystallisation which is confirmed by the X-ray powder diffraction data (Fig. 16.9). Analysis of the TG curves indicates weight losses at these temperatures (Fig. 16.8). Therefore we can conclude that the crystallisation is accompanied by the decomposition of amorphous hydroxides.

XRD analysis of 82 mol.% ZrO_2–10 mol.% HfO_2–8 mol.% Y_2O_3 powders revealed in all the samples lines associated only with a fluorite-type cubic structure. X-ray powder diffraction patterns for samples 1 and 4 calcined at 550 and 1000°C are shown in Figs 16.9 and 16.10, respectively.

As can be seen, they differ only by the half-widths of the diffraction peaks. It is associated with the mean size of crystallites.

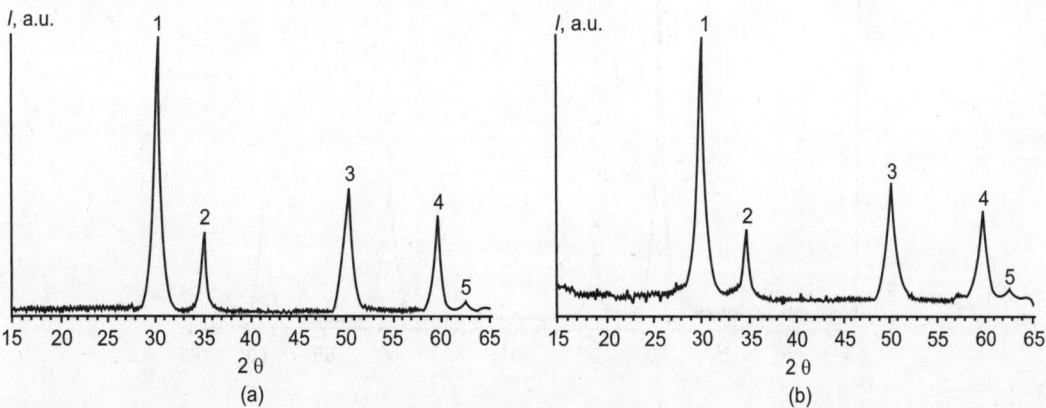

Fig. 16.9. X-ray diffraction patterns for the powders after calcination at 550°C: (a) precipitation temperature 20°C, (b) precipitation temperature 20°C, in an ultrasonic bath.

The calculated mean sizes of crystallites are given in Table 16.4. As the table indicates, the mean crystallite sizes for all the studied samples are in the nanometer range and do not exceed 5 nm for the samples calcined at 550°C and 28 nm for the samples calcined at 1000°C.

Table 16.4. Mean crystallite sizes calculated from the X-ray powder diffraction data.

Sample	Precipitation temperature, °C	The average size of crystallites at 550°C, nm	σ%	The averege size of crystallites at 1000°C, nm	σ%
1	20	3.7	92.1	27.6	98.9
2	0	5.0	93.9	23.3	98.0
3	−5	2.7	90.8	26.5	96.8
4	20	3.1	91.6	19.6	94.5

(*Note*: σ is the confidence of the theoretical description of the experimental X-ray powder diffraction peak).

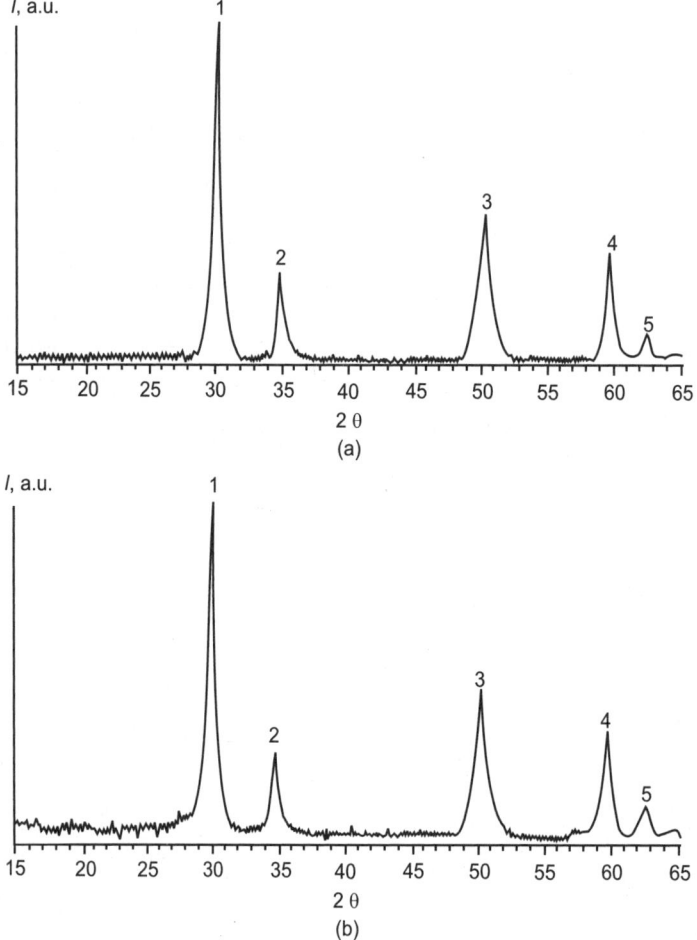

Fig. 16.10. X-ray diffraction patterns for the powders after calcination at 1000°C: (a) precipitation temperature 20°C, (b) precipitation temperature 20°C, in an ultrasonic bath.

The investigation of the agglomeration processes of nanoparticles obtained can be summarised as follows: the mean agglomerate sizes at precipitation temperature of 20°, 0°, and −5°C are equal to

14.41, 5.91, and 5.03 μm, respectively. For the samples precipitated at 20°C in an ultrasonic bath the mean agglomerate size is equal to 600 nm.

Thus, the synthesis of low-agglomerated nanoprecursors in the ZrO_2-HfO_2-Y_2O_3 system are summarised below:

1. Using the back co-precipitation method, 82 mol.% ZrO_2-10 mol.% HfO_2-8 mol.% Y_2O_3 powders were prepared. After calcinations at 550°C the mean crystallite size in the powders doesn't exceed 5 nm. A decrease in the precipitation temperature and the use of ultrasound do not affect the mean crystallite size in the powders.

2. A fluorite-type cubic structure only is observed for all the studied samples after heat treatment at $T > 550°C$.

3. A decrease in the precipitation temperature causes a decrease in the mean agglomerate size in the powders under investigation from 14.4 μm at 20°C to 5 μm at −5°C. The use of ultrasound allows to obtain the precursor powder with the mean agglomerate size of 600 nm.

The results presented above are obtained by using nanoprecursors reveal the technological advantages in various ways. Offering clean separation of n-TPP and its derivatives and enhancement in catalytic efficiency in hydrogen-oxygen combination reaction are clear examples of enhanced surface activity. Synthesis of $YBa_2Cu_4O_8$ under the condition of normal oxygen pressure is an indication of increased chemical reactivity. Lowering of sintering temperature in PZT supports enhancement in densification rate with reduction in precursor size. Growth of continuous thin films of $PbTiO_3$ and $PbZrO_3$ using spin coating is the best illustration of surface coating efficiency. Looking at the technical advantages of using nanoprecursors, there is no wonder if many more laboratories get involved in the research activities related to nanoparticles.

Chapter 17

Core/Shell Nanoparticles

INTRODUCTION

The outstanding potential of core/shell nanoparticles stems from the ability to obtain structures with combinations of properties that neither individual material possesses. Although spherical gold nanoparticles generally have a surface plasmon resonance at a wavelength of about 520 nm, a spherical silica core with a gold shell offers a very highly tunable plasmon wavelength depending on the thickness of the shell and the core diameter.

While silica core/gold shell nanoparticles have been fabricated previously by chemical reduction of gold ions, we attempt to generate these structures by photochemical reduction and by nanosphere lithography. These techniques could provide finer control over the properties of the shell. In addition to using gold, the nanosphere lithography method was also attempted with silver. Optical spectroscopy and electron microscopy were used in the characterisation of these nanoshells. Although both techniques are able to generate nanoparticles on the silica core, the current experimental conditions fail to provide a smooth shell.

Although the inherent nature of core/shell nanoparticles makes them potentially very useful in many areas, it is the biomedical applications that are getting the most attention. Here research is focused on new ways of fabricating nanoparticles with a silica core and a shell of gold or silver, a concept pioneered by Halas. The motivation for developing this structure lies with the surface plasmon resonance, which is the collective oscillation of free electrons in an applied electromagnetic field, resulting in intense absorption and scattering. Using a silica core, the plasmon becomes very sensitive to the shell thickness. Using silica/gold nanoshells with an 800 nm plasmon, Halas found that NIR laser light (808 nm) focused on tumours with nanoshell accumulation results in localised heat delivery that selectively kills the tumours cells. However, the considerable surface roughness of typical core/shell structures leaves room for improvement. This research attempted to apply the methods of photochemical reduction and nanosphere lithography to produce the nanoshells.

EXPERIMENTAL PROCEDURE

The photochemical reduction is a combination of the procedures of Halas and Eustis. Small gold colloid (1–2 nm) was prepared as described by Duff. Silica nanoparticles 110 nm in diameter were suspended in ethanol and functionalised with 3-aminopropyltriethoxysilane (APTES). The APTES covered silica was purified and the pH was adjusted. The negatively charged gold seeds absorbed onto the positively charged amine groups on the silica surface. A solution was prepared of the gold-modified silica and

HAuCl$_4$ in ethylene glycol. Irradiation was performed with a mercury xenon lamp and a band filter selecting wavelengths from 230 nm to 400 nm. Absorbance spectra were measured with a Shimadzu UV-3103-PC spectrophotometer. Nanoshells were analysed using a JEOL100 transmission electron microscope.

The nanosphere lithography method was performed as described by Van Duyne. The unmodified silica nanoparticles were used. A PVD75 filament evaporator was used to deposit 5, 10, and 20 nm of gold and silver on the silica. Without removing the silica layer, absorbance spectra were measured with a Beckman DU 650 Spectrophotometer. Samples were further analysed using a LEO 1530 Scanning Electron Microscope.

RESULTS AND DISCUSSION

Figure 17.1 shows a sample of the photochemical reduction after irradiation. The vast majority of silica nanoparticles appeared as seen here, characterised by incomplete coverage of the silica surface. Some silica particles were observed with complete gold coverage of the silica, shown in the inset. However, these particles had very rough shells, due to the growth of clusters of gold nanoparticles that encased the silica. The low yield of core/shell structures explains the absence of a plasmon peak near 800 nm, and the absorption peak at 540 nm can be attributed to spherical gold nanoparticles formed in solution and on the silica.

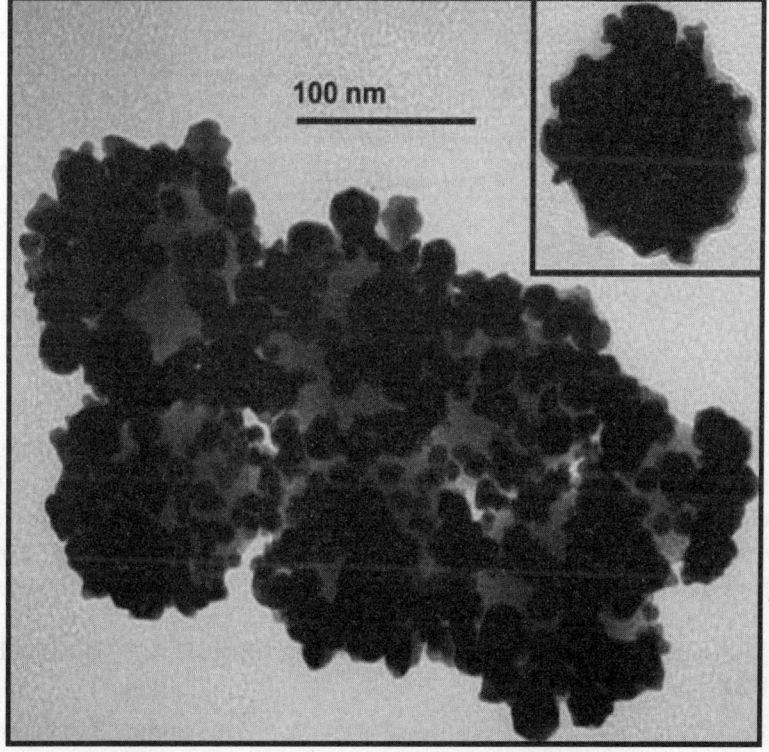

Fig. 17.1. TEM images from photochemical reduction.

Though not yet ideal, this result demonstrates that the photochemical reduction is capable of reducing gold onto the surface of silica nanoparticles. In addition, without the gold seed, no gold is reduced onto the surface of the silica.

With nanosphere lithography, as the deposition thickness was increased for both gold and silver, the plasmon peak red-shifted and the absorbance increased (silver is shown in Fig. 17.2). Both of these facts are consistent with theoretical calculations for the growth of a shell. As seen in Fig. 17.3 from left to right, at 5 nm thickness, quasi-spherical silver nanoparticles pepper the surface of the silica and at 10 nm, almost completely cover the silica surface. However, the silver particles fail to coalesce to form a shell and at 20 nm thickness, the silica is buried under large, irregularly shaped silver particles.

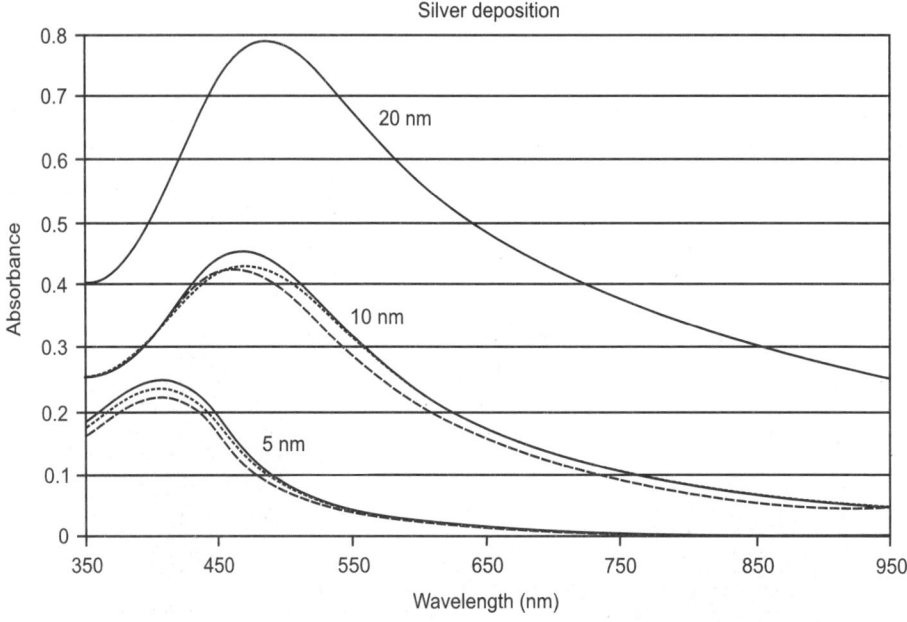

Fig. 17.2. Absorbance spectra for varying deposition thicknesses.

FUTURE WORK

To maximise the amount of gold reduced on the surface of the silica rather than in solution during the photochemical reduction, the initial concentration of $HAuCl_4$ might be reduced or $HAuCl_4$ might be added incrementally. The effects of changing lamp power, using a narrow band filter, or otherwise altering the speed of the reaction, should also be explored. The nanosphere lithography technique may require better monodispersity of the silica nanoparticles or modification of the silica surface. For both methods, it will be necessary to encourage the formation of a smooth shell instead of isolated gold particles.

X-RAY DIFFRACTION ON CORE-SHELL NANOPARTICLES FOR A PRECISE STRUCTURE DETERMINATION

The interest in semiconductor nanoparticles has strongly increased during the last decades. This is partly due to their potential for applications but also to the fact that their properties are of high interest in fundamental research. In particular particles with sizes below 5 nm are striking since they build the bridge from molecular to solid state physics. One aspect for working out a proper picture of nanoparticles is the precise determination of their geometric parameters. A method well suited for this is X-ray diffraction (XRD). Yet, the analysis of the data is a crucial point and requires special care. Many approaches use common solid state techniques to fit the data, like, e.g. the Rietveld refinement. But due to the fact that only several hundred atoms make up one particle, a solid state based analysis must fail in determining detailed structural characteristics. Moreover, core-shell particles cannot be considered at all with the conventional data analysis techniques.

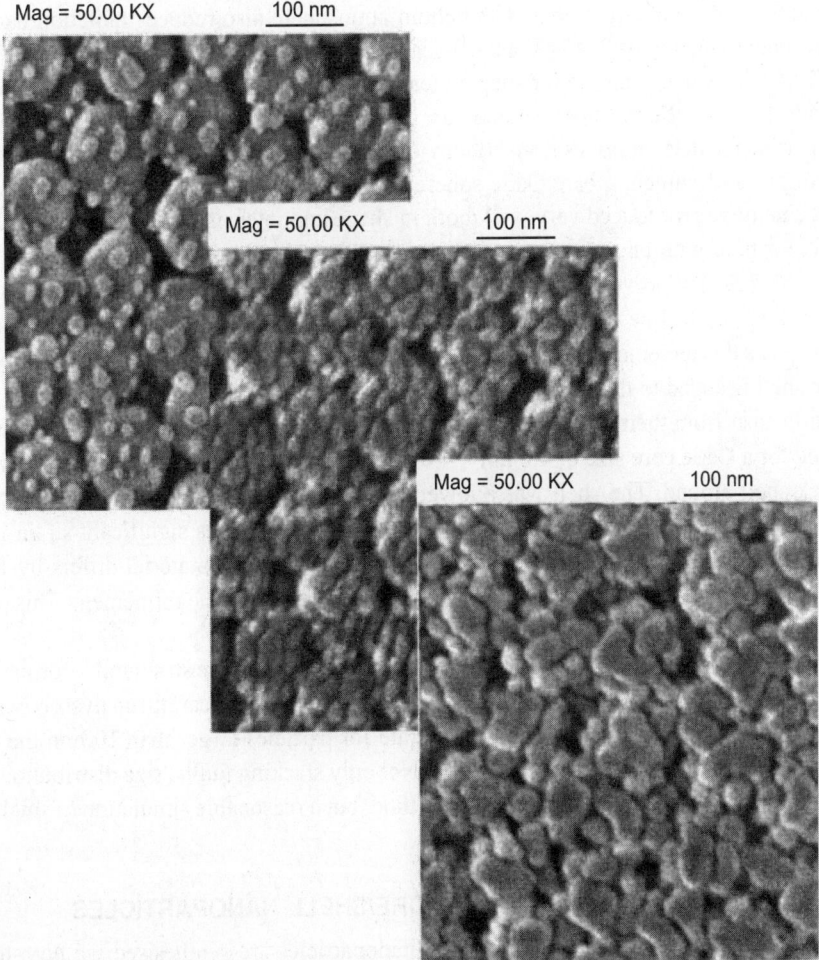

Fig. 17.3. SEM images of silver deposition.

Since techniques that assume an infinite and quasi-fixed crystal structure fail, new approaches must allow for more or different degrees of freedom. In our case, this concerns the shape of the particles, surface strain and relaxation, and stacking sequences differing from those of the bulk structure. With the software package DIFFEV it is possible to directly determine these intrinsic parameters. An entire nanoparticle is modelled and its diffractogram is calculated via the Debye formula. To take into account particle distributions, like a size or stacking fault distribution, the diffraction data of an ensemble of several particles are averaged. The whole procedure is embedded in an evolutionary algorithm to automatically refine the model. As one is independent of any constraints in setting up the atomic models, core-shell particles can easily be realised as well.

Here we show XRD measurements on CdSe/ZnS core-shell particles and several refined diffractograms, which were calculated from different nanoparticle models. The nanoparticles were wet-chemically synthesised and the experiment was performed at beamline BW2 at HASYLAB. To minimise the background scattering, the sample was drop-coated on a silicon wafer and transferred into a helium-flooded chamber for the measurement. The helium atmosphere also reduces radiation damage in the sample. At a photon energy of 9645 eV a θ-2θ-scan from $\theta \approx 0.6$ to 47° was performed.

Figure 17.4 shows one example for such a measurement (empty dots). Additionally, the calculated diffractograms for four different atomic models are presented. The agreement of the different calculations with the experimental data improves from bottom to top. For the bottom-most diffractogram the simplest possible model was assumed, a bare CdSe sphere having bulk-structure and without a ZnS shell. All main peaks cannot be reproduced very well, both in shape and intensity. The fit becomes hardly better when a CdSe sphere with an epitactic ZnS shell is used. Only, the intensity ratio of the triple peak ($q = 2.7$ Å$^{-1} - 3.7$ Å^{-1}) improves a little. Proceeding to the third refined model, a CdSe cylinder without a shell or stacking faults, one can recognise that this fit reproduces the shape of the first peak slightly better but worsens the agreement of the triple peak with the data. However, the situation improves when an epitactic shell is added to this cylindrical model, stacking faults are considered and the shell atoms are allowed to shift from their original position in radial direction. This best fitting particle ensemble was obtained for a CdSe core size of 4.0 nm × 4.0 nm for diameter and height, respectively and with a 0.6 nm thick shell around. The shell was allowed to relax its lattice parameters in the fit and which resulted in values of 95 per cent of those of bulk CdSe. This indicates significant strain in the shell, since the difference of the lattice parameters for ZnS and CdSe bulk material differs by 10 per cent. Moreover, a stacking fault probability of about 2 per cent was found in the refinement. This corresponds to a 20 per cent probability for one stacking fault per particle.

In summary, we could demonstrate that it is possible to obtain precise structural information on very small core-shell nanoparticles from XRD measurements, provided that a careful analysis is being applied. While conventional methods are sufficiently accurate for particles larger than 10 nm, the analysis of very small nanoparticles requires new techniques. Not only stacking faults, size distributions and other ensemble parameters can be addressed with our method, but a reasonable simulation of small core-shell particles is enabled as well.

TAILORING MAGNETIC PROPERTIES OF CORE/SHELL NANOPARTICLES

Bimagnetic FePt/MFe$_2$O$_4$(M = Fe, Co) core/shell nanoparticles are synthesised via high-temperature solution phase coating of 3.5 nm FePt core with MFe$_2$O$_4$ shell. The thickness of the shell is controlled from 0.5 to 3 nm. An assembly of the core/shell nanoparticles shows a smooth magnetisation transition under an external field, indicating effective exchange coupling between the FePt core and the oxide shell. The coercivity of the FePt/Fe$_3$O$_4$ particles depends on the volume ratio of the hard and soft phases, consistent with previous theoretical predictions. These bimagnetic core/shell nanoparticles represent a class of nanostructured magnetic materials with their properties tunable by varying the chemical composition and thickness of the coating materials.

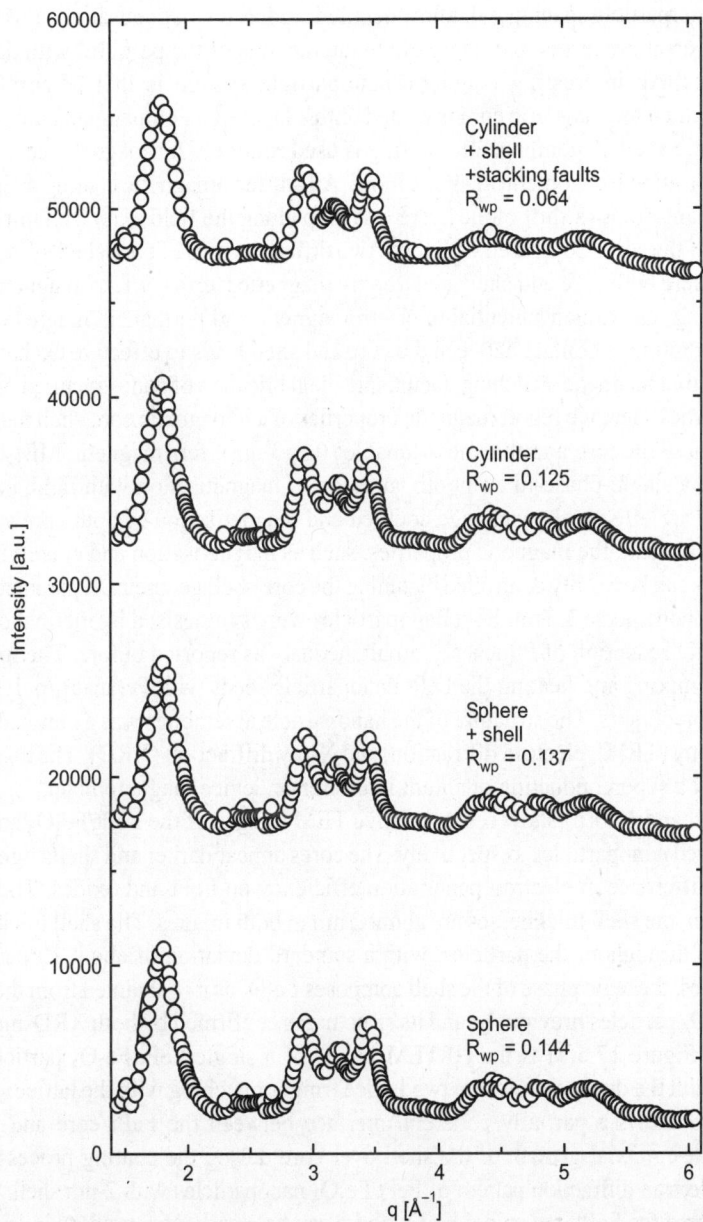

Fig. 17.4. Measured diffraction pattern of CdSe/ZnS core-shell nanoparticles together with calculations of four different simulated atomic models, plotted vs. $q = 4\pi/\lambda^*\sin(\theta)$.

Nanoscale magnetism has stimulated great interest due to its importance in mapping the scaling limits of magnetic information storage technology and understanding spin-dependent transport phenomena. Recent progress in the production of nearly monodisperse magnetic nanoparticles from metallic Fe, Co and Ni, to iron oxides, MFe_2O_4 and FePt and CoPt intermatellic compounds provides various systems suitable for nanomagnetic studies. An assembly of monodisperse magnetic nanoparticles with controlled interparticle spacing will allow detailed studies on magnetisation, anisotropy, as well as magnetisation reversal processes and interparticle interactions of the particles with different sizes and surface properties. An interesting magnetic nanoparticle system is that of core/shell structured nanoparticles in which the magnetic core is coated with a layer of a nonmagnetic, antiferromagnetic, or ferro/ferri-magnetic shell. A nonmagnetic coating is used routinely for magnetic core stabilisation and surface functionalisation for biomedical applications. An antiferromagnetic coating over a ferromagnetic core leads to exchange bias (a shift of the hysteresis loop along the field axis), and improvements in the thermal stability of the core. Compared with these two different types of core/shell systems, a bimagnetic core/shell one, where both core and shell are strongly magnetic (ferro- or ferri-magnetic) is less studied yet more interesting due to their potential in electromagnetic and permanent magnetic applications. In such a system, the intimate contact between the core and shell leads to effective exchange coupling and therefore cooperative magnetic switching, facilitating the fabrication of nanostructured magnetic materials with tunable properties. Here we report magnetic properties of a bimagnetic core/shell nanoparticle system with ferromagnetic FePt core and thickness-tunable (0.5–3 nm), ferrimagnetic MFe_2O_4 (M = Fe, Co) shell. We observe a single-phase-like smooth variation of magnetisation with field, indicating that the core and the shell are effectively exchange coupled and magnetisation of both core and shell reverses cooperatively. As a result, the magnetic properties, such as magnetisation and coercivity, of these core/shell nanoparticles can be readily controlled by tuning the core/shell geometrical parameters and chemical compositions. Monodisperse 3.5 nm FePt nanoparticles were synthesised by thermal decomposition of $Fe(CO)_5$ and polyol reduction of $Pt(acaca)_3$ simultaneously as reported before. The iron oxide coating was achieved via mixing and heating the FePt nanoparticle seeds with $Fe(acac)_3$/polyol, or $Co(acac)_2$/$Fe(acac)_3$/polyol precursors. The structure of the nanoparticle assemblies was examined by transmission electron microscopy (TEM), electron diffraction, and X-ray diffraction (XRD). The magnetic properties were measured by a superconducting quantum interference device magnetometer.

Figures 17.5(a) and 17.5(b) show representative TEM images of the $FePt/Fe_3O_4$ and $FePt/CoFe_2O_4$ core/shell structured nanoparticles, respectively. The cores appear darker and shells lighter in the images due to the large difference in electron penetration efficiency on FePt and oxides. The FePt core has a diameter of 3.5 nm, the shell thicknesses are about 2 nm in both images. The shell thicknesses observed are quite uniform throughout the particles, with a standard deviation of about 10 per cent. For FePt/Fe_3O_4 nanoparticles, the main phase of the shell composes Fe_3O_4 as it is obtained from the same chemistry used to make Fe_3O_4 particles previously, and its structure is confirmed by both XRD and high resolution TEM (HRTEM). Figure 17.6(a) is the HRTEM image of a single FePt/Fe_3O_4 particle. It reveals the crystalline shell with the distance between two lattice fringes matching with the lattice spacing of Fe_3O_4. The HRTEM also shows a partially coherent interface between the FePt core and the Fe_3O_4 shell, indicating possible epitaxial growth of the shell over core during the coating process. Figure 17.6(b) shows a typical electron diffraction pattern of FePt/Fe_3O_4 nanoparticles with 2 nm shell. Diffraction rings from both disordered fcc FePt and spinel Fe_3O_4 phase can be clearly observed, as indexed in the figure.

Both FePt and Fe_3O_4 nanoparticles are ferromagnetic at 10 K. The coercivity for 3.5 nm FePt nanoparticles is 5.5 kOe, while that for 4 nm Fe_3O_4 is only 200 Oe. The large coercivity from disordered fcc FePt nanoparticles likely originates from a uniaxial surface anisotropy. The FePt/Fe_3O_4 core/shell nanoparticle is therefore a two-phase system consisting magnetically of a hard (FePt) and a soft (Fe_3O_4) phase. Figure 17.7(a) shows the 10 K hysteresis loop of the 3.5 nm FePt/1 nm Fe_3O_4 core/shell nanoparticle assembly. Despite consisting of both hard and soft phases, the hysteresis loop shows a single-phase-like behaviour, with the magnetisation changing with the applied field smoothly. The coercivity is determined to be 2.3 kOe, a value in between that of FePt and Fe_3O_4. This indicates that the intimate contact between the FePt core and Fe_3O_4 shell leads to an effective interphase exchange coupling, which results in cooperative magnetisation switching of the two phases.

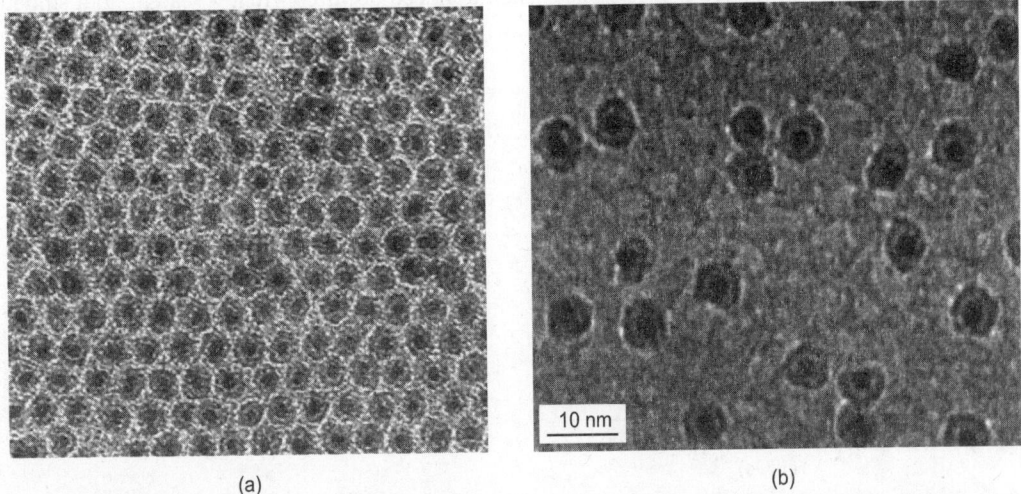

(a)   (b)

Fig. 17.5. TEM images of (a) FePt/Fe_3O_4 and (b) FePt/$CoFe_2O_4$ core/shell structured nanoparticle assembly with shell thickness of 2 nm.

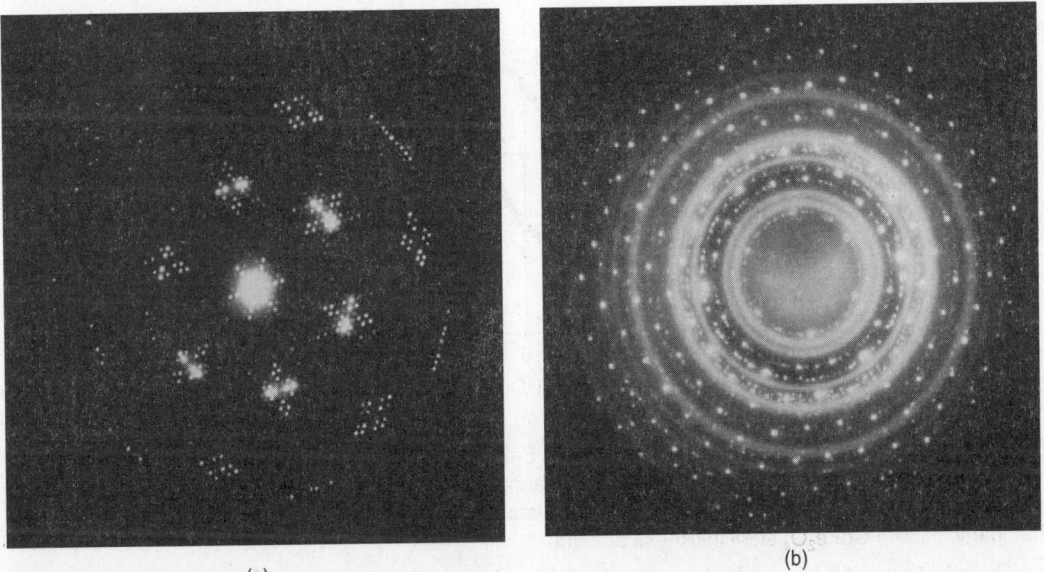

(a)   (b)

Fig. 17.6. (a) HRTEM image of a single FePt/Fe_3O_4 nanoparticle, and (b) electron diffraction pattern of an assembly of FePt/Fe_3O_4 nanoparticles.

Earlier theoretical studies suggest that for hard and soft phases to reverse cooperatively in a hard—soft composite system, the critical dimension of the soft phase (t_s) should be less than twice the domain wall width (δ_W) of the hard phase. Using the measured 10 K H_c of the FePt nanoparticles as an approximation for the effective uniaxial anisotropy field, and $K_u \sim M_s H_K$, the effective anisotropy constant K_u for the FePt nanoparticles is calculated to be on the order of 5×10^6 erg/cm^3. Plugging this K_u value into $\delta_W \sim (A/K_u)^{1/2}$, where A is the exchange constant ($\sim 1 \times 10^{-6}$ erg/cm), we can estimate δ_w to be about 10 nm. The Fe$_3$O$_4$ shell in this study is less than 3 nm and well within the limit $t_s \approx 2\,\delta_W$ of 20 nm. Hence, the switching of the hard and soft phase should indeed occur coherently, leading to a smooth magnetisation transition.

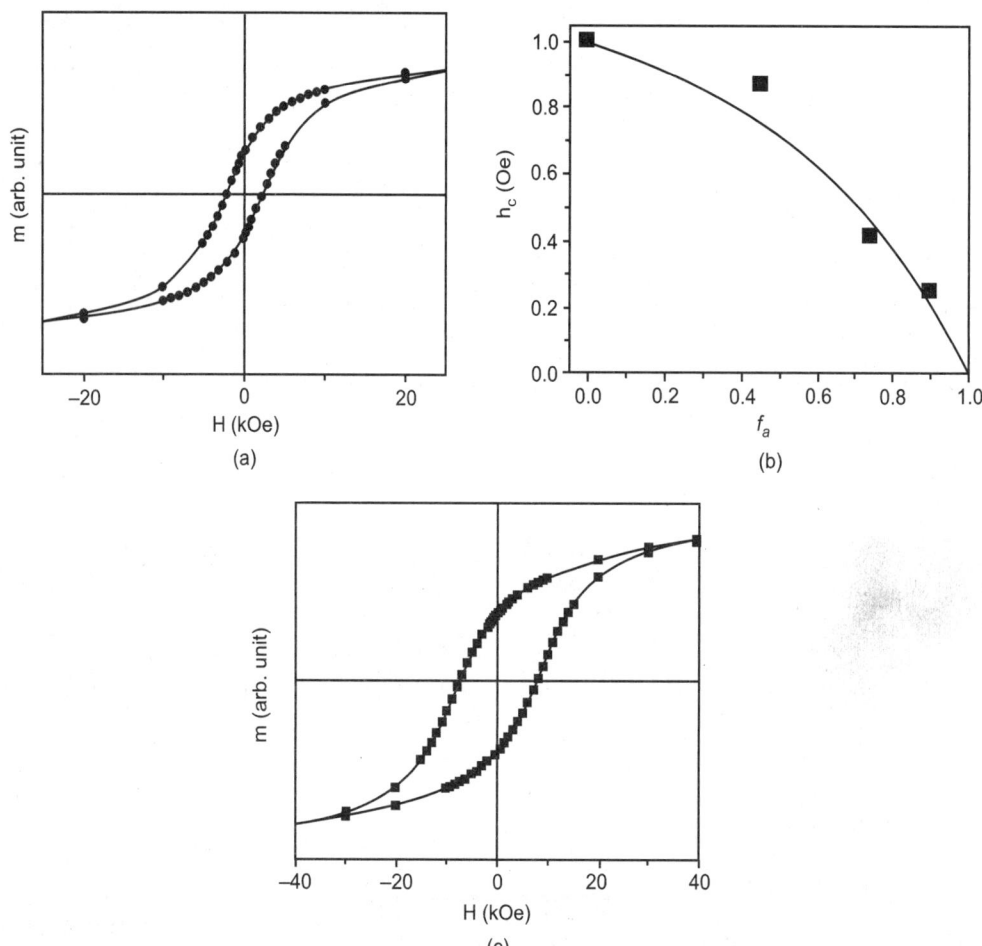

(a)

(b)

(c)

Fig. 17.7. (a) A typical magnetic hysteresis loop of an FePt/Fe$_3$O$_4$ nanoparticle assembly with shell thickness of 1 nm; (b) normalised coercivity h_c of FePt/Fe$_3$O$_4$ nanoparticles as a function of Fe$_3$O$_4$ volume fraction [the curve is calculated from Eq. (17.2), and dots are data points]; and (c) hysteresis loop of FePt/CoFe$_2$O$_4$ nanoparticles with CoFe$_2$O$_4$ shell thickness of 2 nm.

The coercivity of the core/shell particles is observed to decrease with increasing shell thickness. The coercivity of a hard–soft exchange-coupled system with collinear easy axis is

$$H_c = 2\,\frac{K_H f_H + K_S f_S}{M_H f_H + M_S f_S} \qquad \ldots (17.1)$$

where, K is the anisotropy constant, M the saturation magnetisation, f the volume fraction, and the subscripts H and S denote the hard and soft phases, respectively. Equation 17.1 is only strictly valid for uniaxial anisotropy, with the easy axes of the hard and soft phases being collinear. Nevertheless, since K_H of 5×10^6 erg/cm^3 is about 20 times larger than that of Fe$_3$O$_4$, we can ignore the term $K_S f_S$ in Eq. 17.1 for the thickness range we studied without introducing significant error. Based on this, Eq. 17.1 can be transformed into

$$h_c \approx \frac{1}{1 + \dfrac{M_S}{M_H} \dfrac{f_S}{1 - f_S}}, \qquad \qquad ...\ (17.2)$$

where, h_c is the coercivity of the core/shell particles normalised by the coercivity H_{cH} of the hard phase ($h_c = H_c / H_{cH}$). If the above-presented analysis is correct, h_c should decrease monotonically with increasing volume fraction of the soft phase, following Eq. 17.2. Figure 17.7(b) plots h_c of the FePt/Fe$_3$O$_4$ nanoparticles as a function of f_S, in which the curve is calculated from Eq. 17.2 and dots are data points. We can see that they match each other reasonably well. This indicates that the coercivity of the FePt/ Fe$_3$O$_4$ nanoparticles depends only on the volume ratio of core/shell, not on the actual size or thickness of the core and the shell.

The situation of FePt/CoFe$_2$O$_4$ nanoparticles is different from that of FePt/Fe$_3$O$_4$. CoFe$_2$O$_4$ has much larger magnetocrystalline anisotropy than Fe$_3$O$_4$, and exhibits higher coercivity at low temperatures. One study shows that at 10 K, the 8 nm CoFe$_2$O$_4$ nanoparticle assembly has an H_c of 12 kOe and the 18 nm one has H_c of 21 kOe. Compared to FePt/Fe$_3$O$_4$, the hard–soft phases in FePt/Fe$_2$CoO$_4$ core/shell system are therefore reversed, with FePt being magnetically softer and CoFe$_2$O$_4$ harder. Figure 17.7(c) shows the 10 K hysteresis loop of FePt/Fe$_2$CoO$_4$ with 2 nm shell. It can be seen that the H_c increases from 5.5 kOe for FePt to 8 kOe, as expected for such an exchange-coupled system. Since the anisotropy of the hard and soft phase in this system is rather close, Eq. 17.2 cannot be used to describe such systems.

In conclusion, a class of bimagnetic core/shell nanoparticles can be readily synthesised via solution phase chemistry. Magnetic properties of these core/shell nanoparticles can be tailored by controlling the core/shell dimensions, and by tuning the material parameters of both core and shell. Such systems may show interesting nanomagnetism emerging from the exchange-coupling between the core and the shell, and may yield finely tailored materials for various nanomagnetic applications.

Chapter 18

Kinetics and Energetics in Nanolubrication

INTRODUCTION

Nanotribology is a branch of tribology which studies friction phenomenon at the nanometer scale (see nanotechnology, nanomechanics). The distinction between nanotribology and tribology is primarily due to the involvement of atomic forces in the determination of the final behaviour of the system.

Gears, bearings, and liquid lubricants can reduce friction in the macroscopic world, but the origins of friction for small devices such as micro- or nanoelectromechanical systems (NEMS) require other solutions. Despite the unprecedented accuracy by which these devices are now-a-days designed and fabricated, their enormous surface-volume ratio leads to severe friction and wear issues, which dramatically reduce their applicability and lifetime. Traditional liquid lubricants become too viscous when confined in layers of molecular thickness. This situation has led to a number of proposals for ways to reduce friction on the nanoscale, such as superlubricity and thermolubricity.

The micro/nanoelectromechanical systems (MEMS/NEMS) need to be designed to perform expected functions typically in millisecond to picosecond range. Expected life of the devices for high speed contacts can vary from few hundred thousand to many billions of cycles, e.g. over a hundred billion cycles for digital micromirror devices (DMDs), which puts serious requirements on materials. For BioMEMS/BioNEMS, adhesion between biological molecular layers and the substrate, and friction and wear of biological layers may be important. There is a need for development of a fundamental understanding of adhesion, friction/stiction, wear, and the role of surface contamination, and environment. Most mechanical properties are known to be scale dependent. Therefore, the properties of nanoscale structures need to be measured. MEMS/NEMS materials need to exhibit good mechanical and tribological properties on the micro/nanoscale. There is a need to develop lubricants and identify lubrication methods that are suitable for MEMS/NEMS. Methods need to be developed to enhance adhesion between biomolecules and the device substrate. Component-level studies are required to provide a better understanding of the tribological phenomena occurring in MEMS/NEMS. The emergence of micro/nanotribology and atomic force microscopy-based techniques has provided researchers a viable approach to address these problems. This chapter presents a review of micro/nanoscale adhesion, friction, and wear studies of materials and lubrication studies for MEMS/NEMS and BioMEMS/BioNEMS, and component-level studies of stiction phenomena in MEMS/NEMS devices.

Lubrication, one of mankind's oldest engineering disciplines, in the 19th century gained from Reynolds' classical hydrodynamic description a theoretical base unmatched by most of the theories developed in Tribology to date. In the 20th century, however, increasing demands on lubricants shifted the attention from bulk film to ultrathin film lubrication. Finite size limitations imposed constraints on the lubrication process that were not considered in bulk phenomenological treatments introduced by Reynolds. At this point, as is common in many engineering applications, empiricism took over. Functional relationships derived from the classical theories were tweaked to accommodate the new situation of reduced scales by introducing 'effective' or 'apparent' properties.

With the inception of nanorheological tools of complementary nature in the later decades of the 20th century (e.g. the surface forces apparatus and scanning force microscopy) tribology entered the realm of nanoscience. Through an increasing confidence in experimental findings on the nanoscale, kinetic and energetic theories incorporated interfacial and molecular constraints. The very fundamentals have been challenged in recent years. Researchers have realised that bulk perceptions, such as solid and liquid, are defied on the nanoscale. The reduction in dimensionality of the nanoscale imposes constraints that bring into question the use of classical statistical mechanics of decoupled events. The diffusive description of lubrication is failing in a system that is thermodynamically not well-equilibrated.

The challenge any nanotechnological endeavour encounters is the development of a theoretical framework based on an appropriate statistics. In tribology this is met with spectral descriptions of the dynamic sliding process. Statistical kernels are being developed for probability density functions to explain anomalous transport processes that involve long-range spatial or temporal correlations. With such theoretical developments founded in nanorheological experiments, a more realistic foundation will be laid to describe the behaviour of lubricants in the confined geometries of the nanometer length scale.

Since technology is driving lubricant films to molecular thickness, kinetic friction and its dependence on the sliding parameters—especially the sliding velocity—have become of great interest. The complexity of the frictional resistance in lubricated sliding is illustrated in Fig. 18.1 with a Stribeck curve. Various regimes of lubrication can be identified in the Stribeck curve.

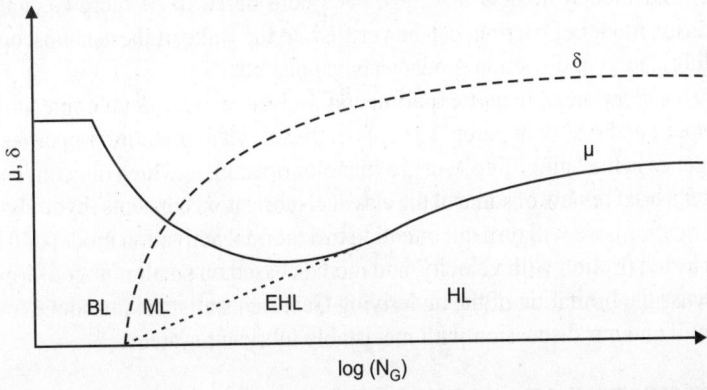

Fig. 18.1. Stribeck curve (schematic) relates the fluid lubricant thickness, δ, and the friction coefficient μ to the Gumbel number $N_G = \eta\,\omega P^{-1}$; i.e. the product of the liquid bulk viscosity, η, the sliding speed (or more precisely the shaft frequency), ω, and the inverse of the normal pressure, P. BL: Boundary lubrication, ML: Mixed lubrication, EHL: Elastohydrodynamic lubrication, HL: Hydrodynamic lubrication.

They express to what degree the hydrodynamic pressure is involved in the lubrication process. In the ultra-low speed regime, called the boundary lubrication regime, no hydrodynamic pressure is built up in the lubricant. Consequently the load is carried by contact asperities coated with adsorbed lubricant molecules. If the speed is raised, a hydrodynamic pressure builds that leads to a mixed lubrication in which the load is carried by both asperities and hydrodynamic pressure. At even higher speeds, elastic contributions of the solid surfaces have to be considered paired with hydrodynamic pressure effects (elastohydrodynamic lubrication), until only hydrodynamic lubrication matters. Hence, the Stribeck curve combines various aspects of lubrication. The curve cannot be discussed without considering the lubricant thickness and the different models of asperity contact sliding.

In one of the first comprehensive physical models of 'dry' friction, Bowden and Tabor introduced a plastic asperity model, in which the material's yield stress and adhesive properties play an important role. Considering this model, which depends on surface energies and mechanical yield properties paired with all the properties that come along with a surface adsorbent lubricant, one can hardly grasp the difficulty level involved in describing the frictional kinetics in lubrication.

Past and current engineering challenges in lubrication have been met with great and complex empiricism. The theoretical modelling of lubrication junctions generally involved only bulk property considerations with inadequately known adsorption mechanisms. The complexity of today's lubricants, most of them, such as motor oil, a product of empirical design over many years, increased exponentially, making it very difficult to meet future challenges. The problem of empiricism is that conventional laws and perceptions are unchallenged. Effective quantities are invented (e.g. effective viscosity), exponential fitting parameters are introduced (e.g. Kohlrausch relaxation parameter), and terminologies such as solid and liquid are taken as granted. Progress based on empiricism is only incremental and rarely revolutionary. One of the reasons for empiricism is a lack of access to a system with fewer and better controlled parameters. In lubrication sliding that challenge has been addressed over the last two decades with the inception of the surface forces apparatus (SFA) by Tabor and scanning force microscopy (SFM) by Binnig. These two instrumental methods allow lubrication studies where roughness effects can be neglected, surface energies controlled, and wear from wearless friction distinguished. Lubricant properties can be studied at nearly mathematically described boundaries, atomistic friction events can be recorded, and fundamental models that have been considered to be mere Gedanken experiments, such as the Tomlinson model of friction, can be verified. In the wake of these nanoscopic tools, exciting new theoretical lubrication and friction models have appeared.

This chapter considers these recent experimental and theoretical developments with a particular focus on sliding speed and real or apparent changes in the lubricant material properties. We will discuss kinetics and energetics in the 'simplified' world of nanolubrication, in which our conventional perception is challenged. After a brief review of some of the classical lubrication concepts (hydrodynamic lubrication and boundary lubrication), we will turn our attention to a thermal activation model of friction, functional behaviour of lubricated friction with velocity, and models based on small non-conforming contacts. We will critically discuss the limitation of the underlying Gaussian statistics, introduce fractal dynamics in lubrication, and will end our discussion with metastable lubricant systems.

FROM BULK TO MOLECULAR LUBRICATION

Hydrodynamic Lubrication and Relaxation

In the classical theories of tribology by da Vinci, Amonton, and Coulomb, not much attention was given to the dependence of kinetic friction on the sliding velocity. This clearly changed in the 19th century during the first industrial revolution, at which time lubricants became increasingly important, for instance, in ball and journal bearings. It was Petrov, Tower, and Reynolds who established that the liquid viscous shear properties determine the frictional kinetics. Reynolds combined the pressure-gradient determined

Poisseuille flow with the bearing surface induced Couette flow assuming, based on Petrov's law, a noslip condition at the interface between lubricant and solid. This led to the widely used linear relationship between friction and velocity. Reynolds' hydrodynamic theory of lubrication can be applied to steady state sliding at constant relative velocity and to transient decay sliding (sliding is stopped from an initial velocity v and a corresponding shear stress τ_o), which leads to the classical Debye exponential relaxation behaviour, i.e.

$$\tau = \tau_o \exp\left(\frac{-D}{A\eta}t\right); \quad \tau_o \propto \frac{\eta}{D} \quad \quad ... (18.1)$$

D is the lubricant thickness, A the area of the slider, and η the viscosity of the fluid. We will later see that this classical exponential relaxation behaviour, obtained in a thermodynamically well mixed three dimensional medium, is distorted when the liquid film thickness is reduced to molecular dimensions.

Boundary Lubrication

Reynolds hydrodynamic description of lubrication was found to work well for thick lubricant films but to break down for thinner films. One manifestation is that for films on the order of ten molecular diameters, the stress in the film does not allow the tension to return to zero. It was also found that the motion in the steady state sliding regime was disrupted, exhibiting a stick-slip-like slider motion. Consequently, this non-Newtonian behaviour was treated with a modified viscosity parameter (effective viscosity), which was composed of the pressure, temperature and rate of shear.

The term boundary lubrication is used to describe a lubricant that is reduced in thickness to molecular dimension and effectively reduces friction between two opposing solid surfaces. Hardy recognised that molecular properties, such as molecular weight and molecular arrangement, are governing the frictional force. This confined concept of lubrication, often visualised by two highly ordered opposing films with shear taking place somewhere in between the two layers, contains many of the rate dependent manifestations of frictional sliding, e.g. stick-slip, ultra-low friction, transitions from high to low friction, phase transitions, dissipation due to dislocations (e.g. gauche and *cis*-transformations), and memory effects.

Boundary lubrication was found to be in many respects unique. In macroscopic experiments, which involved rough surfaces, friction-velocity plots resembled logarithmic functions at moderate speeds. No static stiction force peaks were observed in boundary lubricants close to zero speed. On the contrary, retractive slips could be observed upon halting, constituting a static friction coefficient exceeded by the dynamic friction coefficient. These unique manifestations of boundary lubrication were discussed in terms of a lubricated asperity junction mechanism, which associated 'an increase in the coefficient of friction with a decrease in the adsorptive coverage of the rubbing surfaces by the lubricant substance'. It was argued that in the course of the sliding process of a macroscopic slider, more adsorbed lubricant is expected to exist within the interfacial area than outside. This would lead upon halting to a relaxation process of the elastic restraints on the slider, causing the slider to a retractively slip.

Stick Slip and Collective Phenomena

Based on numerous friction experiments at the initiation of sliding with rough macroscopic contact, it was argued that the distinction between static and kinetic friction is not categorical but rather a manifestation of the apparatus. This was a widely held opinion prior to Briscoe molecularly smooth monolayer SFA experiments of aliphatic carboxylic acids and their soaps. Briscoe found that the character of sliding motion (continuous vs. discontinuous), depends not only on the apparatus but also on the properties (chemistry) of the monolayer. As in the rough boundary layer experiments discussed above, Briscoe's molecularly smooth monolayer experiments exhibited logarithmic-like friction-velocity behaviours. It was Israelachvili who, based on SFA experiments and computer simulations, provided a molecular picture of the stick-slip behaviour caused by the lubricant material. The major achievement of this work was to draw our attention to the molecular structure of the lubricant, which is often different from the bulk, and unstable during the sliding motion.

It was recognised that bulk rheology failed to describe the lubrication process. Finally addressed in the Israelachvili study were in-plane structuring of simple liquids caused by compression forces and 'freezing-melting' transitions due to shear.

The simple concept of a freezing-melting transition is based on a common perception of the two distinctive parts of a stick-slip occurrence: the solid (Hookian)-like sticking part and the liquid (Newtonian)-like slipping part. But a deformation of a solid can be both, coordinated or uncoordinated, and thus can exhibit both solid-like and liquid-like behaviour.

For instance, most of the plastic yielding processes are uncoordinated. On the other hand, slipping within a solid, along a crystal plane in a thermally activated strain-release process for instance, is a highly coordinated molecular process.

Similar arguments can be made for a liquid. For example, stick-slip behaviours were observed in more complex fluidic systems by Reiter and others who compared a molecularly 'wet' lubricant film with a 'dry' self-assembled monolayer lubricant. They concluded that sliding in liquid films is the result of slippage along an interface. In other words, the degree of molecular cooperation determined the frictional resistance. The concept of local-versus-cooperative yield to shear is briefly illustrated here with a frictional-load study of a molecularly entangled polymer melt obtained in a SFM study of Buenviaje. Each of the curves presented in Fig. 18.2(a) represents a polymer film of polyethylene co-propylene of distinctly different degree of entanglement.

Films of thickness above 230 nm exhibit the strongest entanglement strength. Films of 20 nm thickness or thinner are fully disentangled. The reason for the film thickness-dependent entanglement strength is given by the substrate distance-dependent shear strength during the spin coating process of the thin films. For entangled films SFM friction studies exhibit a critical applied load (identified by P_t, and the thickness t) that separates two friction regimes: One identified by a high friction coefficient and the other by a low friction coefficient. At loads below P_t the friction coefficients are high, indicating plastic yielding during sliding.

In these plastic regimes of sliding, molecular cooperation is low, leading to high local shear stresses compensated by local yielding of the material. Above the critical load, the friction coefficient drops, independent of the film thickness, to a low value of 3.0, corresponding to the value obtained from the fully disentangled film.

Note that the polymer molecules in the 20 nm thick film experience high substrate tangential stresses during the spin coating process. Hence the disentangled polymer molecules can be considered to be aligned preferentially along the substrate surface as sketched in Fig. 18.2(b). This leads to a decrease of the structural entropy the closer the material is to the solid substrate surface. Considering the matching friction coefficient of 0.3 above P_t for thicker films, we can assume that any entangled film above a critical load exhibits a similar molecular collective response toward shear as the 20 nm film during spin coating. The critical load and its related pressure represent a barrier that has to be overcome before a collective phenomenon is activated.

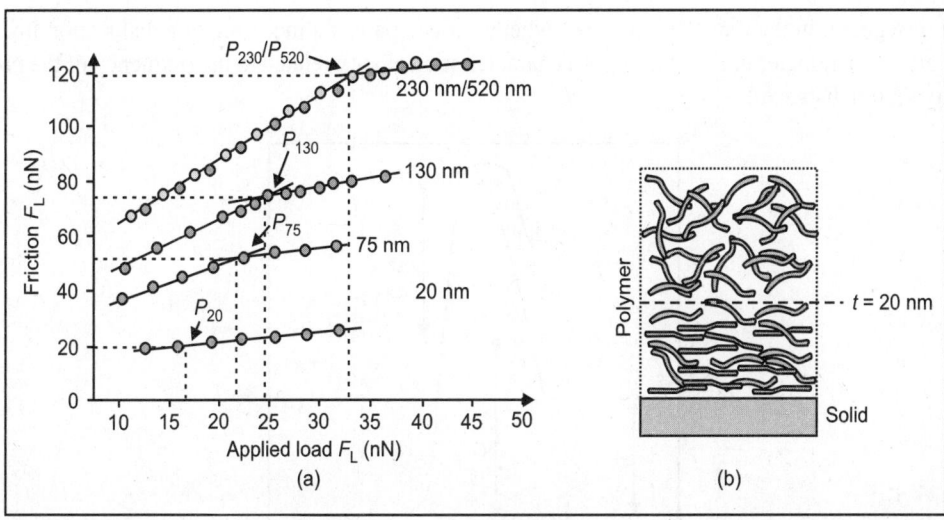

Fig. 18.2. (a) SFM friction measurements at a speed 1 μm/s: cooperative molecular response of polyethylene copropylene to frictional shear forces as a function of the applied load. P_t (t corresponds to the thickness of the polymer film) represents the critical activation load at which collective sliding is energetically more favourable than local plastic yielding. (b) Sketch of the degree of disentanglement in the vicinity to the solid substrate surface.

THERMAL ACTIVATION MODEL OF LUBRICATED FRICTION

With the discussion of shear in entangled polymer systems we have introduced structural entropy as one of the key players that affect frictional resistance in lubricants. We found that the structural entropy was affected by the load of the slider, which introduces an activation barrier in the form of a critical pressure. The terminology used here resembles the one of the Eyring theory of molecular liquid transport.

Eyring discussed a pure liquid at rest in terms of a thermal activation model. The individual liquid molecules experience a 'cagelike' barrier that hinders molecular free motion because of the close packing in liquids. To escape from the cage an activation barrier needs to be surmounted. In Eyring's model, two processes are considered in order to overcome the potential barrier: (i) shear stresses, and (ii) thermal fluctuations. The potential barrier in the thermal activation model is depicted in Fig. 18.3 indicating the barrier modification by the applied pressure force P, and shear stress τ.

Briscoe picked up on this idea to interpret the frictional behaviour observed on molecularly smooth monolayer systems. Starting from the overall barrier height $E = Q + P\Omega - \tau\phi$ that is repeatedly overcome during a discontinuous sliding motion, using a Boltzmann distribution to determine the average time for single molecular barrier-hopping, and assuming a regular series of barriers and a high stress limit ($\tau\Phi/kT > 1$), the following shear strength versus velocity v relationship was derived:

$$\tau = \frac{k_{\mathrm{B}}T}{\phi}\ln\left(\frac{v}{v_o}\right) + \frac{1}{\phi}(Q + P\Omega) \qquad \ldots (18.2)$$

The barrier height, E, is composed of the process activation energy Q, the compression energy $P\Omega$, where P is the pressure acting on the volume of the junction Ω, and the shear energy $\tau\phi$, where, τ is the shear strength acting on the stress activation volume ϕ. T represents the absolute temperature. The stress activation volume ϕ can be conceived as a process coherence volume and interpreted as the size of the moving segment in the unit shear process, whether it is a part of a molecule or a dislocation line. The most critical parameter in Eq. 18.2, v_o, is a characteristic velocity related to the frequency of the process and to a jump distance.

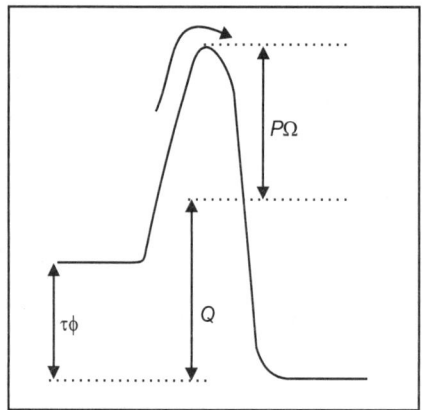

Fig. 18.3. Potential barrier in a lubricant based on Eyring's thermodynamic 'cage-model'. The normal pressure P and the shear stress τ are modifying the barrier height Q.

From Eq. 18.2 the following iso-relationships can be directly deduced:

$$\tau = \tau_o + \alpha P; \quad \tau_o = \frac{1}{\phi}\left(k_B T \ln\left(\frac{v}{v_o}\right) + Q\right); \quad \alpha = \frac{\Omega}{\phi}; \qquad \qquad \dots (18.3a)$$

at constant v, T

$$\tau = \tau_1 - \beta T; \quad \tau_1 = \frac{1}{\phi}(Q + P\Omega); \quad \beta = -\frac{k}{\phi}\ln\left(\frac{v}{v_o}\right); \qquad \qquad \dots (18.3b)$$

at constant P, v

$$\tau = \tau_2 - \theta \ln v; \quad \tau_2 = \frac{1}{\phi}(Q + P\Omega - kT \ln v_o); \quad \theta = \frac{kt}{\phi} \qquad \qquad \dots (18.3c)$$

at constant P, T.

Thus Eyring's model predicts a linear relationship of friction (the product of the shear strength and the active process area) in pressure and temperature and a logarithmic relationship in velocity.

Eyring's model has been verified in lubrication experiments of solid (soap-like) lubricants by Briscoe and liquid lubricants by He and others within three logarithmic decades of velocities. While Briscoe employed a SFA that confines and pressurises the film over several square microns, He and others used a SFM system in which the contact is on the order of the lubricant molecular dimension.

He and others determined the degree of interfacial structuring and its effect on lubrication of n-hexadecane and octamethylcyclotetra-siloxane (OMCTS). For spherically shaped OMCTS molecules, only an interfacial 'monolayer' was found; in contrast, a 2 nm thick entropically cooled layer was detected for n-hexadecane in the boundary regime to an ultra-smooth silicon wafer. SFM measurements of the two lubricants (with similar chemical affinity to silicon) identified the molecular shape of n-hexadecane responsible for augmented interfacial structuring. Consequently, interfacial liquid structuring was found to reduce lubricated friction, (Fig. 18.4). Again as reasoned above, these results can be discussed in terms of a collective phenomenon, i.e. in terms of increased molecular coordination in n-hexadecane versus OMCTS.

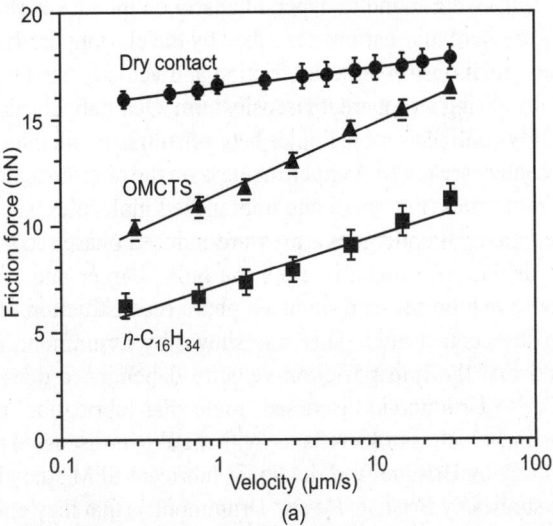

(a)

Fig. 18.4 (a). Logarithmic $F_F(v)$-plots. $F_F(v) = F_0 + \alpha \ln(v[\mu m/s])$: ● 'dry' contact (18 per cent relative humidity) with $F_0 = 16.4$ nN and $\alpha = 0.91$ nN, ▲ OMCTS lubricated with $F_0 = 11.3$ nN and $\alpha = 3.4$ nN, and ■ n-hexadecane (n-$C_{16}H_{34}$) lubricated with $F_0 = 7.1$ nN and $\alpha = 2.5$ nN. The measurements were obtained with rectangular SFM cantilevers (0.4–0.8 N/m) at 100 nN load and 21°C, both feedback-controlled.

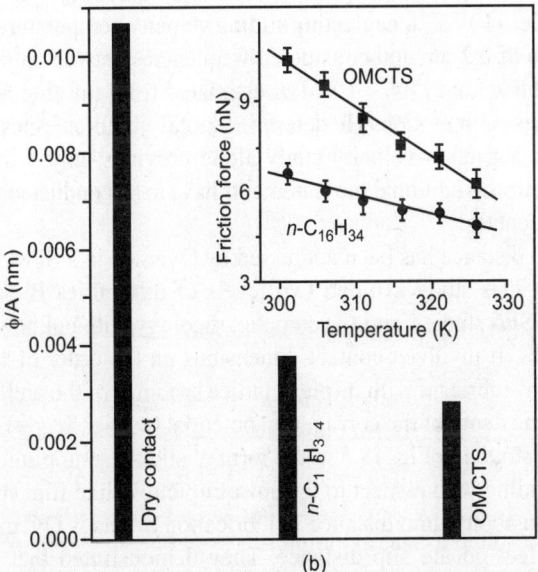

(b)

Fig. 18.4 (b). Stress activation length, ϕ/A. (A area of contact) for OMCTS, n-hexadecane and dry contact. The inset provides a linear relationship between friction and temperature at a velocity of 1 µm/s and a normal load of 100 nN.

FUNCTIONAL BEHAVIOUR OF LUBRICATED FRICTION

Friction-rate experiments are well suited to evaluate the rheological nature of interfacial liquids. In classical theories, such as the Reynolds' hydrodynamic theory discussed above, drag forces in lubricated sliding over thick liquid films were found to depend linearly on the rate of sliding and on the viscosity of the bulk fluid. In high-pressure lubrication, described by the elastohydrodynamic lubrication theory, it was found that the linear relationship between friction and velocity can be retained by adjusting the (apparent) viscosity by introducing an apparent viscosity term. Qualitatively, the same linear relationship has been observed for highly confined simple liquids between ultrasmooth mica surfaces such as alkanes. Note that the lubricated contact area in SFA experiments is on the micron-scale. It significantly exceeds the size of the confined molecules. For small and unbranched molecules, such as simple alkanes, it is possible that the confined material undergoes a pressure-induced phase reconstruction, which leads to material properties that deviate significantly from the bulk. Larger and more complex (branched) molecules are less likely to exhibit pressure-induced phase reconstruction due to internal constraints and poor mixing within the contact area. This was shown by Drummond and others in SFA shear experiments. They found that the linear friction-velocity dependence does not apply for branched hydrocarbon lubricants. Also Drummond discussed 'molecular lubrication' in terms of a logarithmic friction-velocity relationship, which is in accordance with the above-discussed thermal activation model, the solid lubricant SFA study by Briscoe, and the liquid lubricant SFM study by He and others.

Common to the three studies by Briscoe, He and Drummond is that they operate on a single material phase that is disrupted or relaxed over a very specific lateral length scale. In the Eyring model, the length scale is deduced by assuming a regular series of barriers, separated by a virtual jump distance. The distance is embedded in v_o, the characteristic velocity, which is the product of the jump distance and the frequency of the process. Briscoe used the lattice constant of the highly oriented monolayers as the virtual jump distance. It was assumed that the process frequency was related to the vibrational frequency of the molecules (10^{11} s^{-1}), neglecting sliding velocity, temperature, and pressure effects. He assumed a jump distance of 0.2 nm and considered frequencies between a perfectly structured alkane layer (10^{11} Hz) and the bulk fluid (10^{13}–10^{15} Hz, estimated from infrared absorption data for typical covalent bonds). With these assumptions He determined total 'jump-energies' of 4–8 $\times 10^{-20}$ J. Briscoe and He pointed out that a friction-velocity study alone provides only a qualitative measure of the microscopic origin of friction. Additional measurements have to be conducted that quantitatively address jump distances and frequencies.

The issue of the jump distance has been addressed by Overney in a SFM study on a highly ordered lubricant model system. This study avoided two levels of difficulties Briscoe and He encountered: (i) large contact areas of SFA studies, and (ii) complex rheology with unknown structure parameters as in liquid lubricant studies. It involved contact dimensions on the order of 1 nm^2, and the crystalline form a bilayer model-lipid-lubricant with in-plane lattice spacings of 0.6 and 1.1 nm. The study mainly focused on the effect of the depth of the corrugation potential (barrier height) on the static and dynamic friction force. This is illustrated in Fig. 18.5 in the form of stick-slip amplitude plotted as a function of the drag direction (i.e. sliding with respect to the anisotropic row-like film structure).

Relevant to discussion about jump distance in lubrication events is Overney's discussion about the sliding speed and its effect on the slip distance. They demonstrated that within sliding speeds of 36 nm/s to 100 nm/s, the jump distance corresponded to the lattice spacing. At higher velocities, however, they could observe jumps over multiples of lattice distances and found the jump length distribution to become increasingly stochastic at higher velocities. They proposed molecular (or atomistic) friction as a white-noise driven system, which obeys a Gaussian fluctuation-dissipation relation. Hence, based on this finding one should consider discussing kinetic friction in terms of a statistical fluctuation model and understand the jump distance as a statistical quantity.

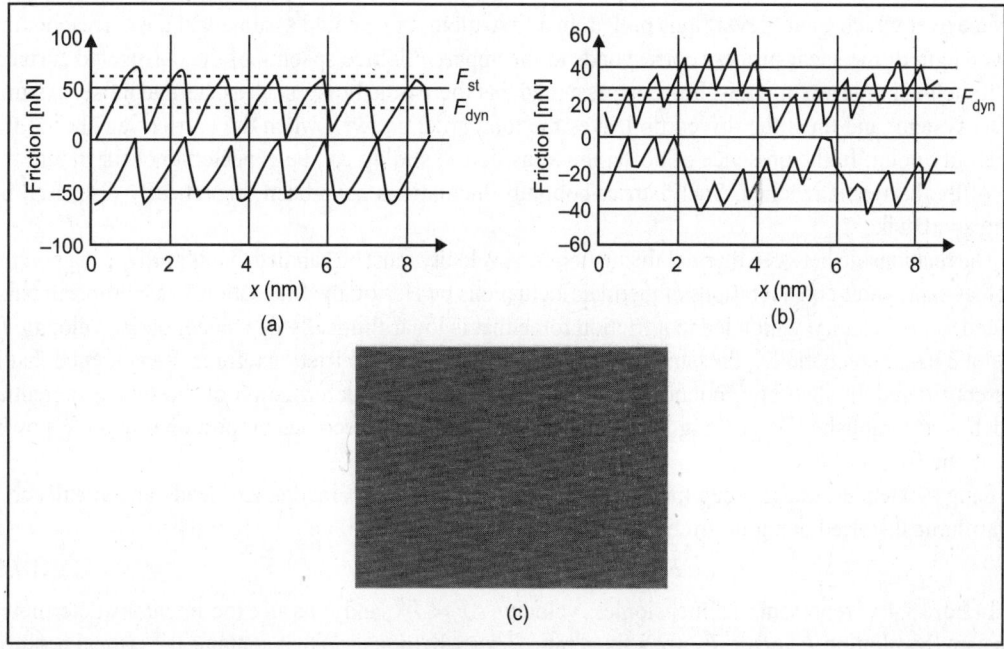

(a)

(b)

(c)

Fig. 18.5. SFM molecular stick-slip measurements of a bilayer lipid system (5-(4′-N,N-dihexadecylamino) benzylidene barbituric acid). (a) High amplitude frictional stick-slip behaviour is observed for scans perpendicular to molecular rows as imaged in 18.5(c). F_{st}, static friction, is assigned to the maximum force occurrence. The average value corresponds to the dynamic friction value, F_{dyn}, determined on large-scale micrometer scans. (b) A 30 degrees out of row direction scan leads to decreased frictional stick-slip behaviour due to smaller molecular corrugations. (c) 12 × 12 nm² SFM lateral force image of a highly structured lipid bilayer. Two crystalline domains with a boundary are imaged. The anisotropic row-like structure is responsible for directional dependent friction forces. The molecular corrugation between the rows is larger than the molecular corrugation in between a single row.

THERMODYNAMICAL MODELS BASED ON SMALL AND NONCONFORMING CONTACTS

The SFM approach simulates a single asperity contact with a very high compliance, provided by a microfabricated and etched ultra-sharp tip and a typically soft cantilever spring. From a realistic, tribological perspective, the SFM approach is targeted toward the study of the intrinsic lubricant properties of a thin film in close vicinity to the solid substrate. The small contact area on the order of the lubricant's molecular dimension allows discussing SFM results in terms of a thermodynamic equilibrium. The area is insufficient in reorganising the lubricant molecules coherently, to cause an apparent material phase-transition, or to generate a metastable situation as observed in SFA experiments. SFM is, therefore, not appropriate to reflect on tribological issues involving large area confinement effects.

In our prior molecular discussion of friction above, we introduced for solid and liquid lubrication a thermal activation model, the Eyring model, which employed a regular series of potential barriers. Note that the concept applies for a solid lubricant of an inherent, highly ordered structure, but also for a liquid system in which the series of potentials is built up and overcome in the course of the shear process. Gnecco showed in a ultrahigh vacuum study on sodium chloride that the concept of the Eyring model also applies for dry SFM friction studies. Thus a molecular theory of lubricated friction involving a molecular contact could be derived from a very simplistic model of an apparent sinusoidal-corrugated surface over which a cantilever tip is pulled. In a first attempt one could assume that the corresponding wavelength of the shear process corresponds to the apparent lattice spacing of the corrugated surface. With such a simple attempt it is, however, assumed that there is no noise, such as thermal noise, existing in the system, and thus, the driven tip leaves the total potential well when the barrier vanishes at the instability point. In the presence of noise, the transition to sliding can be expected to occur before the top of the barrier is reached. Such barrier-hopping fluctuations have been theoretically discussed by Sang and Dudko.

The relationship between thermal fluctuations and velocity must be handled thoughtfully. Sang pointed out that in previous considerations of thermal fluctuations by Heslot, the fluctuations were proportionally related to the velocity, which led to a friction force that is logarithmically dependent on the velocity. In Heslot's linear creep model, the barrier height is proportional to the frictional force. Sang argued that if one considered an absorbing boundary condition (i.e. an elastic deformation of the overall potential which is accomplished by shifting the x-axis) the barrier height becomes proportional to a 3/2-power law in the friction force.

Sang's extended linear creep model resembles a ramped creep model, and leads analytically to a logarithmic distorted dynamic friction-versus-velocity relationship, i.e.

$$F = F_c - \Delta F \ |\ln v^*|^{2/3} \qquad \qquad ... (18.4)$$

In Eq. 18.4 v^* represents a dimensionless velocity, $\Delta F \propto T^{2/3}$, and F_c is an experimentally determined constant (by plotting F versus $T^{2/3}$ for a fixed ratio $T/v = 1$ K/(nm/sec)) that contains the critical position of the cantilever support. The same relationship of friction with velocity was also derived for the maximum spring force by Dudko. ΔF and v^* in Eq. 18.4 were derived as follows by Sang:

$$v^* = 2\left(\frac{v\beta\omega_o^2 U_o}{k_B T\lambda}\right)\frac{\Omega_k^2}{\left(1-\Omega_k^4\right)^{1/2}};$$

$$\Omega_k = \frac{\omega_o}{2\pi\omega_k}; \ \omega_o = \sqrt{\frac{M\lambda^2}{U_o}}; \ \omega_k = \sqrt{\frac{M}{k}} \qquad \qquad ... (18.5a)$$

$$\Delta F = \frac{\pi U_o}{\lambda}\left(\frac{3}{2}\frac{k_B T}{U_o}\right)^{2/3}\left(\frac{\left(1-\Omega_k^4\right)^{1/6}}{1+\Omega_k^2}\right) \qquad \qquad ... (18.5b)$$

In Eqs 18.5a and 18.5b, v is the velocity of the cantilever stage, β is the microscopic friction coefficient or dissipation (damping) factor, ω_o is the frequency of the small oscillations of the tip in the minima of the periodic potential, λ is the lattice constant, U_o is the surface barrier potential height, M and k represent the mass and the spring constant of the cantilever, respectively, and $2\pi\Omega_k$ represents the ratio of ω_o with the intrinsic cantilever resonance frequency ω_k.

Sang's and Dudko's model was experimentally confirmed by Sills and Overney on an unstructured amorphous surface of atactic polystyrene (Fig. 18.6).

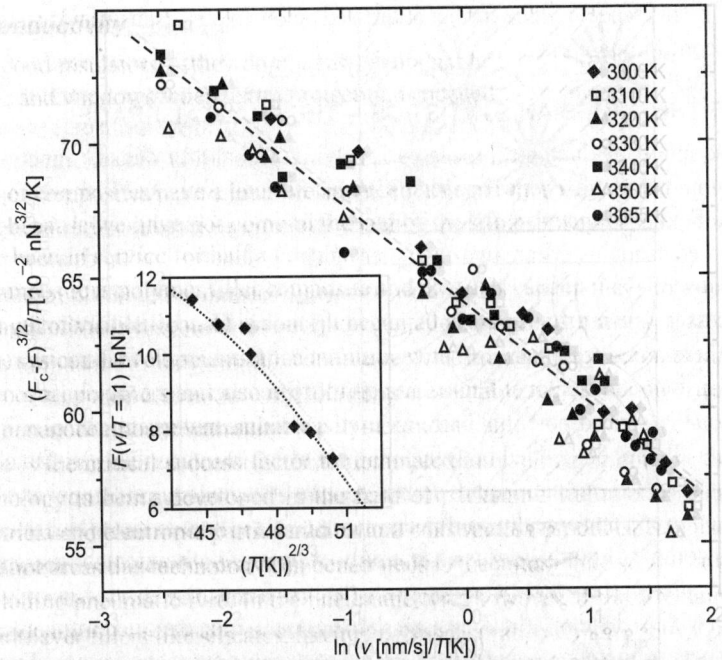

Fig. 18.6. Collapse of SFM friction data obtained on atactic polystyrene using the ramped creep scaling model. The regression parameters from the linear fit (dashed line) are -2.158×10^{-2} N$^{3/2}$·s/m and 40.186×10^{-2} nN$^{-3/2}$/K ($R^2 = 0.9124$). Lower inset: the constant F_c is determined from the intercept of the friction force F versus $T^{2/3}$ for a fixed ratio $T/v = 1$ K/(nm/s): $F_c = 44.5$ nN.

Dudko pointed out that the typically used weak spring constants in SFM measurements are responsible for the more pronounced logarithmic behaviour of friction in velocity as found by Gnecco and He. The ramped creep model is also supported by numerical solutions of the Langevin equation. The Langevin equation combines the equation of motion (including the sinusoidal potential and perfect cantilever oscillator in the total potential energy E) with the thermal noise in the form of the random force, $\xi(t)$, i.e.

$$M\ddot{x} + M\beta\dot{x} + \frac{\partial E(x,t)}{\partial x} = \xi(t) \qquad \ldots (18.6)$$

where

$$E(x,t) = \frac{k}{2}(R(t) - x)^2 - U_0\left(\frac{2\pi x}{\lambda}\right)$$

Equation 18.6 was solved numerically by both Sang and Dudko independently, assuming a Gaussian fluctuation-dissipation relation, $<\xi(t)\xi(t')> = 2M\beta\grave{A}k_bT\delta(t-t')$ to express the random force. Sang confirmed the ramped creep model, and Dudko showed that a force reconstruction approach from the density of states (accumulated from the corresponding Fokker-Planck equation) is equivalent to the Langevin equation. From the dynamic spectral analysis it could be concluded that the locked states (states within the potential wells) contribute mostly to the potential component of the friction force that dominates at low driving velocities, and sliding states contribute to viscous friction dominating at high driving velocities.

It should be noted that all of the above results considered an overdamped SFM system with respect to the driven spring (i.e. $\beta^2 > 4$ kM), and an underdamped system with respect to the periodic potential (i.e. $\beta^2 < 4(M\omega)^2$). This aspect will be further addressed below in our discussion of metastable lubricant systems in large conforming contacts.

LIMITATION OF THE GAUSSIAN STATISTICS—THE FRACTAL SPACE

The spectral description of dynamic processes involving probability density functions has recently been the focus of numerous theoretical papers that treat various statistical kernels, Luedtke, Sokolov, Metzler, Dudko's Fokker-Planck discussion of kinetic friction and Luedtke's Lévy flight model of slip diffusion of adsorbed nanoclusters are two examples in which statistical methods are applied to describe diffusive properties relevant to the kinetics in tribology. Currently most models used to describe tribological processes assume Gaussian statistics, and Dudko. One of the limitations of a Gaussian statistics is that there are no correlations between statistical incidences. In other words, the Gaussian dynamic system is without memory. This is important to remember as the Eyring model discussed above, with its equally spaced potential barriers, used a Gaussian statistics. Simple 'inert' lubricants, such as short chain alkanes embedded between silicon wavers, are described satisfactorily with such a statistics; however, confined complex liquids are not, such as branched molecules, polymers, and generally chemically interactive and entropically confined systems (e.g. perfluropolyether lubricants as discussed below). Confined complex liquids, for instance, easily exhibit strongly interacting glasslike behaviour. The dynamic and stress relaxation behaviours in glasses, frequently discussed only including low interacting system with Arrhenius laws (Gaussian statistics), are often distorted from processes described by independently occurring microscopic processes. For instance, deviations from the Debye exponential relaxation as introduced in Eq. 18.1 are expressed in the form of an extended exponential Kohlrausch relaxation function over time t; i.e.

$$F(t) \equiv \left[\frac{X(t) - X(\infty)}{X(0) - X(\infty)} \right] = e^{-\left(\frac{t}{\tau}\right)^b}; \quad 0 < b < 1 \qquad \qquad \dots (18.7)$$

where, X is the property that is relaxed. The exponent b, the Kohlrausch exponent, can theoretically be determined if one assumes that the process occurs in series, representing a well-determined microscopic origin that correlates the various degrees of freedom. This approach is borrowed from magnetic spin models such as the Ising spin model. The idea is that a given molecular motion is dependent on the availability of other degrees of freedom of mobile neighbouring structural units. Finite relaxation times, t_{max}, are gradually obtained with increasing spin levels (ergodic limit). As mentioned above, the models by Dudko and Sang assumed Gaussian statistics. To illustrate how a diffusion process can deviate from Gaussian statistics, we introduce a simplified version of the Langevin equation, i.e.

$$\ddot{x} = -\eta \dot{x} + \zeta(t) \qquad \qquad \dots (18.8)$$

with the coordinate x, the dissipation (or dampening) parameter η, and the random acceleration $\zeta(t)$. Assuming Gaussian statistics, the mean squared displacement is:

$$\left\langle x^2(t) \right\rangle = 2k_B T \eta t = 2Dt \qquad \qquad \dots (18.9)$$

where, $D = \eta kT$ defines the diffusion constant. It was already realised at the time of Smoluchovski at the beginning of the 20th century that a diffusive description of a dynamic process demands a thermodynamically well-equilibrated or mixed system. Especially in a confined tribological system that involves a third medium (e.g. a lubricant), it can be expected that the Markovian nature of the underlying stochastic process could be disturbed. Consequently, for a monolayer lubricant that is chemically interacting, a nonlinear relationship of the mean squared displacement in time can be expected. Manifestations of anomalous transport are long-range spatial or temporal correlations. Two extreme limits can be distinguished: (i) processes with strong temporal relations (fractal time), and (ii) systems that exhibit long jumps (Lévy flights).

FRACTAL MOBILITY IN REACTIVE LUBRICATION

The importance of the underlying kinetics is illustrated by ultrathin wetting lubricants. The spreading of 'completely wetting' polymer liquids on solid surfaces has revealed unexpected spatial and temporal features when examined at the molecular level. The spreading profile is typically characterised by the appearance of a precursor film of monomolecular thickness extending over macroscopic distances and, in many cases, a terracing (also on the order of molecular dimensions) of the fluid remaining in the reservoir. These spatial features have been shown to be consistent with a Poiseuille-like flow in which the disjoining pressure gradients with film thickness drive the spreading process. The temporal evolution of the spreading profile in this film thickness regime is, however, found to universally scale as $t^{1/2}$ even at short times. That the spreading dynamics are reflective of a diffusive transport mechanism, and not of a pressure driven 'liquid' flow, suggests that interfacial confinement substantially alters the mobility of molecularly thin polymer fluids.

The molecular mobility is of fundamental importance for monolayer lubrication purposes, such as in magnetic storage devices. It has, for instance, been shown that for low surface energy hydroxyl-terminated perfluoropolyether (PFPE-OH) films, the lubricant exhibits spatially terraced flow profiles indicative of film layering and spreading dynamics that are diffusive in nature. In magnetic storage devices the hydroxylated chain ends of molecularly thin PFPE-OH films interact with the solid surface, an amorphous carbon surface, via the formation of hydrogen-bonds with the polar, carbonoxygen functionalities located on the carbon surface. The bonding of the PFPE-OH polymer to carbon is predicated on the ability of the PFPE backbone to deliver spatially the hydroxyl end-group to within a sufficiently close distance to the surface active sites. Kinetic measurements probing the bonding of the PFPE-OH polymer to the carbon reveal two distinctive kinetic behaviours, as illustrated in Fig. 18.7 at two representative temperatures: 50° and 90°C for the two temperature regimes below 56°C and above 85°C. Below 56°C the kinetics are described with a time-dependent (fractal time dependent) rate coefficient of the form

$$k(t) = k_0 t^{1/2} \qquad \qquad \text{... (18.10)}$$

and at temperatures above 85°C with the form

$$k(t) = k_0 t^{1.0} \qquad \qquad \text{... (18.11)}$$

The initial bonding rate constants, k_0, increased abruptly as the temperature rose above 50°C.

The bonding kinetics in the low-temperature regime is characteristic of a diffusion-limited reaction occurring from a glasslike state of the molecularly thin PFPE-OH film. The mobility of the PFPE chain in the glasslike state is limited by the propagation of holes or packets of free volume, which facilitate configurational rearrangements of the chain. The onset of changes in the bonding kinetics at nominally T > 56° signifies a fundamental change in the mobility of the molecularly thin PFPE-OH film. Specifically, the transition in the fractal time dependence suggests that delivery of the hydroxyl moiety to the surface is no longer limited by hole diffusion, and the increase in the initial rate constant indicates an enhancement in the backbone flexibility. These results are consistent with a transition in the film from a glasslike to a liquid-like state in which the enhanced PFPE-OH segmental mobility results from rotations about the ether oxygen linkages in the chain that become increasingly facile. The time dependence observed in the high-temperature rate coefficients, $k(t) = k_B t^{-1.0}$ is characteristic of a process occurring from a confined liquid-like state in which the activation energy increases as the extent of the reaction increases.

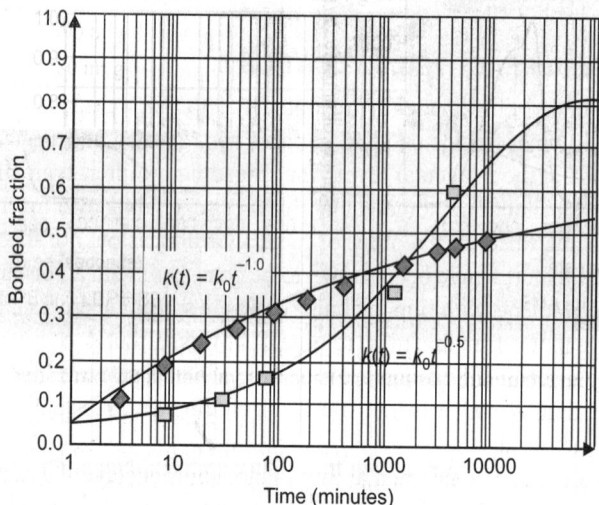

Fig. 18.7. Representative kinetic data for the bonding of PFPE-OH (tradename: Fomblin Zdol) to amorphous carbon at $T = 50°C$ and $T = 90°C$. Solid lines represent fits using a rate coefficient of the form, $k(t) = k_0 t^{-\alpha}$ with $\alpha = 0.5$ for $T = 50°C$ and $\alpha = 1.0$ for $T = 90°C$.

The impact of this transition in the molecular mobility on tribology can be illustrated with sinusoidally modulated shear force experiments. In brief, a molecularly thin (10.7 ± 0.5 Å) PFPE-OH film is subjected to a local shear stress by means of a sinusoidal force applied laterally to a SFM probe (at constant load) where the modulation amplitude is initially set below that required to initiate sliding between the tip and the sample. The amplitude and phase-shift responses are measures of the contact stiffness and the effective viscous dampening, respectively.

From Fig. 18.8 it can be inferred that the SFM-measured nanorheological properties of the PFPE-OH film exhibit the changes discussed above in the molecular mobility. Thus kinetic and rheological data suggest that the thermal transition observed is due to the formation of a two-dimensional (2D) glass. The 'glass transition' results from the preferential 'freezing out' of the out-of-plane torsional motions of the energetically confined PFPE backbone.

The confinement-induced solidification in the molecularly thin precursor film will significantly impact the lubrication properties and challenge thermodynamic, well-equilibrated models of lubrication as introduced above.

METASTABLE LUBRICANT SYSTEMS IN LARGE CONFORMING CONTACTS

It is important to note that in an experiment of thermal activation, the critical time of the experiment t_{exp} decides the system response with its finite relaxation time t_{max}. If $t_{exp} > t_{max}$, the system behaves in an ergodic manner, and thermodynamic laws apply for interpreting lubrication results. On the contrary, if $t_{exp} < t_{max}$ the thermal evolution cannot be described by classical statistical thermodynamics. A metastable configuration is generated. The experimental time depends strongly on the contact area, the parameter that most differs between SFA and SFM measurements. SFA experiments involve large micron-scale contacts while SFM measurements are conducted with contacts on the nanoscale.

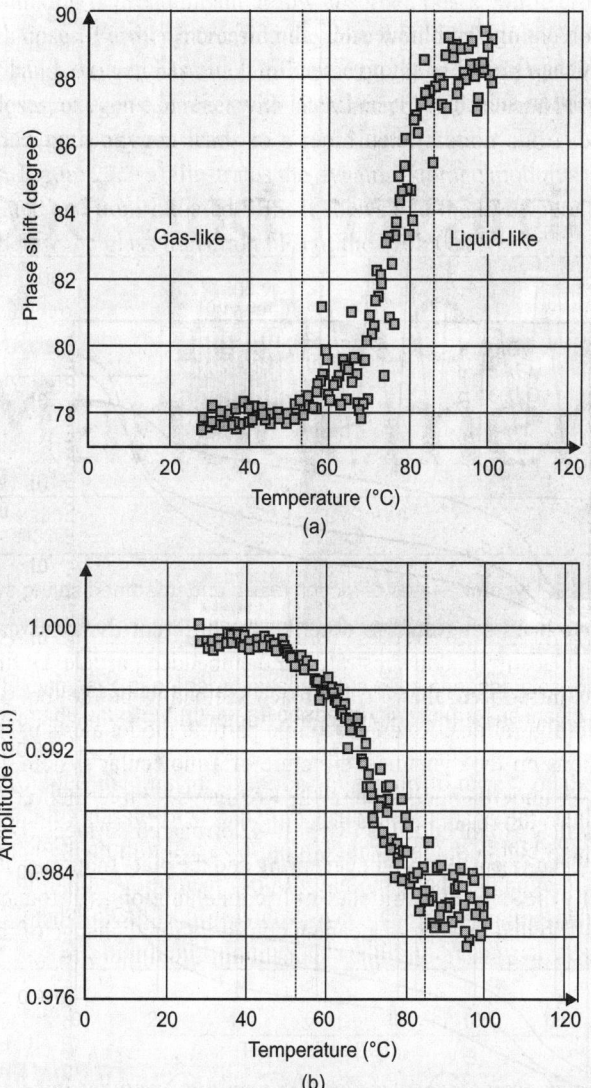

Fig. 18.8. Shear-modulated SFM experiments performed on a 10.7 ± 0.5Å Fomblin Zdol film: (a) phase shift response between disturbance and response, and (b) contact stiffness response vs. temperature.

Because of the large contact area, SFA experiments are very susceptible to generating unequilibrated metastable lubricant configurations. The SFA study by Yoshizawa, in which distinctively different dynamic states of friction were introduced, could be interpreted as such. To date there are three velocity regimes used to describe the dynamic state of friction for a system that is 'underdamped'. In an underdamped system, realised by a stiff spring compared to the friction constant, the characteristic slip time is comparable or smaller than the response time of the mechanical system. One distinguishes three velocity regimes, as depicted in Fig. 18.9. The three regimes distinguish themselves by a single discriminator, v_c, a material, pressure and temperature dependent critical velocity. The regimes are described as:

1. Highly regular with high amplitude stick-slip for $v \ll v_c$.
2. Intermittent stochastic stick-slip for $v < v_c$.
3. Smooth low friction sliding for $v > v_c$.

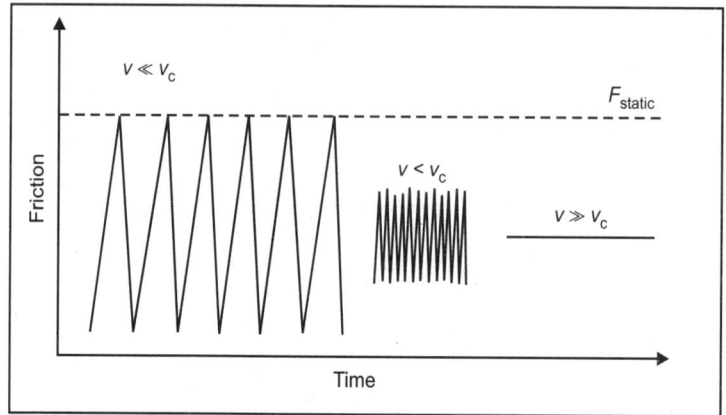

Fig. 18.9. Dynamic states of friction for an underdamped spring system.

Various models have been suggested to describe the different dynamic states, including melting freezing transition, chain adsorption on substrates, and embedded particle. Until the embedded particle model by Rozman was introduced, the SFA approach seemed to be the tool of choice to investigate metastable lubricant configurations. Rozman's single particle model alerts us however about drawing unambiguous conclusions on the dynamical structure of a molecular system embedded between two plates and driven by an underdamped system. In Rozman's simple theoretical model, sketched in Fig. 18.10, a single particle is embedded between two corrugated surfaces (two-wave potential). The top plate is pulled at constant velocity by a linear spring and the plate motion is monitored. Interestingly, the plate exhibits exactly the same dynamic state of frictional motion as introduced above with the three regimes, which were attributed to a rate-dependent configurational change of the lubricant.

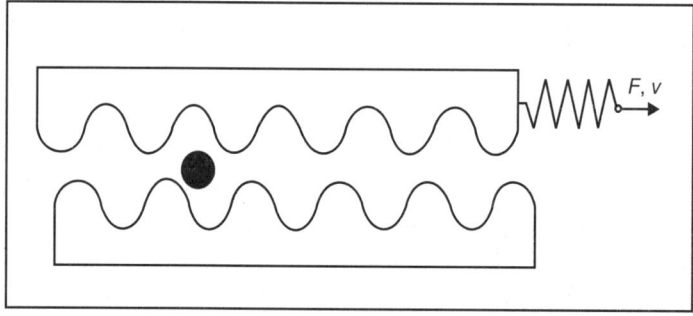

Fig. 18.10. Model of single particle embedded in a two-wave potential driven at constant force, F, and velocity v.

Finally it shall be noted that the slip relaxation pattern depends on the condition under which the stick-slip motion is being studied. In an overdamped system, i.e. a system in which the spring constant is weak compared to the friction constant, Rozman found that one can control the experimentally observed relaxation pattern of the slip by controlling the spring constant.

CONCLUSION

Starting from a classical tribological Master Curve, the Stribeck curve, with its complex description of lubrication for thin lubricants, we launched a discussion of molecular lubrication pulling together disparate approaches to studying friction. We found that phenomenological descriptions of lubrication, such as the Reynolds' theory, were kept alive for ultrathin lubricants by 'adjusting' material properties, such as the viscosity. This is a common engineering approach to introduce 'effective' or 'apparent' properties, if tested models fail to describe new situations. In the case of molecular lubrication, significant progress has been made over the last ten years. Two instrumental techniques have in particular contributed to this progress: the SFA and the SFM. The contributions of these two techniques have been complementary. While the SFA has tested lubricants under pressure constraints with large contacts in respect to the size of the trapped molecules, the SFM has probed the degree of collective mobility with disturbances in the size of the molecules themselves.

One feature common to interpreting lubrication results, and which was discussed here in detail, is the problem of finding the appropriate underlying statistics to describe the lubrication process. Most of the current molecular models that have been used to describe molecular friction and lubrication assumed a Gaussian statistics. Only recently has it been recognised important to consider statistics that embrace long-range spatial or temporal correlations. In the near future, it can be expected that the next leap in an improved fundamental understanding of kinetics and energetics in nanolubrication will come from interpretations that are challenging the Markovian nature of the underlying stochastic process.

Chapter 19

MEMS Packaging

INTRODUCTION

MEMS is a relatively new field which is tied so closely with silicon processing that most of the early packaging technologies will most likely use 'off-the-shelf' packaging 'borrowed' from the semiconductor microelectronics field. Packaging of microelectronics circuits is the science and art of establishing interconnections and an appropriate operating environment for predominantly electrical (and in the case of MEMS, electromechanical) circuits to process and/or store information.

Packaging manifests itself in novel and unique creations that ingeniously reconcile and satisfy what seem to be mutually exclusive application requirements and constraints posed by the laws of nature and the properties of materials and processes. All applications can be summed up in three terms: cost, performance and reliability.

Packaging can span from the consumer to midrange systems to the high performance/reliability applications. It must be noted that no sharp boundaries exist between the classes, only a gradual shift from optimisation for parameters which control performance and cause the cost to increase. All along, the reliability must also be considered. The packaging chapter that follows will summarise the primary package types that will likely apply to MEMS technology and the concerns that traditionally have concerned the microelectronics field.

Webster's dictionary defines package as a group or a number of things, boxed and offered as a unit. MEMS packages can contain many electrical and mechanical components. To be useful to the outside world these components need interconnections. Alone, a MEM die sawed from a wafer is extremely fragile and must be protected from mechanical damage and hostile environments. To function, electrical circuits need to be supplied with electrical energy, which is consumed and transformed into mechanical and thermal (heat) energy. Because the system operates best within a limited temperature range, packaging must offer an adequate means for removal of heat.

FUNCTIONS OF MEMS PACKAGES

The package serves to integrate all of the components required for a system application in a manner that minimises size, cost, mass and complexity. The package provides the interface between the components and the overall system. The following subsections present the three main functions of the MEMS package: mechanical support, protection from the environment, and electrical connection to other system components.

Mechanical Support

Due to the very nature of MEMS being mechanical, the requirement to support and protect the device from thermal and mechanical shock, vibration, high acceleration, particles, and other physical damage (possibly radiation) during storage and operation of the part becomes critical. The mechanical stress endured depends on the mission or application. For example, landing a spacecraft on a planet's surface creates greater mechanical shock than experienced by a communication satellite. There is also a difference between space and terrestrial applications.

The coefficient of thermal expansion (CTE) of the package should be equal to or slightly greater than the CTE of silicon for reliability, since thermal shock or thermal cycling may cause die cracking and delamination if the materials are unmatched or if the silicon is subject to tensile stress. Other important parameters are thermal resistance of the carrier, the material's electrical properties, and its chemical properties, or resistance to corrosion.

Once the MEMS device is supported on a (chip) carrier, the wire bonds or other electrical connections are made, the assembly must be protected from scratches, particulates, and other physical damage. This is accomplished either by adding walls and a cover to the base or by encapsulating the assembly in plastic or other material. Since the electrical connections to the package are usually made through the walls, the walls are typically made from glass or ceramic. The glass or ceramic can also be used to provide electrical insulation of the leads as they exit through a conducting package wall (metal or composite materials). Although the CTE of the package walls and lid do not have to match the CTE of silicon based MEMS as they are not in intimate contact (unless an encapsulating material is used), it should match the CTE of the carrier or base to which they are connected.

Protection from Environment

Simple — mechanical only

Many MEMS devices are designed to measure something in the immediate surrounding environment. These devices range from biological 'sniffers' to chemical MEMS that measure concentrations of certain types of liquids. So the traditional 'hermeticity' that is generally thought of for protecting microelectronic devices may not apply to all MEMS devices. These devices might be directly mounted to a printed circuit board (PCB) or a hybrid-like ceramic substrate and have nothing but a 'housing' to protect it from mechanical damage such as dropping or something as simple as damage from the operator's thumb.

Traditional — hermetic and nonhermetic

Many elements in the environment can cause corrosion or physical damage to the metal lines of the MEMS as well as other components in the package. Although there is no moisture in space, moisture remains a concern for MEMS in space applications since it may be introduced into the package during fabrication and before sealing. The susceptibility of the MEMS to moisture damage is dependent on the materials used in its manufacture. For example, Al lines can corrode quickly in the presence of moisture, whereas Au lines degrade slowly, if at all, in moisture. Also, junctions of dissimilar metals can corrode in the presence of moisture. Moisture is readily absorbed by some materials used in the MEMS fabrication, die attachment, or within the package; this absorption causes swelling, stress, and possibly delamination.

To minimise these failure mechanisms, MEMS packages for high reliability applications may need to be hermetic with the base, sidewalls, and lid constructed from materials that are good barriers to liquids and gases and do not trap gasses that are later released.

Electrical Connection to Other System Components

Because the package is the primary interface between the MEMS and the system, it must be capable of transferring DC power and in some designs, RF signals. In addition, the package may be required to distribute the DC and RF power to other components inside the package. The drive to reduce costs and system size by integrating more MEMS and other components into a single package increases the electrical distribution problems as the number of interconnects within the package increases.

When designs also require high frequency RF signals, the signals can be introduced into the package along metal lines passing through the package walls, or they may be electromagnetically coupled into the package through apertures in the package walls. Ideally, RF energy is coupled between the system and the MEMS without any loss in power, but in practice, this is not possible since perfect conductors and insulators are not available. In addition, power may be lost to radiation, by reflection from components that are not impedance matched, or from discontinuities in the transmission lines. The final connection between the MEMS and the DC and RF lines is usually made with wire bonds, although flip-chip die attachment and multilayer interconnects using thin dielectric may also be possible.

Thermal Considerations

For small signal circuits, the temperature of the device junction does not increase substantially during operation, and thermal dissipation from the MEMS is not a problem. However, with the push to increase the integration of MEMS with power from other circuits such as amplifiers perhaps even within a single package, the temperature rise in the device junctions can be substantial and cause the circuits to operate in an unsafe region. Therefore, thermal dissipation requirements for power amplifiers, other large signal circuits, and highly integrated packages can place severe design constraints on the package design.

The junction temperature of an isolated device can be determined by

$$Tj = Q * R, + T_{case}$$

where,

Q = is the heat generated by the junction and is dependent on the output power of the device and its efficiency.

R = is the thermal resistance between the junction and the case.

T = is the temperature of the case.

Normally, the package designer has no control over Q and the case temperature, and therefore, it is the thermal resistance of the package that must be minimised. Figure 19.1 is a schematic representation of the thermal circuit for a typical package, where it is assumed that the package base is in contact with a heat sink or case.

It is seen that there are three thermal resistances that must be minimised: the resistance through the package substrate, the resistance through the die-attach material, and the resistance through the carrier or package base. Furthermore, the thermal resistance of each is dependent on the thermal resistivity and the thickness of the material. A package base made of metal or metal composites has very low thermal resistance and therefore does not add substantially to the total resistance. When electrically insulating materials are used for bases metal-filled via holes are routinely used, under the MEMS, to provide a thermal path to the heat sink. Although thermal resistance is a consideration in the choice of the die attach material, adhesion and bond strength are even more important. To minimise the thermal resistance through the die-attach material, the material must be thin, there can be no voids, and the two surfaces to be bonded should be smooth.

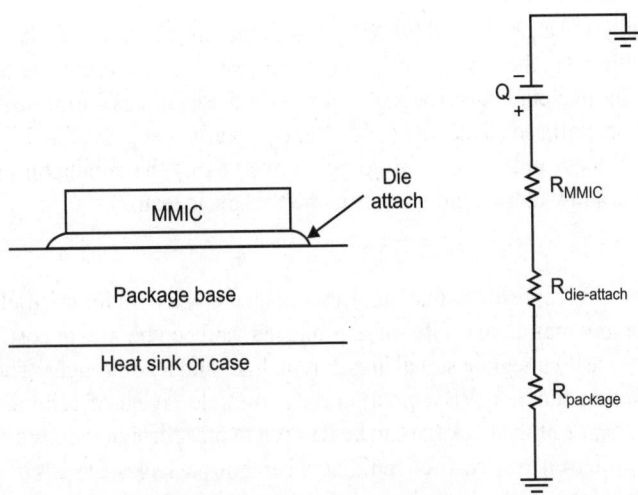

Fig. 19.1. Cross section of MMIC attached to a package and its equivalent thermal circuit.

TYPES OF MEMS PACKAGES

Each MEMS application usually requires a new package design to optimise its performance or to meet the needs of the system. It is possible to loosely group packages into several categories. Four of these categories: (i) all metal packages, (ii) ceramic, (iii) plastic packages, and (iv) thin-film multilayer packages are presented below.

Metal Packages

Metal packages are often used for microwave multichip modules and hybrid circuits because they provide excellent thermal dissipation and excellent electromagnetic shielding. They can have a large internal volume while still maintaining mechanical reliability. The package can use either an integrated base and sidewalls with a lid or it can have a separate base, sidewalls, and lid. Inside the package, ceramic substrates or chip carriers are required for use with the feedthroughs.

The selection of the proper metal can be critical. CuW (10/90), Silvar™ (a Ni-Fe alloy), CuMo (15/85), and CuW (15/85)—all have good thermal conductivity and a higher CTE than silicon, which makes them good choices. Kovar™, a Fe-Ni-Co alloy commonly. All of the above materials, in addition to Alloy-46, may be used for the sidewalls and lid. Cu, Ag, or Au plating of the packages is commonly done.

Before final assembly, a bake is usually performed to drive out any trapped gas or moisture. This reduces the onset of corrosion-related failures. During assembly, the highest temperature curing epoxies or solders should be used first and subsequent processing temperatures should decrease until the final lid seal is done at the lowest temperature to avoid later steps damaging earlier steps. Au-Sn is a commonly used solder that works well when the two materials to be bonded have similar CTEs. Au-Sn solder joints of materials with a large CTE mismatch are susceptible to fatigue failures after temperature cycling. The Au-Sn intermetallics that form tend to be brittle and can accommodate only low amounts of stress.

Welding (using lasers to locally heat the joint between the two parts without raising the temperature of the entire part) is a commonly used alternative to solders. Regardless of the seal technology, no voids or misalignments can be tolerated since they can compromise the package hermeticity. Hermeticity can also be affected by the feedthroughs that are required in metal packages. These feedthroughs are generally made of glass or ceramic and each method (glass seal or aluminium feedthrough) has its weakness. Glass can crack during handling and thermal cycling. The conductor exiting through the ceramic feedthrough may not seal properly due to metallurgical reasons. Generally, these failures are due to processing problems as the ceramic must be metallised so that the conductor (generally metal) may be soldered (or brazed) to it. The metallisation process must allow for complete wetting of the conducting pin to the ceramic. Incomplete wetting can show up as a failure during thermal cycle testing.

Ceramic Packages

Ceramic packages have several features that make them especially useful for microelectronics as well as MEMS. They provide low mass, are easily mass produced, and can be low in cost. They can be made hermetic, and can more easily integrate signal distribution lines and feedthroughs. They can be machined to perform many different functions. By incorporating multiple layers of ceramics and interconnect lines, electrical performance of the package can be tailored to meet design requirements. These types of packages are generally referred to as co-fired multilayer ceramic packages. Details of the co-fired process are outlined below. Multilayer ceramic packages also allow reduced size and cost of the total system by integrating multiple MEMS and/or other components into a single, hermetic package. These multilayer packages offer significant size and mass reduction over metal-walled packages. Most of that advantage is derived by the use of three dimensions instead of two for interconnect lines.

Co-fired ceramic packages are constructed from individual pieces of ceramic in the 'green' or unfired state. These materials are thin, pliable films. During a typical process, the films are stretched across a frame in a way similar to that used by an artist to stretch a canvas across a frame. On each layer, metal lines are deposited using thick-film processing (usually screen printing), and via holes for interlayer interconnects are drilled or punched. After all of the layers have been fabricated, the unfired pieces are stacked and aligned using registration holes and laminated together. Finally, the part is fired at a high temperature. The MEMS and possibly other components are then attached into place [usually organically (epoxy) or metallurgically (solders)], and wire bonds are made the same as those used for metal packages.

Several problems can affect the reliability of this package type. First, the green-state ceramic shrinks during the firing step. The amount of shrinkage is dependent on the number and position of via holes and wells cut into each layer. Therefore, different layers may shrink more than others creating stress in the final package. Second, because ceramic-to-metal adhesion is not as strong as ceramic-to-ceramic adhesion, sufficient ceramic surface area must be available to assure a good bond between layers. This eliminates the possibility of continuous ground planes for power distribution and shielding. Instead, metal grids are used for these purposes. Third, the processing temperature and ceramic properties limit the choice of metal lines. To eliminate warping, the shrinkage rate of the metal and ceramic must be matched. Also, the metal must not react chemically with the ceramic during the firing process. The metals most frequently used are W and Mo. There is a class of low temperature co-fired ceramic (LTCC) packages. The conductors that are generally used are Ag, AgPd, Au, and AuPt. Ag migration has been reported to occur at high temperatures, high humidity, and along faults in the ceramic of LTCC.

Thin-film Multilayer Packages

Within the broad subject of thin-film multilayer packages, two general technologies are used. One uses sheets of polyimide laminated together in a way similar to that used for the LTCC packages described above, except a final firing is not required. Each individual sheet is typically 25 mm and is processed separately using thin-film metal processing. The second technique also uses polyimide, but each layer is spun onto and baked on the carrier or substrate to form 1 to 20 μm thick layers. In this method, via holes are either wet etched or reactive ion etched (RIE). The polyimide for both methods has a relative permittivity of 2.8 to 3.2. Since the permittivity is low and the layers are thin, the same characteristic impedance lines can be fabricated with less line-to-line coupling; therefore, closer spacing of lines is possible. In addition, the low permittivity results in low line capacitance and therefore faster circuits.

Plastic Packages

Plastic packages have been widely used by the electronics industry for many years and for almost every application because of their low manufacturing cost. High reliability applications are an exception because serious reliability questions have been raised. Plastic packages are not hermetic, and hermetic seals are generally required for high reliability applications. The packages are also susceptible to cracking in humid environments during temperature cycling of the surface mount assembly of the package to the mother-board. Plastic packaging for space applications may gain acceptability as time goes on.

Package-to-MEMS Attachment

The method used to attach a MEMS device to a package is a general technology applicable to most integrated circuit (IC) devices. Generally referred to as die attach, the function serves several critical functions. The main function is to provide good mechanical attachment of the MEMS structure to the package base. This ensures that the MEMS chip (or die) does not move relative to the package base. It must survive hot and cold temperatures, moisture, shock and vibration. The attachment may also be required to provide a good thermal path between the MEMS structure and the package base. Should heat be generated by the MEMS structure or by the support circuitry, the attachment material should be able to conduct the heat from the chip to the package base. The heat can be conducted away from the chip and 'spread' to the package base which is larger in size and has more thermal mass. This spreading can keep the device operating in the desired temperature range. If the support circuitry requires good electrical contact from the silicon to the package base, the attachment material should be able to accommodate the task.

The stability and reliability of the attach material is largely dictated by the ability of the material to withstand thermomechanical stresses created by the differences in the coefficient of thermal expansion (CTE) between the MEMS silicon and the package base material (Fig. 19.2). These stresses are concentrated at the interface between the MEMS silicon backside and the attach material and the interface between the die-attach material and the package base. Silicon has a CTE between 2 and 3 ppm/°C while most package bases have higher CTE (6 to 20 ppm/°C). An expression which relates the number of thermal cycles that a die-attach can withstand before failure is based on the Coffin-Manson relationship for strain. The equation below defines the case for die attach,

$$N(f) \propto \gamma^m \frac{2 \times t}{L \times \Delta CTE \times \Delta T}$$

where
γ = shear strain
m = material constant
L = diagonal length of the die
T = die-attach material thickness

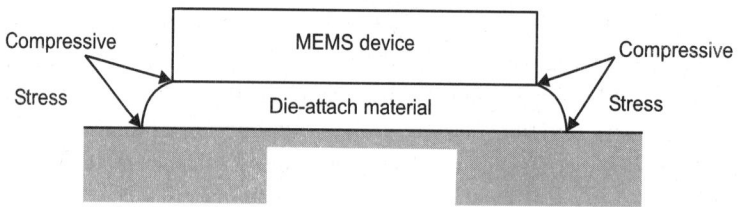

Fig. 19.2. MEMS device in compression.

Voids in the die-attach material cause areas of localised stress concentration that can lead to premature delamination. Presently, MEMS packages use solders, adhesives or epoxies for die-attach. Each method has advantages and disadvantages that affect the overall MEMS reliability. Generally, when a solder is used, the silicon die would have a gold backing. Au-Sn (80–20) solder generally is used and forms an Au-Sn eutectic when the assembly is heated to approximately 250°C in the presence of a forming gas. When this method is applied, a single rigid assembled part with low thermal and electrical resistances between the MEMS device and the package. One problem with this attachment method is that the solder attach is rigid (and brittle) which means it is critical for the MEMS device and the package CTEs match since the solder cannot absorb the stresses.

Adhesives and epoxies are comprised of a bonding material filled with metal flakes as shown in the Fig. 19.3. Typically, Ag flakes are used as the metal filler since it has good electrical conductivity and has been shown not to migrate through the die-attach material. These die-attach materials have the advantage of lower process temperatures. Generally between 100° and 200°C are required to cure the material. They also have a lower built-in stress from the assembly process as compared to solder attachment. Furthermore, since the die attach does not create a rigid assembly, shear stresses caused by thermal cycling and mechanical forces are relieved to some extent. One particular disadvantage of the soft die-attach materials are that they have a significantly higher electrical resistivity which is 10 to 50 times greater than solder and a thermal resistivity which is 5 to 10 times greater than solder. Lastly, humidity has been shown to increase the ageing process of the die-attach material.

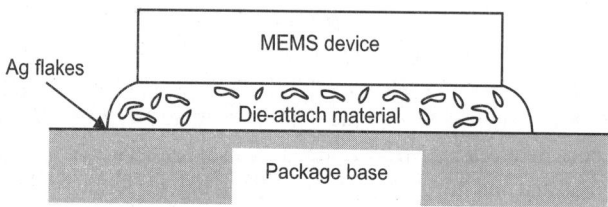

Fig. 19.3. Schematic representation of silver filled epoxy resin.

Chip Scale Packaging

Flip-chip

Controlled collapse chip connection (C4) is an interconnect technology developed by IBM during the 1960s as an alternative to manual wire bonding.

Often called 'flip-chip', C4 attaches a chip with the circuitry facing the substrate. C4 uses solder bumps (C4 Bumps) deposited through a bump mask onto wettable chip pads that connect to matching wettable substrate pads (Fig. 19.4). MEMS technology initially may not use flip-chip packaging but the drive toward miniaturisation may necessitate its incorporation into future designs.

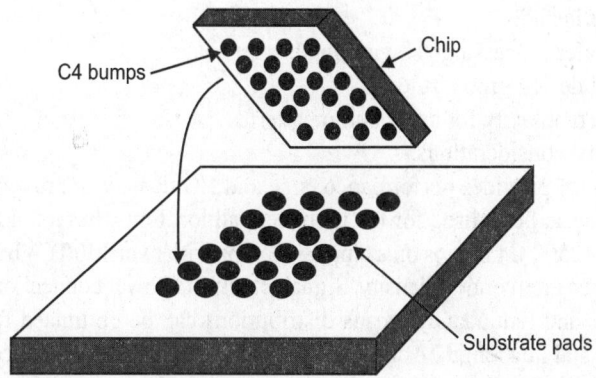

Fig. 19.4. C4 (Controlled collapse chip connection) flip-chip.

'Flipped' chips align to corresponding substrate metal patterns. Electrical and mechanical interconnects are formed simultaneously by reflowing the C4 bumps (Fig. 19.5). The C4 joining process is self-aligning, i.e. the wetting action of the solder will align the chip's bump pattern to the corresponding substrate pads. This action compensates for slight chip-to-substrate misalignment (up to several mils) incurred during chip placement.

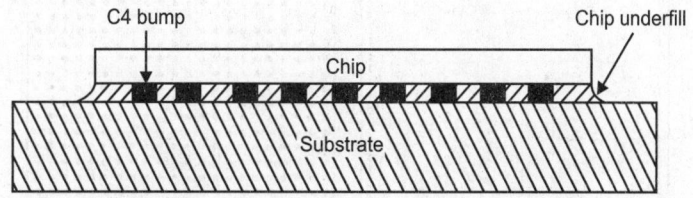

Fig. 19.5. Mechanical and electrical connections.

An added feature of C4 is the ability to rework. Several techniques exist that allow for removal and replacement of C4 chips without scrapping the chip or substrate. In fact, rework can be performed numerous times without degrading quality or reliability.

For improved reliability, chip underfill may be injected between the joined chip and substrate as illustrated in Fig. 19.5. It should be noted that any rework must be performed prior to application of chip underfill.

It is important to recognise certain C4 characteristics when deciding on an interconnect technology. While application, size, performance, reliability and cost all must be factored in the selection process. However, these factors cannot be applied to the chip or product only. The overall impact at the system level must be considered for an equivalent comparison.

The primary advantage of C4 is its enabling characteristics. Specific advantages include:
1. Size and weight reduction.
2. Applicability for existing chip designs.
3. Increased I/O capability.
4. Performance enhancement.
5. Increased production capability.
6. Rework/chip replacement.

Key considerations include:

1. Additional wafer processing vs. wire bond.
2. Supplemental design groundrules.
3. Wafer probe complexity for array bump patterns.
4. Unique thermal considerations.

Most importantly, C4 provides performance, size and I/O density improvements. With C4, nearly the entire chip surface can be utilised for interconnect pad locations. In fact, it has been demonstrated that one can have over 2500 C4 bumps on a chip, and chips with over 1500 C4 bumps are in production.

C4 enables increased interconnect density. Signal, clock and power connections can be placed almost anywhere on the chip and redundancy means distributions can be optimised for minimum noise and skew, current density and line length. Additionally, on-chip wiring can be reduced since z-axis escapes are available where needed.

Figure 19.6 compares single row wirebond and C4 chips. Each chip is 8 mm (200 mil) square. Wirebond pitch is 76 mm (3 mil) pads on 100 mm (4 mil) centres. C4 pitch is 100 mm (4 mil) bumps on 230 mm (9 mil) centres. In this example, interconnect density is increased over 140 per cent using C4.

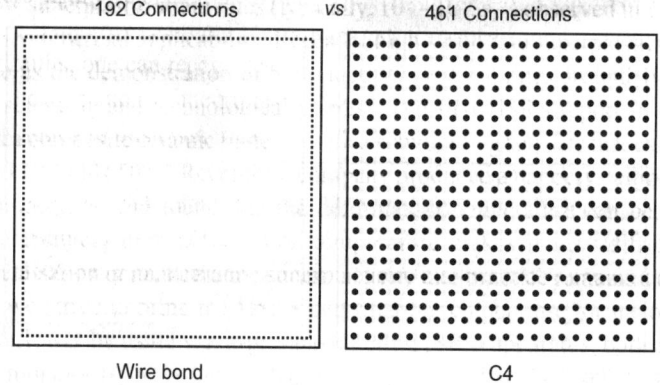

Fig. 19.6. Interconnect density (wire bond *vs.* C4).

The reliability of flip-chip contacts is determined by the difference in the CTE between the chip and the ceramic substrate or the organic printed circuit board (PCB). For example, the CTE for silicon is ~2–3 ppm/°K, for 96 per cent alumina it is 6.4 ppm/°K, and for PCB it is typically 20 to 25 ppm/°K. The CTE mismatch between the chip and the carrier induces high thermal and mechanical stresses and strain at the contact bumps. The highest strain occurs at the corner joints, whose distance is the largest from the distance neutral point (DNP) on the chip. For example, the DNP for a 2.5 × 2.5 mm chip is 1.7 mm. The thermomechanical stress and strain cause the joints to crack. When these cracks become large, the contact resistance increases, and the flow of current is inhibited. This ultimately leads to chip electrical failure. The typical reliability defined failure criterion is an increase in resistance in excess of 30 mW over the zero time value. The trade-off in selecting the bump height is that large bumps introduce a series inductance that degrades high-frequency performance and increases the thermal resistance from the device to the carrier, if that is the primary heat path.

The reliability of the bump joints is improved if, after reflow, a bead of encapsulating epoxy resin is dispensed near the chip and drawn by capillary action into the space between the chip and the carrier. The epoxy is then cured to provide the final flip-chip assembly. Figure 19.5 shows a typical flip-chip package. The epoxy-resin underfill mechanically couples the chip and the carrier and locally constrains the CTE mismatch, thus improving the reliability of the joints. The most essential characteristic of the encapsulant is a good CTE match with the z-expansion of the solder or the bump material. For example, if one uses 95 Pb/5 Sn solder having a CTE of 28 ppm/°K, an encapsulant with a CTE of about 25 ppm/°K is recommended. Underfilling also allows packaging of larger chips by increasing the allowable DNP. In some cases, the encapsulant acts as a protective layer on the active surface of the chip.

Good adhesion among the underfill material, the carrier, and the chip surface is needed for stress compensation. The adhesion between the surfaces can be lost and delamination can take place if contaminants, such as post-reflow flux residue, are present. For this reason, a fluxless process for flip-chip assembly is desirable. Unfortunately, flip-chip bonding on PCB requires the use of flux. However, on ceramic carriers with gold, silver, and palladium-silver thick-film patterns and via metallisations, fluxless flip-chip thermocompression bonding with gold-tin bumps has demonstrated high reliability. The results of reliability testing are summarised in Table 19.1 and may serve as a guideline for future work.

Table 19.1. Summary of reliability test conditions and results for fluxless flip-chip thermocompression bonded bump contacts.

Parameter	Value
Bump height	30 to 70 mm
Chip size	A few mm
Chip carrier	Ceramic
Carrier camber	5 μm per cm
Camber compensation	By bump deformation
Underfill	Yes
Thermal cycling	After 6500 cycles (–55° to +125°C), no contact failure and no change in contact resistance
High-temperature storage	After 1000 hrs no increase in contact resistance
Temperature and humidity	After 1000 hrs (85°C and 85% RH), no change in contact resistance
Pressure-cooker test	After 1000 hrs (121°C and 29.7 psi), contact resistance increased slightly from 3 mW to 4 mW

Finally, care should be taken that the encapsulant or underfill covers the entire underside without air pockets or voids, and forms complete edge fillets around all four sides of the chip. Voids create high-stress concentrations and may lead to early delamination of the encapsulant. After assembly, a scanning acoustic microscope can be used to locate voids in the encapsulant. The encapsulant should also be checked for microcracks or surface flaws, which have a tendency to propagate with thermal cycling and environmental attacks, eventually leading to chip failure.

Ball-grid-array (BGA)

Ball grid array is a surface mount chip package that uses a grid of solder balls as its connectors. It is noted for its compact size, high lead count and low inductance, which allows lower voltages to be used. BGAs come in plastic and ceramic varieties. It essentially has evolved from the C4 technology whereas more I/Os can be utilised in the same area as in a peripherally leaded package (or chip). The CBGA and PBGA are not truly chip scale packaging but the evolution to the μBGA has come out of the experience the industry has gained from the CBGA and PBGA packages.

Ceramic ball-grid-array (CBGA)

Originally designed by IBM, the CBGA was developed to complement their C4 (flip-chip) technology. The package is comprised of a ceramic (alumina) substrate and a C4 chip and an aluminium lid as depicted in Fig. 19.7. The ball-grid spacing is on 50 mil centres with solder balls composed of high melt solder (90/10 Pb/Sn) attached by eutectic solder (63/37 Sn/Pb). Recent designs have concentrated on miniaturisation and have reduced the package size and utilised 40 mil on centre solder balls.

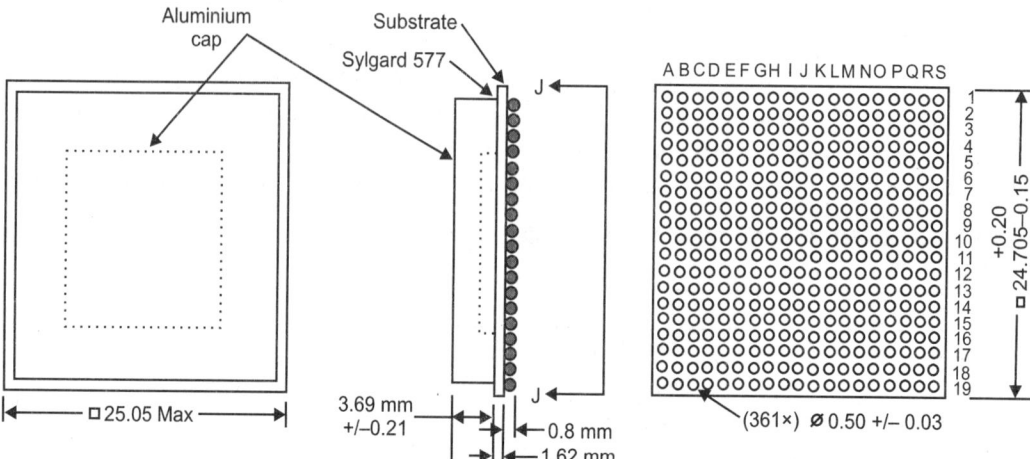

Fig. 19.7. Ceramic ball-grid-array package shown with connections on 50 mil centres with (a) top view, (b) side view and (c) bottom view illustrating the high number of connections.

Aluminum covers that have typically been used with the C4 technology have been bonded with a silicone adhesive (Sylgard 577) to provide a non-hermetic seal. With the flip-chip technology this is usually adequate for most applications. A hermetic seal can be accomplished by designing a seal ring into the ceramic and using a Ni/Fe cover plate for soldering. The package as described above has a cavity which would allow for typical chip-and-wire technology to be utilised. A MEMS device could be utilised in the wire bond package configuration first and migrate to use as a flip-chip in later designs.

Plastic ball-grid-array (PBGA)

The plastic ball-grid-array (PBGA) is very similar to the plastic packaging technology described in section. It is based on the same chip-and-wire technology and has moisture sensitivity (i.e. susceptible to 'popcorn' cracking during solder reflow) issues just like plastic packaging. It is different in that it is built on a printed circuit board substrate rather than a leadframe (metal) material (Fig. 19.8). The attach method (to the motherboard) is accomplished by soldering solder balls or bumps rather than leads.

One advantage this technology has over conventional plastic packaging is that the PC board material (which can vary from FR4 to polyimide to BT resin to name a few materials) can be a simple 2-layer board or be made of multiple layers. Additional layers allow for power and ground planes.

Micro-ball-grid-array (mBGA)

mBGA is a true 'chip scale package' (CSP) solution, only slightly larger than the die itself (die + 0.5 mm). It is the ideal package for all memory devices such as Flash, DRAM and SRAM. uBGA packages enable broad real-estate reductions of typically 50–80 per cent over existing packages. End use applications include cell phones, sub-notebooks, PDAs, camcorders, disk drives, and other space-sensitive applications. This package is also an excellent solution for applications that require a smaller, thinner, lighter or electrically enhanced package. It, therefore, lends itself nicely to space flight applications.

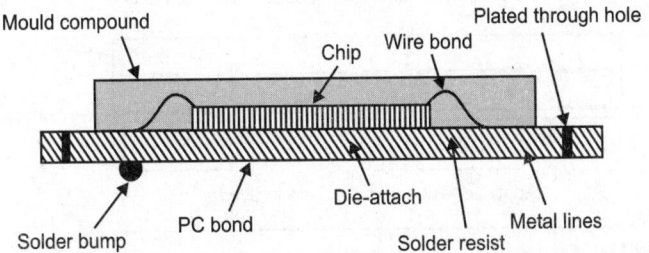

Fig. 19.8. A schematic representaion of a plastic ball-grid-array package.

The uBGA package is constructed utilising a thin, flexible circuit tape for its substrate and low stress elastomer for die attachment. The die is mounted face down and its electrical pads are connected to the substrate in a method similar to TAB bonding. After bonding these leads to the die, the leads are encapsulated with an epoxy material for protection. Solder balls are attached to pads on the bottom of the substrate, in a rectangular matrix similar to other BGA packages. The backside of the die is exposed allowing heat sinking if required for thermal applications. Ball pitches available today are 0.50, 0.75, 0.80, and 1.0 mm. Other features and benefits include: 0.9 mm mounted height, excellent electrical and moisture performance, 63/37 Sn/Pb solder balls, and full in-house design services.

Multichip Packaging

MCM/HDI

Multichip packaging of MEMS can be a viable means of integrating MEMS with other microelectronic technologies such as CMOS. One of the primary advantages of using multichip packaging as a vehicle for MEMS and microelectronics is the ability to efficiently host die from different or incompatible fabrication processes into a common substrate. High performance multichip module (MCM) technology has progressed rapidly in the past decade, which makes it attractive for use with MEMS.

The chip-on-flex (COF) process has been adapted for the packaging of MEMS. One of the primary areas of the work was reducing the potential for heat damage to the MEMS devices during laser ablation. Additional processing has also been added to minimise the impact of incidental residue on the die.

COF/HDI technology

COF is an extension of the HDI technology developed in the late 1980's. The standard HDI 'chips first' process consists of embedding bare die in cavities milled into a ceramic substrate and then fabricating a layered thin-film interconnect structure on top of the components. Each layer in the HDI interconnect overlay is constructed by bonding a dielectric film on the substrate and forming via holes through laser ablation. The metallisation is created through sputtering and photolithography. COF processing retains the interconnect overlay used in HDI, but moulded plastic is used in place of the ceramic substrate. Figure 19.9 shows the COF process flow. Unlike HDI, the interconnect overlay is prefabricated before chip attachment. After the chip(s) have been bonded to the overlay, a substrate is formed around the components using a plastic mould forming process such as transfer, compression, or injection moulding. Vias are then laser drilled to the component bond pads and the metallisation is sputtered and patterned to form the low impedance interconnects.

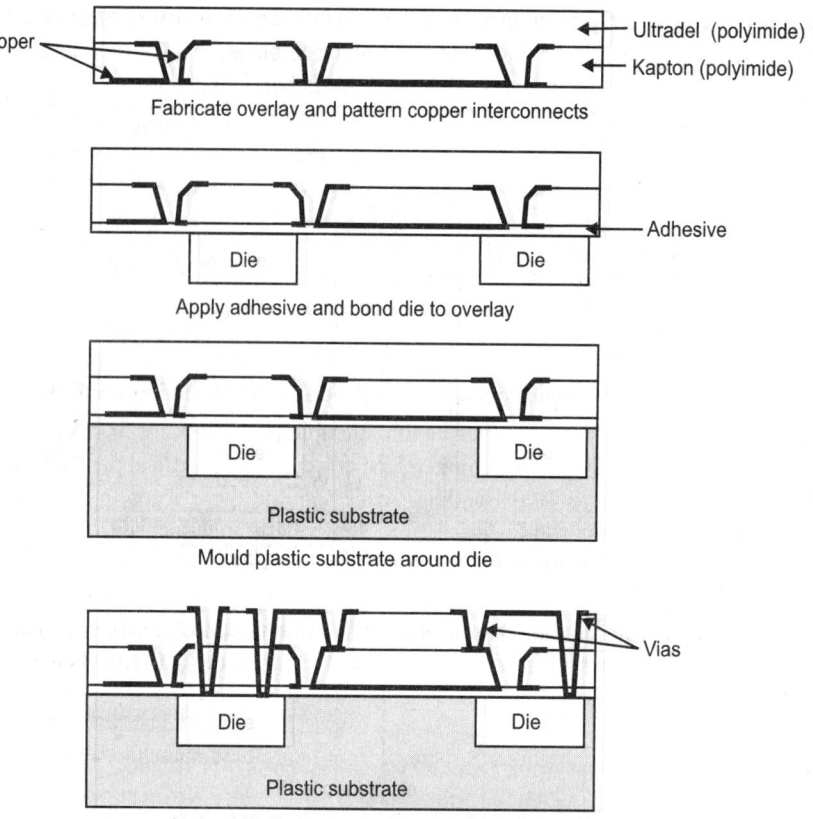

Fig. 19.9. Chip-on-flex (COF) process flow.

For MEMS packaging, the COF process is augmented by adding a processing step for laser ablating large windows in the interconnect overlay to allow physical access to the MEMS devices. Figure 19.10 depicts the additional laser ablation step for MEMS packaging. Additional plasma etching is also included after the via and large area laser ablations to minimise adhesive and polyimide residue which accumulates in the exposed windows.

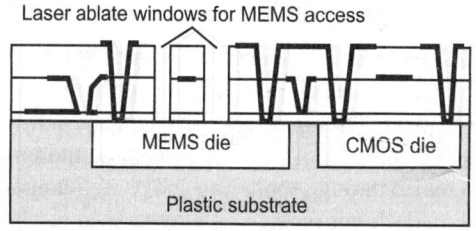

Fig. 19.10. Large area ablation for MEMS access in COF package.

MEMS test chip

MEMS test die can be used in research to assess the impact of various packaging technologies on MEMS. Test die typically contain devices and structures designed to facilitate a structured method of monitoring the health of MEMS devices after packaging. Surface micromachined test die have been available through the multi-user MEMS processes (MUMPs). The MUMPs process has three structural layers of polysilicon which are separated by sacrificial layers of silicon oxide. The substrate is electrically isolated from the polysilicon layers by a silicon nitride barrier. The top layer of the process is gold and is provided to facilitate low-impedance wiring of the MEMS devices but can also be used as a reflective surface for optical devices. Table 19.2 lists nominal thicknesses of the various layers, and Fig. 19.11 shows a cross-sectional view of a notional MUMPs layout.

Table 19.2. MUMPs layer thickness.

Layer	Thickness (mm)
Gold	0.5
Poly 2	1.5
2nd Oxide	0.75
Poly 1	2.0
1st Oxide	2.0
Poly 0	0.5
Nitride	0.6

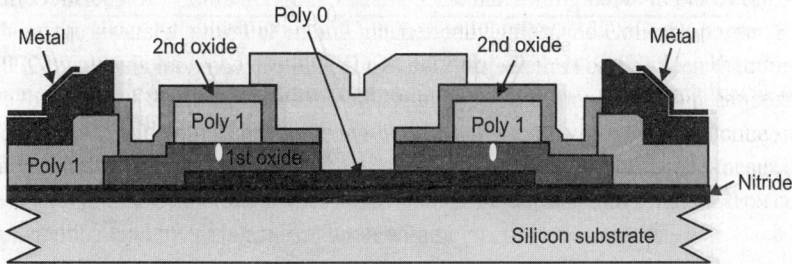

Fig. 19.11. Cross section of MUMPs layout.

Among the test structures on the test die are breakage detectors to monitor excess force and polysilicon resistors to monitor excess heating. Other devices on the die are representative of MEMS structures which might be used in an actual application. Table 19.3 lists general categories of devices on the surface micromachining test die.

Table 19.3. MEMS device categories included on surface micromachining test die.

Device category
Breakage detectors
Polysilicon resistors
Variable capacitors
Flip-up and rotating devices
Thermal actuators
Electrostatic piston mirrors
Electrostatic comb drives

The bulk micromachining test die was fabricated through MOSIS using the orbit CMOS MEMS process. The CMOS MEMS process is based on the standard 2 μm CMOS technology. The CMOS process has two metal and two polysilicon layers. Additional processing is added to allow MEMS fabrication. Provisions are made to specify cuts in the overglass to expose the silicon substrate for bulk micromachining. In addition, regions of boron doping can be specified to form etch steps for anisotropic silicon etchants such as ethylene diamine pyrocatechol (EDP) and potassium hydroxide (KOH). These tools allow for bulk micromachining to be accomplished in the standard CMOS process. Table 19.4 lists some of the device categories represented on our bulk micromachining test die. A sampling of integrated circuits such as ring oscillators for testing package interconnects was also included on the test die (Fig. 19.12).

Table 19.4. MEMS device categories included on bulk micromachining test die.

Device category
Breakage detectors
Polysilicon resistors
Cantilevers
Suspended structures
Thermal bimorphs

Fig. 19.12. Windows laser ablated in COF overly for MEMS access.

System on a chip (SOAC)

System on a chip may not necessarily be classified as a packaging technology. It is derived from the wafer fabrication process where numerous individual functions are processed on a single piece of silicon. These processes, generally CMOS technology, are compatible with the MEMS processing technology. Most SOAC chips are designed with a microprocessor of some type, some memory, some signal processing and others. It is very conceivable that a MEMS device could one day be incorporated on a SOAC. Initially, it may be incorporated by some other packaging technology such as flip-chip or mBGA.

Plastic Packaging (PEMs)

Most MEMS designs either have moving parts or do not allow for intimate contact of an encapsulating material such as in a traditional plastic package. Furthermore, plastic packages have not gained wide acceptance in the field of space applications. However, there are many semiconductor designs that are beginning to be flown in space applications. Programs such as commercial of the shelf (COTS) which include plastic encapsulated microelectronics (PEMs) are gaining wide acceptance. It is, therefore, important to outline the basic issues in PEMs for MEMS applications.

Studies have shown that during the high-temperature soldering process encountered while mounting packaged semiconductor devices on circuit boards, moisture present in a plastic package can vapourise and exert stress on the package. This stress causes the package to crack and also causes delamination between the mould compound and the lead frame or die. This phenomenon is often referred to as 'popcorn' cracking. These effects are most pronounced if the package has greater than 0.23 per cent absorbed moisture before solder reflow. Figure 19.13 shows a typical example of a package crack. The mismatch in thermal expansion coefficients of the package's components also induces stresses. If these combined stresses are greater than the fracture strength of the plastic, cracks will develop. The cracks can provide a path for ionic contaminants to reach the die surface, and/or die delamination can cause wire-bond failure. Hence, these are reliability concerns.

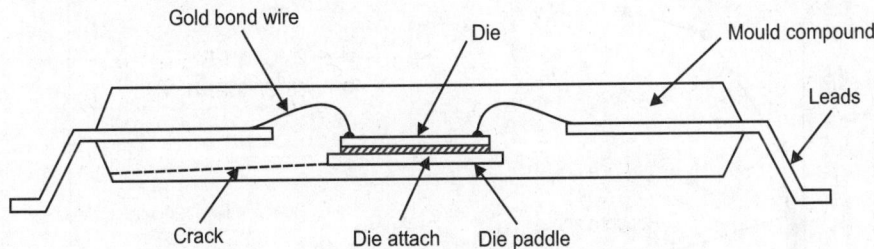

Fig. 19.13. Typical plastic package showing the onset of a crack.

JEDEC defines five classes for moisture resistance of plastic packages and sensitivity to 'popcorning'. Class 1 is defined as unlimited exposure to moisture and the package will still not exhibit delamination during the surface mount operation. Class 5 can tolerate minimal exposure to moisture before it needs to be dried (by baking in an oven set at ~125°C for a duration of 8 to 24 hrs depending on the package). Classes 2 through 4 are defined as somewhere in between the extremes. Most commercial packages are classified as class 3 moisture resistant.

To overcome the delamination problem, results derived from numerical simulation and experimental data can serve as a guide in the selection of suitable moulding compound properties. The properties considered are the adhesion strength, S, and the coefficient of thermal expansion, α. These results are summarised in Fig. 19.14.

The amount of moisture a particular package design can take up prior to delamination and catastrophic popcorning can be empirically determined as shown in Fig. 19.15. As can be seen, a high moisture environment (as well as high temperature) greatly reduces the amount of time on a production floor prior to the surface mount operation.

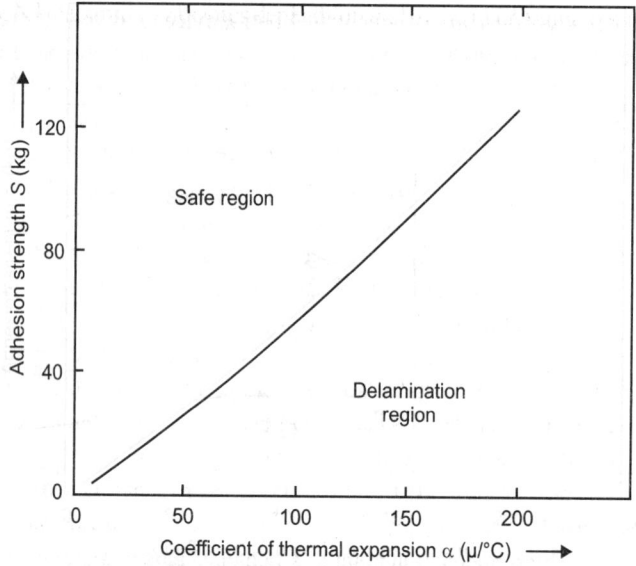

Fig. 19.14. Mould compound properties.

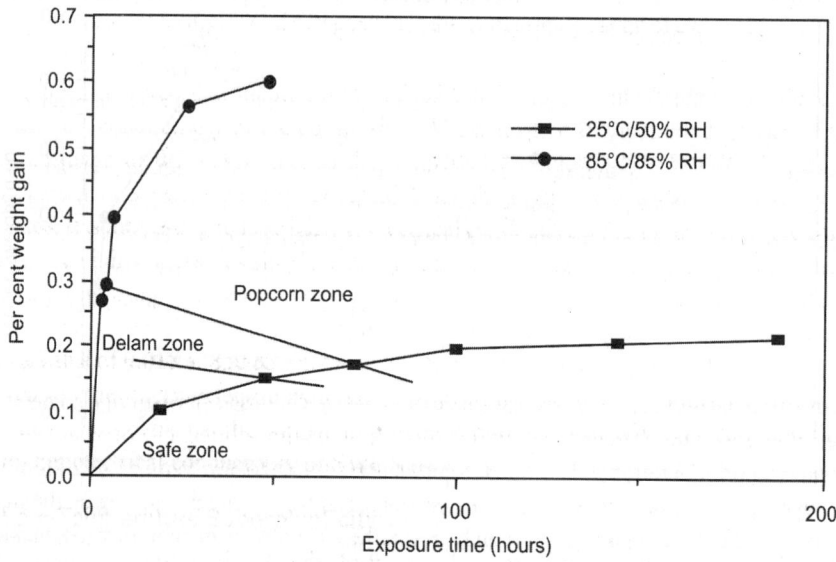

Fig. 19.15. Moisture weight gain of a plastic package exposed to two different moisture conditions.

Also, it has been shown that polyimide die overcoat, or PIX, can reduce the per cent of die or pad delamination by up to 30 per cent on parts subjected to temperature cycling. This PIX coating can mechanically support air bridges during plastic encapsulation, provide a more uniform electrical environment for the die, and provide protection to the surface of the die. Figure 19.16 shows cross-sections of three PIX-treated dies. It has been reported that the PIX shown in Fig. 19.16(a) yields the best improvement in reliability. The PIXs shown in Figs. 19.16(b) and (c) are not as desirable, because, respectively, they cause wire-bond stress and do not protect the die surface.

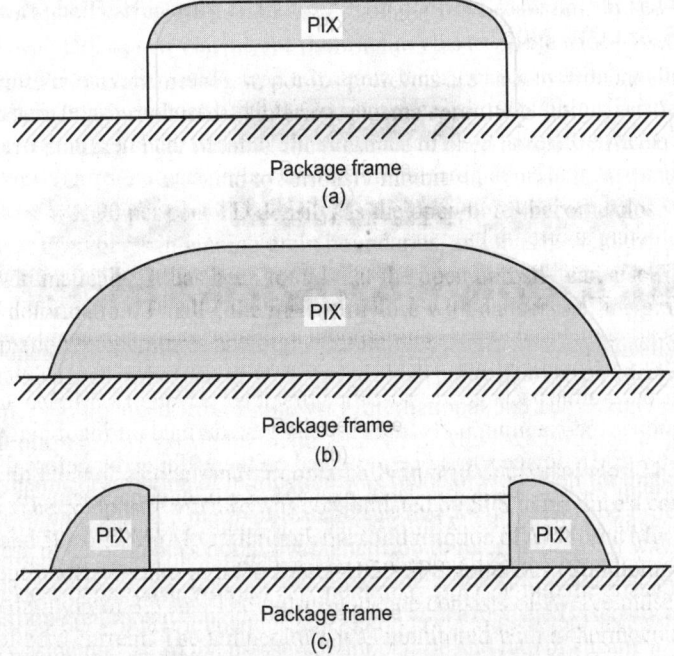

Fig. 19.16. Polyimide die overcoat (PIX) on MEMS die: (a) PIX on MEMS top surface only, (b) PIX on MEMS and package frame, and (c) PIX on package frame and sides of MEMS only.

The last mechanisms by which a chip can fail in a plastic package are caused by bondwire sweep and lift-off, which in turn are caused by the viscous flow of the molten plastic mould compound. The viscosity of the molten plastic is a function of the filler particle size and concentration. Figure 19.17 shows the typical geometries of wire bonds with different die settings. Studies show that of the three wire bonds, the one with the raised die experiences the largest maximum displacement. Further, the raised die and the downset die experience maximum stress at the ball bonds. In these cases, plastic deformation of the ball bonds is a major cause of failure. In contrast, the wire bond for the double-downset die suffers only elastic deformation. Thus, the double downset is the recommended device layout to minimise bond wire sweep.

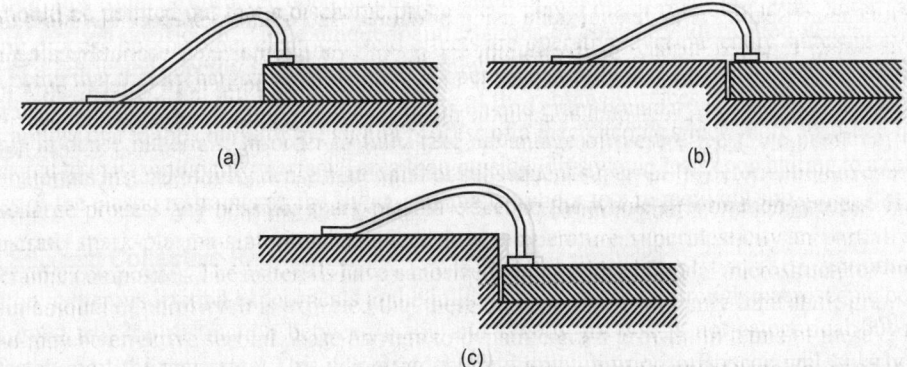

Fig. 19.17. Typical geometry of wire bond with different die settings: (a) raised, (b) downset, (c) double downset.

GENERIC SURFACE MICROMACHINING MODULE FOR MEMS HERMETIC PACKAGING AT TEMPERATURES BELOW 200°C

This section presents the different processing steps of a new generic surface micromachining module for MEMS hermetic packaging at temperatures around 180°C based on nickel plating and photoresist sacrificial layers. The advantages of thin film caps are the reduced thickness and area consumption and the promise of being a low-cost batch process. Moreover, sealing happens by a reflow technique, giving the freedom of choosing the pressure and atmosphere inside the cavity. Sacrificial etch holes are situated above the device allowing shorter release times compared to the state-of-the-art.

With the so-called over-plating process, small etch holes can be created in the membrane without the need of expensive lithography tools. The etch holes in the membrane have been shown to be sufficiently small to block the sealing material to pass through, but still large enough to enable an efficient release.

Unlike CMOS chips, chips containing MEMS (micro-electro-mechanical-system) cannot be directly packaged in a plastic or ceramic package (the so-called first level package) as MEMS are often composed of fragile and/or mobile free-standing parts that can easily be harmed during dicing and assembly. To avoid this damage, a MEMS device should be protected on the wafer level, before dicing.

This is possible with the so-called zero-level package. Zero-level packaging is typically done by bonding a capping wafer or die to the MEMS wafer. One route is to bond a micromachined capping wafer (usually glass or silicon) directly to the device substrate. However, in that case, high process temperature techniques are required. Other common ways are by using anodic or glass frit bonding. These are preferred because they require lower processing temperature (300°–500°C) than fusion bonding (1000°C). In addition, there has been a considerable fast development in packaging technologies using low temperature wafer bonding. However, all the above mentioned techniques increase die area, and thus the cost substantially.

Thin film encapsulation is an alternative wafer-level packaging technique with a minimum amount of wasted area. By this technique hermetic encapsulation can be achieved by the fabrication and sealing of surface micromachined membranes covering the MEMS. By making the membrane thick enough or by using supports, plastic packaging can still be used as first-level packaging technique.

In this work, a new generic surface micromachining module for MEMS thin film packaging with a metallic membrane at temperatures below 200°C based on nickel electroplating is presented. With this technology the MEMS devices can be hermetically sealed and enclosed in a controlled atmosphere and pressure as required for proper operation and ensured lifetime of the MEMS device. The advantages of this integrated packaging technique are the reduced thickness and area consumption, the promise of being a low cost process and the low thermal budget.

State-of-the-art sealing is in general done using horizontal sacrificial etch channels. For example, Stark and Najafi reported surface micromachined caps with horizontal etch channels using nickel electroplating and solder sealing. Release times of several hours were needed. In order to have a high-speed sacrificial release, sacrificial etch holes in the membrane are favourably situated above the device (Fig. 19.18). Recently, Rusu reported a versatile sealing method for sealing vertical access holes by using a two-layer thin film reflow process. Also in this work sealing is accomplished by using a reflow technique, but at much lower temperatures (180°C) compared to Rusu (600°C).

Requirements

Membrane

The membrane layer needs to be rigid, strong and it preferably has a low tensile stress in order to prevent any bending or breaking. A high deposition speed is preferred to be able to deposit thick layers in case plastic moulding is required afterwards. We choose a nickel electroplated membrane on top of a sacrificial photoresist layer because of its high Young's modulus (182 GPa). The membrane layer will be structured in order to allow the sacrificial etching of the underlying material. Sacrificial layer access holes are situated above the device (Fig. 19.18).

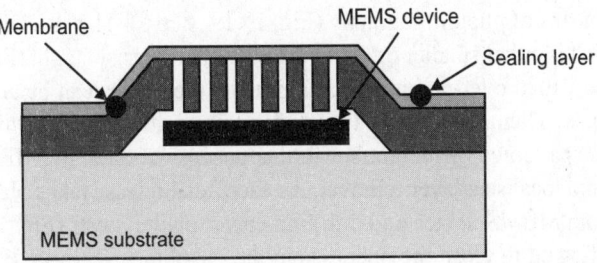

Fig. 19.18. Schematic view of a MEMS package with access holes situated above the device.

Sealing layer

The basic idea of the proposed sealing technique is to deposit a material with low melting temperature on top of openings in the membrane layer (see Fig. 19.19, step 7). After sacrificial layer etching, this material is then reflowed in a furnace with controlled atmosphere and pressure to close the final openings (see Fig. 19.19, steps 8 and 9). An optimisation of the thickness of both membrane and sealing layer as well as an optimisation of the size and shape of the etch holes is required. These holes should be large enough to enable efficient sacrificial etching but also small enough such that no or only negligible deposition inside the MEMS cavity takes place during sealing. The chosen sealing layer is indium with its low melting temperature (156.61°C) because it allows a low thermal budget sealing process. Indium has also an excellent ductility which allows joining materials of different thermal expansion coefficients.

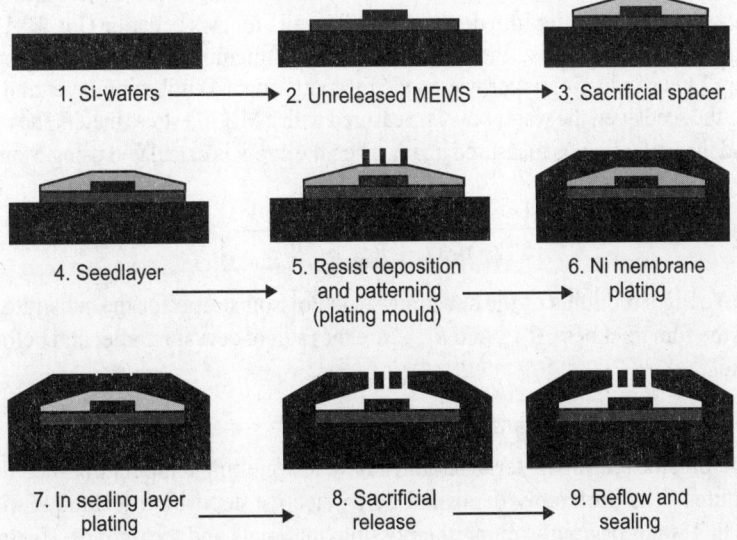

Fig. 19.19. Process in steps for a generic module for MEMS hermetic packaging at T < 200°C.

Experimental Procedure

The process flow in this work starts with unreleased MEMS devices (Fig. 19.19, step 2), followed by the deposition and patterning of the photoresist sacrificial layer (sacrificial spacer) (Fig. 19.19, step 3). Subsequently, the seedlayer for plating is sputtered on (Fig. 19.19, step 4) and the non-conductive plating mould (photoresist) for the membrane is spun on. In order to reduce release times, etch holes are defined above the device by means of photolithography (Fig. 19.19, step 5). Then nickel is electroplated until a certain thickness, which is either smaller or larger than the mould resist height (Fig. 19.19, step 6). In the latter case, with the so-called over-plating process, smaller etch holes can be created in the membrane than in the former case. Then, a wetting layer and a low melting temperature sealing material are electrodeposited on the patterned membrane for further encapsulation of the MEMS device (Fig. 19.19, step 7). After mould and local seedlayer removal, the sacrificial release takes place by etching away the sacrificial layers of both MEMS device and thin film encapsulation layer (Fig. 19.19, step 8). Then the sealing material is reflowed to close the openings in the membrane (Fig. 19.19, step 9). Figure 19.19 shows the proposed packaging process schematically.

Membrane deposition and characterisation

As mentioned previously, electroplated nickel was selected as the membrane material. The nickel is electroplated from a conventional sulphamate plating bath. According to Shipley's specifications, the pH in the solution was kept within the range of 3.8 to 4.5. The plating bath contains nickel sulphamate together with nickel chloride and boric acid. Boric acid buffers pH variations and nickel chloride is used to improve conductivity. The solution, which is held at a fixed temperature, is agitated during plating.

The stress in the membrane layer is a very important parameter for obtaining a stable membrane. Preferably the nickel membrane should have a low tensile stress at the end of the fabrication method. The two types of stress in thin films are thermal and intrinsic stress. Intrinsic stress is generated during film formation and is strongly dependent on process conditions. Thermal stress is caused by the difference in thermal-mechanical properties between the film and substrate and is unavoidable when the films are deposited at elevated temperatures. A common way to determine the stress (σ) in a film is by measuring the wafer curvature before and after film deposition and using Stoney's equation (Eq. 19.1). The substrates used in the experiments to optimise the stress in the nickel membrane were 6'' silicon wafers with a seedlayer on top. This conducting layer was made by sputtering a Ti adhesion layer and a Cu seedlayer. After sputtering the seedlayer, the wafer bow is measured with a MX203 stress meter. The nickel film is then electroplated and the wafer bow is measured again. Then the stress is calculated using Stoney's equation:

$$\sigma = \frac{1}{6} \frac{Y_s}{(1-\upsilon_s)} \frac{t_s^2}{t_f} \left(\frac{1}{R_{\text{after}}} - \frac{1}{R_{\text{before}}} \right) \qquad \text{... (19.1)}$$

where, Y_s is the Young's modulus of the substrate, υ_s is Poisson's ratio for the substrate, t_s the substrate thickness and t_f the film thickness. R_{after} and R_{before} are the radii of curvature after and before, respectively, plating the nickel.

Sealing layer deposition and characterisation

As mentioned before, the sealing material should have a low melting temperature such that it can reflow at low temperatures. It is preferably deposited by a selective deposition technique in order to avoid material where this is not desired. Among the possible materials and techniques, electroplated indium was selected as sealing material. One more requirement for the sealing layer is that the perforation holes in the nickel membrane layer, that are needed for the sacrificial release, are sealed after reflow in such a way that the MEMS device underneath is not damaged by deposition of material through the holes. The sealing layer thickness must be large enough to seal these holes.

Using test wafers with patterned nickel layers, films of different thickness of indium were deposited by electroplating until the trenches were nearly closed. The indium anode had a purity exceeding 99.998 per cent. Normally after mould resist removal, seed layer etch and sacrificial release, the reflow of the indium layer is done. On the test wafers, which are used to characterise and optimise the sealing process, no sacrificial layers were present. Therefore only the mould resist was removed before reflowing the indium. This reflow was performed in a rapid thermal process (RTP) furnace in a N_2 atmosphere at atmospheric pressure and at temperatures above the melting temperature of the material.

Sacrificial layer removal

Etch test structures such as cantilevers were used to determine the etch rate of the photo-resist sacrificial layers in an oxygen plasma. The samples had a 6 µm thick Ni structural layer and a 3 µm thick photo-resist sacrificial layer. The samples were placed in a chamber with downstream oxygen plasma for a specific time at a pressure of 1600 mTorr and a chuck temperature of 25°C.

Results and Discussion

Stress in nickel membrane

Nickel was plated with different thicknesses (6, 8 and 10 µm) at current densities (J) ranging from 1 to 4 A/dm^2 and at temperatures of 50°C, 55°C and 60°C. Figure 19.20 shows the measured stress of nickel layers plated at a constant current density for different temperatures. From the results it was concluded that for the chosen sulphamate bath and titanium-copper seed layer, the best results are observed for a plating current density at 1.41 A/dm^2 and a bath temperature of 55°C. The cathode efficiency was always above 95 per cent. The stress values obtained were always around 50 MPa compressive. Moreover, it was found that stresses are not dependent on the thickness grown. The effect of the reflow process on the internal stress of the plated nickel layers was studied by annealing nickel samples in a furnace with a controlled atmosphere at 180°C for different times. These conditions reflect the real reflow process. The stress in the nickel layer was found to change from low compressive to low tensile after the reflow process (Table 19.5). The low tensile stress after annealing is exactly what is required for the membrane layer.

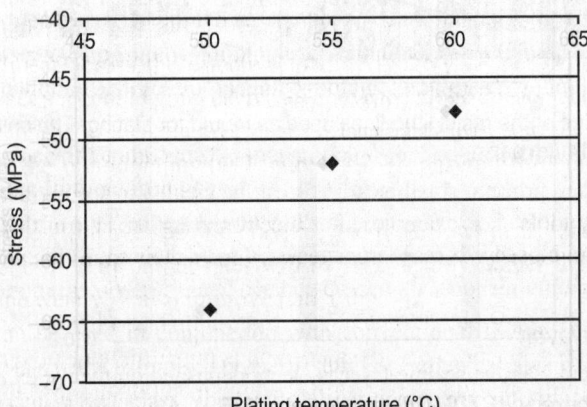

Fig. 19.20. Stress of 6 µm thick Ni layers versus plating temperature. J = 1.41 A/dm^2. Negative readings are compressive.

Table 19.5. Stress in plated Ni layers before and after annealing. The Ni was plated with J = 1.41 A/dm^2, and T = 55°C. Annealing was done at 180°C for 5 min in RTP. Negative values indicate a compressive stress and positive values a tensile stress.

Thickness plated (μm)	Average stress (MPa)	Avg. stress after annealing (MPa)
6 μm	−50.3 ± 1.5	91 ± 2
8 μm	−52.8 ± 0.3	81 ± 1
10 μm	−45 ± 2	85 ± 3

Plating of membrane and sealing layer

Plating of the membrane

Tests on nickel over-plating are done by filling high aspect ratio trenches. The process starts with 6″ silicon wafers on which a titanium/copper seed layer was deposited. To simulate the packaging process, trenches, which are defined by photolithography, mimic the access holes (see Fig. 19.19, step 5 and Fig. 19.20). The photoresist features consist of 6 μm thick and 1 to 15 μm wide lines and equally wide spaces (see Fig. 19.21). Then, nickel is electroplated into the exposed regions up to a thickness, which is larger than the resist thickness. Different thicknesses (8, 9 and 11 μm) were plated.

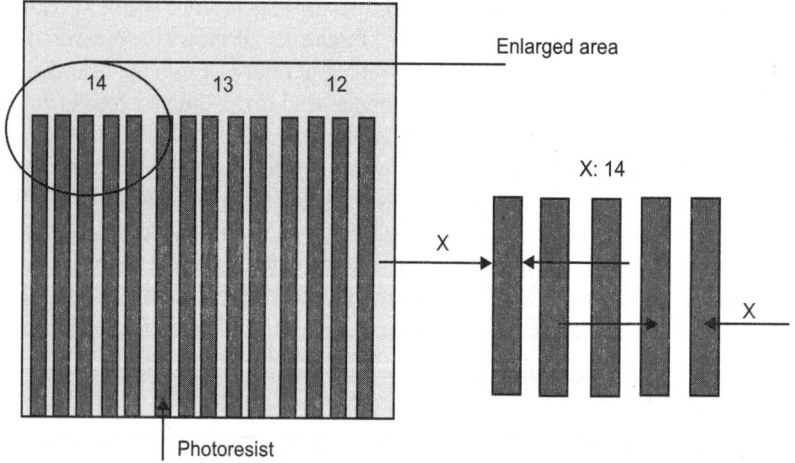

Fig. 19.21. Optical image of the resist litho-lines used as mould for plating. The enlarged feature consists of 14 μm wide lines with 14 μm gaps.

With this over plating process, small etch holes can be created in the membrane without the need of expensive lithography tools. For example, after electroplating an 11 μm thick nickel layer, features smaller than 10 μm are closed whereas larger features have their apertures reduced considerably (see Fig. 19.22).

Plating of the sealing layer

Prior to mould resist removal, the sealing layer is electrodeposited on the nickel trenches (see Fig. 19.19, step 4). Indium was plated with different thicknesses (1, 2, 3 and 4 μm) at current densities (J) ranging from 2 to 5 A/dm^2. Figure 19.23 shows a scheme of the indium plated above the nickel structures before and after reflow.

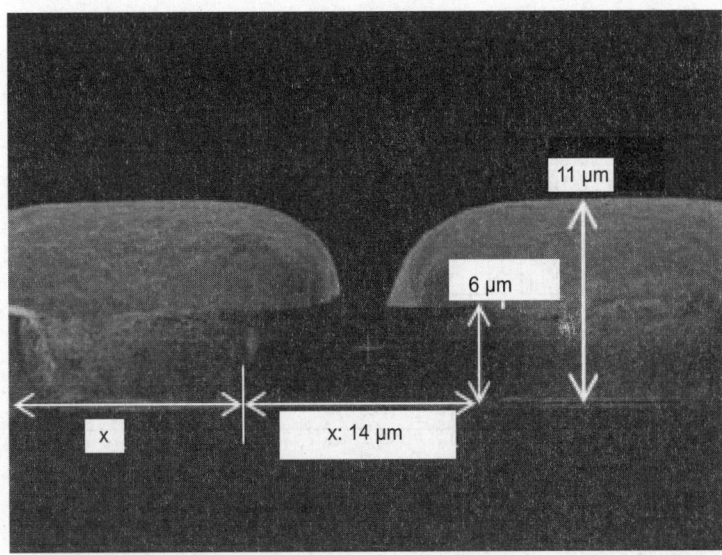

Fig. 19.22. 11 µm high over-plated nickel (mushrooms). Features smaller than 10 µm wide by 6 µm thick are closed.

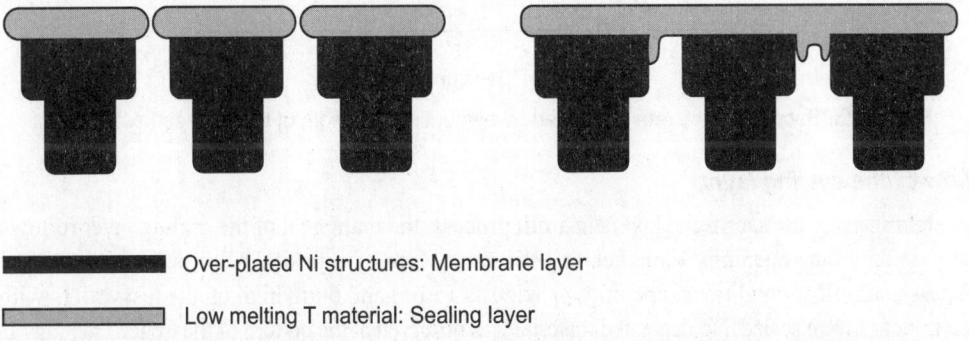

Over-plated Ni structures: Membrane layer

Low melting T material: Sealing layer

Fig. 19.23. One-layer sealing process: (a) Deposition of a low-temperature melting material, and (b) closing of the openings at an arbitrary chosen pressure and atmosphere.

Sacrificial release process

According to the real process, after plating the sealing layer, the mould resist is removed and the seedlayer is etched away. Then the sacrificial resist spacer below the Ni membrane needs to be removed. This can be done using an oxygen plasma.

The sacrificial etch rate was measured on test structures by evaluating the width of the released cantilevers after etching for a certain time. It was found to be 0.58 µm/min (see Figs 19.24 and 19.25). The underetching of the cantilever structures is almost linear as function of the etching time. From the measured etch rate, the release time for a certain design can be estimated. For example, if the distance in between etch holes is 8 µm, we expect a release time less than 10 minutes, while if the distance is 26 µm, the expected release time is less than half an hour.

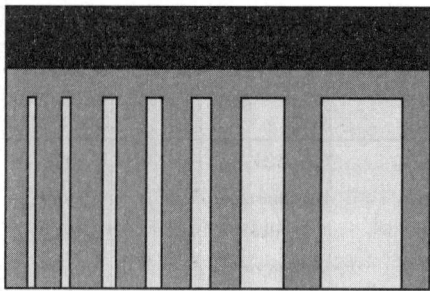

Fig. 19.24. Released cantilevers after 25 min in oxygen plasma: 2, 4, 6, 8, 10 and 25 μm width.

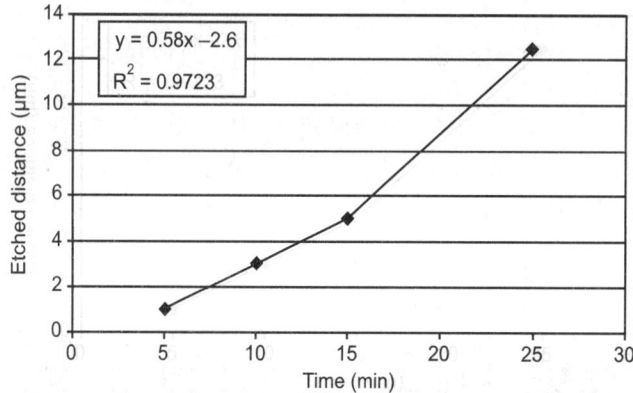

Fig. 19.25. Sacrificial etch rate measured by evaluating the width of the released cantilevers.

Reflow of the sealing layer

After etching away the sacrificial layers in a full process, the main goal of the sealing layer reflow is to close the membrane openings while keeping the cavity free of deposits. After optimising the indium thickness and reflow conditions, openings as wide as 15 μm and 6 μm high on the test wafers with the nickel trenches were sealed. No deposited material was observed at the bottom of the trench (see Figs 19.26 and 19.27).

Fig. 19.26. Overview of the 15, 14 and 13 μm wide plated nickel trenches sealed after indium reflow.

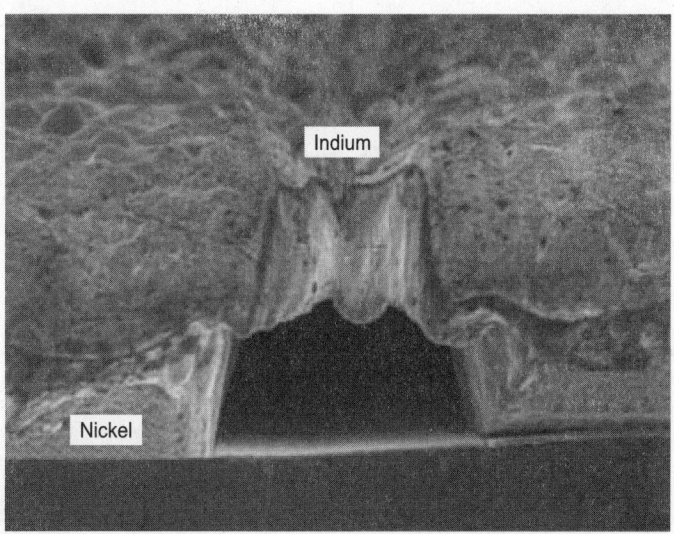

Fig. 19.27. SEM cross-section image of 15 μm wide by 11 μm high Ni plated trenches sealed after indium reflow. No material was deposited in the cavities.

Chapter 20

Nanosensor

INTRODUCTION

Nanosensors are any biological, chemical or surgery sensory points used to convey information about nanoparticles to the macroscopic world. Their use mainly include various medicinal purposes and as gateways to building other nanoproducts, such as computer chips that work at the nanoscale and nanorobots. Presently, there are several ways proposed to make nanosensors, including top-down and bottom-up design, top-down lithography, bottom-up assembly, and molecular self-assembly.

PREDICTED APPLICATIONS

Medicinal uses of nanosensors mainly revolve around the potential of nanosensors to accurately identify particular cells or places in the body in need. By measuring changes in volume, concentration, displacement and velocity, gravitational, electrical, and magnetic forces, pressure, or temperature of cells in a body, nanosensors may be able to distinguish between and recognize certain cells, most notably those of cancer, at the molecular level in order to deliver medicine or monitor development to specific places in the body. In addition, they may be able to detect macroscopic variations from outside the body and communicate these changes to other nanoproducts working within the body.

One example of nanosensors involves using the fluorescence properties of cadmium selenide quantum dots as sensors to uncover tumours within the body. By injecting a body with these quantum dots, a doctor could see where a tumour or cancer cell was by finding the injected quantum dots, an easy process because of their fluorescence. Developed nanosensor quantum dots would be specifically constructed to find only the particular cell for which the body was at risk. A downside to the cadmium selenide dots, however, is that they are highly toxic to the body. As a result, researchers are working on developing alternate dots made out of a different, less toxic material while still retaining some of the fluorescence properties. In particular, they have been investigating the particular benefits of zinc sulphide quantum dots which, though they are not quite as fluorescent as cadmium selenide, can be augmented with other metals including manganese and various lanthanide elements. In addition, these newer quantum dots become more fluorescent when they bond to their target cells. Potential predicted functions may also include sensors used to detect specific DNA in order to recognize explicit genetic defects, especially for individuals at high-risk and implanted sensors that can automatically detect glucose levels for diabetic subjects more simply than current detectors. DNA can also serve as sacrificial layer for manufacturing CMOS IC, integrating a nanodevice with sensing capabilities. Therefore, using proteomic patterns and new hybrid materials, nanobiosensors can also be used to enable components configured into a hybrid semiconductor substrate as part of the circuit assembly. The development and miniaturisation of nanobiosensors should provide interesting new opportunities.

Other projected products most commonly involve using nanosensors to build smaller integrated circuits, as well as incorporating them into various other commodities made using other forms of nanotechnology for use in a variety of situations including transportation, communication, improvements in structural integrity, and robotics. Nanosensors may also eventually be valuable as more accurate monitors of material states for use in systems where size and weight are constrained, such as in satellites and other aeronautic machines.

EXISTING NANOSENSORS

Currently, the most common mass-produced functioning nanosensors exist in the biological world as natural receptors of outside stimulation. For instance, sense of smell, especially in animals in which it is particularly strong, such as dogs, functions using receptors that sense nanosized molecules. Certain plants, too, use nanosensors to detect sunlight; various fish use nanosensors to detect minuscule vibrations in the surrounding water; and many insects detect sex pheromones using nanosensors.

One of the first working examples of a synthetic nanosensor was built by researchers at the Georgia Institute of Technology in 1999. It involved attaching a single particle onto the end of a carbon nanotube and measuring the vibrational frequency of the nanotube both with and without the particle. The discrepancy between the two frequencies allowed the researchers to measure the mass of the attached particle.

Chemical sensors, too, have been built using nanotubes to detect various properties of gaseous molecules. Carbon nanotubes have been used to sense ionisation of gaseous molecules while nanotubes made out of titanium have been employed to detect atmospheric concentrations of hydrogen at the molecular level. Many of these involve a system by which nanosensors are built to have a specific pocket for another molecule. When that particular molecule, and only that specific molecule, fits into the nanosensor, and light is shone upon the nanosensor, it will reflect different wavelengths of light and, thus, be a different colour.

PRODUCTION METHODS

There are currently several hypothesised ways to produce nanosensors. Top-down lithography is the manner in which most integrated circuits are now made. It involves starting out with a larger block of some material and carving out the desired form. These carved out devices, notably put to use in specific microelectromechanical systems used as microsensors, generally only reach the micro size, but the most recent of these have begun to incorporate nanosized components.

Another way to produce nanosensors is through the bottom-up method, which involves assembling the sensors out of even more minuscule components, most likely individual atoms or molecules. This would involve moving atoms of a particular substance one by one into particular positions which, though it has been achieved in laboratory tests using tools such as atomic force microscopes, is still a significant difficulty, especially to do en masse, both for logistic reasons as well as economic ones. Most likely, this process would be used mainly for building starter molecules for self-assembling sensors.

The third way, which promises far faster results, involves self-assembly, or 'growing' particular nanostructures to be used as sensors. This most often entails one of two types of assembly. The first involves using a piece of some previously created or naturally formed nanostructure and immersing it in free atoms of its own kind. After a given period, the structure, having an irregular surface that would make it prone to attracting more molecules as a continuation of its current pattern, would capture some of the free atoms and continue to form more of itself to make larger components of nanosensors.

The second type of self-assembly starts with an already complete set of components that would automatically assemble themselves into a finished product. Though this has been so far successful only in assembling computer chips at the micro size, researchers hope to eventually be able to do it at the nanometer size for multiple products, including nanosensors. Accurately being able to reproduce this effect for a desired sensor in a laboratory would imply that scientists could manufacture nanosensors much more quickly and potentially far more cheaply by letting numerous molecules assemble themselves with little or no outside influence, rather than having to manually assemble each sensor.

ECONOMIC IMPACT

Though nanosensor technology is a relatively new field, global projections for sales of products incorporating nanosensors range from $0.6 billion to $2.7 billion in the next three to four years. They will likely be included in most modern circuitry used in advanced computing systems, since their potential to provide the link between other forms of nanotechnology and the macroscopic world allows developers to fully exploit the potential of nanotechnology to miniaturise computer chips while vastly expanding their storage potential.

First, however, nanosensor developers must overcome the present high costs of production in order to become worthwhile for implementation in consumer products. Additionally, nanosensor reliability is not yet suitable for widespread use, and, because of their scarcity, nanosensors have yet to be marketed and implemented outside of research facilities. Consequently, nanosensors have yet to be made compatible with most consumer technologies for which they have been projected to eventually enhance.

INFRARED TEMPERATURE SENSORS

Measurement of surface temperature is a crucial component of energy transfer. Accurate measurement of the leaf-to-air temperature gradient is essential to the determination of transpiration rate and stomatal conductance in both single leaves and plant canopies.

This gradient is often less than 1°C, which means that leaf temperature should be measured to within 0.2°C. To achieve this accuracy, the Apogee Instruments Infrared Radiometers correct for changes in the sensor body temperature with a subroutine designed for Campbell Scientific dataloggers.

Gold-Flecked Nanosensor Detects Poisonous Mercury

Researchers led by an Indian Australian have pioneered a gold-flecked nanosensor that can precisely measure one of the world's most poisonous substances, mercury.

Developed by Royal Melbourne Institute of Technology (RMIT), the mercury sensor relies on gold flecks that are nano-engineered to make them irresistible to mercury molecules.

In the effort to reduce mercury contamination in the environment and the associated health risks, accurately measuring the toxin has become a priority for mercury-emitting industries like coal-burning power generators and alumina refineries.

Industrial chimneys release a complex concoction of volatile organic compounds, ammonia and water vapour that can interfere with the monitoring systems of mercury sensors.

The mercury sensor was developed with the use of patented electrochemical processes that enabled the RMIT researchers to alter the surface of the gold, forming hundreds of tiny nano-spikes, each one about 1000 times smaller than the width of a human hair. These nanoengineered surfaces are then used with existing technologies such as Quartz Crystal Microbalances—a finely tuned set of scales that measure weight down to molecular levels—to determine the levels of mercury in the atmosphere.

Smarter, Faster Nanosensor

A tiny carbon-nanotube-based chemical sensor can detect low parts-per-billion concentrations of gases. It can also go from detecting one gas to another within half a minute. Typically, carbon-nanotube or nanowire-based sensors, which can be extremely sensitive in detecting gases, take hours to recover and be reused (Fig. 20.1).

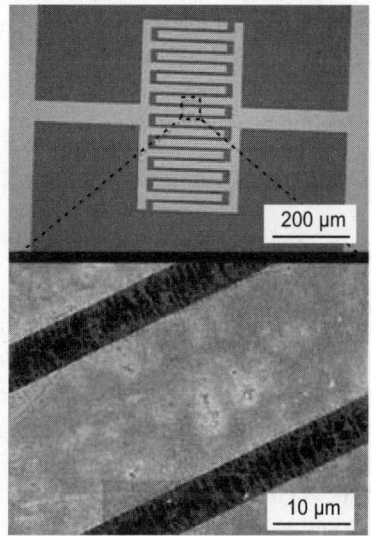

Fig. 20.1. A close-up shows carbon nanotubes (bottom) spanning the space between interlocking gold electrodes in a new type of gas sensor. The nanotubes are coated with an amine through which gases adsorb on the nanotube surface and detach after a few milliseconds. Change in conductivity of the carbon nanotubes specifies which gas was adsorbed.

The researchers coat the carbon nanotubes with chemicals that allow the nanotubes to rapidly switch their response. A network of the sensors could be used to monitor the spread of toxic gases or the movement of various pollutants over a large area. 'Instead of detecting whether a pollutant's there or not, you can detect its motion,' says Michael Strano, a chemical-engineering professor at MIT, who led the work, which was presented in *Angewandte Chemie.* 'Where is the wind moving it? Where is it most toxic?'

The new device is made of two main parts. The first is an ultrasmall gas chromatograph, an instrument commonly used in chemical analysis to separate mixtures of gases. To make a micro version of the instrument, the researchers etch a zigzagging, 35-centimeter-long channel on a silicon chip that is 800 micrometers on each side. As in conventional gas chromatography, different chemicals pass through the column at different rates, depending on their physical and chemical properties, so they exit the column at different times.

The output of the chromatograph feeds into the nanotube sensor. The sensor contains carbon nanotubes spanning the space between tiny gold electrodes. When various gases adsorb on the carbon nanotubes, the nanotubes' electrical conductivity changes by a different amount. By measuring the change in conductivity after the gas binds to the nanotubes, the researchers can identify the gas. 'The idea of incorporating a micro gas chromatograph with a carbon-nanotube sensor is probably the [best] way to go from a practical point of view,' says Pulickel Ajayan, a mechanical-engineering and materials-science professor at Rice University. In a real-world setting, there would be a mixture of gases — air pollutants, say — which would need to be separated before the individual gases can be detected.

Otherwise, even with an extremely sensitive detector, 'you can get very high sensitivity to chemicals, but usually you don't know what chemical it is,' says Ray Baughman, director of the NanoTech Institute at the University of Texas at Dallas. 'By coupling the micro gas chromatograph with the sensor, [researchers] . . . have a reasonable expectation of what the chemical is.'

The researchers test the sensor with a chemical that mimics the nerve toxin sarin. They are able to detect a billion molecules of the gas, corresponding to a concentration of 150 parts per billion.

Increasing sensitivity and selectivity of the nanosensor: If we can increase the sensitivity from parts per billion to parts per trillion, it can detect biochemical markers of lung, breast, and skin cancer. Cancer cells produce different metabolic products which can be sensed by dogs. Their noses have increased sensitivity. Sensing and subsurface technologies are rapidly being used for early cancer detection. If detected early, there is rougly 40 per cent chance of being cured.

CHEMICAL NANOSENSOR DEVELOPMENT AND CHARACTERISATION

Nanoscience and nanotechnology, through the exploration and control of the nanomaterials at the nanometer scale, is considered as one of the key research areas for the future growth of US economy. Many sensor devices are part of our everyday life. Due to the well-organised structure in atomic level of nanomaterials and their large surface-to-volume ratio, nanosensors are becoming very attractive for the next-generation of the sensing devices. Chemical nanosensors (CNS) are fabricated for the space and environmental applications, the safety alert devices, etc. For example, demonstrate a high-resolution CNS that is applied to the rocket fuel hydrazine leak detection. CNS that detects the changes in the electrical conductivity response during the chemical species presence. When the hydrasine is leaked into air, it immediately dissociates into NO_2. Therefore, actually detecting the NO_2 gas in the trace amount from the leakage of the rocket fuel. Detailed works will be briefly reported that includes: sensor chips preparation and process control, detailed studies to the NO_2 response. Finally, the drift of the sensor baseline (stability) and the sensor response.

NANOSENSORS FOR AQUEOUS ENVIRONMENTS

Advances in environmental research will increasingly take advantage of new nanomaterials-based sensing technologies for portable, ultra-sensitive systems for real-time, direct analysis. The extremely high surface-to-volume ratio of nanowires and their synthesis in a myriad of chemical forms (ceramic and polymeric) lend themselves to detection of chemical agents (e.g. pesticides), micro-organisms (e.g. *E. coli*, Giardia), and mineral compounds; the key is to modify interfacial chemistries to achieve selectivity for a specific application. This project will move Idaho's sensor development efforts beyond the proof-of-principle stage into the domain of feasible sensor array development, and will, when integrated with the previously described research, provide insight into research questions at multiple scales.

Research Plan

The research goal of the aqueous nanosensor project is to design, construct, and test nanowire-based sensor devices and integrated arrays for the detection of chemical solutes and biological particulates within water-based mixtures. This team will collaborate with researchers across the state to integrate monitoring requirements into sensor designs; their efforts will address the following questions:

1. What are the best nanowire materials, geometries, and assembly configurations for ultra-high sensitivity and quick response time to the aqueous environment?
2. What is the nanowire electrical behaviour in aqueous media and the electrical contact physics between nanowire material and metal/metal oxide leads?

CHEMICAL SENSORS AND NANOTECHNOLOGY

Nanotechnology can enable sensors to detect very small amounts of chemical vapours. Various types of detecting elements, such as carbon nanotubes, zinc oxide nanowires or palladium nanoparticles can be used in nanotechnology-based sensors. These detecting elements change their electrical characteristics, such as resistance or capacitance, when they absorb a gas molecule.

Because of the small size of nanotubes, nanowires, or nanoparticles, a few gas molecules are sufficient to change the electrical properties of the sensing elements. This allows the detection of a very low concentration of chemical vapours. The goal is to have small, inexpensive sensors that can sniff out chemicals just as dogs are used in airports to smell the vapours given off by explosives or drugs.

The capability of producing small, inexpensive sensors that can quickly identify a chemical vapour provides a kind of nano-bloodhound that doesn't need sleep or exercise which can be useful in a number of ways. An obvious application is to mount these sensors throughout an airport, or any facility with security concerns, to check for vapours given off by explosive devices.

These sensors can also be useful in industrial plants that use chemicals in manufacturing to detect the release of chemical vapours. When hydrogen fuel cells come into use, in cars or other applications, a sensor that detects escaped hydrogen could be very useful in warning of a leak. This technology should also make possible inexpensive networks of air quality monitoring stations to improve the tracking of air pollution sources.

NANOSENSORS AND DEVICES FOR SPACE AND TERRESTRIAL APPLICATIONS

The chemical sensor market has been projected to be $40 billion dollars worldwide in less than 10 years. The development of chemical sensors is to monitor and control environmental pollution, to improve the diagnostics for point care in medical applications, to provide small, low power, fast, and sensitive tools for process and quality control in industrial applications, and to implement or improve detection of warfare threats and security. In all these applications, there is a demand for improved sensitivity, selectivity and stability beyond what is offered by commercially available sensors. It is believed that the emerging field of nanotechnology can play an important role in realising these goals, and the proposed concept represents a significant step in that direction.

Nanotechnology provides the ability to work at the molecular level, atom by atom, to create large structures with fundamentally new molecular organisation. It is essentially concerned with materials, devices, and systems whose structures and components exhibit novel and significantly improved physical, chemical and biological properties, phenomena, and processes due to their nanoscale size.

A nanosensor technology has been developed in NASA Ames using nanostructures, such as single walled carbon nanotubes (SWNTs) and metal oxides nanobelts or nanowires, on a pair of interdigitated electrodes (IDE) processed with a silicon-based microfabrication and micromachining technique. The IDE fingers were fabricated using photolithography and thin film metallisation techniques. Both *in situ* growth of nanostructure materials and casting of the nanostructure dispersions were used to make chemical sensing devices. These sensors have been exposed to nitrogen dioxide, acetone, benzene, nitrotoluene, chlorine, and ammonia in the concentration range of ppm to ppb at room temperature. The electronic molecular sensing of carbon nanotubes in our sensor platform can be understood by electron modulation of the nanostructured devices and analytes in terms of charge transfer mechanisms. As a result of the charge transfer, the conductance of nanostructures will change. The metal oxide nanobelts sensors operate at much lower temperature around 150°C compared with 500°C for conventional metal oxides sensors with same sensing behavior. Due to the large surface area, low surface energy barrier and high thermal and mechanical stability, nanostructured chemical sensors potentially can offer higher sensitivity, lower power consumption and better robustness than the state-of-the-art systems, which make them more attractive for defence and space applications, and for other commercial applications. Combined with MEMS technology, light weight and compact size sensors can be made in wafer scale with high yield and low cost. Additionally, a wireless capability of such a sensor chip can be used for networked mobile and fixed-site detection and warning systems for military bases, facilities and battlefield areas.

Rationale for Recommendation

The nanosensor and nanodevice based on nanostructures are recommended here for possible commercialisation. The research effort on this technology has been funded by NASA and FAA with interagency agreement in past several years. The development of recommended nanosensor and nanodevices has gained NASA's attention for funding in future years to raise the technical readiness level (TRL) for space mission. Currently, NASA is putting $500–$600K in this project to develop a sensor module that has a sensor chip contains 32 sensing channels using different nanostructured materials, a complete electronic system for sensing signal acquisition, and a pneumatic pathway for gas sample delivery. This sensor module will be plugged in a satellite secondary payload in an orbit around 500 km for a flight demonstration of trace chemical detection using nanosensors. A prototype can be easily derived from this flight module.

The research results show that our nanosensors provide high sensitivity (ppm-ppb), low power consumption (μW-mW), compact in size and mass, easy integration to the existing electronic system (e.g. flight demo unit will have power and data communication pins in one connector to interface with satellite's main board). With the optimised combination of different nanostructures in a sensor array, selectivity for detection of complex chemical mixtures or discriminate the different chemical environments will be improved. Nanotechnology allows working at atomic and molecular level, which enables us to miniaturize the sensor device and make a high density sensor array that can greatly increase the detection capability by using the redundant sensing elements, and design an orthogonal sensor array with different nano materials. Fabrication process has been developed for large scale production. The recommended nanosensor is illustrated in Fig. 20.2.

Current State-of-the-Art

Nanotechnology provides the ability to work at the molecular level, atom by atom, to create structures with fundamentally new molecular organisation. Nanostructured materials such as carbon nanotubes (CNT) offer superior performance over conventional approaches due to the remarkable mechanical properties and unique electronic properties as well as the high thermal and structural stability. Single-walled carbon nanotubes (SWNT) have all the atoms on the surface that are exposed to the environment, allowing a change in physiochemical properties sensitively. The lower energy barrier on the surface of

CNT ensures the room temperature sensing, which allow much lower power consumption in μW to mW per sensor. They have great potential for developing a new generation of chemical sensors to detect gas and vapour species. Carbon nanotube-based chemical sensors have the following properties:

1. High sensitivity (potentially single molecule sensitivity) due to large surface to volume ratio.
2. Fast response due to the one-dimensional quantum wire nature that makes its electronic properties very sensitive to gas adsorption.
3. Lower power consumption (at least 100 times less than the current system), which is ideal for persistent surveillance applications.
4. Small size and lightweight.

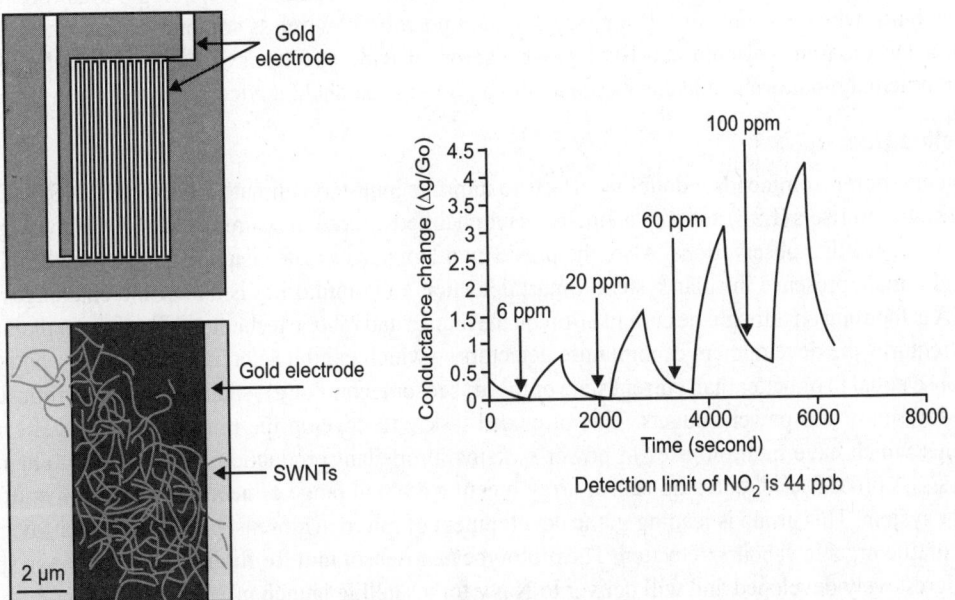

Fig. 20.2. Nanosensor and its response to NO₂ at different concentrations. Image of interdigitated electrode (upper left); Carbon nanotubes across gold electrodes (lower left); sensor response curve to NO₂ (right).

The design of the carbon nanotube-based chemical sensor uses an interdigited electrode platform, and purified single wall carbon nanotubes (SWCNTs) based materials for chemical sensing. Unlike the recent attempts that involved a field effect transistor approach with CNT as the conducting channel, the platform is essentially a chemiresistor. Here, demonstrate a simple SWNT sensor platform that combines the advantages of both single nanotube transistors and the film based nanotube sensors with extended applications to sensitive detection of organic vapours. In this platform, SWNTs form a network or mesh on interdigitated electrodes (IDE) using a solution casting process (Fig. 20.2) providing a large enough density of nanotubes for sensor performance. The IDE configuration enables effective electric contact between SWNTs and the electrodes over large areas while providing good accessibility for gas/vapour adsorption to all SWNTs including semiconducting tubes. This approach provides much simpler fabrication process with high yield of nanosensors that are robust, reliable, and reproducible in their sensor performance. The relatively simple fabrication is important to developing inexpensive sensor systems for the cost-conscious chemical sensor market. Above all, this type of sensor is amenable to integrate into a system. The preliminary work with CNTs indicates a conductivity change with exposure to NO_2, nitrotoulene (stimulant of explosives), a variety of other organic and inorganic vapours and gases. Preliminary results also show sensitivity in the ppb range with reproducibility of sensitivity around 6 per cent sensor to sensor. By loading the CNTs with catalytic metal nano clusters and coating polymers on the surface of CNT, great selectivity was achieved for room temperature detection of methane (combustible) in low ppm and chlorine (toxic) gas in ppb level.

Based on this nanosensor technology, a sensor module with a 32-channel array loaded with different nanostructures is underdevelopment for a space flight demonstration experiment. This sensor module will be in a satellite and fly in the mid-orbit around 500 kM for trace chemical detection.

Combining the carbon nanotube-based chemical sensor with MEMS technology, a lab-on-a-chip can be built, which has onboard data processing and potentially wireless communication capabilities. Such a lab-on-a-chip system can be deployed across a wide spectrum of hardware platforms for environmental monitoring, and can be made into a portable handheld device.

Baseline Use

A recent report recommends redoubling efforts to improve long-term climate modelling. NASA's Earth Science Enterprise's (ESE) action plan includes uninhabited aircraft measurements of greenhouse gases to validate satellite observations. Also, the president announced a new vision in January for NASA to extend human presence in solar system. Smart detection and monitoring is one of the challenges that NASA is focusing on through the current effort on the human and robotic technology for space exploration. This requires the development of miniature detectors—which exhibit selectivity, sensitivity, and time response equal to or better than current state of the art sensor technology—that are compact, robust and have low mass and power budgets. This proposed task will develop the miniature, chemical sensor systems which have multiple uses in power systems, propellant production, and planetary chemical analysis. A broad-based, one-stop-sensor array meeting a broad range of needs is the objective of this sensor system. This group is leading in the development of micro, nanosensor technology for toxic gas and volatile organic vapours detection. The prototype nanosensor unit for flight demonstration in space is progressively developed and will deliver to Navy for a satellite launch in 2012.

The product of proposed work not only has the potential use in space systems, but also has a wide spectrum of terrestrial applications. Some of these include: hydrogen sensors for fuelcell applications; carbon monoxide and NO_x sensors for emissions monitoring; ammonia and perfluoropropane in refrigerant leak monitoring; and carbon monoxide, ozone, and VOC sensors for air quality monitoring. The high performance sensors developed will be suitable for use in hostile environments and are amenable to miniaturisation (both inherent benefits of solid-state sensors over most other gas sensor technologies). The benefits to the commercial market are to solve several key market needs including increased sensitivity, higher reliability, and possibly longer times between calibrations. Sensor users are constantly looking for technologies that are low-cost and enhanced stability, lower detection limits and lower operating power, all potential improvements with the proposed technology.

The immediate use of this product is in defence arena for hazardous gases and vapours detection. The Department of Defence has identified a requirement for 2,70,000 JCAD (Joint Chemical Agent Detector) systems. The JCAD is a small, lightweight, high-performance chemical detection system that offers networked or stand-alone chemical detection. It can be hand-held or operated from a vehicle or from a fixed installation. Our sensing platform fits right there in this application.

Trends Impacting Improvement

Advancement in several areas can directly impact the improvement of our nanosensor technology. They are: (i) nanostructured materials development, (ii) high speed computing capability, (iii) nano printing and inkjetting technologies development, and (iv) wireless and network capability.

Nanostructured materials

Nanostructured materials have unique structural properties, such as high surface area, single crystalline, and well organised molecular structure. These properties are critical to the gas sensing due to the adsorption process prior to the chemical to electrical signal conversion. High surface area ensures high sensitivity, single crystalline provides stability of the nanostructures for sensor stability, and well organised molecular structure can assure the reproducibility of sensors' performance.

High speed computing capability

With the advance of the computing and computer technology, fast data acquisition, transformation for gathering transient chemical signal is possible. This will provide a platform to advance the sensing technology from data collection, signal processing, and information transformation point of view.

Nanoprinting and nanoinkjetting

Currently, there are many efforts on nano printing and nano inkjetting technology to get more precise positioning with less and small size of deposition spots. The development in this area will impact our nanosensors for scale up in mass production with automation capability and easy quality control process.

Wireless and network technology

Taking the advantages of wireless and networking capability, our nanosensors can extend the market from existing point care or measurement to a network mapping measurement. For example, the nanosensors can be used for environmental and safety monitoring each room and hallway in buildings.

Alternative Approaches and Organisations

The challenges in this technology are the selectivity of the sensors and the recovery time for sensor reuse. Alternative approaches can be taken to improve and ensure this technology leading the state of the art. The alternative approaches can be classified as (i) sensing materials, and (ii) sensing platforms.

Sensing materials

Semiconductor nanowires have been the subject of enthusiastic research efforts to enable a large variety of compositions, to incorporate dopants, and to investigate the promise of these structures to field-effect transistor, chemical and biological sensor, and optoelectronic technologies. Although many of these characteristics have been the subject of proof-of-principle experiments, much work remains until semiconductor nanowires can be integrated into a commercially viable technology.

It is within the scope of this concept paper to suggest future integration of this alternative inorganic approach into the fully developed interdigitated electrode array.

Sensing platforms

In analogy to the bulk properties, unadulterated nanowires have been demonstrated to function as switches, diodes, and photonic nanodevices. Related devices have been fabricated to demonstrate the sensitivity of these semiconductor nanostructures to chemical environment. Drawing on the principles of electrostatic gating, polar and charged analytes can be detected as a change in device conductivity, sensitivity to atmospheric gases has been investigated and various examples of biological sensing have been achieved.

Sensing platform, also called transducer, can influence the sensor performance as well. One platform has its own advantages over the others. The effect of sensing platform is also sensing material dependent. A good coupling of sensing material with platform will enable high sensitivity and selectivity with their specific features, such as gating effect of transistors, thermal power of chemically sensitive thermisters, impedence change of capacitors and Shottky diodes. A hybrid sensing platform can be composed to optimise the detection capability using sensor array approach with the advantages of each platform.

Another approach can be used in sensing platform alternation is the high density sensor array. These nanosensor can be self-assembled to a high density (10^3–10^6/cm^2) array with low power consumption (at least 100 times less than the current system), light and compact (at least 10 times smaller than the current system), and with proved high sensitivity (ppb-ppt), accuracy and rapid response. Inputs from high density nanosensor arrays will provide a high discrimination power that can accurately monitor and understand the chemical environment in region and allow appropriate response on the part of the crew or vehicle system. It can also be used for homeland security and battlefield chemical weapon detection.

Leading Aerospace Applications

There are two leading aerospace applications. One is to greatly increase the science measurement capability with compact size, less mass and power requirements for electronics and sensors; to provide a portable device for composition measurements of earth's atmosphere; to make highly miniaturised gas detectors enabling Earth Science Enterprise Plans for *in situ* measurements to validate satellite observations. Another application is to deploy the nanosensors which ensure the proper operation in the microgravity or reduced gravity environment, especially those which must operate in multiphase media, or are strongly impacted by the lack of natural buoyancy, such as combustion and precombustion monitors. The unique microgravity considerations make this deliverable particularly NASA-specific. Microgravity transport modelling capability will be employed to enable interpreration of sensor data, such as localisation of hazards. Sample acquisition for accurate monitoring must account for the complexities of multiphase (gas, liquid, solid) behaviouuur in micro or reduced gravity, and must also require little or no crew time and expendables. Sample handling may be necessary to achieve the necessary sensitivity and selectivity. The recommended nanosensors can be used for external environment monitoring to monitor hazardous conditions in the extra-vehicular environment. Hazards in these regions include but are not limited to reactive chemicals, erosive dust, and radiation.

Potential Nonaerospace Applications

The proposed technology has a specific application in space mission. Meanwhile it has a wide spectrum of application in terrestrial area, such as defense, industry, environmental, and medical and biological applications.

Industry

Nanostructure based chemical sensors possess high sensitivity, small size and low power consumption, which can be used to quickly verify incoming raw materials at the delivery point; to monitor the changing composition of the vapour phase surrounding or contained within the process. Much like vision inspection is used to assess the visual integrity (colour, shape, size) of products, olfactory inspection assesses the chemical integrity (consistency, presence of contaminants). The technology can significantly reduce the amount of time and money spent analysing those materials in a lab, as well as reducing the amount of materials handling.

Environment

Increasing awareness and new regulations for safety and emission control make environmental monitoring one of the most desired amongst the numerous industrial and civil applications for which the development of reliable solid-state gas sensors is demanded. Current methods for air quality control approved by the standards consist of analytical techniques, which need the use of very costly and bulky equipment. For applications in this arena, sensors that are able to selectively detect various gases at a concentration level of a few ppb and in the form of lowcost portable handheld devices for continuous *in situ* monitoring are needed. With unique advantages of high sensitivity, small size and low power consumption, and strong mechanical and thermal stability, carbon nanotube based chemical sensors are best fit for this type of application.

Defence

Chemical sensors are very focused for security and defence applications due to their portability and low power consumption. Carbon nanotube sensors potentially can offer higher sensitivity and lower power consumption than the state-of-the-art systems, which make them more attractive for defence applications. Some examples include monitoring filter breakthrough, personnel badge detectors, embedded suit hermiticity sensors and other applications. Additionally, a wireless capability with the sensor chip can be used for networked mobile and fixed-site detection and warning systems for military bases, facilities and battlefield areas.

Medical/Bio

It is believed that chemical sensors would provide physicians with a quicker and more accurate diagnostic tool. Applications could include obtaining objective information on the identity of certain chemical compounds in exhaled air and excreted urine or body fluids related to specific metabolic conditions, certain skin diseases or bacterial infections, such as those common to leg or burn wounds. Additionally, the chemical sensors may provide more accurate, real-time patient monitoring during anaesthesia administration.

It is note worth that the sensing platform developed in this project can also be used in liquid phase for heavy metals and pH detection, as well as for bio species detection, such as pathogen, bacteria, and enzymes.

Drivers for Change

Chemical sensors currently on the market already demonstrate the necessity for high surface area for analyte capture. Surface roughening is employed in various approaches to enhance surface area. One consequence of this approach is the associated compromise of electronic properties upon the introduction of disorder. The preservation of crystalline order that coexists with the inherent surface area enhancement of organic and inorganic nanowires illustrates one of their major advantages.

The high operating temperatures requisite to conventional metal oxide gas sensors also suggest an area of improvement. In fact, our preliminary investigations into nanostructured metal oxide sensors

(ZnO) suggest that operating temperatures are significantly reduced upon reduction to the nanoscale. The cylindrical configuration of these prototypical devices indicates that surface energy is an important parameter in operating temperature considerations.

Surface energy is intimately tied to surface structure, and further techniques to modify surface structure can be investigated through the concept of functionalisation that can also improve the selectivity.

Recommended Approach(es)

The scope of possibilities that is demonstrated by these early devices suggests a significant payoff for future development. Our extensive knowledge of the surface chemistry of nanostructures combine with simple sensing platform, IDE, suggests the potential for a large assortment of near-term applications. The recommended approaches for advancing our nanosensor technology are:

1. Functionalisation of nanostructures to improve the specificity.
2. Doping the nanostructures with catalysts to improve sensitivity, selectivity, and fast response time.
3. Nanostructures/polymer composites to make low cost, easy fabrication sensor arrays.
4. Hybrid sensing platform to maximise the array performance.
5. Temperature cycling to enhance the recovery time and improve the selectivity.

Rationale for Recommendations

Above recommended approaches are commonly used in commercial sensor technologies. Our design of the sensing platform using interdigitated electrode is mature and easy to work with. The sensing material is the key to the chemical sensing. Nanostructured materials provide not only the unique physiochemical properties in nanoscale, but also provide a well organised structure atom by atom for us to manipulate the surface and bulk structures. This gives us a lot of room to play with and enhance the sensor performance through the modification and stimulation of nanostructured materials.

Nanotechnology applications under development

Sensors using zinc oxide nano-wire detection elements capable of detecting a range of chemical vapours. Sensors using carbon nanotube detection elements capable of detecting a range of chemical vapours. Sensors using pallidium nanoparticle detection elements to detect hydrogen gas.

SECTION II

Nanocomposites and their Applications

21. Nanocomposites   345

22. Clay Based Nanocomposites   362

23. Processing and Characterisation of Nanoceramic Composites   374

24. Polyolefin, Polypropylene and Polystyrene Nanocomposites   388

25. Polyester-clay Nanocomposites   407

26. Nylon-6 Nanocomposites   417

27. Chemical Synthesis of Inorganic Nanomaterials and Composites   423

28. Biologically Derived Synthetic Nanocomposites   431

29. Metal and Oxide Nanocomposites   470

30. Sustainable Flame Retardant Nanocomposites   476

31. Nondestructive Testing of Nanocomposites by Optical Techniques   481

32. Nanoclays for Polymer Nanocomposites, Paints, Cosmetics and Waste-water Treatment   488

33. Hard and Tough Nanocomposite Coatings   503

Chapter 21

Nanocomposites

INTRODUCTION

Composite is normally called to those items, which are made from two or more different materials having a wide difference in physical/mechanical properties. Say, for example, we often talk about rubber fibre composite. Although both rubber and fibre are polymeric origin yet they have wide difference in both physical and mechanical properties.

Solid tyre (i.e. tyre without fabric in it) has failed to provide desired level of: (i) riding comfort, (ii) steering comfort/response, (iii) rolling loss, (iv) traction, and (v) uniformity in mileage.

On the other hand, it is almost impossible to make a tyre with fabric only. For a tyre that rolls on our road today is, therefore, practically made of rubber-fabric composite which gives distinctive advantages on items (i) to (v) mentioned above.

Nanocomposites are materials, which are very very tiny in nature to the level of nanometre (nm) in size, which is even smaller than the wavelength of visible light. A typical nano particle size could be below 100 nm and with their oligoatomic or molecular composition represent a scale of matter where radically different phenomena are manifested. The basic use of nanocomposite is to those products which show many folds of improvement on the physical/mechanical or on the processing properties upon addition very minute quantity of this material.

The basic concept of nanocomposite is as old as more than 20 years now. The system began with finely meshed nylon, with elastomer. Research has been made to utilise this system in various elastomer and plastics. It is understood now that not all polymers as well as not all fillers are suitable for nanocomposite developments.

One of the major finding is such particles increases surface interaction very strongly and reduces the defects in the volume of nano crystals. The effect of nano scale particles are not only for the use of the enhancement of mechanical properties but it has wide potential in the field of electronic, magnetic, optical and chemical field as well. As on today, nanotechnology has been recognised as one of the most major area on the advancement of technology in 21st Century. Polymer nanocomposites or PNC could be polymers of thermoplastics, thermosetting or elastomeric in origin. These materials provide improvement over other known composites in thermal, mechanical, electrical and even air barrier properties.

In this chapter we would review one of the possible future applications on polymer filler nanocomposites in pneumatic tyre.

The major applications on the field of tyre could be on (i) cost reduction, (ii) improvement in rolling loss, (iii) improvement in mechanical properties, and (iv) improvement in air retention properties.

345

APPLIED CONDITION

Normally, silicates or clay are multilayered filler and in some cases the aspect ratio (1/h) of these fillers could be greater than 300. In general condition, when these fillers are loaded in elastomers, the loading volume may increase from 30–50 phr depending on the desired physical/mechanical properties. But under the 'applied condition' in which the multilayer of a filler begins to separate out (Fig. 21.1) from a single particle and each sublayer will show now different particles, the overall reinforcements of the resulting system increases to many folds and in such case only 2–5 phr of filler dose can show similar physical/mechanical properties compared to the one with 30–50 phr of loading.

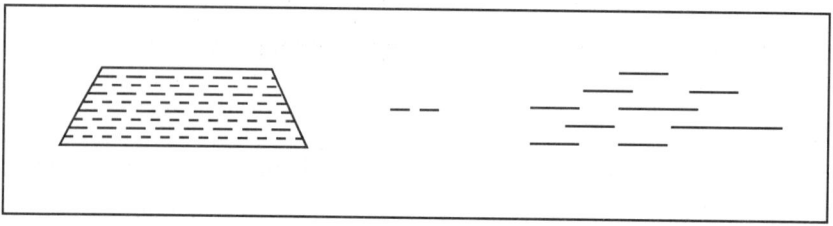

Fig. 21.1. Silicates are multilayered. Nanocomposite is the art of technology to separate the layers from one particle to give multi particles in the polymer interface.

It is true that as of now some producers of nanocomposite polymers have already entered into the commercial applications with their products but they are yet to establish a dominant market positions.

The 'applied condition' could be different for different polymer and different filler type. The ideal condition is how effectively the multilayers of (silicate filler, say) is separated from a single particle to multiparticles and how these multi particles are aligned in the rubber matrix to provide adequate and desired properties. To understand this we can cite one hypothetical example in the development of pneumatic tyre below:

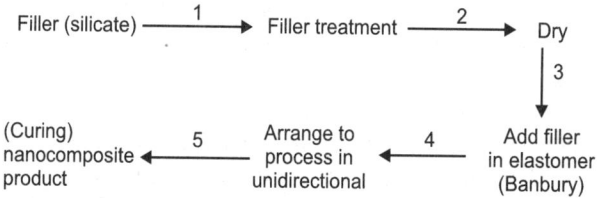

1. Here monomer, initiator, etc. are added under conditions, where monomer liquid can penetrate the multilayer of silicates. Polymerisation of monomer may or may not take place in this stage.
2. It is a simple dry process. Polymerisation may or may not take place in this stage.
3. The monomer may be chosen such that the polymerisation will begin soon the dump temperature (150°–180°C depending on different elastomers) is achieved in the matrix (Fig. 21.2). In fact, it is the polymerisation, force that causes separation of multilayers of silicates.
4. Further polymerisation may continue in the processing stages like milling, calendaring, extrusion, etc. Unidirectional processing will help to align the fillers in the direction of processing. This should further improve filler-rubber interaction because of filler orientation and improve on the crystallinity of composite.
5. Moulding and curing. This is the stage for the formation of nanocomposite product.

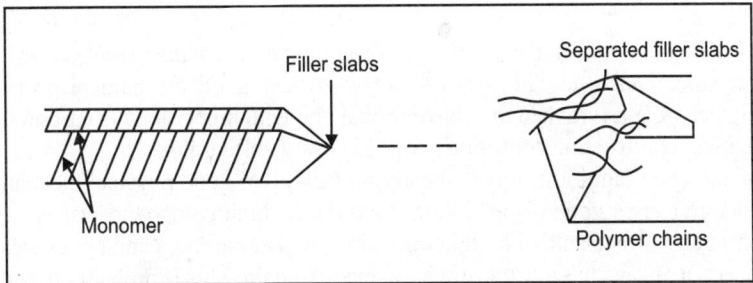

Fig. 21.2. Polymerisation force can be responsible for the separation of multilayered particle of silicates.

PROCESSING

Processing of nanocomposite materials in elastomer, thermo plastic or in thermosetting materials is same as conventional polymer-filler system except in PNC, it is designed to process in unidirectional way to develop better orientation of filler and to increase crystallinity of rubber-filler matrix (Fig. 21.3). Such composites should not show any problem in: (i) mixing, (ii) milling, (iii) calendaring, (iv) extrusion, (v) blow moulding, (vi) injection and micro injection moulding, and (vii) film blowing, etc.

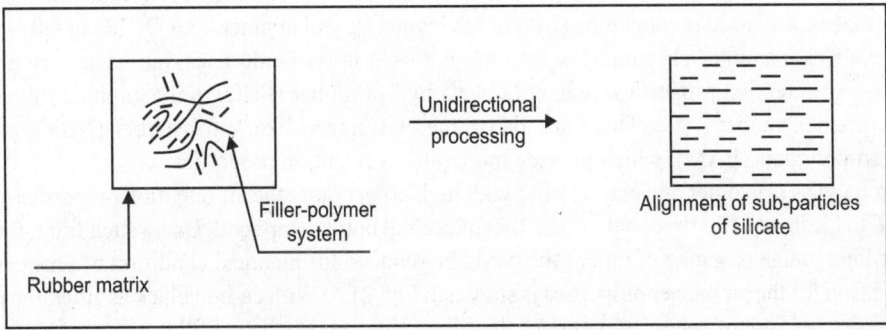

Fig. 21.3. Unidirectional processing of nanocomposite material will tend to provide alignment of nanoparticles.

Formulation of nanocomposite with suitable filler and suitable polymer and the processing technology of the above nano composite is the critical success factor to dominate the product in the market place.

Most companies are still in early stage for the development and processing of nanocomposite products. This gives enough opportunity for new entrants in the market. Nanocomposite technology is being developed for applications in packaging, automotive components, home appliances, electrical parts, electronic parts, building and construction products. Present market expects the demand by end 2010 and a significant demand growth may occur by 2012.

PROPERTIES

It is very interesting to note the enhance properties of nanocomposites. Because of the unique composite nature, nanocomposite may show advantages in pneumatic tyres in the following area.

Mechanical Properties

With low filler : rubber ratio one has further scope to improve mechanical properties by higher dosages of filler without detrimental effect on filler-filler interaction.

Rolling loss

With low filler loading in tread, side wall, ply skim compound, (etc.) the overall weight of the tyre will be on the lower side. Thus one can achieve tyre with very good mechanical properties but with comparatively lower weight tyre than the conventional one with higher load in carbon black. Low tyre weight with identical construction contributes towards lower rolling loss.

Though high structure (agglomeration) of carbon black gives good physical/mechanical properties (high black structure causes good dispersion in the mixture, high compound viscosity, low die-swell with smooth extrudate surface and with higher modulus, higher hardness and better wear resistance of the vulcanisate, yet, it always has adverse effect on heat build up. This is probably because with higher carbon black structure, though the probability of filler-rubber interaction increases, but the probability of filler getting no link with polymer also increases and hence higher heat builds up with high structure carbon black.

With silicates, however, since they cannot exist in high or low structure and otherwise also due to their inherent character, they are low heat build up particles compared to carbon black and hence silicates, with lower dose in elastomer as well as with low heat build up by nature will tend to provide low rolling loss of tyre.

Air Permeability

Excellent air retention properties of butyl/halo butyl is very well-known in tyre industries and these rubber is extensively used in inner tube (IIR) of tube type tyre and in inner liner (XIIR) in tube less tyre. There are also some other elastomers, which are not used in tyre industries, but have very good and even better air retention properties, like, epi chloro-hydrin rubber (ECO), poly sulphide rubber (TM) and poly urethen rubber (AU). There are also some elastomers, like, nitril rubber (NBR), ethylene/acrylate co-polymers (EAM), which provide moderate air retention properties.

Using butyl as elastomer and clay as filler with high aspect ratio, the air retention properties could be improved to the level of 50 times better than that of normal butyl compound. Using such filler, therefore, the inner liner gauge or gauge of inner tube could be reduced for identical condition of services.

The reason for the air barrier properties is shown in Fig. 21.4. With carbon black as filler, the passage of air is faster because of spherical nature of the filler. With needle shaped filler in nano composite the same air will take longer time to travel due to flat shape of the filler.

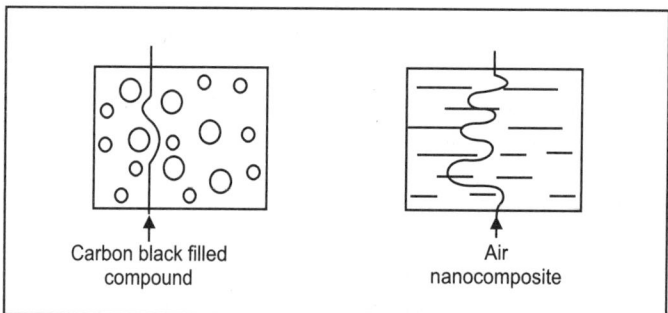

Carbon black filled compound

Air nanocomposite

Fig. 21.4. Air travel time in nanocomposite will take longer time as compared to carbon black filled compound.

Such fillers when used in rubbers (NBR, EAM) having moderate air retention properties can provide substantial increase-in air retention properties. It is known that nanocomposite technology is already commercialised in sports goods (tennis ball) for better durability of the product.

It is not surprising that all future tyre components may be made with nano composite type of filler, which can produce (i) low weight tyre, (ii) low rolling loss tyre, (iii) coloured or transparent tyre although, transparent tyre is very much in use in bicycle and motor cycle categories, and (iv) tubeless tyres without conventional inner liner in it.

It is also not surprising that in future, in place of conventional inner liner as on today, only water based inner paint, based on nanocomposite, may take care of both mould release as well as air retention properties of future tyre.

Reduction in Cost

In fact the items we have covered above are all related to cost saving. Once nanocomposite system is well commercialised the major users will be in the automotive parts and in automobile tyres (PC/TB).

NANOCOMPOSITES: NOVEL POLYMER/INORGANIC HYBRIDS

The idea of reinforcing polymers with other materials is not new and materials like cellulose, clay, calcium carbonate, carbon and various forms of silica have been used in this role for quite some time. Just as in other fields of nanotechnology, however, the novelty of these new materials is the scale on which things happen. The first such clay nanocomposites based on polyamide were synthesised by Toyota group in Japan.

Historically, polymers have been reinforced with rigid inorganic fillers for nearly as long as polymers have been produced. Things like silicates (mica, talc, clays, fumed silica, fibreglass), other metal oxides (titania, alumina), calcium carbonate and carbon black (just to name a few) have a long history of use. Many of these materials are of micrometre dimensions, making systems based on these materials microcomposites, or even macrocomposites, in the case of very large fillers (like fibreglass). From a simple rule of mixtures approach, stiffening is an inherent side effect of the inclusion of these rigid phases; unhappily, lower elongation at break (and often toughness) values are also common. That is, the materials began to take on the properties of the rigid fillers they contained. In many applications, it would be better to avoid this.

Well before the term 'nanocomposite' was coined, researchers, especially those in the rubber industry, were working towards the use of nanoscopic filler particles (carbon black, fumed silica). In the pre nanometer age, their size was measured in either Angstroms or 'millimicrons', but these were some of the first true nanocomposite systems. Even then, some of the basic truths of nanocomposite research had already been explicitly stated: Fine particle size does not necessarily lead to good reinforcement. In practice, the situation is complicated by the fact that very finely divided fillers tend to agglomerate and are extremely difficult to disperse. The use of organic or other coatings in filler surfaces sometimes promotes dispersion, and increases the effective use of fillers of very fine particle sizes. This issue of specific surface area hints at how one might change the nature of reinforcement. In typical micro and macrocomposites, the properties are dictated by bulk properties of both the matrix and the filler. This relationship between the properties of the composite and the properties of the filler is what leads to the stiffening and degraded elongation mentioned earlier.

Clay Structure

The crystal structure of clay is very interesting. Sodium montmorillonite having the molecular formula $Na(Mg_xAl_{2-x})$ $(AlSi_3O_{10})(OH)_2$ has a tetrahedral-octahedral-tetrahedral layered structure. A similar structure is shown in Fig. 21.5. Stacked layers have a regular interlayer gap called gallery. Swelling this hydrophilic clay in water and then exchanging the Na^+ cation of this clay with ammonium ion gives modified clays with increased gallery gap. A polymer containing modified clay gives rise to better properties, due to greater surface interaction through exfoliation of the silicate layered structure.

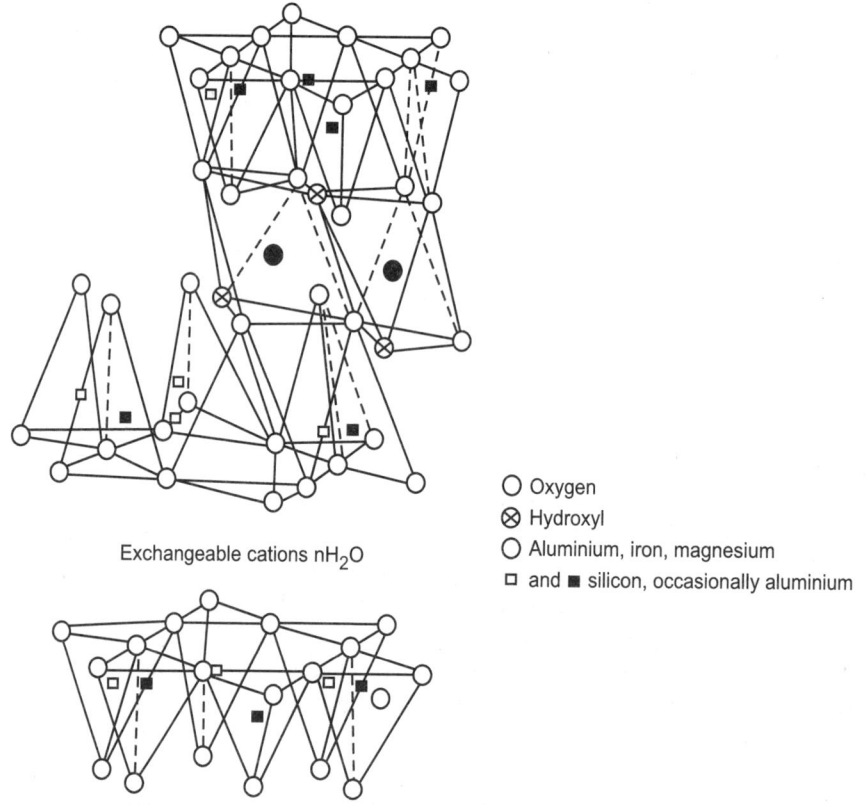

Exchangeable cations nH_2O

○ Oxygen
⊗ Hydroxyl
○ Aluminium, iron, magnesium
□ and ■ silicon, occasionally aluminium

Fig. 21.5. Structure of tetrahedra-octahedral-tetrahedral layered structure of smectite clay.

Bentonite has the chemical formula $(Al_{3.2}Mg_{0.8})Si_8O_{20}(OH)_4Na_{0.2}$. It is basically a mixture of more than one type of clay and includes Na^+-montmorillonite. The basic clay structure is same, but the difference lies in silica and aluminum content.

Nanocomposites

In the case of nanocomposites, the properties of the material are instead tied to the interface. Terms like 'bound polymer', 'bound rubber', and 'interphase' have been used to describe the polymer at or near the interface, but the basic idea is a simple one. Interfacial structure is known to be different from bulk structure, and in polymers filled with fillers with extremely high specific surface areas (that is, hundreds of metres squared per gram), most of the polymer present is near an interface, even with only a small weight fraction of filler. Such fillers are necessarily nanoscopic, as this is the only way to achieve such a high specific surface area. If the interaction at the interface is a strong one, or if the structure of the interfacial polymer is very different from the bulk, one can expect to see markedly different properties in the material as a whole. These changes have a fundamentally different origin than those found in micro and macrocomposites, where the volume of the interphase is only a small fraction of the overall volume of the material, and therefore nanocomposites are often referred to as being 'different' from other reinforced systems.

Nanoscale dispersion is a complex topic, but with respect to layered silicates, three general states of dispersion are often described as existing (or coexisting) in a nanocomposite. As shown in Fig. 21.6, they are the immiscible (or phase separated) state, the intercalated state, and the exfoliated (or dispersed) state.

1. The immiscible state corresponds to the situation where the individual silicate layers do not separate at all, and are confined to multilayer stacks, which may even be agglomerated.

2. The intercalated state is typically described as a situation where the silicate layers expand to accommodate some small number of polymer molecules, but where the layers do not separate and retain a relatively small interlayer spacing and an ordered layered structure, analogous to the swelling of a multilayer stack.

3. Finally, the exfoliated state refers to the situation where the silicate layers are individually dispersed and outside the influence of other silicate layers. This is a gross idealisation, of course, as geometric concerns alone indicate that, for such high-aspect ratio particles, true exfoliation cannot take place except at extremely low inorganic loading levels, even lower than those typically used in nanocomposites.

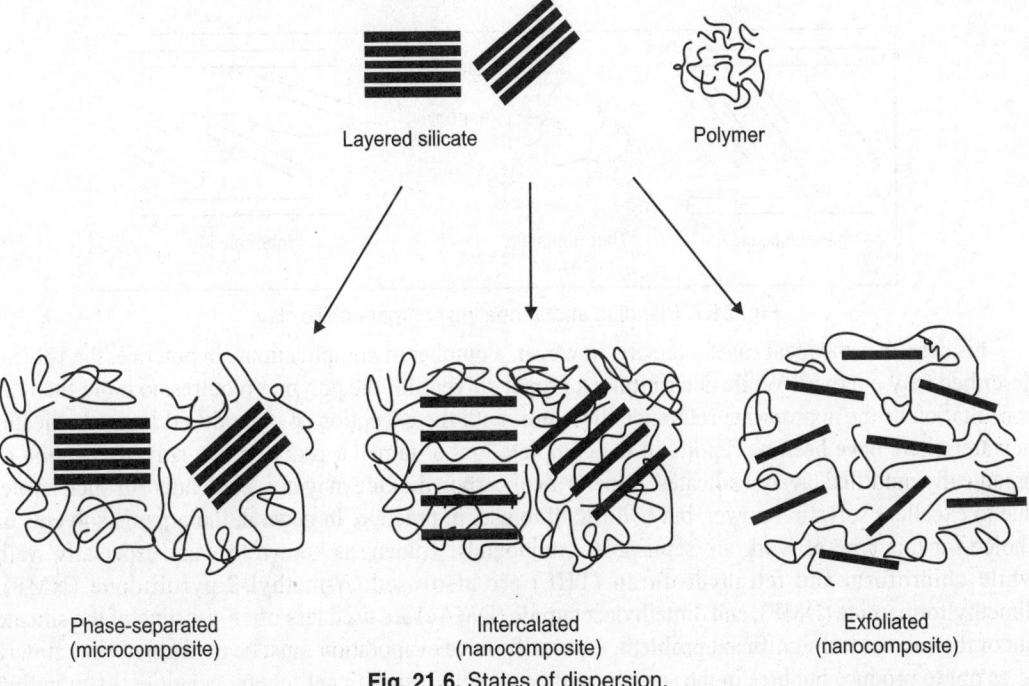

Layered silicate   Polymer

Phase-separated   Intercalated   Exfoliated
(microcomposite)   (nanocomposite)   (nanocomposite)

Fig. 21.6. States of dispersion.

Nanocomposite Synthesis

Methods for nanocomposite production can be classified into two general (and sometimes overlapping) categories:

1. One approach involves enhancing the mobility of the polymer of interest, in the presence of silicate layers, and with thermodynamic compatibility between the two, in order to achieve silicate layer dispersion. This best describes the techniques of melt blending and solvent casting.
2. The other approach attempts to literally force the layers apart through the preferential insertion of material in the interlayer galleries, regardless of thermodynamic compatibility between the silicate and the polymer. This best describes a number of different *in situ* polymerisation techniques.

Both have been used successfully in order to produce polymer/layered silicate nanocomposites, though not surprisingly, there are strong indications that we ignore thermodynamic compatibility at our peril. These techniques are present in many variations, and in some instances they are even combined.

In situ polymerisation, for example, may be used to produce heavily compatibilised layered silicates, to act as concentrated masterbatches and melt blended with pure polymer to produce the final nanocomposite material.

Solvent casting

Solvent casting is one of the simplest techniques by which nanocomposites are produced. A polymer, a silicate, and a solvent are combined and thoroughly mixed, and the solvent is then allowed to evaporate, leaving the nanocomposite behind, typically as a thin film. The solvent imparts the enhanced mobility the polymer needs in order to intercalate between the silicate layers, while thermodynamic compatibility, combined with physical mixing, give rise to a dispersed system. A solvent should be chosen that completely dissolves the polymer and completely disperses the silicate (Fig. 21.7).

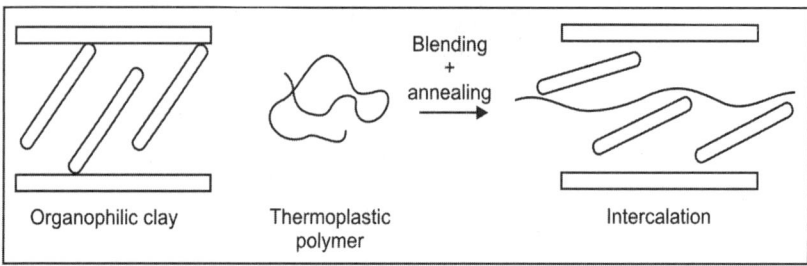

| Organophilic clay | Thermoplastic polymer | Intercalation |

Fig. 21.7. Blending and annealing of organophilic clay.

This describes the ideal case; there are, however, a number of complications. In practice, the solvent described may not exist. While determining a proper solvent for the polymer requires no more than the consultation of the appropriate reference literature, with the exception of unmodified layered silicates in water, there have been no reports of the complete loss of lamellar ordering in a solvent solution of organically modified layered silicates. Supercritical carbon dioxide may be a contender for such a role, due to excellent solvation power, but is difficult to use, in practice. In general, the organic solvents of choice for this type of work are semi-polar compounds. Toluene is known to work especially well, while chloroform and tetrahydrofuran (THF) are also used; N-methyl-2-pyrollidone (NMP), dimethylformamide (DMF), and dimethylacetamide (DMAc) are used less often. Settling of the silicate out of the solvent is a significant problem, especially since evaporation must be performed very slowly so as not to produce bubbles in the sample, and can result in significant inhomogeneities, even in thin films. In general, the clarity of the silicate dispersion in the solvent, as well as the speed at which settling occurs, are good means of telling whether the solvent is effective in dispersing the silicate.

With surfactant-modified Montmorillonite clays, in particular, the layer size is such that agglomerates are very large and scatter light very effectively, but multi-layer stacks and individual layers have lateral dimensions of ~100–500 nm, making them less effective scatterers. As a result, a good solvent dispersion of an organically modified Montmorillonite often appears darker in colour and more transparent than a poor solvent dispersion (which will be opaque and nearly the same colour as the dry material organosilicate).

The origins of the colour change are not precisely known, but have been observed in silicone nanocomposite preparation as well, with liquid prepolymers that wet the organosilicate well changing its colour to a darker shade than those that do not wet well.

Melt blending

Melt blending is akin to solvent casting, in that here too, the polymer is given the enhanced mobility it needs, in combination with physical mixing, in order to disperse compatibilised layered silicates on the nanoscale. There is (typically) no solvent used in this technique, and as the name implies, the enhanced polymer mobility comes simply from thermal energy. Using an extruder or heated mixing chamber of some sort, the polymer and the compatibilised layered silicate are simply physically mixed at high temperature, and a nanocomposite is obtained.

Again, the mixing can be performed via something akin to stirring (i.e. the use of mixing screws in an extruder setup), as well as by using ultrasonics, where a probe is immersed in a polymer melt. The two mixing techniques may again be combined, so long as the polymer viscosity is low enough to allow for effective sonication. Again, however, the possibility of ultrasonic damage, magnified at higher viscosities, is something that must be considered.

The silicate layer orientation may be randomised during melt blending, but extrusion or compression moulding will result in layer alignment along the flow direction or perpendicular to any applied compressive stresses. In addition to these concerns, new problems are created when heat is used to enhance the polymer mobility. In order to achieve good mixing, a long mixing time is preferable, but this must be balanced against the normally undesirable thermal degradation of the polymer and the silicate. Finding an optimum can be difficult, especially for polymers that thermally degrade readily (some biodegradable polyesters, for instance), or for those systems where the processing temperature must be high in order to achieve polymer flow and effective mixing (many polyamides and polycarbonates fall into this category)—in which case it is the organic silicate compatibilisers that degrade. Intuitively, this loss of compatibiliser will result in the loss of thermodynamic compatibility, and thus, nanoscale dispersion, though in at least one example, degradation of the silicate compatibiliser seems to improve mixing.

Advantage over solvent casting

The lack of solvent in this technique solves many problems associated with solvent casting. No impurities or residues are introduced into the sample, settling is generally not an issue due to the much higher viscosities and thus higher shear rates present during mixing, and there are no concerns about complete evaporation or the retention of a small amount of a plasticising solvent. For a given thermal history and mixing procedure, the polymer microstructure is affected only by the composition of the nanocomposite, and pieces of highly varying geometry and size may be produced without problems. As in the case of solvent casting, however, there is still an alignment effect, though it may be at least somewhat less significant.

In situ polymerisation

In situ polymerisation covers any process where the nanocomposite is made by performing some sort of polymerisation reaction in the presence of a layered silicate. There are many variations on this technique, all stemming from the need to disperse the silicate layers. The simplest technique involves mixing a monomer with a layered silicate and polymerising from there.

If all species stay where they are when polymerisation is initiated, i.e. no additional monomer enters the galleries after the chains begin to grow, the amount of layer dispersion will be necessarily minimal and an intercalated system will most likely result. If, however, the polymer forms and uses up the monomer, causing more monomer to come between the layers as the chain grows, the interlayer spacing must necessarily increase. Better yet, if the surface of a silicate layer itself or some silicate-bound functionality catalyses the polymerisation reaction and enhances its rate with respect to material outside of the interlayer galleries, dispersion may be strongly enhanced.

Once the system reaches its gel point, however, further layer separation becomes very difficult, meaning that these systems must typically be carefully tuned, in terms of processing conditions, in order to achieve optimal dispersion.

An alternative approach to layer separation involves the use of emulsion polymerisation. In this case, the issues of intercalating the monomer and ensuring that the reaction rates in the interlayer galleries are reasonable is avoided altogether by dispersing an unmodified silicate in a water/monomer emulsion. High levels of dispersion are virtually guaranteed, then, though any surfactant used must not ion-exchange with the silicate (or it will precipitate, rather than dispersing) and will, in any case, remain in the sample after polymerisation, and may plasticise the sample or affect its properties. In either case, such dispersion will improve the barrier and thermal properties, based on a tortuosity argument alone, but may or may not improve the mechanical performance, due to the lack of polymer/silicate interactions. In addition, without thermodynamic compatibility the silicate layers may even collapse to form multilayer stacks if the nanocomposite is heated. Finally, all of these techniques become somewhat complicated by any changes the silicate layers cause in terms of the polymer's molecular weight distribution, when comparing them with silicate-free controls.

One approach to producing high levels of silicate dispersion while avoiding the aforementioned scenario of complete layer collapse involves the *in situ* polymerisation of such materials using bound initiators or monomers. Specifically, instead of using typical surfactant compatibilisers, it is possible to ion-exchange reactive species into the layered silicates. If a monomer analog is used as the silicate compatibiliser, compatibility with the monomer may be enhanced, and at least some of the polymer chains in the systems should be end-linked to the silicate layers. The level of dispersion in this technique may not be significantly greater than that found in the previously described *in situ* techniques, but the end tethering of some of the chains has two major effects. At least some polymer must stay within the interlayer galleries at all times, preventing collapse to an immiscible state. In addition, the nature of the polymer/silicate interface will change, in that the entanglement density will be lower, but the interaction strength will be higher, at least with regard to this chains bound to the silicate layers.

This technique is amenable to both addition- and condensation-polymerisation, so long as a monomer analog can be found or produced that contains one ionic group that may be bound to a surface without loss of reactivity of the rest of the molecule, and the monomers intercalate into such a silicate modified by such a compound. If, instead, a polymerisation initiator is ion-exchanged into the silicate interlayer, an extreme situation develops. In this case, assuming that some monomer intercalation can still occur and that no other source of initiation is added or present, every single polymer chain must form at the surface of a silicate layer, and must also remain end-tethered to that layer. The aforementioned effects on entanglement density and interaction strength are thus enhanced, dispersion is effectively guaranteed, and layer collapse becomes even less likely. Such materials may be used in their own right, or treated as heavily compatibilised silicates and melt blended with pure polymer.

This technique is amenable only to addition polymerisation, for obvious reasons, but may be applied to any system where an initiator may be bound to a surface and remain active, and where the monomers intercalate into a silicate modified by such a compound.

Overall, then, there are many variations on *in situ* polymerisation. They involve separating the layers either by forcing them apart, due to increasing amounts of new material formed between the layers, or by taking advantage of emulsion polymerisation combined with the high dispersion levels afforded by unmodified silicates in water. They may also produce systems that are untethered (simple and emulsion *in situ* polymerisation), partially end-tethered (*in situ* polymerisation with monomer-modified silicate), or completely end-tethered (*in situ* polymerisation with initiator-modified silicate). In each of these cases, there are differences in materials structure, making it important to be precise with respect to the type of *in situ* polymerised nanocomposite being discussed. All of these techniques may be used with addition polymerised materials, while some of them (notably simple/emulsion *in situ* polymerisation and the use of monomer-modified silicates) can also be applied to condensation-polymerised systems.

Advantages of Nanocomposites Over Conventional Composites

The resultant nanocomposites often have properties superior to conventional composites, and this is achieved at a much lower loading of nanoclay compared to conventional micron-sized clay particles. Nanocomposites have many potential applications in aerospace and automotive structure, food packaging, electronics, biomedical applications, and so on. Improvement of following properties is there in the case of nanocomposites over conventional composites:

1. Mechanical properties, e.g. tensile strength, modulus and dimensional stability.
2. Impermeability to gases.
3. Thermal stability.
4. Flame retardancy.
5. Chemical resistance.
6. Surface appearance.
7. Electrical conductivity.
8. Optical clarity in comparison to conventionally filled polymers.
9. Decreases in thermal expansion coefficient.
10. Weatherability and resistance to ageing.
11. Variations in nanocomposite formulation can allow for precise control of biodegradability.

Industrial Applications

Several potential applications have been identified in the following industrial sectors:

1. Automobile (gasoline tanks, bumpers, interior and exterior panels, etc.).
2. Construction (shaped extrusions, panels).
3. Electronics and electrical (printed circuits, electric components).
4. Food packaging (containers, films).

RESINS FOR COMPOSITES

The general definition of a composite is a combination of different components or elements. Think of a composite as a material made from two or more dissimilar (not alike) materials that, when combined, are stronger than those individual materials by themselves. Composites are defined as 'a combination of plastic resin and a fibre reinforcement'. Another term, used in the past, is reinforced plastics.

Today, the composites industry uses a more specific term: fibre reinforced polymer (FRP) composites. The fibre reinforcement can be glass, carbon, or aramid (*Kevlar*). These fibres are very strong.

The function of the fibres is to provide strength and stiffness to the composite product, while the resin acts to bond and protect the fibres from chemicals and the environment, as well as transfer load between the fibres.

Composites are different from other materials. For example, metals are isotropic, meaning they have equal strength in all directions. Composites are anisotropic, having different properties in different directions. This gives composites an advantage by allowing designers to make efficient use of materials for the design loads.

In composites, the resin (or varnish) is known as the matrix and serves two critical functions: (i) it transfers the load to the reinforcement fibre, and (ii) protects the fibre from environmental effects. There are two family groups of resins that comprise plastic materials: thermosets and thermoplastics. They differ in molecular structure —while thermosets are cross linked and thermoplastics are not.

Thermoset resins are converted from a liquid to a solid using an initiator or heat—the process is irreversible. Thermoplastic resins, on the other hand, are melted and formed and can be remelted and reformed—the process is reversible. Typical household plastics are thermoplastics and consist of nylon, polystyrene, polyethylene, acrylic, and many other plastic compounds. Familiar thermosets are polyester, vinyl ester and epoxy resins. Less common thermoset resins include phenolic, silicone, polyimide and bismaleimide.

Polyester resins are the most commonly used resin systems in FRP because they are low cost and the cured physical properties meet many of the needs in the commercial composite industry. Vinyl ester resins are used where either superior corrosion resistance or toughness are required properties. Vinyl esters are formulated by reacting epoxy resin with methacrylic acid, forming a polymer that has characteristics like both polyester and epoxy. Vinyl ester resins are cured and handle very similarly to polyester resins, but have a higher cost.

Main Components of Resin Systems

Monomer

The monomer serves several purposes in the resin system. First, it co-reacts with the backbone polymer in a resin system when polymerisation (crosslinking) takes place. Second, the monomer also reduces the viscosity of the polymer to provide a workable liquid product, acting as diluents. Styrene is termed a reactive diluent because it takes part in the curing process.

Styrene, vinyl toluene, diallyl phthalate (DAP) and methyl methacrylate (MMA) are the most common monomers used with polyester and vinyl ester resin systems. Most of the styrene monomer in a resin system is captured in the crosslinking reaction. However, a small portion of the monomer may evaporate before curing takes place. This is the characteristic smell of polyester and vinyl ester resins. Lowering styrene emissions is an objective of the composites industry to reduce environmental impact and worker exposure to styrene due to US Environmental Protection Agency (EPA) regulations.

Curing resins

Resin must cure (harden) in a way that is compatible with the fabrication process. Some parts are small and can be laid up quickly. The faster a resin cures, the quicker the turnaround is on the moulds and the greater the production rates. Other parts may involve large lay-ups where more time is required for the lamination process. In compression moulding, pultrusion and sometimes RTM, heated moulds provide rapid curing. Another aspect of curing resin is the physical properties of the cured laminate are determined by the efficiency of the cure.

The hardness of the laminate is affected by the curing process as well as the chemical resistance of the laminate surface. Thick laminates also require special attention. Resin exotherm must be controlled in order to prevent excessive shrinkage and laminate warping.

Initiators/promoters/inhibitors

Initiator is the correct technical term for the product commonly called the catalyst in the composites industry. In technical terms, a catalyst causes a chemical reaction, but does become part of the reaction. An initiator initiates or speeds up a reaction, but becomes consumed in the process. In the case of polymerising polyester resins, the initiator becomes part of the crosslinked polymer. Increasing the amount of initiator added to the resin will increase the rate of cure. An essential factor in maintaining control over the curing process revolves around selecting the correct initiator. There are several types of initiators used to cure polyester and vinyl ester resins: ketone peroxides such as methyl ethyl ketone peroxide (MEKP), acetylacetone peroxides, benzoyl peroxides such as *t*-butyl per benzoate (TBPB), cumine hydroperoxides, etc.

Resin additives

There are a number of additives that are used to modify and enhance resin properties. These include: thixotropes, fillers, pigments, fire retardants, suppressants, UV inhibitors and conductive additives.

Fillers

Adding inert fillers to resin will modify the properties and can reduce cost. Types of fillers include mineral fillers, calcium carbonate, calcium sulphate, talc, mica, organic fillers, wood flour, walnut shells, corn cobs, micro spheres, solid glass spheres, hollow glass spheres, ceramic spheres, thermoplastic spheres, phenolic spheres, etc.

Pigments and colourants

Pigment dispersions and colour pastes can be added to resin or gel coat for cosmetic purposes or to enhance weatherability.

Fire retardants

Most thermoset resins are combustible and create toxic smoke when burned. In critical applications such as aircraft, train interiors or mine equipment, reducing fire hazards is important. Fire retardant additives such as alumina trihydrate and antimony trioxide reduce flame spread and smoke generation of burning composites.

Suppressants/film formers

In order to reduce styrene emissions, suppressant additives can be used to block evaporation. These wax-based materials form a film on the resin surface and reduce the loss of styrene. Additionally, many polyester resins remain tacky on the surface after curing. This is due to air inhibition, which prevents a very thin surface layer from properly curing. The addition of a film former, such as paraffin wax, excludes the air from the surface and allows a non-tacky sandable surface.

UV inhibitors

In the event that a non-gel coated resin will be exposed to sunlight, the addition of a UV inhibitor will slow the surface degradation.

Conductive additives

Composite laminates (except carbon fibre) are inherently nonconductive. In some cases it is necessary to make a laminate conductive to reduce static charge or to enable electrostatic painting. Carbon black, carbon fibres, metallic fibres, or metallised glass can be used to create an electrically conductive laminate.

Gel coat

Gel coat is a specialised polyester resin that is formulated to provide a cosmetic outer surface on a composite product and to provide weatherability for outdoor products. Gel coat is not paint. Paint contains solvents that must evaporate for the paint to dry. The solvent in gel coat is styrene monomer and/or MMA, which cross-links during curing.

Reinforcements

The primary function of fibre reinforcements is to carry load along the length of the fibre to provide strength and stiffness in one direction. Reinforcements can be oriented to provide tailored properties in the direction of the loads imparted on the end product.

Reinforcements can be both natural and man-made. Many materials are capable of reinforcing polymers. Some materials, such as the cellulose in wood, are naturally occurring products. Most commercial reinforcements, however, are man-made. The most common fibre reinforcement is glass fibre. Other fibre reinforcements are carbon and aramid.

Glass fibre

Glass fibre is the least expensive of all reinforcements. Glass fibre (also called fibreglass) is used in more than 90 per cent of manufactured composites. Composites made of polyester resins and glass-fibres are so common, in fact, the term 'fibreglass' is often used for the composite material itself, such as 'fibreglass boat'. Glass fibres, however, are only one part of a composite —they do the reinforcing.

Glassfibres come in several varieties, designated S-, A-, C-, or E-glass. Each variety has special characteristics. S-glass is exceptionally strong. C-glass is extremely resistant to corrosion and chemical attack. A-glass has good resistance to chemicals. E-glass does not conduct electricity. Though economical, glass fibre is relatively heavy. Of the common synthetic reinforcements, it has the least efficient strength-to-weight ratio.

Carbon fibre

Carbon fibre is a very strong fibre and extremely stiff. It is lighter in weight than glassfibre. Carbon fibres come in several varieties and strengths and are the most expensive kind of fibre reinforcements. They are typically used in air planes and spacecraft. Carbon fibre reinforced composites are also used in products such as bicycle frames, tennis rackets, skis and golf club shafts.

Aramid fibre

Aramid fibre resists impact. It is used extensively in bulletproof vests and body armour. Racing drivers wear aramid suits that help protect them from burns in fiery, high-speed crashes. Aramid is commonly known as *Kevlar*, produced by Dupont. Aramid fibres cost is between glass and carbon. Aramid is more difficult to work with than glass and has a tendency to absorb moisture.

Other fibres

Different fibres can be combined to make a composite cost less or perform better. Composites that are made of more than one fibre are called hybrid composites. Fibres with special characteristics are used when a composite must be exceptionally strong or heat-resistant—for high-performance military aircraft, for instance, or aerospace applications. These materials are quite expensive. Examples include boron (an extremely hard natural element) and ceramics (hard, manufactured materials that can withstand high heat and harsh chemicals).

In most composite products, the fibre reinforcements are bundled together for strength. Fibres are assembled in various patterns called fabrics. Typical forms include: unidirectional, woven, mat, knit, stitched, braid and veil. Because each pattern carries loads differently, how the fibres are placed or assembled is important to engineers and designers. The cost of each of these forms also varies, depending on the amount and the quality of the fibre used.

Core reinforcement

Core materials are widely used in composites to make stiff, lightweight products. Typical core materials include balsa (wood from the balsa tree), polyurethane foam and PVC (polyvinyl chloride) foam (both manufactured in chemical processes), and honeycomb.

These materials are light and strong. Core materials are used to make sandwich construction. Using sandwich construction, a core material is placed between two outside surfaces (called 'face skins') of fibre reinforcements. This sandwich is bonded together with an adhesive or glue. As the core is made thicker, the stiffer is the sandwich panel. Sandwich construction is used in commercial aircraft flooring because it is lightweight, strong, stiff, and economical.

Advantages of Composites

Light weight

Composites are light in weight, compared to most woods and metals. Their lightness is important in automobiles and aircraft, for example, where less weight means better fuel efficiency (more miles to the gallon). People who design airplanes are greatly concerned with weight, since reducing a craft's weight reduces the amount of fuel it needs and increases the speeds it can reach. Some modern airplanes are built with more composites than metal, including the new Boeing 787, Dreamliner.

High strength

Composites can be designed to be far stronger than aluminum or steel. Metals are equally strong in all directions. But composites can be engineered and designed to be strong in a specific direction.

Strength related to weight

Strength-to-weight ratio is a material's strength in relation to how much it weighs. Some materials are very strong and heavy, such as steel. Other materials can be strong and light, such as bamboo poles. Composite materials can be designed to be both strong and light. This property is why composites are used to build airplanes—which need a very high strength material at the lowest possible weight.

A composite can be made to resist bending in one direction, for example. When something is built with metal, and greater strength is needed in one direction, the material usually must be made thicker, which adds weight. Composites can be strong without being heavy. Composites have the highest strength to weight ratios in structures today.

Corrosion resistance

Composites resist damage from the weather and from harsh chemicals that can eat away at other materials. Composites are good choices where chemicals are handled or stored. Outdoors, they stand up to severe weather and wide changes in temperature.

High-impact strength

Composites can be made to absorb impacts—the sudden force of a bullet, for instance, or the blast from an explosion. Because of this property, composites are used in bulletproof vests and panels, and to shield airplanes, buildings, and military vehicles from explosions.

Design flexibility

Composites can be moulded into complicated shapes more easily than most other materials. This gives designers the freedom to create almost any shape or form. Most recreational boats today, for example, are built from fibreglass composites because these materials can easily be moulded into complex shapes, which improve boat design while lowering costs. The surface of composites can also be moulded to mimic any surface finish or texture, from smooth to pebbly.

Part consolidation

A single piece made of composite materials can replace an entire assembly of metal parts. Reducing the number of parts in a machine or a structure saves time and cuts down on the maintenance needed over the life of the item.

Dimensional stability

Composites retain their shape and size when they are hot or cool, wet or dry. Wood, on the other hand, swells and shrinks as the humidity changes. Composites can be a better choice in situations demanding tight fits that do not vary. They are used in aircraft wings, for example, so that the wing shape and size do not change as the plane gains or loses altitude.

Nonconductive

Composites are nonconductive, meaning they do not conduct electricity. This property makes them suitable for such items as electrical utility poles and the circuit boards in electronics. If electrical conductivity is needed, it is possible to make some composites conductive.

Nonmagnetic

Composites contain no metals; therefore, they are not magnetic. They can be used around sensitive electronic equipment. The lack of magnetic interference allows large magnets used in MRI (magnetic resonance imaging) equipment to perform better. Composites are used in both the equipment housing and table. In addition, the construction of the room uses composites rebar to reinforced the concrete walls and floors in the hospital.

Radar transparent

Radar signals pass right through composites, a property that makes composites ideal materials for use anywhere radar equipment is operating, whether on the ground or in the air. Composites play a key role in stealth aircraft, such as the U.S. Air Force's B-2 stealth bomber, which is nearly invisible to radar.

Low thermal conductivity

Composites are good insulators—they do not easily conduct heat or cold. They are used in buildings for doors, panels, and windows where extra protection is needed from severe weather.

Durable

Structures made of composites have a long life and need little maintenance. We do not know how long composites last, because we have not come to the end of the life of many original composites. Many composites have been in service for half a century.

Thus, nanocomposite is polymer filler composite and the filler used in it is very small, even smaller than the wavelength of visible light. This concept began 20 years back from now with elastomer-nylon (chopped, nylon) system. Developments are continuing with different polymers and fillers system. It is understood that not all polymers and also not all fillers are suitable for nanocomposite developments.

Formulation of nanocomposite with suitable polymer and suitable filler and the processing technology of the composite is the critical success factor to dominate the resulting product in the market. Nano composite technology is being developed in the field of packaging, automotive components, home appliances, electrical and electronic parts, building and construction products, etc.

One of the major areas this technology can penetrate is in the automobile sectors, which includes all rubber parts including pneumatic tyre. In the pneumatic tyre, however, the key technology lies on the fact that how multilayer fillers like silicates, having high aspect ratio (l/h), are activated in the polymer system, such that, the multilayer filler will now break into number of particles which give rise to higher surface area to be reinforced in the polymer interface.

It is also not surprising that in future, in place of conventional inner liner as on today, only water based inner paint, based on nano composite, may take care of both mould release as well as air retention properties of future tyre. In the elastomers, when such systems are used they may give rise to distinct improved properties such as low weight, low rolling loss, low air permeability, low tyre cost, better physical/mechanical properties.

It appears that polymer nanocomposites have very bright future. Nanotechnology is recognised as one of the most promising avenues of technology development for the 21st century. In the materials industry, the development of polymer nanocomposites is a rapidly expanding multidisciplinary research activity. There has been enormous interest in the commercialisation of nanocomposites for a variety of applications, and several of these applications will be successful in the near future. To generate the information necessary to construct a reasonable future market for polymer nanocomposites, it is necessary to take a hard headed look at the potential advantages and pitfalls of the current crop of these materials, as compared with conventionally-filled polymers.

Chapter 22

Clay Based Nanocomposites

INTRODUCTION

Nanocomposites are recently finding application in numerous fields—space research, automobile industry and gas barriers, amongst others. Nanometric size distribution allows remarkable improvement in thermal and mechanical properties. These include decreased gas permeability, increased mechanical robustness and increased storage modulus.

These improvements are obtained without any compromise or penalties like brittleness, high density and loss of transparency. Thus, nanocomposites have gained an edge over other materials. This chapter discusses the improvements in properties due to the nanocomposites.

Flame retardancy is one of the most deeply studied properties of nanocomposites. The flammability properties of nanocomposites are discussed with reference to their degradation temperature and heat release rate. The chapter also discusses the usefulness of nanocomposites in radiation resistance. XRD and ESR patterns reveal that addition of clay into polymer blocks at nanometer scale improves the radiation resistance of nanocomposites to UV and gamma irradiation.

Dispersion at the nanometric scale allows nanocomposites to exhibit markedly improved properties, including increased modulus, decreased gas permeability, increased solvent resistance and, above all, decreased flammability.

Heat distortion temperature for nanocomposites shows an increase of upto 100°C, allowing their use in higher temperature environments, such as automotive engine applications.

Numerous physical and chemical properties of polymer materials may be altered in the presence of UV and gamma radiations and even electron beams. Layers of organophilic substances absorb a large amount of radiation to protect the chain of polymers from irradiation and allow grafting of broken chains onto the surface of the layers of clay.

The synthetic approach achieves molecular level incorporation of layered silicates, such as montmorillonite (MMT), into the polymer matrix, by addition of a modified silicate during the polymerisation (*in situ*) or to a solvent-swollen polymer or to the polymer melt.

The final composite is either an intercalated or an exfoliated hybrid. In intercalated structures, the organic component is inserted within the layers of clay, causing the interlayer spacing to expand, but the layers still bear a well-defined spatial relationship to each other. In delaminated (or exfoliated) structures, the layers of clay are completely separated and individual layers are dispersed throughout the organic matrix. In this chapter, we discuss the flammability and radiation resistance properties of clay-based nanocomposites.

FLAMMABILITY PROPERTIES

Fire is one of the major hazards. Hence, it is necessary to have fire-resistant and flame-retardant materials. Lately, the desire to develop halogen-free fire retardant (HFFR) polymer systems has accelerated, due to the environmental problems associated with halogenated compounds. In addition, toxicity, corrosiveness and excessive smoke generation of halogenated fire-retardant compounds, when burnt, also pose problems.

Nanotechnology provides the much-required solution. There are several proposed mechanisms as to how layered silicates affect the flame-retardants properties of polymers. One of them refers to the increased char layer that forms when nanocomposites are exposed to flames.

Interestingly, it has been reported that the flame retardance obtained in nanocomposites is at least as good as when mere intercalation has occurred, compared with complete exfoliation. In fact, excellent performance has been observed when clay layers have remained separated by only approximately 3 nm, which is considered to be well within the 'intercalation' zone.

Transmission electron microscopy (TEM) analysis suggests that the intercalated arrangement of the layered silicates breaks down during combustion, giving a large array of even platelet coverage. This results in very thermally stable char, which inhibits oxygen transport to the flame front and therefore reduces the heat release rate (HRR) of the burning polymer.

In the study by C.M.L Preston the following polymers were used: low-density polyethylene (2.5 dg min^{-1} Ml, 0.921 gram cm^{-3}) (LDPE); ethylene-methyl acrylate copolymer (6.0 dg min^{-1} Ml, 21.5 per cent MA, 0.942 gram cm^{-3}) (EMA); ethylene-methyl acrylate-acrylic acid terpolymer (5.0 dg min^{-1} Ml, 18 per cent MA, 6 per cent AA, 0.953 gram cm^{-3}) (EMAAA); ethylene-vinyl acetate copolymer (1.7 dg min^{-1} Ml, 18 per cent VA, 0.939 gram cm^{-3}) (EVA-18); maleated ethylene-propylene copolymer (9.0 dg min^{-1} Ml, 0.87 gram cm^{-3}) (PE-g-MAH); and ethylene-vinyl acetate copolymer (6.0 dg min^{-1} Ml, 28 per cent VA, 0.951 gram cm^{-3}) (EVA-28). Hydrogenated tallow quaternary amine modified bentonite clay, Bentone 34 (B34, Rheox Inc.), was used as the organoclay for nanocomposite preparation.

TEM analysis suggests that the intercalated arrangement of the layered silicates breaks down during combustion, giving a large array of even platelet coverage. This leads to the formation of a thermally very stable char, which is effectively reinforced and performs as an excellent insulator and mass transport barrier.

It is found that the organoclay effectively stabilises the matrix. The marked effect that the modified bentonite has on the shape of the heat-release curves (obtained by cone calorimetry) of all the polymer clay composites is clearly seen in Fig. 22.1. A remarkable protection effect is noted for all polymers upon addition of the organoclay.

The reduction in peak energy output of the burning composite, as compared with its unfilled host matrix, is apparent. Also the addition of the organoclay has caused a decrease in the ignitability of the PE derivatives. Similar results were observed for all the polymer systems under consideration.

RADIATION RESISTANCE PROPERTIES

The presence of radiation such as UV light, γ-radiation and electron beams can adversely affect many physical and chemical properties of polymer materials.

Incorporation of clay into the polymer matrix helps resolve the problem to an extent. Zhang reported the results obtained on incorporating clay into the tri-block copolymer styrene-butadiene-styrene (SBS) matrix to form SBS/clay nanocomposites.

SBS is a well-known thermoplastic elastomer widely used in plastics, coatings, etc. Like other polymers, SBS is also susceptible to ageing. The effects of γ-radiation on SBS/clay nanocomposites under atmospheric oxygen include a decrease of the dynamic storage modulus of the nanocomposites at high temperatures relative to SBS.

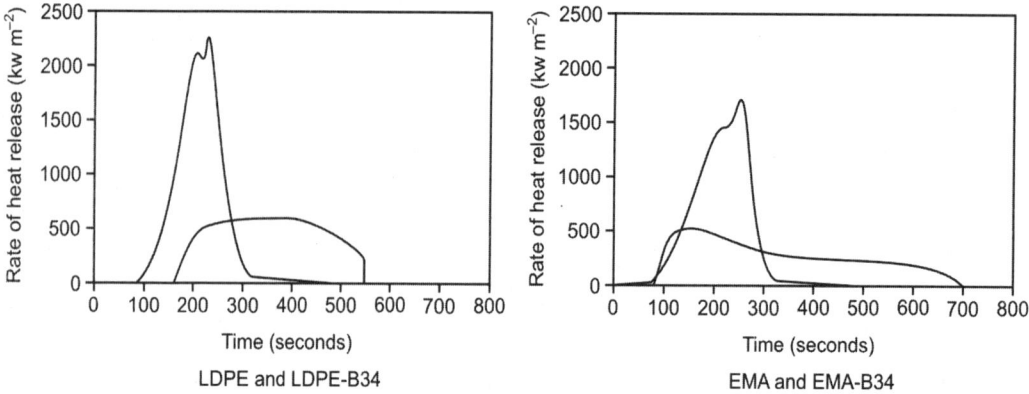

LDPE and LDPE-B34   EMA and EMA-B34

Fig. 22.1. Comparison of the heat-release rate curves of unfilled matrices and composites. Reduction in peak energy output and prolonged reduced intensity combustion are seen due to the presence of the organoclay.

These nanocomposites have intercalated structure due to polystyrene (PS) blocks intercalating into the clay layers, as characterised by X-ray diffraction. When nanocomposites are exposed to γ-radiation under atmospheric oxygen, the decrease of the storage modulus related to SBS can be efficiently restrained by the clay at temperatures above the glass transition temperature of the polybutadiene (PB) block.

The layers of organophilic clay cannot only absorb a large amount of radiation to protect the chains of SBS from being irradiated, but also produce a large number of free radicals so that the broken chains of SBS can be grafted on to the surface of the layers of the organophilic clay, as confirmed by electron spin resonance (ESR), as well as gel fraction measurements. Hence, the formation of SBS/clay nanocomposites can remarkably improve the radiation stability of SBS.

The effect of radiation doses on the structure of SBS/clay nanocomposites is shown in the Fig. 22.2. The intensity of the peak is slightly influenced by the radiation dose.

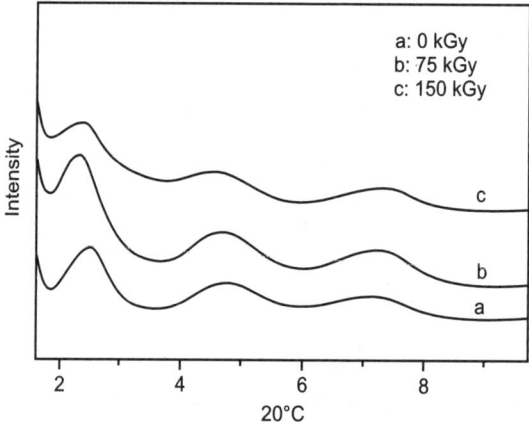

Fig. 22.2. X-ray diffraction curves of an SBS/clay nanocomposite at different radiation doses.

When the dose is 75 kGy, the diffraction peak of SBS/clay nanocomposites is a little more ordered than that without radiation. When the dose is further increased to 150 kGy, the diffraction peak becomes less orderly. This is because SBS/clay nanocomposites undergo both cross-linking and main-chain scission when exposed to γ-ray radiation. Obviously, SBS should be favourable to undergo cross-linking. For most polymers, cross-linking is predominant at low absorbed doses, while cross-linking and scission are comparable at high doses. Further increasing the dose would lead to the dominance of main chain scission. On the other hand, oxygen has much influence on the radiation reaction process.

Even at very low doses, oxygen can react with lateral macro radicals and prevent crosslinking. γ-ray radiation in conjunction with oxygen leads to a rapid deterioration and subsequent breakdown of mechanical properties. Figure 22.3(a) illustrates the dynamic storage modulus (E′) as a function of the temperature for irradiated and nonirradiated SBS. It shows that the E′ of pure SBS increases with the dose at temperatures below the glass transition (T_g) of the PB block.

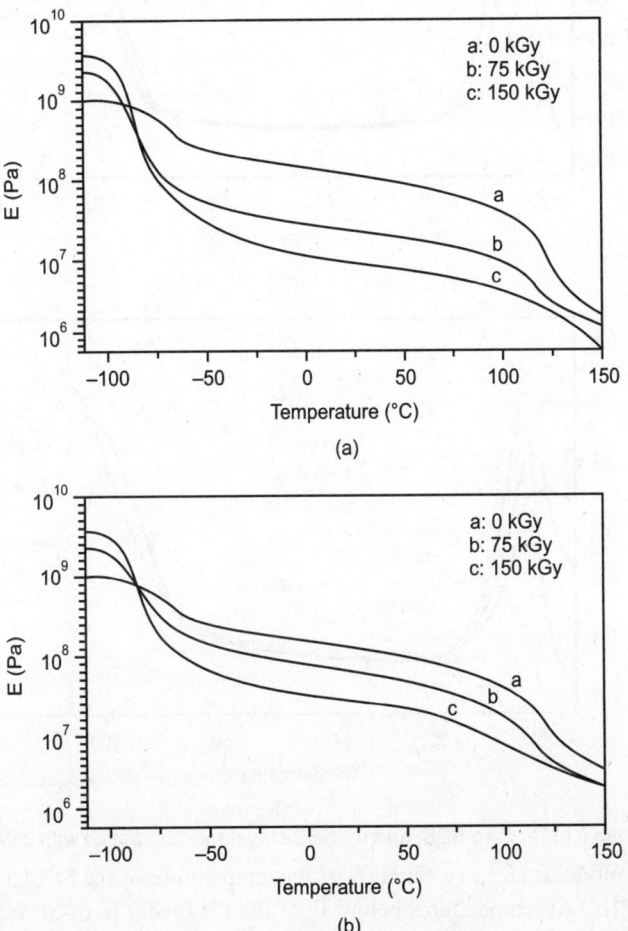

Fig. 22.3. The storage modulus of: (a) pure SBS with different radiation doses; and (b) SBS/clay nanocomposite at different radiation doses.

However, when the temperature is higher than T_g of the PB block, E′ of pure SBS dramatically decreases with increase of the γ-ray dose. This is because both cross-linking and main-chain scission of pure SBS occur when exposed to γ-ray irradiation. The loss tangent tan δ of pure SBS is shown in Fig. 22.4(a). The peak of the tan δ curve corresponds to the main relaxation processes, and the temperature at the main relaxation is taken as T_g. The loss tangent, tan δ, of SBS/clay nanocomposites is similar to that of pure SBS [Fig. 22.4(b)].

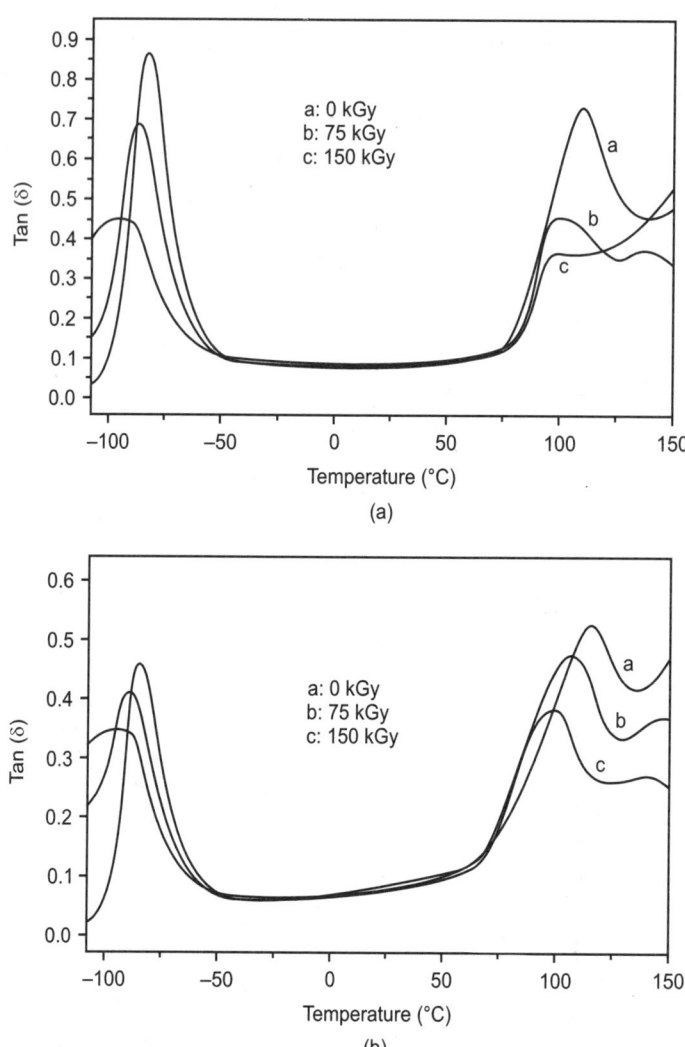

Fig. 22.4. Variation of tan δ of (a) pure SBS; and (b) SBS/clay nano-composite with different radiation doses.

Dynamic storage modulus (E′) as a function of the temperature of the SBS/clay nanocomposite is shown in Figure 22.3(b). At temperatures below T_g of the PB block, E′ of SBS/clay nanocomposites increases with the dose increasing, which is almost the same as for SBS.

However, it is found that the E′ of SBS/clay nanocomposites decreases much more slowly than that of the pure SBS with the dose increasing at temperatures beyond the T_g of the PB block, though both SBS and SBS/clay nanocomposites show almost the same reaction under irradiation. After OMMT is exposed to γ-ray irradiation, it was found that a large amount of free radicals resulted, as characterised by ESR.

Obviously the layers of OMMT absorbed some irradiation energy to produce free radicals. Because the chains of SBS alternate with the layers of OMMT, these layers of OMMT can effectively protect the chains of SBS from being irradiated. The ESR peak of SBS/clay nanocomposites after irradiation is much more intense than that of pure SBS, which is due to the cooperating effect of the radicals of SBS with those of OMMT. All these radicals can combine with each other, which leads the chain radicals of SBS to be grafted onto the surface of the layers of OMMT. The gel fraction of SBS/clay nanocomposites with different OMMT contents is illustrated in Fig. 22.5.

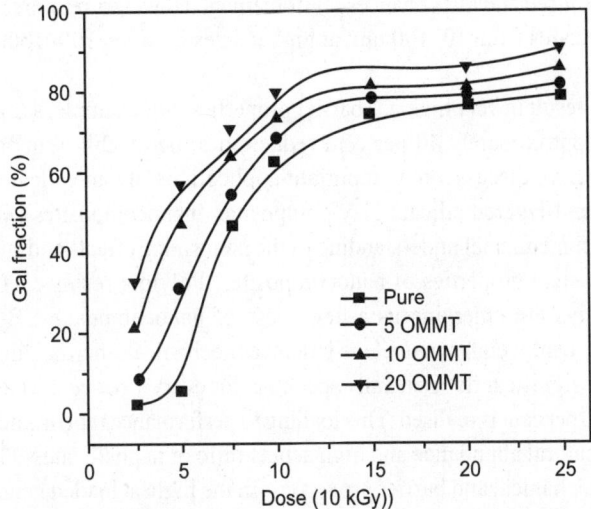

Fig. 22.5. Gel fraction v/s. the radiation dose for pure SBS and SBS/clay nanocomposites with different amounts of OMMT.

It can be seen, as expected, that the gel fraction increase with increasing the radiation doses. Moreover, the higher the content of OMMT, more quickly the gel forms. In particular, initially with a dose of about 25 kGy, pure SBS scarcely forms a gel, while about 28 per cent gel was formed for SBS/clay nanocomposites with 20 per cent OMMT. This further indicates the grafting reaction occurs between PS blocks and OMMT; moreover, the probability of the reaction increases with OMMT content.

NANOSILICATES

A scientific and technical revolution has just begun based upon our ability to systematically organise and manipulate matter at the nanoscale and 'Nanotechnology' is a 'catch-all' phrase for materials and devices that operate at the nanoscale and nanocomposite technology has been described as the next great frontier of material science.

Nanomaterials, nanocomposites, nanotechnology are now common language in the 'nano' world. Applications of nanocomposites are as varied as plastics or elastomers themselves and they potentially could influence a multitude of products used by consumers and industry every day.

Nano fillers are not new, only the applications are becoming new now-a-days. A number of nano fillers, precisely used in elastomers and plastics, were already available with us. These include: carbon blacks, fumed silica, precipitated silica and precipitated calcium carbonates. For example, nanoscale zinc oxide is being used for its UV absorbing properties to create sunscreens. The small size of the particles makes them invisible to the naked eye, so the lotion or the ointment containing nano zinc oxide appears to be clear. Zinc oxide nanoparticles are also being utilised in bacteria-resistant fabrics and surfaces. Yet another example is nanocarbon. Its use in elastomers has high potential for reduction of material weight, besides resulting in better durability and mechanical strength.

Rubber-clay nanocomposites, with clay particles between 0.1–10 nm in size, are increasingly being used both for reinforcement, as well as a gas barrier. The intercalation of a polymer into silicate inter-layers increases the active surface area of the filler. In addition, polymer chains confined between silicate layers are immobilised and this enhances reinforcement. However, reinforcement is only witnessed when the particle size ranges from 10–100 nm; at higher sizes between 100–1000 nm there is partial or semi-reinforcement.

Use of nanosilicates result in very high gas barrier properties. For example, adding 2 per cent nanoclay into nylon film gives approximately 40 per cent reduction in permeability in commercial gas barrier products. In this section, we discuss a few important applications of nanosilicates.

Recently, polymer and layered silicate clay composites (nanocomposites), have been extensively investigated to obtain a fundamental understanding of the nanosizing effects and subsequently to enhance the mechanical and physical properties of nanocomposites. Polymer resins containing well-dispersed layered silicate nanoclays are emerging as a new class of nanocomposites. By employing minimal addition levels (<10 per cent weight) nanoclays enhance mechanical, thermal, dimensional and barrier performance properties significantly. Generally speaking, for every 1 per cent weight addition, a property increase on the order of 10 per cent is realised. This loading to performance ratio is known as the 'nanoeffect' and is basically due to natural abundance and high aspect ratio of nanosilicates. The nano-effect is quite evident judging from mechanical and barrier properties. In the highest loaded nanocomposites scientists have already achieved a 110 per cent increase in flexural and tensile moduli and a 175 per cent increase in heat distortion temperature under load.

Nylon-6 is the most studied plastic in the nanocomposite field, largely due the pioneering work of Toyota Research and Development Laboratory and the popularity of nylon as a film for food packaging, and an engineering plastic for injection moulded parts. Organic treatment on clay surface serves to enhance compatibility between hydrophilic montmorillonite and hydrophobic nylon. In general, two methods have been employed to disperse montmorillonite nanolayers into nylon 6 matrix. One is *in situ* polymerisation, in which polymerisation takes place after mixing monomer or oligomer with organically modified montmorillonite.

The second method is melt compounding, which adds an organically modified montmorillonite into a polymer melt. Based on current literature, *in situ* formed nanocomposites out-perform melt compounded ones by a significant margin. By creating a nanocomposite via *in situ* polymerisation, mechanical properties and gas barrier can be significantly improved. The degree of property enhancement meets the criteria of the nano effect and, in some instances, even exceeds those criteria.

Nanosilicate Treatments for Fabric

The nanosilicate treatment for fabric is based on the creation of nanosized particles of silicates, which are then added to a matrix. The solution can then be manipulated in order that it becomes suitable for coating almost any surface. When treated, the surface of the fabric becomes coated with a durable nanolayer of silicate and the fabric become stain resistant, fire resistant and waterproof. It retains these properties even when washed (up to 50 wash cycles).

The main advantages of such treatment are given below:

1. It is environmentally beneficial as there is less, if any, need for cleaning products.
2. A minimal amount of water is required
3. The treatments are very easy to apply; no special tools or expertise is required.
4. The coating protects the substrate and prevents corrosion, staining and wear, even as the substrate continues to breathe.

Gas Barrier

Gas barrier is very important for air retention in tyres. General-purpose rubber (GPR) is too permeable to gases to permit use in many applications. Tyres hence use approximately $1-bn worth of butyl rubber just to help hold air. Air retention is also important for sports balls for trade off feel and bounce.

Nanocomposites have better barrier properties to gases such as air, oxygen, carbon dioxide etc. It also helps to protect migration of moisture and odour. Greater barrier also impedes absorption of flavours and vitamins by the plastic packaging itself. Excellent air retention properties of butyl and halo butyl rubbers are very well known in the tyre industries and these rubbers are extensively used in inner tube (IIR) of tube-type tyres and in inner liner (CIIR, BIIR) in tube less tyres.

There are also some elastomers like nitrile rubber (NBR), ethylene/acrylate copolymers (EAM), which provide moderate air retention properties. The air retention properties of these elastomers could be enhanced by nanofillers like silicates.

With carbon black as filler in a rubber compound such as a tyre, the passage of air through the material is faster because of spherical nature of the filler (Fig. 22.6). With needle shaped fillers, like silicates, for example in a nanocomposite, the same air will take longer time to travel due to the flat shape of the filler (Fig. 22.7).

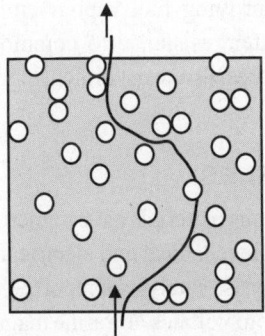

Fig. 22.6. Air passes faster in black filled compound.

Using butyl as elastomer and clay with high aspect ratio as filler, the air retention properties could be improved to the level of 50 times that of the normal butyl compound. Using such fillers, therefore, the inner gauge of a tubeless tyre or the gauge of the inner tube on a tube type tyre could be reduced for identical condition of services.

Such fillers when used in rubbers (NBR, EAM) having moderate air retention properties can provide substantial increase in air retention properties. It is known that nano composite technology is already commercialised in sports goods (tennis ball) for better durability of the product.

Barrier enhancement is also important for longer product shelf-life. Where barrier is a limiting factor in existing packaging, its improvement can lead to lower weight packages and reduce package cost.

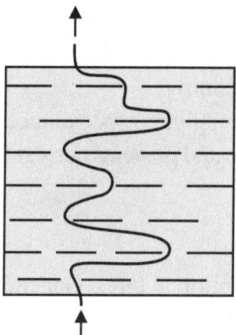

Fig. 22.7. Air takes longer time to pass through a nanocomposite.

In the future, all tyre components may be made with nano type of fillers. This strategy will result in production of:

1. Low weight tyre with low rolling loss.
2. Coloured or transparent tyres (such tyres are very much in use in bicycle and motor cycles).
3. Tubeless tyres without conventional inner liner in it.

The transportation sector, especially automotive, is a heavy user of performance plastics. An average car has over 350 lbs of plastic components. Performance plastics contain reinforcements, thermal stabilisers, chemical resistors and flame resistant additives, each increasing weight and therefore decreasing fuel efficiency.

Since nanocomposites contain low nanoclay addition levels, they offer the automotive engineer an opportunity to minimise this weight penalty. The nanocomposite's barrier enhancement property also plays a role in transportation by improving fuel vapour emissions in fuel tanks and distribution components. It also contributes greater resistance to common corrosive agents such as gasoline, anti freeze and road salt. Nanocomposites also resist dimensional changes by reinforcing plastics in all directions and at a submicron level.

CERAMIC–METAL NANOCOMPOSITES

Nanosize (10^{-9} metre), powders of various materials have numerous applications in our industrial sector. Due to their unique physico-chemical, mechanical and electrical properties, these materials are used as catalysts, photo-catalysts, electrocatalysts, catalysts-support, electrodes, fabrication of dense bodies, semiconductors for energy storage. Photo voltaics, ultrafine magnetic materials for information storage, as absorbent for toxic and hazardous materials, water purifier, optical computers and many more. In view of many attractive and potential applications of nanomaterials, a lot of research and development are being carried out in recent years, to produce various nano materials including composites, alloys and coated materials. The processes developed and being applied to produce the nano materials include gas phage condensation, mechanical milling, thermal crystallisation, chemical precipitation, sol-gel technique, and chemical decomposition, etc.

Nanosized particles reinforced ceramic represents a new class of materials with improved mechanical properties, even at high temperature compared to monolithic ceramics. One notable system is aluminium oxide ceramic containing five volume per cent of silicon carbide nano particles. In this system, an increase in strength from 350 MPa for pure aluminium oxide up to 1.5 GPa coupled with an annealing treatment at 1300°C, has been reported. Following the development of such a wonderful nanocomposite, in recent years, a lot of research and development are taking place to produce various types of metalceramic nanocomposites.

During the last two decades, a large number of metal ceramics composites are being developed for their use as special engineering materials. One of these, that is the metal-metal oxide composites can be more cheaply and casily produced by mixing the metal with the same or another metal oxide in the desired proportions. In the type of metal-metal oxide composites, aluminium oxide is used in many cases, because of its cheaper price, stability, excellent oxidation resistance and superior high temperature mechanical properties.

Since about two decades, 'Glidcop' a copper metal dispersion strengthened with ultra fine aluminium oxide, has been developed and commercially available in three grades containing 0.3 to 1.1, weight per cent of Aluminium oxide. This composite is reported to have good electrical and thermal conductivities with high strength and excellent resistance to high temperature annealing.

These low alumina, copper alumina composites are produced by internal oxidation powder metallurgy process. In this process, aluminium oxide is formed in the copper matrix. These fine aluminium oxide particles harden the metal matrix without very much affecting the electrical and thermal conductivities of the metal.

High density aluminium oxide-nickel nanocomposite with superior mechanical and magnetic properties, has been developed by hydrogen reduction of nickel oxide to nickel and subsequent hot pressing of aluminium oxide and nanosized nickel particles. The nanocomposite thus produced with a small amount of homogeneously dispersed nanonickel particles, has exhibited very high fracture strength. Further, the nanocomposite has been found to show ferro magnetic properties due to dispersion of nano particles of nickel, which points out the possibility of adding a functional value to the structural compostes.

The photocatalytic detoxification of organic and inorganic compounds is very promising for the purification of polluted air and liquid industrial effluents. Titanium dioxide, due to its stability, non toxicity and low cost, has been found to be the most investigated photocatalyst. A photocatalyst is generally suspended in waste-water for the degradation of pollutants. In this process, an additional separation step to remove the catalysts from the treated water, is necessary. However, removing the photocatalyst from large volumes of water, is very difficult because of their small particle size. In order to minimise this problem, research and development work have been undertaken by immobilising titania on to various porous substrates, such as active carbon, silica, or zeolite. While this approach provides a solution to the solid liquid separation problem, the porous materials are found to be not cheap, because these are systhesised through various complicated processes. To treat the huge amounts of industrial effluents by utilising photocatalyst, a cheaper photocatalyst should be developed. In this regard, recently, a process has been developed successfully by coating nanoparticle of titanium dioxides on flyash powders to proved a cheaper tiatania mobilised photocatalysts.

Recently, an attractive process for the preparation of nanosize oxide powders has been developed involving molecular decomposition. This process involves the synthesis of a compound, the precursor, which contains in a chemically combined form the desired final product and a fugitive constituent. The precursor is then allowed to react selectively with a suitable liquid reagent, which leads to the formation of nanosize product.

There is a large volume change when the precursor converts into the product. This leads to the formation of cracks and fissures at the nanolevel. In this manner, nanosize powder is formed. As neither the precursor nor the product is soluble in the liquid, particle growth, which requires dissolution reprecipitation, cannot occur. Using this technique, several nano-oxides such as curium oxide, ceramic oxide, titanium oxide, iron oxide, yttrium oxide, etc. have been successfully prepared.

Recently, a novel chemical processes has been developed to produce copper-aluminium oxide, nickel-aluminium oxide and copper-nickel-aluminium oxide nanocomposites through chemical decomposition of their nitrates in aqueous phase. In a similar manner, they have also produce copper-nickel-nano alloys from the mixed aqueous solution of the nitrates of copper and nickel through coformation of their ultrafine mixed oxide by heating around 375°C followed by reduction with hydrogen at a very low temperature of 350°C.

The alloying of copper and nickel has taken place during the reduction of nano-sized oxide particles of copper and nickel, prepared by the above mentioned chemical routes. The alloy powder has been sintered at 1000°C in hydrogen atmosphere. The density and hardness of the consolidated alloy, have been measured and found to be close to the theoretical values.

Recent R & D studies on nano powders or composites of metals and ceramics have revealed their great potentiality for special application due to their extraordinary properties. In this regard, the industries should come forward to participate and sponsor various R & D activities for developing nano materials. This will go a long way in developing may novel nano materials, which will find numerous special applications in various fields.

Thus a nanocomposite is polymer-filler composite in which the filler used is very small in size, even smaller than the wavelength of visible light. This concept began few years back with elastomer nylon system, but developments are continuing with different polymers and filler systems. Nanocomposite technology is a young material science and plastic nanocomposites are among its first commercial applications.

One of the major areas this technology can penetrate is the automobile sector, which includes rubber parts including pneumatic tyres. Over the past decade patented technologies have been developed for producing nanoscale clays suitable for incorporation into plastics and technologies for making nanocomposites. Commercial nanoclay production began in 1998.

This technology can improve not only air barrier as described earlier, but also flame resistance, thermal resistance and structural properties of many plastics. They are not only used to improve existing products, but also are extending their reach into areas formerly dominated by other materials such as metal, glass and wood. Presently, the main focus is in the appliance, construction, electrical, food packaging and transportation sectors, but unique uses surfaces almost daily.

The nanocomposite approach improves air retention properties without weight, rolling resistance and cost penalties. Following benefits of the technology are envisaged by tyre companies:

1. Reduce annual fuel usage by 2.5 bn gallons in the US alone.
2. Reduce annual solid waste by ~1 bn lbs/year.
3. Improve tyre safety with reduced maintenance requirements.

Several leading tyre companies are evaluating this technology and seeking additional commercialisation partners and government support. It is also not surprising that in future, in place of conventional inner liner as on today, only water-based inner paint coating, based on nanocomposite, may take care of both mould release as well as air retention properties of future tyres.

A reduction in peak energy output of the burning composite compared to its unfilled host matrix pair is clear, while the combustion is prolonged at lower heat intensity, in all instances. This indicates the sheltering effect of layered silicates on combustion of the polymers.

A decrease in the storage modulus of SBS/clay nanocomposites can effectively be retarded by introduction of clay into the SBS matrix, when exposed to γ-radiation in air. The layers of OMMT allow absorption of a large amount of the radiation dose and hence broken SBS chains are grafted onto the surface of the layers of OMMT, as indicated by ESR and gel fraction measurements.

A similar process should be expected to occur for other polymer/clay nanocomposites when they are exposed to radiation such as UV light, γ-radiation or electronic beams in air.

Chapter 23

Processing and Characterisation of Nanoceramic Composites

INTRODUCTION

Nanocrystalline materials have demonstrated substantial changes in physical, chemical, and mechanical properties at severely diminished length scales. The extraordinary mechanical, electrical, and thermal properties of single-wall carbon nanotubes (SWCNTs), have prompted intense research into a wide range of applications in materials, electronics, chemical processing, and energy management. Attempts have been made to develop advanced engineering materials with improved or novel properties through the incorporation of carbon nanotubes in various matrices (polymers, metals, and ceramics). However, the potential application of carbon nanotubes in the reinforcement of ceramic composites has not yet been successfully demonstrated. Recently, we have successfully realised this possibility in reinforcing nanocrystalline ceramics through the use of a fast, comparably lower temperature, sintering technique, e.g. spark plasma sintering (SPS). SWCNTs were also successfully used to convert insulating nanoceramics to metallically conductive composites. Additionally, SWCNTs have been demonstrated as outstanding thermal barrier materials in ceramic systems for the first time. Novel thermoelectric properties have also been found recently in these nanocomposites. Such multifunctional carbon nanotube/ceramic composites are envisaged for a wide range of applications. These results will be discussed in the context of microstructural investigations and mechanistic interpretation. The current status and the challenges for the development of carbon nanotube composites are also outlined. In addition, recent findings on novel processing of transparent nanoceramic composites and new forming method for high-strain-rate and low-temperature superplasticity of nanocrystalline ceramic composites will be presented.

Nanostructured materials is an exciting area of materials research because such bulk materials with grain size less than 100 nm exhibit novel properties as compared to their microcrystalline counterparts. However, the brittleness of nanocrystalline ceramics has limited their potential and promise for use in structural and functional applications. Many strategies have been proposed to improve the mechanical properties of nanocrystalline ceramics by using reinforced second phases to develop nanometer-scale composite materials. Carbon nanotubes (CNT), originally discovered as a by-product of fullerene research, are attracting increasing interest as constituents of novel nanostructured materials for a wide range of applications. There are many predictions for the potential applications of nanotubes, such as field emission displays, radiation sources, sensors, probes, interconnects, energy storage and conversion devices, hydrogen storage media, nanometer-sized semiconductor devices, and high-strength conductive composites. CNT should be ideal reinforcing fibres for composites. There are two main types of carbon nanotubes, single-wall carbon nanotubes (SWCNT) and multi-wall carbon nanotubes (MWCNT). Both of these types can have high structural perfection. SWCNT have a particularly desirable combination of mechanical, thermal, and electrical properties.

Specifically, they have an elastic stiffness comparable to that of diamond (1.4 TPa), but they are several times as strong (yield strength 50 GPa). Moreover, carbon nanotubes conduct electricity along their length with very little resistance. Depending on bonding orientation within the nanotube, SWCNT may theoretically be either metallically conducting or semiconducting. The size, shape, and properties of SWCNT make them prime candidates for use in the development of potentially revolutionary composite materials.

The development of advanced engineering composites incorporating carbon nanotubes has become an interesting concept. Attempts have been made to develop advanced engineering materials with improved mechanical properties through the incorporation of CNT in various matrices (polymers, metals, and ceramics) by taking advantage of the exceptional strength of the nanotubes. Most of the investigations on carbon nanotube containing composites have so far focused on polymer-based composites with improved electrical and mechanical properties. For example, their addition to a polymer matrix leads to a very low electrical percolation threshold and improved electrical conductivity. Work on carbon nanotubes in metals and ceramics has been much less focused. To date, the utilisation of the extraordinary mechanical properties of carbon nanotubes in composites has not been successfully realised.

Additionally, theory predicts an extremely high value (6000 W/mK) for the room-temperature thermal conductivity of an individual SWCNT, suggesting that SWCNT should be ideal for high-performance thermal management. Although this speculation has not yet been proven for SWCNT, a recent measurement of 3000 W/mK for the room temperature thermal conductivity of an individual MWCNT has been reported. However, the experimental measurements indicated that aligned bundles of SWCNT show a measured thermal conductivity of only 250 W/mK at room temperature and surprisingly, only 2.3 W/mK for the sintered sample. These results suggest that ropes of SWCNT would be ideal thermal barrier materials for thermal management application. However, no researcher has yet been concerned with thermal properties of carbon nanotube composites in ceramic systems. This study will report for the first time that ropes of SWCNT have been shown to successfully confer anisotropic thermal conductivity to nanoceramic materials.

Among nanostructures, carbon nanotubes have generated special interest due to their unusual properties, including both their thermal and electrical behaviour. While the results of several investigations have been published, these have been directed to individual carbon nanotubes or to 'ropes', 'mats', and films of carbon nanotubes, in attempts to understand the thermoelectric properties and behaviour of these materials. Typically, thermoelectric power of metallic carbon nanotubes is in the range of $-50 \sim +65$ μV/K at 300K. Greatly enhanced values (~260 μV/K) were discovered in semiconducting SWCNT devices.

This opens up the possibility of using SWCNT for thermoelectric applications. Despite these investigations, an effective way to utilise carbon nanotubes in a thermoelectric application has been elusive since the high electrical conductivity and also high thermal conductivity of carbon nanotubes, the calculated ZT of pure CNTs can be expected to be very low (~10^{-4}). Recently, it was discovered that incorporation of single-wall carbon nanotubes into nanoceramics leads to a dramatically improved electrical conductivity of the composites combined with a significant decrease in thermal conductivity, suggesting that the carbon nanotube reinforced nanoceramic composites might make promising thermoelectric materials.

Superplasticity can be defined as the ability of a polycrystalline material to undergo large elongations prior to failure. It can be described by the phenomenological equation,

$$\dot{\varepsilon} = \frac{A}{k_B T} \left(\frac{b}{d} \right)^P \left(\frac{\sigma}{E} \right)^n D_0 \exp\left(-\frac{Q}{k_B T} \right) \qquad \text{... (23.1)}$$

In this equation, $\dot{\varepsilon}$ is the strain rate, A is a material constant, k_B is Boltzmann's constant, b is the Burger's vector, d is the grain size, σ is the applied stress, E is Young's modulus, D_0 is the pre-exponential factor for diffusion, Q is the activation energy for superplastic flow, and T is the temperature in degrees Kelvin. The constant n is the stress exponent, and p is the grain size exponent. The phenomenon usually occurs in fine-grained materials under conditions of moderate temperatures ($T > 0.5 \ T_m$) and moderate-to-slow strain rates (10^{-6} to $10^{-2} \ \mathrm{s}^{-1}$). It is expected that nanocrystalline ceramics, which are characterised by a grain size in the range of 10–100 nm, can deform at low temperatures and at high strain rates. However, evidence for really low temperature superplasticity in nanostructured materials is either very limited or not very convincing. To date, compressive superplasticity has been observed in several studies on nanocrystalline ceramics but elevated temperatures are required for this phenomenon to take place. In addition, the low superplastic strain rates (typically, 10^{-5}–$10^{-4} \ \mathrm{s}^{-1}$) observed in fine-grained ceramics have limited their commercial applicability. Recently, high strain rate superplasticity (HSRS), which is usually referred to as the demonstration of high ductility at strain rates around $10^{-2} \ \mathrm{s}^{-1}$ or greater, has stimulated much scientific and technological interest. The first HSRS report in ceramic materials is Kim who reported a composite ceramic material with a submicron grain size that exhibits superplasticity at strain rates up to 1 s^{-1} at 1650°C. Recently, a company produced a nanocrystalline ceramic composite with the same composition and found that the deformation temperature can be reduced to 1400°C. However, the prohibitively high deformation temperature makes it very difficult for commercial application. The realisation of nanoceramic superplasticity into practice remains a challenging issue. In the present study we strive to bring the HSRS deformation temperature for nanoceramic composites down to 1050°C or lower by using spark-plasma-sintering (SPS) apparatus, both as a sintering tool as well as a forming tool.

EXPERIMENTAL PROCEDURE

The composites in this study were produced from consolidation of two types of alumina powders and SWCNT. The first type of powder was a mixture of α-alumina and γ-alumina, consisting of 80 per cent α-Al_2O_3 and 20 per cent γ-Al_2O_3 with particle sizes of 300 nm (40 nm crystallite size) and 20 nm respectively (Baikowski International Corporation, Charlotte, North Carolina). The second type was gas condensation synthesised γ-Al_2O_3 with average particle sizes of 15 nm and 32 nm (Nanophase Technologies Corporation, Austin, Texas, USA and Nanotechnologies Corporation, Darien, Illinois, respectively). Purified single-wall carbon nanotubes (SWCNT) in paper form, produced by the HiPco process with more than 90 per cent of the catalyst particles removed were obtained from Carbon Nanotechnologies.

SWCNT tend to self-organise into 'ropes' that consist in many (typically, 10–100 nm) tubes running together along their length in van der Waals bonding with one another. Due to their high surface area and high aspect ratio, the ability to homogeneously disperse the nanotubes into the matrix is a challenge in the processing. The SWCNT were received in the form of a pressed sheet or 'paper'. Care has been taken to mix the nanopowders and SWCNT. First, the as-received SWCNT in the 'paper' form must be dispersed into ethanol using an ultrasonic bath (40 Hz, ~50°C, 2 hours, Branson Ultrasonic Cleaner B2510, American Airworks, Sophia, WV 25921). Second, alumina nanopowder was mixed with the dispersed SWCNT alcohol media. Finally, the composite powders were wet-sieved through 200 mesh, ball-milled for 24 hr in ethanol using zirconia ball media, and then dried. Using these techniques three contents of SWCN at 5.7 vol.%, 10 vol.%, and 15 vol.% were produced. For certain experiments, the γ-Al_2O_3 nanopowder with particle size of 15 nm was mechanically activated by high-energy ball milling (HEBM) prior to being combined with the SWCNT. The high-energy ball milling was performed with 1 weight per cent polyvinyl alcohol on a Spex 8000 mixer mill (Spex Industries, Metuchen, New Jersey, USA) in a zirconia vial. The polyvinyl alcohol was included to prevent severe agglomeration of the powder. Milling was performed at room temperature for 24 hours, after which time the polyvinyl alcohol was removed by heating the powder at 350°C for 3 hrs under vacuum. The alumina was then mixed with the SWCNT dispersion, and the combined dispersion was sieved, ball-milled, and dried as described above. An alumina-SWCNT mixture at 15 per cent SWCNT, by volume, was prepared in this manner.

In order to obtain fully dense nanocomposites, retain nanocrystalline alumina grain size, and not damage the carbon nanotubes during sintering, spark plasma sintering (SPS) was employed in the present study. SPS is a moderate pressure sintering method based on the conjecture of a high temperature plasma momentarily generated in the gaps between powder materials by electrical discharge during DC pulsing. It has been suggested that the ON-OFF DC pulse energising method could generate: (i) spark plasma, (ii) spark impact pressure, (iii) Joule heating, and (iv) an electrical field diffusion effect. In this process, powders are loaded into a graphite die and were heated by passing an electric current through the assembly. The low heat capacity of the graphite die allows rapid heating. Therefore, SPS can rapidly consolidate powders to near theoretical density through the combined actions of a rapid heating rate, pressure application, and proposed powder surface cleaning. In recent research, spark plasma sintering is carried out under vacuum in a Dr. Sinter 1050 apparatus (Sumitomo Coal Mining Co., Japan). The powder mixtures were placed into a graphite die (20 mm in inner diameter) and cold-pressed at 200 MPa, giving rise to a sample size with a diameter of ~19 mm. Typical SPS processing parameters are: (i) an applied pressure between 50 and 100 MPa, (ii) pulse cycles with a period of 2.5 ms and follow the pattern of 12 cycles on and 2 cycles off, and (iii) maximum pulse parameters of 10,000 A and 10 V. After applying the given pressure, samples are heated to 600°C in 2 minutes and then ramped to their sintering temperature at rates of 150°C/min to 500°C/min.

The final densities of the sintered compacts were determined by the Archimedes' method with deionised water as the immersion medium. The theoretical densities (TD) of the specimens were calculated according to the rule of mixtures. It is to be noted that the density of graphite (2.25 g/cm^3) was usually used for SWCNT but the estimated density for the present SWCNT is 1.8 g/cm^3. The density of carbon nanotubes is a function both of their diameter and number of shells. Microstructural observation was carried out using an FEI XL30-SFEG high-resolution scanning electron microscope (SEM) with a resolution better than 2 nm. Grain sizes were estimated from XRD and SEM images. Additional characterisation by analytical electron microscopy and high-resolution transmission electron microscopy (HRTEM) was performed on a Philips CM-200 with a field emission gun operating at 200 kV. This instrument has a Link energy dispersive X-ray detector with energy resolution of 1.36 eV and a Gatan imaging filter with 1K by 1K CCD camera and energy resolution of 0.9 eV with 1 nm spatial resolution. Indentation tests were performed on a Wilson Tukon hardness tester with a diamond Vickers indenter. The indentation parameters for fracture toughness (KIC) measurements were a 2.5 kg load with a dwell of 15 s. The fracture toughness was calculated by using the Antis equation.

Electrical conductivity measurements using a four-wire-probe technique were carried out on an Agilent 34420A nano Volt/micro Ohm meter. Even at ambient temperatures, voltages can arise due to thermal EMF generated by dissimilar material junctions in the testing circuit. To remove the effect of these, or any other extraneous voltages, two readings are made for each measurement of conductivity. One reading is made with the current on and the other with the current off. The resulting change in voltage is used to calculate the conductivity of the specimens. Using this configuration, the meter has a resolution of 0.1 μΩ. The four-point probe electrical conductivity (σ) of the composites was measured at four temperatures –196°C, –61°C, 25°C, and 77°C.

Thermal diffusivity measurements were performed by the use of a Xenon flash thermal diffusivity system both on the sintered disks and on rectangular bars cut from the sintered disks. The measurements were taken in two orientations, i.e. the orientation along the uniaxial compression axis, which is referred to herein as the 'transverse' orientation and the orientation perpendicular to the hot-pressing direction, which is referred to herein as the 'in-plane' direction. For each orientation, the measurement was performed by applying a short heat pulse (less than 1 millisecond in duration) to one surface of the disk or bar using a xenon flash lamp, while using an InSb infrared detector to measure the temperature change at the surface across from the surface to which the pulse was applied, the direction normal to the two surfaces defining the orientation being measured. Thermal conductivity is calculated from measurements of thermal diffusivity, specific heat, and bulk density. The room temperature specific heat for alumina of 0.79 J/gK and ropes of SWCNT of 0.65 J/gK is used to calculate the thermal conductivity of the composites using the rule of mixtures.

Specimens for thermoelectric measurements were prepared by cutting the sintered compacts into rectangular bars whose dimensions were 14 mm × 5 mm × 1–3 mm. The heads of two Pt-Pt 13 per cent Rh thermocouples were embedded at the two ends, respectively, of each specimen and fixed with platinum wires. Thermoelectric properties of the specimens were then measured either in air or argon atmospheres by use of an automatic thermoelectric measuring apparatus (RZ2001 K, Ozawa Scientific Co., Japan). Electrical conductivity measurements were performed at 300–900 K by a DC four-probe technique.

RESULTS AND DISCUSSION

Processing and Microstructure

In the open literature all the other carbon nanotubes reinforced ceramic composites have been consolidated by hot-pressing methods that require higher temperatures and longer duration. These sintering parameters damage the carbon nanotubes in the composites, leading to decreases in, or total loss in, reinforcing effects and electrical conductivity. For example, in carbon nanotubes-metal-ceramic composite systems some of the hot-pressing temperatures were as high as 1600°C. That damaged most of the carbon nanotubes and decreased the quantity and quality of carbon nanotubes in the sintered composites. Fully dense nanocomposites could still not be obtained. Siegel used the hot-pressing method at 1300°C/1 hr to obtain fully dense multiple-wall carbon nanotubes filled alumina nanocomposites but the matrix grain size was in submicron range. Spark-plasma-sintering technique allows much lower sintering temperatures and shorter times for obtaining dense nanocrystalline ceramics as compared to conventional sintering techniques. It can be seen that the pure alumina nanopowders can be consolidated by SPS at 1150°C for 180 seconds to get full density. The microstructure of the pure Al_2O_3 consisted of equiaxed grains with an average value of 350 nm. Note that both 5.7 vol.% $SWCNT/Al_2O_3$ and 10 vol.% $SWCNT/Al_2O_3$ nanocomposites can also be successfully consolidated to their theoretical densities at the same sintering conditions as that for pure alumina, suggesting that addition of SWCNT to the alumina matrix was not detrimental to the sintering process. The density of 10 vol.% SWCNT/Al_2O_3 nanocomposite was decreased to 95.2 per cent TD when the sintering duration time was shortened to two minutes at 1150°C. Only 86 per cent TD could be obtained when the sintering temperature was decreased to 1100°C. These results show that spark plasma sintering is an effective sintering technique to obtain fully dense nanocrystalline ceramic composites without damaging carbon nanotubes. When γ-Al_2O_3 nanopowders were used, the sintering temperature to get dense materials is 50°C higher than that for α-Al_2O_3 nanopowders. When high-energy ball milling was applied to the γ-Al_2O_3 nanopowders, the sintering conditions are the same as pure alumina while the obtained grain size is smaller than that for using α-Al_2O_3 nanopowders, suggesting the HEBM along with using finer alumina powders is beneficial to the consolidation process.

SEM and TEM analysis of the microstructures of the consolidated specimens was performed. Some interesting features can be noted. First, most of the alumina grains were less than 200 nm, indicating that the introduction of carbon nanotubes leads to refinement of grain size. When using much finer nanopowder such as γ-Al_2O_3 nanopowder and the use of HEBM processing technique, truly nanocrystalline grains less than 100 nm have been obtained. Second, it can be seen that carbon nanotubes are fairly homogenously dispersed in the matrix. A minor level of agglomeration was observed when the content of SWCNT was increased to 15 vol.%. Third, ropes of carbon nanotubes were effectively entangled with the alumina grains to form a network structure. These unique features observed in the present nanocomposites are quite different from *in situ* carbon nanotubes reinforced alumina nanocomposites where the cohesion between carbon nanotubes and the matrix was rather poor and pullouts of carbon nanotubes were observed. Figure 23.1(a) shows a typical bright field TEM image of the 10 vol.% SWCNT/Al_2O_3 nanocomposite. This Fig. 23.1 shows that ropes of single-wall carbon nanotubes are distributed along grain boundaries to develop an intertwining network microstructure. Some of catalyst particle such as iron were found in the ropes of carbon nanotubes. This microstructure simultaneously provides stiffness, toughness, and strength to the ceramic and also leads to continuity of the nanotube phase providing an interlinked electrical and thermal pathway for improved electrical conductivity and novel thermal properties. Good bonding between carbon nanotubes and alumina was observed in this material at HRTEM [Fig. 23.1(b)]. XDS profile and spot scans were performed to analyse the chemical composition of the different grains, grain boundaries, and particles in the samples. The results showed that no other carbon forms such as graphite were detected along the grain boundary, suggesting carbon nanotubes were not damaged during the consolidation process by SPS.

Mechanical Properties

Measured Vickers hardness and fracture toughness for pure alumina are 20.3 GPa and 3.3 MPam$^{1/2}$ respectively. Fracture toughness of 5.7 vol.% SWCNT/Al_2O_3 nanocomposite is over two times higher than that of pure alumina and there is almost no decrease in hardness. A toughness of nearly three times pure alumina was achieved in the 10 vol.% SWCNT/Al_2O_3 nanocomposite in the present study. No reinforcing effect was noted in the *in situ* carbon nanotubes—Fe-Al_2O_3 nanocomposites in the investigation of Peigney. Moreover, only a marginal increase in fracture toughness can be obtained by improving the quality and quantity of carbon nanotubes in the latter work. The reasons given were mainly related to the damage of carbon nanotubes during hot-pressing. Smith noted that volume contents of carbon nanotubes in the sintered products are lower than those in the starting powders. So far, the best reported result by Siegel was a 24 per cent increase in toughness in 10 vol.% MWCNT/Al_2O_3

nanocomposites. However, it can be seen that fracture toughness increases significantly with the introduction of single-wall carbon nanotubes for the present nanocomposites.

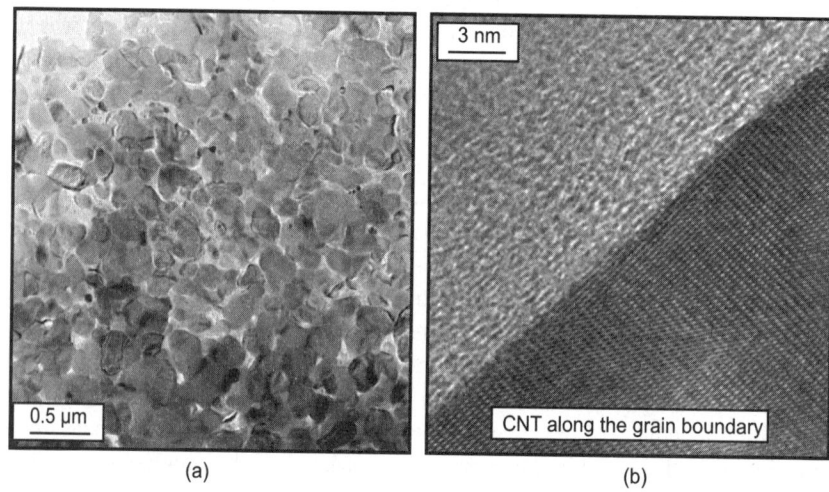

(a)   (b)

Fig. 23.1. TEM images of the nanocomposites showing the intertwining network structure of carbon nanotubes in the matrix. Figure 23.1(a) is an image of the 10 vol.% SWCNT/Al$_2$O$_3$ composite. Light regions are filled with SWCNT bundles. Figure 23.1(b) is a higher magnification image of the 10 vol.% SWCNT/Al$_2$O$_3$ composite that shows the SWCNT concentrated at the boundaries between alumina grains.

More interestingly it was found that there is a positive dependence of fracture toughness on the density in the 10 vol.% SWCNT/Al$_2$O$_3$ nanocomposites in addition to an increase of hardness with density. The dependence of hardness on density is reasonable, but the positive dependence of toughness is contrary to conventional expectation. This may be related to the extent of bonding of carbon nanotubes in the matrix as a function of density. As mentioned earlier, a network structure was developed in the fully dense nanocomposites where ropes of SWCNT were strongly entangled with alumina matrix. It can be noted that some of carbon nanotubes were entangled with alumina grains and some of them encapsulated alumina nano-scale grains. However, the situation is quite different for non-fully-dense nanocomposites. Depending on the density, SEM observation indicates that the entangling network structure appears to disappear in the 86 per cent TD nanocomposite where a loose network with a poor interfacial bonding can be noted. Stronger bonding of ropes with the matrix can be noted in the 95 per cent TD nanocomposite. These results suggest that the extent of interfacial bonding might be a factor in increasing the toughness of the composites.

Other toughening effects may be likely related to the following factors. First, it is due to extraordinary mechanical properties and the more perfect structure of SWCNT. MWCNT are similar to SWCNT, but contain more defects, which limit their properties. Furthermore, there are differences in the ability to transfer load from the matrix to the nanotubes between SWCNT and MWCNT. This may account for the improvement in the toughening effect between SWCNT in the present study and that observed in 10 vol.% MWCNT/Al$_2$O$_3$. Second, the toughening effect may be related to the unique entangling network structure. Therefore, crack deflection along the continuous interface between carbon nanotubes and nanocrystalline matrix grains is possibly one of the toughening mechanisms. Third, it is related to fast sintering technique that allows lower sintering temperatures and short duration. Therefore, the high quality ropes of single-wall carbon nanotubes can be retained in the sintered compacts. This is also consistent with the results by Flahaut where the increase in the quality and quantity of SWCNT may result in an easier transfer of the stress and, thus, can account for the increase in the toughness in the *in situ* SWCNT/Fe/Al$_2$O$_3$ nanocomposites. In order to be effective as reinforcing elements, high quality carbon nanotubes, without damage during consolidation, must be effectively bonded to the matrix so that they can actually carry the loads. The dependence of toughness on density directly supports this statement. The present work gives a promising future for application of carbon nanotubes in reinforcing structural ceramic composites and other materials systems as well. Ongoing investigation to optimise the microstructure and to examine the toughening mechanisms is currently underway.

Electrical Properties

It is well known that pure alumina is an insulator with extremely low electrical conductivity (10^{-10}–10^{-12} S/m). It is interesting to note that the SWCNT/Al$_2$O$_3$ composites become much more electrically conductive when small amounts of carbon nanotubes were incorporated into alumina. It was found that the electrical conductivity increases with carbon nanotube content. The room-temperature electrical conductivity is up to 1050 S/m in 5.7 vol.% SWCNT/Al$_2$O$_3$ nanocomposite. More than three times the best reported conductivity of the CNT-Fe-Al$_2$O$_3$ nanocomposites (~400 S/m) has been obtained in our investigation in the 10 vol.% SWCNT/Al$_2$O$_3$ nanocomposite. An additional increase in conductivity (up to 3345 S/m) has been achieved in the 15 vol.% SWCNT/Al$_2$O$_3$ nanocomposite. This value is 735 per cent higher than that of the hot-pressed CNT-Fe-Al$_2$O$_3$ nanocomposites. These values lie in the semiconductor range of conductivity but are very close to the metallic conductor threshold (10^4 S/m), as shown in Fig. 23.2. This dependence of the electrical conductivity on carbon nanotube content in the present ceramic nanocomposites is in contrast to the carbon nanotube metal matrix composites where the electrical conductivity decreases with carbon nanotube content.

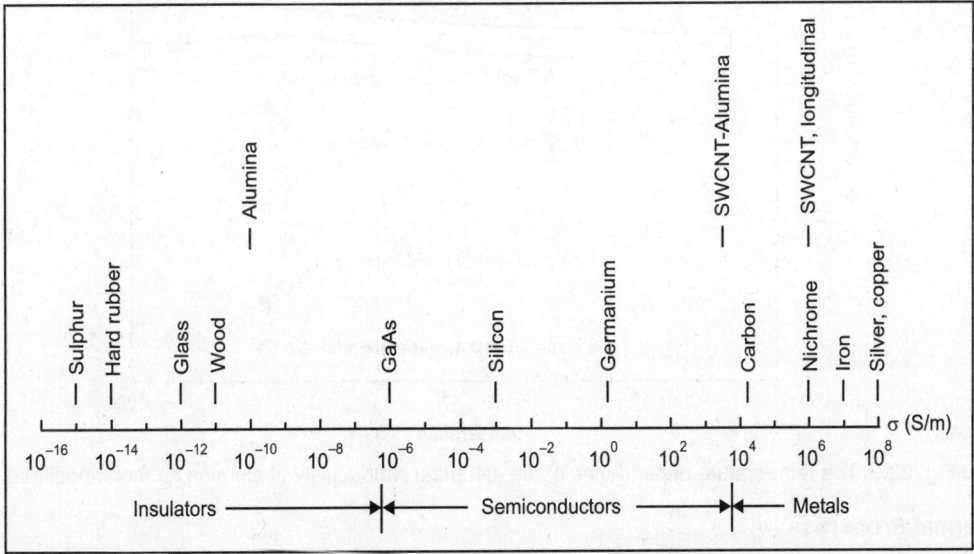

Fig. 23.2. The electrical conductivity of various representative materials at room temperature. Note more than 13 orders of magnitude increase in conductivity of the 15 vol.% SWCNT-alumina composite compared to pure alumina.

The temperature dependence of the electrical conductivity of the alumina nanocomposites is shown in Fig. 23.3. The conductivity of the present nanocomposites increases with increasing temperature. The largest conductivity increase, from 2060 S/m at –194°C to 3375 S/m at 77°C, is in the 15 vol.% SWCNT/Al_2O_3 nanocomposite. This behaviour is also opposite to the trend in metal matrix composites containing carbon nanotubes. It has also been reported that carbon nanotube-metal-oxide nanocomposites can be aligned to improve the electrical conductivity by high temperature extrusion. The best reported conductivity measured along the extrusion direction is 2000 S/m whereas much lower values are measured in the transverse direction (~60 S/m) in CNT-Fe/Co-$MgAl_2O_4$ systems. Note that the best conductivity in the aligned $MgAl_2O_4$-based systems could only be obtained in the centre of the extrusion while other regions were quite low due to the damage of the nanotubes by exposure to high temperature extrusion. The significant increase in electrical conductivity may be related to the use of high quality ropes of SWCNT that were distributed along grain boundaries to develop an intertwining network of electrically conducting pathways. In order to directly compare the effects of the network structure an additional composite was produced. The experimental procedure and materials were identical to the 5.7 vol.% SWCNT/Al_2O_3 composite except carbon black was substituted for the SWCNT. The carbon black is composed of mixed fullerenes and has an average particle size of 42 nm. The 5.7 vol.% carbon black/Al_2O_3 composite had a measured conductivity of 15 S/m, nearly two orders of magnitude less than the corresponding SWCNT containing composite.

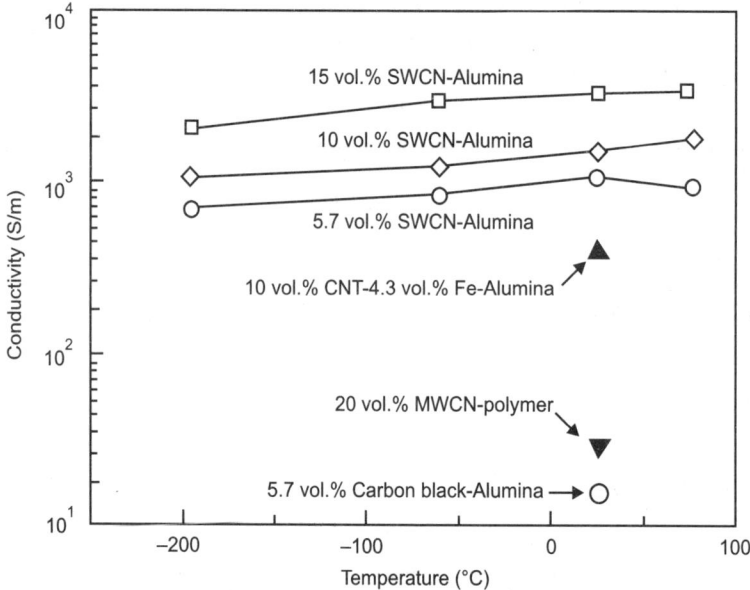

Fig. 23.3. The temperature dependence of the electrical conductivity of alumina nanocomposites.

Thermal Properties

Figure 23.4 is a bar graph showing the thermal diffusivities of various samples. The first three sets of bars each represent measurements taken directly on sintered disks in the transverse direction, while the last two sets represent measurements in transverse and in-plane directions taken on bars cut from sintered disks. It is very interesting to note that the incorporation of ropes of SWCNT does not change the in-plane thermal diffusivity of the matrix. By contrast, the transverse thermal diffusivities are significantly decreased when the carbon nanotubes are present, and that an increase in the level of carbon nanotubes produces a greater drop in thermal diffusivity in the transverse direction. These findings are directly in contrast to the results observed for SWCNT/polymer composites.

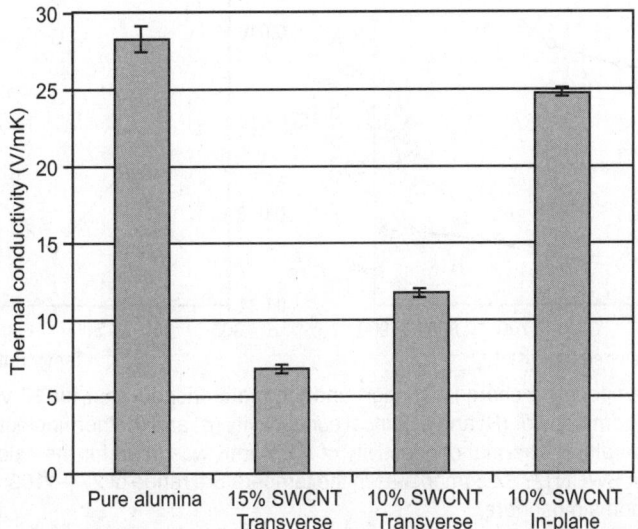

Fig. 23.4. Bar graph comparing thermal diffusitivites of various materials in different materials and orientations.

The temperature dependence of the transverse thermal diffusivity for pure alumina and composites has been measured in the temperature ranges from 25° to 500°C. The thermal diffusivity of the pure alumina and carbon nanotube composites decreases with increasing temperature. This is consistent with the results observed for other carbon materials. Moreover, it is interesting to note that the transverse thermal diffusivity of 15 vol.% SWCNT/Al$_2$O$_3$ composite increases slightly from room temperature to ~100°C and then decreases with increasing temperature until 300°C. Above that the curve coincides with that of 10 vol.%SWCNT composite. The observed reduction in thermal diffusivity with increasing temperature can be attributed to the dominant effect of Umklapp scattering (phonon-phonon scattering) in reducing phonon mean-free path length.

Thermoelectric Properties

The thermopower measurement of the 10 vol.% SWCNT/20 vol.% 3Y-TZP/Al$_2$O$_3$ composite was carried out in air atmosphere. It was found that thermoelectric power increases linearly with increasing temperature. The measured values are ranging from 28.5 µV/K at 345K to 50.4 µV/K at 644 K. The lowest temperature data is comparable to that of aligned ropes of SWCNT (~27 µV/K at 300 K). The positive value of thermopower is consistent with the oxygen-doping p-type semiconducting behaviour of SWCNT. It can be seen (Fig. 23.5) that the electrical conductivity of the composite decreases with increasing temperature, indicating the metallic conducting behaviour. It was found that the residual catalyst remaining in the SWCNT plays a critical role in the charge and thermal transport of the composites. This is different from that of purified SWCNT/ceramic composites. Also, the electrical conductivity of the composite is extremely low, which is more than two orders of magnitude lower than that of purified SWCNT/ceramic composites. Previous studies found that the in-plane thermal conductivity of the CNT/ceramic composites depends on that of the matrix and it decreases with increasing temperatures.

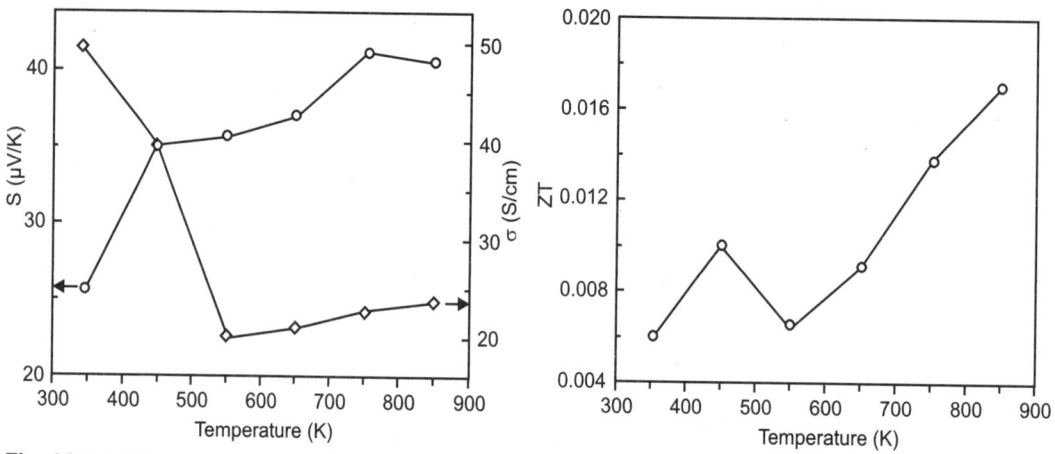

Fig. 23.5. Variable-temperature charge transport and thermal transport data for 10 vol.%SWCNT/3Y-TZP nanocomposite: (a) Thermopower (S) and electrical conductivity (σ) and (b) thermoelectric figure of merit, ZT, as a function of temperature. Thermal conductivity of 0.2 W/mK was used for the calculation of ZT. This is referred to the data for SWCNT/YSZ composites in the temperature range of 27°–1100°C. The measurement was carried out in argon atmosphere.

It has been reported that incorporation of SWCNT into zirconia leads to the development of the lowest thermal conductivity of the composites. The measured thermal conductivity of the composites was in the range of 0.2–0.4 W/mK in the temperature range of 27°–1100°C. It was found that the present 10 vol.% SWCNT/3Y-TZP composite has the highest electrical conductivity among all the composites. It is interesting to note that the electrical conductivity decreases with increasing temperature until 545 K and then increases slowly with increasing temperature above that value. The thermopower also has a positive sign and it increases with increasing temperature. Based on this data, the calculated ZT of the composite is shown in Fig. 23.5. It was found that the ZT increases with increasing temperature and has a value of 0.018 at 850 K. This is two orders of magnitude higher than that of pure SWCNT. As compared to other thermoelectric materials, however, the electrical conductivity of the CNT/ceramic composites is low. This can be further improved if pure metallic SWCNT can be utilised due to the much higher electrical conductivity of SWCNTs.

Ultra-Low-Temperature Superplasticity

SPS has been demonstrated not only to be an effective sintering process to fabricate fully dense nanocrystalline ceramics and composites but also to be a new forming method to enhance ceramic ductility. The first attempt to apply the SPS approach to speed up superplastic forming was by Shen who started with fully dense ceramics that sinter via either transient or permanent liquid phase modes. The observed enhanced ductility is thought to be associated with the enhanced grain sliding at the boundary of the glassy/liquid phase resulting from the electric-field induced motion of charged species. Despite remarkable success (rapid superplastic deformation with high strain rates in the range 10^{-2} to 10^{-3} s^{-1}), the deformation temperatures were still extremely high (1500°C is typical). In the present study, completely different strategies and ideas were utilised. Here we started with a porous preform instead of fully dense blanks and simultaneously consolidated and superplastically formed the specimens in the SPS equipment. The use of concurrent deformation and consolidation provides nanoceramic superplasticity with many advantages. First, porous preforms are easy to produce (75–90 per cent TD) by different methods such as pressureless sintering, plasma spray, SPS, etc.

Second, in term of effects of pore pinning, the presence of open porosity network in the materials at densities below 90 per cent TD was found to seriously inhibit static grain growth during the sintering. However, at densities over 90 per cent TD density, as the open pores become closed, the closed pores are not as effective as open pores in pinning grain boundaries, and thus beyond this point the grain size begins to increase dramatically. It has been found that the open porosity can also limit dynamic grain growth during the deformation. Finally, the microstructure with nanocrystalline grain size can lead to much lower deformation temperatures and higher strain rates.

The starting materials used in the present study include high purity alumina powder doped with 500 ppm MgO and 300 ppm Y_2O_3 obtained from Baikowski International and cubic MgO nanopowder with a particle size of 40 nm. 86.7 wt.% Alumina and 13.3 wt.% MgO were mixed by conventional ball-milling method with ethanol alcohol and zirconia balls in a plastic vessel for 24 hours to prepare a composite powder. The composite mixture was consolidated by SPS to produce a composite consisting of 50 vol.%Al_2O_3 and 50 vol.%$MgAl_2O_4$ (through the solid reaction of Al_2O_3 and MgO during sintering). SPS was carried out under vacuum in a Dr. Sinter 1050 SPS apparatus (Sumitomo Coal Mining Co., Japan) with pulse duration of 3.3 ms. The pulse sequence consists of twelve pulses followed by two periods (6.6 ms) of zero current. The temperature was monitored with a thermocouple inserted into a 'non-through' hole (2 mm in diameter and 5 mm in depth) in the graphite die. The sample was a disk shape with a diameter of ~20 mm and a thickness up to 4 mm. The compressive deformation tests of SPS-prepared, partially dense nanoceramic composites were performed via the same SPS apparatus. Before the desired temperature was reached, a constant load of about 3–4 kN was applied to the sample, which corresponds to an initial stress of 40-50 MPa. When the desired temperature was reached, constant load rates of 45–125 N/s were applied. The compressive strain is defined here as $Ln\,(1-\Delta L/L_0)$, where ΔL and L_0 represent the shrinkage of sample height and the original height of the sample before deformation, respectively. The use of pulsed direct current may generate spark discharges between the powder particles and plasma generation may also occur. Although the generation of a plasma has not been confirmed yet by direct experiment, it is likely that the electric field generates internal localised heating, impact pressure, and ionisation, which promote mass transfer and accelerate localised reactions, and as a result lead to enhanced densification.

It should be pointed out that a discharge process can play a major role only in the initial part of a sintering process but the other electrical field effects are operative over the entire sintering cycle, the reason being that the discharge process cannot be operative in fully dense samples. The field-enhanced mass transfer effects such as grain-boundary diffusion and grain-boundary migration processes can be effective in dense materials. In order to fully take advantage of these effects, we prepared partially dense materials instead of fully dense materials for subsequent superplastic deformation to enable both the discharge process and possible spark plasma effect to the whole deformation process. Here, we demonstrate spark-plasma-sintering, enhanced low temperature superplasticity in partially dense nanoceramic composites. The materials have nanosized grains with a bimodal microstructure containing a certain amount of porosity. It is expected that these open porosities not only limit static grain growth but also may be effective second phase barriers to dynamic grain growth. In terms of these combined effects, it may be easily understood why such high deformation temperatures were observed in Shen's work since the only electrical field effect could be operative for the fully dense materials during the deformation. Figures 23.6(a) and 23.6(b) show the spark-plasma-enhanced superplastically formed samples at 1000° and 1050°C, respectively. It can be seen that the strain rates up to 10^{-2}/s could be achieved.

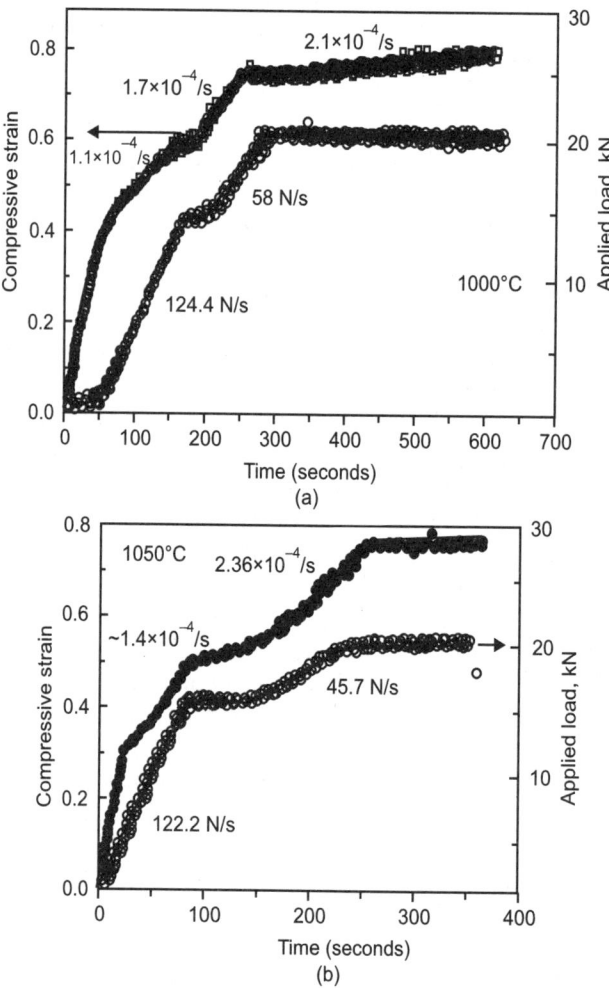

Fig. 23.6. Spark-plasma-enhanced superplastic deformation behaviour of 50 vol.% Al_2O_3/$MgAl_2O_4$ composite. The compressive strain data obtained plotted vs. time for the composite at 1000°C (a) and at 1050°C (b).

A compressive strain of –0.8, equivalent to a reduction to half the original height, can be achieved without cracking. Surprisingly, the process deformation temperature is as low as 1000°C—remarkably lower than those found for conventional superplastic ceramics. These record temperatures are comparable to that of Ni-based superalloys (typically, 950°C), suggesting that an existing metallic superplastic shape tooling might be applied to nanoceramic composites. Importantly, the present study demonstrated that the more practical use of ceramic superplasticity is not in the shaping of already-dense materials but in the near net shape of dense parts from less-than-fully dense preforms. It should be noted that the same composites did not exhibit superplasticity by conventional deformation methods since both static grain growth during the slow heating and dynamic grain growth during high temperature deformation occur.

It is generally accepted that application of mechanical pressure is helpful in removing pores from compacts and enhancing diffusion. The increasing applied pressure during deformation is expected to provide extra strain energy to promote rapid densification and grain boundary sliding. Therefore, applying a high pressure at low temperature that allows the grain-boundary sliding to become kinetically favourable can enhance deformation rates. This is consistent with our findings where the strain rate increases with increasing loading rate. When a constant load instead of a constant stress is applied, the strain rates are significantly decreased. The latter is similar to the observations of Shen work where they applied a constant load during the deformation, indicating a decreasing applied stress. It can also be noted that the strain rates are significantly increased when a little higher deformation temperature is applied [Fig. 23.6(b)] even though the loading rates are slightly lower, suggesting the temperature has profound effect on deformation. Finally, it is very interesting to note that nearly nanosized microstructure has been obtained in the deformed composites whereas much grain growth can be seen in the non-deformed samples even at the same SPS temperature. These results suggest that SPS approach may be a new way for fabricating truly nanocrystalline ceramic composites with novel properties. Moreover, nearly fully dense materials can be achieved in the final products through SPS superplastic deformation. Interestingly, the deformed nanoceramic composites exhibit excellent infrared transparency (Fig. 23.7). The infrared transparent nanoceramic composites instead of polycrystalline monolithic ceramic materials have never been reported in the literature. Such materials are highly desirable and have potential applications. In conclusion, this new SPS approach, starting with partially dense materials and concurrent deformation and densification in the SPS apparatus, provides a new route for low temperature and high-strain-rate superplasticity for nanostructured materials and should impact and interest a broad range of scientists in materials research and superplastic forming technology.

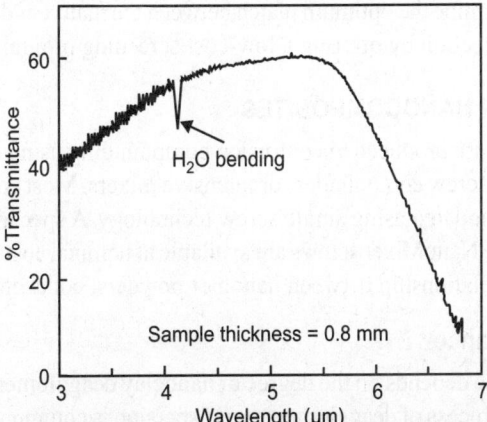

Fig. 23.7. Infrared spectrum of spark-plasma-enhanced superplastic formed sample at 1050°C. The infrared spectra of the alumina-spinel composite samples were collected on a FTIR instrument (Mattson Galaxy Series FTIR 3000). The spectrometer was set to collect 16 scans in transmittance mode with a resolution of 4 cm^{-1} over a range from 650 nm to 16 μm. Over 40 per cent transmittances have been obtained in the wavelength range of 3–6 μm, showing a maximum value of 65 per cent at 5 μm. Note that there is an absorbing peak in the 4.4 mm due to the water bending effect that comes from environmental condition.

To sum up SPS technique has been found to not only be a very effective processing method for consolidation of composites containing carbon nanotubes with novel mechanical, electrical, thermal, and thermoelectric properties but also as a tooling for ultralow-temperature superplasticity.

Polyolefin, Polypropylene and Polystyrene Nanocomposites

INTRODUCTION

Polyolefin nanocomposites are useful for a wide variety of applications in packaging, engineering and fire retardant markets. These nanocomposites are near molecular blends of montmorillonite nanoclay and a polyolefin matrix. Nanocor offers nanoclay concentrates suitable for polyolefin nanocomposites under the Nanomer® concentrate tradename. This chapter outlines performance properties for typical polyolefins and directs the reader to additional Tech Data Sheets with explanations of options or creating nanocomposites. Polyolefins represent a wide range of homo and copolymers, principally containing propylene and ethylene.

Also included are polyolefin/rubber alloys usually called TPO's and polyolefin/EVA blends. Because propylene and ethylene contents as well as polydispersities vary widely, some nanomer concentrate screening is required to determine the optimum match between the matrix and nanoclay surface treatment. Nanocor assists with the selection by offering a low-cost screening program.

CREATING POLYOLEFIN NANOCOMPOSITES

Polyolefin nanocomposites are produced via extrusion compounding using concentrates. Film or bottle grade products require twin screw compounders or intensive mixers. Most engineering and fire retardant applications can be accommodated using single screw technology. A specially designed screw, called a NanoMixer™, must be used. NanoMixer screws are available at nominal cost from New Castle Industries. Figure 24.1 illustrates the relationship between nanomer powders, concentrates and nanocomposites.

Nanocomposite Performance

Nanocomposite performance depends on the degree of nanoclay deagglomeration and dispersion within the polyolefin matrix. The process of deagglomeration/dispersion is commonly referred to as exfoliation. Three factors influence exfoliation: the nanoclay surface treatment, nanocomposite loading, and the amount of dispersive energy used.

The optimum nanoclay surface treatment is readily determined through screening, as described earlier. Nanomer loading will not hinder exfoliation within the 2–8 per cent wt/wt range and is optimised at 6 per cent wt/wt. The proper choice of compounding equipment and use of Nanocor guidelines assists in achieving the dispersive energy requirement. However, another factor impacting dispersive energy is matrix resin melt flow. Resin melt flow influences the level of shear resulting from dispersive energy input.

The lower the melt flow, the greater the shear on nanomer particles. It follows that exfoliation is maximised in lower melt flow resins, where shear is the highest.

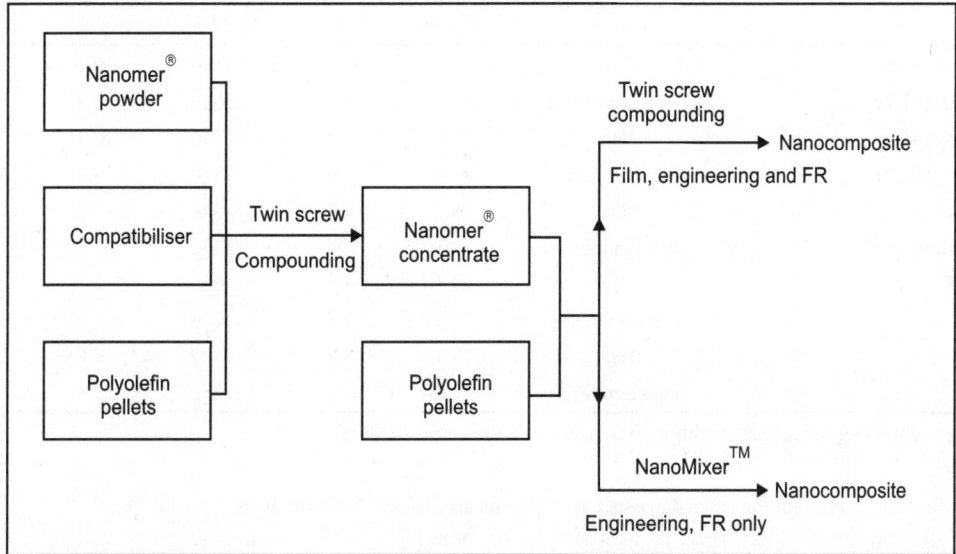

Fig. 24.1. Relationship between nanomer powders, concentrates and nanocomposites.

Properties

Mechanical properties

Polyolefin nanocomposites exhibit mechanical properties similar to highly filled systems, but with the added advantage of low density. The density (and resulting part weight) is just slightly higher than unfilled resin. Nanocompositing also provides a high degree of reinforcement without significant loss of impact. In most cases notched Izod impact remains the same as neat (unfilled) resin.

Table 24.1 provides typical mechanical performance data for a variety of polyolefins and melt flows. The concentrate addition level is 12 per cent wt/wt in all cases. This corresponds to a final nanomer loading of 6 per cent wt/wt.

Reinforcement is linear across the effective loading range (2–8 per cent wt/wt) nanomer. The magnitude of increase per per cent loading decreases above the range due to nanoclay packing. Thermal improvement is curve linear in the effective range because nanoclay surface treatment exerts some plasticising effect on the nanocomposite. Figure 24.2 graphs reinforcement for homopolymer polypropylene (MF 4).

Polyolefin nanocomposites have specific gravities and melt flows nearly the same as neat resin, so they process like neat resin. Part output rates will be significantly higher than conventionally filled composites and machine wear will also improve noticeably.

Nanocomposites can be used in conjunction with conventional fillers, i.e. glass fibre or talc, to create synergistic materials with customised property sets. For example a 6 per cent wt/wt nanocomposite containing 10–12 per cent wt/wt glass fibre will deliver a property set similar to 30 per cent wt/wt straight glass fibre, but lower specific gravity and part weight.

Table 24.1. Mechanical properties.

Resin	Melt flow (gms/10 min.)	Condition	Tensile strength (Mpa)	Elongation @break (%)	Flexural modulus (Mpa)	Notched izod (ft-lb/in)	HDT[2] (°C)
Homopolymer	4	Neat	32	280	1150	0.7	86
Polypropylene		Nanocomposite[1]	38	20	2040	0.7	115
Homopolymer	35	Neat	35.2	105	1590	0.4	113
Polypropylene		Nanocomposite[1]	39.9	11	2310	0.4	121
Copolymer	6.5	Neat	26.3	420	840	0.7	71
Polypropylene		Nanocomposite[1]	30.3	93	1350	1	81
HDPE	4	Neat	23.8	460	680	3	71
		Nanocomposite[1]	26.5	200	1170	1.1	77
TPO	12	Neat	19.5	320	780	9.8	71
		Nanocomposite[1]	21.8	42	1230	9.8	85

[1] 12 per cent wt/wt concentrate addition (6 per cent wt/wt nanomer loading).

[2] 66 psi.

Note: Specific gravity for the nanocomposites is 0.93 compared to 0.90 for neat resins.

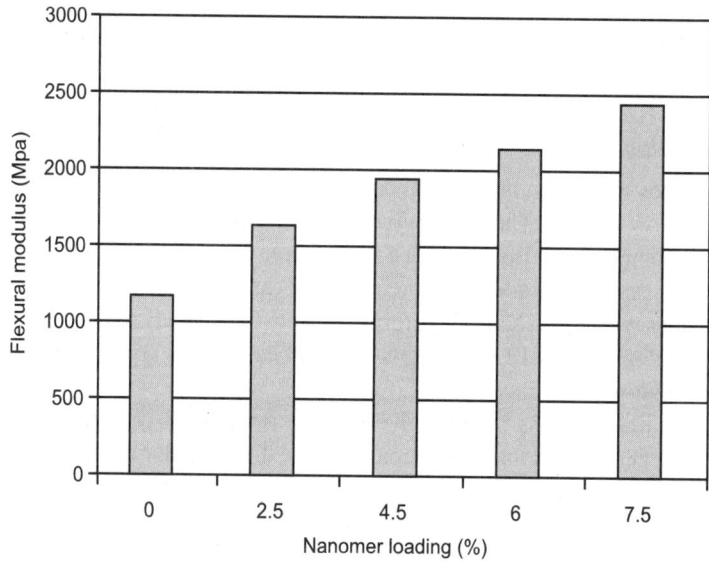

Fig. 24.2. Nanomer loading and reinforcement.

Fire resistant properties

Polyolefin nanocomposites are potent char formers. In fire retardant (FR) applications they function by insulating the resin substrate from the air-surface interface, starving the flame by reducing the fuel needed to continue the burning process. The result is a dramatically lower peak heat release rate (HRR) (Fig. 24.3).

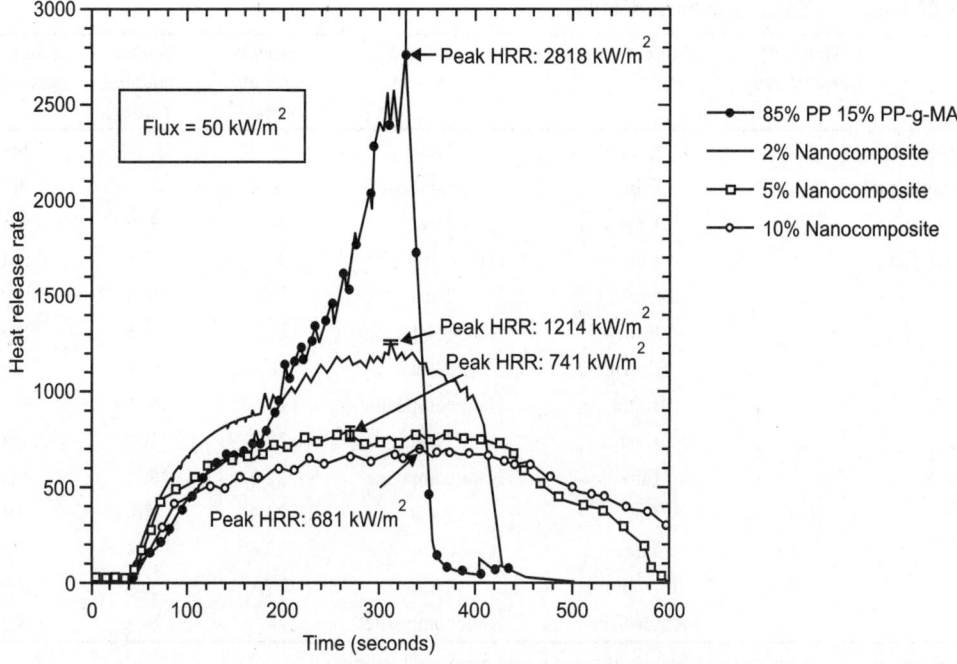

Fig. 24.3. HRR for polypropylene nanocomposites.

Nanocomposites alone do not conform to a UL V0 rating. But in conjunction with reduced levels of conventional FR additives they act synergistically to achieve V0. Often traditional FR additives can be reduced by 50 per cent wt/wt. With less FR additive dilution and the enhanced reinforcing effect of nanocompositing, mechanical properties are restored, usually at lower cost than would be the case without nanocompositing.

Another benefit of nanocomposite/FR additive synergy is lower density and part weight, since a lot of high density additive has been removed from the formulation.

Gas Barrier Enhancement

As polyolefins take an increased role in packaging applications, gas barrier becomes an important requirement. Nanocomposites improve oxygen transmission rates (OTR) between 1.2–2X. Carbon dioxide transmission rates (CO_2TR) improve 1.4–1.8X. There is also some improvement in water vapour transmission (WVTR), around 10–15 per cent. Improved gas barrier follows the tortuous path model and is valid for cast, blown and moulded products. Tables 24.2(a) and 24.2(b) present performance data for a variety of polyolefins.

Predictably, nanocomposite films possess greater tensile strength and modulus, although elongation at break is reduced. Simultaneous improvement in tensile properties and gas barrier provides the formulator with opportunities to downgauge films and containers.

Tear strengths in both the machine and transverse directions remain the same as neat polyolefin films of the same mil thickness. Tear strength of downgauged nanocomposite film drops in proportion to the reduction in film thickness.

Table 24.2(a). Film/Bottle strength properties.

Resin	Melt flow (gms/10 min.)	Product form	Condition	Tensile strength (Mpa)	Tensile modulus (Mpa)	Elongation @break (%)
Homopolymer Polypropylene	1.3	Cast Film	Neat Nanocomposite[1]	36.4 33.4	1230 1710	660 480
Copolymer Polypropylene		Cast Film	Neat Nanocomposite[1]	38.7 31.7	910 1500	650 330
	2	Stretched Film	Neat Nanocomposite[1]	N/A N/A	N/A N/A	N/A N/A
		Blown Bottle	Neat Nanocomposite[1]	N/A N/A	N/A N/A	N/A N/A
HDPE	4	Cast Film	Neat Nanocomposite[1]	22.0 22.5	700 854	800 690
TPO	0.9	Cast Film	Neat Nanocomposite[1]	23.4 27.6	56 178	730 780
TPE	N/A	Injection Moulded liner	Neat Nanocomposite[1]	N/A N/A	N/A N/A	N/A N/A

[1]12 per cent wt/wt concentrate addition (6 per cent wt/wt nanomer loadings).

Table 24.2(b). Gas barrier properties.

Resin	Melt flow (gms/10 min.)	Product form	Condition	OTR (cc-mil/ 100 in²·day)	CO₂TR (cc-mil/ 100 in²·day)	WVTR (cc-mil/ 100 in²·day)
Homopolymer Polypropylene	1.3	Cast Film	Neat Nanocomposite[1]	142 108	N/A N/A	N/A N/A
Copolymer Polypropylene		Cast Film	Neat Nanocomposite[1]	205 108	890 460	0.64 0.55
	2	Stretched Film	Neat Nanocomposite[1]	110 70	N/A N/A	N/A N/A
		Blown Bottle	Neat Nanocomposite[1]	191 119	N/A N/A	N/A N/A
HDPE	4	Cast Film	Neat Nanocomposite[1]	160 126	N/A N/A	N/A N/A
TPO		Cast Film	Neat Nanocomposite[1]	1820 1270	N/A N/A	N/A N/A
	0.9	Blown Film	Neat Nanocomposite[1]	2270 1020	N/A N/A	N/A N/A
TPE	N/A	Injection Moulded liner	Neat Nanocomposite[1]	808 354	N/A N/A	N/A N/A

[1]12 per cent wt/wt concentrate addition (6 per cent wt/wt nanomer loading).

PREPARATION AND PROPERTIES OF POLYOLEFIN NANOCOMPOSITES

Polyolefin nanocomposites have been successfully manufactured using commercially available surface modified montmorillonite (Nanomer®). These nanocomposites are notable for their improved mechanical properties, such as tensile strength and modulus and for their dimensional stability. They also demonstrate enhanced barrier to gas permeation, particularly oxygen. Nanocomposite processing uses twin screw technology, coupled with standard processes for injection moulding, film casting and blow moulding. In addition to packaging applications, polyolefin nanocomposites are of particular interest for automotive and industrial applications where wall downgauging is desirable. Nanomer® chemistry, nanocomposite processing, and properties will be discussed in this section. Polyolefins are one of the fastest growing classes of thermoplastics due to their good balance of physical and chemical properties, their low cost, lightweight, favourable processing and recycle characteristics. Some polyolefins, such as polypropylene, find extensive use in both packaging and engineering applications, especially automotive.

Like all plastics, polyolefins demonstrate some performance shortcomings. In packaging for example, they are poor gas barriers. In automotive on the other hand, dimensional and thermal stability limit their range of application. Most schemes to improve polyolefin gas barrier involve either addition of higher barrier plastics via a multilayer structure or high barrier surface coatings. Although effective, the increased cost of these approaches negates one big attraction for using polyolefins in the first place-economy.

Currently, automotive and appliance applications employ glass or mineral-filled systems with loading levels ranging from 15–50 per cent by weight. This approach improves most mechanicals, i.e. dimensional and thermal stability, but it carries two drawbacks. First, polyolefin's easy processing is compromised and second, high filler loadings inevitably lead to heavier products.

The emerging field of polymer-layered nanocomposites is unique in that it addresses shortcomings of polyolefins for both packaging and engineered applications, and it does so with favourable cost, processing and weight profiles. Polymer-layered silicate nanocomposites are plastics containing low levels of dispersed platey minerals with at least one dimension in the nanometer range. The most common mineral is montmorillonite clay. Its aspect ratio exceeds 300, giving rise to enhanced barrier and mechanical properties. In general, every one weight-per-cent of these 'nanoclays' creates a 10 per cent property improvement. Their interaction with resin molecules alters the morphology and crystallinity of the matrix polymer, leading to improved processability in addition to the benefits to barrier, strength and stability.

Polyolefin nanocomposites have attracted significant research interest. Nanocomposite research activities are mainly focused on the use of chemically modified montmorillonite. Recently, companies have successfully synthesised polypropylene-layered silicate nanocomposites employing a masterbatch route. Improved gas barrier and mechanical properties were achieved for different grades of polymers at low nanoclay loading (typically 4~6 wt%). The findings provide one opportunity for end users to downgauge finished products and maintain equivalent or enhanced performance properties.

Experimental Procedure

Materials

Nanomers® I.30 P, I.31 PS, and I.44 PA are commercial products manufactured by Nanocor, Inc. Each is an onium ion modifed montmorillonite, designed for maximum compatibility and dispersion in a polyolefin matrix. Available as free-flowing powders with a mean size of 15–25 microns, they are capable of dispersing to nanoscale in twin screw compounders. In addition to the typical onium treatment, Nanomer I.31PS contains a silane-coupling agent to promote higher tensile properties. I.30P and I.31PS are film grade Nanomers and I.44 PA is an engineering grade.

Formation of polyolefin nanocomposites

Polyolefin nanocomposite was formed using a two-step process: Nanomer® masterbatching and subsequent letdown into neat polyolefin. A typical masterbatch contains 50~60 wt% Nanomer® and the remainder a combination of compatibiliser (such PP-g-MA) and standard polyolefin. Masterbatching predisperses nanomers, promoting full nanocomposite formation while minimising heat history during letdown. The final nanomer loading after letdown is typically 2–7 per cent by weight.

For our work a Lestritz co-rotating twin screw extruder was used to produce both the masterbatches and nanocomposites. The extruder has a diameter of 27 mm and a L/D ratio of 36:1. There are three shear-mixing zones in the extruder set-up to maximise dispersive energy. Screw speeds ranged from 300–500 rpm and extrusion temperatures ranged from 170° to 190°C.

A detailed screw configuration is shown in Fig. 24.4.

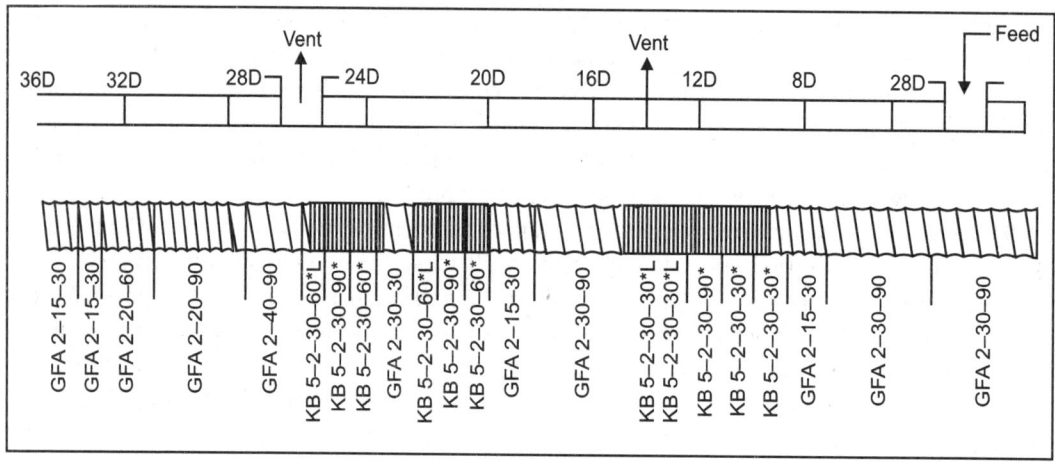

Fig. 24.4. Extruder screw element.

Materials testing

Nanocomposite formation and the degree of nanomer dispersion can be monitored using wide-angle X-ray diffraction (XRD) and transmission electronic microscopy (TEM). Both techniques were used in this study. Polyolefin nanocomposite films were made using cast and blown processes. Oxygen transmission rates (OTR) of nanocomposite film samples were measured using a Mocon 2/60 at room temperature and 65 per cent relative humidity. Mechanical properties were determined on injection moulded specimens according to ASTM testing methods.

Results and Discussion

Nanoclay dispersion

Both XRD and TEM confirm good Nanomer® dispersion in the polyolefin matrices. XRD measures the degree of dispersion by estimating the distance between individual platelets after compounding. Table 24.3

presents *d*-spacing data for Nanomers® before and after dispersion. The increases confirm that polyolefin host polymers have successfully interacted with Nanomer® surface treatments, promoting nanocomposite formation. TEM confirms the results (Fig. 24.5).

Table 24.3. Formation of polyolefin nanocomposite: compositions and dispersion.

Polyolefin	MFI (gm/10 min)	d-Spacing (Å)	
		Nanomer®	Nanocomposite
PP-PE random copolymer	2.0	22	29
Homopolymer PP	0.45	22	28
Homopolymer PP	4.0	22	32
TPO	12	22	32

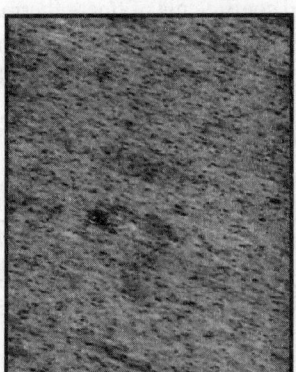

Fig. 24.5. TEM images of HPP nanocomposite (Homopolymer PP- 6 wt% I.31PS).

Film properties

Table 24.4 provides tensile properties for the PP/PE copolymer nanocomposites cast into monolayer films. Reinforcing effects can be easily observed in modulus with minimal impact on strength. Assuming gas barrier improves commensurate with tensile modulus, the film can be downgauged without compromising integrity. Barrier for this film improved by 45 per cent, indicating that significant down-gauging can be done.

Barrier improvement is explained using tortuous path theory, as it relates to alignment of the nanoclay platelets. It follows, then, that film processing conditions can influence nanocomposite barrier improvement. Table 24.5 compares cast and blown films using homopolymer polypropylene as the matrix. Blown film, with its greater stretching ratio, creates more uniform nanoclay platelet orientation, leading to a more tortuous path and better barrier.

Table 24.4. Properties of PP copolymer nanocomposite films.

Nanomer	Loading level (wt %)	Tensile modulus (Mpa)	Improv.	OTR (cc.mil/100 in².day)	Improv.
Control	0	909	–	205	–
I.30 P	6.0	1751	+ 93%	137	1.5x
I.31 PS	6.0	1495	+ 64%	108	2x

Table 24.5. OTR of HPP nanocomposite blown and cast films.

Processing	Nanomer grade	Loading level (wt %)	OTR (cc.mil/100 in^2·day)	Improv.
Blow film	Control	0	2520	–
Blow film	I.31 PS	7.0	1019	2.5x
Cast film	Control	0	1823	–
Cast film	I.31 PS	7.0	1272	1.4x

Mechanical properties

We have previously reported the mechanical properties of homopolymer PP (HPP) and TPO nanocomposites using film grade nanomers I.30 P and I.31 PS. These nanocomposites showed improved mechanical properties, thermal properties (HDT), and coefficients of linear thermal expansion (CLTE). Recently nanocor developed nanomer I.44 PA, a grade designed specifically for non-film applications. Tables 24.6 and 24.7 summarise the performance of this new grade for HPP and TPO nanocomposites. A standard grade HPP with a melt flow of 4 g/10 min. and a standard grade TPO with a melt flow of 12 g/10 min. were used as representative matrix polymers.

Table 24.6. Tensile properties of PP nanocomposites.

Nanomer	Loading level (wt %)	Resin	Tensile modulus (MPa)	Improv.	Tensile modulus (Mpa)	Improv.
Control	0	TPO	19.6	–	957	–
I.44 PA	6.0	TPO	23.5	+20%	1458	+53%
Control	0	HPP	31.3	–	1388	–
I.44 PA	6.0	HPP	35.5	+13%	2180	+57%

Table 24.7. Flexural properties of PP nanocomposites.

Nanomer	Loading level (wt %)	Resin	Flexural strength (MPa)	Improv.	Flexural strength (Mpa)	Improv.	HDT (°C)	Improv.
Control	0	TPO	22.4	–	811	–	72.8	–
I.44 PA	6.0	TPO	29.5	+32%	1295	+60%	93.3	+28%
Control	0	HPP	34.6	–	1181	–	88.3	–
I.44 PA	6.0	HPP	46.0	+ 33%	1777	+50%	109.1	+24%

Grade I.44PA creates the 'nanoeffect' in both HPP and TPO as evidenced by improved mechanical properties and HDT.

Thus, nanocomposites have been successfully prepared in a range of polyolefins for both film and engineering applications. Significantly improved gas barrier properties, as well as mechanicals, are obtained at low Nanomer® additions.

This provides downgauging opportunities for films and rigid containers. A new Nanomer® grade expands options for improving performance in the engineered plastics arena, especially for automotive and industrial applications.

This more cost-effective grade improves mechanical, thermal and dimensional properties, while preserving two inherent advantages of the technology—low weight and easy processing.

PREPARATION AND CHARACTERISATION OF POLYPROPYLENE NANOCOMPOSITES CONTAINING POLYSTYRENE-GRAFTED ALUMINA NANOPARTICLES

Traditional polymer nanocomposites have improved mechanical properties, such as toughness, as a result of the incorporation of inorganic particulate fillers. However, high filler loadings (up to 20 per cent by volume) are required for such an enhancement of performance, leading to a loss of the easy processability of the polymers.

Consequently, polymer-based nanocomposites are attracting considerable attention because of the unique properties that result from their nanoscale microstructures. They are much lighter in weight, more transparent, and easier to process than conventional inorganic particle-reinforced polymers, in addition to displaying improved mechanical properties. However, the homogeneous distribution of inorganic nanoparticles into the polymer matrix is required to obtain the desired polymer-based nanocomposites because agglomeration of inorganic nanoparticles caused by immiscibility between the inorganic nanoparticles and the polymer matrix leads to a reduction, rather than an improvement, of the material's properties. To improve the mechanical properties through uniform dispersion of inorganic nanoparticles into the matrix, we synthesised polystyrene-grafted γ-Al$_2$O$_3$ through a high energy irradiation method after surface modification of nano-γ-Al$_2$O$_3$. The polypropylene nanocomposites were fabricated by blending surface-modified γ-Al$_2$O$_3$ nanoparticles and cross-linking agents, followed by e-beam irradiation. Their characterisation is described.

Experimental Procedure

Materials

Polypropylene (PP; B310; MW: 5,23,000; Honam Petrochemical Co., Ltd.) was employed as a polymer matrix. Micropolished γ-Al$_2$O$_3$, possessing an average diameter of 50 nm, was purchased from Buehler Company. 3-(Trimethoxysilyl)propyl methacrylate (TMSPM), xylene, methanol, and styrene were purchased from Aldrich Chemical Company and used as received. 1,4-Butanediol dimethacrylate (1,4-BDDA) and trimethylopropane triacrylate (TMPTA) were supplied by the Aldrich Chemical Company and used as cross-linking agents without further purification (Table 24.8).

Table 24.8. Graft polymerisations onto TMSPM-modified Al$_2$O$_3$.

Styrene concentration[a] (vol%)	Total dose (kGy)	Dose rate (kGy/hr)	Graft yield (%)[b]
50	2.5	2.5	2.1
50	5	2.5	4.8
50	7.5	2.5	7.5
50	10	2.5	8.1
70	2.5	2.5	1.9
70	5	2.5	4.4
70	7.5	2.5	6.2
70	10	2.5	7.1

[a]Methanol used as solvent. [b]Weight of grafting polymer/weight of TMSPM-treated γ-Al$_2$O$_3$ measured by TGA.

Graft polymerisation onto γ-Al$_2$O$_3$ nanoparticles

Prior to a silylation, γ-Al$_2$O$_3$ nanoparticles (10 g) were dried in a vacuum oven (195°C) for 24 hrs and then dispersed in 250 ml of dry xylene with the aid of an ultrasonic probe. The nanoparticle/xylene mixture was added to a 500 ml round-bottom flask containing a stirrer a bar, and then a 10 per cent v/v TMSPM/xylene solution was added with stirring. The mixture was heated under reflux for 20 hrs under a nitrogen atmosphere. The mixture was cooled, filtered, and then dried in a vacuum oven at room temperature for 24 hrs. The dried TMSPM-modified γ-Al$_2$O$_3$ (0.5 g) was immersed into various concentrations of styrene in methanol. The resulting solutions were flushed for 15 min. with nitrogen and then irradiated using γ-rays from a ^{60}Co source at a dose rate of 2.5 kGy/hr at room temperature. The resulting polystyrene grafted γ-Al$_2$O$_3$ (γ-Al$_2$O$_3$-g-PS) was washed thoroughly with hot benzene in a Soxhlet extractor to remove any residual monomer and homopolymer. The γ-Al$_2$O$_3$-g-PS was dried in a vacuum oven at 80°C for 24 hrs. The overall synthetic scheme is shown in Fig. 24.6. The weight loss of the grafting polymer was calculated using a TA thermogravimeter.

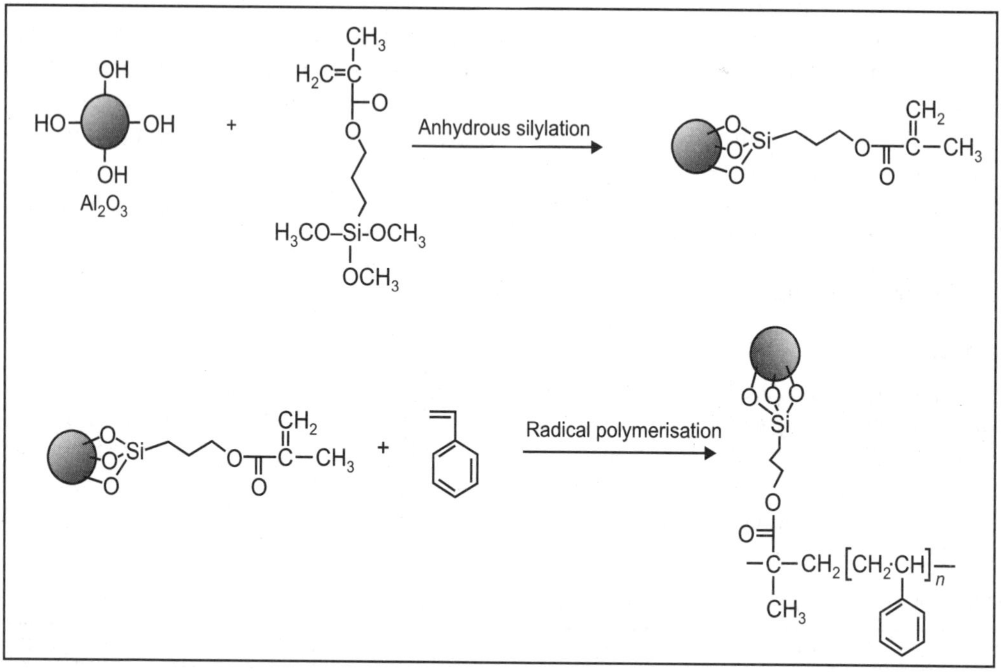

Fig. 24.6. Surface modification of alumina particles: (a) silylation with TMSPM and (b) graft polymerisation with styrene.

Preparation and characterisation of polypropylene nanocomposites

Polypropylene nanocomposites were fabricated by blending polypropylene pellets, γ-Al$_2$O$_3$-g-PS, and cross-linking agents using a lab-scale Brabender instrument, followed by E-beam irradiation. E-beam irradiation of the mixed samples was performed in the EB-tech using an ELV-4 electron beam accelerator with an energy of 1.0 MeV.

The integral irradiation dose levels were conducted at 4 kGy. The stress-strain properties of the prepared nanocomposite were determined using an Instron model 4411 testing machine according to ASTM D 638. The test procedure was performed at a cross-head speed of 50 mm/min at room temperature. The dispersion of γ-Al$_2$O$_3$-g-PS in the nanocomposites was investigated using a scanning electron microscope (SEM; XL30S FEG, Philips Co.).

Results and Discussion

Graft polymerisation onto γ-Al$_2$O$_3$ nanoparticles

The FT-IR spectra of neat γ-Al$_2$O$_3$, TMSPM-modified γ-Al$_2$O$_3$, and γ-Al$_2$O$_3$-g-PS are shown in Fig. 24.7. After surface modification with TMSPM, a carbonyl peak at 1730 cm^{-1} and an aliphatic C–H band at 2950 cm^{-1} appeared in TMSPM. After graft polymerisation, aromatic C–H bands (at 3010, 1600, and 1475 cm^{-1}) resulting from the polystyrene were newly generated. The surface modification was also confirmed by performing a floating test on water. The surface-modified γ-Al$_2$O$_3$ nanoparticles did not wet because the hydrophilic surface of the alumina particles had become hydrophobic. Quantitative results of the graft polymerisation on γ-Al$_2$O$_3$ are given in Table 24.8. The graft yield increased upon increasing the irradiation doses. The graft yield was higher at a 50 vol% monomer concentration than at 70 vol%. The highest graft yield was obtained when 50 vol% of styrene was irradiated at 10 kGy. In addition, neat γ-Al$_2$O$_3$, TMSPM-treated γ-Al$_2$O$_3$, and Al$_2$O$_3$-g-PS were characterised using an SEM.

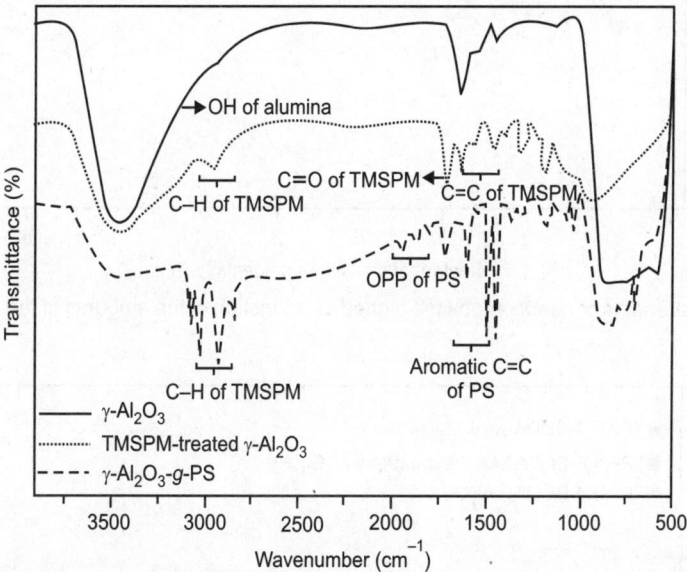

Fig. 24.7. FT-IR spectra of pure γ-Al$_2$O$_3$, TMSPM-treated γ-Al$_2$O$_3$, and Al$_2$O$_3$-g-PS.

Characterisation of polypropylene nanocomposites

The tensile strengths of nanocomposites prepared with different cross-linking agents and 5 phr of TMSPM-treated γ-Al$_2$O$_3$ are shown in Fig. 24.8. The tensile strengths of pure PP and 4 kGy-irradiated PP were 2.74 and 2.78 kgf/mm^2, respectively. Both the 1,4-BDDA- and TMPTA-cross-linked nanocomposites displayed similar tendencies. The tensile strengths were enhanced upon increasing the content of the cross-linking agents up to 3 phr, but they gradually decreased at contents of cross-linking agents over 3 phr. At higher concentration more than 3 phr the monomers may lead to the production of homopolymers, rather than reaction with PP, because so much monomer surrounds the radicals formed during irradiation. The tensile strength of the 1,4-BDDA-cross-linked nanocomposite was higher than that of the TMPTA-cross-linked one. The reason may be that the rate of cross-linking of TMPTA is faster because of its higher reactivity for BDDA; as the cross-linking density of the matrix increases, however, the chain radical and TMPTA diffusion become restricted and, therefore, lead to a lower cross-linking efficiency. The nanocomposites fabricated with 3 phr of 1,4-BDDA showed the highest tensile strength. From these results, we fixed the content of 1,4-BDDA at 3 phr and prepared nanocomposites according to various contents of the three kinds of fillers.

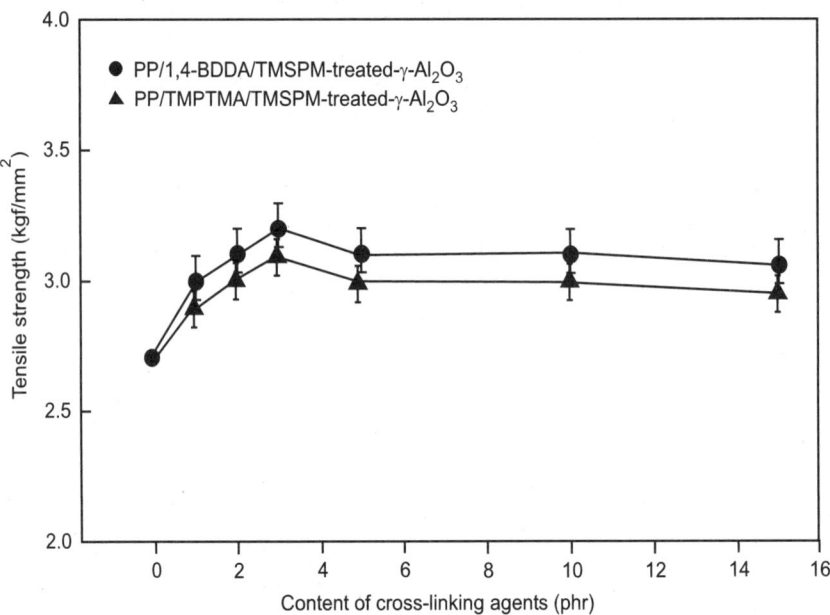

Fig. 24.8. Tensile strengths of nanocomposites plotted as a function of the amounts of the two kinds of cross-linking agents.

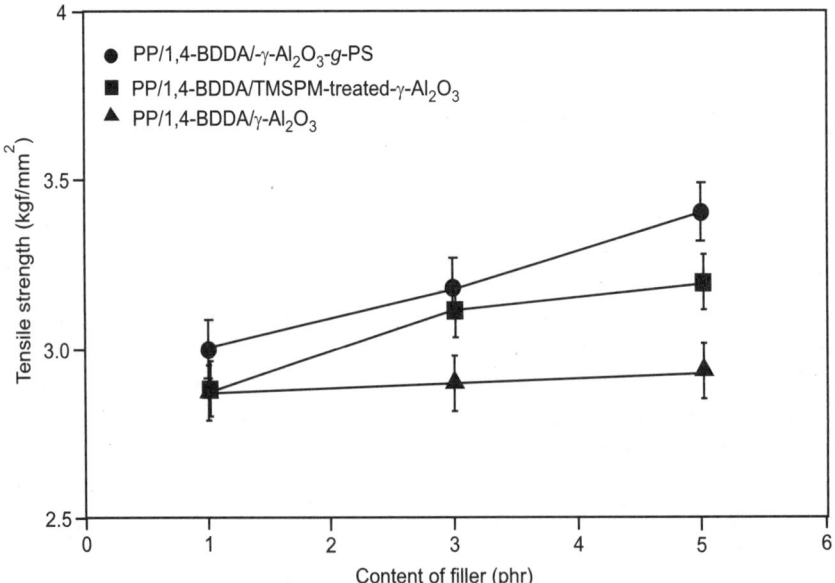

Fig. 24.9. Tensile strengths of nanocomposites plotted as a function of the filler content.

The tensile strengths of the prepared nanocomposites (Fig. 24.9) increased upon increasing the γ-Al$_2$O$_3$-g-PS content up to 5 phr. The nanocomposite containing 5 phr of γ-Al$_2$O$_3$-g-PS exhibited the highest tensile strength. The SEM micrographs of the fracture surfaces of the composites are shown in Fig. 24.10. γ-Al$_2$O$_3$-g-PS formed comparatively smaller agglomerates than did the untreated γ-Al$_2$O$_3$ or TMSPM-treated γ-Al$_2$O$_3$.

The reason for this phenomenon is that the polystyrene chain grafted onto the γ-Al$_2$O$_3$ interfered with the agglomerisation of the nanoparticles. This result is in agreement with the results of the tensile strength measurements above.

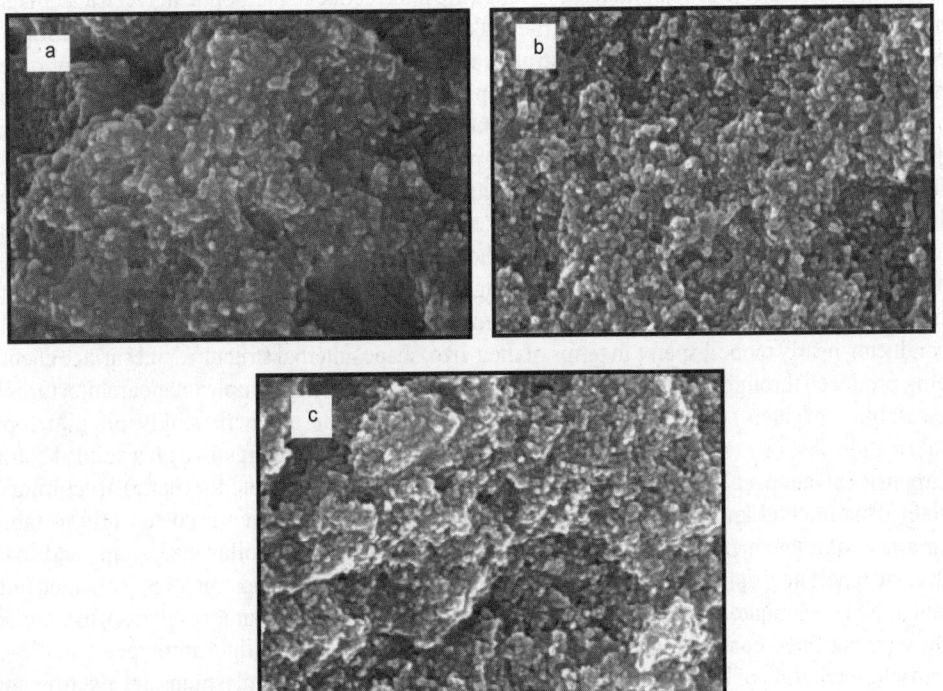

Fig. 24.10. Fracture surfaces of nanocomposites: (a) PP/5 phr of untreated γ-Al$_2$O$_3$, (b) PP/1,4-BDDA/5 phr of γ-Al$_2$O$_3$, and (c) PP/1,4-BDDA/5 phr of Al$_2$O$_3$-g-PS.

Thus, in this section, we performed surface modification of neat γ-Al$_2$O$_3$ with TMSPM. The graft polymerisation of styrene onto TMSPM-modified γ-Al$_2$O$_3$ was performed using the simultaneous irradiation polymerisation technique under various conditions. We found that the graft yields increased upon increasing the absorbed dose; the graft yield at 50 vol% of monomer was higher than that at 70 vol%.

The highest graft yield was obtained when 50 vol% of styrene was irradiated at 10 kGy. The nanocomposite fabricated with 5 phr of γ-Al$_2$O$_3$-g-PS and 3 phr of 1,4-BDDA showed the highest tensile strength. The homogeneous dispersion of γ-Al$_2$O$_3$-g-PS into the polypropylene matrix and the cross-linking by e-beam irradiation improved the mechanical properties of the nanocomposites.

Additional studies are underway to incorporate various polymer grafted γ-Al$_2$O$_3$ samples into polymer matrixes for the preparation of nanocomposites with improved mechanical properties.

OPTICAL AND STRUCTURAL CHARACTERISATION OF PERIODIC SILVER-POLYSTYRENE NANOCOMPOSITES

Ordered arrays of polystyrene nanospheres were self-assembled on glass substrate and subsequently used as templates for thin silver film deposition. Silver-polystyrene nanocomposites with periodicity given by the polystyrene spheres diameter (400 nm) were obtained and investigated by optical measurements and electron microscopy. The resonant interaction of light with surface plasmons supported by periodically structured silver-polystyrene nanocomposite leads to important spectroscopic applications like molecular sensing and enhanced Raman scattering.

Nanostructured materials are of increasing scientific interest due to their optical, electrical, magnetic, thermal properties, which are not present in the bulk phase. The main parameters which determine their properties are divided in two categories: (i) structural parameters (nanometric size, periodicity, ordering, and interparticle distances), and (ii) compositional parameters (chemical or biological functions attached to the surface of nanostructures). In order to implement novel applications in the field of nanotechnology is of paramount importance to control as much as possible the above parameters.

There are two possible approaches to fabricate nanostructures in a controllable manner: the so-called top-down and bottom-up methods. The first category includes photolithography, holographic interferometry, electron beam lithography, X-ray lithography. Recent developments of scanning tunnelling microscopy, atomic force microscopy and scanning probe lithographic techniques show great promise. However, during the past decades, various nanoparticles of polystyrene, silica, noble-metal and semiconductor, nearly monodisperse in terms of their size, shape, internal structure, and surface chemistry, are being produced through a reliable manufacturing process. Building complex nanoarchitectures from self-assemblies of such stable building blocks has also become an increasingly popular topic in nanofabrication. As for example polystyrene (PS) nanospheres can self-organize on a solid substrate in two-dimensional compact hexagonal arrays and, then, can serve as template for (nano)structuring other materials. This method known as nanosphere lithography (NSL) has been used recently to fabricate regular arrays of noble metal nanoparticles and metallic holes with controlled size, shape and spacing.

Here, we report new optical properties of periodically arrayed silver nanostructures fabricated through a variant of NSL technique. The optical properties of noble-metal nanostructures result from the interaction of light with confined conduction electrons. For instance, when visible light impinges a noble-metal nanoparticle, excitation of localised surface plasmons occurs. The surface plasmons are electromagnetic waves coupled to the oscillations of conduction electrons at the metal-dielectric interface. A few of the consequences of the surface plasmons excitation are strong selective photon absorption, scattering and local electromagnetic field enhancement. The strongly enhanced fields (10^3–10^5 times) decay exponentially with the distance (on the order of 10 nm in the metal and ~100 nm in the dielectric).

Noble-metal nanostructures are highly desirable in several technological applications including optical energy and information transport, photonic crystals, near-field scanning optical microscopy, surface enhanced spectroscopy, chemical and biological sensors.

Sample Preparation

As solid substrates we used glass microslides of 24×24 mm size. The slides were sonicated in isopropylic alcohol for degreasing, then treated in piranha solution (mixture of 95 per cent H_2SO_4 and 30 per cent H_2O_2) and sonicated again in a mixture of $H_2O/NH_4OH/H_2O_2$ for at least 30 min. The substrates were stored in deionised water until nanospheres deposition. A water suspension of polystyrene nanospheres of 400 nm diameter was dropped onto the cleaned substrate [Fig. 24.11(a)]. We deposited the right volume of suspension to cover the substrate taking in account the substrate area, diameter of the spheres and concentration of the PS spheres in solution. The wet substrate was placed in an oven at constant temperature between 40°–80°C in order to evaporate the water and get self-assembled nanoparticles [Fig. 24.11(b)]. In the next step of sample preparation, a silver film of 40 nm thickness was deposited over the PS nanospheres layer by means of an electron-beam evaporator. A schematic representation of the evaporation system is shown in Fig. 24.11(c). The accelerated electron beam from a wolfram filament is directed onto a solid piece of silver, the metal heats up and silver atoms arrive on the PS array substrate mounted above.

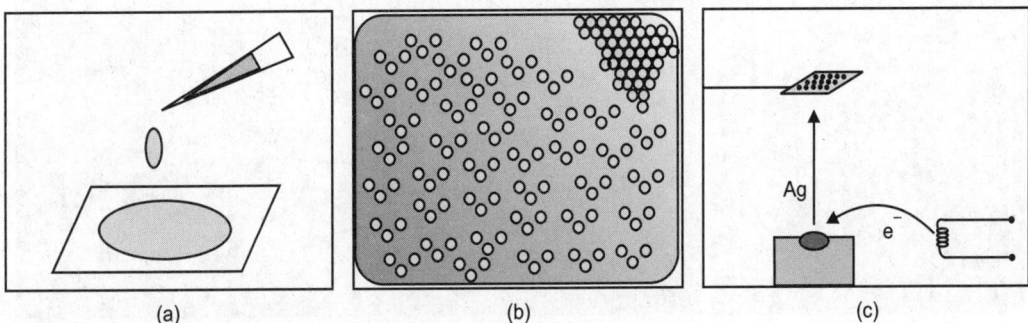

(a)   (b)   (c)

Fig. 24.11. Schematic representations of (a) drop-coating of glass slide with nanosphere solution, (b) self-organisation of the nanospheres and (c) the vacuum deposition chamber.

Results and Discussion

The microstructure of the prepared sample was mapped by scanning electron microscopy using a JEOL JSM 5510 LV electronic microscope. Figure 24.12(a) is a representative SEM picture of regular hexagonal close packed nanospheres on the substrate. Most of sample surface exhibits a single layer 'polycrystalline' structure with ordered domains ranging from 10 to 50 µm lateral size. Typically the crystallisation process leads to the formation of two-dimensional (2D) colloidal crystals along with multilayered structures and randomly self-assembled nanospheres. Small domains where the spheres form a rectangular lattice can also be observed (not shown here). In the literature it is generally accepted that the mechanism and the driving forces of assembling process involve: (i) formation of a nucleus, under the action of attractive capillary immersion forces, and (ii) crystal growth, through convective particle flux caused by the water evaporation from the already ordered array. This implies that the 2D crystal nuclei are formed under the capillary attraction arising when the tops of the spheres protrude from the water layer. In the schematic representation of Fig. 24.11(b) the spheres from the darker region of the picture move towards the ordered array by passing through the brighter region. The origin of the lateral capillary forces is the overlap of perturbations in the shape of the liquid surface surrounding the PS nanospheres. The dynamics of ordering is strongly dependent on the evaporation rate. A spring-block stick-slip model was recently introduced for simulating the phenomenon of nanosphere self-assembling and to study the influence of several controllable parameters on the sample polycrystallinity.

Many different silver-polystyrene nanocomposites can be easily fabricated by depositing silver on top of 'polycrystalline' structure. As for example, Fig. 24.12(b) shows ordered array of triangular silver particles left on glass substrate after removing the spheres mask, and Fig. 24.12(c) shows randomly distributed nanocavities made in metal film. PS spheres can be removed from the substrate by dissolution in toluene [Fig. 24.12(b)] or happens that the spheres fall down accidentally [Fig. 24.12(c)]. A periodic array of smaller metallic particles than those shown in Fig. 24.12(b) can be obtained by depositing metal through a mask consisting of two layers of PS spheres. On the substrate, along with crystalline domains, there are various types of defects of crystallisation too. The so-called 'line defects' (gaps between domains) and 'point defects' (missing spheres) can serve as optical (nano) microcavities to control the fluorescence rate of emitters via density of localised optical modes. The metallic nanoapertures as those fabricated in this work [see Fig. 24.12(c)] were recently employed as nanocavities for studying the luminescence of single quantum dots.

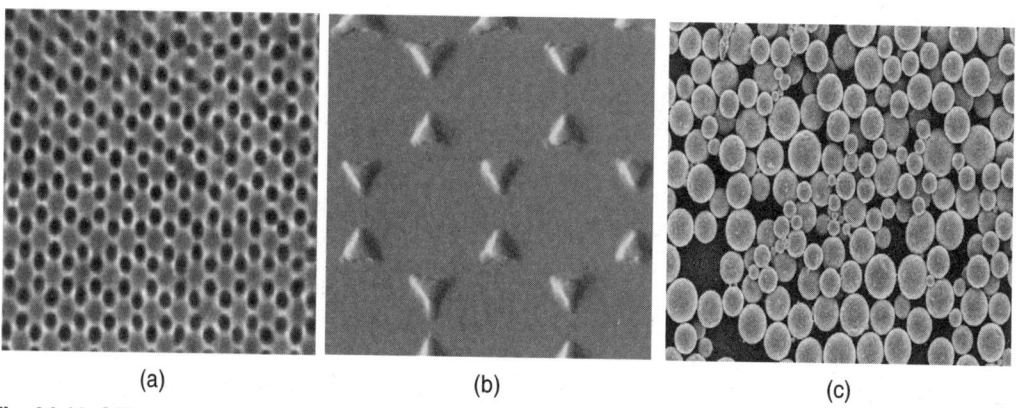

(a)   (b)   (c)

Fig. 24.12. SEM images of: (a) PS spheres array; (b) array of triangular silver nanoparticles, and (c) perforated silver film area.

The first proof of interesting optical properties of fabricated nanostructures is given even by naked eye. If observed in reflection, the substrates show beautifully iridescent green, blue, violet colours, depending on angle of observation. Contrary, the samples appear to be brownish-red coloured in transmission. In the first approximation, the colloidal lattice surface can be considered a two-dimensional diffraction grating of close packed spheres and the appearance of such colours can be explained in terms of light diffraction and scattering. The phenomenon is rather complex as many different local 2D gratings (domains) and structural defects (points, lines and gaps) co-exist. In order to characterise the optical properties of as fabricated nanostructures, optical extinction measurements were recorded for three distinct cases: (i) bare layer of PS nanospheres (before silver deposition); (ii) silver-coated polystyrene layer; (iii) silver nanoparticles on the substrate (after removal of the spheres). The extinction spectra were recorded using a JASCO UV-530 spectrometer with a probe beam of approximately 9 mm^2 size and unpolarised light.

Figure 24.13(a) shows the grating extinction recorded at 0°, 20° and 40° angles of incidence. Resonant phenomenon between the incident light and the eigenmodes of the regular array of nanospheres are responsible for the peak (transmission dip) observed in the normal incidence spectrum around 480 nm, bulk polystyrene film not being an absorber at these wavelengths.

The peak is a specific optical response of ordered arrays of nanospheres and its existence demonstrates the degree of ordering on the substrate. It is conceivable that the individual responses of different single domains are very locally sensitive. However, the probe light beam of about 9 mm^2 cross-section integrates many domains and the measured optical spectra give finally only an area-averaging extinction behaviour.

The peak moves towards longer wavelengths as the incidence angle is increased as a result of crossover phenomenon due to the interaction between eigenstates of the monolayer and those of the substrate. Light interacts not only with surrounding spheres but also with the substrate and the optical interaction between spheres as well as interaction between spheres and substrates changes as function of the incidence angle. This grating resonance at wavelength comparable with the size of the spheres is likely associated with Mie resonances and cannot be explained by Bragg diffraction as in the case of colloidal photonic crystals.

The diameter of nanospheres and refractive indices of polystyrene and substrate determine the position of the peak, while the degree of polycrystallinity (defects) determine its width.

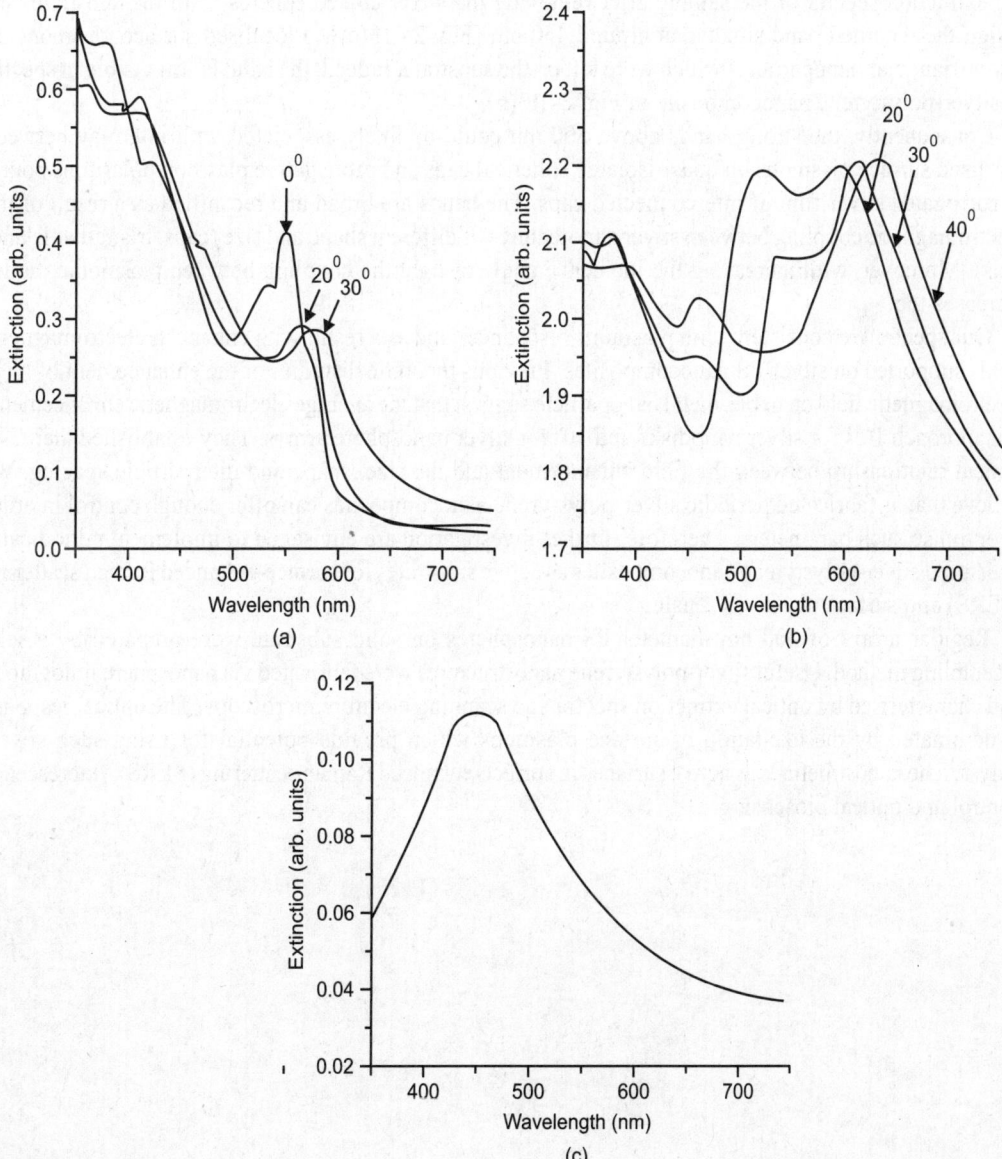

Fig. 24.13. Extinction spectra of: (a) array of PS spheres; (b) composite silver/PS nanostructure; (c) array of silver nanoparticles.

The spectra in Fig. 24.13(b) look drastically different as compared with spectra in Fig. 24.13(a). On the contrary to eigenmode resonances observed in the grating of dielectric spheres, the main features of optical spectra in Fig. 24.13(b) are connected with surface plasmons resonances. Both propagative plasmons supported by corrugated silver film on polystyrene hemispheres (interconnected caps) and localised plasmons supported by quasi-isolated caps and nanoparticles on glass have to be involved.

In order to get insight into the origin of spectra shown in Fig. 24.13(b) we subsequently measured the extinction spectra of the sample after removing the silver coated spheres from the substrate. We assign the recorded band situated at around 450 nm [Fig. 24.13(c)] to localised surface plasmons on silver triangular nanoparticles which were left on the substrate. Indeed, the band is still visible in spectra of silver-polystyrene nanocomposite in Fig. 24.13(b).

Consequently, the strong bands above 500 nm could be likely associated with interplay between localised surface plasmons on quasi-isolated spherical caps and propagative plasmon polaritons bound to corrugated silver film of interconnected caps. The bands are broad and redshifted as a result of the electromagnetic coupling between silver nanofeatures of different shape and size (caps, triangular island, rods). Moreover, with increasing the incidence angle of light the coupling between plasmonic modes increases too.

Our spectra are consistent with plasmons resonances and as a result with enhanced electromagnetic fields supported on silver/PS nanocomposites. Previous theoretical studies of the enhancement of local electromagnetic field on noble-metal nanoparticle suggest that the average electromagnetic enhancements can approach 10^{11} for silver nanodisks and 10^9 for silver nanosphere arrays. They established there is a critical relationship between the field enhancement and the size, shape and interparticle spacing. We believe that as fabricated periodic silver-polystyrene nanocomposites can offer enough control in order to optimise such parameters. Therefore, further investigation are envisaged to implement periodically structured silver-polystyrene nanocomposites as active substrates for surface-enhanced Raman scattering (SERS) and surface plasmons sensing.

Regular arrays of 400 nm diameter PS nanospheres on solid substrate were prepared by a self-assembling method. Useful silver-polystyrene nanostructures were fabricated via nanosphere lithography and characterised by optical extinction spectra and scanning electron microscopy. The optical response is dominated by the excitation of surface plasmons which provide potential for using such silver-polystyrene nanostructures as active surfaces in surface-enhanced Raman scattering (SERS), fluorescence control and optical biosensing.

Chapter 25

Polyester-clay Nanocomposites

INTRODUCTION

It has been proven in recent years that polymer based nanocomposites reinforced with a small amount of nanosize clay particles (<5 per cent) significantly improve the mechanical, thermal, and barrier properties of the pure polymer matrix. Experiments on these materials have shown that virtually all type of nanocomposites lead to new type with improved material properties.

Nanocomposites are a new class of composites, which are particulate filled polymer in which at least one dimension of the dispersed particles are in the nanometer range. One can distinguish three types of nanocomposites, depending on how the dimensions of the dispersed particles are in the nanometer range. When the three dimensions are in the nanometers, they are known as the isodimensional nanoparticles, such as spherical silica nanoparticles. When two dimensions are in the nanometer range and third is larger, they form an elongated structure, such as nanotubes/whiskers. The third type of nanocomposites is characterised by only one dimension in the nanometer range. In this case the filler is present in the form of sheets of one to few nanometer thick to hundreds to thousands crystal nanometer long. This family of composites can be gathered under the name of polymer-layered crystal nanocomposites. These materials are almost exclusively obtained by the intercalation of the polymer (or a monomer subsequently polymerised) inside the galleries of host crystal.

Recent studies reveal that nanoclay composites dispersed in polymer matrix exhibit dramatic improvement in stiffness, strength, increased dimensional stability, improved flame retardancy, improved solvent and UV resistance, and reduction in permeability to gases, improved thermal stability, and ablative properties, have been reported in a wide range of polymers. These property improvements resulting from the formation of nanocomposites occur at extremely low counteractions of the aluminosilicates (1–5 vol.%) compared to conventional phase-separated composites of a filler material in a polymer (20–30 per cent).

Nanoscale layered clays with high aspect ratio and high strength play all important role in forming effective polymer nanocomposites owing to their intercalation/exfoliation chemistry. Montmorillonite has been particularly important in polymer clay nanocomposites. It is a crystalline material consisting of 1 nm thick layers (or sheets) which are made of an octahedral sheet of alumina fused into two tetrahedral sheets of silica. The layers are continuous in the 'a' and 'b' direction and are stacked one over the other in the 'c' direction. When the nanoclay is dispersed in the polymer matrix, two types of *t* nanocomposite structures are obtained namely intercalated and exfoliated structure. Intercalated structure is one in which a single (and sometimes more than one) extended polymer chain is intercalated between the silicate layers resulting in a well-ordered multi-layer morphology built up with alternating polymeric and inorganic layers. When the silicate layers are completely and uniformly dispersed in a continuous polymer matrix, an exfoliated or delaminated structure is obtained.

Kornmann reported the dispersion of montmorillonite in unsaturated polyester resin and found that the fracture toughness is doubled by dispersing 1.5 per cent (by volume) of aluminosilicate. Smith demonstrated that the resulting properties of polyester clay nanocomposites were mainly dependent upon mixing of the clay, polyester resin promoters as well as the curing conditions. Smith also showed that the tensile modulus and oxygen permeability rate has decreased with increasing clay content.

Engineering structures are generally designed based on stress sustaining capacity of the structural components. Stability and vibration are also very important, especially when the structural elements are thin and subjected to dynamic loads. Since weight is a crucial factor in aircraft structures, the use or conventional isotropic material gives very little room for weight savings. Unlike the isotropic materials, the properties of composite materials can be tailored to have very high strength and yet being very light. As the strength-to-weight ratio of composite materials is high, structures made of composites often become very thin. In case of a structure made of isotropic material, the natural frequency is high if the element is thin. Adjustment in the natural frequency or such an element can be made by either changing the thickness or adjusting the boundary conditions. Nevertheless, in case of composite materials the natural frequency can be changed by designing lamination scheme.

Therefore, it is important to study the dynamic characteristics of polyester clay nanocomposites. In this work, we report the preparation or polyester clay nanocomposites and the study of dynamic characteristics such as natural frequency, damping factor and loss factor for different concentration or clay in addition to the characterisation of composites.

EXPERIMENTAL DETAILS

The organically modified montmorillonite was mixed with the low viscosity polyester resin to prepare the nanocomposites. Several day concentrations in the cross-linked polyester resin were investigated (1,2,3 and 5 wt.%). An appropriate amount of the organically modified clay was added to the resin and mechanically stirred by a shear mixer at ~1000 rpm. This solution resulted in well-dispersed, stable suspension of the clay in the polyester resin. Approximately 1.5 vol% of MEKP catalyst was added to premixed clay in polyester resin at room temperature to initiate the cross-linking process. Samples were allowed to cure for 24 hrs at room temperatures.

CHARACTERISATION

Characterisation and Property Evaluation

The X-ray diffraction was performed on both clay and cured samples of nanocomposites to evaluate the degree of intercalation and the d-spacing between the platelets. It is carried out with a scanning rate of $2°/min.$ and CuKα radiation at 30 KV and 15 mA ($\lambda = 1.5406$ A$°$). Tensile test was performed on the Instron machine with cross head speed of 1 mm/min. according to the ASTM 0638. The Izod impact test was conducted to study the impact energy of polyester clay according to the ASTM D256. Heat deflection temperature was obtained by ASTM D648.

The dynamic mechanical analysis (DMA) results were obtained using NETZSCH DMA242 C thermal analyser/dynamic mechanical analyser (DMA) system with frequency rate of 10 Hz. This system measures the modulus and damping or a material as it is deformed under periodic resonant stress at low strain.

Modal Testing

Modal testing is a process of constructing mathematical test object for suitable measurements or its vibrational characteristics. This study is used for comparison of modal properties, identification of damping, structural modification and optimisation and force determinations.

Polyester clay nanocomposite plates were prepared for different concentration or clay having physical dimension of $200 \times 100 \times 3$ mm. The first four modes of natural frequencies were obtained for polyester clay cantilever plates by hammer impact test.

The accelerometer was placed over the free end of the plate and connected to FFT analyser through the signal-conditioning amplifier. The hammer was used to generate the impact force at different locations of the plate.

The natural frequency of the plates was computed from the frequency response function (FRF) which was recorded from the FFT analyser. The FFT analyser was tuned in averaging mode and maximum frequency range obtained was 800 Hz. In order to verify and fine tuning of natural frequencies the function generator excites the electrodynamic exciter to excite the plate with computed natural frequencies at resonant condition.

The damping factor was computed from the frequency response functions of the nanocomposite plates by half-power method. In half-power method, the FRF is reduced to $1/\sqrt{2}$ times the amplitude and corresponding bandwidth is used to calculate the damping factor.

Results and Discussion

Structure and morphology

The X-ray scattering intensities for the organically modified clay and the polyester-clay nanocomposites are shown in Fig. 25.1. In the scattering curve for the pure clay, a prominent peak corresponding to basal spacing of organically modified montmorillonite clay occurs at a d-spacing or 17.8 A. This reflection is absent in the scattering curves for all the polyester-clay nanocomposites, irrespective of the clay concentrations, confirming the formation of nanocomposites.

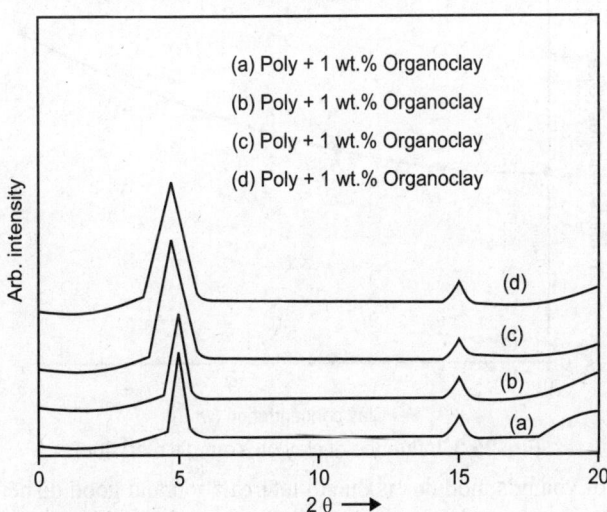

Fig. 25.1. XRD pattern of polyester/organo clay series.

Mechanical properties

The stress strain behaviours or all nanocomposites under uni-axial tension are shown in Fig. 25.2. From the Fig. 25.2 it is clear that there is enhancement or young's modulus. It is also seen that, increasing the clay concentration in polyester matrix increases the tensile modulus. From Fig. 25.3, the formation of nanocomposite yields the 43 per cent improvement in tensile modulus at 5 per cent clay concentrations.

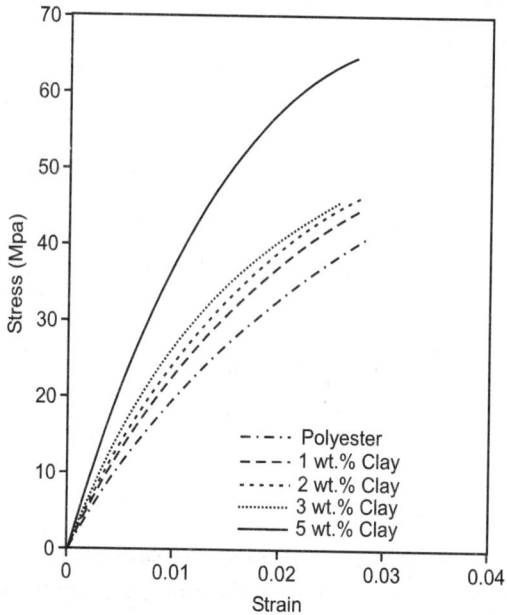

Fig. 25.2. Stress/Strain curves for polyester/organo clay series.

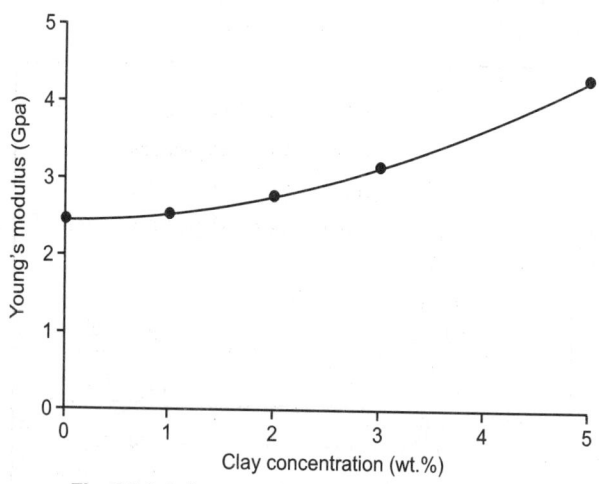

Fig. 25.3. Influence of clay on Young's modulus.

The improvement of young's modulus is due to intercalation and good dispersion of clay particles that restrict the mobility or the polymer chains under loading as well as good interfacial adhesion between the particles and the polyester matrix. The orientation or the clay particles and the polymer chains with respect to the loading direction can also contribute the reinforcing effects, which requires more investigation. The 55 per cent enhancement in the tensile strength of nanocomposites is seen to increase with increasing clay content as shown in Fig. 25.4.

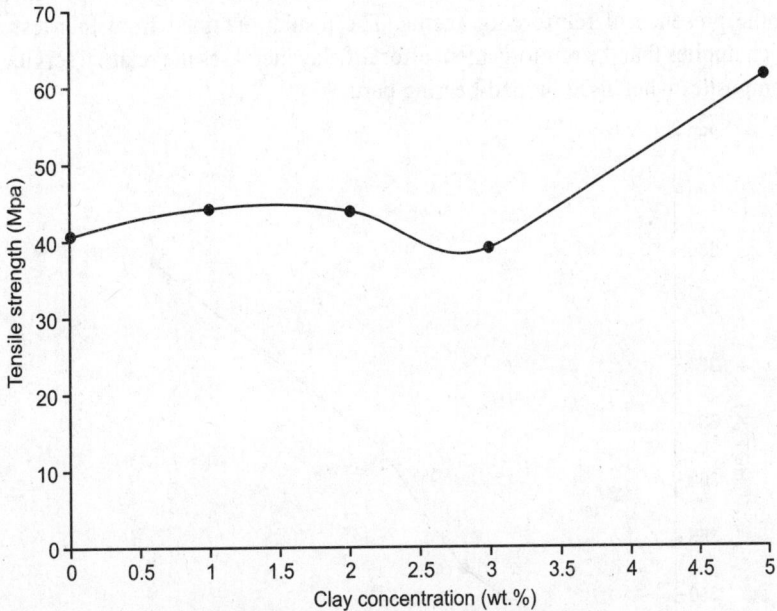

Fig. 25.4. Influence of clay on tensile strength.

A combination of the morphology and the extent of cross-linking in the polyester clay nanocomposites can be used to understand this phenomenon which requires further investigation. The charpy impact test results are shown in Fig. 25.5.

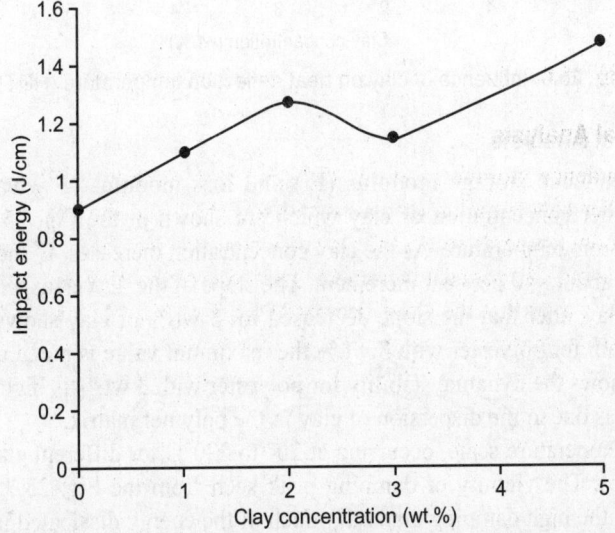

Fig. 25.5. Influence of clay on impact energy of polyester clay nanocomposite.

It shows that the reinforcement effect of the clay is reflected in higher impact energy as compared to polymer. It improves the impact energy by 68.27 per cent for 5 wt. per cent of clay concentration. The Heat deflection temperature (HDT) is a measure or a polymer's resistance to distortion under a given load at elevated temperatures. The value obtained for a specific polymer grade will depend on the base resin and on the presence of reinforcing agents. The results obtained from this test are shown in Fig. 25.6, which implies that the reinforcement effect of clay increases the relative service temperatures of the nanocomposites when used in load-bearing parts.

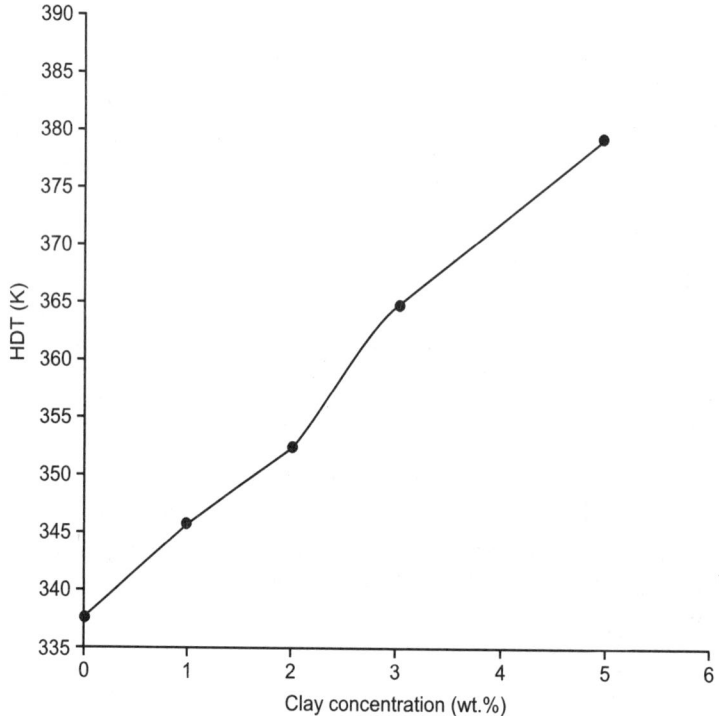

Fig. 25.6. Influence of clay on heat deflection temperature (HDT).

Dynamic Mechanical Analysis

The frequency dependence storage modulus (E′) and loss modulus (E″) values are increasing monotonically at higher concentration of clay which are shown in the Fig. 25.7(a). Pure polyester shows E′ of 5500 at room temperature. As the clay concentration increases, E′ increases upto 7500 for 5 wt.% clay showing about ~40 per cent increment. The slope of the E′ versus temperature is constant upto the 2 wt.% of clay, alter that the slope decreased for 3 wt.% of clay shows that the stability of storage modulus. Again for polyester with 3 wt.% the maximum value is noted at the glass transition temperature which shows the dynamic stability for polyester with 3 wt.% is high at T_g. The increased storage modulus (E′) is due to the dispersion of clay in the polymer matrix.

The tanδ versus temperature scale, occurring at 20° to 220°C for different clay concentrations are shown in Fig. 25.7(b). The vicinity or damping peak seen from the Fig. 25.7 shows that the clay reinforcement results the high damping material, much of the energy dissipated into a heat. There is a clear peak at ~100°C for pure polyester which is glass transition temperature. After the glass transition temperature the variation or storage modulus is constant for all wt.% of clay. The clay concentration broadens the peak for all polyester clay nanocomposites. Also the secondary damping peak observed for 2 wt.% of clay indicates that the high impact stability or nanocomposite.

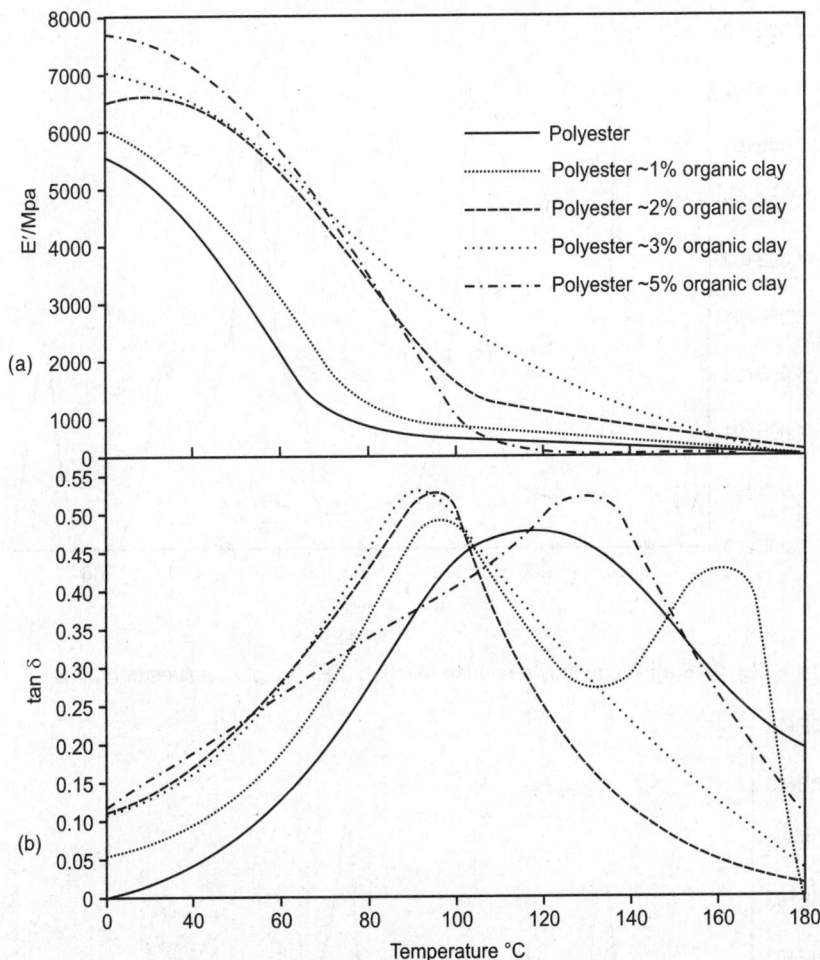

Fig. 25.7 (a) and (b). Dynamic mechanical analysis of polyester/organo clay series.

Natural Frequency

The frequency response function (FRF) of polyester and with 5 wt.% of clay is shown in Fig. 25.8(a) and (b). It is seen from the Fig. 25.8 that the second and third modes are predominant in the 800 Hz bandwidth.

The natural frequency of the polyester clay nanocomposite cantilever plates with different concentration of the clay is shown in Fig. 25.9. It is seen that the reinforcement effect of clay in the polyester matrix shifts the natural frequencies. In first bending mode the 60 per cent increasing the natural frequency is obtained whereas the 7.8 and 8.1 per cent improvements are obtained for third and fourth mode respectively. This improvement in first bending mode is due to the dispersion of clay having large bending stiffness. In higher mode, there is a small increase in the natural frequency. It is evident from the experiments that the clay reinforcement in the polymer matrix increases the natural frequencies.

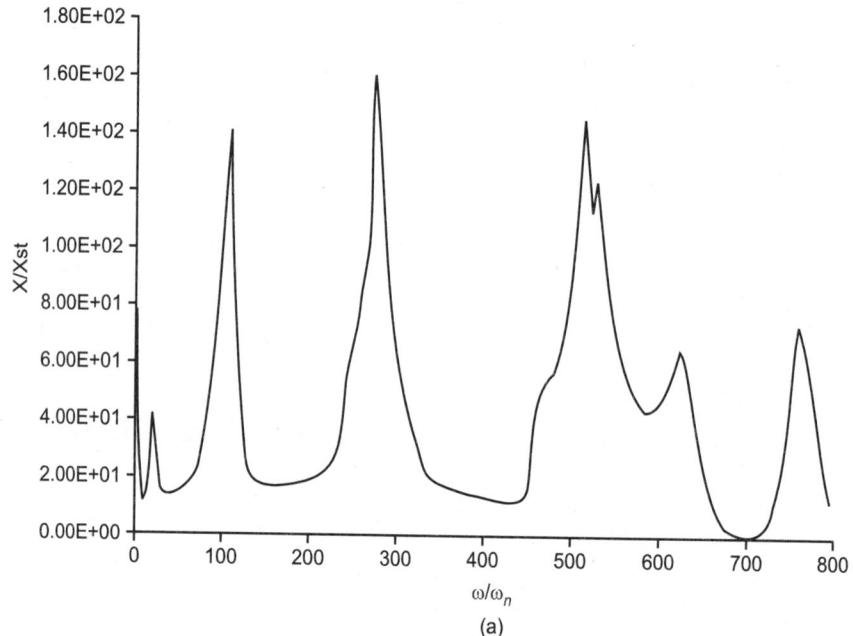

Fig. 25.8(a). Frequency response function (FRF) of pure polyester matrix.

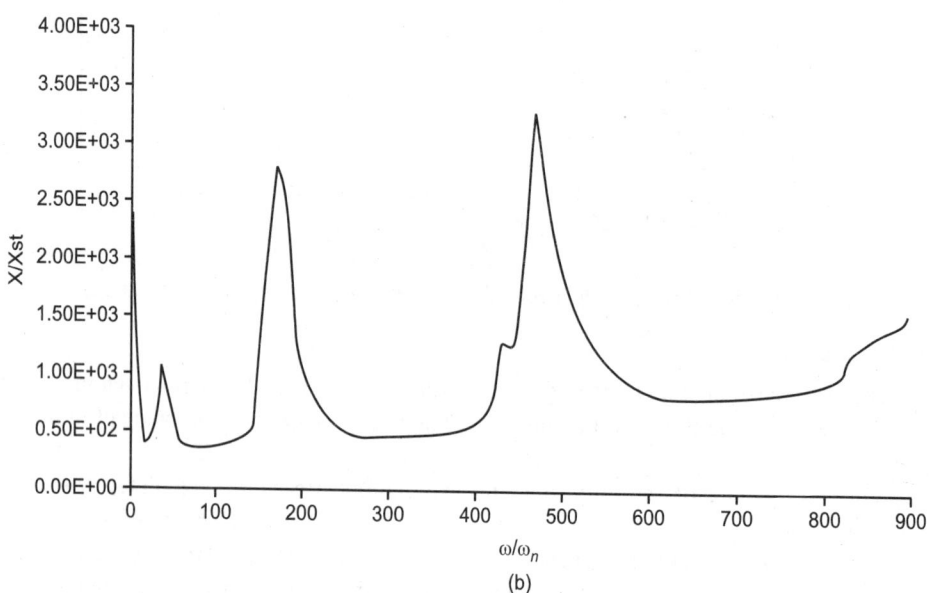

Fig. 25.8(b). Frequency response function (FRF) of polyester with 5 wt.% clay.

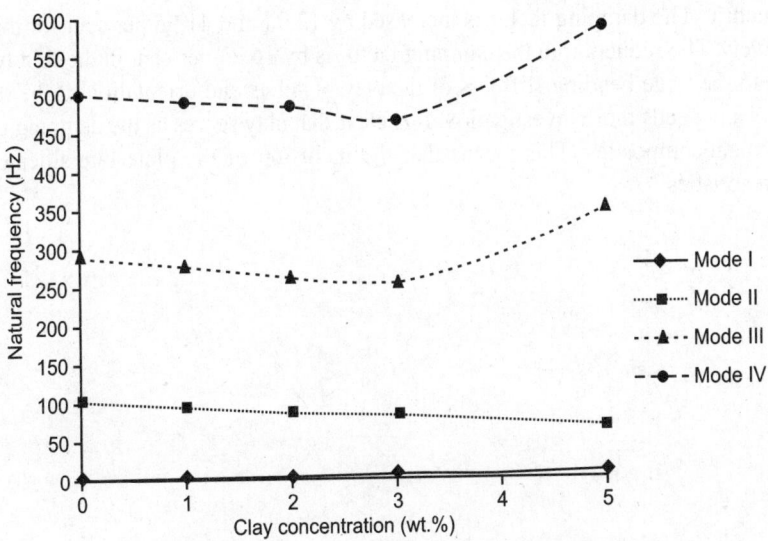

Fig. 25.9. Influence of clay on natural frequency.

Damping Factor

The series of the curves in Fig. 25.10 show that the influence of the clay in the polyester clay nanocomposites. As the clay concentration increases, the damping factor is increased for all modes of nanoclay composites.

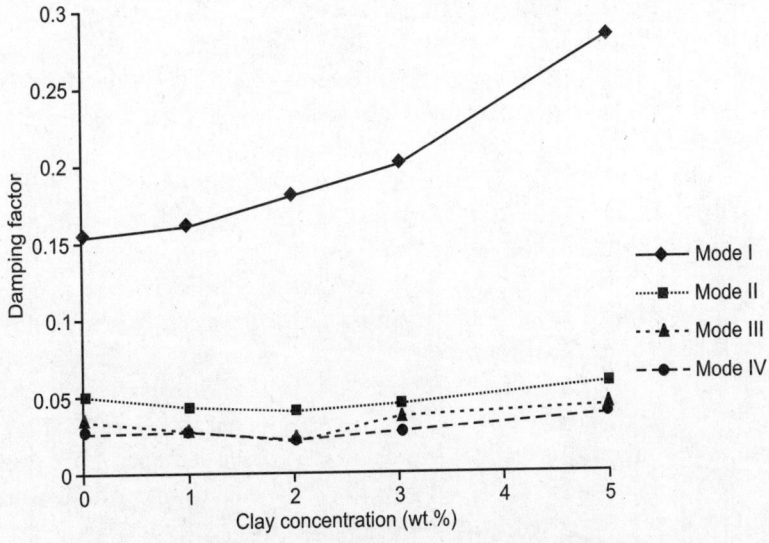

Fig. 25.10. Influence of clay on damping factor.

The platelet structure in the polymer matrix results in the improvement or first mode bending damping factor by 89 per cent. That is for 60 per cent improvement in first bending mode natural frequency, the corresponding damping factor is increased by 89 per cent. This improvement is due to large surface area of platelet structure. The damping factor is increased by 12.94 and 41.96 per cent for third and fourth mode respectively. The reduction in the damping factor is by 16.23 per cent in the first twisting mode. This may due to the large bending stiffness of the clay platelets and orientation of the platelets in the polymer matrix and needs more investigation. It is clear that clay serves as the damping medium in the polyester clay nanocomposites. This means that the inclusion of the plate like silicates have better dynamic characteristics.

Nylon-6 Nanocomposites

INTRODUCTION

The study of polymer nanocomposites is currently an expanding field of research because it exhibits a wide range of improved properties over their unmodified starting polymers due to small size, high surface area and unique properties or nanoparticles. Polymer nanocomposites combine these two concepts, i.e. composites and nanometer size materials.

The purpose of this chapter is to present an overview of basics of polymer clay nanocomposites, their superiority over conventional composites and its applications, with a brief study of clay chemistry, different routes of nanocomposite synthesis, and their characterisation.

'Using clay is a green alternative to current practices and reduces flammability in the wide range of nylon-6 polymer.' Natural clays are currently the most used because they are the same clays already used in many products. 'However, synthetic clays, because of their tailored properties, may prove essential for high added value products, such as in biomedical devices and space applications. The clay chemistry plays vital role in determining the properties of nanocomposites.

This chapter also focuses on different properties of nylon-6 nanocomposites and their comparison with unmodified nylon-6 polymer. Thermoplastics filled with nanometer size materials have different properties than thermoplastics filled with conventional materials. Nanocomposites show drastic improvement in the properties derived from the addition of a few per cent of clay in the polymer matrix. This improvement in the properties is the result of the ultrafine phase dimension of the filler.

A composite consists of two phases: an organic phase, and an inorganic phase. These phases, when combined together, synergistically reinforce each other; this enhances various properties such as tensile strength, flexural strength.

A nanocomposite is a type of composite in which one of the phases is of nanodimension. The system of polymer-clay nanocomposites has clay as the inorganic phase, and polymer as the organic phase. This clay consists of slacks of clay platelets, where the individual dimension of these clay platelets is of nano (10 exp–9 m) range, and their thickness is of the order of micron range. This imparts a very unique aspect ratio (LID) ratio of about a 1000 order. This high aspect ratio makes these types of nanocomposites very special in their properties.

CLAY PLATELETS

Clay is highly hydrophilic in nature and polymer is highly hydrophobic. These two are, therefore, not naturally compatible with each other. But, the clay can be made a bit hydrophobic (and thus more organophilic) by treating their surface with modifiers containing long molecules having polar groups, which can attach themselves to clay surface, and an organic part, which increases interaction with organic phase (polymer matrix). Similarly, the polymer phase is made to approach the polar end of spectrum by converse method. This increases interaction between the two phases, and results in better composite materials. These substances added to increase interaction between the two are called compatibilisers.

Due to their small size, surface interactions, such as hydrogen bonding, become magnified. What this means is that the particles tend to agglomerate and dispersion in resins is quite difficult, even in polymers that should be relatively compatible. The problem is solved by using compatibilisation agents that will bond to the montmorillonite surface and also interact with the resin to form a more miscible system. Each resin requires a compatibilisation agent suited to it. Among the various types of clays available, such as smectic clays, hydrotalcites, montmorillonites, etc. the montmorillonite clays have a layered silicate structure and are used most widely for preparing polymer-clay nanocomposites. Amongst the polymers in use for these types of nanocomposites are mainly nylon and polypropylene.

An example of compatibilised nylon-6 nanocomposite with montmorillonite clay and surface modified with ammonium tallow is shown in Fig. 26.1.

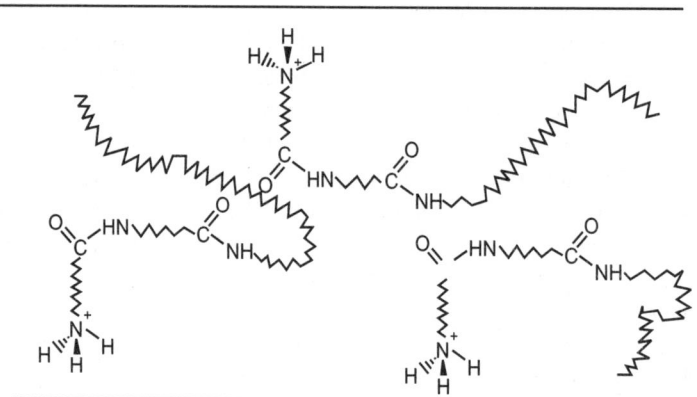

Fig. 26.1. Long chains are organic polymer chains.

Montmorillonite

The approximate formula for montmorillonite is [R+ 0.33 (Al,Mg)2Si4O10(OHhnH20], where R in natural mineral can be one or more of the Na^+, K^+, Ca^{2+}, cations. Montmorillonite has a dimension below the visible wavelength, properly oriented and transparent; this is a critical requirement in barrier packing. Na-montmorillonite is the major mineral constituent of bentonites and has high swelling capacity, as compared to bentonites having Ca-montmorillonite as major mineral. Na-Montmorillonite is a smectite in which Na and water molecules are the interlayer material. The largest and highest quality Na-bentonite deposits in the world are situated in South Dakota, Wyoming and Montana.

Montmorillonite are the preferred materials in nanocomposites because of:

1. Low level of loading.
2. High aspect ratio (as high as 100).
3. Good transparency safe handling.

4. Natural abundance.
5. Economic and good chemical resistance.

Montmorillonite shows hydrophillic nature hence it cannot be homogeneously dispersed into the organic polymeric phase. The higher aspect ratio of silicate particles can be maximised by dispersing individual silicate layers into the polymer matrix. Therefore, to incorporate Montmorillonite into the polymer phase, it is organically modified by a cation exchange reaction. This improves compatibility of montmorillonite with the polymer owing to its cation exchange properties the clay gallery ions can be replaced by organic cationic species such a alkyl ammonium ions. As the gallery cation changes from inorganic to organic, the surface property changes from hydrophilic to organophilic.

Compatibilising Agents

Clay do not disperse well in organic polymer matrix. The differences in surface tension force between the clay and polymer matrix produces an obstacle in the dispersion of clay particles in the polymer. The direct blending results in clustering tendency of the silicate layers, thus posing a limitation on its use as a nanoclays, hence chemical modification is essential for these clay before using it in a nanocomposite. The inorganic cations present in the clay intergallery provides the site for the water molecules to form a monolayer or multilayer structure. The layered spacing is governed by the amount of water present. This makes clay hydrophilic in nature. To mix the clay with the polymer matrix its hydrophilicity has to be decreased and organophilicity has to be improved. As smectite clays exhibit cation exchange properties, the clay gallery ions can be replaced by organic species such as alkyl ammonium ions. Compatibilising agents are typically a molecule having one hydrophilic and one organophilicity functional group. Nonpolar nature of the alkyl chain reduces the electrostatic interactions between silicate layers to facilitate the diffusion of the polymer into the galleries of layered silicate.

Complete destruction of layered structure of clay: These structures can be determined by transmission electron microscopy (TEM) imaging of these systems. One such image is provided in Fig. 26.2.

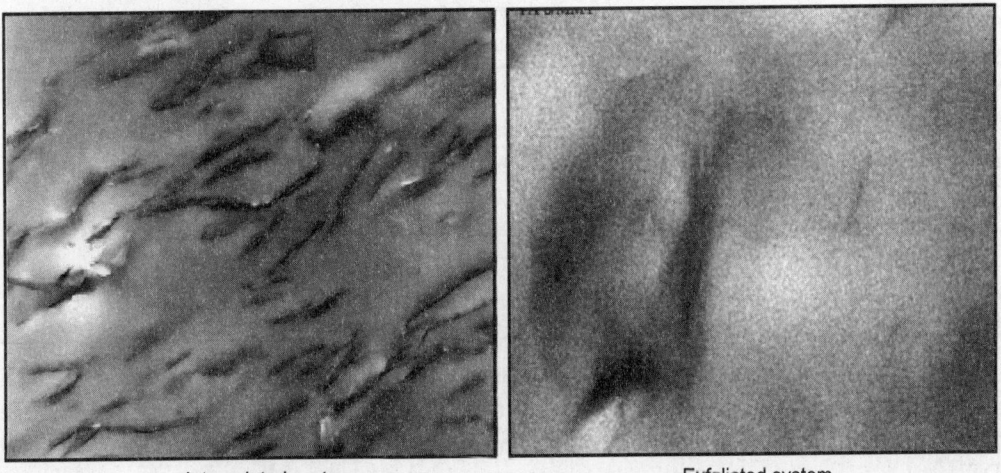

Intercalated system   Exfoliated system

Fig. 26.2. Complete destruction of layered structure of clay.

The image on the left shows clusters of clay platelets sticking together and hence a preservation of layered structure. The image on right shows single dark strands of delaminated clay platelets.

MECHANICAL PROPERTIES

The nanoeffect is quite evident judging from mechanical and barrier properties. As for mechanical properties, in the highest loaded nanocomposite we achieved a 110 per cent increase in flexural and tensile moduli (Fig. 26.3), and a 175 per cent in heat distortion temperature under load (DTUL) (Table 26.1)

On the other hand, there is no negative effect on notched Izod impact strength from nanocomposite formation.

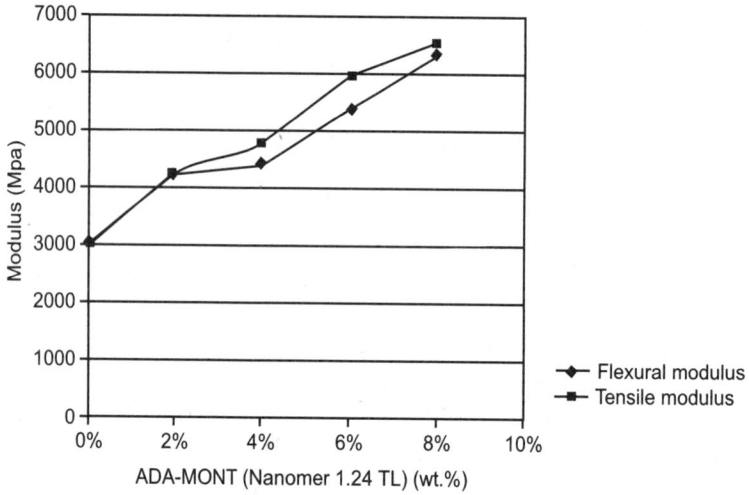

Fig. 26.3. Dependence of modulus on nanoclay loading.

Table 26.1. Comparison of properties of nylon-6 and its nanocomposite.

	Nylon-6	*Cloisite®* *Nanocomposite (5%)*
Tensile strength (Mpa)	82	101
Tensile modulus (Mpa)	2756	4657
Flexural modulus (Mpa)	2431	3780
Notched Izod, J/m	38	27
HDT, °C	57	96
% Elongation	12	8
Shore D	83	83

Nylon-6 nanocomposites containing 2 wt% nanoclay are currently available from two commercial sources. Currently, available products feature dry-as-moulded (DAM) strength improvements of 30 per cent and heat distortion increases double those of neat nylon. As loading increases, so too do strengths and HDT's.

Since real world uses of nylon require performance at moderate-to-high humidity, dry-as-moulded values are not indicative of actual service requirements. Fortunately, comparisons of nanocomposite performance vis-*a*-vis neat nylon are even better under humid conditions.

GAS PERMEABILITY

Gas permeability also improves with nanoclay loading. Current commercial products deliver about 50 per cent (2X) improvement in barrier to oxygen. At higher loadings the reduction exceeds 3X. Because nanoclay promote rapid crystallisation, clarity is better than neat nylon, making nanocomposites ideal for films.

Taking into account their improved strength, nanocomposites can be run at higher line speeds. Add to this the benefit of better print hold-out, and they become a superior, low cost film material.

Properties

Properties of the nanocomposite material are governed to a greater extent by complex interactions between different phase and the interfaces between them.

Mechanical properties

Surface modified silicates used in extrusion melt compounding when properly mixed with melt enhances the mechanical properties. For nylon-6 clay nanocomposite the flexural modulus increases rapidly with increasing clay content form 0–3.5 wt%.

Further increase in the clay content has little effect on the flexural modulus. Flexural strength approaches a peak at 3 wt % and decreases with further increase of the clay content. The tensile modulus increases in the range of 0–15 wt% put little effect was observed for clay content higher than 15 wt%. Notch impact strength remains almost constant in range 0–17 wt%. For low filler content the properties were found to be far superior to those of conventional counterpart composites. There is a strong interaction, between the matrix and the clay layers, which is attributed to the nanoscale size and uniform dispersion of the clay layers in the nylon-6 matrix.

Barrier properties

The barrier properties of the polymer clay nanocomposite were enhanced by the addition of nanoclays in the polymer matrix. If silicate layers are dispersed parallel in the polymer reaction the tortuous factor ($''t$) is given by:

$$t = d'/d$$
$$t = 1 + (L/2 \, W) \times Vf$$

where, Vf = volume fraction of clay.
d = thickness of a film, L = length of clay, W = width of clay.

$$d' = d + (d \times L \times Vf)2W$$

Therefore, relative permeability coefficient is given by:

$$Pc/Pp = 1/[1 + (L/2 \, W) \times Vf]$$

This equation roughly estimates the mechanism of 'Tortuous path' in polymer nanocomposites.

Thermal properties

The HDT increases rapidly for nylon-6 nanocomposites with an increase in clay content from 0–5 wt%. The density of nanocomposite was found to be 65–75 per cent of that of general composite. Nanocomposites offer effective flame retardancy without affecting environment.

Ablative property

A relative tough, inorganic char formed during ablation of nylon-6 nanocomposite. An order of magnitude decrease in the mass loss rate was observed as compared to the pure nylon-6. The formation of char depends on specific interaction between the polymer and the clay surface.

Biomedical Nanocomposites

Polyurethane multiblock copolymer are used in a variety of applications including as blood containing components in ventricular assist systems. One of the drawbacks of these materials is, their relative high permeability to air and moisture. Conventional biopolymers PUU elastomers are composed of soft materials mostly polytetramethyl oxide. It is the low T_g that pertains to its high permeability. This problem is resolved by manipulating the backbone of the polymer chain.

Chapter 27

Chemical Synthesis of Inorganic Nanomaterials and Composites

INTRODUCTION

Materials with features on the scale of nanometers (10^{-9} meters) often have properties dramatically different from their bulk scale counterparts. For example, nanocrystalline copper is five time harder than ordinary copper with its microsized crystalline structure. The development of such materials is currently a research of great interest.

Important among these nanoscale materials are nanocomposites, in which the constituents are mixed in a nanometer length scale. They often have properties that are superior to the conventional microscale composites and can be synthesised using surprisingly simple and inexpensive techniques.

In recent years, a large number of investigations are being carried out to develop special nanomaterial and composites. The interest in these materials have increased tremendously due to the fact that, owing to the very small size and high source to volume ratio these materials are expected to exhibit unique mechanical, optical and magnetic properties. The properties of nanomaterials and composites largely depend on the following four-microstructure feature:

1. Fine grain size and size distribution (<100 nm).
2. The chemical composition of the constituent phases.
3. The presence of interfaces, more specifically grain boundaries, heterophases interfaces, or free surfaces.
4. Interaction between the constituent domains.

The presence and interplay of these four features mostly determine special properties of nanomaterials and composites. In these materials, a variety or size related properties can be incorporated by controlling the size or the component, for example, nanostructure materials and ceramics has been found to better mechanical properties to that of conventional materials and this is due to ultrafine microstructure. Further, nanomaterial and composites has got a very important capabilities particularly, while preparing complicated parts through powder metallurgy techniques. These materials can be sintered at much lowered temperature than the conventional powders. Magnetic application of nanomaterials include fabrication of devices with giant magneto-resistance effects, the property used by magnetic heads to read data on computer hard drives, as well as the development magnetic refrigerators that use solid magnets as refrigerants rather than compressed ozone destroying chlorofloro carbon. In addition, nanostructure metal and ceramics are high application as catalyst.

In view of special applications or nanoparticles and composites in various sophisticated industries, it is very necessary to develop suitable economic process to produce nanomaterials and composites. In this chapter some general methods of preparing nanomaterials and composites are described.

SYNTHESIS OF NANOMATERIALS AND COMPOSITES

The synthesis of nanomaterials and composites generally follow the four methods as described below:
1. The production of isolated, ultrafine crystallites having uncontaminated free surfaces followed by a consolidation processes either at room or at elevated temperature. The specific processes used to isolate the nanomaterials are inert gas condensation, decomposition of the starting chemicals or the precursors and precipitation from solutions.
2. Chemical vapour deposition, physical vapour deposition and electrochemical methods.
3. By introducing defects in formerly perfect crystal such as dislocation or grain boundaries, new classes of nanomaterials can be synthesised either by high energy ball milling, extrusion, shear or high energy irradiation.
4. The final approach used to make nanomaterials is based on crystallisation or precipitation from unstable states or condensed matter such as crystallisation from glasses or precipitation from supersaturated solid or liquid solutions.

Although these are the general methodologies employed in the synthesis of nanomaterials, several variants of the process have been developed to produce compounds or alloys with specific composition and properties.

These methods can broadly classified in two areas as: (i) physical methods, and (ii) chemical methods.

Physical Methods

Several different physical methods are being used to produce nanomaterial and composites. The most widely used technique involves, the synthesis or single phase metallic and ceramics by inert gas evaporation technique. The generation of atom clusters by gas phase condensation proceeds by evaporating or precursor material either a single metal or a compound generally in a gas at a low pressure.

The sputtering process is another technique, where it is possible to produce nanomaterials and composites. In this method the ejection of atoms or clusters of desired materials by subjecting them to an accelerated and highly focused beam of insert gas such as argon and hellium.

The nanomaterials and composites can be prepared through mechanical processes. In this process, the particles are grounded to very fine size by suitable grinding media. The structural degradation of coarser-grained structures induced by high mechanical energy. The nanometer size grains nucleate within shear bands of the deformed materials by reducing coarse to ultrafine structure. This is effected by means of a high energy ball mill or high energy shear process. In this method though it is possible to produce large quantity of ultrafine powders, it is a energy intensive process also there is a chance of contamination of material with the grinding media.

Chemical Method

Chemical synthesis of nanomaterials and composites is more flexible and many time easy to operate. The nanomaterials production through chemical process over the physical and mechanical methods is good chemical homogeneity, because of chemical synthesis offers the advantage of mixing at the molecular level.

Precipitation of a solid from solution is a common technique for the synthesis fine particles. The general procedures involve in the aqueous or nonaqueous solution containing the soluble or suspended salts. Once the solution became supersaturated with the product, the desired material is precipitated by homogenous or heterogeneous precipitation. The growth of nuclei after formation usually proceeds by diffusion, in which the reaction temperature and concentration gradient are the determining factor for the growth of the particles.

Preparation of nanometals and alloys by chemical routes, by decomposition of some metal salts to the metal or to their oxide followed by hydrogen reduction of oxides could result in the formation or the nanosize metal particles. Smith and co-worker successfully prepared Cu-Ni and their alloys from their nitrates by their salt solution at low temperature reduction to their respective oxides followed by low temperature hydrogen reduction. Some metal chloride like $CrCl_3$, $MoCl_3$, $MoCl_4$ and WCl_4 have been successfully reduced in toluene solution with $NaEt_3H$ at room temperature to form corresponding metal colloids in high yield. When the same metal chlorides were reduced in tetrahydrofuran (THF) solution with $LiBEt_3H$ and $NaBEt_3H$, metal carbide were formed.

Another commonly used reduction procedure is called the polyol process. In this method, liquid polyols such as ethylene glycol or diethylene glycol are used both as a solvent and as a reducing agent for the chemical preparation of metallic powders from various inorganic precursors. The basic reaction scheme or the synthesis of these metal powders by the polyol process involves the dissolution of the solid precursor, the reduction of the dissolved metallic species by the polyol itself, nucleation of the metallic phase, and growth of the nuclei. To obtain metal powders with a narrow size distribution, two conditions must be fulfilled: (i) a complete separation of the nucleation and growth steps is required, and (ii) the aggregation of metal particles must be avoided during the nucleation and growth steps.

Soluble compounds of transition metals and post-transition metals in dimethyl ether or THF were rapidly reduced at $-50°C$ by dissolved alkalides or electrides to produce metal particles with crystallite sizes from less than 3 to 15 nm. Salts of Au, Cu, Te, and Pt formed metallic particles with little or no oxidation even when washed with degassed methanol. The reduction of salts of Ni, Zn, Ga, Mo, Sn and Sb yielded surface oxidation over a metallic core. Stoichiometric amounts of the alkalide or electride were used and these were prepared either separately or *in situ*.

A typical reduction reaction follows the scheme.

$$2AuCl_3 + 3K^+ (15C5)_2 e^- \rightarrow 2Au + 3K^4 (15C5)_2 Cl^- + 3KCl$$

or

$$AuCl_3 + 3K^+ (15C5)_2 e^- \rightarrow Au + 3K^+ (15C5)_2 Cl^-$$

where, C5 = 15-crown-5 ether.

The intermetallic TiAl, $TiAl_3$, $NiAl_3$, $NiAl_3$, can he prepared by the reduction or Ti or $NiCl_2$ with $LiAlH_4$ in a mesitylene slurry followed by heating in the solid state. In the course of reaction Al and Ti Ni are initially precipitated in segregated component phase, which on subsequent heating, result in nanocrystalline intermetallics. The microstructure of the intermetallic indicates that the particle size in between 25–35 nm. $MoSi_2$ has also been prepared by the treatment of their chloride with sodium potassium alloys as per the following reactions.

$$MoCl_5 + 2SiCl_4 + (13/2) NaK \rightarrow MoSi_2 + (13/2)NaCl + (13/2) KCl$$

Smith and others have produced metal–metal oxide based nanocomposite such as $Cu-Al_2O_3$, $NiAl_2O_3$, and Cu-Ni Al_2O_3 through their nitrate solution route. The low temperature reduction resulted in the nanooxides of copper–nickel aluminium by low temperature hydrogen reduction at $300°C$. These nanocomposites have surprisingly high mechanical properties as compared to their parent and conventional metal oxides. Thus, as the chemical methods are flexible and versatile and the future of nanomaterials and composites will largely depend upon our ability to modified the various processes to develop better and desired nanomaterials with superior properties.

INDUSTRIAL APPLICATION OF NANOCOMPOSITE FILLERS BASED ON ORGANIC INTERCALATED BENTONITES

Nanotechnology appears to become a key technology of the 21st century. At present new products are intensively developed with the help of nanotechnology in various industrial and university research laboratories. Whereas various fields of this technology are still in their infancy stage, a first commercial success could be achieved in the field of materials technology. Most progresses could be achieved in the development of new polymer material with nanometer architecture.

Nowadays, plastic material and rubber are applied in almost all areas of our life and still have a stimulating function for the development of future technologies.

In contrast to metallic ceramic materials, plastic is relatively cheap, can easily be processed and has a variety of fields of application for textiles, packaging, automotive parts, furniture and household appliances. Plastic needs clearly less energy than ceramic or even metal for production and shaping. On the other hand, the isolation effect of plastics reduce the consumption of fuel oil natural gas and—as a result—the emission of carbon dioxide.

In the automotive industry—in comparison to metals—the very lightweight plastics are saving weight and are very effectively contributing to a reduction in the need for fuel and in the emission of exhaust gases. About 100 kg of plastic substitutes about 350 kg of other materials. Per 100 kg reduction in weight half a litre of fuel is expected to be saved. If the project '3-litre-car' is to be realised, next to the car industry the plastic industry will be concerned with regard to the production of lightweight plastic car parts. The high requirements for plastic materials in the automotive industry are based hereon. Facing competition with metals, plastics have to show a high thermostability, i.e. at high temperatures the dimensional stability of plastic moulding compounds should exactly remain the same. Car parts and bumpers are not allowed to deform under solar radiation and motor heat. Further on, the resistance to mechanical stress is important.

In order to meet the multi-fold requirements to plastic products, composite material has been developed from the beginning of the plastic technology area. The plastics are reinforced by incorporating mineral fillers or fibres. However, in order to achieve suitable properties, 20 to 60 per cent of mineral fillers such as talc, kaolin or calcium carbonate have to be added, which also increases the weight of the material. Nanotechnology offers the chance to realise a high mechanical stability at a very low filler concentration when incorporating highly effective nanofillers.

Although nanotechnology is subject to a revival one should not forget that nanofilling material has a very long tradition in the plastic industry. For many centuries carbon black has been applied for the improvement of rubber. Worldwide some hundred of thousands of tons of carbon black are used for the production of tyres. At the end of the eighties, the Toyota Research and Development department developed a completely new concept for the improvement of plastic parts for the automotive industry with the help of nanocomposite fillers.

The Toyota researchers hereby applied a clay mineral which has been used in industry for more than 100 years: bentonite. Bentonite consists of the mineral montmorillonite (60–95 per cent). Montmorillonite belongs to the so-called layered silicates representing the main component of clay minerals. Its very good swelling capacity in water is a special characteristic of montmorillonite. Water molecules can easily diffuse in the layers of montmorillonite which is homogeneously dispersing in water in the shape of extremely thin nm-sized plates.

Now a very similar concept is followed-up with regard to polymer nanocomposites based on these natural layered silicates. Instead of water, the polymer should diffuse between the silicate layers, disperse them and finally be homogeneously distributed in a plastic matrix in the shape of nm-sized thin plates. However, to achieve this, it is necessary to organically modify the surface of bentonite and thus guarantee the complete dispersion in the plastic matrix (Fig. 27.1). The Toyota researchers aiming at the production of an improved polyamide material used a water soluble monomer of the polyamide for the organic modification of bentonite. The entire available silicate surface was thus modified with amino dodecanoic acid in an environmentally friendly process.

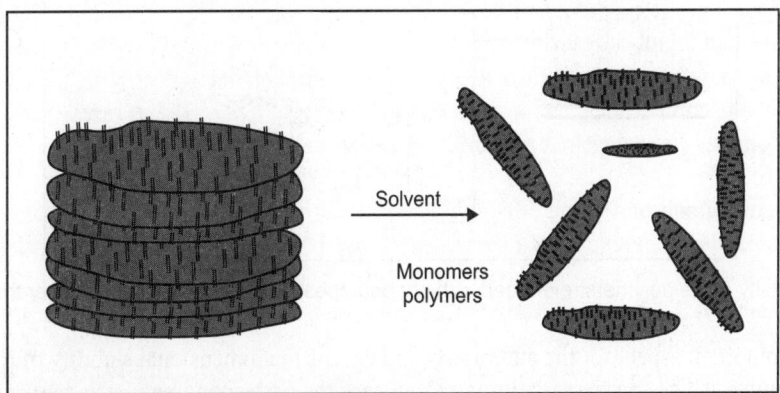

Fig. 27.1. Exfoliation of organic modified montmorillonite in monomers and polymers.

Considering the fact that bentonite shows a specific surface of about 700 m^2/g after a complete dispersion, it is becoming clear that it is necessary to modify bentonite with 20–40 per cent of these organic 'intercalation components'. Mostly, conventional fillers such as calcium carbonate or kaolin are coated with less than 2–3 per cent of a suitable additive in order to achieve a better workability during compounding.

In the meantime, the development of (in addition to polyamide) suitable nanocomposite fillers on the basis of layered silicate for a variety of technical important polymers has been successful.

When starting the polymer nanocomposite development, nanocomposite fillers were incorporated into the polymer matrix only during polymer synthesis. It speaks for itself that this technology was originally only used by polymer producers. In the meantime, the development of suitable process technologies has also been successful. According to these technologies, nanocomposite fillers can be incorporated into the plastic material like conventional fillers via compounding processes. This technology has thus become available to all compounders and even plastic part processors.

Contrary to conventional layered fillers like kaolin, talc or mica, the typical thickness of nanofillers is 10 to 50 times smaller. The diameter of the fully exfoliated nanofillers varies between 100 and 500 nm at a layer thickness of only 1 nm. This special structure of the layers results in an extraordinarily high aspect ratio of more than 100.

The high aspect ratio makes nanofillers superior to all other conventional layered fillers or short glass fibres. The very low particle size and the high aspect ratio yield an extraordinary improvement of the properties in a wide variety of polymer materials. This improvement of the properties may be reached with a very low concentration of nanofillers.

The density of the polymers reinforced with nanocomposite filler is only slightly higher than the unfilled polymers. This leads to a definite weight advantage especially in the area of automotive applications (Fig. 27.2).

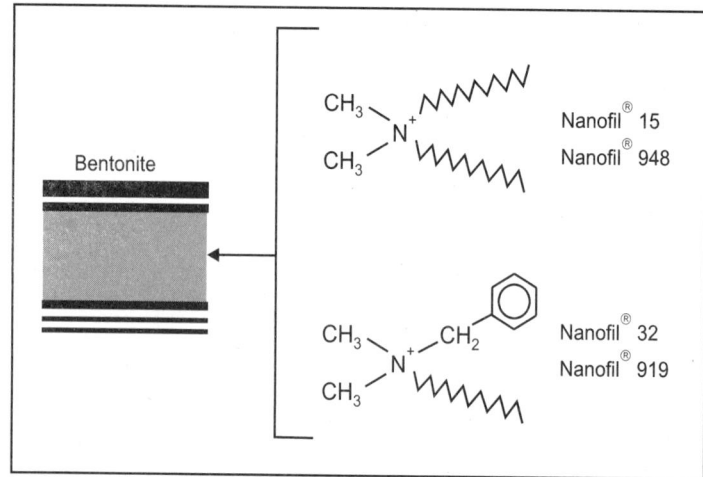

Fig. 27.2. Density of the polymers reinforced with nanocomposite filler is only slightly higher than the unfilled polymers.

Another important aspect for the automotive industry is the dimensional stability of plastic parts at high temperatures. This is particularly important in case the corresponding plastic parts are used in the direct environment of the motor. In contrast to conventional fillers, it is possible to increase the thermostability of polyamide by about 50°C with nanofillers. Only a few per cent of nanofillers are sufficient.

One of the most remarkable characteristics of plastic nanocomposites is the unusual barrier behaviour towards gases and liquids. In case the nanosized silicate layers, which are finely dispersed in the plastic are parallely orientated — which is successfully done in conventional film production, they will form a barrier for that material that is having an effect on the composite.

As a result, the permeability for oxygen, carbon dioxide, or even water steam can be reduced by more than 50 per cent by adding only 2 weight per cent of nanocomposite fillers. A lot of applications are therefore expected for food packaging (Fig. 27.3).

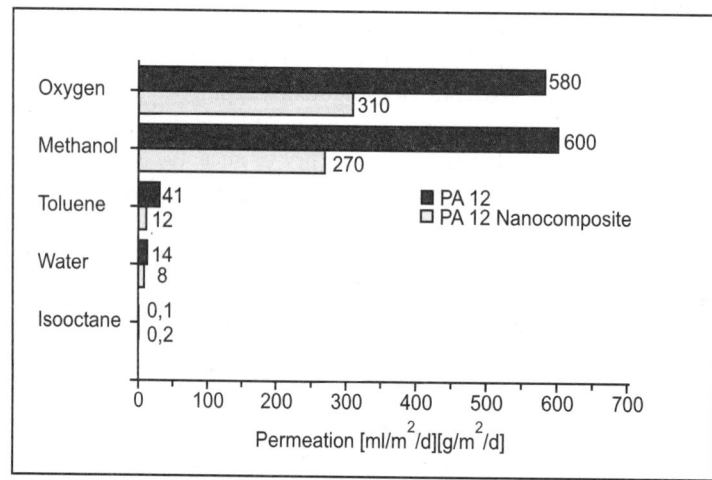

Fig. 27.3. Permeation of different gases by polyamide 12.

Even in the automotive industry nanocomposite technology can be used to develop a zero-emission-car. Nowadays, a car is loosing a part of its fuel via diffusion through fuel lines and plastic tanks. It is the object of present co-operations between additives producers, plastic producers and the automotive industry to improve the fuel barrier of tanks and fuel lines by special plastic materials applying nanotechnology. The first results are very promising.

The high inflammability of plastic material contrary to metallic or ceramic working material is a disadvantage. Even by adding special flame retardants it is not possible to prevent various plastics from inflammability.

In general, these flame retardant additives help to lower the inflammability, prevent the flames from spreading and limit smoke development. This makes it possible to apply plastics even in those fields where inflammability of plastic cannot be excluded.

However, some recent large fires have shown that flame retardant agents are subject to criticism. In case of fire, the highly effective halogen containing organic flame retardant agents generate caustic or toxic gases. In addition, these flame retardants lead to a higher smoke density.

Nowadays, nontoxic flame retardants, such as Mg- or Al-hydroxide are available and meet all toxicological requirements as well as those corresponding to the environment. Unfortunately, these hydroxide flame retardants have only a comparably low effectivity. As a result, plastic material, which is to be correspondingly a flame retardant, has to be filled with about 50–65 per cent of these hydroxide flame retardant agents.

However, mechanical properties are reduced while specific weight is increased. Nanocomposite fillers offer the chance to clearly reduce this amount of hydroxide flame retardant agents, still meeting the flame retardant requirements.

The extremely fine particle size and the extraordinarily high aspect ratio of bentonite nanocomposite fillers have the effect that a three-dimensional network of inorganic silicate layers can be built up in the plastic matrix with only a few weight per cent of nanofiller. In case of fire, this inorganic network assists the formation of a strong carbon crust. This crust prevents the burning polymer from dropping and the fire can thus not spread.

Moreover, it has been found out that smoke density can be reduced when adding nanocomposite fillers; which could be very important for survival in case of fire. In many years of co-operation in research and development, Kabelwerk Eupen AG and Süd-Chemie AG succeeded in using the technology for the production of flame retardant electrical cables. Since the beginning of 2002 these flame retardant cables have been marketed.

The objective of another research co-operation between Siemens AG, Prof. Mülhaupt from the Freiburger Material-forschungszentrum and Süd-Chemie AG is to reduce the inflammability of this plastic material, which is applied in electrical engineering. Possibilities are being traced to find a halogene free flame retardant agent for important thermoplastic material which industry is demanding (Fig. 27.4).

This project is financially supported by the German Federal Ministry for Education and Research. In the framework of European cooperative projects, new fields of application for plastic nanocomposite technology are intensively developed. In one of the cooperative projects financed by the European Communities in which producers of polymer additives, compounders and end users of plastic are participating, new processing technologies are developed making it possible to produce cheap plastic nanocomposite materials. Those cooperative projects are helping the European industry to compete on the international market with regard to nanotechnology. Especially in the United States and in Japan, a lot of money is spent on nanotechnology research.

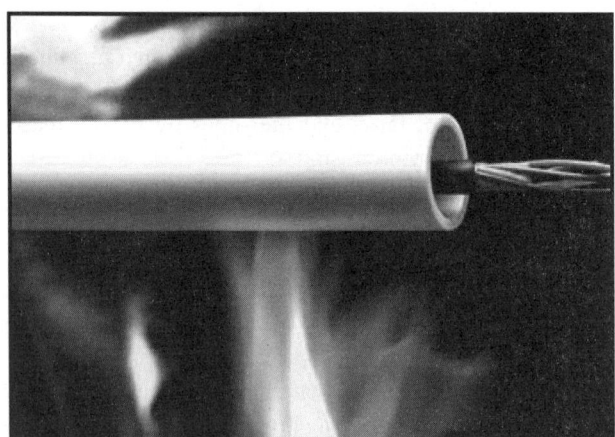

Fig. 27.4. Flame retardancy test UL 1666 for cables.

First 'nanofilms' have become commercially available for packaging technology. Few years ago, Bayer AG already produced a polyamide film for packaging purposes; in doing so, they succeeded in improving the oxygen barrier by 50 per cent. In addition to that, this polyamide film shows an improved transparency in comparison with the unfilled film. Even this characteristic, which was not expected to appear, is a result of the extremely low particle size of nanocomposite fillers. These particles are so small that they cannot interact with light; as a result, they are invisible to the eye. In Japan and in the United States similar packaging films are commercially offered with nanocomposite fillers based on bentonite, too.

Since we are only at the beginning of the industrial commercialisation of nanocomposite technology it is extremely difficult at the moment to estimate future market chances for this technology. However, considering the fact that the total plastic fillers market shows a potential of more than two billion Euro one cane estimate a market potential based on this figure. In case we are able to enter only a few per cent of this market we are talking about sales amounting to a few million Euro.

Nanocomposite fillers based on natural bentonites are even nowadays a very good example for solving the problem how to find completely new markets for very old products—bentonites have been used in industry for more than 100 years—with the help of consequent research. However, there is a clear need in doing research and development in order to be able to use plastic nanocomposite technology in many fields of application.

The market for nanocomposite fillers has to be developed via cooperations between producers of polymer additives, compounders and endusers of plastic. It is necessary to establish a cooperation between public research institutes and industry as soon as possible to achieve a rapid market realisation. For the first time nanocomposite technology is opening the door to the growing market 'plastic material' for the traditional European bentonite industry.

Chapter 28

Biologically Derived Synthetic Nanocomposites

INTRODUCTION

Because of the limited variety of materials available by purely biological routes, there is great interest in utilising the sophistication offered by biological systems in concert with synthetic procedures to create materials with otherwise unobtainable nanostructures and thus, potentially, unique properties. Biology offers a unique selection of building blocks that would be difficult or impossible to synthesise in the laboratory. These include proteins, DNA, RNA, and small but highly functional molecules.

PROTEIN-BASED NANOSTRUCTURE FORMATION

The protein S-layer on the surface of bacteria create complex nanostructures. The protein S-layer present on the surface of some bacteria has now also been used to create entirely synthetic nanostructures. The diversity of S-layer structures, coupled with the potential for chemical functionalisation, make them ideal starting points for nanostructure synthesis. To review, S-layers are 2D protein crystals that have oblique, square, or hexagonal lattice symmetry with lattice constants between 3 and 30 nm. Interesting for potential nanostructure formation, and to enable bacterial respiration, S-layers almost universally contain pores of identical size and identical surface chemistry—properties that make S-layers useful for nanostructure and nanocomposite fabrication. In an early example, an S-layer was used to template a periodic structure into a thin metal film. Specifically, a 1 nm-thick metal (Ta/W) film containing 15 nm holes periodically arranged in a triangular lattice with a lattice constant of 22 nm was created by S-layer templating. This was accomplished by first depositing a suspension of S-layer fragments onto an amorphous carbon support film. After S-layer deposition, a 1.2 nm thick film of Ta/W was evaporated onto the S-layer at an angle of 40° from the normal of the substrate surface. When examined in the transmission electron microscope (TEM), the resulting film shows contrast indicating some thickness variation in the metal, but the contrast is not significant. To improve the contrast and to open up holes in the metal film, the mm was argon-ion-milled for a short time. The result was a metal film that contained a periodic array of holes with the same symmetry as in the S-layer template. This work was done more than 15 years ago, yet creating such a nanostructure today would still be a challenge, even with a modern e-beam lithography system.

Much more recently, an S-layer was used to create nanostructured semiconductor films. The S-layer in this study had an oblique 2D lattice (space group p_1; $a = 9.8$ nm, $b = 7.5$ nm, $\theta = 80°$). Depending on the mechanism of S-layer deposition, either its negatively charged inner face or its charge-neutral outer face was exposed to a 10 mM $CdCl_2$ solution. After drying, the cadmium-ion-doped S-layer was exposed to H_2S gas, resulting in formation of a nanocrystalline film of the semiconductor CdS on the S-layer. In the TEM, the CdS film exhibited a superlattice structure that was a direct copy of the structure of the S-layer. The S-layer may have survived the entire process, and thus the actual mineral structure formed may be a nanocomposite of CdS and protein. It is still unclear what applications might be found for such materials, although robust metal or semiconductor nanoporous structures may find application as filter membranes, sensors, and optoelectronic devices.

Another protein assembly that has been used to form nanocomposite materials is ferritin (Fig. 28.1). In its native form, ferritin consists of a supermolecular arrangement of proteins around an iron oxide core. The iron oxide core can be selectively dissolved without damaging the structure of the protein shell, yielding a hollow ball of protein about 10 nm in diameter. The demineralised protein (apoferritin) can be refilled with iron oxide, demonstrating that its structure was not greatly affected by the demineralisation process. Interestingly, apoferritin can also be filled with other mineral nanoparticles. In one example, the iron oxide core of ferritin was converted to FeS by treating the ferritin with H_2S. Nanoparticles of MnOOH, UO_3, Fe_3O_4, and CdS can also be formed inside apoferritin through the appropriate chemical treatment. Because the protein shell is not disrupted by the mineralisation process, the final product is truly a nanocomposite of protein and inorganic material.

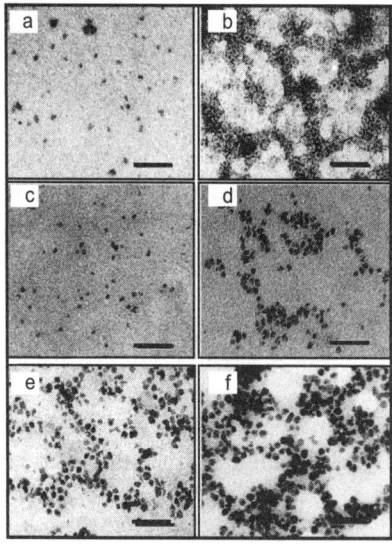

Fig. 28.1. Transmission electron micrographs of magnetite (Fe_3O_4), and maghemite (γ-Fe_2O_3)-filled ferritins. (a) 260 Fe atoms/molecule, unstained; only the discrete electron-dense inorganic cores can be seen. (b) 260 Fe atoms/molecule, after staining with uranyl acetate, showing encapsulation of inorganic cores by intact protein shell (white halo around each particle). (c) 530 Fe atoms/molecule, unstained. (d) 1000 Fe atoms/molecule, unstained. (e) 2040 Fe atoms/molecule, unstained. (f) 3150 Fe atoms/molecule, unstained. Scale bars in all figures = 50 nm.

DNA-TEMPLATED NANOSTRUCTURE FORMATION

DNA offers great potential as a building block for nanocomposite materials. It can be tethered to a wide range of substrates, can direct assembly with specificities that greatly exceed that of any synthetic molecule, is relatively robust, can be synthesised in relatively large quantities, and can be functionalised with tags such as fluorescent molecules to enable rapid detection of binding events.

The use of DNA to assemble and create nanostructures and nanocomposite materials is only in its infancy, but even the preliminary work done so far indicates the great potential of DNA-based assembly technique. A few of the approaches that have been studied to date include the mineralisation of DNA, the use of DNA to assemble nanoparticles, and the use of DNA to assemble much larger colloidal particles.

These three approaches are outlined in this chapter; however, the number of possibilities is vast, and significant work on DNA-mediated assembly of nanostructures is currently being done.

The possibility of using plasmid DNA as a template for mineralisation was first explored in 1996. In this work, single strands of a 3455-basepair circular plasmid DNA were mineralised with CdS nanoparticles, by mixing the plasmid and cadmium perchlorate in solution, followed by spin coating this solution onto a polylysine-coated glass slide. The DNA/cadmium perchlorate-coated glass slide was then exposed to H_2S, converting the cadmium perchlorate to CdS, which preferentially mineralised the DNA. This procedure results in ring-like structures consisting of DNA embedded within a 5–10 nm thick CdS strand that can be directly observed in the TEM (Fig. 28.2). The diameter of the ring formed by the CdS-encrusted DNA is directly related to the diameter of the DNA plasmid, and the thickness of the CdS/DNA composite strand is about 10 nm. This represents the first example of DNA templating of semiconductors but, given the vast array of structures that can be formed from DNA, should not be the last.

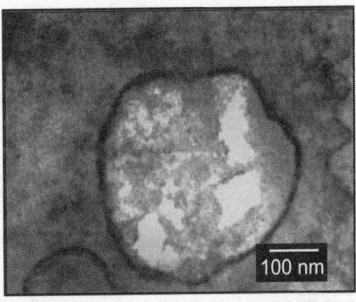

Fig. 28.2. Transmission electron micrograph of a nanostructure obtained by mineralisation of a circular plasmid DNA with CdS nanoparticles. The ring diameter closely matches the predicted diameter of the plasmid DNA.

Another example of DNA-based nanostructure development can be seen in the work on DNA-based nanoparticle assembly. Several approaches have been explored, with the majority based on the Watson–Crick base-pairing of DNA strands attached to various nanoparticles. The first approach involved the functionalisation of gold nanoparticles with multiple strands of thiol-terminated type-A or -C DNA in separate reactions (where 'type' is used as a label for different nucleotide sequences). The thiol terminal group is important, because it enables covalent attachment of the DNA strand to the gold nanoparticles. Importantly, the A and C strands were designed so that they would not hybridise. The result, after mixing the solutions of each kind of DNA-functionalised nanoparticles, is a simple mixture of DNA-functionalised gold of type A and DNA-functionalised gold of type C. Then a single strand of DNA with one end complimentary to the type-A DNA and the other end complimentary to the type-C DNA was added. We label this strand B-D. Upon hybridisation of the B-D strand with both nanoparticles of type A and nanoparticles of type C, a DNA-linked aggregate of nanoparticles was formed. Aggregate formation is thermally reversible by increasing the temperature above the melting point of the DNA, at which point the strands dehybridise. The assembly could be monitored by UV/visible spectroscopy, because the UV/visible absorption of gold nanoparticles changes greatly when they are close together, due to changes in the surface plasmon.

The formation of controlled aggregates, although interesting, is really a starting point for the creation of much more complex structure through a marriage of biological macromolecules and synthetic materials. In addition to placing multiple DNA molecules on a nanoparticle, it is possible to place a single or a finite number of DNA molecules on a nanoparticle. However, this requires a significant degree of synthetic effort, because one usually starts with a population of nanoparticles with various numbers of attached DNA molecules. Through careful separation, nanoparticles with exactly one or exactly two single-stranded DNA molecules on their surface can be recovered. These DNA-functionalised particles can then be assembled in highly controlled fashions through DNA hybridisation reactions, by adding single-stranded DNA that is complimentary to the single-stranded DNA on the nanoparticles (Fig. 28.3). This is similar to the approach used to make DNA-linked aggregates; however, because each nanoparticle contains only one DNA strand, not multiple strands, the result after hybridisation is not the formation of an aggregated structure, but rather, individual dimers of gold nanoparticles. Trimers of gold nanoparticles were also created by a similar approach, except that nanoparticles with one strand of DNA of three different sequences was mixed with a strand of DNA complimentary to all three of the DNA strands on the surface of the nanoparticles.

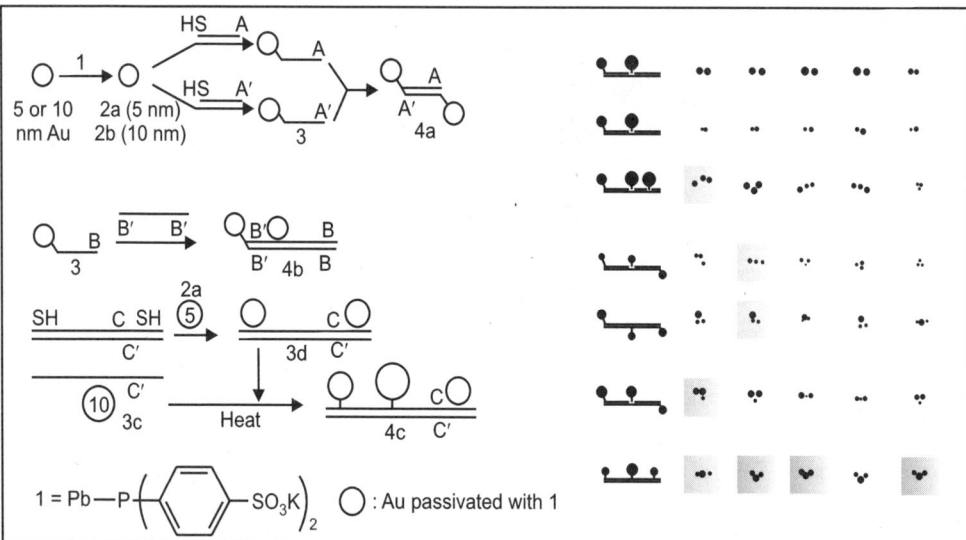

Fig. 28.3. DNA allows for precise manipulation of the order and spatial arrangement of nanoparticles. The large and small nanoparticles can be designed to hybridise or react with a single site on the DNA backbone. See examples A, B, and C on the left and the resulting nanostructures on the right.

Recently, it was demonstrated that DNA can assemble much larger colloidal particles into 3D assemblies of controlled shape. The basic approach is very similar to the assembly of nanoparticles; however, here the results after hybridisation are assemblies of colloidal particles with defined connectivity.

The overall power of DNA-based assembly of nanostructures is based on several powerful aspects of DNA hybridisation. First, hybridisation is reversible with temperature; thus it is possible to anneal structures that at least partially eliminate kinetic traps that may be present in the assembly process. For example, if during nanostructure assembly, a particle does not hybridise in the correct location, the structure can be gently heated, releasing the particle, and then the particle can be retrapped by cooling below the melting point of the DNA. Second, DNA hybridisation is very specific. Unlike most organic linking chemistries, which cannot distinguish small changes in molecular structure. DNA can be designed to hybridise only to its exact complimentary strand. Thus, it may be possible to combine many DNA-functionalised nanoparticles into very specific arrays by hybridising to a long strand of DNA that is encoded so as to react with each DNA functionalised nanoparticle at only a very specific location.

PROTEIN ASSEMBLY

To achieve a level of complexity similar to what can be obtained by DNA-based assembly, one can engineer or select biological organisms that recognise and bind with high specificity to specific minerals, semiconductors, and metals. In the first example, *Escherichia coli* containing genetic sequences specific for recognising and binding iron oxide but not other metal surfaces was identified and multiplied. Bacteria that would specifically adhere to iron oxide were identified via serial enrichment from a population of bacteria. The experimental procedure was as follows: A population of genetically diverse *E. coli* was exposed to iron oxide particles. The bacteria that bound to the iron oxide were collected, and the remainder were discarded. After repeating this procedure for several generation, bacteria with high degrees of specificity for iron oxide were obtained. The reason only some of the bacteria bind iron oxide is that *E. coli* like many bacteria, express proteins on their outer surface whose sequences are a function of the unique genetic make-up of the bacteria. By starting with a library of $\sim 10^7$ *E. coli* that express surface proteins of slightly different sequences, the probability that at least one of the *E. coli* has a highly specific protein is reasonable. Through enrichment, it is possible to extract and multiply just those bacteria that bind to the surface of interest.

It is also possible to select for protein sequences that attach to specific metal surfaces, and presumably also to metal nanoparticles. Through such biological experiments, a library of $\sim 10^7$ different polypeptides 14 or 28 amino acids long was created.

Then, the polypeptides that adhere to the metal surface were isolated, and their sequences were determined. This approach is exceeding powerful because of the vast number of different molecular sequences that can be studied in parallel in a single experiment.

From a technological standpoint, it is interesting to consider the possibility of recognising semiconductor surfaces and semiconductor nanoparticles with this procedure. If an appropriate genetic sequence can be identified, it would be possible to assemble nanoparticles via biological organisms that are designed to present the appropriate biological macromolecules on their surface. Phage display enables the simultaneous testing of many peptide sequences for specificity to a given surface. A phage library containing 10^9 polypeptide sequences was exposed to a surface, and all the phages that did not bind to the surface were washed away. The phages that stuck were then removed from the surface by lowering the pH and were amplified by infecting *E. coli* bacteria with them. This process was repeated until only phages that stuck strongly to the surface were present. The DNA of these phages was then sequenced to determine the specific peptide sequence(s) that bind with such high affinities to the specific surface of interest.

A phage that identifies and binds to a surface of a specific material also binds strongly to nanoparticles of this material. The result of mixing a nanoparticle-binding phage with nanoparticles is 'decoration' of the phage with nanoparticles, which occurs only at the and of the phage that contains the specific binding polypeptide, and nowhere else on the phage. Depending on the experimental conditions and phage design, it may even be possible to bind single nanoparticles to an individual phage or to assemble the nanoparticles into defined structures, although in the present state of the art, multiple nanoparticle are bound in a fairly random fashion to each phage.

BIOLOGICALLY INSPIRED NANOCOMPOSITES

The properties of biocomposites and synthetic pathways for their formation have inspired wide ranging research. However, early on it was recognised that it is not always necessary or even desirable to use biologically derived materials far many applications and that it may be possible simply to use biology as an inspiration for totally synthetic nanocomposite systems. It is interesting to consider biological systems as an inspiration for nanocomposite materials, because biological systems exhibit many characteristics that would be attractive in synthetic materials, but it is also clear that direct mimicking of biology will be limited to a specific small subset of materials and to specific nanostructures. However, many lessons can be learned from biology on how to form complicated nanostructures and on the potential properties of these synthetic nanostructured materials, should one be successful in synthesising them. Of course, just because a synthetic material resembles a natural process and the process of forming the materials resembles a natural synthesis does not always mean that the scientists and engineers who performed the work were inspired by biology. Often it is not stated whether biology was an inspiration for the work, and thus some care must be taken before assuming that, just because something appears biologically inspired, it is biologically inspired. In this chapter, there is no attempt to make this distinction and by necessity assume that, if the work has a biological analogue, then it can be considered 'biologically inspired', although certainly in some cases this assumption may be wrong.

Biological Systems for Complex Inorganic Structures

Much can be learned from biological systems to further the development of synthetic approaches to the formation of complex inorganic structures. Each of the routes to nanostructure formation that is discussed in this section — liquid crystal templating, colloidal particle templating, block copolymer templating, and surfactant-inorganic self-assembly (mesoporous silica being the most famous of this approach to nanostructure formation) — invokes many of the tenets of biologically directed mineral growth. As already discussed, biological systems rely on self-assembly and mineralisation in the synthesis of hard inorganic structures such as shells, teeth, and bone, and their approaches to materials fabrication can provide guidance and direction to synthetic systems. Often, the term 'biomimetic' has been applied to any approach using self-assembly in the synthesis of nanocomposite materials. As mentioned above, biological systems do indeed use self-assembling molecules, and high levels of molecular organisation are a very important part of an organism's inorganic structure development. However, it would be naive to expect to simulate this process in the laboratory except on the most basic level. Biological processes are extremely dynamic, involving huge numbers of very specific proteins and other molecules being generated and transported to very specific locations, with temporal control. The best synthetic systems are indeed very simple approximations of living systems and generally are much too simple to be considered to be truly mimicking biology. Although the synthetic systems are only simple approximations of life, still much can be learned by attempting to mimic living systems, even if biomimetic is perhaps not the best term. For example, many attempts centred around the synthesis of mineral phases in a self-organised matrix have indeed mimicked the mineralisation processes of many biological systems. Examples of matrix-mediated biological mineralisation processes include the reliance of bacteria, plants, shells, and even mammals on organically mediated growth of mineral phases to eliminate by-products (bacteria and plants), create exoskeletons (shells), and grow teeth (mammals). If during the synthetic process, organic molecules are incorporated into the mineral phase, the resulting material may resemble the spicule of a sea urchin, which contains much less than 1 per cent protein intercalated into the crystal lattice.

Despite the low degree of sophistication of synthetic systems, they do have very distinct advantages over biologically based schemes. Biological systems operate with only a limited subset of elements and compounds, but synthetic systems can be designed to use a wider range of elements and compounds, many of which would be toxic to most living organisms. Biological systems form and operate near room temperature in the presence of water and oxygen (for the most part), but synthetic systems can be formed and operate under a wide range of temperature and conditions. Finally, biological materials generally take days to years to form, but synthetic systems may be formed rapidly. For biology, systems that form slowly in response to external stimuli have significant advantages, for example, bone remodels to meet the demands of applied loads. However, in general, the longtime scales required for formation of biological nanocomposites limit the application of direct biological synthesis of engineering materials to a few very specific cases. Through self-assembly of organic molecules, biological systems have succeeded in synthesising a wide range of composite and inorganic nanostructures. Biological systems contain large quantities of lipids, or soap-like molecules, which self-assemble (along with many other biological macromolecules, including proteins) to form the external membranes of cells, as well as smaller vesicles within the cell. These membranes serve to protect the interior contents, as well as to provide synthetic microreactors for biological processes. The concept of performing syntheses in self-assembled micro reactors is in this sense biologically inspired, has been exploited by many researchers to create nanoparticles and composite nanostructures, and is the topic of this section.

Nanoparticles

The simplest examples of nanomaterials are zero-dimensional materials, known as 'nanoparticles'. The study of small semiconductor particles in general is of significant interest, because the electronic and optical properties of semiconductor nanoparticles change drastically as the characteristic dimension of a particle is reduced to the nanoscale regime, largely due to the quantum confinement of electrons within the particle, although surface-area effects may also play a role (Figs 28.4 and 28.5). The most popular nanoparticles for basic studies have been II–VI semiconductor particles, because of their scientifically interesting and technologically important properties. In this context, synthetic methodologies for the formation of metal sulphide and selenide quantum dots and their assembly into higher-order structures have been widely studied. For nanostructuring to dominate the properties of metal sulphides and other semiconductors, it is generally necessary for the characteristic dimension of the nanoparticle to be <10 nm. Semiconducting nanoparticles are generally synthesised through one of three routes. The first, grinding of large chunks is rarely done for nanoparticle preparation, because the grinding process is poorly regulated, generally generating a very polydisperse population of particles, and grinding introduces too many contaminates for most applications. The other two methods, gas-phase and solution-phase synthesis, are much more common. Gas-phase synthesis is essentially a vapourisation and condensation process—a crucible containing the desired semiconductor (or other material) is heated until it starts to sublime, and then an inert carrier gas is flowed over the material. The carrier gas then heads into a cool region where the gaseous semiconductor atoms or molecules condense into nanoparticles and are collected. Although this method is fairly versatile, it operates only under conditions of high temperature and vacuum and generally produces solid spherical particles.

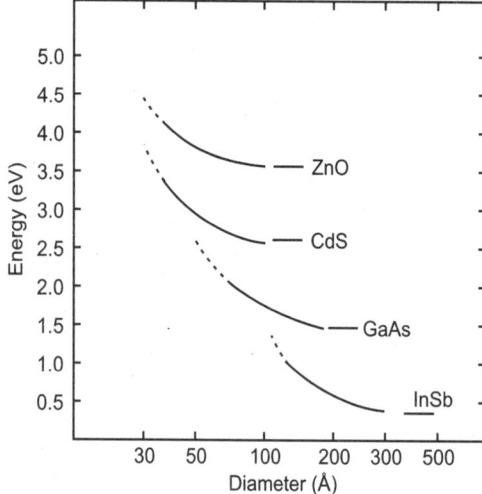

Fig. 28.4. Theoretical calculation of band-gap energy as a function of particle diameter for several different semiconductors.

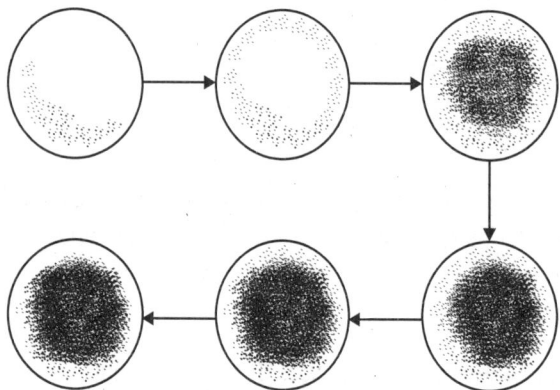

Fig. 28.5. Cd_3P_2 quantum dots. Particle size increases in the direction of the arrows. The white particles are about 1.5 nm, and the black particles are larger than 5 nm.

Solution based route for nanoparticles

Solution-based synthetic routes for nanoparticle formation have ranged from simple precipitation reactions to much more complex self-assembly-based routes. In general, simple precipitation results in agglomerates of nanoparticles, and the size distribution generally varies widely. These problems led to research into synthetic procedures that would result in nanoparticles that are stable against aggregation and have narrow size distributions. A primary route to preventing both these problems is to use self-assembly-based techniques, which in many respects resemble nanostructure development in biological systems, including biomineralisation, cell membrane development, and other biological structure formation. Solution-phase synthesis of semiconductors is often preferred over other techniques, because it is generally mild (even being carried out at room temperature and pressures) and can be used to create reasonable volumes of materials. Solution-phase synthesis has been widely used to grow semiconductor quantum dots, yielding particles with low polydispersities and novel optical properties. Solution-based chemical syntheses are very attractive because they allow for direct control over the actual concentrations of the chemical precursors.

Depending on actual conditions, it is even possible to cap the surface of the particles with organic molecules, which allows for further solution-based processing.

Conventional route to creating nanostructured materials

The more conventional route to creating nanostructured materials is of course through top-down lithographic methods. Some examples of top-down techniques for generating very small features are extreme UV ($\lambda \ll 200$ nm) lithography, electron beam writing, focused ion-beam lithography, X-ray-lithography, scanning probe lithography, and microcontact printing. In general, all these can form structures on the scale of tens to hundreds of nanometers, although only on very flat substrates, and they can be quite slow and often very expensive. The self-assembly-based route to nanostructure formation has the significant advantage that it is not limited to feature generation on flat surfaces, and it can be massively parallel.

Of course, the general problem with self-assembly is that it is not possible to regulate the exact spatial position of the nanostructure, and thus we are still many years away from creating highly functional self-assembled electronic circuits.

In the micellar routes to nanoparticle formation, micelles are self-assembled from surfactant molecules and a solvent that contains at least one of the precursors for the inorganic nanoparticles in solution. The result is a solution that contains vast numbers of discrete nanoreactors which individually contain only a finite number of precursor species for the inorganic phase. When these ions are converted to mineral, generally by reduction or oxidation, the result can be one nanoparticle per micelle. The polydispersity in nanoparticle diameter is thus directly related to the polydispersity in the initial micelle size. If it were possible to create a suspension of monodisperse micelles, it would be a straightforward process to create nanoparticles with a very narrow size distribution.

Taking a lesson from biology, this might be possible through the use of complex macromolecules that organise into particles of only a specific size. A very good biological example of a potential nanoreactor with a tight size distribution is a virus particle. It may be possible to use virus particles to synthesise nanoparticles with tight size distributions if a way to load the interior of the virus particles with precursors for nanoparticles can be developed.

Nanoparticles synthesis routes

Most of the early nanoparticle synthesis routes produced solid semiconductor particles with morphologies never far from spherical. The creation of nanoparticles with complex morphologies is most interesting, because even such sophisticated patterning techniques as e-beam lithography are limited to ~10 nm features, which are often too large to result in the desired quantum confinement and other properties. Nanoparticles having certain complex morphologies cannot be created through e-beam patterning or other top-down processing routes.

The possibility of generating complicated morphologies was examined in studies of CdS synthesised under Langmuir monolayers, in which dendritic structures were generated (Fig. 28.6). Subsequently, a range of strategies have resulted in rod-like and even complex nanoparticles with complex morphologies that are based on self-assembly regulating the growth of semiconductor particles.

Fig. 28.6. Transmission electron micrograph of nanocrystalline CdS structure grown under an arachidic acid monolayer at room temperature. Scale bar = 200 nm.

Self-assembly processes

As examples of self-assembly processes which create nanoparticles with complex shapes, Alivisatos and coworkers demonstrated the formation of CdSe nanorods with aspect ratios of 30:1, as well as arrow-, teardrop-, tetrapod-, and branched tetrapod-shaped nanocrystals of CdSe (Fig. 28.7). These highly shaped nanoparticles result from using a mixture of hexylphosphonic acid and trioctylphosphine oxide as passivating agents in the synthesis.

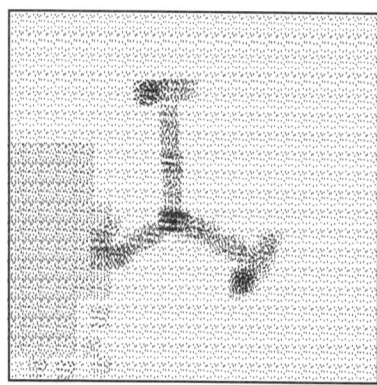 

Fig. 28.7. Examples of CdSe nanoparticles with complex shape and form that can be created by solution synthesis from a mixture of surface passivating agents.

Apparently, it is possible to block the growth of specific crystallographic faces while encouraging the growth of other faces. Thus, particles that have one crystallographic direction as their long axis, and their short axis as another crystallographic direction can be formed. The advantage of high aspect ratio particles over normal, spherical particles has not yet been proven, but the electronic or physical properties of the particles might be impacted by the change in shape. In addition, the particles might be able to self-assemble into higher-order structures because of their high aspect ratios: for example, they may form liquid crystalline phases, much like molecules with high aspect ratios.

Semiconductor nanostructures

Semiconductor nanostructures may also have potential nonquantum confinement-based properties not found in the bulk equivalent. Besides quantum confinement, one way to greatly modify the properties of nanostructured semiconductors may be to design synthetic methodologies that result in dispersion of organic molecules within the inorganic phase at the molecular level. These composite materials could exhibit novel properties significantly enhanced over those of either the inorganic or organic phase alone, as has actually been observed in a wide range of materials. Composite materials that are tougher, have increased thermal stability, are electronically more sophisticated, or have enhanced chemical selectivity than either of the constituent parts have been created. Even without the incorporation of organic material, periodically nanostructured semiconductors have great potential in solid-state science and technology, for example, because of their potential for both electronic and catalytic activity. For example, a periodically nanostructured semiconductor might behave as an array of antidots (a material with a regular array of scattering centres spaced closer than the mean free path of electrons travelling through them). A nanostructural material is necessary because, if the mean free path of the electrons is shorter than the spacing between the scattering centres, the antidot lattice does not operate. At high magnetic fields (>2 Tesla), quantum steps in the conductivity as a function of electric field may be observed if the lattice spacing of the antidots is on the order of the cyclotron diameter of the electrons. In contrast to forming a solid of quantum dots, the formation of an antidot lattice requires the semiconducting structures to be continuous, with a periodic array of nanocavities.

Nanoparticles with narrow size distribution

Early on, the greatest emphasis was on creating nanoparticles with narrow size distributions, not on creating superlattice structures. However, through careful control of size distribution and chemical functionality, CdSe nanocrystals, and now many other nanocrystals, have been observed to order into superlattice structures. These structures may present properties beyond simple quantum effects. The individual crystallites in this system do not form a continuous mineral structure, but are actually separated by thin layers of organic molecules, which are composed of the self-assembling molecules used in the synthesis to regulate the diameter and polydispersity of the particles. During synthesis, the organic self-assemble into a shell around the nanoparticles, imparting organic solubility to the nanoparticles, which enables them to be processed similar to organic compounds. Because of the high degree of regularity in size and shape, these organically coated nanoparticles assemble into a crystal of nanoparticles with long range periodicity, much in the way that organic molecules and atoms can crystallise.

Nanostructures with complex predefined morphologies

The next level of complexity in nanostructure formation is the creation of nanostructures with complex, predefined morphologies. Here, the biological concepts of self-assembly and nanostructure formation become most applicable. For example, in biology, it is common to have complex predefined structures on the nanometer scale, yet this length scale is exceedingly difficult to regulate in synthetic materials. However, if the power of self-assembly is coupled with the materials synthesis strategies known today, there is great potential for the formation of complex composite nanostructures.

Liquid crystal templating of inorganic nanostructures, an approach in which the periodic structure of liquid crystals is imparted to a mineral phase, is one such route. Liquid crystals present an ideal matrix for the creation of nanoscale composite materials, because the characteristic 1–10 nm length scales most often expressed in liquid crystals are similar to the size scale of interest for semiconductor nanostructures.

Furthermore, the periodic structure in liquid crystals can be quite long-range, and thus the periodic nanostructure also has the potential for long-range order, which is of exceptional interest for many applications. A very important goal of several research groups is in fact the creation of long-range nanoperiodic order in materials such as semiconductors through liquid-crystal templating.

Semiconductors with long-range periodic composite structures could come in several forms, for example as a particle containing a periodic array of embedded second-phase material, as a thin film with a periodic topography, or as a periodically porous material. Highly porous, periodically nanostructured semiconductor particles could be quite interesting for solution-based chemistry. For example, the photochemical nature of their semiconducting phase and zeolite-like pore structure could make them highly applicable for photochemical degradation of toxic compounds or for performing shape-selective chemistry, that is chemistry that operates only on molecules of specific shape and size. Thin films may be even more technologically important, given the wide range of potential uses for both supported and freestanding thin films. The ability to predefine a nanoperiodic array of features in semiconductor thin films may open up many applications, including electronic devices, sensors, and filter membranes. Three-dimensional semiconductor structures may have unique optical or electronic properties, depending on the characteristic length scale of the structure. As the length approaches hundreds of nanometers, the materials may even exhibit photonic band-gap effects. Because direct top-down patterning of <10 nm-long nanostructures is difficult or impossible, and the patterning of 3D nanostructures of almost any length scale is difficult, templating of nanostructures through self-assembly processes has significant promise. Essentially, self-assembly-based templating can take place in either 2 or 3D, and the templating agent mayor may not be removed, as desired. Here, we discuss several templating methodologies with potential for nanocomposite formation, the properties of such templated materials, and their potential applications.

LYOTROPIC LIQUID-CRYSTAL TEMPLATING

Lyotropic liquid-crystal templating is an approach for nanostructure and nanocomposite formation that utilises the self-assembled structure of a liquid crystal to regulate the structure of a growing inorganic material. When processed correctly, the structure of the inorganic phase directly replicates the structure of the liquid crystal; thus the liquid crystal is a 'template' for the inorganic. The most important aspect of liquid-crystal templating of inorganic material is the lyotropic liquid-crystal; thus, we should review some of the basics of lyotropic liquid crystals.

Lyotropic liquid crystals are composed of at least two covalently linked components, one of which is usually an amphiphile, which is a molecule that has two or more physically distinct components, and the other a solvent. Typically, one of the components making up the amphiphile is hydrophobic, and the other component is hydrophilic, as in common soaps, although this is not the necessary distinction. The dual solvent properties of an amphiphile lead to the interesting self-assembly of these molecules in solution by means that include surface segregation, formation of micelles and vesicles, as well as the formation of a wide range of LC structures.

Amphiphilic Molecules and Segments

Most amphiphilic molecules contain two or occasionally three segments, of which at least one is water soluble. Clearly, if half a molecule dissolves in a solvent and the other half does not, interesting self-assembled structures may result. Within the general class of amphiphilic molecules are four distinct subclasses; cationic, anionic, zwitterionic, and nonionic. Cationic, anionic, and zwitterionic amphiphiles all contain a formally charged polar moiety, typically called the head group, and a nonpolar moiety, typically termed the tail. As the names imply, cationic amphiphiles contain a cationic head group, such as a quaternary ammonium salt, and anionic amphiphiles, an anionic head group such as a sulphonate salt. Zwitterionic amphiphiles contain a head group having both positive and negative charge, for example trimethylammonium phosphonate. Nonionic amphiphiles do not contain charged functional groups, but instead contain polar segments such as oligo(ethylene oxide) or oligo(vinyl alcohol). Examples of anionic and non ionic amphiphiles, sodium octanoate and oligo(ethylene oxide) oleyl ether, respectively, are shown below.

Amphiphiles exhibit very rich, complex phase behaviour as a function of solvent concentration. In the dilute amphiphile limit, structures such as micelles form, and in the concentrated amphiphile limit (even solvent free), some amphiphiles show liquid crystalline or crystalline phases. Between these endpoints, a wide range of phases and structures can form. A typical nonionic amphiphile, oligoethylene oxide (10) oleyl ether [$(EO)\overline{_{10}}$ oleyl], forms micelles, micellar rods, and hexagonal, cubic, and lamellar liquid crystals as the amphiphile:water ratio increases. This rich self-assembling behaviour occurs because the polar segment is readily solubilised by water, and the nonpolar tail is not. For a single molecule in solution, other than curling up on its self (which is entropically very unfavourable), there is no physical way to reduce the unfavourable interactions, but if multiple molecules are allowed to associate or self-assemble, the unfavourable interactions can be reduced. Determination of the structure of this minimum-energy aggregate is beyond the scope of this chapter but essentially one can calculate the free energies of each possible phase for a particular concentration and temperature and then select the lowest-energy phase.

Phases Observed in Mixtures of Water and Amphiphile

The three most common phases observed in mixtures of water and amphiphile are hexagonal, lamellar and cubic. Correctly stated, the cubic phase encompasses a wide range of potential phases, all exhibiting cubic symmetry but with differing degrees of continuity in the hydrophilic or hydrophobic phases. Similar to the structures observed for block copolymers, the structures observed in lyotropic liquid crystals can be rationalised as a function of the volume fraction of the various components. Figure 28.8 shows representative phase diagrams for two nonionic amphiphiles; the first liquid crystalline phase observed as the amphiphile concentration is increased is the hexagonal phase, which is in essence the hexagonal close packing of rod-like micelles (Fig. 28.9).

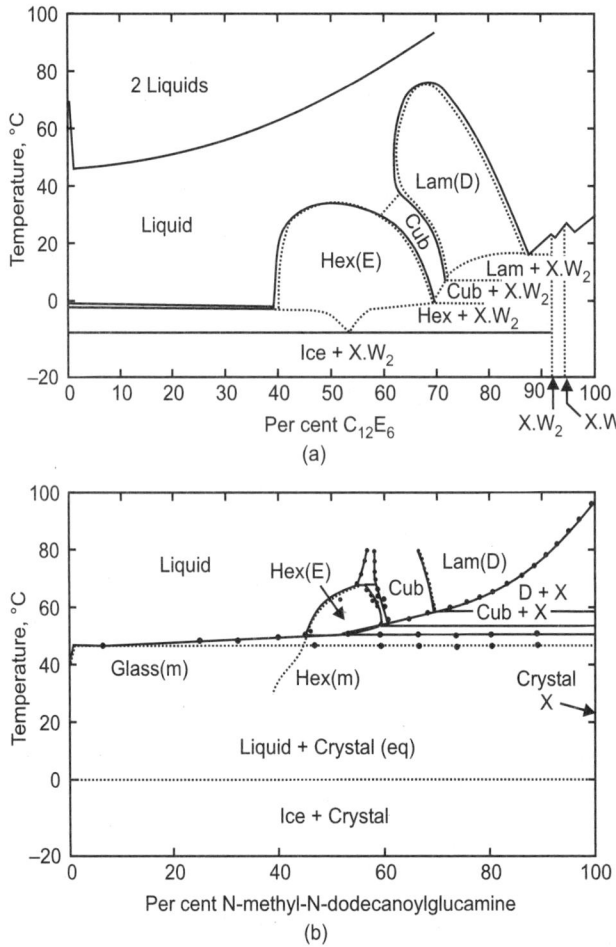

Fig. 28.8. Phase diagrams for nonionic amphiphile/water systems. (a) $C_{12}E_6$ (C_{12} represents the number of carbons in the tail, and E_6 represents the number of ethylene oxide groups in the head), (b) N-Methyl-N-dodecanoylglucamine.

These rod-like micelles develop in solution as the concentration of the amphiphile increases, but below a critical concentration do not pack closely. As the concentration of amphiphile is increased further, the spacing between the rod centres decreases, and eventually a critical point is reached where a bicontinuous cubic phase is formed (Fig. 28.10). Finally, at high enough concentration of amphiphile, a lamellar phase forms (Fig. 28.11). In some systems, especially those composed of nonionic triblock amphiphiles, a close-packed cubic phase is observed, usually at lower concentrations of amphiphile than the hexagonal phase (Fig 28.12).

Rather than relying on coassembly, which is the process by which nanostructures develop in many mesoporous oxide systems, many biological processes utilise the order present in a preformed structure to form nanostructured inorganics. Substantial efforts have been made to utilise the order present in an organic mesophase to directly template the growth of an inorganic phase; liquid crystal templating has been one of the most successful of these approaches.

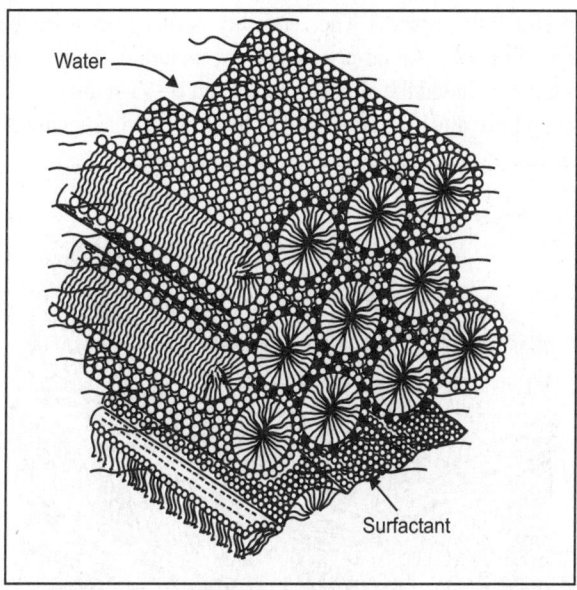

Fig. 28.9. Schematic illustration of the hexagonal phase. The polar head group of the surfactant points out into the water phase, and the hydrophobic tail inserts into the centre of the rod-like micelles.

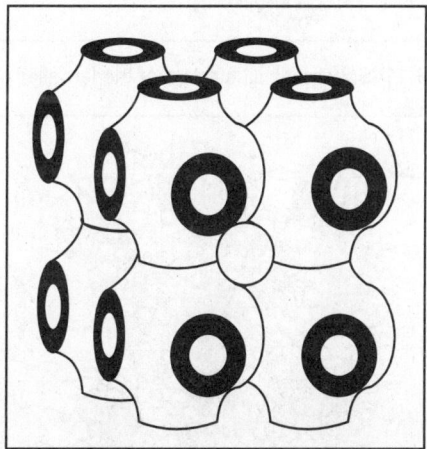

Fig. 28.10. Schematic illustration of a bicontinuous phase.

Early efforts in this area resulted only in oblong or cubic crystallites or microporous reticulated structures. More recently, through liquid-crystal templating, the successful synthesis of periodically nanostructured semiconductors that copied directly the symmetry and dimensionality of the precursor liquid crystal was demonstrated. Liquid-crystal templating appears to be a general route to the synthesis of semiconductor nanostructures. The general concept of liquid-crystal templating is to first form a liquid crystal that contains at least one of the precursors of the mineral phase, and then to induce a mineral phase to precipitate in only one chemical region of the liquid crystal by applying an outside perturbation. Obviously, it is very important to select the correct synthetic conditions, mesophase, and mineral phase to be successful in this process. The versatility of this process has been demonstrated by the fact that materials have been formed in liquid-crystal phases, including the hexagonal, lamellar, and cubic phases, and the materials have included the already mentioned II–VI semiconductors; periodically nano-structured metals, both as thin films and in bulk; as well as films of the chacogenides selenium and tellurium.

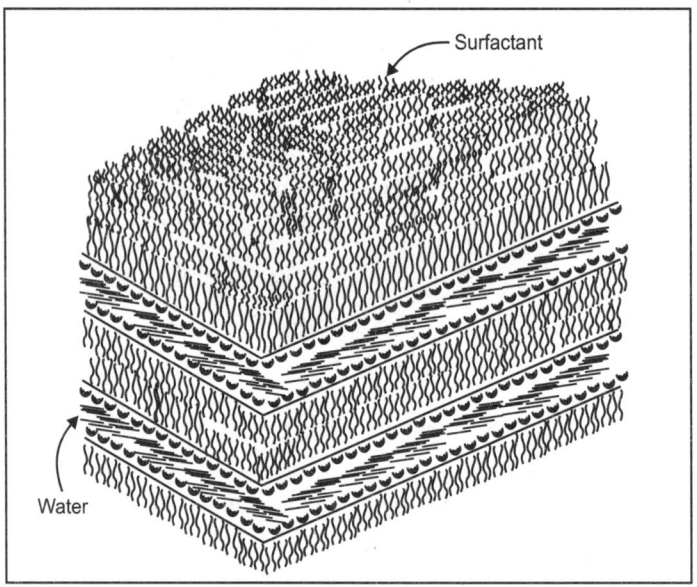

Fig. 28.11. Schematic illustration of the lamellar phase.

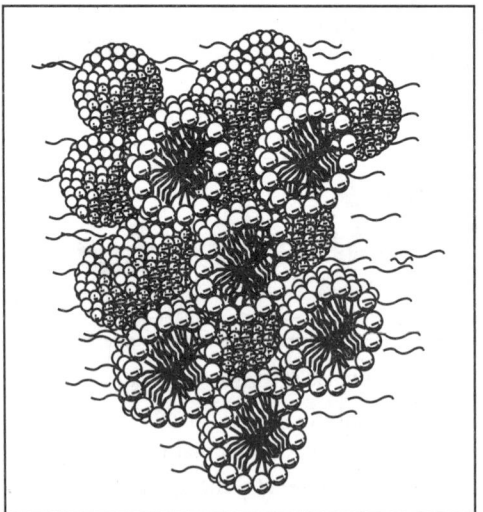

Fig. 28.12. Schematic illustration of the simple cubic phase.

Not only are the characteristic dimensions of the materials synthesised by liquid-crystal templating smaller than those obtainable by lithographic techniques, but they are often attainable through bulk synthesis, which is obviously not possible via lithography.

Direct Templating of Materials

The direct templating of materials in the preordered environment of a nonionic amphiphilic mesophase generates semiconductor/organic superlattices containing both the symmetry and the long-range order of the precursor liquid crystal. In the templated growth of II-VI semiconductors, the semiconductor is grown in a water-containing liquid crystal by reaction of H_2S or H_2Se with a dissolved salt such as $Cd(NO_3)_2$. Both the chemical nature and the structure of the amphiphile are important for direct templating. For example, the order obtained in the nanostructured systems was even observed to be dependent on the counterion for the metal.

One advantage of direct templating with liquid crystals as a route to nanostructure formation, as was already discussed, is that there are a large number of amphiphilic liquid crystals, with lattice constants ranging from a few nanometers to tens of nanometers, and which include lamellar, hexagonal, cubic, and bicontinuous phases. Potentially, many of these systems can be mineralised, generating materials with an array of novel structures and properties.

Specifically, II–VI semiconductors have been directly templated by hexagonal liquid crystals based on $(EO)_{\overline{10}}$ oleyl, and a lamellar liquid crystal formed from oligo(vinyl alcohol) oleyl ester. As expected, the hexagonal liquid crystal yielded product with a hexagonal nanostructure, and the lamellar liquid crystal yielded a lamellar structure. $(EO)_{\overline{10}}$ oleyl can also form a lamellar liquid crystal, which also was successful in templating a lamellar CdS product. Other lyotropic phases, such as a variety of bicontinuous and cubic liquid crystals, may also yield interesting mineral nanostructures. In this regard, CdS has been grown in a body-centred cubic phase by using a triblock copolymer of poly(oxyethylene)-poly(oxypropylene)-poly(oxyethylene) $[(EO)_{\overline{160}} (PO)_{\overline{70}} (EO)_{160}]$ as the amphiphile. When mixed with water, the PO segment is only weakly solvated, but the EO is highly solvated. Thus, when this molecule is hydrated it forms micelles which pack closely, forming a cubic phase. The result of precipitation in this cubic phase was the formation of hollow nanospheres of CdS, which are covered later in this chapter.

It is remarkable that a soft organic liquid crystal can directly template a hard covalent mineral phase. As a specific example, a hexagonal mesophase consisting of 50 vol.% aqueous 0.1 M $Cd(OAc)_2$ and 50 vol.% $(EO)_{\overline{10}}$ oleyl templated an inorganic/organic nanocomposite of CdS and amphiphile when exposed to H_2S gas (Fig. 28.13). The composite material contained an internal nanostructure that replicated the symmetry and dimensions of the liquid crystal in which it was grown. Interestingly, CdS and ZnS exhibited a superlattice morphology when formed in a liquid crystal from their respective nitrate salts and H_2S, but Ag_2S, CuS, HgS, and PbS did not (Fig. 28.14). CdSe is also nanostructured by this method (Fig 28.15). An additional virtue of the $(EO)_{\overline{10}}$ oleyl system is that, when combined with H_2O, it forms a hexagonal mesophase at 25°C over the range of ~35 to ~65 vol.% amphiphile and a lamellar mesophase from ~70 to ~85 vol.% amphiphile, and it can directly template the growth of mineral over almost this entire range.

As shown in Fig. 28.14, the nanostructures of the semiconductors CdS and ZnS synthesised by precipitation in hexagonal mesophases doped with their respective nitrate salts have hexagonal symmetry with a periodicity and dimensionality commensurate with that of the template. The hexagonal nanostructure is not always evident in TEM micrographs, due to random orientation of particles in the field of view.

Presumably, if properly oriented, all the particles would show hexagonal symmetry. This was partially observed by tilting the samples on the TEM stage, revealing many more particles with hexagonal nanostructures.

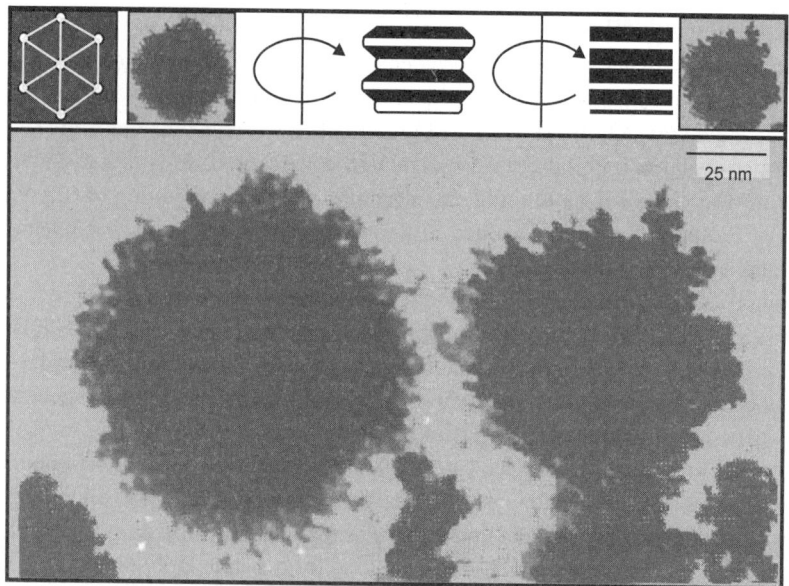

Fig. 28.13. The structure on the left in the transmission electron micrograph (bottom) is nanostructured CdS viewed with the hexagonally packed cylindrical templated pores parallel to the electron beam; the one on the right has the cylindrical structures perpendicular to the electron beam. Top: model of the nanostructure as a hexagonal arrangement of cylindrical pores of low electron density, corresponding to organic material in a solid matrix of semiconductor. The left and right schematic representations correspond to the adjacent micrographs; the central view shows the cylindrical assemblies in an intermediate state of rotation.

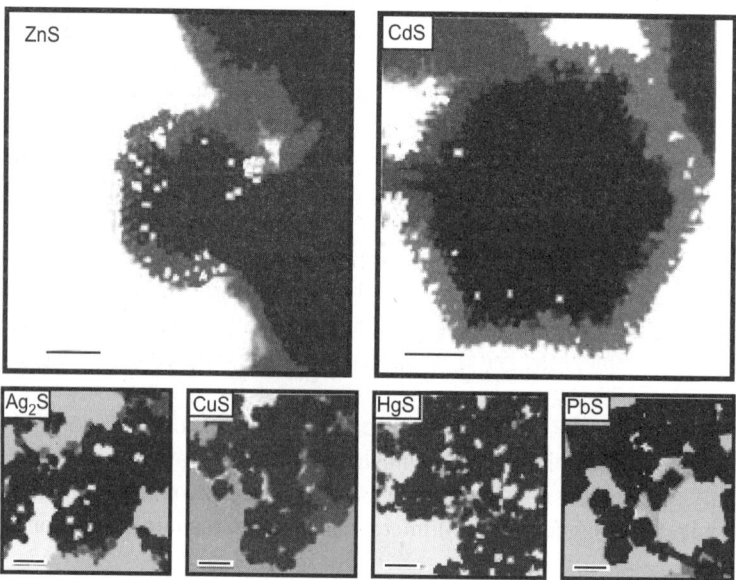

Fig. 28.14. Transmission electron micrographs of mineralised structures grown in hexagonal mesophases from their respective nitrate salt and H_2S.

(a)   (b)

Fig. 28.15. Transmission electron micrograph of CdSe mineralised in hexagonal mesophases doped with (a) cadmium nitrate and (b) cadmium acetate.

CdS is also templated when grown from its acetate salt, although the order of its nanostructures is not nearly as well defined as in the nitrate systems (Fig. 28.16). When ZnS is generated from its acetate salt, only spherical polycrystalline particles with a porous appearance are formed. Another difference between the product obtained from the acetate and nitrate salts is the average particle diameters of the semiconductor product: both CdS and ZnS grown from their respective nitrate salts are approximately five times larger than when grown from their acetate salts.

$$Cd(OAc)_2 + H_2S \rightleftharpoons CdS + 2HOAc$$

$$Cd(NO_3)_2 + H_2S \rightleftharpoons CdS + 2HNO_3$$

Fig. 28.16. Transmission electron micrographs of CdS grown in identical hexagonal mesophases except for precursor salt. Note the significantly improved order when cadmium nitrate is used as the precursor over the order obtained with cadmium acetate.

This size difference can be clearly observed in low magnification electron micrographs. As controls, CdS was also grown in aqueous environments from both the nitrate and acetate precursors, and as expected, no nanostructure was generated. The counterion of the metal did not affect the templating of the other mineral systems studied: Ag_2S, CuS, HgS, and PbS were not generated with a superlattice morphology when grown in a hexagonal mesophase, irrespective of whether the acetate or nitrate salts were used. The reason may be that the by-product of the synthesis from the nitrate salt is nitric acid, whereas the by-product of the synthesis from the acetate salt is acetic acid. Nitric acid is a much stronger acid, and apparently enables the mineral phase to reform around the template during growth so as to remove any structural defects.

As already mentioned, in addition to the hexagonal mesophase, a lamellar mesophase of $(EO)_{\overline{10}}$ oleyl also can template a precipitated mineral (Fig. 28.17). The lamellar periodicity in the resulting CdS is ~7 nm, which agrees very well with the periodicity of the lamellar template. The lamellar morphology can be confirmed by careful tilting of particles within the TEM. If a particle is tilted about an axis perpendicular to the stripes, no change is observed in the pattern. However, if it is tilted on an axis parallel to the stripes (and perpendicular to the electron beam), the pattern quickly disappears.

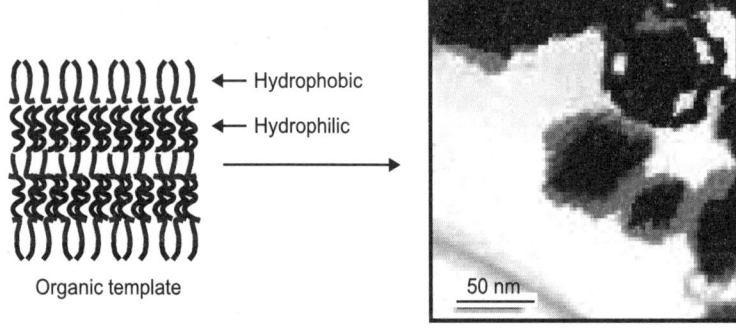

Organic template — Hydrophobic — Hydrophilic

Mineralised product

Fig. 28.17. Schematic representation of the lamellar organic template and transmission electron micrograph of the resulting product after mineralisation. The dark bands are CdS that mineralised in the hydrophilic region of the template liquid crystal; mineral growth is prohibited in the hydrophobic regions.

These observations constitute strong evidence for a lamellar morphology within the particles and agree closely with results obtained from lamellar nanocomposites formed in a poly(vinyl alcohol)-based liquid crystal. As shown in Fig. 28.17, the mineralised product consists of disk-like particles with a long axis in the plane of the layers that is ~1.5 times the maximum width perpendicular to the layers. Very interestingly, particles having lamellar morphology do not disperse, even with repeated ultrasonication, perhaps due to mineral or organic bridging between the CdS layers. Presumably, CdS nucleates within a hydrophilic layer of the mesophase and grows rapidly in the plane, but concurrently an occasional finger forms perpendicular to this layer, piercing the hydrophobic region. This finger then nucleates another layer of CdS, resulting in a mineral bridge between layers. Another possibility is tethering of organic molecules in the mineral phase, tying the layers together. Of course, growth in the plane of the layers is faster than growth perpendicular to the layers, generating the disk-like habit observed. Not surprisingly, since it seems to enhance the templating effect, lamellar-nanostructured CdS was seen only when the nitrate salt was used, while the acetate salt afforded only small particles.

The layered structure formed by templating with a lamellar liquid crystal is in fact reminiscent of the layered abalone shell structure; however, in the templated material the characteristic dimension is just a few nanometers, not hundreds of nanometers to micrometers. It still remains to be seen if the properties of the layered material are significantly improved over those of the solid equivalent, but if the abalone shell is any guide, advanced mechanical properties are a possibility.

Mineral Growth in Hexagonal and Lamellar Phases

As described, mineral growth in the hexagonal and lamellar phases yields interesting, controlled nanostructures. Mineral growth in the cubic phase formed from $(EO)_{106} (PO)_{70} (EO)_{106}$ was also done, using cadmium acetate and H_2S as precursors. Although the mechanism is not entirely understood, the result was hollow spheres 20–200 nm in diameter (Fig. 28.18), which can be observed by both TEM and SEM. When the sample is tilted in the stage, the shape and observed structure do not change, as expected for a hollow sphere (Fig. 28.19). If the particles had been corpuscular in shape, their appearance would change as a function of sample tilt. The strongest evidence for their hollow nature is the dark edges of the spheres observed in TEM micrographs. If each sphere were solid. TEM would show greater scattering from the centre than the edges, making the centre appear darker. In the SEM, the spheres also scatter the most electrons from their edges and appear somewhat transparent in their centres, providing further proof of their hollow nature (Fig. 28.18). Unlike materials produced by direct templating in hexagonal and lamellar liquid crystals, the hollow spheres are not of a size commensurate with the structure of the liquid crystal, but are rather 1–10 times the size of the characteristic dimension of the liquid crystal in which they were formed.

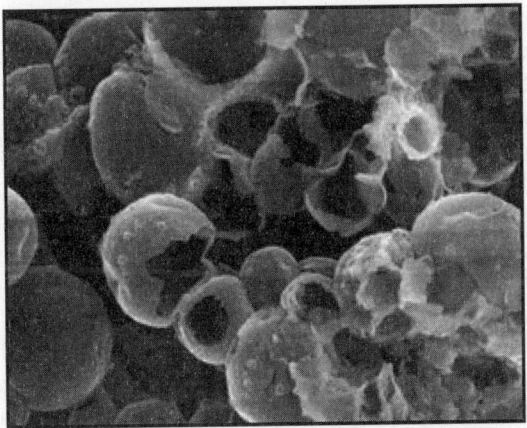

Fig. 28.18. Scanning electron micrograph of hollow spheres of CdS mineralised within a cubic mesophase.

Because the characteristic dimension (diameter) for hollow CdS spheres obtained from the $Cd(CH_3CO_2)_2$-doped cubic phase is 20–200 nm, (1–10 times the diameter of the micelles making up the cubic phase), it does not appear that the mineral nanostructure is directly templated by the liquid crystal. In addition, this nanostructure is entirely absent from the CdS when the cubic phase is doped with $Cd(NO_3)_2$ as the precursor salt. This result is important when taken in the context of previous results in which the use of the nitrate salt led to a nanostructure with enhanced order. In essence, the nitrate salt allowed the growing mineral to access a thermodynamically lower-energy morphology, which was a nanostructure commensurate with the structure of the liquid crystalline matrix. The inability of the nitrate salt to 'sharpen' the order in the cubic system is not surprising, because, as observed in previous studies, the nitrate salt only improves the registry between the nanostructure and the liquid crystal and does not result in a new nanostructure. In the cubic system, the spheres are not a copy of the liquid crystal, so there is no registry to improve. The mineralisation of the cubic phase must lead to local rearrangements of the liquid crystal, leading to the hollow-sphere morphology observed; however, the detailed mechanism is still not understood.

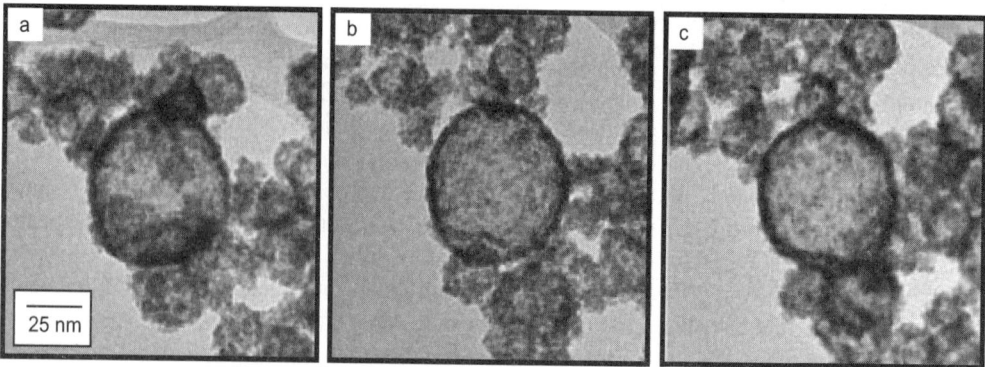

Fig. 28.19. Transmission electron microscope tilt series of hollow CdS spheres grown in an acetate-doped cubic phase. The tilt axis was diagonal running from upper left to lower right of each micrograph. (a) −45°, (b) 0°, (c) 45°.

Cubic Phase and Other Phases

A potentially significant difference between the cubic phase and the ether phases studied (hexagonal and lamellar) is the connectivity of the hydrophilic and hydrophobic portions of the liquid crystal. In the lamellar and hexagonal phases, both these regions are continuous in at least one direction. In the cubic phase, however, the hydrophobic regions are confined to discrete micelles. This confinement may cause a difference in templating ability relative to the other phases. For all templating phases, as the mineral nucleates and grows, it is necessary to expel some molecules from the volume occupied by the mineral. For both the lamellar and hexagonal phases, a molecule can diffuse away from the growing mineral without ever exposing its hydrophobic (oleyl) or hydrophilic (EO) segments to domains of the opposite nature. In contrast, in the cubic phase, when a molecule is forced away from the growing mineral it must leave its micelle and expose its hydrophobic (PO) segment to the polar surroundings (EO + H$_2$O), a high-energy situation. As already stated, the nitrate salt allows the mineral phase to access a lower-energy configuration, which in the lamellar and hexagonal systems results in a high degree of fidelity between the template and the semiconductor nanostructure. That a hollow-sphere morphology is not observed in the cubic system when the nitrate salt is used is not surprising, given that the hollow morphology is not directly templated by the liquid crystal and that formation of the spheres requires the mesophase to go through a high-energy intermediate state. The full reason for the spherical morphology when Cd(CH$_3$CO$_2$)$_2$ is used as the semiconductor precursor must be due to a subtle kinetic balance that is not yet understood. At the very least, the energy difference between the lamellar, hexagonal, and cubic systems is rather small, and thus the fact that templating is successful in the first two systems but unsuccessful in the last indicates that some fairly specific interactions are necessary for direct templating. This is similar to the action of many biological systems, in which very specific interactions between proteins and other macromolecules and growing inorganic phases are exceedingly important for structural development.

In addition to simple one-component systems, it is interesting to consider the result of templating binary mixtures of the precursor salts within a liquid crystal. In one system, a hexagonal mesophase containing 0.05 M Cd(NO$_3$)$_2$ and 0.05 M Zn(NO$_3$)$_2$ was mineralised. The resulting semiconductor (Cd$_x$Zn$_{1-x}$S; $x \sim 0.5$) product's nanostructure exactly matched that of the template (Fig. 28.20). In a system of 0.05 M Pb(NO$_3$)$_2$ and 0.05 M Zn(NO$_3$)$_2$, the result was a very different nanostructure, which consisted of a single-crystal core of PbS surrounded by a shell of nanostructured ZnS (Fig. 28.21).

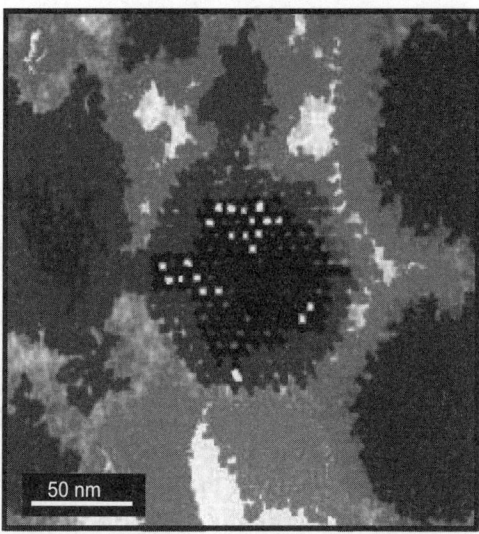

Fig. 28.20. Transmission electron micrograph of the product obtained from precipitation in a hexagonal mesophase doped with 0.05 M $Cd(NO_3)_2$ and 0.05 M $Zn(NO_3)_2$. The templated product is a solid solution of CdS and ZnS.

Fig. 28.21. Transmission electron micrograph of a composite product of PbS and ZnS grown in a hexagonal mesophase doped with 0.05 M $Pb(NO_3)_2$ and 0.05 M $Zn(NO_3)_2$. The single crystal cube at the core of the particle is PbS, and the shell is a periodically nanostructured solid consisting of ZnS.

From this it appears that, when mixed metal products are synthesised, there is a direct correspondence between the behaviour of the constituent solids and the mixed solid. $Cd_xZn_{1-x}S$ was nanostructured (as were CdS and ZnS, Fig. 28.14), but precipitation from Pb^{2+} and Zn^{2+} ions resulted in a single crystal of PbS surrounded by nanostructured polycrystalline ZnS, corresponding with the result seen for PbS and ZnS when grown discretely (Fig. 28.14). The formation of mixed metal precipitates gives an insight into the growth processes and also opens the possibility of engineering a property. The system of $Cd_xZn_{1-x}S$ offers the possibility of band-gap engineering, although, due to the somewhat lower solubility of CdS than ZnS in water, the particles may be cadmium-rich in their centre and zinc-rich on their exterior. The differing solubilities of PbS and ZnS in water play a very important role in the structure of the particles formed from this mixed system. PbS has a much lower solubility than ZnS and, as a result, upon exposure of the doped mesophase to H_2S, single-crystal PbS cubes nucleate and grow. Then, after most of the Pb^{2+} ions are locally exhausted, ZnS heterogeneously nucleates on the PbS particles. As expected, because ZnS is templated by liquid crystals, the shell of ZnS around the PbS single crystal contains a periodic superlattice structure.

Direct Templating of an Inorganic by an Organic Liquid Crystal

Direct templating of an inorganic by an organic liquid crystal may depend on many factors, the most important of which is probably the thermodynamic stability of the mesophase throughout the mineral growth process. The mesophase must be stable to the addition of mineral precursors, and the mineral precipitation process must not disrupt the order of the liquid crystal. In studying direct templating, researchers have observed that the textures observed by polarised optical microscopy are the same for the pure mesophase and for a mesophase that contains the precursor salt, indicating that the doping did not lead to radical disruption of the order in the mesophase. Nuclear magnetic resonance (NMR) can also be utilised to verify the structure of liquid crystalline mesophases. To verify that the characteristic molecular order of the mesophase was not disrupted by ion doping, broadline ^2H NMR spectra were obtained from both cadmium ion-doped and undoped mesophases. For both mesophases, the same quadrupole splitting was observed (Fig. 28.22). If ionic doping had perturbed the structure of the mesophase, the splitting would have decreased. As additional proof of molecular order in the mesophase, X-ray diffractograms were collected to characterise both the mesophase's long period and symmetry. For systems containing 35, 40, 50, and 60 per cent amphiphile, the 100, 110, and 200 reflections are clearly observed, indicating that the liquid crystalline structure is hexagonal (Fig. 28.23). A mesophase containing 78 per cent amphiphile forms, a lamellar liquid crystal, as indicated by the presence of 100 and 200 reflections and the absence of a 110 reflection (Fig. 28.24). As expected, a strong correlation was found between the phase diagrams as determined by optical analyses and the X-ray data. Similar experiments have been performed for other liquid-crystal templating systems, as well as for the formation of mesoporous silica.

In addition to the metal sulphides, which can be successfully templated as already mentioned, several sulphide materials, including Ag_2S, CuS, HgS, and PbS, were not templated by the liquid crystal in which they were grown, irrespective of the counterion. Design of amphiphiles with proper structures and binding constants for both the inorganic precursors and the inorganic product may enable a wide variety of inorganic and organic compounds to be templated in the future.

To better understand the scientific underpinnings of liquid-crystal templating, it is instructive to consider a few additional experiments. For example, liquid-crystal templating at elevated temperature sheds some light on the mesophase-ion-product interactions. As already mentioned, it appears the periodically nanostructured materials are thermodynamically stable with respect to their solid equivalent. This implies that there is a critical energy balance between the energy gained by reducing the surface area of the mineral phase and the energy lost due to disruption of the mesophase structure. To study this further, CdS was precipitated in mesophases at both 35°C and 50°C. These temperatures are both below the isotropisation temperature of the doped mesophases, so the reactions were carried out in a self-assembled medium. The CdS produced from the reaction at 35°C did express the order of the mesophase, albeit poorly when compared with the order obtained at room temperature (22°C). Precipitation at 50°C resulted in mineral with no periodic order, clearly indicating that the energetic difference between periodically nanostructured and disordered product is small and that it only takes a small perturbation to result in nonnanostructured product.

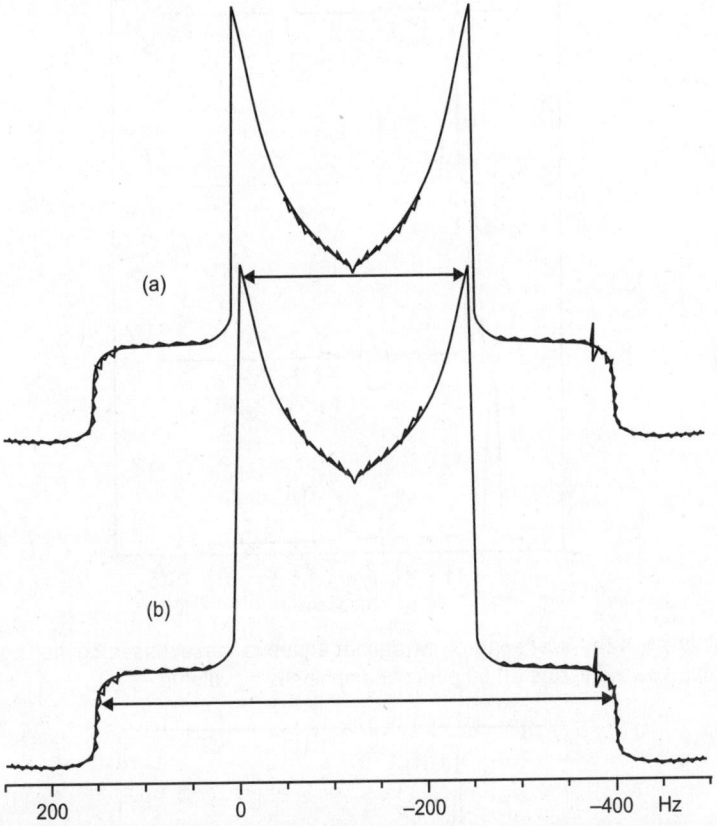

Fig. 28.22. Quadrupole splitting for a hexagonal mesophase (a) doped with 0.1 M cadmium acetate, and (b) undoped. The splitting indicated by an arrow is the same for both samples.

As additional evidence for the direct templating mechanism, samples composed of 35, 40, 50, and 60 per cent by weight $(EO)_{\overline{10}}$ oleyl were mineralised with CdS. By varying the amphiphile content of the mesophase, the spacing between the cylindrical aggregates of amphiphilic molecules making up the hexagonal mesophase was varied. Assuming that CdS is directly templated by the liquid crystal in which it is grown, the hexagonal symmetry and associated length scale should be nearly identical to that found in the precursor hexagonal mesophase, which is what indeed happened. The result (Fig. 28.25), demonstrating that the superlattice dimension in the precipitate can be varied by changing the lattice constant of the mesophase, is very strong evidence for direct templating. If the nanostructure present in the templated inorganic phase had been the result of a cooperative self-assembly process, much as occurs for most mesoporous silica systems, the result would have been that the periodicity of the nanostructure would not have changed with varying water content.

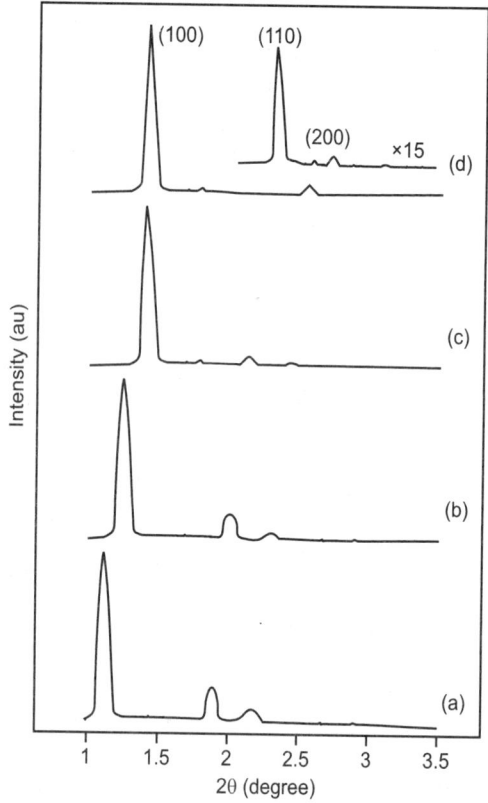

Fig. 28.23. Small-angle X-ray scattering of hexagonal aqueous mesophases containing (a) 35 per cent, (b) 40 per cent, (c) 50 per cent, and (d) 60 per cent amphiphile by volume.

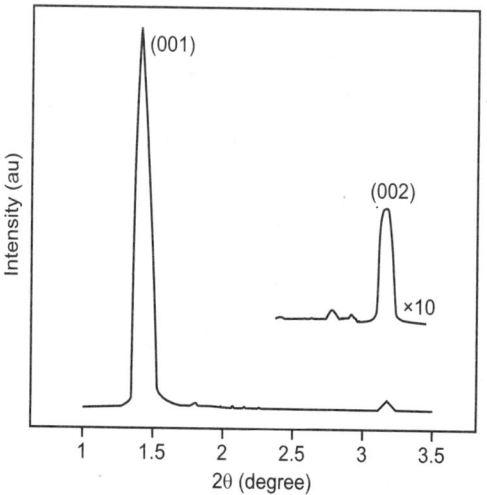

Fig. 28.24. Small-angle X-ray scattering of a lamellar mesophase containing 78 per cent amphiphile and 22 per cent water by volume.

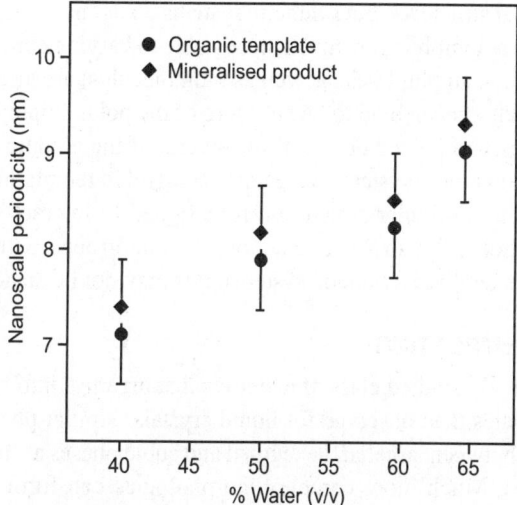

Fig. 28.25. Center-to-center spacing in a cylindrical assembly of amphiphilic molecules of hexagonal mesophases as determined by X-ray diffraction (●), and center-to-center pore spacing in templated CdS as measured by transmission electron microscopy (◆). Both are plotted as a function of the water content in the mesophase.

LIQUID-CRYSTAL TEMPLATING OF THIN FILMS

The templating of thin films by self-organised organic structures should find significant application in both technology and scientific study. Thin-film templating is structurally related to the bulk templating of inorganic materials, which generally results in periodically structured particles; however, because the result is a periodically structured thin film, the potential for application is clearer. Templating with organic structures is especially intriguing because of the potential to create features much smaller than those feasible by almost any top-down technique, because it utilises the nanoscale molecular order inherent in self-assembled organic structures to define the structure of the resulting thin film. A few key points must hold true for successful thin-film templating by liquid crystals. First, clearly, for templating to be successful, the self-assembled matrix must be compatible with the substrate. Then, via some process, the inorganic material must be deposited or grown on the substrate. Realistically, only chemical and electrochemical routes for materials deposition can meet these requirements. Other conventional methods of thin-film deposition require high vacuum, which is incompatible with lyotropic liquid crystals and furthermore cannot operate through a thick overlying layer of liquid crystal.

The synthetic routes for liquid-crystal templating of thin films are relatively straightforward. To date, most studies have used electrochemical techniques to drive the material deposition. First, a precursor containing lyotropic liquid crystal is interlaced with the substrate. Then, under an applied potential, material is electrochemically deposited at the liquid-crystal/substrate interface. Nanostructured materials that have been created through this process include a variety of metals, selenium, and tellurium. It may also be possible to electrodeposit other interesting materials including semiconductors; however, no publications have yet appeared demonstrating success.

Throughout, it is quite interesting that all the successful templating experiments have relied on liquid crystals formed of nonionic amphiphiles. However, in biology, most preformed, matrices are formed via ionic macromolecules. In part, this is likely because biology makes use of very specific interactions to create mineral structures, but synthetic systems do not have this degree of sophistication. Thus, the fact that nonionic amphiphiles are much more stable to varying concentrations of soluble salts is actually an advantage. Ionic amphiphiles (as well as biomolecules) are affected much more strongly by salts, because a single salt ion can bind to one or more of the polar amphiphile head groups, greatly reducing their polarity. This was indeed observed for several of the anionic amphiphiles studied, and most likely was the reason that ionic systems were not successful in templating the growth of a mineral phase. In biology, where the molecular structures are designed to interact specifically with one salt under very specific conditions, the strong interaction of ionic groups with dissolved species is an advantage. However, for generalised synthetic systems, this may not be an advantage.

BLOCK COPOLYMER TEMPLATING

Block copolymers are a widely studied class of materials that organise into both 2D and 3D structures at slightly longer length scales than observed for liquid crystals. Similar phase behaviour is observed, with systems transitioning between lamellar, hexagonal and cubic phases as the relative volume fraction of the two blocks changes. Much more complex morphologies can form in triblock systems. The characteristic periodicity in block copolymer systems ranges from ~5 nm to hundreds of nanometers and primarily depends on the molecular weight of the block copolymer. As expected, as the molecular weight of the block copolymer increases, the characteristic length scale increases. It was realised that if there was a way to impart this nanoscale order into a substrate, one might have a powerful technique for patterning materials with a periodic array of nanometer-sized structures. Unlike lyotropic liquid crystals block copolymers are generally solvent-free and can be taken to elevated temperatures and under vacuum without destabilising the self-assembled structure; thus high-vacuum material deposition and processing approaches can be used. Usually the chemistry of the block copolymer is designed so that one of the two blocks can be removed via a dry etch with ozone or other reactive compound to generate the porous structure, which will subsequently serve as a template for nanostructure formation.

Procedure for Block Copolymer

The procedure for block copolymer templating of nanostructures usually is as follows. A thin film of some block copolymer is spun-coated from solvent onto a substrate and allowed to self-assemble. After this, one of the blocks of the polymer is removed by ozone etching. The result of this etching procedure is a substrate coated with a thin polymer film containing a periodic nanoscale void structure. After removal of any solid polymer film that overlies the void structure, the polymer film is used as a mask. Material can be evaporated through the polymer film onto the substrate, material can be electrochemically grown from the substrate up through the polymer film, or the polymer film can be used as an etch mask. In all cases, the result is material structured to be a replica of either the polymer film (when the polymer is used as an etch mask) or the pore structure of the polymer film (when material is deposited in the pores). Usually, at the end, the polymer film is removed with solvent or reactive ion etch, leaving behind nanostructured templated features on a substrate. With this approach it is possible to create features as small as ~20 nm holes or dots in a periodic array on a substrate of a wide range of materials, including oxides, semiconductors, magnetic materials, and of course the polymer itself (Fig. 28.26). The power of block copolymer templating is further enhanced due to the facts that both the size and spacing of the feature can be modulated simply by varying the molecular weight and composition of the polymer and that the lattice structure can be modulated by varying the relative length of the two blocks. For example, both the lamellar and hexagonal phases can template nanometer-scale lines if they are oriented properly on the substrate. The periodic arrays of dots have attracted the greatest attention for applications such as magnetic storage media.

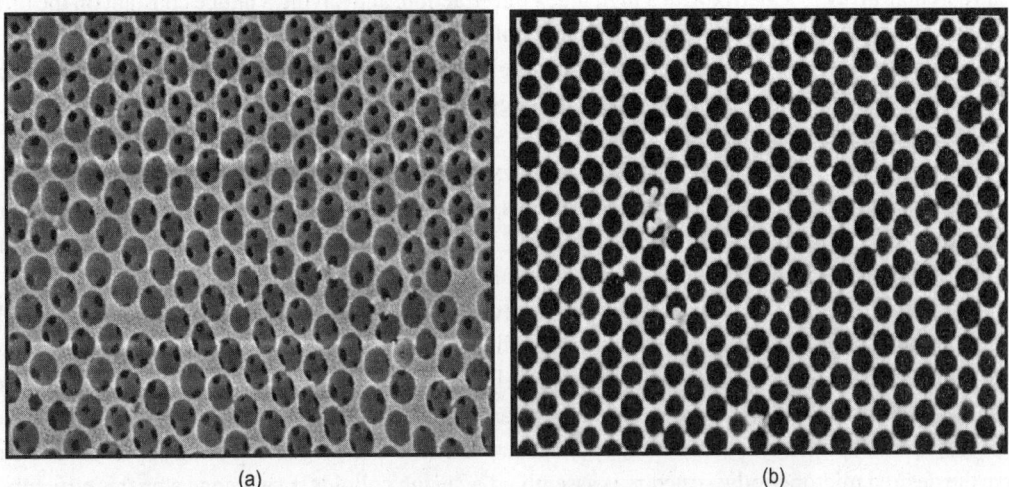

(a)   (b)

Fig. 28.26. (a) Scanning electron micrograph of a polystyrene/polyisoprene block copolymer template after it has been partially etched with ozone and the continuous polystyrene layer on the top has been removed. The now empty PI domains are now holes and are darker in the micrograph. (b) Scanning electron micrograph of hexagonally ordered arrays of holes in silicon nitride on a thick silicon wafer. This pattern is formed by using as a template a copolymer film such shown in (a). The darker regions are ~20 nm-deep holes in the silicon nitride layer, which are formed by etching through the overlying template.

COLLOIDAL TEMPLATING

In any discussion of biologically inspired nanocomposite materials, one must include recent developments on colloidal crystal templating of photonic materials. The basic premise behind this approach is to use the 3D periodic structure of synthetic opals to direct the structure of a second phase material. This approach is not biologically inspired, but should more accurately be described as 'naturally inspired', because opals, although natural, are geological, not biological, in origin. Furthermore, although the lower limit to the characteristic length scale of the material generated is 10–20 nm, the characteristic dimension is often relatively large, on the order of 500 nm. This is in fact intentional, since most of the applications for these materials are optical, and thus the characteristic length should not be much smaller than the wavelength of the light that one desires to modulate. Commonly, however, much smaller features are embedded within the templated structure, and these features may be as small as a few nanometers.

Microperiodic Structures and Photonics

The interest in microperiodic 3D structures has grown tremendously due to the exciting potential of such materials, particularly in the area of photonics. Such 3D structures, often termed photonic crystals, are the extension of the well-known dielectric stack into three dimensions. Although the colours that occur in opals, which stem from diffraction of white light by planes of highly ordered submicrometer silica spheres, are our inspiration; for practical application, synthetic approaches are needed to create materials and structures with the necessary refractive index and periodicity to meet the requirements for most optical applications, which opals simply do not have.

A particularly interesting class of optical structures are the so-called photonic band-gap materials. For example, a microperiodic material consisting of low-refractive-index spheres arranged in a face-centred-cubic array in a matrix with a high index of refraction, and having a lattice constant on the order of the wavelength of light (visible or infrared), could be such a photonic band-gap material. Similar to how a dielectric stack has a stop-band for light in a given frequency range, this material would not allow light in a given frequency range to travel through it in any direction. In essence, it would be an omnidirectional, perfectly lossless, mirror.

The synthesis of these structures however is exceedingly difficult. Layer-by-layer fabrication of photonic crystals using state-of-the-art VLSI tools, e.g. deep UV photolithography; chemical vapour deposition (CVD), chemical-mechanical polishing, has been demonstrated, but formidable processing difficulties limit the formation of large area and truly 3D structures.

When appropriately formed, self-assembled colloidal crystals are natural candidates for the construction of photonic crystals. Good crystal quality is achieved only with colloids that have very low size polydispersity (<5 per cent), which currently limits the choice or materials to SiO_2 or polymers, both of which have a fairly low index of refraction around 1.5, which is much smaller than that required for most optical applications. This has led researchers to take a two-stage templating approach. In a first step the desired microperiodic structure is assembled by using colloids. In a second step this structure is used as a template to build a complementary structure with a material having a higher index of refraction.

Colloidal Crystal Templating

Colloidal crystal templating is a very promising approach for production of high-resolution, micrometer-scale, 3D periodic photonic crystals, but as conventionally applied to the fabrication of photonic band-gap structures it has serious optical limitations, unless materials with the necessary optical properties can be used. Typically, the most important point is to infill with a material with a sufficiently high refractive index to generate an optically interesting material. A range of approaches have been suggested to maximise the index contrast, including sol-gel, chemical vapour deposition, imbibing of nanoparticles, reduction of GeO_2 to Ge, electroless and electrochemical deposition, and melt imbibing. In addition, polymers have been used to infill colloidal crystals, and in one report, the colloidal particles were less than 100 nm in diameter, which, although perhaps not interesting from an optical standpoint, may have potential for separation membranes and confined chemical reactor spaces. Although they are not the focus of the following discussion, the colloidal templates used in these attempts are commonly polycrystalline and can contain unacceptably high numbers of defects; thus, substantial effort has also gone into creating colloidal crystals with low defect densities. Each of these infilling techniques has various advantages and disadvantages, which are discussed below. In general, they all consist of approaches to infill the interstitial space of the colloidal template, after which the colloidal template is generally removed.

Sol-Gel Infilling Colloidal Crystal Templates

The sol-gel infilling of colloidal crystal templates is intriguing to consider as a route to 3D porous materials, although it is somewhat limited in application for photonic materials for several reasons. First, the refractive index of most materials that can be formed via sol-gel is <2 (with the exception of TiO_2, which can have a refractive index of ~2.5), second, there is considerable reduction in volume during the conversion of the sol to solid material, and third, the refractive index of most sol-gel-derived material is substantially less than that of a single crystal of the same material. The net effect is that most sol-gel-derived macroporous materials have relatively low refractive index contrast, and their long-range order is somewhat disrupted due to the uneven contraction of the matrix. However, if one is not interested in photonic materials, but rather is attempting to make a ceramic macroporous material, sol-gel infilling of colloidal crystals may be a very good route. The contraction of the matrix may in fact be an advantage, in that it may be possible to make structures with pore diameters on the order of 50 per cent of the diameter of the template. Because it is difficult to make colloidal crystal templates from spheres smaller than a few hundred nanometers, this may be valuable for creating nanoperiodic structures.

Another pathway to macroporous materials is to fill the interstitial space of a colloidal crystal with nanoparticles, followed by removal of the colloidal template. This has some advantages over sol-gel infilling, in that the contraction of the structure upon removal of the template from a nanoparticle filled colloidal crystal is significantly less than that seen upon removal of the template from a sol-gel filled system, and a much larger subset of materials can be prepared as nanoparticles, including semiconductors, metals, and ceramics. The first example of semiconductor nanoparticle infilling of colloidal template used II–VI semiconductor nanoparticles; since then, Er-doped TiO_2 nanoparticles, for example, have been filled into a colloidal template, followed by removal of the template to generate a macroporous solid (Fig. 28.27).

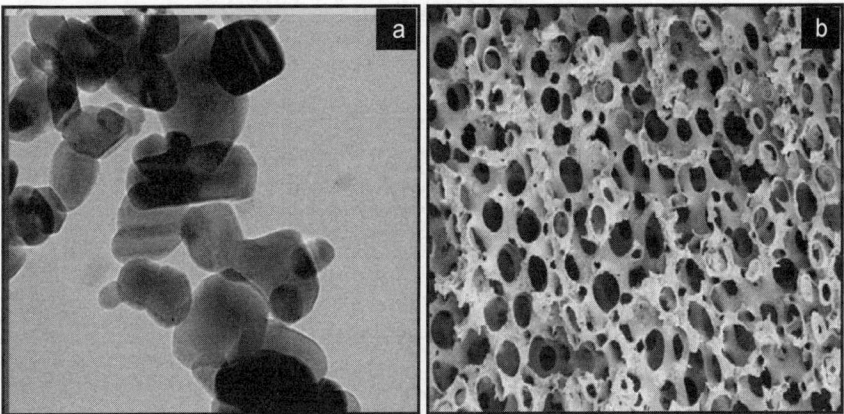

Fig. 28.27. (a) Transmission electron micrographs of 3 per cent erbium-doped hydrothermally synthesised titania nanoparticles. (b) Scanning electron micrographs of macroporous titania structure formed by imbibing these nanoparticles into a colloidal template formed from 466 nm polystyrene colloidal particles. The colloidal template was removed by calcination at 300°C for one hour in air.

The use of CVD as a pathway to filling colloidal crystals at first may seem counter-intuitive. After all, CVD generally is most efficient at coating planar surfaces, and it would seem almost impossible to fill structures with deep pores, such as the interstitial space of a 3D colloidal crystal. However, significant strides have been made in the past few years, and now virtually complete infilling of colloidal structures with both Si and Ge via CVD has been demonstrated. After dissolution of the colloidal template, the result is an inverse structure with the necessary refractive index and structural conditions to exhibit a complete photonic band-gap.

Electrodeposition-based infilling is intriguing for several reasons and has the potential to be general with respect to both characteristic lattice constant and material (Fig. 28.28).

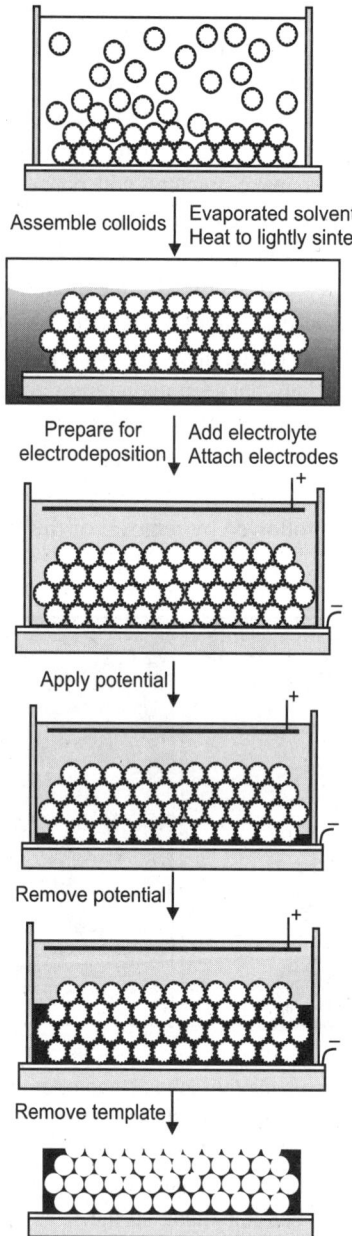

Fig. 28.28. Generalised procedure for creating 3D periodic macroporous materials by colloidal templating and electrodeposition. Monodisperse colloids sediment onto a conducting substrate, self-assembling into a crystal. The sample may be dried and sintered before electrolyte is added. A counter electrode allows electrodeposition of the desired material (semiconductor, polymer, metal) into the interstitial spaces. In a final step the electrolyte and the templating colloid are removed. For polymeric colloids this can be done by treatment at elevated temperature or by dissolution with a solvent. For silica colloids, aqueous HF is effective for dissolving the template.

The potential for high-refractive-index materials, large area structures, and the complete infilling of thick 3D colloidal templates, as well as the low cost of electrodeposition have led to interest in this area. To date, three different classes of materials that have been electrodeposited into self-assembled colloidal crystals, namely semiconductors, polymers, and metals.

Semiconductors for Photonic Crystals

Semiconductors are interesting candidates for photonic crystals, primarily because of their high refractive indices and generally robust nature. For example, CdS has a refractive index of 2.5, and materials such as GaP, Si, and Ge have indices of 3.4, 3.5, and 4.0, respectively. However, routes to creating periodic macroporous structures from such materials are limited because of their very high melting points and low solubility in common solvents.

To date, the II–VI semiconductors CdS and CdSe, and ZnO, have been electrochemically grown through colloidal templates, resulting, after dissolution of the template, in macroporous semiconductor films. For all systems, a conducting oxide film on glass was used as the substrate. Macroporous CdS films were generated by galvanostatic deposition through the interstitial spaces of a colloidal crystal formed from 1 µm SiO_2 spheres, and CdS and CdSe macroporous films were generated by potentiostatic deposition through a colloidal template generated from 466 nm polystyrene spheres (Fig. 28.29). After electrodeposition, the SiO_2 and polystyrene colloidal templates were removed with aqueous HF and toluene, respectively. Because of the high rigidity of the semiconductor network, contraction upon removal of the template was limited to a few percent at most. The fine and gross morphologies of the electrodeposited semiconductors are shown in Fig. 28.30.

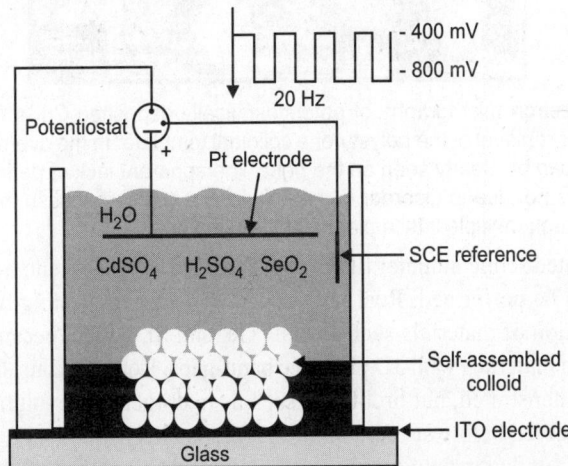

Fig. 28.29. Schematic representation of the experimental setup for potentiostatic deposition of CdSe through the interstitial space of a colloidal crystal.

Macroporous ZnO films were formed by potentiostatic deposition through a colloidal crystal formed from 368 nm polystyrene spheres, and the spheres were removed with toluene. Careful control of the electrodeposition conditions was necessary: if electrodeposition was done at a potential less negative than –1.0 V vs. Ag/AgCl, large crystalline grains of ZnO formed, which disrupted the structure of the conoidal template. Using a deposition potential more negative than –1.0 V suppressed the formation of large-grain ZnO, and the colloidal template was not disrupted.

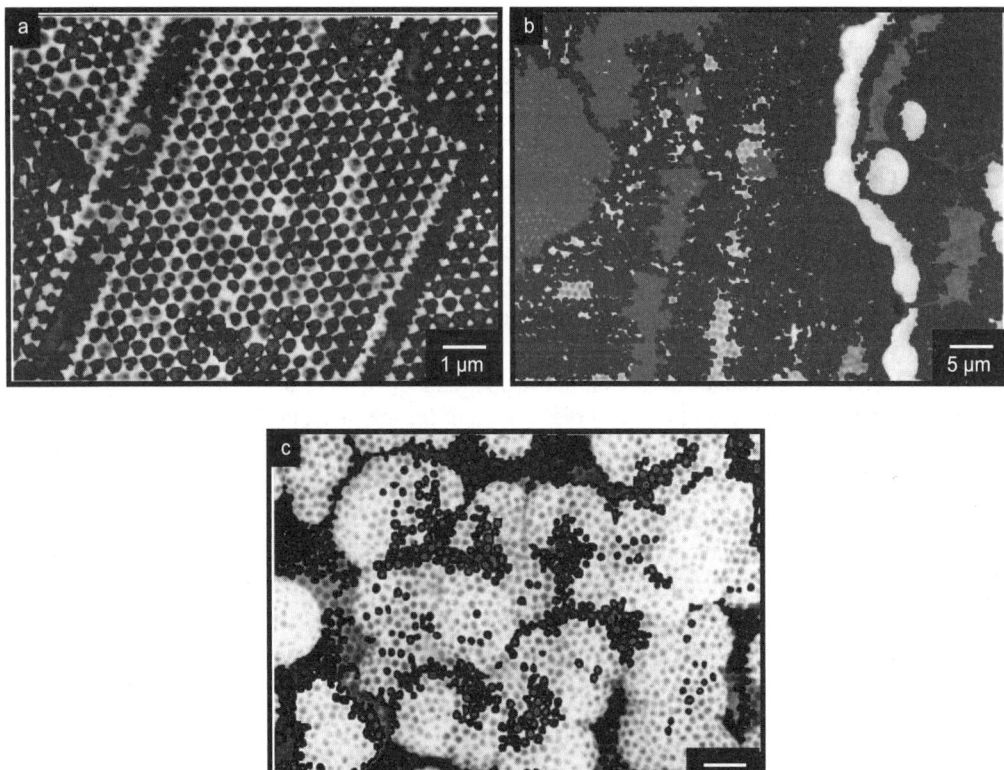

Fig. 28.30. Scanning electron micrographs of potentiostatically deposited CdSe (a) and galvanostatically deposited CdS (b, c) after removal of the polystyrene colloidal template. In the overdeposited system (b), the overlying solid CdS film can be clearly seen on the right. The apparent lack of periodic pore structure in the underdeposited system is not due to disorder in the colloid, but occurs because the nodular surface of the semi-conductor cuts through multiple lattice planes of the template.

All the electrodeposited semiconductor films are reported to be opalescent; however, detailed optical spectroscopy has yet to be performed. Real progress in optically interesting materials may await the electrochemical deposition of materials such as GaP, Ge, and Si, which, because they have refractive indices >3, may result in materials with 3D photonic band-gaps. Routes to the electrodeposition of such materials have been demonstrated, but problems, such as hydrogen gas evolution and generally harsh conditions, need to be solved before success in these areas is likely.

Electrodeposition of conducting polymers (electropolymerisation) through self-assembled colloidal crystals, followed by removal of the colloidal template, is a promising route to achieving active macroporous materials. Several significant advancements over the past few years have begun to demonstrate the potential of conducting polymer-based microperiodic photonic structures. Inherently, because of the low refractive index of polymeric materials, it is quite unlikely that a 3D photonic band-gap material will result from a polymer-based photonic crystal; however, conducting polymers have advantageous properties as compared to conventional polymers or inorganic materials: their optical properties can he electrochemically modulated, fine control over properties can be obtained through organic chemistry, and they are often mechanically flexible.

Electrochemical Growth of Conducting Polymers

Electrochemical growth of conducting polymers is a fairly well developed field, and many procedures for growing solid films have been published. There are, however, only a few reports on the growth of porous conducting polymer films. Fibres of polypyrrole, poly(3-methylthiophene), and polyaniline were formed in the early 1990s by electrodeposition from the appropriate monomer solution through a porous membrane. The first example of electrochemical deposition of a conducting polymer around a colloidal template was in 1992, when polypyrrole was grown around latex particles. However, no attempt was made to remove the colloidal particles, and the optical properties of the resulting films were not measured. Only in the past few years have researchers been exploring the possibility of the templated growth of conducting polymers for photonic applications.

Colloidal templating of conducting polymers

To date, three reports on colloidal templating of conducting polymers have appeared, all of which followed the general procedure of: (i) colloidal crystal formation on a conducting substrate, (ii) electrochemical deposition from solution, and (iii) dissolution of the colloidal template with an appropriate solvent. In the first example, polypyrrole was grown potentiostatically from a solution of pyrrole in acetonitrile through a colloidal crystal composed of SiO_2 spheres with a mean diameter of 238 nm assembled on F-doped SnO_2-coated glass, followed by removal of the colloidal template with aqueous HF. Macroporous polypyrrole, polyaniline, and polybithiophene films have been potentiostatically polymerised through a colloidal crystal assembled from 500 nm and 750 nm polystyrene spheres, on a substrate of gold-coated glass. The polystyrene template was then removed with toluene. In the most recent example, polypyrrole and polythiophene macroporous films were potentiostatically grown through colloidal crystals assembled from 150 nm and 925 nm polystyrene spheres, respectively, on glass coated with indium tin oxide: the polystyrene was removed with tetrahydrofuran. A preliminary optical characterisation showed a weak dip in transmittance that appeared to be correlated with the periodic structure.

Removal of solvent

One significant issue is the contraction of the period structure upon removal of solvent for electrodeposited macroporous polymers. This was most clearly observed in polystyrene-templated systems, in which significant contraction, ranging from 13 to 40 per cent, was observed for the macroporous polypyrrole and polyaniline.

However, very little contraction was observed in polystyrene-templated macroporous polybithiophene or when SiO_2 spheres were used as the template. For example, poly(ethylenedioxythiophene) was templated by silica colloidal particles (Fig. 28.31). The primary difference is that organic solvents are used to remove the polystyrene spheres and an aqueous HF solution is used to remove the SiO_2 spheres. This suggests that the organic solvent softens the electrodeposited polymer, allowing it to contract; however, there may be other systems similar to polybithiophene, in which contraction of the macroporous matrix does not occur. This is less of a problem for macroporous metals and semiconductors.

Metallic macroporous ordered replicas of colloidal assemblies are of potential interest for a wide range of applications including filtration, separation, and catalysis. In addition, they might have interesting electrical, magnetic, or optical properties. The tools and techniques for electrochemically plating metals have been well established for thin films and even bulk materials. It is thus fairly straightforward to develop recipes to backfill the interstitial space of a colloidal self-assembled crystal with almost any metal.

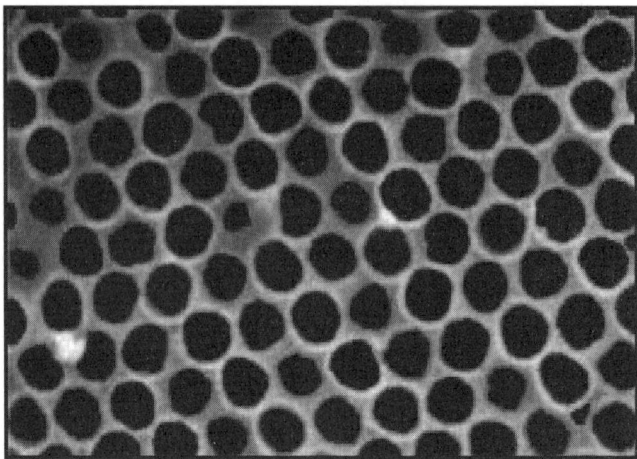

Fig. 28.31. Scanning electron micrograph of poly(ethylenedioxythiophene) electrodeposited around a colloidal crystal template, after dissolution of the template.

From a photonic standpoint, the properties of bulk metals are very poor, although of course templated structures may have many other applications. The imaginary components of the dielectric constants of bulk metals are large, hence they readily absorb light. When the metallic structures become small enough, however, strong optical resonances associated with plasmon frequencies of the conduction electron in the metals can lead to qualitatively new phenomena. A well-known example of this is the red colour of a nanosized dispersion of gold colloid. A more recent manifestation of unexpected behaviour is the anomalously highlight transmission through small holes (<200 nm) in thin metallic films.

Theoretical calculations on ordered 3D arrays of metallodielectric spheres show that these are promising for the construction of materials with full photonic band-gap in the visible part of the optical spectrum. The advantage of metallo-dielectric structures over purely dielectric structures is that it should be easier to achieve a full band-gap in the visible. A full band-gap in the visible is exceedingly difficult if not impossible, to create with purely dielectric structures, because very few dielectric materials have an index of refraction >3 and very low absorption in the visible. This has led to the development of synthesis routes to produce metallo-dielectric colloidal core-shell particles with sizes in the submicrometer range and metallic shell thicknesses or cores that are small enough to show resonance effects.

Gold replica of colloidal crystals

Smith made gold replicas of colloidal crystals made of silica (radius 113 nm) and polystyrene (radius 322 nm). Prior to electrodeposition of the gold, the silica spheres were sintered by heat treatment at 600°C. After electrodeposition of the gold, the silica template was removed by etching with aqueous HF, and the polystyrene spheres were removed by combustion at 450°C (Fig. 28.32). There are no dimensional changes between the dried, sintered colloid and the final replica, although some cracking is observed during the original drying and sintering process, indicating that the electrochemically formed gold is dense and structurally robust. This is a definite improvement over other methods of infilling macroporous structures with high-dielectric materials, e.g. liquid-phase or sol-gel chemistry, and infiltration with nanosized particles, in which considerable contraction of the matrix is observed, which leads to serious crack formation and warping of the colloidal structure. Other electrodeposited materials include Ni, Pt, and a SnCo alloy, Pd, Pt, and Co; electroless deposition has also been attempted.

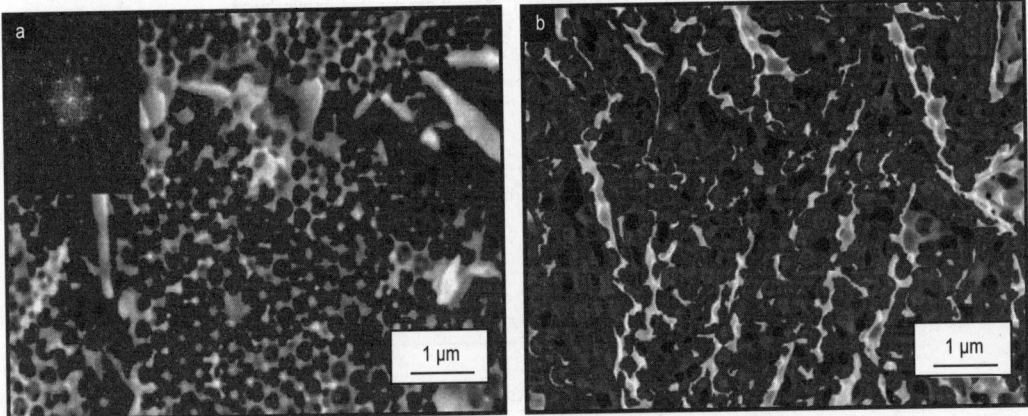

Fig. 28.32. (a) Scanning electron micrograph of a crystal of air spheres (radius 111 nm) in gold, made with a silica template. The inset is a Fourier transform of the image. (b) Scanning electron micrograph of macropores (radius 322 nm) in gold, made with a latex template. The structure has short-range order, but no long-range order.

As just described, recent work has demonstrated that templating of the interstitial space of highly ordered colloidal crystals has promise for creating macroporous photonic crystals from a diverse set of materials including oxides, semiconductors, metals, polymers, and glasses. The resulting 3D macroperiodic materials have been formed with close-packed macropores ranging in diameter from 100 nm to a few micrometers, giving the potential to modulate light ranging from deep UV to the infrared. However, problems with the infilling process still need to be overcome before this approach to photonic structures comes to fruition. As outlined, filling the 3D interstitial space of a colloidal crystal with a high-index material has been problematic, because many techniques either only deposit material in the top few layers of a colloidal crystal or do not fully fill the colloidal crystal with a material of high enough refractive index.

Route to infilling of colloidal crystals

Another route to infilling of colloidal crystals to generate a high-refractive-index structure is melt-imbibing of a chalcogenide glass such as selenium, followed by dissolution of the silica template. Selenium was selected because it has a high refractive index of 2.5, and thus can provide a nearly complete 3D photonic gap (Fig. 28.33), a very low optical loss coefficient between 1 and 10 μm, low melting point (217°C), and relatively low surface tension (~100 dynes cm^{-1}), which reduces the force necessary to in fill the structure. Importantly, selenium vitrifies easily, forming an optically isotropic glass. Other chalcogenide glasses certainly could be used to infill colloidal crystals; however, they have higher softening points and thus are not suitable for initial investigation. Through melt imbibing, essentially complete infilling of a colloidal crystal was demonstrated (Fig. 28.34).

Infilling was accomplished by imbibing molten selenium under high pressure into the colloidal crystal, followed by quenching to vitrify the selenium. Subsequently, the colloidal crystal template was removed with HF, resulting in a macroporous selenium/air structure with a high contrast in refractive index.

Thus, the confluence of nanoscience, biotechnology, and materials chemistry offers great potential for discovery and fabrication of advanced composite materials.

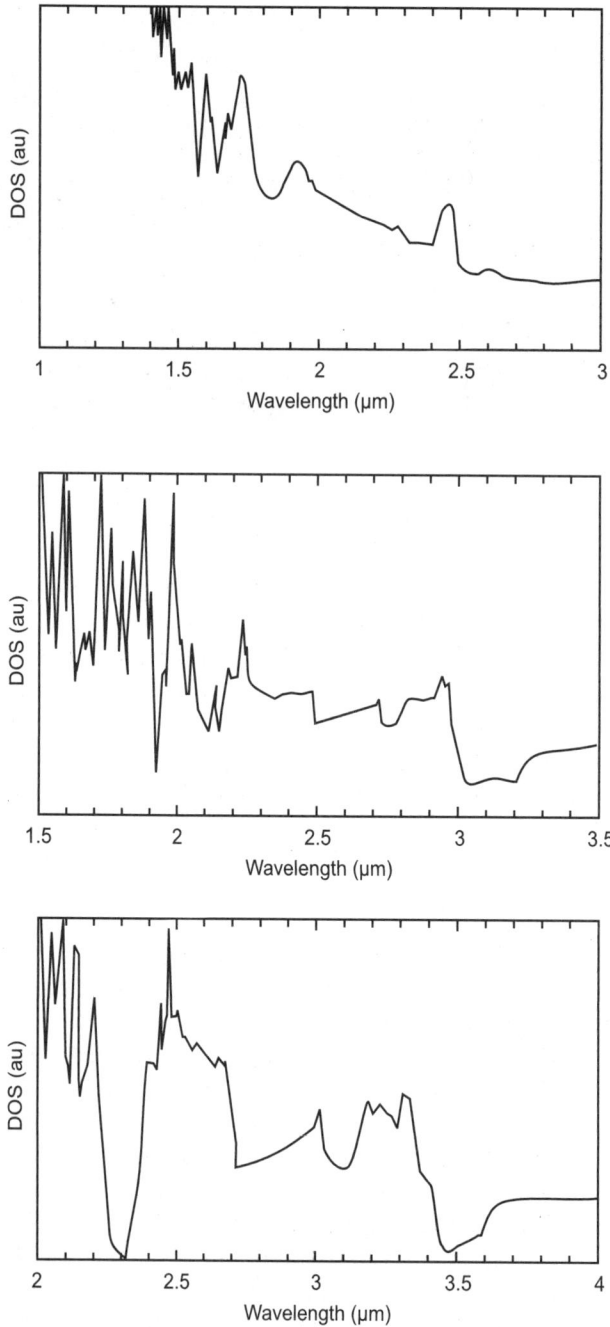

Fig. 28.33. Photonic density of states for inverse fcc structure for refractive index contrasts of (a) 1.45, (b) 2.5, and (c) 2.95; 1.45 corresponds to filling with a simple oxide such as silica, 2.5 to filling with selenium, and 2.95 to filling with a high-index chalcogenide glass such as Ge_{25}, $As_{20}Se_{25}T_{30}$. Note the deep photonic band-gap for an index contrast of 2.5 at 2 µm and a complete gap at 2.3 µm for the system with an index contrast of 2.95.

Great amounts of information still need to be gleaned from the study of biological systems, but we have now reached a point where the current body of knowledge on how biological systems can create highly functional nanocomposites is starting to enable the creation of advanced materials. For example, natural systems widely exploit self-assembly to create a great diversity of interesting and highly functional materials, and today we are beginning to also create synthetic systems by similar processes.

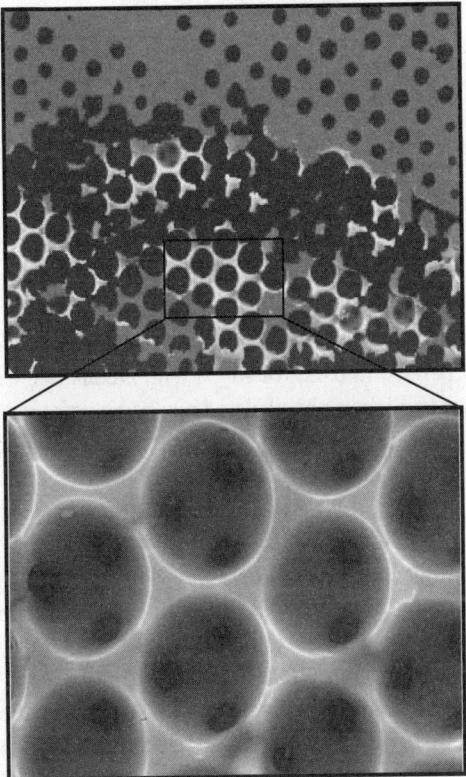

Fig. 28.34. SEM micrograph of the selenium photonic crystal before removal of the silica colloids. The top image was taken of the surface that was once adjacent to the glass slide. While this sample had not been chemically etched in HF, several layers or selenium/colloids were removed during the preparation of this sample as shown in the top image. This is included to demonstrate the nature of the three-dimensional infiltration. The magnified bottom image shows the quality of the infiltration and the tetrahedral coordination of the silica and the sintered necks.

We must always remember that biological organisms work with a limited subset of materials and take up to years to create nanocomposite structures, so, rather than attempting to create materials by direct mimicking of biology, it will likely be much more productive to create materials by exploiting the design rules expressed by biological systems and applying them to synthetic systems. The dividends for research into biologically inspired nanocomposite materials are great, and much progress is expected in the near future.

Chapter 29

Metal and Oxide Nanocomposites

INTRODUCTION

Composites are considered to be preferable materials with superior physical and mechanical properties compared to their individual constituents. Further, the extensive studies made in recent years have established the fact that, nanocomposites are far superior to the conventional composites. In this chapter the methods or preparation of nanocomposites of Al_2O_3 with metals like Cu, Ni, and Co with the oxide as the major or minor constituent and their physical and mechanical properties, have been described and discussed. The preparations and applications of some special nanocomposites containing TiO_2 have been described briefly. The novel chemical route developed by Smith and his co-workers, on the preparation of $Cu-Al_2O_3$, $Ni-Al_2O_3$, $Cu-Ni-Al_2O_3$, and Cu-Ni nanocomposites, has been highlighted with specific findings.

Ceramics are used in many high temperature operations. However, the use of ceramic as engineering material is critically determined by its brittleness. Therefore, the toughening of ceramic materials is very necessary. The dispersion of a ductile metallic phase like Cu, Ni, W, Cr, etc. in a brittle ceramic matrix, is found to have a very promising effect. Reinforcement models show the importance of the homogeneity and the small size of the metallic inclusion. The control of the microstructure of ceramic–metal composites is generally difficult to achieve through traditional techniques involving mechanical mixing of ceramic and metallic powders followed by hot pressing.

In recent years, a lot of efforts are being made to develop suitable metal or ceramic base composites for various special applications due to their superior mechanical and physical properties. Traditionally, these composites have been produced by techniques such as squeeze casting, powder metallurgy and rapid solidification. In such techniques of producing the composites, the reinforcing phase is fast mixed with the matrix material. The size of reinforcing phase is constrained by that of the starting material, which is typically in order of microns to a few tens of microns and rarely below one micron. However, with an objective to have more liner size of the constituents in the composite, *in situ* technologies have emerged for fabricating such composites and have received more attention in recent years. These technologies include self-propagating high temperature synthesis, direct metal oxidation method, exothermic dispersion, mechanical alloying and reactive infiltration or reaction powder metallurgy. The fineness and thermodynamic stability of the reinforcing phase in the *in situ* composites, offer excellent dispersion of line reinforcing particles and active interface, resulting in better mechanical properties and high temperature performance. In addition to this, *in situ* techniques can be more conveniently applied to produce composites that could be difficult to obtain by other conventional methods.

DEVELOPMENT OF ALUMINA-METAL NANOCOMPOSITES

In this section, development or some alumina-metal nanocomposites of industrial importance, have been described and discussed.

Strengthening or Al_2O_3 due to microstructural refinement by incorporating Ni has been reported. The dispersion of nickel particles may lead to toughening of alumina by the plasticity of the metal. Another interesting feature of this composite is that, it shows magnetic properties due to Ni dispersion. It is well-known that, ferro magnetic properties depend on the crystalline size of materials. Therefore, particular magnetic properties are due to the nanometer size nickel dispersed in the composites. Superior magnetic and mechanical properties could be realised in the Al_2O_3–Ni nanocomposite, which was developed by reducing the NiO by H_2 and hot pressing of Al_2O_3 and nanosized Ni mixed powder. The nickel nitrate was selected as the starting material or NiO.

Microscopic observations revealed that, the Ni particles of about 100 nm size were dispersed homogeneously at the matrix grain boundary. The composite containing a small amount of Ni (5 vol%) exhibited an increased fracture strength of 1090.0 MPa from 682.8 MPa for only Al_2O_3.

Alumina nickel nanocomposite was developed by Kaplan by another process where the Al_2O_3 preforms were slip casted and dried in air at 900°C for 5 hours to reach 68 per cent of the theoretical density. The fired preforms were infiltrated with a nickel nitrate solution in distilled water. The pressure less sintering was applied to the preform. The sintering process was carried out in two stages, i.e. reduction process at 900°C followed by sintering at 1400°C. The reduction was suggested to occur within the open pores of the ceramic preform after evaporation of the metal salts in to the gas phase. By controlling the partial pressure of oxygen during the sintering process, a reaction between the nickel particles and the alumina matrix was suggested to form Ni spinal ($NiAl_2O_4$) particles.

$$Ni + Al_2O_3 + \tfrac{1}{2}O_2 \rightarrow NiAl_2O_4$$

Three-point bending experiment showed an increase in strength for the nickel reinforced alumina and a signification increase for the spinal reinforced alumina. A similar work as reported by Niihara on the preparation and study of properties of nickel dispersed alumina composites. The mixture was prepared by using NiO or $Ni(NO_3)_2 \cdot nH_2O$ as dispersion source of nickel metal. The composites in both cases were prepared by hydrogen reduction of the nickel oxide. Microstructural investigations of the composite fabricated using nitrate powder compared to that with nickel oxide, revealed that, fine nickel particles about 100 nm in diameter were dispersed homogeneously at the matrix grain boundaries forming the intergranular nanocomposites. The microstructure was found to affect the crack propagation process very significantly. The toughness increased for the composites that contained a large amount of nickel dispersion. However, the plasticity of the metal contributed slightly to the toughening of the materials. The fracture strength, of more the >1 GPa, was measured for the Al_2O_3/5 vol% nickel composite prepared from the $Ni(NO_3) \cdot nH_2O$ powder. The strengthening was caused by the refinement of the microstructure. It indicated the advantage of the present method adopted for fabricating Al_2O_3/metal composites. The ferromagnetism of nanometer sized nickel that possessed a high coercive force, contributed to the magnetic properties of Al_2O_3-based, composites.

Niihara have prepared copper dispersed alumina matrix nanocomposites with sound microstructure and improved mechanical properties. The α-Al_2O_3 with particles size or 0.2 μm and copper oxide (CuO) powder with particles size of 1–2 μm were ball—milled for 24 hours in ethanol with high purity Al_2O_3 balls as grinding media with desired proportion of both the oxides. Slurries of the powder mixture were dried and then again ball milled in a polypropylene bottle for 24 hours to avoid agglomeration.

The powder mixtures thus prepared, were placed in a rectangular graphite die and heated in flowing hydrogen gas at 350°C for 30–120 minutes to convert the CuO to the metal. The hydrogen atmosphere was maintained to a temperature of 1100°C and then changed to argon gas. Sintering was carried out at 1450°C for one hour in argon atmosphere with an applied pressure of 30 MPa. The composite thus prepared, revealed that, the nanosized copper particles were mostly intergranularly dispersed in the Al_2O_3 matrix. The addition or copper particles resulted in improved fractured strength from 536 in absence of copper to 820 MPa with 2.5 and 5 vol% of copper. The toughness increase was explained by crack bridging. The strengthening was mainly attributed to the refinement of Al_2O_3 matrix grains by the nanosized Cu dispersion. With increasing Cu content, the composite showed inhomogeneous microstructure due to the agglomeration of Cu particles.

In a latter period, Niihara prepared copper dispersed $Cu-Al_2O_3$ nanocomposites using Al_2O_3 and Cu $(NO_3)_2$ mixture and compared its properties with those of the nanocomposite prepared by using Al_2O_3 and CuO. The Al_2O_3-Cu composite prepared through nitrate, was obtained by mixing α-Al_2O_3 powder in the copper nitrate solution in ethanol and ball milled for 24 hrs. with high purity alumina balls. The dried mixtures were calcined at 300°C for 2 hrs in air to obtained the mixed oxides powder. Rest of the steps were similar to that used for Al_2O_3-Cu powder mixture to obtain the Al_2O_3-Cu composite as mentioned above. The relative density and mechanical properties for monolithic Al_2O_3 and Al_2O_3 with 5 vol% copper composite produced through the above two routes, were measured.

In comparison with monolithic Al_2O_3 the fracture strength of the composite showed an enhanced value of 890 MPa as produced by using the mixture of Al_2O_3-CuO and 953 MPa through Al_2O_3-Cu nitrate. Though both the composites showed identical fracture toughness, the one produced through the nitrate route exhibited improved fracture strength which could be ascribed for the refinement of matrix grain size. Microstructural investigation for the composites through nitrate route showed that, fine copper particles, about 150 nm in diameter were homogeneously distributed within the Al_2O_3 matrix grains and also at the grain boundaries.

Smith and others have also prepared high density Nickel-cobalt alloy dispersed Al_2O_3 composites by using both nickel and cobalt nitrates in a similar manner as mentioned above. Microstructural investigations revealed that, nanometer-sized alloy particles were dispersed homogeneously at the matrix grain boundaries, forming the inter granular-type nanocomposite. High strength (>1 Gpa) was registered for the Al_2O_3/10 wt% Ni-Co alloy composite. The strengthening was caused by the refinement of the microstructure. The Ni-Co alloy that possessed a high magneto mechanical coupling factor, contributed to the inverse magnetostrictive response of the Al_2O_3 based composites.

A heterogeneous precipitation method was developed for producing nanocomposite powders containing nickel nanoparticles homogeneously dispersed within γ-Al_2O_3. Prepared by using nanoparticles of NiO, $Al(NO_3)_3 \cdot 9H_2O$ and NH_4OH as the starting materials. The amorphus aluminium hydroxide nucleated on the surface of NiO nanoparticles and crystallised to γ-Al_2O_3 at 900°C. After the calcination, the nickel oxide was selectively reduced at 700°C in a hydrogen atmosphere to nickel metal with the size of 25–35 nm. The TEM and AES analysis showed that, nickel nanoparticles were uniformly dispersed in the alumina matrix.

IMPROVING PHYSICAL AND MECHANICAL PROPERTIES OF METALS BY INCORPORATING NANOPARTICLES OF AN OXIDE

In the previous section, it has been found that, alumina based nanocomposites with small amounts of nanoparticles of metals like nickel, copper, cobalt, etc. could significantly improve the physical and mechanical properties of the oxide. In a similar manner, studies have been made to improve the physical and mechanical properties of some metals by incorporating nanoparticles of an oxide particularly Al_2O_3.

Conventionally, metal-metal oxide composites are produced by the internal oxidation method. One of the main defects of internal oxidation method is the nonhomogeneous distribution of the oxide particles which negatively influence the mechanical and electrical characteristics or the composites. In view of this, the thermochemical process to produce the nanocomposite powders with a homogenous distribution of the nanosize oxide particles in the metal matrix, is suggested to solve this problem in order to improve electrical and mechanical properties.

To improve the electrical and mechanical properties of electrodes for contact welding, the Cu-Al_2O_3 nanocomposite powders were developed by thermochemical process. This method used water soluble copper and aluminium nitrates which were decomposed at a low temperature to their respective oxide and finally at a temperature of 150°–200°C and hydrogen reduction of the copper oxide to metallic copper was carried out. In the Cu-Al_2O_3 nanocomposite powders prepared by the thermochemical method, the Al_2O_3 particles were found to be around 20 nm and it was uniformly distributed inside the matrix or copper aggregate.

Copper powders with a median particle diameter of 60 μm were mixed with different amounts of nanosized alumina powder with a median particle diameter of 14 μm in an asymmetric moved mixture followed by hot extrusion. The nanoparticles of alumina was produced by evaporation of corundum with a pulsed radiation laser and subsequent condensation of the laser induced vapour in a controlled aggregation gas. Copper composites containing one and three vol% nanoalumina powder, pure copper sample and copper sample with 1 vol% microscale alumina powder addition were produced. The investigation or the micro structure or the copper nanocomposites reveal considerable less recrystallisation and grain growth up to temperatures near to the melting point of copper compared to hot extruded samples made of copper powder without alumina and composites with microsized alumina powder. In the nanocomposites, the nanosized alumina particle were found to be located at the grain boundaries of the copper grains. The hardness of the nanocomposite samples of the nanoparticles of alumina, were found to be much higher than the values for pure copper or copper alumina microcomposite. Even at a temperature of 1065°C due to nanopowder addition an increase in hardness and yield stress were observed. The improved properties of the composite containing nanoparticles of alumina, was attributed to inhibited recrystallisation and grain growth of the matrix copper. Smith developed a novel method to produce nanocomposites of Cu-Al_2O_3 by chemical routes. Copper–alumina composite briquettes were prepared by hydrogen (pH_2 = 0.1 atm.) reduction of a homogenous mixture of finely divided CuO and Al_2O_3 formed by two chemical processes. In these processes, the Al_2O_3 of desired per cent was formed *in situ* by: (i) addition of $Al(NO_3)_3$ solution to CuO, and (ii) by addition of stoichiometric amount of ammonia to a slurry of CuO in $Al(NO_3)_3$ solution, followed by filtration and then heating the mixture in both the cases, at 850°C for 2 hr. The Cu-Al_2O_3 composites thus formed were sintered at 975°C for 2 hrs in H_2 atmosphere. Composites samples were also prepared by H_2 reduction of the briquettes made out of the mixture of Al_2O_3 and CuO powders or by direct addition of Al_2O_3 to Cu powder, followed by sintering at 975°C for 2 hrs in order to compare their homogeneity and microstructures with those made by the above two chemical routes.

The characterisation of composite samples were carried out by metallography, SEM and TEM techniques. The distribution of Al_2O_3 particles in the Cu matrix, was found to be improved when the Cu was produced *in situ* by reduction of CuO by H_2 and very significant improvements in this regard were noticed in case of composites prepared by the chemical methods, particularly the one through $Al(NO_3)_3$

decomposition. In this case besides the grain refinement of Cu (50–250 nm), and homogeneous distribution of Al_2O_3 in it (about 10 nm in size), an appreciable amount of a third phase was found to appear. The existence of the third phase has been indicated by SEM, TEM, and EDS analyses. The third phase is suggested to be possibly Cu-Al_2O_3 through the following reaction:

$$2CuO + H_2 \rightarrow Cu_2O + H_2O$$
$$Cu_2O + Al_2O_3 \rightarrow 2\ CuAlO_2$$

In a subsequent investigation Smith established the presence of the third phase along with Cu and Al_2O_3 phases and suggested a major role of the third phase for imparting superior physical and mechanical properties to the nanocomposites. Giannelis prepared Ni-α-Al_2O_3 metal-ceramic composites with metal loading ranging from 33 to 67 vol% via a sol-gel (both polymeric and colloidal) technique. The polymeric route involved introducing nickel formate during the hydrolysis and condensation of aluminium isopropoxide, while the colloidal route involved precipitating $Ni(OH)_2$ on to boehmite particles. Transformation of Ni-α-Al_2O_3 composites was accomplished by calcination in a reducing environment followed by hot pressing at about 1400°C in an oxygen partial pressure of 10^{-12} atm. The composites exhibited a highly interconnected microstructure, composed of continuous Ni and alumina networks with enforced fracture properties compared with pure alumina.

Recently, Smith and others have prepared Ni-Al_2O_3 and CuNi-Al_2O_3 nanocomposites through a chemical route and studied their microstructures. Nickel-alumina nanocomposites were prepared by coformation of their oxides from an aqueous solution of their nitrates followed by preferential reduction of the nickel oxide with hydrogen at a low temperature (400°C). This Ni-Al_2O_3 nanocomposite powder, thus formed, was briquetted and sintered at 1000°C for two hours. The consolidated composite or Ni-Al_2O_3 with 0.5 wt% Al_2O_3 was characterised by XRD, SEM, EDS and TEM. The SEM image and EDS composition maps showed the homogeneous dispersion of Ni, Al and O through out the microstructure. Studies of the Ni-Al_2O_3 (0.5 wt%) nanocomposite by TEM had shown the formation of nanosized Ni matrix grains and had also conformed the homogeneity of the nanocomposite. As the novel chemical method developed by the various scientists had been successfully applied to obtain homogeneous nanocomposites or Cu-Al_2O_3 and Ni-Al_2O_3, the same process was also applied successfully to obtain homogenous nanocomposite of Cu Ni-Al_2O_3.

SOME OTHER NANOCOMPOSITES AND NANOALLOYS (BESIDES Al_2O_3)

Some other nanocomposites and nanoalloys besides Al_2O_3 metal composites, were developed in recent years. For example, some composites containing TiO_2 were found to have a lot of applications in various fields. The photo catalytic detoxification of organic and inorganic compounds is very promising for the purification of polluted air and industrial waste-water. The TiO_2 is an ideal photocatalytic material due to its stability, nontoxicity and low cost. Smith developed a nanosize TiO_2 coated coal fly-ash photo catalyst by precipitation method. The titanium hydroxide was precipitated on coal fly-ash by hydrolysis of titanium chloride and the hydroxide was treated in the temperature range of 300°–400°C. The crystalline TiO_2 was formed on the fly-ash having a particle size of 9 nm. This TiO_2 photocatalyst was reported to be very efficient for removal of 63–67.5 per cent of NO gas in the air. Such photocatalyst also can be used to treat industrial effluents to remove the toxic materials before the industrial waste is discharged to the environment.

Now-a-days, the metal nanoparticles are being used under extreme conditions such as optical limiters and three-dimensional optical memories applying laser with high intensity. Gold and silver nanoparticles are some of the best optical limiters known so far. However, at high light intensities, these are susceptible to damage, leading to photo fragmentation, legand destruction, etc. To make them stable at extreme conditions, it is necessary to protect them with stable and chemically inert shells such as an oxide. The coating of this metal nanoparticles with TiO_2 shell was suggested to be very suitable. Yu and Mulvancy developed Au/TiO_2 core-shell structure nanoparticles by sol-gel process. In this process, the thickness of TiO_2 shell on the surface of gold particle was claimed to be 1 nm.

Using chemical-route, it is also possible to produce nanoalloy powders for various applications and fabrication of intricate components by powder metallurgy techniques at a relatively lower pressure and temperature. Recently, Smith developed the new chemical route for making nanoalloys of copper-nickel. The alloy was prepared from the mixed aqueous solution of the nitrates of copper and nickel, through coformation of their ultrafine mixed oxides by heating around 375°C followed by reduction with hydrogen at a very low temperature (below 350°C). It was possible to get high pure Cu-Ni alloy powder (50 per cent each) free from any detectable oxygen, from their coformed oxides, by hydrogen reduction at 350°C in less than 20 minutes. The alloying of the two metals took place during the H_2 reduction of nanosize oxide particles of copper and nickel, prepared by the above-mentioned chemical route. The nanoalloy powder was cold pressed and sintered at 1000°C. The density and hardness of the consolidated alloy were measured and found to be close to theoretical values. The SEM characterisation or the sintered sample confirmed the homogeneity and purity of Cu-Ni alloy.

Chapter 30

Sustainable Flame Retardant Nanocomposites

INTRODUCTION

This chapter examines the current state of research into sustainable flame retardants with the work on nanocomposites highlighted. The motivations to move away from halogen-based flame retardants are discussed and a number of life-cycle-assessments are mentioned which set the stage for a similar LCA study of nanocomposite flame retardant products. Additives, such as hydrotalcite and cellulose nanofibrils, are proposed as components of potential future sustainable flame retardant nanocomposites (Fig. 30.1).

Fig. 30.1. A Pop art sculpture outside the National Arboretum in Washington DC depicting the challenges which our current non-sustainable plastics present to the world.

CURRENT ECONOMIC AND ENVIRONMENTAL CLIMATE

In the current economic and environmental climate it is more critical than ever to develop the tools and information that enable quantitative evaluation of the sustainability of utilising nanotechnology in products. Environmental concerns over the potential risks that halogenated chemicals pose have been a reality for decades. This is rooted in the persistence, bioaccumulation and toxicity (PBT) associated with specific brominated organic compounds. To respond to this, the flame retardant research community and others began developing non-halogenated flame retardants. Initial nonhalogenated research focused on developing new phosphorus based flame retardants. Numerous publications and patents were issued in this area based on phosphorus, aluminium trihydroxide and magnesium dihydroxide, boron, siloxane and silica. A more recent class of flame retardants based on nanoadditives was also developed in response to the nonhalogen FR issue. This later class of FR additives utilise naturally occurring smectite clays (layered silicates), such as Montmorillonite (Mt), Hectorite (Hc), or Laponite (Lp). Incorporation of clay in polymers has been reported to have as much as a 75 per cent decrease in the peak heat release rate (PHRR), as measured in the cone calorimeter. These materials exhibit enhancements in a variety of physical properties at one tenth the loading required as compared to when micrometre size additives are used. However, in practice, i.e. in the patent and archival journal literature, the publications show that the best advantage is found when the clay FR is combined with another non-halogenated FR.

STUDIES OF POLYMERS WITH LAYERED SILICATES

The first studies of polymers combined with layered silicates at the nanoscale to form 'nanocomposites' was work by Carter in 1950, which was followed by *in situ* polymerisation of vinyl monomers in the interlayer space of Mt by a series or researchers in the early 1960s. Most of this early work involved intercalated clay polymer nanocomposites (CPN) comprised of much higher loadings of clay mineral (50 per cent mass fraction) than are used today in nanocomposites (5 per cent mass fraction).

Nanocomposites with lower loadings (1 mass fraction % to 10 mass fraction %) characterise the type of materials that are the focus of more recent studies. Examples include those disclosed in initial patents in the 1970's and mid-1980s from General Motors (GM), imperial chemical industries (ICI) and DuPont. The GM patent primarily claims the use of clay minerals as substitutes for antimony oxides, while the ICI patent teaches the use of 'delaminated vermiculite' to impart self-extinguishing and charring properties to expanded polystyrene beads. The DuPont patents also discuss the flame retardant properties of CPN, but only as anti-drip additives to formulations heavily filled with conventional flame retardants. The inventors note an increase in char formation, which they attribute to the polyester. Kamigato, at Toyota also filed patents on the *in situ* polymerisation of styrene, isoprene, vinyl acetate and caprolactam. Although some of these patents indicate that clays nanocomposites enable self-extinguishing properties, or a V0 rating (self-extinguishing in under 10 s) in the UL 94 test, or they may pass other large scale fire tests such as the UL 910, no other study of the char forming flame retardant properties of nanocomposites appeared in the literature until the mid-1990s.

Groups at NIST and Cornell both reported that polymer nanocomposites, with no other flame retardant, reduced the parent polymer's flammability and enhanced char formation. Giannelis found self-extinguishing properties for nanocomposites when they were exposed to small open flame tests. Researchers at NIST used cone calorimetry and radiative gasification to show that Mt nanocomposites had enhanced char formation and gave up to 75 per cent lower flammability, as measured by reduction in the peak heat release rate (PHRR) or peak mass loss rate. In most cases, the carbonaceous char yield was limited to (2 to 5) mass fraction %; consequently, the total heat release (THR) was not significantly affected. In addition, ignition times were either minimally or not all improved. However, the unique character of this new approach to flame retardant polymeric materials was the dual benefit of reduced peak heat release rate and improved physical properties, a combination not usually found with conventional flame retardants. A significant number of papers have since been published on this topic, with many shedding light on the flame retardant mechanism.

DISCUSSION

The focus of this chapter is to bring to light the need to develop and evaluate sustainable approaches to flame retardancy. In 2008, the Environmental Protection Agency (EPA) invited a number of research groups to present work where nanotechnologies were preventing pollution. The nanocomposite flame retardant area was identified and incorporated into a special session on case studies. While it is reasonable to propose that substituting clay for polybrominated diphenyl ether (PBDE) flame retardants might reduce pollution and be called sustainable, the fact is that no quantitative study has ever been done to support this assertion. Life-cycle-assessment (LCA) methods are the tools of choice for such an analysis. According to a report from a similar conference in 2006, from the Woodrow Wilson International Centre for Scholars, only 2 LCAs, which meet the full scope of an LCA as defined in the ISO standards (ISO14040:2006, ISO14044:2006) have been published on nanotechnology based products as of 2005.

A LCA study of nanocomposites used for automotive applications has appeared, but the issue of flame retardancy was not addressed. However, this LCA study did provide an illustration of how one might approach this type of analysis. This LCA was performed by Lloyd and Lave from Carnegie Mellon University, and was motivated by the view within the auto industry that the use of polymeric components, instead of metal components, in body panels will reduce the mass of an automobile and improve the fuel consumption. Whereas improved fuel consumption is one of the main selection criteria for automotive customers, the manufactures must be able to provide this in a cost competitive manner and without any additional environmental consequences. In addition, the authors of the LCA study acknowledged the additional drivers associated with nanotechnology are the potential reduction in the energy and materials needed to manufacture products, 'while improving environmental performance and sustainability'. But, they caution that a broad spectrum of issues must be examined to make a responsible assessment, and 'it is important not to compromise safety, cost, or other desired attributes'.

They estimated potential economic and environmental impacts for the use of claypolypropylene nanocomposites, or aluminium, instead of steel in light duty vehicle body panels. As the data in Fig. 30.2 show, although the manufacturing costs for the nanocomposite body panel are currently higher, a significant potential benefit of this approach is in reducing energy use and environmental discharges during manufacturing.

Smith concluded the use of nanocomposites would increase fuel economy at a low cost, which potentially leads to large economic and environmental benefits, primarily through reduction in the production of CO_2 during the lifetime of the vehicle. However, since the study was published before the recent worldwide fuel cost crisis they also assert that 'US consumers have little interest in greater fuel economy, and so this technology is unlikely to be developed and employed in this application without government intervention'. Obviously, changes in the economic situation can radically change the potential that a more sustainable approach will be utilised. The same can be said of how a nanocomposite, or any other new non-halogenated flame retardant products could be approached, i.e. both economic conditions and governmental regulations can be strongly coupled, and can be equally important factors as the environmental realities of the analysis in determining the feasibility of the new approach.

The issues associated with performing a LCA of a product flame retarded using a nanocomposite, or another non-halogenated flame retardant, as compared to a halogen based flame retardant, are somewhat complex. However, the LCA reported by Lloyd and Lave provides an example of how the evaluation of a nanocomposite flame retarded products might be performed.

In addition, several LCAs of various flame retardant products (television sets, wire and cable, and sofas), performed at Swedish National Testing and Research Institute (SP), can also be used to provide insight as to how to structure such an LCA. The unique information that these LCAs offer is the inclusion of the effect of accidental fires on the LCA, something not usually included in most LCAs. In the 2005 SP study of halogenated flame retarded versus non-flame retarded television sets made of high impact polystyrene (HIPS), a similar approach to that taken by Lloyd and Lave was used, i.e. the incorporation of the effect of a different additive on the gasses released into the environment. However, instead of CO_2 emission savings, the SP researchers found reduced emissions from incineration of recycled TV sets with halogenated flame retardants as compared to those without any flame retardant. Furthermore, they pointed out that since TV sets with halogenated flame retardants had a V0 rating (self-extinguishing in 10 s), which 'essentially removes the risk of TV fires' the societal cost of using no flame retardant in the TV sets is 165 TV fires per million TVs or 160 deaths and 2000 injuries per year in the European Union. Additional distinctions from the SP study include the fact that the halogenated FR HIPS performed better in ageing and recycling studies than the non-FP HIPS, and the FR in the HIPS did not bloom (phase separate to the surface) during tests.

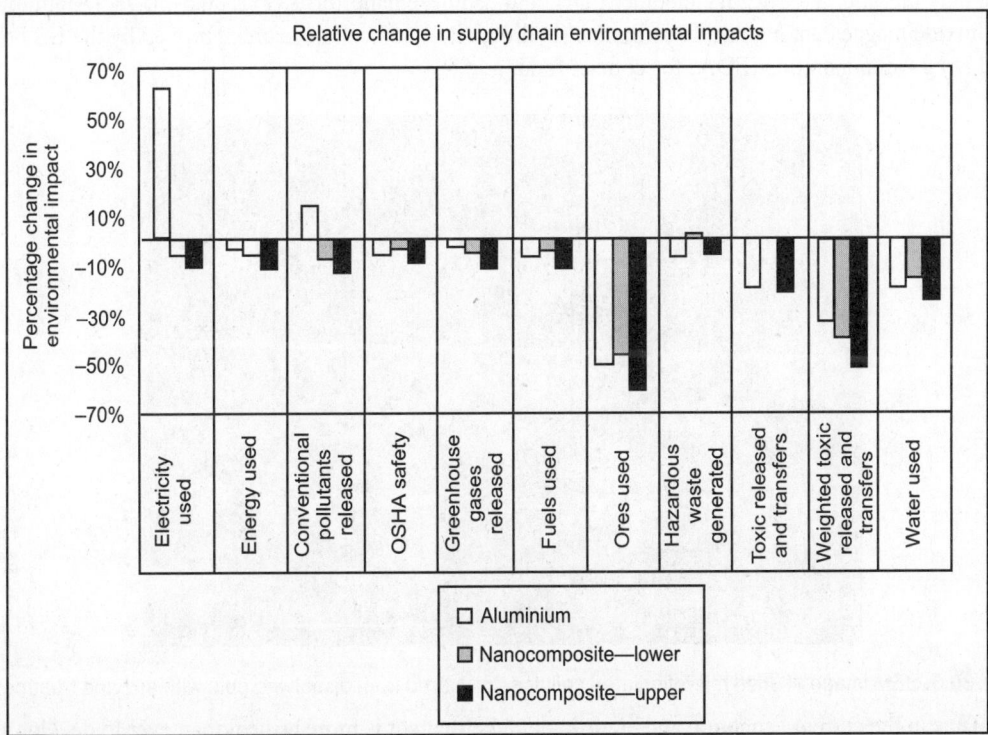

Fig. 30.2. Table from LCA study by Lloyd and Lave showing the higher amount of energy used for aluminium as compared to the range of values expected for energy use during manufacture for the nanocomposite, and the higher pollutants released for aluminium as compared to the range of values expected for that for the nanocomposite.

Nanocomposites have been found to prevent blooming, which may reduce environmental release of any additive present in the nanocomposite product. This raises another unique aspect of nanocomposite based FRs. Specifically, there is a lack of environmental health and safety (EH&S) data on nanoparticles, which is required information for many of the inputs of an LCA. Mechanism of release into the environment over the life of the product, toxicity, effect nanoparticles have on the combustion gasses formed during accidental burning or during incineration, are some of the areas where research is needed so the necessary data can be produced and made available to enable companies to perform meaningful LCAs. This challenge is particularly daunting not only when natural nanoparticles, such as clays, are considered, but when engineered nanoparticles are included. Engineered nanoparticles are man-made, using multiple techniques, and often post processed so the number of different varieties of a given nanoparticle can be huge. Furthermore, since many engineered nanoparticles have only been prepared recently very little is known about their EH&S properties, in addition to the economic feasibility of manufacturing them. This is relevant to the nanocomposite FR approach since in the last several years many nanoparticles have been found to have flame retardant properties, such as, layered double hydroxides (LDH), carbon nanofibres, and carbon nanotubes.

One approach which may simplify developing sustainable FR additives is to utilise nanoparticles where a significant amount of favourable EH&S data is already available. Some of the nanoparticles that may fall into this category include: LDH, and cellulose nanofibrils (Fig. 30.3). LDH (Aluminium hydroxide magnesium hydroxide carbonate (hydrotalcite)) and cellulose are approved by the US Food and Drug Administration (FDA) for contact food.

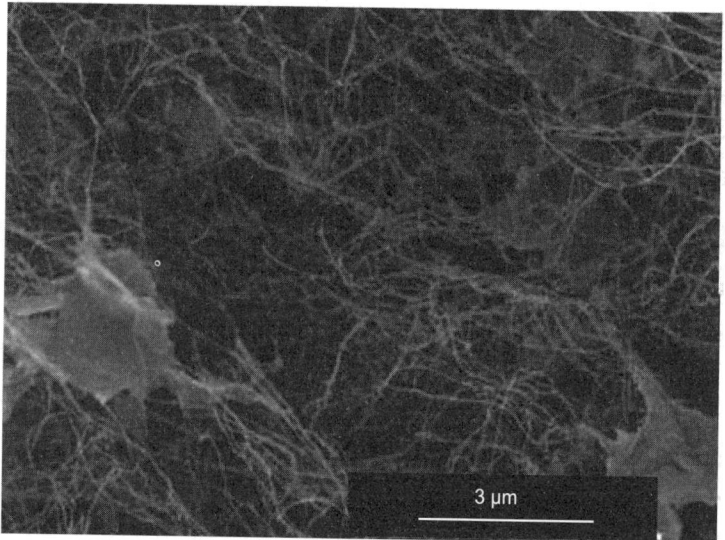

3 μm

Fig. 30.3. SEM image of dried microfibrilated cellulose prepared from dissolving pulp with enzyme treatment.

Thus in the current economic and environmental climate it is more critical than ever to develop the tools and information to enable quantitative evaluation of the sustainability of utilising nanotechnology in products.

This is true for nanoparticle based flame retardant products as well. However, only a handful of LCAs has appeared that address some of the issues which must be included. A lack of EH&S data on many nanoparticles will continue to hamper the effort to perform complete LCAs. This suggests that the international research community must gather the necessary EH&S data, while at the same time investigate approaches utilising materials currently known to have favourable EH&S attributes if nanotechnology is going to actually be successfully used in commercial products.

Chapter 31

Nondestructive Testing of Nanocomposites by Optical Techniques

INTRODUCTION

Increased complexity and sophistication of nanocomposites and structures, together with the demand for lighter, stronger and more reliable materials, naturally requires a more detailed knowledge of any flaws which may be present in the components. New manufacturing processes and fabrication techniques to realise such structures need to address the problems caused by flaws which might be encountered in production or developed in service.

The main objective of this chapter is to discuss the three optical methods:

1. Holographic interferometry (HI).
2. Electronic speckle pattern interferometry (ESPI).
3. Shearography.

These methods are well suited for identifying the flaws in nanocomposites developed during machining or during course of operation. Different stressing techniques which are essential parts of the above have been outlined. The stressing techniques are able to induce extremely small stresses without damaging the nanocomposite structures. Excitation with low level vibration from suitable sources generate and reveal modal geometry, resonant behaviour and other characteristics in addition to structural and material defects that may be present.

These techniques have the advantages of: (i) noncontact and nonintrusive method, (ii) resolution is very high of the order of 0.3 μ when He-Ne laser is used, (iii) moving objects can be recorded using pulsed lasers, (iv) specimen preparation is generally not required, and (v) applicable to any solid material.

Nanocomposites are the emerging area and a lot of work is being carried in some developed countries and a few organisations in India. Nanocomposites are finding various applications in space, defense, nuclear, electronics, environment, medicine, etc. While a lot of work is going in this area and many new nanocomposite materials are being evolved, at the same time the characteristics and performance of these new materials are very important under particular conditions. The permanent demand for increasing product quality requires the use of reliable, fast and economical techniques for quality assurance and flaw detection. To encounter such problems, the measuring techniques required to be used, should be on-line, full field, nondestructive and nonintrusive. Optical techniques have become an interesting alternative for nondestructive testing over conventional methods because of the following advantages:

1. Noncontact and nonintrusive method making inspection distance flexible and there is nondamage to the specimen.

2. Full-field: specimens ranging from few mm to few square meters can be studied in a single exposure enabling huge time saving or high inspection rate.
3. Moving objects can be recorded using pulsed lasers.
4. Results can be processed using an intelligent software.
5. Specimen preparation is generally not required.
6. It is applicable to any solid material.

Most of the optical techniques have the above characteristics. They are a fundamental tool to save costs because time to perform quality control and inspection personnel can be reduced substantially. In this chapter holographic interferometry (HI), electronic speckle pattern interferometry (ESPI) and shearography techniques are discussed with special reference to their applications in nanocomposites. Though photoelasticity is also a well-known optical technique, it is not discussed here, because it always need specimen preparations and consumables and so it is not suitable for fast inspection but rather for research.

The above techniques have extensive applications in nondestructive testing, quality control, inspection and design optimisation. Some of the important industrial applications are:
1. Displacement measurement.
2. Stress intensity measurement.
3. Crack detection.
4. Defects identification.
5. Determination of elastic constants.
6. Thermal expansion coefficients, etc.

The response of a structure depends upon the type of stressing. Stressing techniques usually employed for the above applications are:
1. Acoustic vibration.
2. Thermal excitation.
3. Pressurisation.
4. Mechanical stressing.

The choice of method depends on:
1. Component itself.
2. Defect to be identified.
3. Accessibility of the components, i.e. whether the components can be tested by itself or must remain an integral part of a more complex system.

HOLOGRAPHIC INTERFEROMETRY

HI and speckle interferometry have been used to extend the methods or classical interferometry to investigate a much wider range of applications.

Holography is three-dimensional photography. Dr. Denis Gabor, who is known as father of holography, coined the term which is derived from the Greek 'holos' meaning 'the whole' and 'graphy' which means 'writing'. Holography means recording the entire wave, i.e. both phase and amplitude in the form of an interference pattern. Light waves scattered or diffracted by an object carry all information about the shape, intensity distribution and surface texture of the object. A photographic plate records intensity information by a well-understood simple process. Recording of phase information is slightly complicated and is achieved by superimposing a reference wave coming from the source with the scattered light from the object on the film. Hence, an interference pattern is recorded in holography, unlike recording of only intensity distribution in conventional photography.

The simple optical schematic and components required for recording holograms are shown in Fig. 31.1. A laser beam is split into two components, one of which is directed towards the object and the other directly to the photographic plate. Light waves scattered from the object combine with those from the direct beam to form a complex interference pattern, which is recorded on a high-resolution photographic plate. Both the amplitude and phase information of the scattered object waves is contained in the interference pattern.

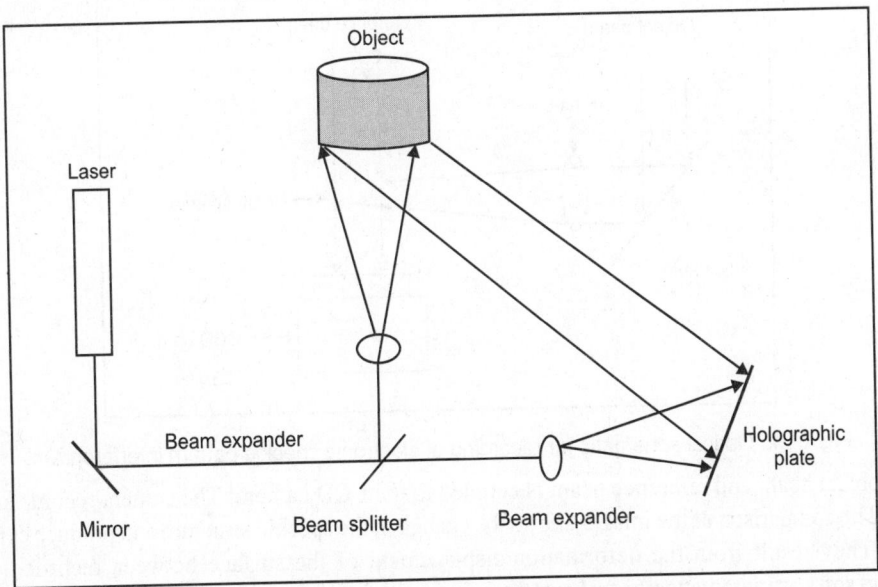

Fig. 31.1. Optical schematic for recording hologram.

After developing the plate, it is illuminated with a coherent light source of same wavelength with which it was recorded. The complex interference pattern will diffract this light to produce a reconstructed wave pattern indistinguishable from the original object wave.

To the eye, these diffracted waves appear to form the original object itself in three dimensions behind the hologram.

Holometry combines both holography and interferometry making it a powerful nondestructive and noncontact technique. This technique not only detects material defects but also measures the strength of materials. Double exposure, real time and time average holography are three holometric methods of making an interferogram. Each type of interferometry has its own advantages in a particular test situation. Double exposure holography is the process of recording a hologram by superimposing the hologram of the object under stress with the hologram of the same object in normal condition on the same recording medium. When the processed double exposed hologram is positioned for reconstruction, a fringe pattern is observed upon the virtual image of the object. The fringe pattern represents the contour map of the displacement field which is weighed for quantitative and qualitative analysis of the object under stressed conditions.

ELECTRONIC SPECKLE PATTERN INTERFEROMETRY

This technique is also called TV-holography or digital holography. It was first demonstrated by Butters and Leendertz in 1971 as a method of producing interferometric data without using traditional holographic recording technique. The premise was to use a CCD camera in place of film to record a low spatial frequency hologram. A schematic arrangement for recording of electronic speckle pattern interferogram is shown in Fig. 31.2. The main differences between ESPI and holography are primarily in the optical set up and data processing. Since ESPI records images at video rate it is thousands of times faster than holographic technique making it suitable for online inspection.

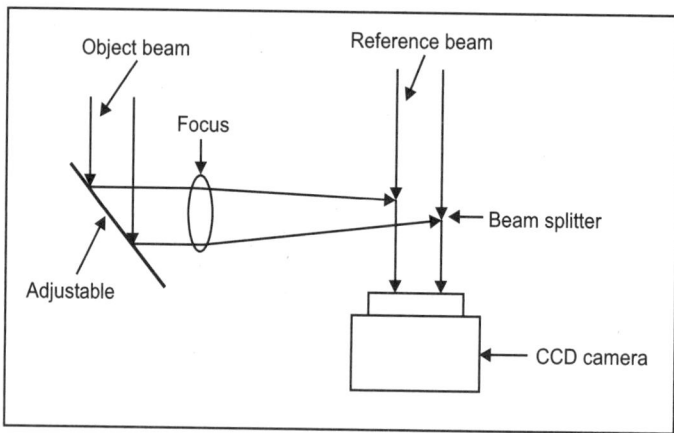

Fig. 31.2. Optical schematic for recording of electronic speckle pattern interferogram.

The object beam and reference beam is coupled to the CCD camera. The camera records a speckle images. The comparison of the images shows the change in the speckle structure and produce correlation fringes. They result from the deformation/displacement of the surface between recorded images. Intelligent software automatically analyses these fringes and calculates quantitative displacement values. From this data, strains, stresses, vibration modes, etc. can be derived. Material industry uses such technique to measure Young Modulus, Poisson's ratio, crack growth, true strain/true stress functions and other material parameters which are required to characterise new materials. High speed measuring system can also deliver dynamic material curve which are used for crash tests and crash simulation.

SHEAROGRAPHY

Shearography is also a speckle interferometry technique, but the optical setup is slightly modified. An optical schematic is shown in Fig. 31.3 for recording a shearogram. In this technique instead of a reference beam, the image of the object is doubled, laterally sheared and superimposed in the CCD camera. A speckle pattern is produced which shows the deformation gradient of the surface being tested or analysed. Shearography measures only the gradient of deformation, it is relatively insensitive to rigid body motions. Typical optical setup for recording a shearogram is shown in Fig. 31.3.

This technique is widely used in NDT and NDI. During production process of modern materials, many different components are bonded together. Usually, parts of this assembly process are carried out manually. Therefore, it is of utmost importance for product reliability and quality control carry out NDT at certain steps of production line.

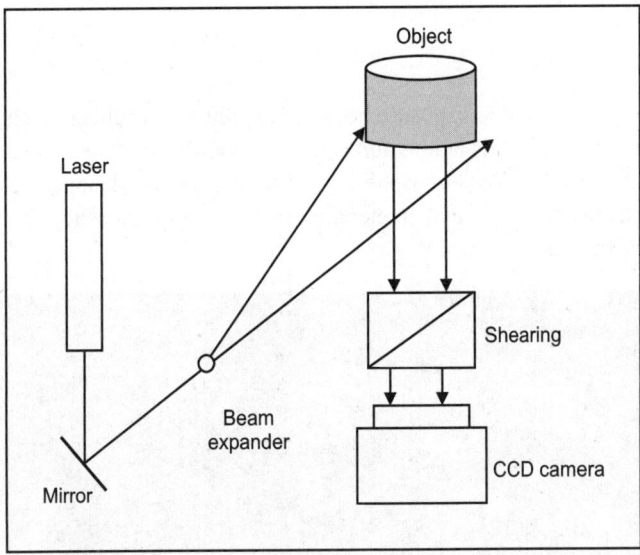

Fig. 31.3. Optical schematic for recording shearogram.

Aerospace industries use shearography to test composite materials of glass fibre reinforced plastics, carbon fibre reinforced plastics (CFRP), GLARE, foam and aluminium composites, etc.

Hand portable shearography is now a reality. These portable shearography systems are currently in production use in the Airbus A 380 commercial aircraft program.

Typical uses include inspection of debond, detection or delamination, and characterisation of bonds between composites and metal parts. Advanced shearography technique is readily adaptable to a wide range of applications. Recently shearography has been validated for maintenance repair of concord parts.

COMPARISON OF HOLOGRAPHY, ESPI AND SHEAROGRAPHY

The main differences among three techniques are primarily in the optical setup and data processing as mentioned above. The above three methods have been compared with special reference to the defect identification of composites.

Since the composite material is a combination of two or more materials, the likelihood of having flaws in composite materials is generally higher than in metals. Consequently, there is a need to monitor the integrity of composite structures during and after fabrication. Since flaws and damage may develop during service, nondestructive inspections are also required in service.

All the three methods will reveal imperfections in composite materials. Shearography is more practical as it employs a very simple setup and does not require special vibration isolation. Furthermore, shearography generally reveals defects more prominently. This is because shearography measures displacement derivatives whereas holography and ESPI measure displacements. Since defects normally create strain concentrations, it is easier to correlate defects with strain anomalies rather than displacement anomalies. In particular, shearography, which measures displacement gradients, is not sensitive to rigid body motions.

EXAMPLES

Example 1

Shows an illustration of the double exposure hologram applied to a rubber-to-aluminium laminate. A uniform vacuum was applied to this lamination and then returned to atmospheric pressure. The rubber continues to creep after the return to atmospheric pressure. The small debonds between the rubber and the aluminium, which are about 1/4″ in diameter are easily detected since they are contoured by a series of concentric circular fringes (Fig. 31.4).

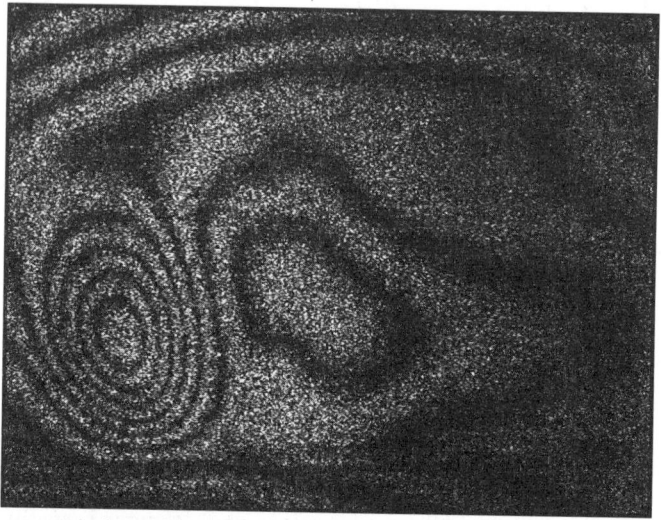

Fig. 31.4. Double exposure holographic fringe pattern showing debonds in a rubber-to-aluminium laminate composite structure under uniform vacuum load.

Example 2

Fringe pattern of both holographic and ESPI images of a honeycomb panel with a debonded surface (Fig. 31.5).

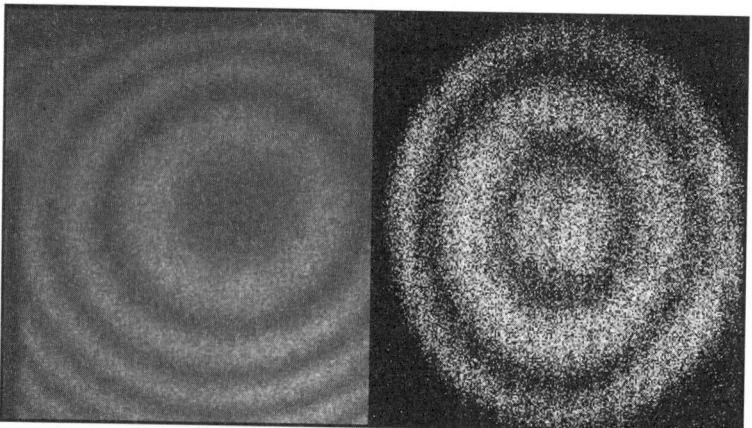

Fig. 31.5. Holographic and ESPI fringe pattern of a honeycomb panel showing debonds.

Example 3

Real time holographic fringe pattern shows internal structural flaw in an advanced graphite-epoxy composite material when local thermal stress of 15°C temperature was applied locally on the back of the composite material. The non-uniform displacement distribution indicates the location of the internal flaw (Fig. 31.6).

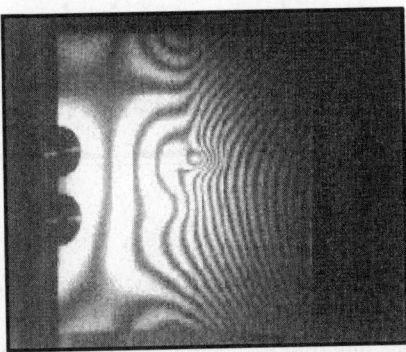

Fig. 31.6. Real time holographic fringe pattern showing internal structural flaw in an advanced graphite-epoxy composite structure under nonuniform vaccum load.

Example 4

Inspection of a sandwich component (carbon fibre and foam) shows delaminations at –70 mbar using shearography (Fig. 31.7).

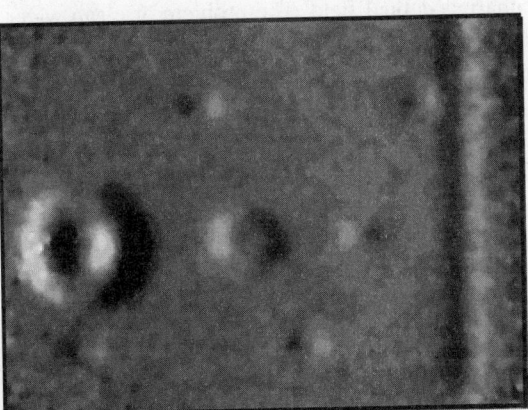

Fig. 31.7. Inspection of a sandwich component showing delaminations.

Thus, all techniques discussed in this chapter are having their own advantages and disadvantages for a particular test situation. With special reference to the development of new materials like nanocomposites a lot of work is going on. In developing such materials various combination of two or more different materials and their performance is required to be studied frequently, quickly and effectively. For such development and production shearographic NDT could be the most suitable candidate for defects identifications a to cut costs of inspection. It is generally used qualitatively, less susceptible to environment noise and operating environment and typically requires less technical understanding.

Chapter 32

Nanoclays for Polymer Nanocomposites, Paints, Cosmetics and Waste-water Treatment

INTRODUCTION

An overview of nanoclays or organically modified layered silicates (organoclays) is presented with emphasis placed on the use of nanoclays as the reinforcement phase in polymer matrices for preparation of polymer/layered silicates nanocomposites, rheological modifier for paints, inks and greases, drug delivery vehicle for controlled release of therapeutic agents, and nanoclays for industrial waste-water as well as potable water treatment to make further step into green environment. A little amount of nanoclay can alter the entire properties of polymers, paints, inks and greases to a great extent by dispersing 1 nm thick layered silicate throughout the matrices. The flexibility of interlayer spacing of layered silicates accommodates therapeutic agents which can later on be released to damaged cell. Because the release of drugs in drug-intercalated layered materials is controllable, these new materials have a great potential as a delivery host in the pharmaceutical field. The problem of clean water can be solved by treating industrial and municipal waste water with organoclays in combination with other sorbents like activated carbon and alum. Organoclays have proven to be superior to any other water treatment technology in applications where the water to be treated contains substantial amounts of oil and grease or humic acid.

Organically modified layered-silicates or nanoclays have become an attractive class of organic–inorganic hybrid materials because of their potential use in wide range of applications such as in polymer nanocomposites, rheological modifier in paints, inks, greases and cosmetics, adsorbent for toxic gases, effluent treatment and drug delivery carrier. The generic term, layered silicates, refers to natural clays as well as synthesised layered silicates such as montmorillonite, laponite and hectorite.

The most commonly used clay in the synthesis of polymer nanocomposites is montmorillonite (MMT) which is the major constituent of bentonite. It is well-known that filler anisotropy, i.e. large length to diameter ratio (aspect ratio), is especially favourable in matrix reinforcement. Due to unique structure of montmorillonite, the mineral platelet thickness is only one nanometer, although its dimensions in length and width can be measured in hundreds of nanometers, with a majority of platelets in 200–400 nm range after purification. Due to very small size and thickness of the platelets, a single gram of clay contains over a million individual platelets. The term polymer layered silicate nanocomposites describes a class where the reinforcing phase, in the shape of platelets, has only nanolevel dimensions. There is substantial improvement in mechanical and physical properties of nanocomposites and this too at a very low silicate content (3–6 wt per cent). Improved mechanical and thermal properties are of interest for under-the-hood applications in the automotive industry. Excellent barrier properties combined with good transparency make these materials ideal for packaging applications. The era of polymer nanocomposites received an impetus after the work of a researcher from Toyota in 1987. Toyota discovered the possibility of synthesising polymer nanocomposites based on nylon-6/organophilic montmorillonite clay that showed dramatic improvements in mechanical and physical properties and heat distortion temperature at very low content of layered silicate. The intercalation chemistry of polymers when mixed with appropriately modified layered silicate and synthetic layered silicates has long been known. The field of polymer/layered silicate nanocomposites has gained momentum by the observation of Vaia wherein they have shown that it is possible to melt-mix polymers with layered silicates, without the use of organic solvents. Today, efforts are being made globally for using nanoclays in almost all types of polymer matrices.

In addition to organically modified natural montmorillonite, synthetic layered silicates such as laponite, hectorite and saponite have been used as rheological modifiers in paints, inks, greases and cosmetics. Organoclays obtained by interaction of these layered silicates with ammonium or phosphonium salts act as thixotropic agent in the above applications. A small addition of nanoclays can greatly enhance the rheological properties of the paint system. These properties prevent pigment settling and sagging on vertical surfaces and gloss is minimally affected due to the low levels of addition. Thermal stability of grease is greatly enhanced by the addition of small amount of organoclays. Nanoclays provide colour retention as well as good coverage in cosmetics and inks. The organic binds to the ionic surfaces of layered silicates and converts it from a hydrophilic form to an oilwet, a hydrocarbon adsorbent material, ideal for water treatment applications. When used for water treatment, organoclays are commonly utilised in the upstream sector of the petroleum industry for removing hydrocarbons from refinery process water, but it has seen little use in petroleum production. Organoclays have also been tested for treating ground and surface water and for other toxic organic chemicals from pharmaceuticals and pesticides industries. Organoclays can offer dramatic performance improvements in many other adsorption applications, including removing oil, grease, heavy metals, and polychlorinated biphenyl; organic matter, such as humic and fulvic acids; polynuclear and polycyclic aromatics; and sparingly soluble hydrophobic, chlorinated organics. Removing radionuclides, including pertechnetate, from water is another application with tremendous potential. Nanoclays are potentially useful materials in the field of controlled release of therapeutic agent to patients, where it acts as a drug vehicle. MMT could adsorb dietary toxins, bacterial toxins associated with gastrointestinal disturbance, hydrogen ions in acidosis, and metabolic toxins such as steroidal metabolites associated with pregnancy. All these conditions result in a host of common symptoms, including nausea, vomiting and diarrhea, most of which are typical symptoms of the side effects caused by anticancer drugs.

The focus of discussion of this chapter is on nanoclays which includes polymer-layered silicate nanocomposites, rheological modifier in paints, inks, greases and cosmetics, nanoclays as drug carrier in medicinal application and treating industrial waste-water. The replacement of the inorganic exchangeable cations in the interlayer spacing of the layered silicates by alkyl ammonium surfactants can compatibilise the surface chemistry of the clay and the hydrophobic polymer matrix or liquor which is the key property of nanoclays suitable for above applications.

CLAYS AND THEIR MODIFICATION

Clays are naturally occurring minerals with variability in their constitution depending on their groups and sources. The clays used for the preparation of nanoclays belong to smectite group clays which are also known as 2:1 phyllosilicates, the most common of which are montmorillonite $\{Si_4[Al_{1.67}Mg_{0.33}]O_{10}(OH)_2 \cdot nH_2O \cdot X_{0.33} = Na, K \text{ or } Ca\}$ and hectorite $\{Si_4[Mg_{2.7}Li_{0.3}]O_{10}(OH)_2 \cdot X_{0.4} = Na\}$, where octahedral site is isomorphically substituted. Other smectite group clays are beidillite $\{[Si_{3.67}Al_{0.33}]Al_2O_{10}(OH)_2 \cdot nH_2O \cdot X_{0.33} = Na, K \text{ or } Ca\}$, nontronite $\{[Si_{3.67}Al_{10.33}]Fe_2O_{10}(OH)_2 \cdot X_{0.33} = Na, K \text{ or } Ca\}$ and saponite $\{[Si_{3.67}A_{0.33}]Mg_3O_{10}(OH)_2 \cdot X_{0.33} = Na, K \text{ or } Ca\}$ in which tetrahedral site is isomorphically substituted. Crystal lattice of smectite group clay consists of a two-dimensional, 1 nm thick layers which are made up of two tetrahedral sheets of silica (SiO_2) fused to an edge-shaped octahedral sheet of alumina.

The lateral dimensions of these layers vary from 30 nm to several microns depending on the particular silicate. Stacking of the layers leads to a regular van der Waals gap between them called the interlayer or gallery. Isomorphic substitution within the layer by Mg^{2+}, Fe^{3+}/Fe^{2+} or Al^{3+} generates negative charges that are normally counterbalanced by hydrated alkali or alkaline earth cations (Na^+, K^+, Ca^+, etc.) residing in the interlayer as shown in Fig. 32.1. Because of the relatively weak forces existing between the layers (due to the layered structure), intercalation of various molecules, and even polymer, is facile.

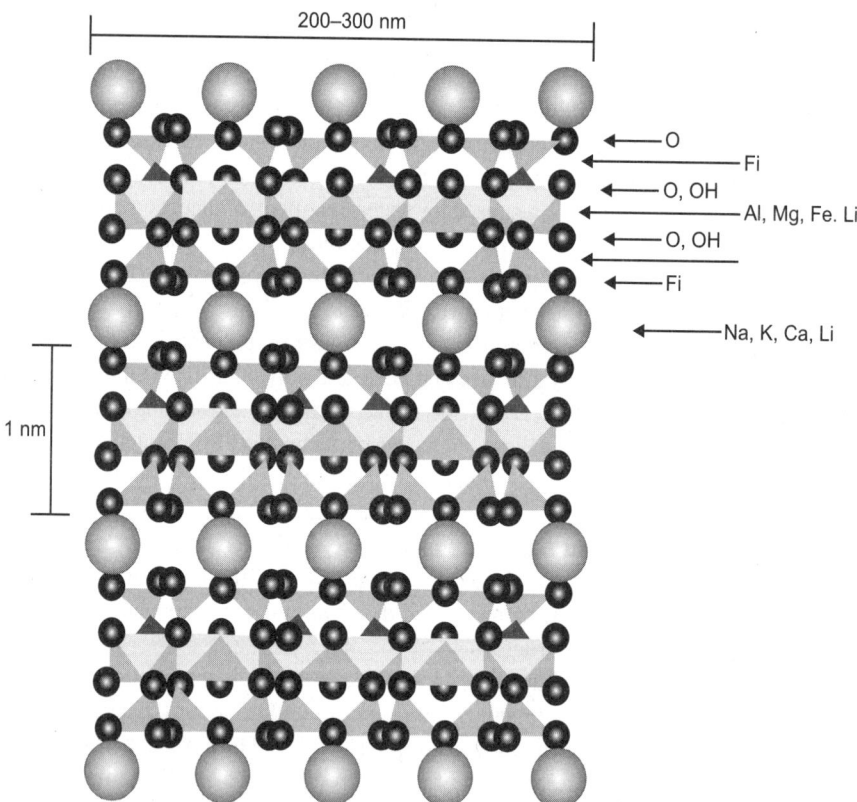

Fig. 32.1. The structure of 2 : 1 layered silicates.

The clay platelets are truly nano-particulate. In the context of nanocomposites, it is important to note that the molecular weight of the platelets (*ca.* 1.3×10^8 Dalton) is considerably greater than that of typical commercial polymers, a feature which is often misrepresented in schematic diagrams of clay-based nanocomposites. In addition, platelets are not totally rigid, but have a degree of flexibility. The clays often have very high surface areas, up to hundreds of m^2/g of clay. The clays are also characterised by their cation exchange capacities, which can vary widely and depends on source and type of clay. The purity of the clay can affect the final nanocomposite properties; due to this it is very important to have montmorillonite with minimum impurities of crystalline silica (quartz), amorphous silica, calcite, kaolin etc. The technique mainly used for purification of clays includes hydrocyclone, centrifugation, sedimentation method and chemical treatment. Clays are inexpensive materials, which can be modified by ion exchange, metal/metal complex impregnation, pillaring and acid treatment to develop catalysts with desired functionality.

Hectorite is a clay mineral similar in structure to MMT but with more negative charges on its surface. It is also hydrophilic swelling clay composed of silicate sheets which delaminate in water to provide an open three dimensional structure. Hectorite clays have the ability to thicken water and are widely used as rheological additives in waterborne coatings and inks.

One important consequence of the charged nature of the clays is that they are generally highly hydrophilic species and therefore, naturally incompatible with a wide range of non-polar systems. Organophilic clay (also known as nanoclay) can be obtained by simply the ionexchange reaction of hydrophilic clay with an organic cation such as an alkyl ammonium or phosphonium ion. The inorganic ions, relatively small (sodium), are exchanged with more voluminous organic onium cations.

This ion-exchange reaction has two consequences; first, the gap between the single sheets is widened, enabling organic cations chain to move in between them and second, the surface properties of each single sheet are changed from being hydrophilic to hydrophobic or organophilic as shown in Fig. 32.2.

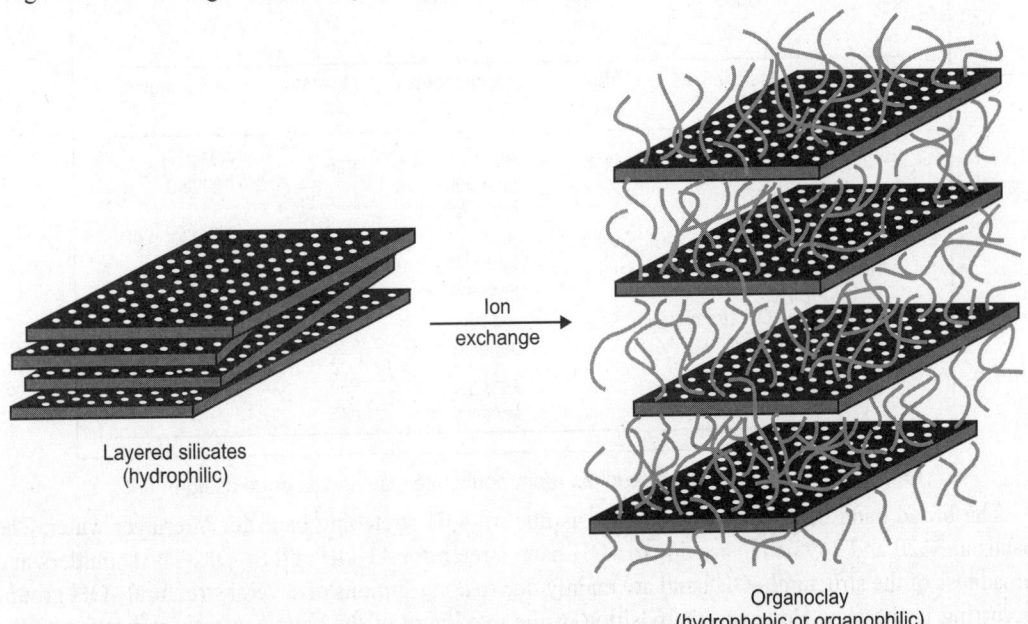

Layered silicates
(hydrophilic)

Ion
exchange

Organoclay
(hydrophobic or organophilic)

Fig. 32.2. Schematic of an ion-exchange reaction in layered silicates.

Clays and their modified organic derivatives are characterised using simple as well as modern characterisation tools which include determination of chemical compositions by gravimetric analysis, inductively coupled plasma (ICP) or XRF, cation exchange capacity (CEC) using standard ammonium acetate method, surface area measurement, Fourier transform infrared spectroscopy (FT-IR), powdered X-ray diffraction (PXRD) and others. Generally, ionic formula is computed on the basis of its chemical compositions, charge density and cation exchange capacity of clays which provide information about the types of layered silicates. The instrumental techniques mainly, FT-IR and PXRD are basic methods for identification of clay structure. The FTIR spectrum for montmorillonite clay recorded as KBr pellet and its band assignments are shown in Fig. 32.3.

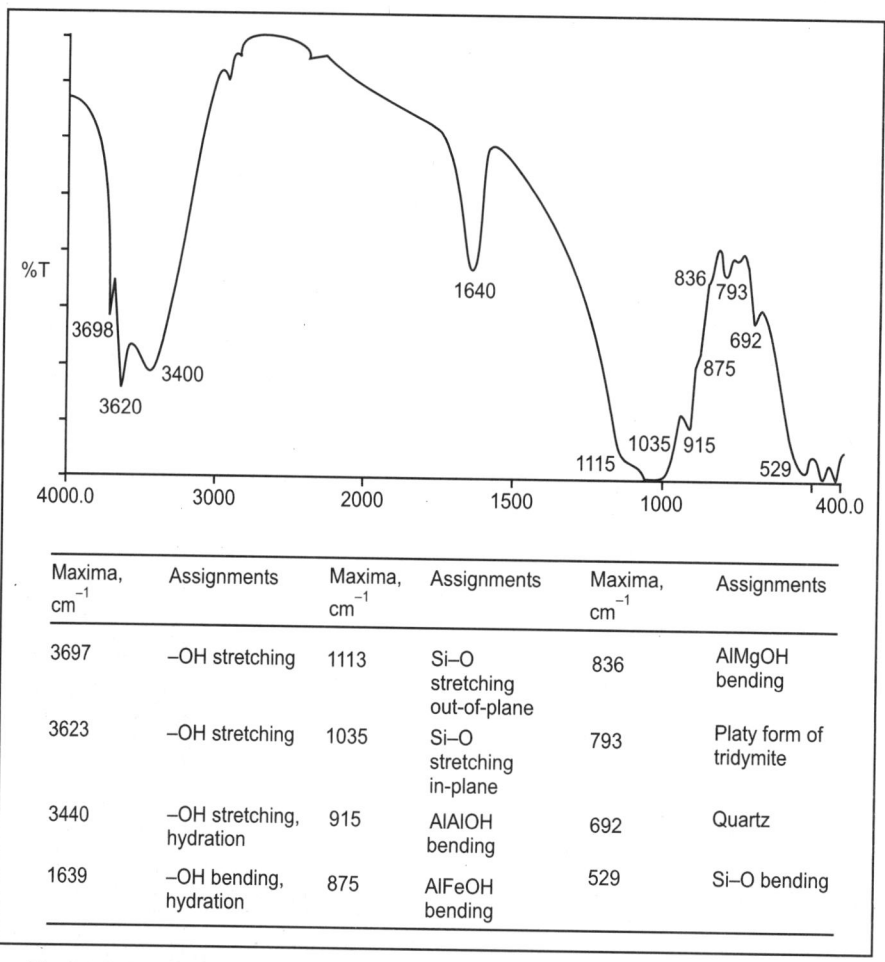

Maxima, cm^{-1}	Assignments	Maxima, cm^{-1}	Assignments	Maxima, cm^{-1}	Assignments
3697	–OH stretching	1113	Si–O stretching out-of-plane	836	AlMgOH bending
3623	–OH stretching	1035	Si–O stretching in-plane	793	Platy form of tridymite
3440	–OH stretching, hydration	915	AlAlOH bending	692	Quartz
1639	–OH bending, hydration	875	AlFeOH bending	529	Si–O bending

Fig. 32.3. FT-IR spectrum of purified montmorillonite clay and its band assignments.

The broad band centred near 3400 cm^{-1} is due to –OH stretching band for interlayer water. The bands at 3620 and 3698 cm^{-1} are due to –OH band stretch for Al–OH and Si–OH. The shoulders and broadness of the structural –OH band are mainly due to contributions of several structural –OH groups occurring in the clay. However, the position of the maximum of the band is clearly indicative of the chemical composition of montmorillonite. The overlaid absorption peaks in the region of 1640 cm^{-1} in the FTIR spectrum of purified clay (montmorillonite) is attributed to –OH bending mode in water (adsorbed water). The characteristic peak at 1115 cm^{-1} is due to Si–O stretching (out-of-plane) for montmorillonite. The peak at 1035 cm^{-1} is attributed to Si–O stretching (in-plane) vibration for layered silicates. The IR peaks at 915, 875 and 836 cm^{-1} are attributed to AlAlOH, AlFeOH and AlMgOH bending vibrations, respectively. An organically modified layered silicate indicates vibrational bands of organic modifier without causing any distortion of structure of clay.

PXRD is one of the most important techniques to determine the structural geometry, texture and also to illustrate impurities (kaolin, quartz, calcite, etc.) in layered silicates which are present in clays.

Generally, the PXRD pattern (as shown in Fig. 32.4) indicates that there is presence of impurities such as kaolin (K) and quartz (Q) in raw montmorillonite which are partly or fully removed on further purification by sedimentation. The reflections relative to the planes [001] and [002] confirmed the presence of montmorillonite as main phase.

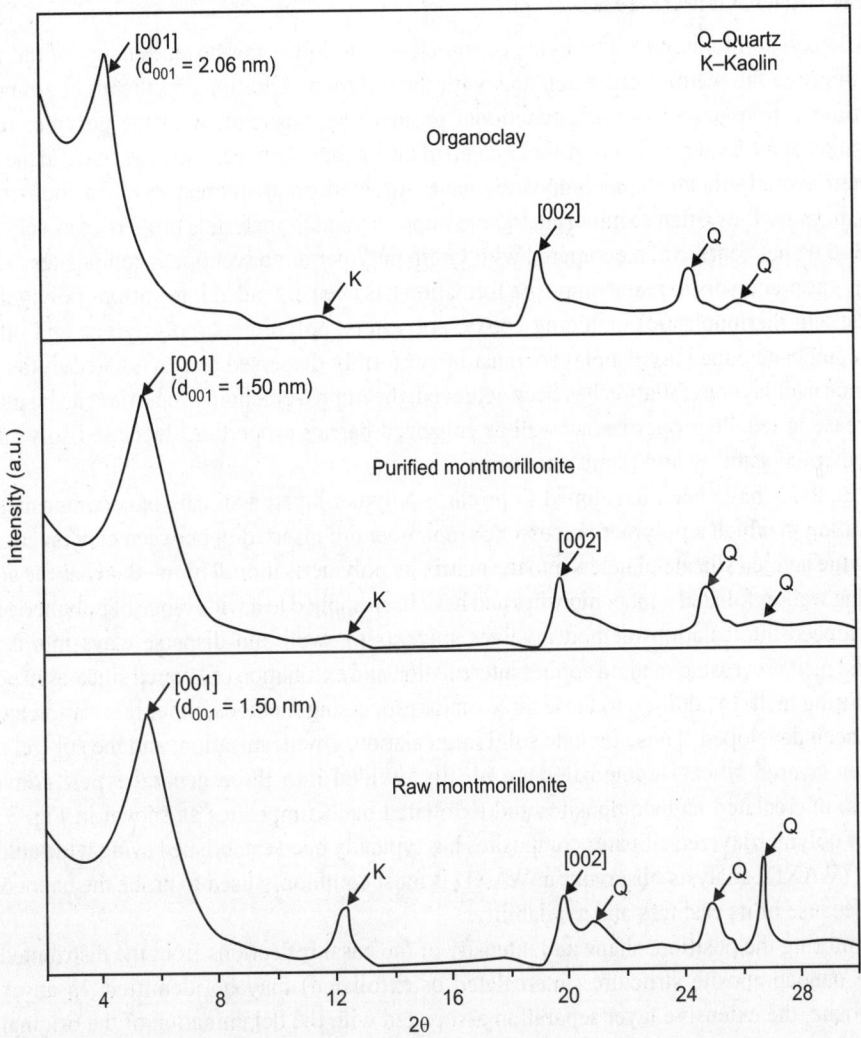

Fig. 32.4. PXRD pattern for raw montmorillonite, purified montmorillonite and organoclay.

The X-ray diffractograms of the organoclay revealed a shift in the position of [001] planes (2θ changed from 5·7°–4·32°C), meaning an increase in the basal spacing of these planes. The increase was relatively large, from 1·5–2·05 nm and confirms the occurrence of organic molecule intercalation between silicate platelets. From the PXRD of Fig. 32.4, it is also observed that the peaks from [002] planes of montmorillonite, were not affected during treatment.

This observation highlights that the unique effect of organic modifier in layered silicates structure is the intercalation of [001] planes of montmorillonite. The basal spacing of layered silicates depends on the kind of organic modifier, with bulkier as well as high concentration of organic modifier and higher interlayer spacing results. The effect of types and concentration of organic modifier on properties of organoclay was studied by FT-IR, XRD and NMR spectroscopy and are well reported.

POLYMER NANOCOMPOSITES

The role of alkyl ammonium cations in the organoclays is to lower the surface energy of the inorganic host and improve the wetting characteristics with the polymer. Additionally, the alkyl ammonium or phosphonium cations could provide functional groups that can react with the polymer or initiate polymerisation of monomers to improve the strength of the interface between the inorganic and the polymer.

Polymer/layered silicate nanocomposites have attracted great interest, both in industry and in academia, because they often exhibit remarkable improvement in materials properties at very low clay content (3–6 wt per cent), when compared with virgin polymer or conventional composites. The use of organoclays as precursors to nanocomposite formation has been extended into various polymer systems (thermostat and thermoplastic) including epoxy, polyesters, polyolefins, polystyrene and others. For true nanocomposites, the clay nanolayers must be uniformly dispersed and exfoliated in the polymer matrix. Once nanolayer exfoliation has been achieved, the improvement in properties can be manifested as an increase in tensile properties, as well as enhanced barrier properties, decreased solvent uptake, increased thermal stability and flame retardance.

Three methods have been developed to produce polymer/layered silicate nanocomposites: *in situ* polymerisation in which a polymer precursor or monomer are inserted in between clay layers and then expanding the layered silicate platelets into the matrix by polymerisation. This method has the advantage of producing well-exfoliated nanocomposites and have been applied to a wide range of polymeric systems; solution-induced intercalation method involves solvents to swell and disperse clays into a polymer solution and melt processing method applies intercalation and exfoliation of layered silicates in polymeric matrices during melt. In addition to these three major processing methods, other fabrication techniques have also been developed. These include solid intercalation, covulcanisation, and the sol-gel method.

Polymer layered silicate composites are ideally divided into three general types: conventional composites, intercalated nanocomposites and exfoliated nanocomposites as shown in Fig. 32.5. The structure of polymer/layered silicates composites has typically been established using wide angle X-ray diffraction (WAXD) analysis observation. WAXD is most commonly used to probe the nanocomposite structure because of its easiness and availability.

By monitoring the position, shape and intensity of the basal reflections from the distributed silicate layers, the nanocomposite structure (intercalated or exfoliated) may be identified. In an exfoliated nanocomposite, the extensive layer separation associated with the delamination of the original silicate layers in the polymer matrix results in the eventual disappearance of any coherent X-ray diffraction from the distributed silicate layers. On the other hand, for intercalated nanocomposites, the finite layer expansion associated with the polymer intercalation results in the appearance of a new basal reflection corresponding to the larger gallery height.

Any physical mixture of a polymer and an inorganic material (such as clay) does not form a nanocomposite. Conventional polymer composites that are prepared by reinforcing a polymer matrix with inorganic materials like reinforcing fibres and minerals have poor interaction between the organic and the inorganic components, which leads to separation into discrete phases. Therefore, the inorganic fillers are required to be added in higher concentrations to achieve enhancements in the thermomechanical properties of the polymer. Table 32.1 shows a comparison of the physical properties of nanocomposites and conventional composites of polyamide. The reasons for the greater effectiveness of the nanoclay in polymer/layered silicates nanocomposites are two-fold: first, the nanoclay can be dispersed to the level of individual platelets. This nano scale dispersion of silicates provides very high surface area for polymer clay interaction. Secondly, the lamellar surfaces of the nanoclay can be modified through an ion exchange reaction to make them compatible with the polymer matrix.

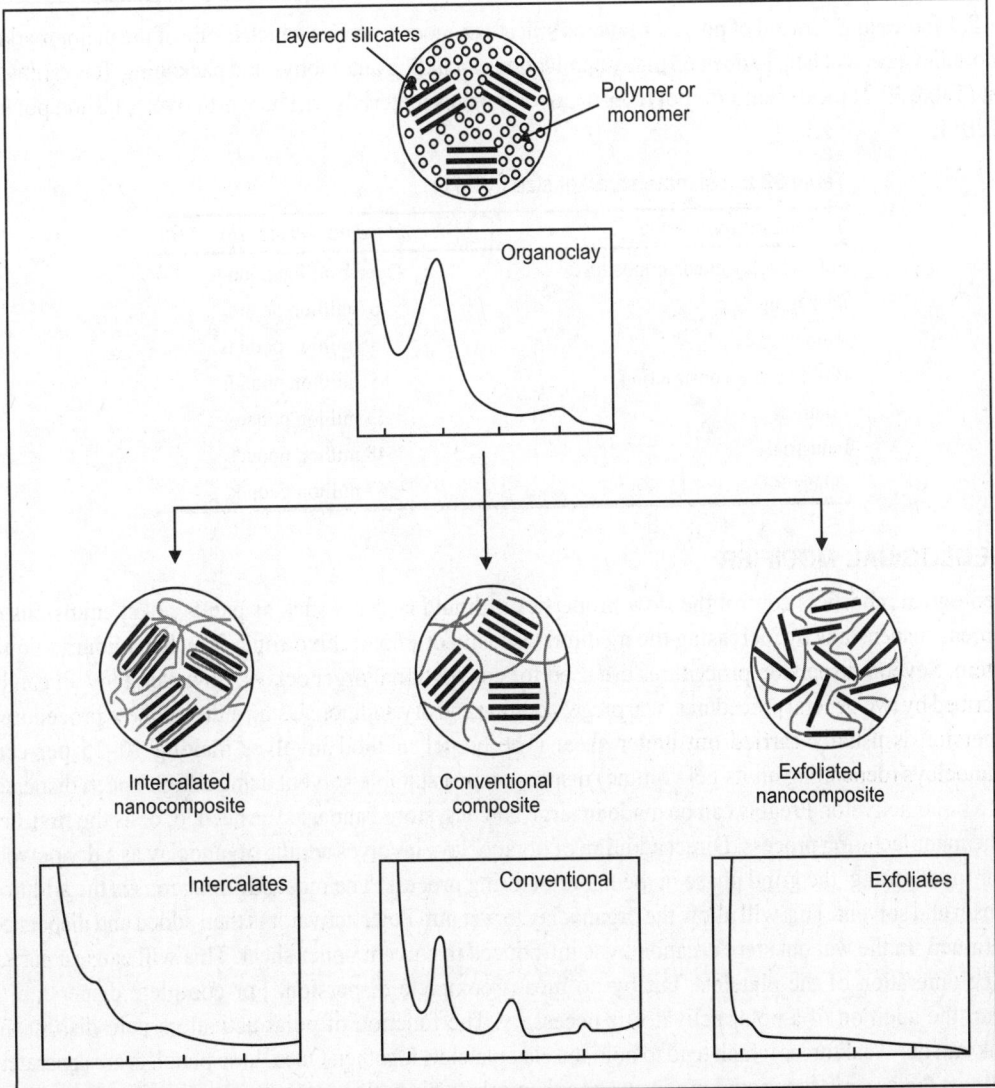

Fig. 32.5. Three idealised structures of polymer/layered silicates composites with their WAXD patterns.

Table 32.1. Comparison of mechanical and physical properties of nylon-6 nanocomposites and conventional composites.

Materials properties	Pristine nylon-6	3–6% Organoclay	30% Mineral	30% Glass fibre
Tensile strength (psi)	7250	11800	8000	23000
Flexural modulus (psi)	120	500	650	1000
Notched izod impact strength (J)	1·2	1·2	1·6	1·8
Heat deflection temperature (°C)	66	110	120	194
Specific gravity (g/cc)	1·3	1·14	1·36	1·35

The increasing demand of polymer/layered silicates nanocomposites which is one of the major markets for organoclays will help to open up many applications including automotive and packaging. It is estimated that (Table 32.2) the demand for polymer nanocomposites materials will be worth over 1 billion pounds by 2011.

Table 32.2. Estimated market size by 2011.

Technology/application	Estimated market size (by 2011)
Polymer/clay nanocomposites	Over 1 billion pounds
Packaging	367 million pounds
Automotive	345 million pounds
Building and construction	151 million pounds
Coatings	63 million pounds
Industrial	48 million pounds
Others	67 million pounds

RHEOLOGICAL MODIFIER

Rheological modifiers control the flow properties of liquid systems such as paints, inks, emulsions or pigment suspensions by increasing the medium viscosity or impart thixotropic flow behaviour to liquid system. Several dispersion procedures are used for conventional organoclays; however, they all can be described by two general procedures, viz. pregel addition and dry addition. Using either of these procedures, dispersion is usually carried out under shear. The pregel method involves making 10–15 per cent organoclays (depending on its gel volume) dispersion in a suitable solvent using a high-speed disperser and a polar activator. Pregels can be made in large batches, stored and used as needed, or as the first step in the manufacturing process. Direct addition of organoclays involves adding organoclay as a dry powder prior to, or during, the grind phase in the manufacturing process. The most advantageous is the addition of resin and solvent. This will allow the organoclay to wet out. Polar activator is then added and dispersion continued. In the wet out step, organoclay is introduced to solvent under shear. This will cause a partial deagglomeration of the platelets, but by no means complete dispersion. For complete dispersion to occur, the addition of a polar activator is necessary. The function of polar activators is to disrupt the weak van der Waal forces which tend to hold the clay platelets together. Once these platelets are separated, it allows the organic functional groups to free themselves from close association with the clay surface. These functional groups are now free to solvate in the organic liquid; i.e. they have a much higher affinity for the organic solvent than the inorganic clay surface (Fig. 32.6). These functional groups are part of the organic modifier that is attracted to the clay surface through electrostatic forces.

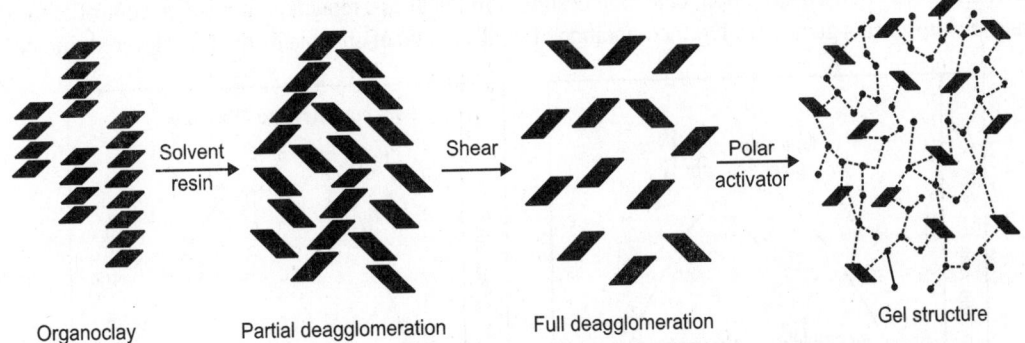

Organoclay   Partial deagglomeration   Full deagglomeration   Gel structure

Fig. 32.6. Mechanism of gel formation.

The typical structure of organoclay consists of layered silicate platelet having a long-chain organic compound bonded to its two faces. In a system containing the fully dispersed and activated organoclay additives, a gel structure is developed by edge-to-edge hydrogen bonding between hydroxyl groups on the organoclay platelet edges. Solvation of the long-chain organic molecule tails makes them stand away from the clay platelet faces. In most cases a chemical activator (also known as polar activator) is added to ensure complete delamination, dispersion, and full activation of the organoclay; however, polar activator free organically modified layered silicates had also been developed by subsequently ion exchanged organic modifier with polar functional groups. Polar activators are defined as low molecular weight compounds of a polar nature. The most commonly used polar activators are propylene carbonate, methanol/water and ethanol/water mixture. Acetone is an excellent polar activator but it is seldom used today due to safety and environmental concerns. Several other low molecular alcohols are also used as polar activator in the industry. While all the polar activators are highly efficient, the methanol/water or ethanol/water combinations are most frequently used due to cost considerations in paint formulation, however, propylene carbonate/water mixture is best suitable for formulating high temperature resistant greases. An optimum amount of polar (chemical) activator must be used to avoid problem of reduction in gel strength. If not enough polar activator is used, then even with the application of shear not all the platelets will be wedged apart. This will result in partial delamination and inadequate gel strength as shown in Fig. 32.7(a), (b). An excessive level of polar activator interferes with hydrogen bonding weakens the gelation forces leading to a reduction in gel strength. It should also be noted that when using an alcohol as a polar activator, it must contain at least 5 per cent water. If water is absent, the polar activator will not function efficiently, thus reducing final product performance. A little is gained by omitting the 5 per cent water addition. Poor gel strength development without the water addition indicates that not enough water molecules were available to form a bridge between the hydroxyls on organoclay platelet edges. There are three main types of mechanisms that occur while adding organoclay into solvent based system: hydrogen bridging or OH-bonds formation, associativity space orientation which results into gel structure as shown in Fig. 32.6. The rheological properties of the paint system are enhanced by small addition of organoclays either by pregel or dry addition as discussed above. The gel formation prevents pigment settling and sagging on vertical surfaces to ensure that the proper thickness of the coating is applied as shown in Fig. 32.8. They also ensure good levelling for the removal of brush marks and storage stability even with high temperatures. Organoclay is used in the ink formulation. It helps to adjust the consistency of printing inks to the desired values, avoiding pigment sedimentation, providing good colour distribution, obtaining desired film thickness, reduction in misting, control of tack, water pickup and dot gain control by incorporation of small amount of organically modified layered silicates.

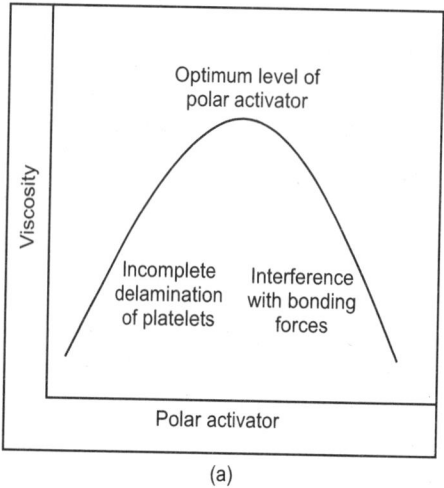 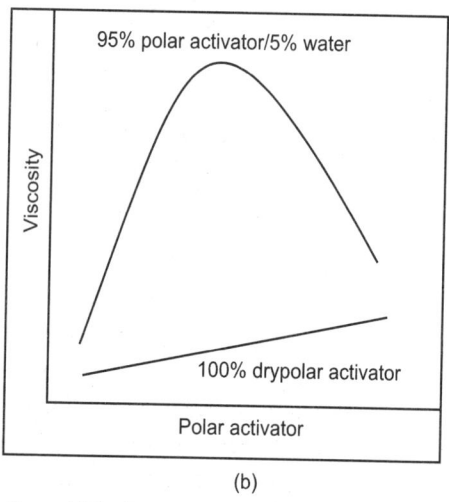

(a) (b)

Fig. 32.7. (a) Effect of amount of polar activator on viscosity and (b) effect of polar activator/water on viscosity of fluid system.

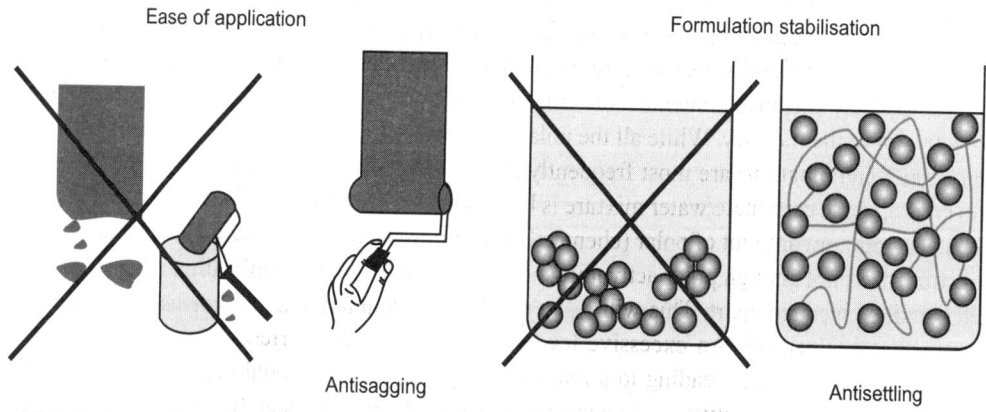

Fig. 32.8. Incorporation of small amount of organoclay improves antisagging and antisettling properties in paints and inks.

Thickening lubricating oils with organoclays can produce specially high temperature resistant lubricating greases. Organoclay also gives good working stability and water resistance to the greases. Such greases are typically used for lubrication in foundries, mills and on high-speed conveyors, agriculture, automotive and mining. The performance of cosmetics is enhanced by the use of organoclays and they allow good colour retention and coverage for nail lacquers, lipsticks and eye shadows. They have been tested to be nonirritant for both skin and eye contact. The applicability of organoclays as rheological modifiers in paints, inks, grease and cosmetics are shown in Fig. 32.9.

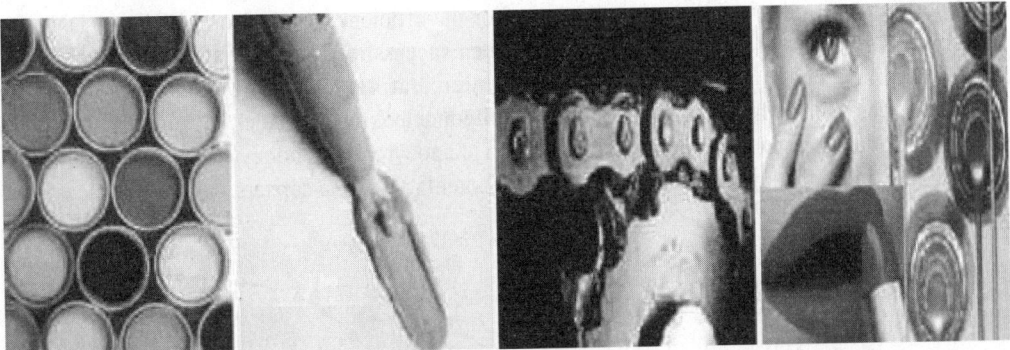

Fig. 32.9. Fields of application (paints, inks, greases and cosmetics).

NANOCLAY AS DRUG VEHICLE

The continuous development of new controlled drug delivery systems is driven by the need to maximise therapeutic activity while minimising negative side effects. One class of drug delivery vehicle that has received more attention in recent years is layered materials which can accommodate polar organic compounds between their layers and form a variety of intercalated compounds. Because the release of drugs in drug-intercalated layered materials is potentially controllable, these new materials have a great potential as a delivery host in the pharmaceutical field. Calcium montmorillonite has also been used extensively in the treatment of pain, open wounds, colitis, diarrhea, hemorrhoids, stomach ulcers, intestinal problems, acne, anemia, and a variety of other health issues. Not only does montmorillonite cure minor problems such as diarrhea and constipation through local application, it also acts on all organs as well.

Yuancai and Si-Shen described the novel poly (d,l-lactide-co-glycolide)/montmorillonite nanoparticle drug delivery system, formulating the drug carrier from a material, which can also have therapeutic effects, either synergistic with or capable to mediate the side effects of the encapsulated drug. Paclitaxel (anticancer drug)-loaded poly (d,l-lactide-co-glycolide)/montmorillonite nanoparticles were prepared by the emulsion/solvent evaporation method and was tested for *in vitro* drug release. The initial burst of 22 per cent on the first day can be observed for sample. After that, the release of paclitaxel was at a slow constant rate. In three weeks, about 36 per cent drug was released with a slightly reduced initial burst and speed release as shown in Fig. 32.10(a). The adsorption and desorption of organic molecules and surfactants on layered silicates indicates that these materials can be used for drug delivery. The release of buformin from buformin/montmorillonite complex and pure buformin hydrochloride in artificial intestinal juice over 360 minutes is presented in Fig. 32.10(b).

Buformin/montmorillonite complex released 70 per cent of buformin with lower rate as compared to pure compound in 360 minutes. Medical devices such as a drug delivery patch, implantable or insertable medical device comprise of polymer carrier (as matrix) and drug intercalated layered silicates (as reinforcement) provides controlled release of therapeutic agent to damaged cell of a patient.

In addition to surface unmodified and modified montmorillonite, layered double hydroxides are also used as drug carrier in various applications. Intercalation of fenbufen in a layered double hydroxide followed by coating with Eudragit® S 100 gives a composite material which shows controlled release of the drug under *in vitro* conditions which model the passage of a material through the gastrointestinal tract. Intercalations of anti-inflammatory drug in layered double hydroxide have the advantage of gradual release over a longer period of time. Gene therapy is gaining growing attention for the treatment of genetic deficiencies and life-threatening diseases. For the efficient introduction of foreign DNA into cells, a carrier system is required. Recently, it has been successfully demonstrated that novel layered double hydroxide could form a nanohybrid by intercalating with bimolecular anion such as mononucleotides, DNA which shows that antisense oligonucleotide molecules packaged in the layered double hydroxide can enter cells, presumably through phagocytosis or endocytosis. The leukemia cells were used to explore the layered double hydroxide's potential as gene carriers.

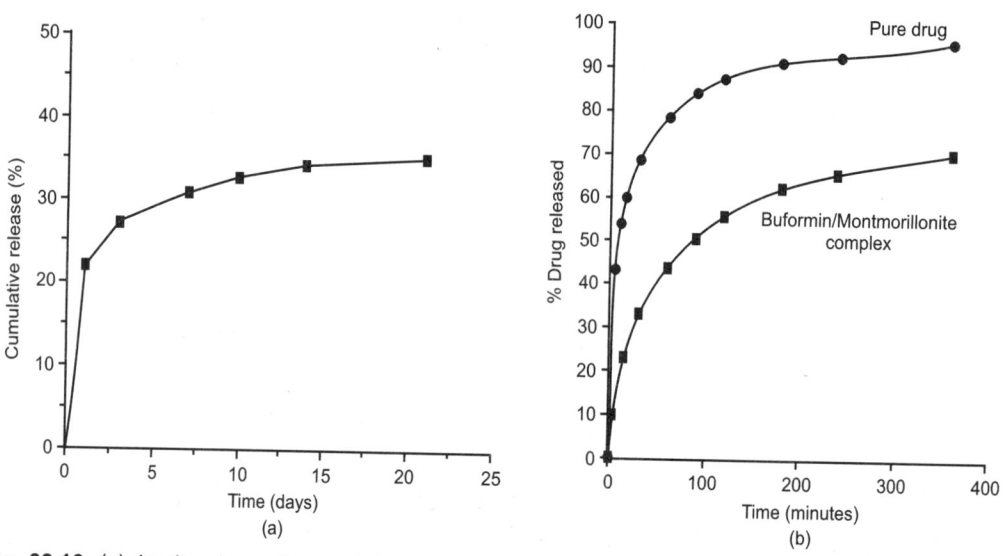

Fig. 32.10. (a) *In vitro* drug release of the paclitaxel-loaded poly (d,l-lactide-co-glycolide)/montmorillonite nanoparticles, and (b) rate of desorption of buformin from the buformin hydrochloride and buformin/montmorillonite complexes in artificial intestinal juice.

WASTE-WATER TREATMENT

The use of organoclays in waste-water treatment has become common in industry today. Organoclays exhibit a synergistic effect with many commonly utilised water treatment unit processes including granular-activated charcoal, reverse osmosis, and air strippers. Granular-activated carbon is particularly effective at removing a large range of organic molecules from water, however, is very poor for removing large molecules such as humic acid and waste-waters containing emulsified oil and grease. Organoclays have proven to be the technology of choice for treating oily waste-waters. Humic acid is one of the common contaminant in potable water and is difficult to remove with conventional flocculation techniques commonly used for drinking water treatment and activated carbon is very ineffective due to its weak interaction with humic acid. The comparative studies for removal efficiencies of humic acid from groundwater using different sorbents are shown in Fig. 32.11.

If humic acid is not removed from drinking water, subsequent chlorination produces unacceptable levels of trihalomethanes which are known carcinogens.

The ability of organoclays to sorbs organics was studied by McBride and Kokai. They concluded that 1 litre of water containing 12,000 ppm of oil reduced to 12 ppm by addition of 10 g of organoclays.

Partitioning is the mechanism responsible for the sorption of organics by organoclays. The organoclay contains alternating organic and inorganic layers. The organic layer is comprised of the quaternary ammonium compounds ion exchanged on the surface. This hydrophobic layer acts as an organic phase into which organic substances that are dissolved in water can partition. The partitioning efficiency should be a function of water solubility of the organic substance since the two phases available to it are water and the quaternary compound in the gallery of the clay.

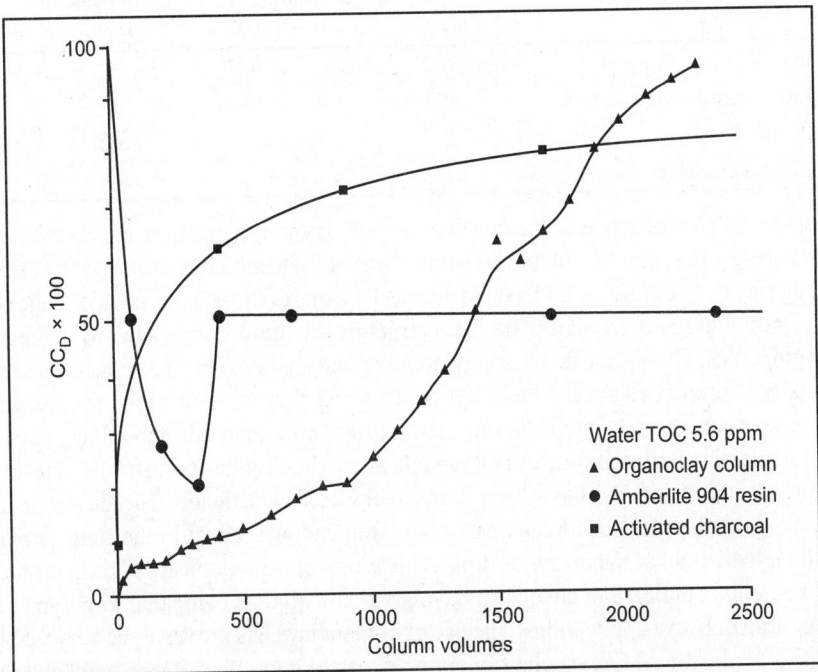

Fig. 32.11. Sorbents (organoclay, amberlite 904 resin and activated charcoal) removal efficiencies for humic acid from groundwater.

Organoclay is one of the ideal materials for treating industrial waste-water. In the process of manufacturing heater coils, as part of their quality control (QC), the company leak tests each unit. The leak testing is conducted by pressurising the coils while they are submersed in large tanks of water. Since the leaks are spotted visually, the water in the tanks must be as clear as possible. As more and more coils are tested, residual Parco88 oil utilised in the manufacturing process builds up in the water, causing the turbidity to climb rapidly. In the past, the test tanks required frequent dumping and refilling, causing problems with discharge of the oily water and the expense of additional water to refill the test tanks. In the initial testing of treatment alternatives for this water, three processes were studied which included pH adjustment, alum flocculation and organoclay column treatment. It was found that pH adjustment was totally ineffective. The alum flocculation test worked reasonably well; however, this process required substantial outlays for equipment and allocation of space, plus a solid waste stream would be produced requiring disposal.

The disposal of such sludges is becoming increasingly more difficult and expensive. The disposal of such waste also leaves the waste generator with a potential long-term liability. The best alternative tested was column treatment of the waste with organoclay. A comparison of the results obtained from the alum/NaOH/polymer flocculation and organoclay column test is given in Table 32.3. As can be seen in the results, the organoclay treatment yielded superior clarity and oil and grease removal when compared to the alum treatment. In addition to leak test tank waste-water treatment, organoclays can be extensively used to treat military base effluent, oil well acid returns, boiler feed water, steam condensate among others.

Table 32.3. Comparison of alum flocculation and organoclay treatment.

Type sample	pH	Turbidity (NTU)	Oil and grease (ppm)	Appearance
Untreated water (a)	6·6	18·0	52	Turbid
(a) Treated with alum-NaOH-polymer	8·0	0·4	32	Clear
Untreated water (b)	7·1	15·0	62	Turbid
(b) Treated with organoclay	7·1	<0·4	<2	Clear

To sum up significant progress in the development of clay/polymer nanocomposites has been made over the past twenty years. The advantages and limitations of the technology have become clear. However, we have a long way to go before we understand the mechanisms of the enhancement of major engineering properties of polymers and can tailor the nanostructure of these composites to achieve particular engineering properties. Finally, the fact that clay/polymer nanocomposites show concurrent improvement in various material properties at very low filler content, together with the ease of preparation through simple processes such as melt intercalation, melt extrusion or injection moulding, opens up a new dimension for plastics and composites. Organoclays as rheological modifier is one of the oldest methodologies in industries and have been extensively used worldwide. The development of polar activator free organoclays in last fifteen years made tremendous impact in the field of paint, ink and greases. Although the field of nanoclays as drug vehicle for controlled release of drug is one of the born age area in medicinal application, nanoclays have great potential as compared to polymer and carbon nanotubes for drug delivery applications. The use of organoclays has proven to be very viable for many water treatment applications. Organoclays operate via partitioning phenomena and have a synergistic effect with activated carbon and other unit processes such as reverse osmosis. They have proven to be superior to any other water treatment technology in applications where the water to be treated contains substantial amounts of oil and grease or humic acid. The commercial application of organoclays to trihalomethane control in drinking water has not yet occurred. With increasing concerns about the carcinogenic effects of trihalomethanes, the commercialisation of this technology could be around the corner.

Chapter 33

Hard and Tough Nanocomposite Coatings

INTRODUCTION

Toughness and hardness are important aspects for coating applications in manufacturing industry. Extensive theoretical and experimental efforts have been made to synthesise and study nanocomposite coatings with super hardness and high toughness. The materials can be hardened through various or combined hardening mechanisms. However, for engineering applications, coating toughness is as important as, if not more than, super hardness. At present, there is neither a standard test procedure, nor a standalone methodology for the assessment of thin film toughness. The determination of the toughness is still a difficult task, and very much a fully open problem. In this chapter, we review the hardening and toughening mechanisms of nanocomposite films, and the toughness characterisation techniques. Based on these reviews, an outlook will be presented in the concluding remarks.

Nanostructured or nanocomposite coatings are a new branch of materials that possess unique physical and mechanical properties. A nanocomposite coating comprises of at least two phases: a nanocrystalline phase and an amorphous phase, or two different nanocrystalline phases. These nanocomposite coatings represent a new class of materials, whose mechanical and tribological properties are not subjected to volume mixture rules, but depend on grain boundary effects, and on synergetic interactions of the composite constituents owing to the size effect. Now, it is accepted that materials can be classified as hard, superhard or ultrahard for hardness over 20, 40 or 80 GPa, respectively. Hard, superhard or even ultrahard nanocomposites have been a hot research topic recently. Hard coatings are used in many applications, for example, cutting and polishing tools, moulds, dies, hard disk and other wear-resistant applications. However, for engineering applications, hardness must be complimented with high toughness, which is a property of equal importance as hardness. Therefore, it is vital to master the formation of hard films with high toughness (Fig. 33.1). Toughness is an important mechanical property related to the materials resistance against shock loads, and it describes the resistance against the formation of cracks resulting from stress accumulation in the vicinity of structure imperfections. In an energetic context, toughness is the ability of a material to absorb energy during deformation up to fracture. According to this definition, toughness encompasses the energy required to create a crack and to enable the crack to propagate until fracture.

Therefore, a high toughness coating has high resistance to formation of cracks under stress, and in the meantime, high energy absorbance to deter crack propagation, whereby preventing chipping, flaking or catastrophic failure. Creating tough, fracture-resistant ceramics has been a focal point of ceramics research for decades. Ceramics have inherently low fracture toughness, and are thus subject to brittle fracture (in contrast with ductile metals, which are much tougher because of the plasticity induced by dislocation motion). It is known that mechanical properties of a solid depend strongly on the density of dislocations, interface-to-volume ratio and grain size.

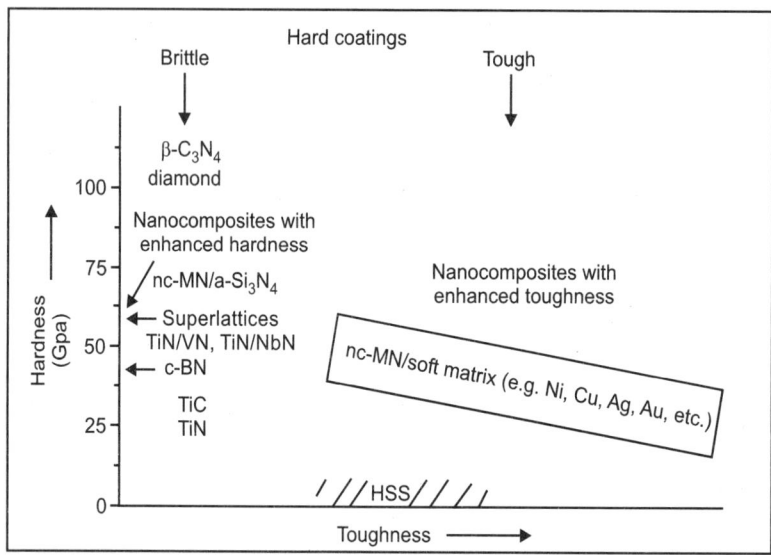

Fig. 33.1. Classification of nanocomposites according to their hardness and toughness.

An enhancement in damping capacity of a nanocomposite solid may be associated with grain-boundary sliding, or with energy dissipation mechanism localised at interfaces. A decrease in the grain size significantly affects the yield strength and hardness. The grain boundary structure, boundary angle, boundary sliding and movement of dislocations are important factors that determine the mechanical properties of the nanocomposites. In most cases, coatings are applied to protect surfaces from the consequences of mechanical loads. A greatly simplified examination of the abrasive wear resistance in a variety of materials as a function of the fracture toughness (as applied to bulk materials) reveals a dependence, namely the maximum wear resistance results from a specific, favourable combination of hardness and toughness.

DESIGN METHODS FOR HARD YET TOUGH NANOCOMPOSITES

Hardening Mechanisms in Nanocomposite Coatings

Hardness is defined as the resistance of a material to plastic deformation. Plastic deformation of crystalline materials occurs predominantly by dislocation movement under applied load. It means that a material with enhanced hardness has a higher resistance to dislocation movement. Very often, several of the classical hardening or strengthening mechanisms are active in hard coatings deposited by plasma-assisted vapour deposition techniques. Such hardening mechanisms work by providing obstacles for the dislocation movements, which can also be applied to some extent to hard films. Hindering of dislocation movements can be achieved by: (i) grain boundary hardening, (ii) solid solution hardening, (iii) age hardening and (iv) compressive stress hardening.

Grain boundary hardening

In a recent review, Veprek has analysed the design criteria to produce superhard or even ultrahard nanocomposite coatings. From material-selection point of view, combinations of nanocrystalline transition-metal nitrides, for example TiN, W_2N and VN, with amorphous Si_3N_4 or BN as the grain boundary phase, have a potential to achieve coating hardness exceeding 40–50 GPa. This hardening behaviour, especially in the regime where the crystallite size d is less than 10 nm, is because the tiny crystallites will restrict the grain boundary sliding in case of a thin amorphous grain boundary. One of the best known theories based on dislocation pile-up is described by the Hall-Petch equation. The relationship indicates that the hardness of the material is inversely proportional to the square root of the grain size. However, as grain sizes are reduced to the nanometer scale, and the percentage of grain boundary atoms increases correspondingly, this traditional view of dislocation-driven plasticity in polycrystalline metals needs to be reconsidered. In a sample with grain diameters of 20 nm, 10 per cent of atoms are located at grain boundaries. Dislocation sources and pile-ups are hardly expected to exist in such a material, and deformation is believed to be carried mostly by the grain boundaries via a certain accommodation mechanism.

For very small grains, the numbers of atoms at inter-granular boundaries and inside grains become comparable. In this case, softening may happen via inter-granular slips, as the inter-granular boundaries cease to be barriers for the motion of dislocations. The search for very hard materials is coupled with the study of low-compressibility solids, which have high values of the bulk (K_b) or shear (G) modulus. Transition-metals-containing compounds have large cohesive energies and high K_b, associated with the distribution of their valence electrons between bonding and anti-bonding states within the partly filled electronic bands. Transition metal carbides, nitrides, borides, or oxides have very low compressibility and also often possess high hardness. Based on thermodynamically driven spinodal segregation in a binary or ternary system, Veprek proposed a design to achieve superhard or even ultrahard coatings; a highly thermal stable nanocomposite structure with ≤10 nm size crystallites of the transition metal nitride separated by about one monolayer of a-Si_3N_4 was synthesised. The approximately one monolayer thin tissue of Si_3N_4 acts as a 'glue' for the MeN nanocrystals to achieve ultrahardness, thus avoiding the reverse Hall-Petch effect. Though the original value given was 103 GPa, more recently studies show that the Fischerscope used normally gives too high hardness values for superhard coatings.

Nevertheless, the design concept should be on the right track. Subsequently, many superhard nanocomposites with similar design have been synthesised, but none approaches the hardness of diamond. Figure 33.2 shows the hardness of nc-TiN/a-Si_3N_4 coatings deposited by close-field unbalanced magnetron sputtering as a function of the silicon nitride fraction. Silicon-free TiN forms elongated crystallites that are several hundreds of nanometers long and tens of nanometers wide (inset [a] in Fig. 33.2). TiN crystallites become very small at increased Si_3N_4 contents. At 15–20 at.-% Si_3N_4, the mean grain size of TiN does not exceed 7 nm; this size is too small for dislocation activities. Therefore, under mechanical loading, such a material can react only by grain-boundary sliding (i.e. by moving single, undeformed TiN nanocrystallites against one another). This process requires more energy than deformation by dislocation movement; hence a higher hardness can be achieved for such coating structures. Estimations of the mean grain separation at this composition show that only a few monolayers of silicon nitride separate the nanocrystallites (inset [b] in Fig. 33.2). At high silicon nitride fractions, the mean grain separation becomes so wide, that ordinary crack propagation in Si_3N_4 takes place, and the hardness approaches that of Si_3N_4 (inset [c] in Fig. 33.2).

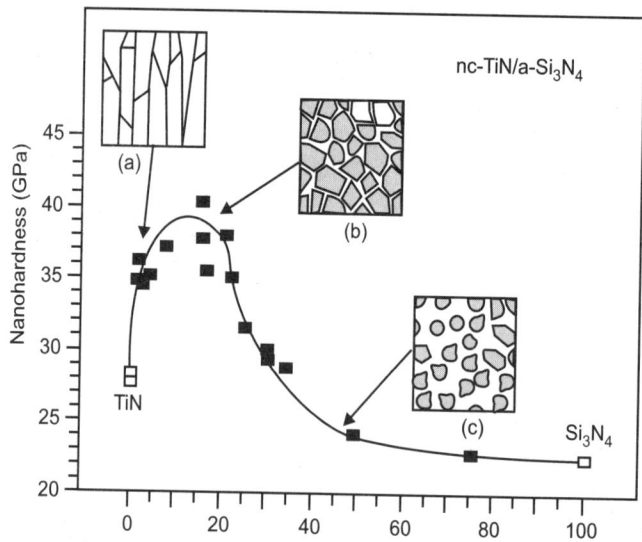

Fig. 33.2. Hardness of nc-TiN/a-Si$_3$N$_4$ as a function of silicon nitride fraction. Insets illustrate the schematic nanostructure for different compositions.

Solid solution hardening

Solid solution hardening comes from the lattice distortion as a result of insertion of atoms of an alloying element in the interstitial location or substation for some of the host atoms. This is, perhaps, one of the oldest hardening methods used in bulk materials. In thin films or coatings, the same principle works too. TiCN, TiAlN, CrAl-N, and CrZrN are a few popular examples. Formation of a non-equilibrium supersaturated solid solution of B in TiN may contribute to further strengthening of the material, since the gliding of the dislocations eventually formed inside the crystallites is hindered by the strain exerted by the insertion of B. To increase the solubility of B in TiN, non-equilibrium growth processes, such as physical vapour deposition (PVD), or plasma-assisted chemical vapour deposition (PACVD) have been used, where a maximum B solubility of 17.4 per cent has been reported for PACVD with hardness up to 43 GPa. For reactive arc evaporation from Ti-B compound targets, the highly ionised flux of film forming species was utilised to synthesise a promising nanocrystalline metastable supersaturated solid solution of B in TiN at lower nitrogen fractions.

The maximum in hardness (34.5 GPa) is obtained when the crystallites are 6–8 nm in size, and are strained by the formation of a substitutional solid solution of B in TiN. The formation of a metastable solid solution was confirmed by XRD-observed widening of the lattice upon B substitution of N as compared to TiN. Therefore, the mobility of deformation-induced dislocations is hindered in these small crystallites that are initially dislocation-free. At high N$_2$ fractions, the substitutional B sites in the TiN crystallites are replaced with additional N to form energetically favourable unstrained TiN crystals. This unstrained TiN crystals combined with high fraction of less strongly bonded BN amorphous phase lead to a hardness drop (Fig. 33.3). Another example is the superhard Ti-Al-Si-N film (maximal hardness ≈55 GPa) synthesised by a hybrid coating system of arc ion plating and sputtering method. Based on XPS and XRD results, it was suggested that the Ti-Al-Si-N films were a composite consisting of solid-solution (Ti, Al, Si)N crystallites and amorphous Si$_3$N$_4$.

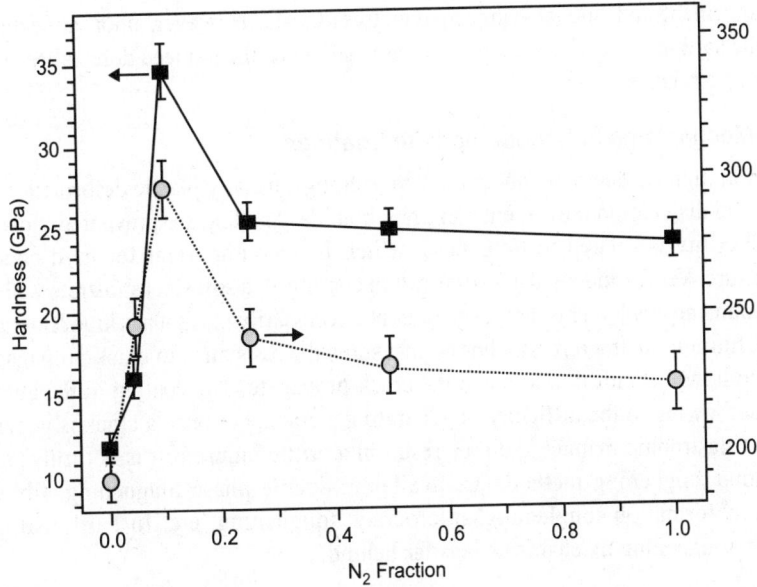

Fig. 33.3. Indentation hardness and reduced modulus values for Ti-B-N coatings for different N_2 fractions.

Age hardening

Age hardening is yet another 'aged' hardening mechanism in strengthening bulk materials (for example Al alloys and steels). Age hardening achieves higher hardness through precipitation over time of a small and uniformly distributed strengthening phase in the form of particles from a supersaturated solid solution. This works equally well in hard coatings. Supersaturated phases can be easily obtained by non-equilibrium deposition techniques such as PVD or PACVD due to limited atomic kinetics during deposition. As-deposited (TiAl)N and Ti(BN) films show a dense columnar microstructure of a supersaturated NaCl structured phased based on fcc TiN whose Ti location is partly substituted by Al to form (TiAl)N, and N is partly substituted by B to form Ti(BN). During annealing, (TiAl)N and Ti(BN) coatings undergo spinodal decomposition and transform into coherent cubic-phase nanometer-size domains, resulting in hardness increase.

Compressive stress hardening

Ion bombardment during deposition at low temperatures has been utilised to increase the density and modify the morphology of the films. During growth, stresses are generally induced by energetic particle bombardment. The lattice defect arrangements induced by impinging energetic ions are responsible for the stress. As the incoming ion or knock-on atoms possess enough kinetic energy, they knock atoms out of their lattice positions and create secondary collisions. In this way, collision cascades. The atom first knocked absorbs most of the energy. The resulting strong atomic motion along the trajectory of the ions leads to a rearrangement of the lattice atoms.

The hardness enhancement results from a complex synergistic effect of the decrease of crystallite size, densification of grain boundaries, and compressive stress upon ion bombardment. Musil reported enhancement of hardness of TiN (up to 80 GPa) and (TiAlV)N (up to 100 GPa) during deposition by means of unbalanced magnetron sputtering at negative substrate bias, where a correlation between the hardness enhancement and biaxial compressive stress exists. However, upon annealing at a certain temperature, the hardness may decrease back to 'normal' when the induced defects are annealed and the compressive stress relaxes.

Toughening Mechanisms in Nanocomposite Coatings

In order to obtain high hardness in nanocomposite coatings, usually plastic deformation is designed to be prohibited, and dislocation movement and grain boundary sliding are prevented, thus causing a loss in ductility. Ductility is related to toughness, which is very important for hard coatings to avoid catastrophic failure. Veprek indented a 6–10 µm-thin coating on a soft steel substrate with a high load of 1000 mN and found no cracks. This, however, does not necessarily mean it has high 'fracture toughness': the classical definition of 'fracture toughness' measures the resistance to crack propagation—a crack exists first, then how difficult it is to make the crack propagate. The coating in the above example is 'damage tolerant' owing to the difficulty in generating a crack, but once a crack is generated, it might propagate in a catastrophic manner and thus result in a brittle failure if it is not truly 'tough'. In bulk ceramics, various toughening methods are available: ductile phase toughening, fibre and whisker toughening, transformation toughening, microcrack toughening, etc. In hard coatings, however, investigation on toughening mechanisms lags far behind.

Ductile phase toughening

In order to toughen hard ceramic films, a straightforward method is to incorporate a ductile phase. Ductile phase toughening arises from two major mechanisms: the relaxation of the strain field around the crack tip through the ductile phase, and the yielding and bridging of cracks by ligaments of the ductile phase. These two mechanisms can increase the work for plastic deformation. Zhang doped Al into a-C films by co-sputtering of graphite and Al. The composite films exhibit a reduction in the residual stress and also an improvement in toughness. However, this came at huge expense of hardness, where the hardness determined by nanoindentation dropped from 31.5 to 8.8 GPa. In order to restore the hardness, nanocrystalline TiC (or nc-TiC) was embedded into the a-C matrix doped with Al to form nc-TiC/a-C(Al) nanocomposite film. The hardness is restored to $\approx$20 GPa, while toughness remained high (indentation plasticity of 55 per cent) and residual stress low (only 0.5 GPa). Although toughness is not measured due to lack of proper measurement facilities for such a high toughness coating, the nc-TiC/a-C(Al) nanocomposite film is evidently very tough: the optical comparison of scanning scratch tracks between nc-TiC/a-C [Fig. 33.4(a)] film and nc-TiC/a-C(Al) [Fig. 33.4(b)] shows that without doping of Al, the coating basically fails in a brittle way [Fig. 33.4(a)]; with doping of Al, however, the scratching tip ploughs into the coating causing the scanning amplitude to reduce along the scratch direction [Fig 33.4(b)]. Co-sputtering of Ti, TiNi and Si_3N_4 targets produced nanocomposite coatings where the amorphous metallic Ni increased the toughness of nc-TiN/a-SiN$_x$ coating. Doping from 0 to $\approx$40 at.% Ni brings about an increase in toughness from 1.15 to 2.60 MPa·m$^{1/2}$, but at the expense of hardness, which diminishes from 30 to 14 GPa.

Phase transformation toughening

Some phase transformation occurs under stress and is accompanied by absorption of a large quantity of energy, which could be used to toughen a material. Transformation from tetragonal to the monoclinic phase takes place under stress. The process occurs in the stress field around the tip of the crack and is accompanied by a volume increase of about 4 per cent. The resultant strain during the transformation relieves the stress field and absorbs the fracture energy, whereby toughens the material. One example is ZrO_2-toughened ZrB_2 composite. In order to facilitate the phase transformation, retention of the high-temperature tetragonal phase is the key, which can be easily realised during deposition. Sprio deposited thin zirconium oxide films by radio-frequency (RF) magnetron sputtering. The films have a mixture of tetragonal and monoclinic zirconia phases. It was shown that increasing the substrate bias power disrupted the columnar grain growth. TEM confirmed a reduction in the intergranular porosity, but also an increase in lateral defects. These defects are hypothesised to be stress-induced microcracks caused by a tetragonal to monoclinic phase transformation. Tetrahedral zirconia has a high toughness and is proposed to increase properties of brittle substrates. Trinh prepared γ-Al_2O_3-ZrO_2 thin films by RF magnetron sputtering with hardness up to 30 GPa. However, owing to the lack of proper toughness measurement instrumentation, toughness was not measured.

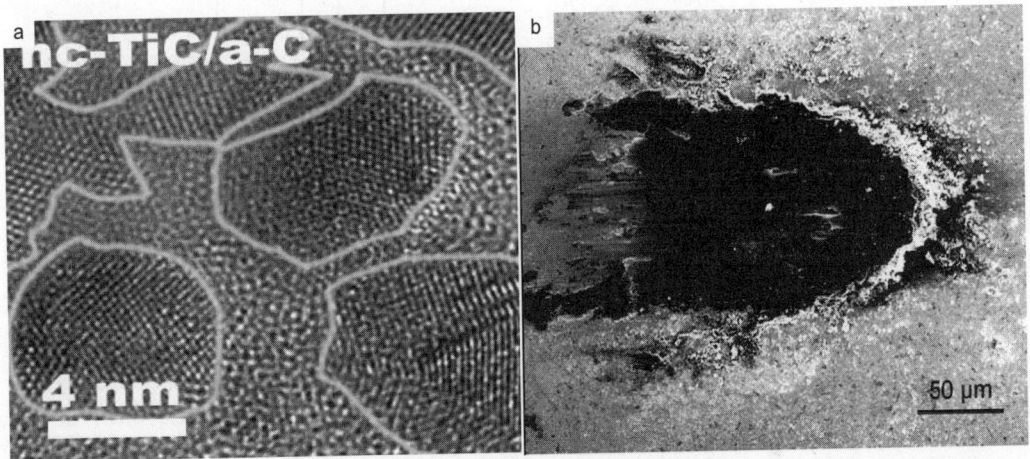

Fig. 33.4. Optical image of scratch tracks on: (a) nc-TiC/a-C, and (b) nc-TiC/a-C(Al) coatings.

Compressive stress toughening

Generally cracking is initiated by tensile stress, thus compressive residual stress has to be overcome first before a crack is initiated in tension. When a coating has a high compressive residual stress to start with, the coating will be able to take more tensile strain before fracture, as is demonstrated in Zr-Cu-O coatings (Fig. 33.5). Although a certain level of compressive stress can increase toughness, excessively large compressive stress can cause film delamination or cracking.

Nanotube toughening

The discovery of carbon nanotube (CNTs) has opened up new avenues in producing unique carbon-based materials. Novel mechanical tests on individual CNTs and atomistic calculations suggest that CNTs have ultra-high elastic modulus approaching 1 TPa, and exceptional tensile strengths, in the range of 20–100 GPa. Due to their outstanding properties, carbon nanotubes have attracted a growing interest and are considered to be the most promising materials for applications in nanoengineering. Toughness enhancement can be obtained through crack deflection, crack bridging and CNT pull-out. CNT toughened bulk ceramics are attractive, and outstanding results have been successfully achieved, for instance, in Al_2O_3 matrix, and $BaTiO_3$ matrix. However, nanotube toughened thin films is yet to be realised, due to difficulties in controlled alignment, interface reaction and reasonable volume fractions. Xia anodised aluminium in order to obtain an amorphous alumina matrix having pores with diameters in the 30–40 nm range.

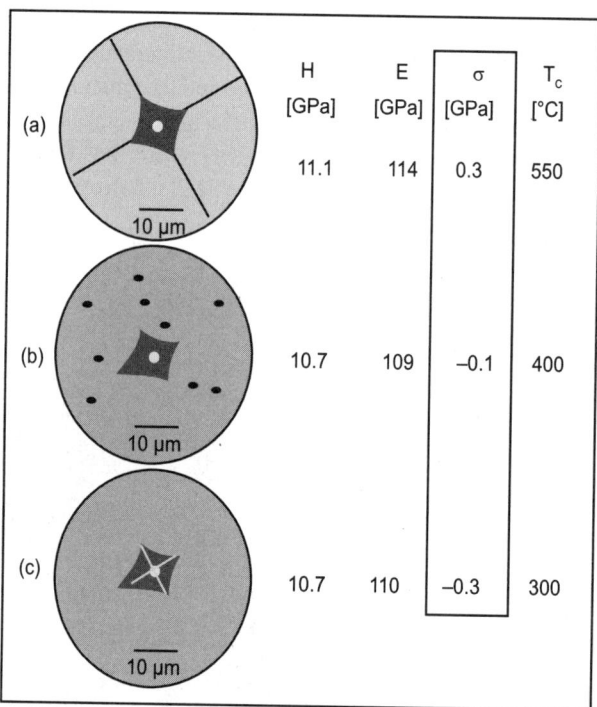

Fig. 33.5. Comparison of Vicker's indenter impressions into Zr-Cu-O coatings with different residual stress: (a) 0.3 GPa; (b) –0.1 GPa; and (c) –0.3 GPa.

Multiwall carbon nanotubes were grown by CVD in this matrix to realise CT/alumina composite layers of 20 to 90 μm in thickness. Crack deflection at the CNT/matrix interface [Fig. 33.6(a)], crack bridging by CNT [Fig. 33.6(b)], CNT pull-out [Fig. 33.6(c)] are all observed.

Composition or structure grading toughening

Graded inter-layering is a known effective method to reduce cracks concentration and enhance adhesion between coating and substrate. In the gradient design, the substrate is first coated with a high adhesion layer, and then the coating constituents are allowed to vary homogeneously or heterogeneously while the coating thickness builds up. With biased graded (varying substrate bias voltage while coating thickness builds up) in magnetron sputtering, Zhang prepared a 1.5 μm a-C gradient coating on tool steels with moderately high hardness (25 GPa), but very high toughness (plasticity 57.6 per cent). During deposition, substrate bias gradually increased from –20 to –150 V, and a graded sp^2/sp^3 was achieved through coating thickness. The bottom layer has the highest sp^2 fraction for high adhesion, whereas the top layer possess the highest sp^3 fraction to render high hardness. Bias-graded coating design enhances the toughness and the adhesion of the coating on tool steel (Fig. 33.7). The adhesion strength increased more than two times as compared to the same coating deposited at constant bias (–150 V). Pei changed the substrate bias in order to change the microstructure of the nc-TiC/a-C:H coatings prepared by close field unbalanced magnetron sputtering. Increasing the bias resulted in a clear transition from columnar to glassy microstructure in the coating.

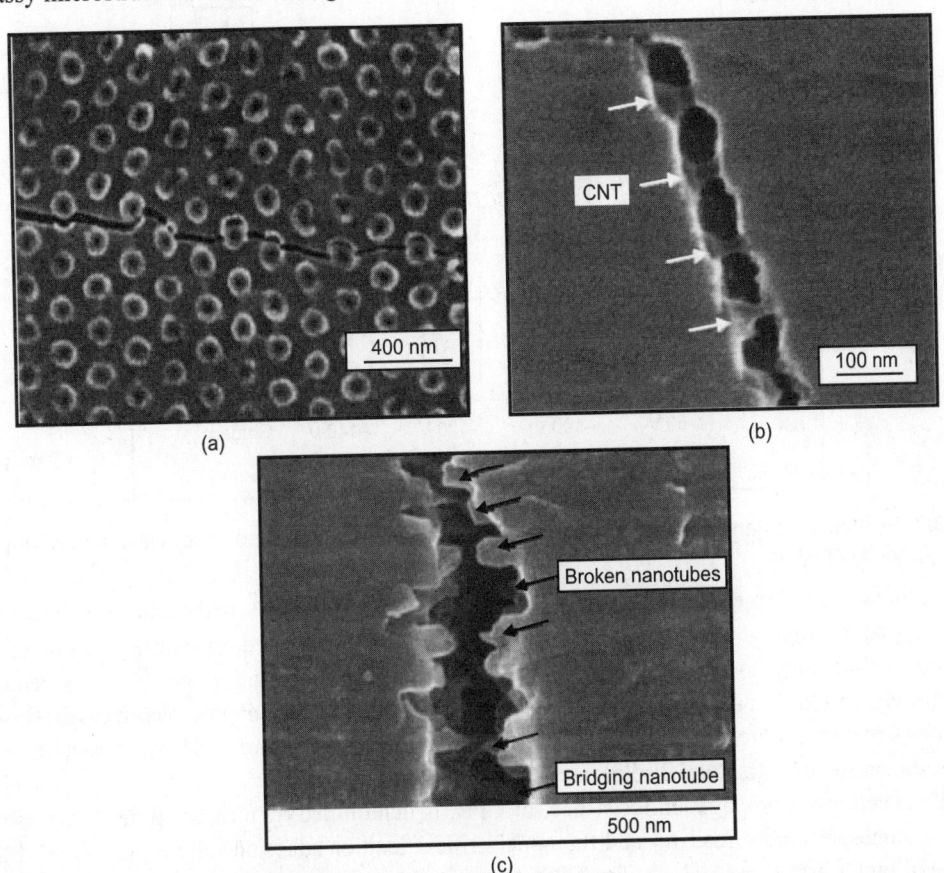

(a)

(b)

(c)

Fig. 33.6. Micrographs of deformations in CNT-alumina composites fabricated using ordered porous alumina templates: (a) transverse crack deflection at the CNT/matrix interface; (b) CNTs bridging longitudinal matrix cracks; (c) nanotube pull-out, and subsurface nanotube bridging.

Substrate bias also simultaneously and greatly enhanced the coating hardness, toughness and adhesive strength. The coating with glassy microstructure exhibits a substantial toughening. However, a word of caution on the actual toughness value given in the chpater—the value of few tens of MPa·m$^{1/2}$ is unbelievably high, as also pointed out by Smith, probably due to experimental limitations. Ion bombardment can densify the structure and prevent the formation of columns which are detrimental in terms of microcrack initiation and propagation under load. The induced residual stresses grow as bias increasing.

Toughness Characterisation

In a classical definition, toughness measures the ability of a material to resist the crack propagation until fracture. For bulk materials and very thick films, toughness can be easily measured according to ASTM standards. For thin films, there is no proper definition for toughness, since in thin films or coatings, the cracks do not necessarily pre-exist, and thus energy has to be used to initiate the crack.

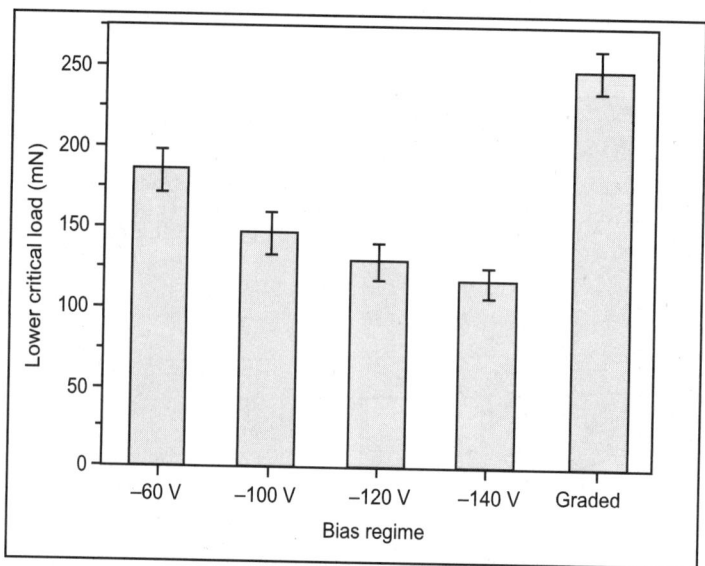

Fig. 33.7. Scratch adhesion strength of the non-hydrogenated DLCs deposited under various constant bias or bias-graded conditions.

Thin film toughness measurements are difficult or tedious in practice due to the thickness limitations. Although many proposed methods can be used in some cases, a standard procedure or a universally accepted methodology does not exist yet for ceramic thin film or coating toughness measurements. Qualitatively, toughness can be simply estimated by the plasticity, the MDP (microhardness dissipation parameter), and the scratch crack propagation resistance. The plasticity or MDP are measured as the ratio of the plastic over the total displacement or work.

The scratch crack propagation resistance, or CPR, is determined by measuring the lower critical load, i.e. cohesive failure load L_{c1} and the upper critical load or adhesion failure load L_{c2} and then calculated by CPR = L_{c1} (L_{c2}–L_{c1}). The CPR measures the scratch resistance to crack initiation and propagation. The most widely method used in the assessment of thin film toughness is indentation. When the stress exceeds a critical value during indentation, a crack or spallation is generated. Failure of the coating is manifested by the formation of a kink or plateau in the load-displacement curve, or by the crack formation in the indent impression. The length, c, of radial cracks is related to the fracture toughness K_{IC} through the following:

$$K_{IC} = \delta \left(\frac{E}{H} \right)^{1/2} \left(\frac{P}{c^{3/2}} \right) \qquad \text{... (33.1)}$$

where, P is the applied indentation load; E and H are the elastic modulus and the hardness of the coating, respectively. δ is an empirical constant which depends on the geometry of the indenter. Major uncertainties in this method come from the crack formation threshold, the substrate effect and the crack length measurement.

Another approach based on indentation is through determining the fracture energy. The energy difference before and after through-thickness cracking is believed to be responsible for the coating fracture.

The fracture toughness of the film is calculated by:

$$K_{IC} = \left(\frac{E}{(1 - v_f^2) 2\pi C_R} \frac{\Delta U}{t} \right)^{1/2} \qquad \dots (33.2)$$

where, E is the elastic modulus of the coating, v_f is Poisson's ratio of the coating, $2\pi\ C_R$ the planar crack perimeter, t the effective coating thickness, and ΔU is the strain energy difference responsible for the cracking. This method also inherited the uncertainties for the substrate effect on the deformation fields, the crack size measurement and negligent of the interface debonding.

Tensile testing of freestanding thin films with a precrack induced by indentation was recently used to measure the toughness of 0.52 and 1 μm tetrahedral amorphous diamond-like carbon (ta-C). During preparation of the standalone film, wet etching of SiO_2 interlayer between the coating and silicon substrate has been employed. This method has difficulties in obtaining freestanding film without curling, precrack sharpness and accurate measurement of the precrack dimension. A simple two-step tensile method, based on the strain energy difference between two tensile processes is proposed by Zhang. The 'two-step' tensile method extends a very elastic substrate coated with the ceramic film in question until the coating fracture. A kink is observed in load-extension curve. The load is then released to zero but reloaded to the previous extension. The energy difference is considered to be the energy used in creating the fracture. In this method, nc-TiN/a-SiN$_x$ nanocomposite layer with thickness of 3.0 μm gives a toughness value of 2.6 MPa·m$^{1/2}$. The method requires very a good elasticity for the substrate (so that when the coating is fractured the substrate is still in the elastic deformation regime), and that the thickness of the substrate to be small enough (for the kink to be visible); plus, there should not be a coating delamination during tension. Currently, there are relatively few toughness data available for thin nanocomposite films due to the difficulties in forming a sensible crack while reducing the substrate effect in the thin films. Toughness values of some nanocomposite are listed in Table 33.1.

Table 33.1. Toughness values for some hard coatings.

Materials	Toughness MPa·m$^{1/2}$	Evaluation method
TiC$_x$N$_y$/SiCN	1.0	Indentation
TiN/SiN$_x$	1.15	Indentation
Ni-doped nc-TiN/a-SiN$_x$	2.60	Two-step tensile
YSZ/Au	1.6	Indentation
Nanolayered Cr/a-C	1.81–3.49	Indentation
Ni/Al$_2$O$_3$	2.4	Indentation
ta-C	4.25–4.4	Tensile with precrack

SUPERHARD NANOCOMPOSITE COATING

A variety of superhard coatings with Vickers plastic hardness exceeding 40 GPa have been reported by several research groups during the last five years. However, one has to distinguish between superhard nanocomposites, such as nc-TiN/a-Si$_3$N$_4$, nc-TiN/a-Si$_3$N$_4$/a- and nc-TiSi$_2$, nc-(Ti$_{1-x}$Al$_x$)N/a-Si$_3$N$_4$, nc-TiN/TiB$_2$, nc-TiN/BN, etc. where the high hardness originates from the nanostrucutre and, therefore, remains stable upon annealing to high temperatures, and coatings, such as CrN/Ni, ZrN/Ni, and others in which the measured high hardness is due to a high compressive stress that is induced in the coatings due to energetic ion bombardment during their deposition (e.g. by magnetron sputtering). We also summarise the recent progress in the industrial applications of the superhard nanocomposite coatings on machining tools. There are many examples which show that compressive stress in the coatings or in a bulk material results in an apparent increase of the plastic hardness measured by the load-depth sensing indentation technique. In the case of relatively soft ductile material that showed pile-up during the indentation, this apparent enhancement of the measured hardness is an artifact of that technique because no change of the hardness was found when the hardness was evaluated from the projected area of the remaining plastic deformation. Because of the pile-up, the load-depth sensing technique underestimates the indentation depth and consequently overestimates the values of hardness and elastic modulus.

A strong enhancement of the hardness due to a high biaxial compressive stress in the coatings was reported, e.g. for HfB_2 (70 GPa), (TiAlV)N, TiN (100 GPa and 80 GPa, respectively), and TiB_2 (68 GPa). However, when the coatings were annealed at $\geq 400°C$, the stress decreased below 2 GPa and the hardness decreased to the value of bulk materials (34 GPa for TiB_2 and ≤ 20 GPa for the others).

Obviously, the enhancement of the hardness due to high compressive stress is of little interest for applications, such as cutting tools for dry machining where the coatings reach a high temperature of $600°–800°C$. Recent studies showed that for the ZrN/Ni and CrN/Ni coatings whose high hardness of ≥ 40 GPa is solely due to a high compressive stress of ≥ 5 GPa, as well as for superhard nanocomposites, such as $nc\text{-}TiN/a\text{-}Si_3N_4$ and $nc\text{-}(Ti_{1-x}Al_x)N/a\text{-}Si_3N_4$ with either a high (≥ 5 GPa) or low (≤ 1 GPa) compressive stress, the values of plastic hardness measured by the load-depth sensing indentation technique agree fairly well with the Vickers hardness evaluated from the projected area of the remaining plastic deformation. Also, the hardness values of a variety of superhard nanocomposites measured by the load-depth sensing technique at sufficiently large loads of 50 to 200 mN (in the case of 15–20 μm-thick coatings even up to 1000 mN) agree fairly well with the Vickers hardness calculated from the projected area of the plastic deformation within the broad range of hardness between 20 and 100 GPa. These nanocomposites were deposited by plasma CVD and, therefore, have a low compressive stress of ≤ 1 GPa. It can be concluded that the very high values of the plastic hardness of these nanocomposites were carefully checked and can be considered as correct. All the superhard nanocomposites are deposited by means of plasma induced processing, such as plasma-induced CVD and PVD including magnetron sputtering and vacuum arc evaporation. Alternatively, combined plasma PVD and CVD are used in which the metals (Ti, Al, ...) are evaporated either by means of vacuum arc or by sputtering and the nonmetals are introduced as gaseous reactants (e.g. SiH_4, BCl_3, $B_3N_3H_6$). Voevodin used pulsed laser ablation for the deposition of hard wear resistant nanocomposite coatings.

Generic Design Concept and Thermal Stability of Superhard Nanocomposites

The superhard nanocomposites are prepared according to the generic design principle that is based on a strong spinodal decomposition, which results in a formation of a nanostructure. The thermodynamic criterion for spinodal decomposition to occur is negative second derivative of the Gibbs free energy of the mixed phase (e.g. $Ti_xSi_yN_z$) with a change of the composition (e.g. toward $TiN + Si_3N_4$).

It means that any spontaneous infinitesimal, local fluctuation of the composition of the mixed phase leads to a decrease of the Gibbs free energy of the system. Consequently, the phase segregation occurs spontaneously without any need for nucleation of either phase. In such a way, a nanocomposite is formed by a self-organisation with a characteristic length scale (crystallite size), which is determined by the balance between the decrease of the Gibbs free energy due to phase segregation, chemical gradients, and incoherency strain at the interface. When formed, such nanocomposites are stable against coarsening within the temperature range where the second derivative of the Gibbs free energy of the segregated system remains negative. Thus, an appropriate selection of binary or ternary refractory hard materials allows one to prepare superhard nanocomposites whose nanostructure and the resultant hardness remain stable up to high temperatures of 1100°C. In these nanocomposites, the nanocrystals of the transition metal nitride are imbedded within about 1 monolayer thin amorphous tissue which provides the materials with a high hardness, resistance against crack formation and propagation and a high thermal stability up to 1100°C. Examples of such systems are nc-MnN/a-Si_3N_4 {M = Ti, W, V, $(Ti_{1-x}Al_x)N$/a-Si_3N_4 and other hard transition-metal nitrides}, TiN/TiB_2, nc-TiN/BN and others. In the case of the nc-TiN/a-Si_3N_4, nc-$(Ti_{1-x}Al_x)N$/a-Si_3N_4, nc-TiN/a-Si_3N_4/a-$TiSi_2$, and TiN/TiB_2 it was shown that the hardness, measured after the annealing at room temperature, remains unchanged up to the recrystallisation temperature of 900°–1100°C. The highest recrystallisation temperature is achieved for nanocomposites with the optimal Si_3N_4 content corresponding to the percolation and lowest crystallite size of about 3 nm. Figure 33.8 shows a typical example of the thermal stability of such superhard nanocomposites. This is an unambiguous evidence that the superhardness is due to the nanostructure.

Another evidence of the superhardness being due to the formation of the nanostructure is the spontaneous formation of the nanocomposite due to self-organisation. For example, Hammer deposited amorphous Ti-B-N films by magnetron sputtering at room temperature with a hardness between 27 and 29 GPa, depending on the composition. Upon annealing to ≥600°C, a nanocomposite structure was formed and the hardness increase to about 40 GPa.

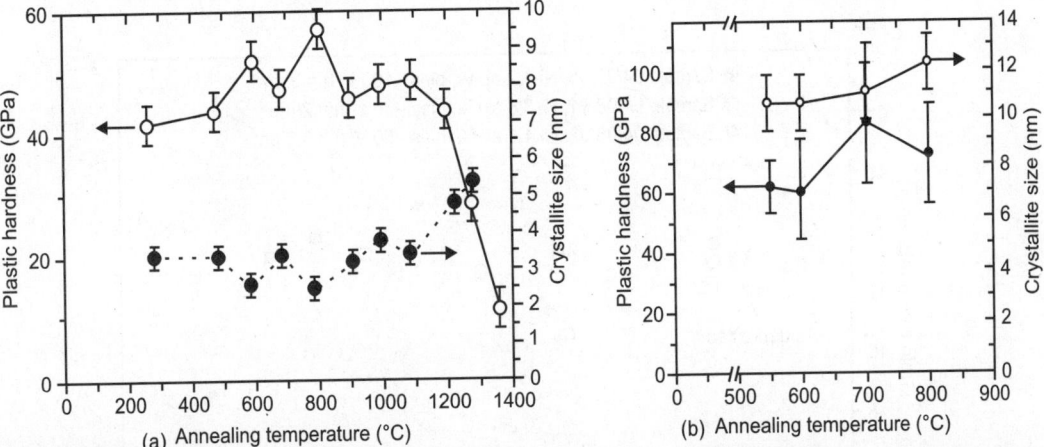

Fig. 33.8. (a) Example of annealing behaviour of a $(Ti_{1-x}Al_x)N$/a-Si_3N_4 sample deposited on cemented carbide substrate at a temperature of about 300°C; (b) ternary nc-TiN/a-Si_3N_4/a-$TiSi_2$ nanocomposite deposited on steel substrate; total Si-content 8 at. %, 3.5% Si_3N_4, 4.1% $TiSi_x$, thickness of the coating 20 μm.

In many cases reported in the literature, the high measured hardness was attributed to 'nanocomposites', although it was predominantly or solely due to the high biaxial compressive stress. For example, Musil reported superhardness of ≥40 GPa for a series of coatings consisting of hard transition-metal nitride (e.g. ZrN, CrN) and a soft metal that does not form nitride (e.g. Ni, Cu). More recently, it was shown that the high hardness of these coatings is due predominantly to the high biaxial compressive stress induced in the coatings during their deposition by means of unbalanced magnetron sputtering. The high hardness strongly decreases to the ordinary value for the bulk material of ≤20 GPa upon annealing to ≥400°C when the stress relaxes as shown in Fig. 33.9.

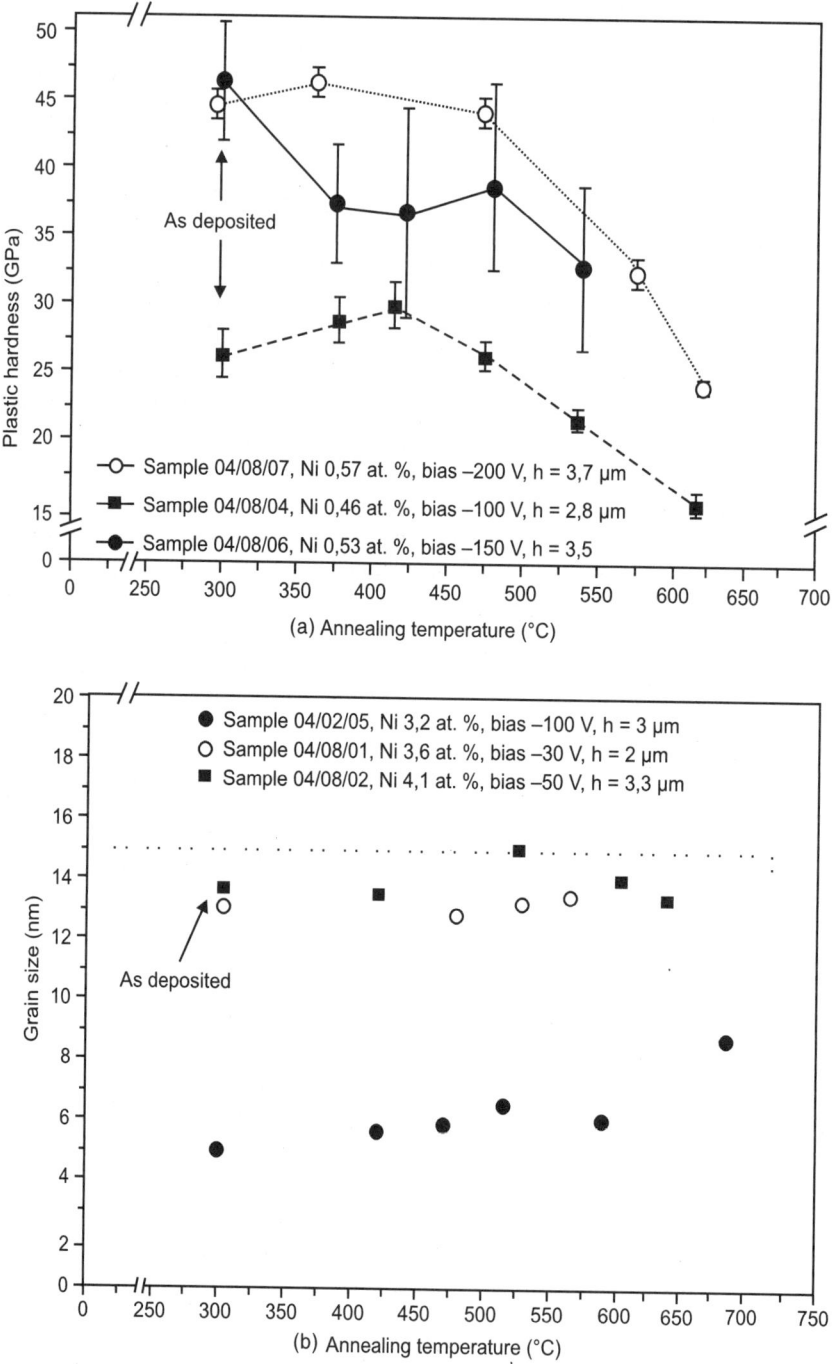

Fig. 33.9. (a) Decrease of the hardness, and (b) no change of the crystallite size is observed upon annealing of the ZrN/Ni coatings above 400°C. Simultaneously, a decrease of the high compressive stress in the coatings is observed.

The experimental data available so far strongly suggest that the reported superhardness in these coatings (called also nanocomposites) is a simple consequence of the high biaxial stress in a similar way as reported for HfB_2, (TiAlV)N, TiN, and TiB_2 mentioned above. No effect of the nanostructure on the reported enhancement of the superhardness could be found because no changes of the microstructure occurred upon the annealing. An interesting question is also the extent of the hardness enhancement due to the formation of a nanostructure and to compressive stress in the TiN-TiB_2 coatings prepared by magnetron sputtering at low pressure. The data reported in showed an enhancement of the hardness to about 68 GPa for pure TiB_2 (bulk value of 34 GPa) due to the compressive stress and its decrease to ≤55 GPa for Ti-B-N nanocomposites with nitrogen content of 25–30 per cent. With reference to the above mentioned results of Hammer regarding hardness increase up to about 40 GPa upon annealing of such films deposited at room temperature one would expect that the high hardness of 55 GPa is due partially to the formation of nanostructure and to the high compressive stress. A lack of changes of the crystallite size upon annealing of these films is not any guarantee of the stability of their hardness, because it does not reflect the relaxation of the compressive stress, as shown in Fig. 33.9. Therefore, more detailed studies of this system are needed.

Mechanical Properties of the Superhard Nanocomposites

Conventional hard materials are brittle and undergo fracture at a strain of ≤ 0.1 per cent. It was therefore, surprising to find in the nanocomposites an unusual combination of very high hardness of 40–100 GPa, high elastic recovery of 80–94 per cent, and a high resistance against crack formation. A recent theoretical analysis of the experimental data showed that they can be understood in terms of the conventional fracture physics and mechanics when this is scaled down to a crystallite size of 3–5 nm. As already discussed, the superhardness is a simple consequence of the formation of stable nanocomposites with a strong interface which avoids grain boundary sliding. The high resistance against crack formation is due to the small stress concentration factor of 2–4 in the nanocomposites, which is orders of magnitude smaller than that of ordinary microcrystalline materials. For the same reasons, the yield stress needed to propagate a nanocrack is much higher in the nanocomposite than in a microcrystalline material even if they have the same stress intensity factor. The high elastic recovery of up to 94 per cent and high range of predominantly elastic deformation of ≥10 per cent which results in a very high elastic energy density upon indentation can be understood in terms of reversible flexing. The very high elastic modulus as measured by the indentation technique is due most probably the high pressure under the indenter.

Industrial Applications

The industrial-scale coating technology by means of vacuum arc evaporation from a central cathode consisting of segments of the different metals was developed for the superhard nc-$(Ti_{1-x}Al_x)N$/a-Si_3N_4 coatings and is now available for a large-scale production. The coatings show a significantly improved cutting performance as compared with the conventional ones. Figure 33.10 shows schematics of the equipment. It consist of a central cathode with two segments made of titanium and an aluminium/silicon eutectic alloy. The movement of the cathode spot of the arc is controlled by magnetic field, which is induced by a combination of permanent magnets and electric coils. The axial component of the field drives the motion of the spot around the axis of symmetry, whereas the axial shift of the magnetic field, which is controlled by the coils, determines the fraction of the duty time of the cathodic spots on the individual segments. This spot is moved at a time scale of several tens of millisecond per segment, which assures atomic mixing of the metallic components during the deposition on tools. The latter are mounted on the substrate holders, which are fixed in an axially symmetric manner and rotate along their own axes. The typical deposition conditions are: substrate temperature: 550°C; substrate bias 200 V; total pressure 0.2 Pa; arc voltage 32 V; arc current 100 A; substrate bias 200 V; substrate current 6 A. Meanwhile, the arrangement of the central cathode and the control of the movement of the cathodic spot were further developed and improved, which resulted in a significant decrease of the emission of the macroparticles and a much better smoothness of the surface, which in turn led to a further improvement of the cutting performance of the coated tools.

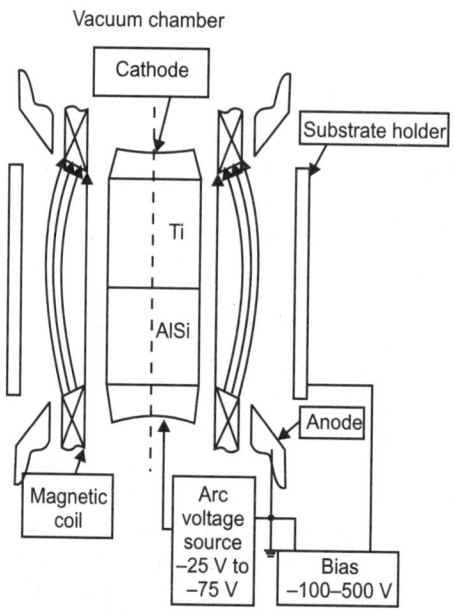

Fig. 33.10. Schematics of the industrial coating equipment based on vacuum arc evaporation from segmented central cathode consisting of titanium and aluminium/silicon eutectic alloy. The movement of the cathode spot is controlled by magnetic field, which is induced by a combination of permanent magnets and electrical coils.

A variety of tools for dry and fast turning, milling, drilling are coated and successfully used by the customers. Under the conditions of the dry machining, which saves the environmentally risky coolants and saves costs, the temperature of the cutting edge and of the rake reaches 600°–800°C. This illustrates the need of thermally stable and oxidation resistant coatings. Figure 33.11 shows an example of the performance of the recently developed superhard nanocomposite coating nc-$(Ti_{1-x}Al_x)$N/a-Si_3N_4 in comparison with the conventional TiN and $(Ti_{1-x}Al_x)$N coatings in dry milling at a relatively high cutting speed. One can clearly see the progress in the development and improvement of these coatings. The performance of the indexable inserts coated with the 3rd generation of multilayer nc-$(Ti_{1-x}Al_x)$N/a-Si_3N_4 coatings MARWIN® MT is improved by a factor of 9 as compared to the uncoated tool and more than 2 as compared to the standard $(Ti_{1-x}Al_x)$N coating. To sum up a significant increase of the hardness measured by the load-depth sensing indentation technique can be achieved either by a high biaxial compressive stress induced in the coatings by energetic ion bombardment during their deposition or by the formation of a stable nanostructure due to spinodal decomposition and phase segregation.

Whereas in the former case, the hardness decreases upon annealing to ≥400°C, the latter superhard coatings remain stable up to high temperatures of 1100°C. This is a significant advantage of these coatings in industrial applications, such as dry machining, where high temperatures are reached at the cutting edge of the tool.

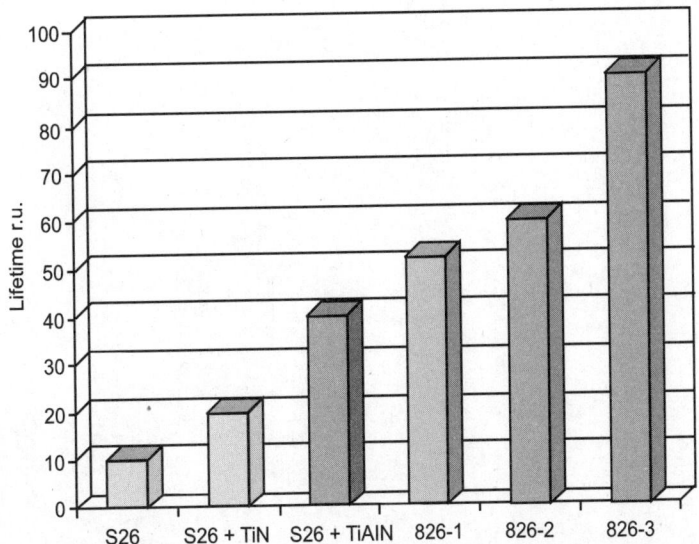

Fig. 33.11. Example of the cutting performance of various coatings. Symmetric milling of steel CK45, feed 0.23 mm/tooth, depth of cut 2 mm, cutting speed 179 m/min. The meaning of the symbols: S26 indexable insert SPCN 1203EDSR (P20-P30) without any coating, S26+TiN TiN-coated, S26+TiAlN (TiAl)N-coated, 826-1 and 826-2 the same insert coated with nc-$(Ti_{1-x}Al_x)N$/a-Si_3N_4 nanocomposite of the 1st and 2nd generation, respectively, 8026 nc-$(Ti_{1-x}Al_x)N$/a-Si_3N_4 multilayer nanocomposite coatings 3rd generation 2000.

FUTURE TRENDS AND DIRECTIONS

For engineering applications, the zest for sole superhardness will subside, while a combination of hardness and toughness will be pursued more by the academics driven by the industrial needs. A sensible route is to toughen superhard coatings in order to achieve what we call 'hard yet tough' coatings. In principle, all toughening mechanisms can be applied to the toughening of ceramic thin films or coatings, the real challenge lies in the implementation. *In situ* synthesis of carbon nanotube-imbedded nanocomposite (such as nc-TiC/a-C(CNT) or nc-TiN/a-Si_3N_4(CNT)) or nanocrystals imbedded in a complete matrix of carbon nanotubes (for instance, nc-TiC/CNT, or even nc-Diamond/CNT) may hold the key (at least, one of the keys) to this ideal 'hard yet tough' coating, where the CNT-imbedded matrix or the CNT matrix provides the ultimate 'bulk' elasticity and toughness to stop or block crack propagations, and the nanocrystalline phase provides the superior hardness. In respect to the toughness measurements, the lack of a universally accepted toughness measurement methodology and instrumentation for thin films or coating has effectively hindered the progress of the research. It is hoped that in a not too far future, a universally accepted methodology will be born, and that this will accompany the birth of a toughness measurement apparatus for ceramic thin films or coatings; a standalone machine, or perhaps, as an integral part of a hardness testing system. To catalyse the birth of the universal toughness measurement for ceramic thin films or coatings, the community needs first to have a consensus on the definition of the toughness or 'fracture toughness' for thin films or coatings (which we believe it should take into account the energy needed to generate the cracks), and regular international seminars should be organised especially on this topic.

References

Allen, K.M., *Nanostructured Systems*, Butterworths, London.

Benaim Pinto, C., *Introduction to Superconductivity*, Prentice-Hall, London.

Budyko, M.I., *Solid State Theory*, D. Van Nostrand, New York.

Chang, J.C., *Fundamental Theory of Metals*, John Wiley & Sons, New York.

Coolingwood, R.W., *Electronics Microstructure Science*, John Wiley & Sons, New York.

Downe, S.A., *Sensors and Actuators*, John Wiley & Sons, New York.

Dugan, P.R., *Silicon Sensors*, Plenum Publishing Corporation, London.

Feynman, R.P., *Superconductivity of Metal and Alloys*, Wiley, New York.

Gould, G.W., *Micromachining and Microfabrication*, D. Van Nostrand, New York.

Gregory, T., *Nanotechnology*, Springer (India) Pvt. Ltd., New Delhi.

Harding, G., *Nanostructures*, Prentice-Hall, John Wiley & Sons, New York.

Hidy, G.M., *Material Science and Engineering*, Wiley, New York.

Jackwerth, E., *Carbon Nanotubes and Fullerenes*, Academic Press, London.

Jarvis, B., *Atoms, Molecules and Clusters*, John Wiley & Sons, New York.

Jencks, W.P., *Nanotechnology and Its Applications*, John Wiley & Sons, New York.

Kim, C.K., *Nanostructures and Mesoscopic Systems*, Marcel Dekker, New York.

Kittel, C., *Introduction of Solid State Physics*, Wiley, New York.

Lewis, S.T., *Nanotribology*, Reston Publishing Co., Reston, Virginia.

Lowman, R.L., *Supramolecular Chemistry*, Chilton Book Company, Radnor, Pennsylvania.

Mark, R., *Introduction to Nanotechnology*, Pearson Education, South Asia.

Mason, R., *Nanosensors*, McGraw-Hill, New York.

McCaull, J. and Crossland, J., *Biosensors and Bioelectronics*, Harcourt Brace Jovanovich, New York.

Mitchell, R., *Chemistry of Nanomaterials*, McGraw-Hill, New York.

Odum, P.L., *Nanoscale Materials*, W.B. Saunders and Co., New York.

Olsen, S.K., *Chemistry of Fullerene*, W.B. Saunders and Co., New York.

Phillips, D.J.H., *Nanotechnology*, Applied Science Publishers, London.

Sawyer, C.N. and McCarty, P.L., *Nanosensors and Nanotribology*, McGraw-Hill, Tokyo.

Sax, N.I., *Carbon Nanotubes*, Van Nostrand Reinhold, New York.

Schrowebel, J., *Fullerenes and Carbon Nanotubes,* Pergamon Press, New York.

Smith, H.S., *Sensors and Silicon Sensors,* Chilton Book Company, Radnor, Pennsylvania.

Tyler, E.G. and Jack, D., *Nanostructures Systems,* McGraw-Hill, New York.

Wyatt, G.M., *Application of Nanotechnology*, Reston Publishing Co., Reston, Virginia.

Index

A

Ab initio methods, 28
Ablative property, 422
Acoustic droplet ejection (ADE), 161
Acoustic waveguides, 220
Adiabatic chemical dynamics, 24
Advantages of composites, 359
Advantages of LOCs, 164
Advantages of nanocomposites over conventional
  composites, 355
Aerospace, 21
Affinity for cations, 132
Age hardening, 507
Agriculture, 22
Air permeability, 348
Algorithmic self-assembly, 109
Amphiphiles, 139
Amphiphilic molecules and segments, 443
Anti-stokes scattering, 81
Applications of clusters in catalysis, 206
Applications of nanoshell particles, 254
Applications of nanotechnology, 16
Approaches to miniaturisation, 120
Aqueous-sol derived thin films, 269
Aramid fibre, 358
Arc discharge, 66
Aromaticity, 54
Atom chains, the ultimate nanowires, 98
Aza-crowns, 132

B

Back-end processing, 46
Ball milling, 262
Ball-grid-array (BGA), 313
Barrier properties, 421
Baseline use, 338
Basic concepts of nanotechnology, 3
Biological folding, 140
Biological macrocycles, 142
Biological systems for complex inorganic structures, 436

Biological systems, 150
Biologically derived synthetic nanocomposites, 431
Biologically inspired nanocomposites, 436
Biologically-derived units, 128
Biomaterials, 8
Biomedical nanocomposites, 422
Biomimetics, 126
Block copolymer templating, 458
Boron buckyball, 52
Boron nitride nanotubes, 71
Bottom-up approaches, 6
Boundary lubrication, 289
Buckyball, 51
Building blocks of supramolecular chemistry, 127
Bulk micromachining, 117
Bundling mode, 80

C

Calixarene, 129
Carbon allotropes, 120
Carbon fibre, 358
Carbon nanotube chemistry, 86
Carbon nanotube composites in the photoactive layer, 72
Carbon nanotubes as a transparent electrode, 73
Carbon nanotubes in photovoltaics, 72
Carbon nanotubes, 53, 59
Carcerand, 130
Catalysis, 18, 128
Catalytic residues in enzymatic reaction, 204
Categories and sub-categories of supramolecular
  chemistry, 129
Catenane, 131
Cathodoluminescence, 80
Cavitand, 131
Ceramic packages, 308
Ceramic–metal nanocomposites, 370
Characterisation of polypropylene nanocomposites, 399
Characterisation of SAMs, 170
Chemical dynamics, 24, 30
Chemical libraries, 254

Chemical nanosensor development and characterisation, 334
Chemical sensors and nanotechnology, 335
Chemical separation of metallic and semiconducting SWNTs, 86
Chemical synthesis of inorganic nanomaterials and composites, 423
Chemical vapour deposition (CVD), 66
Chemistry and environment, 18
Chip materials and fabrication technologies, 163
Chip scale packaging, 310
Clay based nanocomposites, 362
Clay plateletes, 417
Clay structure, 350
Clays and their modification, 489
Cluster chemistry, 206
CNTs in dye-sensitised solar cells, 74
Coarse WDM, 224
Colloidal crystal templating, 460
Colloidal stability, 254
Colloidal templating of conducting polymers, 465
Colloidal templating, 459
Colloids, 237
Colorimetry and biosensing, 255
Colossal carbon tube, 88
Common host molecules, 138
Comparison of holography, ESPI and shearography, 485
Compatibilising agents, 419
Components of a nanophotonic system, 219
Composites containing BN, 71
Composition or structure grading toughening, 510
Compressive stress hardening, 507
Compressive stress toughening, 509
Computational chemistry, 25
Concepts in supramolecular chemistry, 125
Conduction of electrons, 94
Conductivity of nanowires, 92
Continuous-flow microfluidics, 159
Control of supramolecular chemistry, 125
Conventional route to creating nanostructured materials, 339
Core reinforcement, 359
Core/shell nanoparticles, 276
Coulomb blockade, 192
Coupler, 220
Covalent reactivity via acid-oxidation, 86, 88
Creating polyolefin nanocomposites, 388
Crown ether, 131
Cryptand, 132
Cryptophane, 132
Crystal engineering, 133
Crystal growth, 183
Crystal structure and properties, 34
Cubic phase and other phases, 452

Cucurbituril, 133
Curing resins, 357
Cyclodextrin inclusion compounds, 140
Cyclodextrin, 135

D

D mode, 81
Damping factor, 415
Data storage and processing, 128
Densification of lead zirconium titanate (PZT), 269
Dendrimer, 136
Dense WDM, 224
Density functional methods, 29
Density functional theory, 24
Design and development of functional molecules, nanomaterials, and nanodevices, 202
Design methods for hard yet tough nanocomposites, 504
Development of alumina-metal nanocomposites, 471
Device test, 47
Diagnostics, 17
Diamond patterning, 115
Diazonium chemistry, 83
Die preparation, 48
Digital (droplet-based) microfluidics, 160
Digital microfluidics, 161
Dimensional stability, 360
Direct templating of an inorganic by an organic liquid crystals, 454
Direct templating of materials, 447
Directed assembly of materials on the surface of a patterned SAM, 180
Disadvantages of LOCs, 164
DNA chips (microarrays), 160
DNA intercalation, 141
DNA machine, 106
DNA nanomechanical devices, 109
DNA nanotechnology, 107
DNA nanotubes, 108
DNA origami, 108
DNA polyhedra, 109
DNA-templated nanostructure formation, 432
Drug delivery, 17
Dry etching, 116
Ductile phase toughening, 508
DX arrays, 108
Dynamic covalent chemistry, 126, 136

E

Efficiency of Pd impregnated sol-gel derived γ-alumina porous spheres as catalyst, 266

Efficiency of sol-gel derived nanostructured γ-alumina porous spheres as an adsorbent in liquid chromatography, 266

Electrical connection to other system components, 306

Electrochemical growth of conducting polymers, 465

Electroluminescence, 80

Electromagnetic waveguides, 219

Electronic properties, 35

Electronic speckle pattern interferometry, 484

Electronic structure of carbon nanotube, 75

Electronic structure, 23, 207

Electro-optic modulator, 222

Elementary properties of superconductors, 40

Endohedral fullerene, 137

Endohedral metallofullerenes, 137

Energy applications of nanotechnology, 14

Energy, 18

Enhancement of luminescence, 252

Enhancement of thermal stability, 253

Environment, 125

Etchant etching, 115

Etching processes, 115

Examples of geometric phases, 111

Examples of global LOC application, 165

Examples of molecular machines, 105, 148

Excitation mechanism, 78

Excursions in cluster science, 208

Existing nanosensors, 331

Experimental investigation on thermal conductivity of nanofluids, 166

F

Fabrication of colloids, 180

Fabrication of nanowires at surfaces, 95

Fabrication on curved surfaces, 179

Fabrication using patterned SAMs as resists, 178

Families of catenanes, 131

Filtration, 18

Fire resistant properties, 390

Flammability properties, 363

Flip-chip, 310

Fluorination, 82

Foldamer, 138

Folding (chemistry), 138

Formation of polyolefin nanocomposites, 394

Foucault pendulum, 111

Fractal mobility in reactive lubrication, 299

From bulk to molecular lubrication, 288

From fission to coulomb explosion, 209

Front-end processing, 46

Fuel cells, 161

Fullerene, 50

Fullerite (solid state), 58

Functional behaviour of lubricated friction, 294

Functions of MEMS packages, 304

Future of NEMS, 121

G

G mode, 80

Gas barrier enhancement, 391

Gas barrier, 369

Gas cluster ion beam, 205

Gas permeability, 421

Gas phase cluster, 200

Generic design concept and thermal stability of superhard nanocomposites, 514

Generic surface micromachining module for MEMS hermetic packaging at temperatures below 200°C, 322

Geometric phase, 110

Global challenges, 165

Gold replica of colloidal crystals, 466

Gold-flecked nanosensor detects poisonous mercury, 332

Graft polymerisation onto γ-Al_2O_3 nanoparticles, 397, 399

Grain boundary hardening, 505

Green chemistry, 129

G' mode, 81

H

Hard and tough nanocomposite coatings, 503

Hardening mechanisms in nanocomposite coatings, 504

Hazardous materials, 49

Health and environmental concerns, 13

High aspect ratio (HAR) silicon micromachining, 117

High temperature superconductivity, 44

Holographic interferometry, 482

Host guest interactions, 129

Host-guest chemistry, 138

Hydrodynamic lubrication and relaxation, 288

Hydrogen bond, 139

Hydrophobic effect, 139

I

Importance for AFM, 120

Improving physical and mechanical properties of metals by incorporating nanoparticles of an oxide, 472

In situ polymerisation, 354

Industrial application of nanocomposite fillers based on organic intercalated bentonites, 426

Initiators/promoters/inhibitors, 357

Intercalation (chemistry), 140

Interpreting molecular wave functions, 30

Inverse/reverse micelles, 144
Ionophores/receptors, 203

K

Kataura plot, 77
Kinetics and energetics in nanolubrication, 286
Kinetics, 170

L

Lab-on-a-chip, 162
Large carcerands, 130
Laser ablation, 66
Laser, 225
Lasers as weapons, 231
Left-handed helix of polypeptides, 204
Limitation of the Gaussian statistics—the fractal space, 298
Liquid-crystal templating of thin films, 457
Lithium ion batteries, 89
Lithography, 115
LOCs and global health, 164
Low thermal conductivity, 361
Luminescence, 78
Lyotropic liquid-crystal templating, 442

M

Macrocycle, 141
Macrocycle effect, 141
Macrocycles, 127
Magnetic behaviour, 38
Magnetic nanoparticles, 239
Magnetic properties, 253
Main components of resin systems, 356
Materials for MEMS manufacturing, 113
Materials science of quasicrystals, 37
Materials technology, 128
Mathematical description, 36
MCM/HDI, 315
Mechanical bond, 142
Mechanical properties of the superhard
    nanocomposites, 517
Mechanical properties, 347, 421
Mechanically-interlocked molecular architectures, 126, 142
Medical diagnostics and treatments, 240
Medicine, 16, 128
Meissner effect, 42
Melt blending, 353
Membrane deposition and characterisation, 324
Membrane, 322
Memory storage, 19
MEMS basic processes, 114

MEMS manufacturing technologies, 117
MEMS packaging, 304
MEMS thermal actuator, 119
Metal and oxide nanocomposites, 470
Metal packages, 307
Metallacrown, 142
Metal–nonmetal transition in finite systems, 191
Metastable lubricant systems in large conforming
    contacts, 300
Micelle, 143
Micro- and nano-electromechanical system, 113
Micro-emulsion, 264
Microfluidics, 158
Micro-opto-electromechanical systems, 119
Microperiodic structures and photonics, 459
Microscale behaviour of fluids, 158
Mineral growth in hexagonal and lamellar phases, 451
Mixed monolayer coverage on gold nanoparticles for
    interfacial stabilisation of immiscible fluids, 174
Model of Sherrington and Kirkpatrick, 38
Molecular biology, 160
Molecular borromean rings, 144
Molecular clusters and interaction forces, 202
Molecular dynamics, 30
Molecular electronics, 182
Molecular encapsulation, 145
Molecular imprinting, 146
Molecular knot, 146
Molecular machine, 103, 147
Molecular machinery, 127
Molecular mechanics, 28
Molecular nanotechnology, 5
Molecular orbital, 24
Molecular recognition and complexation, 126
Molecular self-assembly, 125, 129, 149
Molecular sensor, 150
Molecular shuttle, 151
Molecular tweezers, 145
Molecular wires, 93
Monomer, 356
Montmorillonite, 418
Morphology, 238
Multichip packaging, 315
Multiwalled carbon nanotubes, 86
Multiwalled, 60

N

Nanoarchitecture, 109
Nanobud, 62
Nanoclay as drug vehicle, 499
Nanoclay dispersion, 394

Nanoclays for polymer nanocomposites, paints, cosmetics and waste-water treatment, 488

Nanocomposite performance, 388

Nanocomposite synthesis, 352

Nanocomposites, 350

Nanocomposites: novel polymer/inorganic hybrids, 349

Nanoelectromechanical systems, 119

Nanoelectronics, nanooptoelectronics, and information nanoprocessing, 196

Nanofluid: engineering the fluid, 165

Nanofoods, 22

Nanomaterial synthesis and application, 232

Nanomaterials, 6

Nanomechanical device, 205

Nanomechanics, 102

Nanoparticle properties, 170

Nanoparticles synthesis routes, 439

Nanoparticles with narrow size distribution, 441

Nanoparticles, 183, 437

Nanophotonics, 218

Nanoprinting and nanoinkjetting, 339

Nanoscale iron particles, 239

Nanoscale regime, 196

Nanosensor, 330

Nanosensors and devices for space and terrestrial applications, 335

Nanosensors for aqueous environments, 334

Nanoshell particles: synthesis, properties and applications, 241

Nanosilicates, 367

Nanosilicate treatments for fabric, 369

Nanostructured advanced materials, 186

Nanostructured materials, 339

Nanostructures with complex predefined morphologies, 441

Nanotechnology applications under development, 342

Nanotechnology, 150

Nanotoxicity: threat posed by nanoparticles, 257

Nanotube toughening, 509

Nanotubes, 183

Nanowire battery, 101

Nanowire, 90

Nanoworld is different, 189

Natural, incidental, and controlled flame environments, 67

Non-adiabatic chemical dynamics, 25

Nonconductive, 360

Noncovalent bonding, 151

Nondestructive testing of nanocomposites by optical techniques, 481

Non-ergodic behaviour, and applications, 39

Nonmagnetic, 360

Non-metal doped fullerenes, 137

Novel optoelectronic devices, 20

Novel semiconductor devices, 20

Nuclear fusion driven by cluster coulomb explosion, 213

Nylon-6 nanocomposites, 417

O

Optical absorption, 77

Optical and structural characterisation of periodic silver-polystyrene nanocomposites, 402

Optical circulator, 231

Optical properties of carbon nanotubes, 75

Optical switch, 221

Optical Systems, 180

Optical waveguides, 219

Organic chemistry, 180

Organic nanotubes and nanowires, 203

Other devices and functions, 129

Other fibres, 359

Other tile arrays, 108

Oxide nanoprecursors: a technological perspective, 261

P

Package-to-MEMS attachment, 309

Packaging, 48

Passivation of surfaces: protection from corrosion, 178

Patterning, 115

Perspectives in nanoscience and nanotechnology, 198

Pharmaceuticals, 152

Phase transformation toughening, 508

Phases observed in mixtures of water and amphiphile, 443

Photo-/Electro-chemically active units, 128

Photodetector, 221

Photonic band gap materials, 254

Physics of nanowires, 92

Physics of quasicrystals, 37

Plasma etching, 116

Plastic packages, 309

Plastic packaging (PEMs), 319

Plating of membrane and sealing layer, 326

Polarised light in an optical fibre, 111

Polyester-clay nanocomposites, 407

Polymer nanocomposites, 494

Polymers, 114

Polymorphism, 133

Polyolefin, polypropylene and polystyrene nanocomposites, 388

Porphyrin, 152

Preparation and characterisation of polypropylene nanocomposites containing polystyrene-grafted alumina nanoparticles, 397

Preparation and characterisation of polypropylene nanocomposites, 398

Preparation and properties of polyolefin nanocomposites, 393
Preparation of core shell assemblies, 246
Preparation of SAMs, 169
Probing selective chemistry via optical absorption, 84
Procedure for block copolymer, 458
Processing and characterisation of nanoceramic composites, 374
Prologue, 186
Properties of nanoshell particles, 251
Protein assembly, 435
Protein purification, 140
Protein-based nanostructure formation, 431

Q

Quantum chemistry and quantum field theory, 25
Quantum chemistry computer programs, 30
Quantum chemistry, 23
Quantum computers, 20
Quantum electrochemistry, 33
Quantum mechanics, 56
Quantum structures: synthesis, characterisation, manipulation, and assembly, 187
Quasicrystal, 36

R

Radar transparent, 361
Radial breathing mode, 80
Radiation resistance properties, 363
Raman scattering, 80
Raman spectroscopy, 84
Rationale for recommendation, 336
Rayleigh scattering, 82
Reaction mechanism, 85
Reactivity of bound guests, 130
Reactivity of guests, 145
Recycling of batteries, 19
Reduction of energy consumption, 18
Refineries, 21
Reflow of the sealing layer, 328
Removal of solvent, 465
Resin additives, 357
Resins for composites, 356
Resorcinarene, 153
Reversibility of diazonium chemistry, 86
Rheological modifier, 496
Rolling loss, 348
Rotaxane, 153
Route to infilling of colloidal crystals, 467

S

Sacrificial release process, 327
Safety and toxicity, 56
Sealing layer deposition and characterisation, 324
Selective chemistry of single-walled nanotubes, 82
Selective reaction and Raman features, 84
Selective reaction conditions, 83
Self-assembled monolayer, 168
Self-assembled monolayers in organic chemistry, 170
Self-assembly processes, 440
Semiconductor device fabrication, 45
Semiconductor nanostructures, 441
Semiconductors for photonic crystals, 463
Semi-empirical and empirical methods, 29
Sensing materials, 340
Sensing platforms, 340
Sensitisation, 79
Shape effects, 195
Shearography, 484
Sidewall functionalisation, 82
Silicon nanotubes, 88
Simple to complex: a molecular perspective, 5
Single-walled carbon nanotubes, 88
Single-walled, 59
Size determination of nanoparticles, 265
Size effects for transport in nanostructures, 195
Size effects, 189
Smarter, faster nanosensor, 333
Software packages, 30
Solar cells, 70
Solar nanowires promise efficient, low-cost solar power, 100
Sol-gel infilling colloidal crystal templates, 460
Sol-gel, 236, 263
Solid solution hardening, 506
Solid-solid phase transition, 262
Solid-state physics, 34
Solution based route for nanoparticles, 438
Solution-phase synthesis, 92
Solvation, 143
Solvent casting, 352
Some other nanocomposites and nanoalloys (besides Al_2O_3), 474
Sound synthesis, 220
Spectroscopy and functionalisation, 83
Spin glass, 38
Stacking (chemistry), 155
Stacking in biology, 156
Stacking within supramolecular chemistry, 155
Stem loop controllers, 109
Stick slip and collective phenomena, 289

Stochastic pump effect, 112
Stress in nickel membrane, 325
Studies of polymers with layered silicates, 477
Superconducting phase transition, 41
Superconductivity, 39, 57
Superhard nanocomposite coating, 513
Supermolecule, 156
Supramolecular assembly, 156
Supramolecular chemistry, 122, 153
Supramolecular polymers, 157
Supramolecular systems, 149
Surface chemical and catalytic properties, 253
Surface micromachining, 117
Surface plasmon resonance of metal nanoparticles and metallic nanoshells, 249
Sustainable flame retardant nanocomposites, 476
Synthesis of dielectric cores, 243
Synthesis of high T_c superconductor $YBa_2Cu_4O_8$ (1–2–4) under the condition of normal oxygen pressure, 268
Synthesis of low-agglomerated nanoprecursors in the ZrO_2-HfO_2-Y_2O_3 system, 270
Synthesis of metal nanoparticles, 244
Synthesis of nanomaterials and composites, 424
Synthesis of nanoshell particles, 242
Synthesis of nanowires, 91
Synthesis of oxide nanoparticles, 261
Synthesis of semiconductor nanoparticles, 244
Synthetic recognition motifs, 127

T

Tailoring magnetic properties of core/shell nanoparticles, 280
Template-directed synthesis, 126
Templates, 183
Textiles, 22
Theories of superconductivity, 43
Therapeutic applications and drug delivery, 256
Thermal activation model of lubricated friction, 291
Thermal considerations, 306
Thermal properties, 421
Thermodynamical models based on small and nonconforming contacts, 295
Thermodynamics, 125
Thin-film multilayer packages, 308
Threshold size effects from a single particle to collective phenomena, 191

Tile-based arrays, 108
Tissue engineering, 17
Tools and techniques, 11
Top-down approaches, 7
Topoisomer, 157
Torus, 62
Toughening mechanisms in nanocomposite coatings, 508
Toughness characterisation, 511
Traditional—hermetic and nonhermetic, 305
Types of carbon nanotubes and related structures, 59
Types of EOMs, 222
Types of MEMS packages, 307
Types of SAMs, 169

U

Ultracapacitors, 70
Ultraintense laser—cluster interactions, 212
Ultra-low-temperature superplasticity, 384
Use of more environmentally friendly energy systems, 19
Use of nanowires in molecular electronics, 95
Uses of nanowires, 93

V

Valence bond, 24
van Hove singularities, 76
VLS Growth, 91

W

Wafer test, 47
Wafers, 45
Waste-water treatment, 500
Wave model, 24
Waveguide, 219
Wavelength-division multiplexing, 223
WDM systems, 223
Welding nanowires, 93
Wet etching, 115
Wetting control, 182
Wireless and network technology, 339
Working principle, 161

X

X-ray diffraction on core-shell nanoparticles for a precise structure determination, 278

Z

Zero electrical 'dc' resistance, 40